Dynamics of the Norwegian Margin

It is recommended that reference to all or part of this book should be made in one of the following ways:

NØTTVEDT, A. *et al.* (eds) 2000. *Dynamics of the Norwegian Margin.* Geological Society, London, Special Publications, **167**.

SUNDVOR, E., ELDHOLM, O., GLADCZENKO, T. P. & PLANKE, S. 2000. Norwegian–Greenland Sea thermal field. *In:* NOTTVEDT, A. *et al.* (eds) *Dynamics of the Norwegian Margin.* Geological Society, London, Special Publications, **167**, 397–410.

GEOLOGICAL SOCIETY SPECIAL PUBLICATION NO. 167

Dynamics of the Norwegian Margin

EDITED BY

A. NØTTVEDT
Norsk Hydro Research Centre, Bergen, Norway

Co-editorial Board

BJØRN T. LARSEN
Norsk Hydro Exploration, Oslo

ROY H. GABRIELSEN
University of Bergen

SNORRE OLAUSSEN
Saga Petroleum, Oslo

HARALD BREKKE
Norwegian Petroleum Directorate, Stavanger

BJØRN TØRUDBAKKEN
Saga Petroleum, Oslo

ØRJAN BIRKELAND
Statoil, Stavanger

JAKOB SKOGSEID
University of Oslo

2000

Published by

The Geological Society

London

THE GEOLOGICAL SOCIETY

The Geological Society of London was founded in 1807 and is the oldest geological society in the world. It received its Royal Charter in 1825 for the purpose of 'investigating the mineral structure of the Earth' and is now Britain's national society for geology.

Both a learned society and a professional body, the Geological Society is recognized by the Department of Trade and Industry (DTI) as the chartering authority for geoscience, able to award Chartered Geologist status upon appropriately qualified Fellows. The Society has a membership of 8600, of whom about 1500 live outside the UK.

Fellowship of the Society is open to those holding a recognized honours degree in geology or cognate subject and who have at least two years' relevant postgraduate experience, or who have not less than six years' relevant experience in geology or a cognate subject. A Fellow with a minimum of five years' relevant postgraduate experience in the practice of geology may apply for chartered status. Successful applicants are entitled to use the designatory postnominal CGeol (Chartered Geologist). Fellows of the Society may use the letters FGS. Other grades of membership are available to members not yet qualifying for Fellowship.

The Society has its own publishing house based in Bath, UK. It produces the Society's international journals, books and maps, and is the European distributor for publications of the American Association of Petroleum Geologists, (AAPG), the Society for Sedimentary Geology (SEPM) and the Geological Society of America (GSA). Members of the Society can buy books at considerable discounts. The publishing House has an online bookshop (*http://bookshop.geolsoc.org.uk*).

Further information on Society membership may be obtained from the Membership Services Manager, The Geological Society, Burlington House, Piccadilly, London W1V 0JU, UK. (Email: *enquiries@geolsoc.org.uk*: tel: +44 (0)171 434 9944).

The Society's Web Site can be found at *http://www.geolsoc.org.uk/*. The Society is a Registered Charity, number 210161.

Published by The Geological Society from:
The Geological Society Publishing House
Unit 7, Brassmill Enterprise Centre
Brassmill Lane
Bath BA1 3JN, UK

(*Orders*: Tel. +44 (0)1225 445046
Fax +44 (0)1225 442836)
Online bookshop: *http://bookshop.geolsoc.org.uk*

First published 2000

British Library Cataloguing in Publication Data
A catalogue record for this book is available from the British Library.

ISBN 1-86239-056-8
ISSN 0305-8719

Typeset by Aarontype Ltd, Bristol, UK

Printed by The Alden Press, Oxford

Distributors

USA
AAPG Bookstore
PO Box 979
Tulsa
OK 74101-0979
USA
Orders: Tel. +1 918 584-2555
Fax +1 918 560-2652
Email: *bookstore@aapg.org*

Australia
Australian Mineral Foundation Bookshop
63 Conyngham Street
Glenside
South Australia 5065
Australia
Orders: Tel. +61 88 379-0444
Fax +61 88 379-4634
Email: *bookshop@amf.com.au*

India
Affiliated East-West Press PVT Ltd
G-1/16 Ansari Road, Daryaganj,
New Delhi 110 002
India
Orders: Tel. +91 11 327-9113
Fax +91 11 326-0538

Japan
Kanda Book Trading Co.
Cityhouse Tama 204
Tsurumaki 1-3-10
Tama-shi
Tokyo 206-0034
Japan
Orders: Tel. +81 (0)423 57-7650
Fax +81 (0)423 57-7651

Contents

Present stress

Integrated Basin Studies – Dynamics of the Norwegian Margin: an introduction

ARVID NØTTVEDT

Norsk Hydro Research Centre, N-5020 Bergen, Norway
Present address: Norsk Hydro Canada, 111–5th Avenue SW, Calgary AB, T2P 3Y6, Canada

Several lithospheric and upper crustal processes interact in the formation of rift basins and evolution of such basins into passive margins. Similarly, the fillings of rift basins depend on a variety of tectonic, morphological and sedimentary processes. Areas that have passed through a complete evolution from intra-cratonic rifting through breakup and passive margin formation offer particular opportunities to study and link the different processes of basin formation and filling. One such area, that has a unique combination of a well-preserved rock record and abundant data, is the Norwegian North Sea–North Atlantic margin, herein referred to as the Norwegian margin.

During post-Caledonian times the North Sea and mid-Norwegian margin underwent several episodes of lithospheric extension (multi-phase rifting), of which the latest led to crustal break-up and accretion of oceanic crust between Norway and Greenland near the Paleocene–Eocene transition. Prominent pre-breakup extensional episodes, in late Permian–Triassic, late Jurassic, early Cretaceous and mid-Cretaceous time, led to the development of the North Sea rift system, the large Cretaceous Vøring and Møre sedimentary basins off Norway, and conjugate equivalents off eastern Greenland. Latest Cretaceous–Paleocene rifting and breakup were accompanied by large-scale igneous activity, developing the present conjugate volcanic margins of the North Atlantic.

The North Sea, in particular, is covered with an exceptionally good geological and geophysical industry and academic database, comprising both high-quality geophysical profiles and a large number of industrial wells. About 30 years of active exploration in the North Sea has led to an advanced level of understanding of the geological evolution and complexity of the basin. Numerous papers have been published on the formation and filling of the North Sea intra-cratonic rift structure.

The mid-Norwegian margin has a less extensive, but increasing, industry database. Nevertheless, the Vøring margin is among the best studied volcanic margins globally due to a regional coverage of multichannel seismic lines, expanded spread profiles and other geophysical and geological data. In addition, the successful scientific drilling through a sequence of seaward dipping reflectors on the Vøring Plateau has greatly contributed to the understanding of the margin.

This setting makes the study region an excellent laboratory in which to study progressive rift evolution and its inter-relationship with basin formation and filling, including the interplays between structural evolution, erosion, sedimentation and magmatism, on both regional and local scales. Consequently, the region allows research into fundamental earth processes, which also have direct implications for hydrocarbon exploration and assessment.

The Integrated Basin Studies (IBS) project

The papers and research results presented here have been prepared as part of the Integrated Basin Studies project (Fig. 1).This project was funded under the DG IIX, JOULE II programme, with the objective to study the lithospheric and upper crustal processes governing the formation and evolution of extensional and foreland basins and to decipher the role of tectonics, sea-level and sedimentary processes in the filling of such basins. As part of this task, the project also aimed at studying the physical laws of compaction in fine-grained sediments. Based on these results, a new generation of descriptive as well as numerical basin formation and basin fill models has been derived.

Results from the Integrated Basin Studies project have been reported in an extensive Final Report to the DG IIX, and selected papers prepared within Modules 1 and 2 have been published in two previous Geological Society Special Publications, nos 134 and 156.

From: NØTTVEDT, A. *et al.* (eds) *Dynamics of the Norwegian Margin*. Geological Society, London, Special Publications, **167**, 1–14. 1-86239-056-8/00/$15.00

This book has been prepared with the intention to give a representative picture of the scientific span and results of the project Module 3: Dynamics of the Norwegian Margin. It includes papers prepared by PhD students funded directly through the project, as well as papers written by researchers in academia and industry that have been working in, or closely associated with, the project.

IBS Module 3: Dynamics of the Norwegian Margin (IBS-DNM)

The Integrated Basin Studies project Module 3, Dynamics of the Norwegian Margin (Fig. 2), was technically part of the IBS project, but received funding directly from the Research Council of Norway (RCN). The module was coordinated by Norsk Hydro, and contractual arrangements were made between Norsk Hydro and EU, Norsk Hydro and RCN, and Norsk Hydro and partners. The module was organized into three themes and subordinate topics (Fig. 3).

Research objectives

The scope of the IBS-DNM project module was to analyse and model the dynamics of the basins off mid-Norway and in the northern North Sea, in order to establish a better understanding of the processes controlling basin formation and filling and to develop new models for multiphase, intraplate rifting and volcanic margin formation.

Thematically, the project module focused on three related themes: (i) Intra-plate rifting and basin formation, (ii) Basin infill and (iii) Conjugate volcanic margins. The data coverage and geological setting resulted in a topical focus on two geological provinces, the northern North Sea and the Norwegian-Greenland Sea rifted volcanic margins. The Møre Basin, on the volcanic Møre margin, links these two provinces. *Geographically*, the two former research themes (Intra-plate rifting and basin formation and Basin infill) focused on the northern North Sea and the Møre Basin, while the latter (Conjugate volcanic margins) primarily dealt with the Møre-Vøring margins and their conjugates.

Of particular interest in the intra-plate setting is the kinematics and relative importance of deeper versus shallow crustal processes and how these control overall subsidence patterns; the erosional and provenance history of sediment supply areas; and the architecture, composition and history of basin fill. The work concentrated on syn-rift basin fill, but included also post-rift strata. It has been attempted to establish general and coupled models for rift basin formation and sediment filling in the northern North Sea.

The Norwegian-Greenland Sea represents complete extension and thinning of the lithosphere. Therefore, the margin basins bear an imprint not only of the same events as the intra-plate basins, but have also undergone a structural, magmatic and depositional history reflecting the formation and subsequent

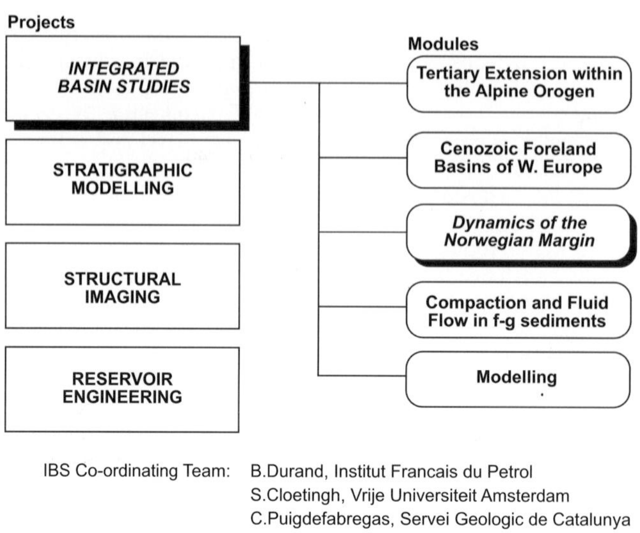

Fig. 1. DG IIX: Joule II Geoscience II Programme

development of the present volcanic passive margin. To understand the basin evolution and tectono-magmatic history it is necessary to assess the timing and magnitude of the different tectonic, magmatic and subsidence episodes for the various basins. In order to determine the style of deformation for each tectonic episode and to quantify the magnitude of crustal and lithospheric extension, restoration of the structurally defined extensional deformation through time was particularly emphasized.

Database and methods

The project database (Fig. 4) consisted of scientific deep reflection and refraction seismic data, large volumes of industry regional 2D reflection seismic data, 3D seismic data, well data, potential field data, industry special studies and reports, in addition to published literature.

It is important, therefore, that the project had access to much of the existing seismic, structural and stratigraphic framework in the study region to define and select the optimal target areas for the detailed studies described in the various project topics. The industrial partners were instrumental in achieving this objective.

The work on crustal structure and basin formation concentrated around seven transects from the northern North Sea to the Vøring margin (Fig. 4), whereas the topic on stress field covers the entire Norwegian continental shelf. The syn-rift infill studies focused on selected fault-blocks in the northern North Sea, whereas the post-rift (Cenozoic) infill studies integrated data from the Danish and Norwegian sectors of the North Sea, up to 62°N. The work on erosion and provenance included the northern North Sea. The studies on volcanic margins concentrated on the mid-Norwegian margin, but comparative studies were made with other volcanic margins around the world.

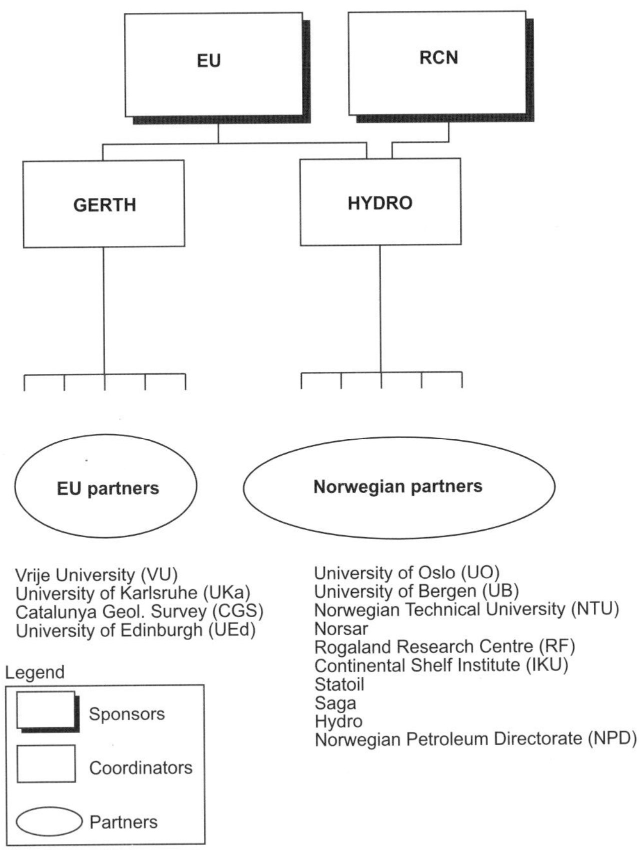

Fig. 2. IBS-DNM contractual framework and involved partners. RCN, Research Council of Norway.

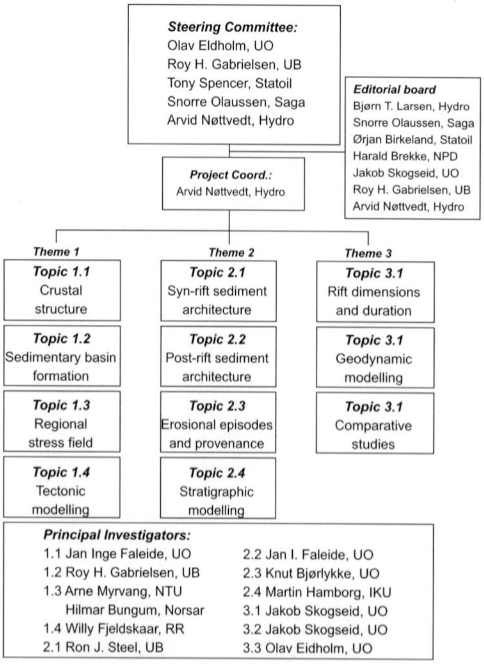

Fig. 3. IBS-DNM project structure and organization. Past members of the Steering Committee include Alv Orheim (Statoil), Jan Volset (Statoil) and Ron J. Steel (UB).

Theme 1: Intra-plate Rifting and Basin Formation

Topics 1.1: Crustal structure, 1.2: Sedimentary basin formation and 1.4: Tectonic modelling

The dynamic framework for Mesozoic and Cenozoic basin development in the northern North Sea and on the mid-Norwegian margin has been well documented over the last decade. However, there is major uncertainty as to the nature of the deep crustal structure and its relation to the sedimentary cover. In addition, the detailed nature, significance and areal extent of the faulting and differential subsidence in the post-rift intervals are poorly known.

For a complete understanding of evolution of the northern North Sea and Møre Basin it is essential to consider the earlier Permo-Triassic rift, which generated the structural framework on which the later Jurassic and Cretaceous rifts acted (Fig. 1.3). Compaction of Permo-Triassic basin fill and residual Permo-Triassic thermal anomalies may enhance Cretaceous–Cenozoic subsidence and may therefore contribute significantly to the stretching factor as calculated from subsidence analysis of later rift episodes. A major question is the relative contribution to final extension contributed by each of the main episodes of extension.

Key results. The deep seismic reflection survey NSDP84 has formed the basis for numerous papers on the crustal structure and basin evolution of the northern North Sea rift system. However integration with all available geophysical and geological data from both offshore and onshore areas provided some interesting new results.

Reprocessing of the deep reflection seismic data resulted in enhanced data quality. By integrating seismic refraction data, ESP data and gravity/magnetic data, and combining these transects with high-quality commercial reflection seismics, it has been possible to map out the structural outline of the graben geometry of the northern North Sea (**Christiansson *et al.*, Odinsen *et al.* (a)**), including linkage between upper and mid crustal faults, fault geometries of the deep crustal levels and structure of the lowermost crust. Of particular interest is the improved identification of top basement, the confirmation of intra upper mantle dipping reflectors and identification of a high-velocity lower crustal rock body.

New information on the levels of detachment and general geometries of the different fault systems has also been gained (**Fossen *et al.***). These investigations reveal a more complex interaction between fault systems than anticipated earlier. The kinematics of the faulting was studied by the use of analogue models (Fossen & Gabrielsen 1995). In addition, the confirmation of the existence of 'graben units,' which are characterized by shifting polarities along the basin axis of the Permo-Triassic basin axis, is important (Færseth *et al.* 1995*a*).

Forward/backward numerical modelling of crustal geometry, palinspastic graben topography and thermal development have resulted in better insight in the formation of the North Sea rift (**Odinsen *et al.* (b), ter Voorde *et al.***). These investigations conclude that the Permo-Triassic stretching event has been underestimated in many previous works in the area. The Permo-Triassic event is responsible for a more extensive extension than the Late Jurassic event and the former event also resulted in extension of a wider area than did the latter.

The results from the structural mapping of the Møre Basin particularly emphasize the complexity of the basin, where large basement-involved, rotated fault-blocks have contributed to the internal compartmentalization of the basin. As for the North Sea, the Møre Basin was subject to multi-phase extension (Grunnaleite & Gabrielsen 1995). Lithospheric stretching probably commenced in the late Permian-early Triassic, followed by a second episode in the late Jurassic. In contrast to the North Sea, however, early-mid-Cretaceous rifting and successive Palaeogene breakup strongly influenced further development of the basin, resulting in a cumulative beta-factor in excess of 3. Regional-scale antiforms along the northeastern basin margin, basin inversion in the Slørebotn Sub-basin and the northeastern Møre Basin, compressional reactivation of faults as seen on the near base Cretaceous level, and reverse drag and folding associated with faults of primarily extensional origin, suggest that pronounced multi-phase late Mesozoic–Cenozoic inversion has taken place (**Mogensen *et al.***, Gabrielsen *et al.* 1999).

Relevance to the petroleum industry. The improved models for basin development and fault geometries obtained within this topic provide new concepts to the industry that are relevant for the understanding of sand distribution in general, and in search for the subtle trap in particular. It is also expected that the results will be of importance in basin modelling, both on the regional (crustal) scale, and for maturation and migration studies on the sub-regional scale. In addition, the focus on the pre-Jurassic basin development has opened new perspectives that may be of importance for future exploration, particularly in the deeper parts of the North Sea Basin.

Topic 1.3: Present regional stress field

Some rift episodes in the northern North Sea–Møre Basin–Vøring Basin correlate with breakup and initial sea floor spreading in other basins within the North Atlantic rift system and the inter-relationships between pre-rift structures (structural fabric) and the regional stress field probably played an important role in structural evolution, by rejuvenation of inherited structures.

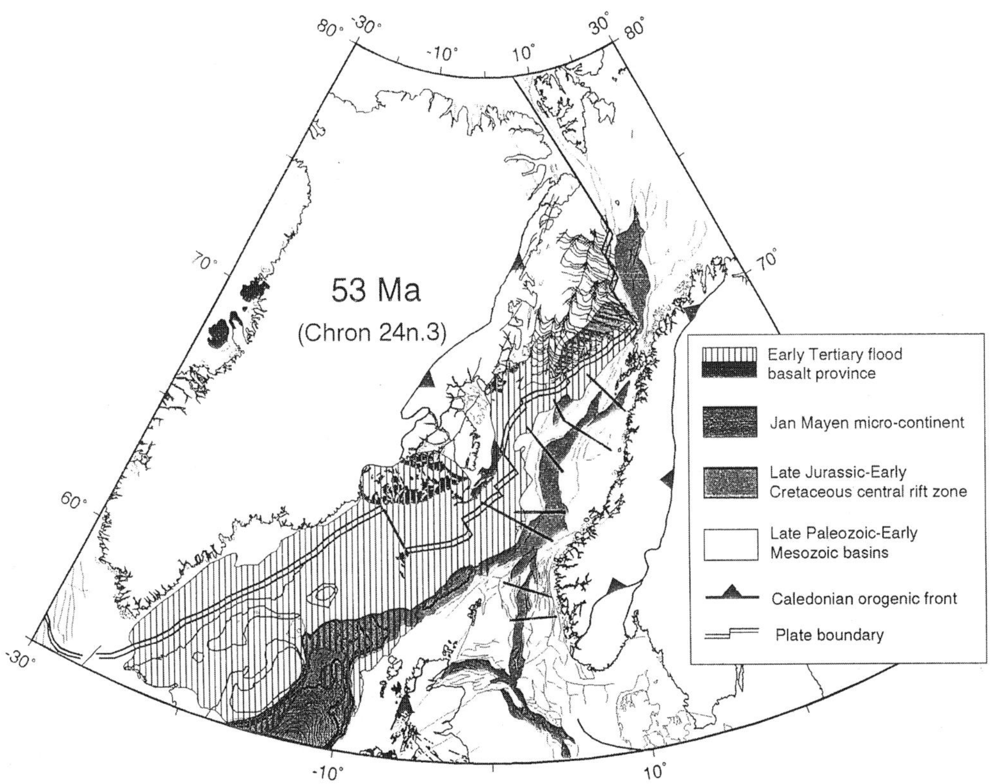

Fig. 4. Schematic outline of the IBS-DNM study area and database (after Skogseid *et al.*, this volume).

The orientation and magnitude of the present stress field are reflected in both well data and earthquake focal mechanism solutions. The use of both *in situ* measurements and earthquake based data enables mapping of crustal stress as a function of depth. Most important are strain markers for individual time intervals.

Regional and local variations in stress directions and magnitude yield information on the intra-plate forces, plate tectonics and recent geological history. Thus, it provides important input to the tectonic modelling. Coupling of non-linear lithosphere rheology to a model involving stress changes acting on a basin offers the prospect of understanding non-thermal subsidence and of solving discrepancies in present crustal extension estimates.

Key results. A database of unprecedented quality and quantity for the region has been established, comprising stress information from a variety of sources. The project has contributed to the compilation of an earthquake focal mechanism database that is constantly being expanded, and which currently comprises 109 solutions for northwestern Fennoscandia and Svalbard (**Lindholm *et al.***, Lindholm *et al.* 1995). In addition, a total of 345 borehole breakout observations and 104 overcoring measurements have been collected and quality-assessed into a high-quality database (**Fejerskov & Lindholm (a)**, Borgerud & Suave 1995, Fejerskov *et al.* 1995, Gølke *et al.* 1995).

All these data were used in a joint interpretation of the crustal stress field in Norway and adjacent offshore regions (**Fejerskov & Lindholm (b)**). Regionally, the analysis has revealed a crustal horizontal compressive stress with a dominating WNW–ESE direction(s_{Hmax}) in southern Norway, which gradually rotates into N–S compression in the Barents Sea region. The observed stress is largely consistent throughout the upper crust, indicating that large-scale tectonic mechanisms are the main source. This can, with a high degree of confidence, be attributed to crustal spreading along the mid-Atlantic Ridge. The consistent stress pattern shows regional variations probably caused by regional stress generating mechanisms like the continental margin–ridge push and sediment loading. Faults that perturb and deviate the regional stress locally also have been observed.

With important exceptions in local areas, the dominant type of faulting is found to be contractional and, to some extent, strike-slip, indicating that the continental margin is subject to compression, as also indicated by *in situ* measurements onshore. Compressive differential horizontal stresses are found in all parts of the margin analysed, and the data indicate that the stress regime in the upper crust often alters with depth. A stress homogeneity throughout the brittle crust is, however, well documented by the homogeneous stress directions obtained from complementary data (deep earthquakes and shallow boreholes).

Relevance to the petroleum industry. On a short-term basis, this topic has provided important information to the general understanding of the stress field along the Norwegian Margin, which is of considerable importance when entering into new areas like the Vøring Basin. In addition, detailed information on *in situ* stresses in several petroleum fields is expected to be directly relevant to activities in those fields. *In situ* rock stress data are particular important for the planning and drilling of stable and safe wells. The state of stress in and around the hydrocarbon reservoir is also important for field development and reservoir management, and the success of hydraulic fracturing and injection depends on knowing how fractures and fluid fronts propagate under different stress systems. The use of such data may on a long-term basis reduce costs and help optimize drilling and production.

Theme 2: Basin Infill

Topic 2.1: Syn-rift sediment architecture

The stratigraphy and architecture of the Triassic–lower Jurassic and Cretaceous–Cenozoic post-rift basinal infill of the North Sea are relatively well documented, whereas the Permo-Triassic and upper Jurassic syn-rift infill are known only in broadest outline. Details of timing, style and rates of the events that make up the syn-rift episodes, and of the sedimentary response to these events, have only recently started to emerge. Extension across the northern Viking Graben is known to vary between the Tampen Spur, Viking Graben proper, Horda Platform and Sogn Graben segments. This variability in space, together with temporal variations (Permo-Triassic, Late Jurassic) has considerable consequences for the composition of syn-rift infill. These lateral changes and related provenance and sediment transport issues have been addressed by analysis of different patterns of syn-rift stacking and by studies of the relation between rates and style of structural events.

Key results. A database of seismic examples as well as field analogues has been built to illustrate the variability of syn-rift architecture, involving integration of data on rift topography, erosional and drainage patterns, relative sea level and its changing position, sediment transport processes and facies and resultant sand body geometry and distribution (Marjanac 1994). In several such examples, the spatial and temporal evolution of reservoir sand facies has been semi-quantitatively linked to the structural evolution of the parent half-graben and neighbouring footwall sediment source areas (Farseth *et al.* 1995*b*; Ravnås & Bondevik 1997; **Nøttvedt *et al.***).

A synthesis has been compiled of various aspects of the three-dimensional geometry of syn-rift stratigraphic architectural elements and their stacking pattern, and the spatial organization of potential facies tracts and stratigraphic surfaces present in marine syn-rift basin fills (Gabrielsen *et al.* 1995; Nøttvedt *et al.* 1995; Ravnås *et al.* 1997*a*; Ravnås & Steel 1998). This includes an analysis and comparison of a number of half-graben sub-basins in the northern North Sea (Færseth & Ravnås 1998; **Ravnås *et al.***).

The studies have led to a classification of marine rift-basins and syn-rift successions in terms of sediment supply into overfilled, sediment-balanced, sediment-underfilled and sediment-starved basins (Ravnås & Steel 1997). Dependent on whether the rift-basin was overfilled/sediment-balanced, sediment-underfilled or sediment-starved, a three-fold, sand–clay–sand package, two-fold sand–clay package or one-fold mud-prone package constitutes the syn-rift succession, respectively. In cases where the rift episode was characterized by repetitive rift phases, results show that the successive rotational tilt events are often separated by periods characterized by less intense faulting, the so-called intra-rift quiescence or relaxation stages.

The project also included a study of the syn-rift infill of the Lusitanian Basin in Portugal (Ravnås *et al.* 1997*b*).

Relevance to the petroleum industry. The results and interpretations obtained within this topic provide new insight and ideas for analysis of rift-basin development and syn-rift infill. The recognition of the different types of syn-rift sedimentary architectures provides a step forward in erecting predictive models for the analysis of half-graben syn-rift sediment fill. In addition, integrated detailed analysis of rift basin structural evolution, high-resolution biostratigraphic zonation, reworked biozonation, basin-fill facies and architecture and seismic facies offer a sensitive tool for analysing the detailed sequential basin development and predicting reservoir sand distribution. The models erected may serve as important references to the conceptual training of explorationists.

Topics 2.2: Post-rift sediment architecture and 2.3: Erosional episodes and provenance area

Post-rift intervals commonly have a duration of 50–100 Ma. Thermal relaxation following the late Jurassic rift event had almost ceased and the North Sea basin was in a state close to thermal equilibrium during the transition to the Cenozoic. Subsidence and deposition during the Cenozoic, therefore, were controlled by other factors. The Cenozoic succession in the Norwegian North Sea can be divided into genetic sequences of 8–16 Ma duration, representing major periods of clastic wedge progradation and retreat with respect to the Norwegian hinterlands. The clastic wedges are bounded by major flooding surfaces and reflect variable subsidence rates. The character of these variations and of the resulting genetic sequences have been investigated and related to the uplift and erosion of the surrounding land areas.

The sand/clay ratio in sedimentary basins is commonly viewed as a function of depositional energy. A factor that tends to be overlooked is the primary provenance area composition. A better understanding of the processes that cause changes in the composition of the sedimentary sequence in the North Sea Basin therefore is needed. The relative contributions of the Norwegian mainland and East Shetland Platform as source areas has been evaluated, based on analysis and description of basin fill composition and characteristics, mapping of palaeodrainage directions and transport routes, estimations of palaeo-provenance areas and source rock lithologies, and correlation of basin fill to provenance area.

Key results. The project has integrated seismic data from the Danish sector with the Norwegian North Sea (Jordt *et al.* 1995). It also includes mineralogical and geochemical analyses of 1600 samples of cuttings and cores from about 40 wells in the Norwegian sector of the North Sea, and thus represents the most comprehensive database currently available. These data are important in terms of understanding the provenance and provide a new basis on which to interpret directions of sediment transport and areas of tectonic uplift.

The seismostratigraphic study shows that changes in seismic sequence geometry occurred in-phase with intra-continental stress variations on the European Platform and that sediment supply and differential tectonic movements in the basin and in the provenance areas controlled Cenozoic deposition in the central and northern North Sea (**Jordt *et al.***). Seismic sequences were generated largely independent of marked global glacio-eustatic sea level falls. Generation of seismic onlap appears to have been controlled by sediment supply and basin floor topography. Thinning of sedimentary strata towards inclined surfaces results in marked seismic onlap and apparent unconformable relationships.

It is further shown that seismic velocities in the North Sea Cenozoic mudstones vary systematically as a function of mineralogical composition. A commonly observed velocity inversion at the base of the Pliocene and Pleistocene sequence (Reemst *et al.* 1996) has been found to relate to a low content of smectite and therefore much faster compaction of these sediments compared to the underlying smectite-rich mudstones (**Thyberg *et al.***). Silica cementation from biogenic silica, as well as alteration of opal-A to opal-CT, have been shown to cause abrupt increases in seismic velocities.

It has been demonstrated that pore pressure in the Tertiary succession is governed by mineralogical composition. Thick sequences of mudstones with high smectite content, which typically have high specific surface and low permeability, typically give rise to overpressure, providing there are no sandy beds causing lateral drainage. Moreover, because of their high water content smectite rich mudstones usually have lower density than the overlying mudstones and sandstones. This density inversion frequently caused diapirism.

Finally, the project has provided results contributing to the understanding of the Cenozoic uplift of Norway (Fjeldskaar 1994; Stuevold & Eldholm 1996).

Relevance to the petroleum industry. This topic has contributed to the understanding of the Cenozoic basin formation and filling in relation to the uplift and erosion history of Southern Norway. This uplift may strongly influence secondary and tertiary migration and trapping of petroleum, particularly in areas close to the coast of Norway.

The present project is one of the first in the North Sea where seismic stratigraphy has been related to mineralogical composition in a systematic way. Understanding the regional variations in the velocity/depth function and the processes responsible for these trends is important when depth converting seismic profiles. The data on mineralogy and diagenesis is also of value to calibration of seismic response to lithology, in order to understand better the information that can be extracted from seismic attributes.

The demonstrated relationship between mineralogical composition, diagenesis and overpressure is expected to be of great interest to drilling, particularly horizontal drilling, and for handling of rock mechanical problems during production. Moreover, the present study has shown that the North Sea Cenozoic rock properties cannot be realistically represented as a simple function (i.e. linear or exponential) of burial depth. This has important consequences to basin modelling, as porosity reduction (compaction) is usually assumed to be a function of overburden or effective stress.

Topic 2.4: Stratigraphic modelling

Numerical modelling is becoming increasingly important in the understanding of sedimentary basin deposition and filling. In particular, it helps evaluate the complex interplay between tectonics, eustacy, climate, erosion and sediment transport, and better constrain the boundary conditions of geological interpretations and models.

The overall sedimentary architecture of the upper Jurassic syn-rift infill in the Oseberg area as well as the Cenozoic infill along one of the North Sea regional profiles have been modelled, as a function of sediment input, subsidence and sea level fluctuations to obtain additional information about their interplay and relative importance. The modelling has used forward process-based simulation programs of dynamic-slope type. Comparisons have been made between observed stratigraphic architecture and synthetic stratigraphic models where the most important controlling factors have been considered.

Key results. The numerical modelling of the syn-rift infill of the Oseberg area demonstrate the interaction between structural evolution, sedimentation and resulting stratigraphic pattern across the rotating fault-blocks (terVoorde *et al.* 1997). The modelling further confirms that the sedimentary architecture of the North Sea Cenozoic infill cannot be explained by sedimentary processes and/or eustacy alone (**Kyrkjebø *et al.***). Periods of anomalous subsidence, deviating from the post-rift thermal subsidence, are required in the late Paleocene and late Miocene. However, some of the Paleocene subsidence may be related to the initial basin form e.g excess water depth after late Cretaceous. The Miocene event, on the other hand, includes basinal subsidence in the northern North Sea, as well as source area uplift in the Norwegian mainland, and is believed to represent an intraplate effect coupled to the opening of the North Atlantic.

Relevance to the petroleum industry. The iterative process between interpretation and modelling allows the geologist to better constrain the possible geological models and to narrow in on a less number of likely interpretations. This type of modelling is also an excellent tool to visualize the complex interaction between sediment supply, sea-level and tectonics in filling of sedimentary basins.

Theme 3: Conjugate Volcanic Margins

Topics 3.1: Rift dimensions and duration of rifting, 3.2: Geodynamic modelling and 3.3: Comparative studies

It has long been recognized that the formation of the northern North Atlantic continental margins was accompanied by excess magmatic activity. This caused a sub-division of rifted margins into volcanic and non-volcanic types. Until recently, the volcanic margins were thought to be an exceptional case. Recent comparative studies involving other rifted margins now suggest that excess magmatism during breakup is far more common than previously thought. In fact, the volcanic margin might represent the normal evolutionary case rather than being anomalous.

The volcanic signature of the North Atlantic margin has been explored by seismic reflection data and scientific drilling, whereas the tectonic signature, and thus development, have to a large extent been hidden below the volcanic rocks.

The project has involved geophysical–geological mapping of the Vøring, Møre and conjugate margins, in order to obtain key dimensions and timing of the late Cretaceous–Paleocene rift episode, by focusing on: (1) width, style and timing of syn-rift lithospheric extension; (2) extent and timing of syn-rift regional uplift; (3) extent, character, time of emplacement and dimensions of igneous units (extrusives, intrusives, lower crustal high-velocity bodies); (4) location of continent-ocean transition; (5) extent of post-rift thermal margin subsidence. The availability of data also off Greenland offers a possibility to map and model the entire rift and to compare the late Cretaceous–Paleocene rift dimensions with the previous rift episodes.

In addition, comparative studies of the volcanic margin formation in the North Atlantic, as well as globally, have been undertaken, with the objective to improve our understanding of tectono-magmatic volcanic margin setting and the processes governing volcanic margin initiation and development. It has also helped in providing a framework for analysis of the implications of volcanic margins for erosion and sedimentation on local and regional scales, and for the environment (palaeoceanography, palaeoclimate) on local, regional and global scales.

Key results. A number of crustal transects has been constructed across the North Atlantic margin by use of deep seismic data across the margin. It is estimated that the conjugate margins experienced some 140 km of crustal stretching during the Maastrichtian–Paleocene rifting and breakup, and about 50–70 km of lateral displacement during Late Jurassic–Cretaceous rifting (Skogseid 1994; Skogseid & Eldholm 1995; **Skogseid *et al.***). The Rockall Trough has, however, experienced far more stretching, which is interpreted to be related to separate rifting in the mid-Cretaceous. By using the stretching estimates, palinspastic maps have been constructed based on restorations to 53 Ma pre-drift, 75 Ma pre-Cenozoic breakup and 170 Ma Late Jurassic pre-rift plate configurations. The results also include an evaluation of the tectono-magmatic events associated with plume–lithosphere interaction during rifting, with particular focus on relative vertical movements and provenance development.

The tectonic development of the Vøring and Møre basins is controlled by two structural trends, NE–SW and NW–SE. It is suggested that the NE–SW trend was established in the Paleozoic and was active during all subsequent tectonic phases, whereas the NW–SE trend, probably reflecting the old Precambrian grain of the basement, controlled the tectonic activity throughout the Cretaceous and Cenozoic (**Brekke**). The results show that during the Cretaceous and Cenozoic the Vøring Basin was tectonically active, with repeated phases of normal faulting and contraction causing large-scale folding. The Møre Basin, particularly towards the south, was overall more tectonically quiet and experienced mainly continuous subsidence.

Comparative studies between the NE Atlantic, the Namibian, the US Atlantic and the NW Australian margins show that all margins seem to have had a protracted rift development prior to continental separation, that significant crustal thinning is observed adjacent to the continent–ocean boundary, and that the main pulse of igneous activity coincides with the time of breakup (Eldholm *et al.* 1994; **Eldholm *et al.***; Planke & Eldholm 1994). On the other hand, it is also recognized that excess volcanism may occur without a direct link to an active mantle plume, which may explain the large variety in tectono-magmatic development of volcanic margins worldwide.

An extensive Norwegian–Greenland Sea thermal field data base named HEAT has been compiled, containing all public domain data in the region; i.e. 436 heat flow values (**Sundvor *et al.***). The heat flow on oceanic crust reveals a clear, first-order heat flow–crustal age relationship, whereas continental slope maxima on the Møre and Barents Sea margins contrast greatly with the typical low heat flow of old oceanic and thinned continental crust.

Relevance to the petroleum industry. The understanding of basin geometries, rift dimensions and subsidence history is important to exploration companies working offshore mid-Norway. New information on magnitude and temporal development of intra-basinal vertical movements in combination with palinspastic reconstruction provides a tool for predicting reservoir facies in the outer margin basins. Such vertical movements, including the formation of a land bridge from the Charlie Gibbs Fracture Zone to the SW Barents Sea margin, may have led to the establishment of large drainage systems and the probability of prospective well-sorted sediment sequences of generally Late Cretaceous Paleocene age in the adjacent subsiding parts of the basins.

In addition, the geometrical definition of units of igneous materials at crustal levels has relevance for calculation of heatflow history, as bodies of underplated materials at the base of the crust can spike heatflow and may have caused maturation of organic matter that cannot be predicted by present heatflow measurements. The compilation of the heat flow database is an important tool in this respect.

The comparative work between volcanic margins worldwide is important for the understanding of basic geodynamic processes and tectono-magmatic development in general.

Concluding remarks

The results of the Integrated Basin Studies–Dynamics of the Norwegian Margin project emphasize the value of academia and industry working together in order to make significant scientific progress. It shows how public scientific grants and industry funding can be focused to leverage investments put into research. It also presents a model for how academia and industry research can be effectively organized into a single project.

The achievements of the Dynamics of the Norwegian Margin project have brought our understanding of this region forward, and it is our hope that the results presented in this book will serve as an important reference for the area in the years to come. The project has also established a database that will serve as a good basis for continued research on the Norwegian margin and several new research projects have already been initiated that build on the results of this project. In addition, it has highlighted some new avenues of research that may further increase our understanding of this margin and of multiphase rift evolution in general.

Project staff and participants

Direct project participation

Principal Investigators (PI), Doctorate Students (PhD), Participating Scientists (PS), Master Students (MS) and Research Associates (RA)

Theme 1: Intra-plate Rifting and Basin Formation

Topics 1.1: Crustal structure, 1.2: Sedimentary basin formation, 1.4: Tectoniic modelling

PI:	Prof. J. I. Faleide (UO), Prof. R. H. Gabrielsen (UB), Dr. W. Fjeldskaar (RF)
PhD:	M.sci. P. Christiansson (UO), C.sci. T. Odinsen (UB), C.sci. I. Grunnaleite (UB)
RA:	C.sci. K. Lokna (UB)
Industry:	Dr. A. M. Berge (Hydro), C.real. B. T. Larsen (Hydro), Prof. R.B. Færseth (Hydro), Dr. A. Nøttvedt (Hydro), Dr. H. Fossen (Statoil)
EU:	Dr. P. Reemst (VU), Dr. P. van der Beek (VU), M.sci. M. ter Voorde (VU), Prof. S. Cloetingh (VU)

Topic 1.3: Present regional stress field

PI:	Prof. A. Myrvang (NTH), Prof. H. Bungum (Norsar)

PhD:	Siv.ing. M. Fejerskov (NTH)
PS:	Dr. C. Lindholm (Norsar)
MS:	T. Jørgensen (NTH), L. Borgerud (NTH), E. Svare (NTH), M. Villgran (Norsar/UB), E. Hicks (Norsar/UO)
Industry:	Dr. T. H. Hanssen (Hydro), C.real. B. T. Larsen (Hydro), C.real. R. K. Bratli (Saga), Dr. L. N. Jensen (Statoil)
EU:	Dr. M. Golke (UKa)

Theme 2: Basin Infill

Topic 2.1: Syn-rift sediment architecture

PI:	Prof. R. J. Steel (UB), Dr. J. Underhill (UEd)
PhD:	C.sci. R. Ravnås (UB)
PS:	Dr. P. Theriault (UB/Statoil), D. Mellere (UB/Statoil)
Industry:	C.sci. K. Bondevik (Hydro), Prof. R. B. Færseth (Hydro), Dr. A. Nøttvedt (Hydro), J. Windelstad (Statoil)

Topic 2.2: Post-rift sediment architecture, 2.3: Erosional episodes and provenance area

PI:	Prof. J. I. Faleide (UO), Prof. K. Bjørlykke (UO)
PhD:	M.sci. H. Jordt (UO), C.sci. B.I. Thyberg (UO)
RA:	
Industry:	Dr. P. van Veen (Hydro), Dr. L. J. Skjold (Hydro), Dr. A. Ryseth (Hydro), Dr. M. Ramm (Hydro) Dr. A. N. Nøttvedt (Hydro)

Topics 2.4: Stratigraphic modelling

PI:	C.real. M. Hamborg (IKU)
MS:	R. Kyrkjebø (NTH)
PhD:	C.sci. R. Ravnås (UB)
Industry:	C.sci. K. Bondevik (Hydro)
EU:	M.sci. M. ter Voorde (VU), Prof. S. Cloetingh (VU)

Theme 3: Conjugate Volcanic Margins

Topics 3.1; Rift dimensions and duration, 3.2: Geodynamic modelling, 3.3: Comparative studies

PI:	Dr. J. Skogseid (UO), Prof. O. Eldholm (UO)
MS:	B. Flakstad (UO), U. Byrkjeland (UO), F. Neverdal (UO), S. Ren (UO), E. Alvestad (UO)
PhD:	C.sci. T. Gladszenko (UO)
PS:	Dr. S. Planke (UO), Dr. T. Pedersen (UO), Prof. A. M. Myhre
Industry:	C.real B. T. Larsen (Hydro)

Referees. The following persons kindly served as technical and linguistic referees for the book:

John Akselsen, Arild Andresen, Kuvet Atakan, W. Scott Baldridge, Giovanni Bertotti, Ørjan Birkland, Eric Bogoslowski, Lars Boldreel, Ross Boutilier, Nicolas Chamot-Rooke, Sierd Cloetingh, Denis Couturier, Tom Dreyer, Richard England, Doug Gardner, Rob Gawthorpe, Felix Gradstein, Pål Haremo, Jens Havskov, William Helland-Hansen, Helge Hjelmeland, Chuck Hurich, Erik P. Johannesen, Reidar Kanestrøm, Ridvan Karpuz, Oddbjørn Kløvjan, John Knight, John Korstgård, Yngve Kristoffersen, Axel Makurat, Ole J. Martinsen, Alain Mascle, Jock McCracken, Wojtec Nemec, Johan P. Nystuen, Arvid Nøttvedt, Lars Nørgård-Jensen, Nigel Platt, John Palmer, Sarah Prosser, Garry Quinlan, Phil D. Rice, Jan Rivenæs, Ellen Roaldset, Alan Roberts, Yngve Rundberg, Alf Ryseth, William Sassi, Ken Saunders, Roger Scrutton, Michel Seranne, Morten Sparre Andersen, Gerald Sullivan, Tore Torske, Bjørn Tørudbakken, Jan Volset, Erling Vågnes, John Walsh, Marjorie Wilson, Graham Yielding and four anonymous referees.

Acknowledgements. The project was funded by the European Union and of Norway Research Council under the JOULE II research programme (contract No. JOU2-CT 92–0110). Norsk Hydro, Statoil and Saga provided additional funding. Are B. Carlsson at the Research Council of Norway is gratefully acknowledged for supporting the project. As project leader, I am very grateful to Bjørn T. Larsen for his involvement in the project on an industry client basis and to Vigdis Michelsen for her secretarial efforts. I also owe many thanks to past and recent members of the project Steering Committee, Olav Eldholm, Roy H. Gabrielsen, Snorre Olaussen, Alu Orheim, Tony Spencer, Ron J. Steel and Jan Volset and to my fellow colleagues on the Editorial Board, Bjørn. T. Larsen, Ørjan Birkeland, Harald Brekke, Roy H. Gabrielsen, Snorre Olaussen, Jakob Skogseid and Bjørn Tørudbakken, without whom this book would never have come to light. Some extended thanks also go to the many scientists who reviewed the manuscripts. However, the principal investigators, participating researchers and students, who showed great efforts and a remarkable spirit of co-operation throughout, of course are the key to the success of the project. Finally, I would like to pass a word of appreciation to Bernard Durand, Sierd Cloetingh, Cai Puigdefabregas and all other scientists in the Integrated Basin Studies project, for 3 years of stimulating co-operation.

Research contributions

Publications

BLYSTAD, P., BREKKE, H., FÆRSETH, R. B., LARSEN, B. T., SKOGSEID, J. & TØRUDBAKKEN, B. 1995. Structural elements of the Norwegian continental shelf. Part II: The Norwegian Sea region. Norwegian Petroleum Directorate Bulletin, **8**.

BORGERUD, L. & SVARE, E. 1995. In-situ stress field on the Norwegian margin. *In*: FEJERSKOV, M. & MYRVANG, A. (eds) *Workshop on Rock Stresses in the North Sea*. NTU Trondheim, 13-14.02.95, pp. 165–178.

CLOETINGH, S., SASSI, W. & TASK FORCE TEAM 1994*a*. The origin of sedimentary basins: a status report from the task force of the International Lithosphere Program. *Marine and Petroleum Geology*, **11**, 659–683.

CLOETINGH, S., ELDHOLM, O., LARSEN, B. T., GABRIELSEN, R. H. & SASSI, W. (eds) 1994*b*. Current state and perspectives of models for extensional and inverted basins. *Tectonophysics*, Special Volume, **240**.

ELDHOLM, O. & THOMAS, E. 1993. Scratching the surface: Environmental impact of volcanic margin formation. *Earth and Planetary Science Letters*, **117**, 319–329.

ELDHOLM, O., MYHRE, A. M. & THIEDE, J. 1994. Cenozoic tectono-magmatic events in the North Atlantic: potential paleoenvironmental implications. *In*: BOULTER, M. C. & FISHER, H. C. (eds) *Cenozoic Plants and Climates of the Arctic*. NATO, ASI Series, **127**, Springer, Heidelberg, 35–55.

ELDHOLM, O., SKOGSEID, J., PLANKE, S. & GLADCZENKO, T. P. 1995. Volcanic margin concepts. *In*: BANDA, E., TALWANI, M. & TORNE, M. (eds) *Rifted Ocean. Continent Boundaries*. NATO ASI Series Volume. Kluwer, Dordrecht, 1–16.

FEJERSKOV, M., MYRVANG, A. M., LINDHOLM, C. & BUNGUM, H. 1995. In-situ rock stress pattern on the Norwegian continental shelf and mainland. *In*: FEJERSKOV, M. & MYRVANG, A. (eds) *Workshop on Rock Stresses in the North Sea*. NTU Trondheim, 13-14.02.95, pp. 191–201.

FJELDSKAAR, W. 1994. The amplitude and decay of the glacial forebulge in Fennoscandia. *Norsk Geologisk Tidsskrift*, **74**, 2–8.

FOSSEN, H. & GABRIELSEN, R. H. 1995. Experimental modelling of extensional fault systems. *Journal of Structural Geology*, **18**(5), 673–687.

FROSTICK, L. E. & STEEL, R. J. 1993*a*. Tectonic signatures in sedimentary basin fills: an overview. *International Association of Sedimentologists Special Publication*, **20**, 1–9.

FROSTICK, L. E. & STEEL, R. J. 1993*b*. Sedimentation in divergent plate-margin basins. *International Association of Sedimentologists Special Publication*, **20**, 111–128.

FÆRSETH, R. B. & RAVNÅS, R. 1998. The structural configuration of the Oseberg Fault Block in the context of the northern North Sea structural framework. *Marine and Petroleum Geology*, **15**, 467–490.

FÆRSETH, R. B., GABRIELSEN, R. H. & HURICH, C. A. 1995*a*. The influence of basement in structuring of the North Sea Basin offshore west Norway. *Norsk Geologisk Tidsskrift*, **75**, 2/3, 105–119.

FÆRSETH, R. B., SJØBLOM, T. S., STEEL, R. J., LILJEDAHL, T., SAUAR, B. E. & TJELLAND, T. 1995*b*. Tectonic controls on Bathonian-Volgian syn-rift successions on the Visund Fault Block, northern North Sea. *In*: STEEL, R. J., FELT, V. L., JOHANNESSEN, E. P. & MATHIEU, C. (eds) *Sequence Stratigraphy on the Northwest European Margin*. Norwegian Petroleum Society Special Publication, **5**, 325–346.

GABRIELSEN, R. H. & STRANDENES, S. 1994. Dynamic Basin Development – A complete geoscientific tool for basin analysis. Proceedings World Petroleum Congress 1994, 13–21

GABRIELSEN, R. H., GRUNNALEITE, I. & RASMUSSEN, E. 1997. Cretaceous and Tertiary inversion in the Bjørnøyrenna Fault Complex, south-western Barents Sea. *Marine and Petroleum Geology*, **14**(2), 165–178.

GABRIELSEN, R. H., ODINSEN, T. & GRUNNALEITE, I. 1999. Structuring of the northern Viking Graben and the Møre Basin; the influence of basement structural grain, and the particular role of the Møre–Trøndelag Fault Complex. *Marine and Petroleum Geology*, **16**, 443–465.

GABRIELSEN, R. H., STEEL, R. J. & NØTTVEDT, A. 1995. Subtle traps in extensional terranes: A model with reference to the North Sea. *Petroleum Geoscience*, **1**, 223–235.

GØLKE, M., COBLENTZ, S., CLOETINGH, S. & FEJERSKOV, M. 1995. Stress system of the Norwegian Continental Margin – Part I, In-situ rock stress pattern on the Norwegian continental shalf and mainland. *In*: FEJERSKOV, M. & MYRVANG, A. (eds) *Workshop on Rock Stresses in the North Sea*. NTU Trondheim, 13-14.02.95, pp. 250–274.

GRUNNALEITE, I. & GABRIELSEN, R. H. 1995. The structure of the Møre Basin. *Tectonophysics*, **252**, 221–251.

JORDT, H., FALEIDE, J. I., BJØRLYKKE, K. & IBRAHIM, M. T. 1995. Cenozoic stratigraphy of the central and northern North Sea Basin: tectonic development, sediment distribution and provenance areas. *Marine and Petroleum Geology*, **12**, 845–879.

LINDHOLM, C. D., BUNGUM, H., VILLAGRAN, M. & HICKS, E. 1995. Crustal stress and tectonics in Norwegian regions determined from earthquake focal mechanisms. *In*: FEJERSKOV, M. & MYRVANG, A. (eds) *Workshop on Rock Stresses in the North Sea*. NTU Trondheim, 13-14.02.95, pp. 77–91.

MELLERE, D. & Steel, R. J. 1996. Tidal sedimentation in Inner Hebrides half-grabens, Scotland: the mid-Jurassic Bearreraig Sandstone Formation. *In*: DE BATISTE, M. & JACOBS, P. (eds) *Geology of Siliciclastic Shelf Seas*. Geological Society, London, Special Publications, **117**, 49–79.

NØTTVEDT, A., GABRIELSEN, R. H. & STEEL, R. J. 1995. Tectonostratigraphy and sedimentary architecture of rift basins, with reference to the northern North Sea. *Marine and Petroleum Geology*, **12**, 881–901.

PLANKE, S. & ELDHOLM, O. 1994. Seismic response and construction of seaward dipping wedges of flood basalts: Vøring volcanic margin. *Journal of Geophysical Research*, **99**, 9263–9278.

RAVNÅS, R. & BONDEVIK, K. 1997. Architecture and controls on the Bathonian–Kimmeridgian shallow-marine syn-rift wedges of the Oseberg-Brage area, northern North Sea. *Basin Research*, **9**, 197–226.

RAVNÅS, R. & STEEL, R. J. 1997. Contrasting styles of late Jurassic syn-rift turbidite sedimentation: a comparable study of the Magnus and Oseberg areas, northern North Sea. *Marine and Petroleum Geology*, **14**, 417–449.

RAVNÅS, R. & STEEL, R. J. 1998. Architecture of marine rift-basin successions. *AAPG Bulletin*, **82**, 110–146.

RAVNÅS, R., BONDEVIK, K., HELLAND-HANSEN, W., LØMO, L., RYSETH, A. & STEEL, R. J. 1997*a*. Sedimentation history as an indicator of rift initiation and development: The late Bajocian-Bathonian evolution of the Oseberg-Brage area, northern North Sea. *Norsk Geologisk Tidsskrift*, **77**, 205–232.

RAVNÅS, R., HANSEN, J. W., MELLERE, D., NØTTVEDT, A., SJØBLOM, T. S., STEEL, R. J. & WILSON, R. C. L. 1997*b*. A marine late Jurassic syn-rift succession in the Lusitanian Basin, western Portugal – tectonic significance of stratigraphic signature. *Sedimentary Geology*, **114**, 237–266.

REEMST, P., SKOGSEID, J. & LARSEN, B. T. 1996. Base Pliocene velocity inversion on the eastern Vøring margin – causes and implications. *Global and Planetary Change*, **12**, 201–211.

SKOGSEID, J. 1994. Dimensions of Late Cretaceous–Paleocene Northeast Atlantic rift derived from Cenozoic subsidence. *Tectonophysics*, **240**, 225–247.

SKOGSEID, J. & ELDHOLM, O. 1995. Rifted continental margins off mid-Norway. *In*: BANDA, E., TALWANI, M. & TORNE, M. (eds) *Rifted Ocean. Continent Boundaries*. NATO ASI Series Volume, Kluwer Academic PUBLISHERS, pp. 147–153.

SKOGSEID, J., ELDHOLM, O. & PLANKE, S. 1994. Mesozoisk kontinental rifting og Kenozoisk margindannelse: dypseismikk og skorpestruktur på Vøringmarginen. *Geonytt*, **21**, 3–18.

STEEL, R. J. 1993. Triassic–Jurassic megasequence stratigraphy in the northern North Sea: rift to post-rift evolution. *In:* PARKER, J. R. (ed.) *Petroleum Geology of Northwest Europe*. Geological Society, London, 299–315.

STUEVOLD, L. M. & ELDHOLM, O. 1996. Cenozoic uplift of Fennoscandia inferred from a study of the mid-Norwegian margin. *Global and Planetary Change*, **12**, 359–386.

TERVOORDE, M., RAVNÅS, R., FÆRSETH, R. B. & CLOETINGH, S. 1997. Tectonic modelling of middle Jurassic syn-rift stratigraphy in the Oseberg-Brage area, northern North Sea. *Basin Research*, **9**, 133–150.

THERIAULT, P. & STEEL, R. J. 1995. Aspects of synrift sedimentation in the Upper Jurassic (Helmsdale Boulder Beds) of the Inner Moray Firth Basin. *In*: STEEL, R. J., FELT, V. L., JOHANNESSEN, E. P. & MATHIEU, C. (eds) *Sequence Stratigraphy on the Northwest European Margin*. Norwegian Petroleum Society Special Publication, **5**, 365–387.

Reports

Reports listed herein are public and can be made available through the Research Council of Norway (RCN), University of Bergen (UB), University of Oslo (UO) and Norwegian Technical University in Trondheim (NTU = NTH).

FEJERSKOV, M. 1993. Bergspenninger i Norge og på den norske sokkel. *In*: MYRVANG, A., JOHANSEN, T., HANSEN, Å. & BERG, K. R. (eds) *Fjellsprengingsteknikk, bergmekanikk, geoteknikk*. Oslo 1993, 17pp.

FEJERSKOV, M. 1994. *Breakout as a tool for stress determination in deep wellbores*. NTH Report No. 1, 15pp.

FEJERSKOV, M. 1994. *Breakout interpretation. Methods and software used at NTH*. NTH Report No. 3, 14pp.

FEJERSKOV, M. 1994. *Breakout identification in 7 wells near the Troll Field on the eastern flank of the northern Viking Graben*. NTU Report No. 4, 39pp.

FEJERSKOV, M. 1995. *Criteria for breakout identification based on 4-arm oriented caliper logs*. NTU Report No. 2, 17pp.

FEJERSKOV, M. 1995. *Breakout interpretation – 11 wells on the Visund Field, northern Viking Graben*. NTU Report No. 5, 39pp.

FEJERSKOV, M., in prep 1995. *Breakout interpretation in the Tampen Spur area*. NTU Report No. 6.

GLADCZENKO, T. P. & ELDHOLM, O. 1995. *LIP Database: Large Igneous Provinces – distribution and references*. Geophys. Res. Group, Dep. Geol., Univ. Oslo, Computer Pgm./Database Doc. Ser. No. 15, 5pp.

HAMBORG, M., KYRKJEBØ, R. & RAVNÅS, R. 1995. *Syn- and post-rift depositional modelling, northern North Sea*. IKU report xxxxx.

MARJANAC, L. T. 1994. *Reference data base for rift basins (syn-rift)*. UiB Report, 41pp.

NØTTVEDT, A. & IBS-DNM working group 1993. *IBS Module 3 – Dynamics of the Norwegian Margin. First Periodical Report – Project Description, June 1993*, 28pp.

NØTTVEDT, A. & IBS-DNM working group 1993. *Minutes of Meeting, IBS-DNM Project Seminar*. Geilo, November 1993, 57pp.

NØTTVEDT, A. & IBS-DNM working group 1993. *IBS Module 3 – Dynamics of the Norwegian Margin. Second Periodical Report, December 1993*, 24pp.

NØTTVEDT, A. & IBS-DNM working group 1994. *Minutes of Meeting, IBS-DNM Project Seminar*. Stavanger, May 1994, 81pp.

NØTTVEDT, A. & IBS-DNM working group 1994. *IBS-DNM Project Status Report*, May 1994.

NØTTVEDT, A. & IBS-DNM working group 1994. *IBS Module 3 – Dynamics of the Norwegian Margin. Third Periodical Report, June 1994*, 32pp.

NØTTVEDT, A. & IBS-DNM working group 1994. *IBS Module 3 – Dynamics of the Norwegian Margin. Fourth Periodical Report, December 1994*, 49pp.NØTTVEDT, A. & IBS-DNM working group 1995. *IBS-DNM Project Status Report, February 1995*. Three volumes.

NØTTVEDT, A. & IBS-DNM working group 1995. *IBS Module 3 – Dynamics of the Norwegian Margin. Fifth Periodical Report, June 1995*, 37pp.

NØTTVEDT, A. & IBS-DNM working group 1994. *IBS Module 3 – Dynamics of the Norwegian Margin. Final Report, December 1995*.

PLANKE, S. 1993. *HEAT – Heat flow data base program*. Geophys. Res. Group, Dept. Geol.,

Univ. Oslo, Computer program/Data base Doc. Ser. No. 2 version 3.1, 10pp.

PLANKE, S. 1993. *VELO – Seismic refraction/wide-angle reflection velocity data base program.* Geophys. Res. Group, Dept. Geol., Univ. Oslo, Computer program/Data base Doc. Ser. No. 1 version 3.1, 14pp.

RAVNAAS, R., HANSEN, J. W., MELLERE, D., NØTTVEDT, A., SJØBLOM, T. S. & STEEL, R. J. 1995. *A marine to continental syn-rift succession: the Kimmeridgien Abadia and Lourinha Formations of the Santa Cruz area, Lusitanian Basin, Portugal.* University of Bergen Report.

VANVEEN, P., SKJOLD, L. J. & RYSETH, A. E. 1994. *A High-Resolution Stratigraphic Framework for the Paleogene in the Northern North Sea.* Norsk Hydro Report R-055419.

Theses

BORGERUD, L. 1995. *Relations between Rock Stresses and Pore Pressure on the Norwegian Margin, 62–67° north – a study based on borehole breakouts.* Diploma thesis, Dep. of Geology and Mineral Resources Engineering, Norwegian Technical University, Trondheim.

FEJERSKOV, M. 1996. *Determination of in-situ rock stresses related to petroleum activities on the Norwegian continental shelf.* Dr. ing. thesis, Department of Geology and Mineral Resources Engineering, Norwegian Technical University, Trondheim, Norway, 162pp.

FINSTAD, A. G. 1995. *Main structural elements on the Norwegian continental margin, 68–72°N and adjacent land areas: temporal and spatial development.* Cand. scient thesis, Dep. of Geology, University of Oslo.

HOLMSEN, C. 1994. *Undersøkelse av kenozoiske og mesozoiske sedimenter i Vikinggraben i relasjon til den tertiære landhevingen.* Cand. scient. thesis, Dep. of Geology, University of Oslo.

IBRAHIM, M. T. 1993. *Post-Jurassic basin fill in the northern North Sea.* Cand. scient thesis, Dep. of Geology, University of Oslo.

JORDT, H. 1995. *The Cenozoic geological evolution of the central and northern North Sea based on seismic sequence stratigraphy.* Dr. scient thesis, Department of Geology, University of Oslo.

JØRGENSEN, T. 1994. *Determination and Evaluation of In-situ Rock Stresses at the Snorre Field.* Diploma thesis, Dep. of Geology and Mineral Resources Engineering, Norwegian Technical University, Trondheim.

KARLBERG, T. 1995. *En geofysisk undersøkelse av Hovgaardryggen.* Cand. scient thesis, Dep. of Geology, University of Oslo.

RAVNÅS, R. 1996. *Variability of syn-rift sedimentary architecture in marine rift-basins: examples from the middle-late Jurassic of the northern Viking Graben, North Sea, and the Lusitanian Basin, western Portugal.* Dr. scient. thesis, Department of Geology, University of Bergen, 210pp.

SVARE, E. 1995. *Relations between Rock Stresses and Pore Pressure on the Norwegian Margin, 62–67° north – a study based on leak-off tests and formation pressure.* Diploma thesis, Dep. of Geology and Mineral Resources Engineering, Norwegian Technical University, Trondheim.

Crustal structure in the northern North Sea: an integrated geophysical study

P. CHRISTIANSSON,[1,2] J. I. FALEIDE[1] & A. M. BERGE[3]

[1] *Department of Geology, P.O. Box 1047 Blindern, N-0316 Oslo, Norway*
[2] *Present address: Norsk Hydro ASA, P.O. Box 200, 1321 Stabekk, Norway*
[3] *Norsk Hydro Research Centre, P.O. Box 646, 5020 Sandsli, Norway*

Abstract: This study focuses on the deep structure of the Viking Graben and adjacent areas of the northern North Sea (60–62°N), and its implications for the amount, timing and nature of lithospheric extension. Two regional transects have been constructed based on an integrated analysis of deep seismic reflection and refraction data, gravity and magnetic data, and correlations between offshore and onshore geology. The shallow interpretation is based on high-quality conventional seismic reflection data calibrated against a large number of exploration wells. The new and partly reprocessed seismic data, combined with the other geophysical data, make possible a better documentation of the crustal configuration, such as the pre-Jurassic sediment distribution, basement and Moho relief, and deep fault geometries. A lower-crustal body characterized by an 8+ km s^{-1} velocity and an average bulk density of 2.95 g cm^{-3} is present beneath the Horda Platform. This body probably represents a deep crustal root of partially eclogitized rocks that formed during the Caledonian orogeny. Heterogeneities within this body give rise to the non-typical velocity–density relation. The crust–mantle boundary is located at the base of this body at a depth of 30–35 km and does not coincide with the seismically defined Moho. The geometry of crustal thinning reflects the cumulative effect of several post-Caledonian rift phases. Results show that Permian rifting affected a wide area, from the Øygarden Fault Complex to the Hutton Fault.

Deep seismic reflection data have formed the basis for numerous papers through the last decade, focusing on the crustal structure and basin evolution in the northern North Sea rift system. However, poor data quality, especially the low S/N ratio at depth, has led to many model-driven interpretations. The NSDP84-lines were first described by Gibbs (1987*a*, *b*) and Klemperer (1988). Additional interpretations have been presented by Harrison (1987), Kusznir & Matthews (1988), White & McKenzie (1988), Klemperer & White (1989), Pinet (1989), White (1989, 1990), Klemperer & Hurich (1990), Reston (1990) and Brun & Tron (1993). A commercial deep seismic profile, Britoil NNS83-22, that is located close to NSDP84-1 (Fig. 1) was described by Beach (1985, 1986) and Beach *et al.* (1987). The eastern ends of the NSDP84 lines are tied by two north–south profiles, ILP-10 and -11 (Fig. 1), recorded by the Norwegian International Lithosphere Project (Hurich & Kristoffersen 1988; Klemperer & Hurich 1990; Færseth *et al.* 1995; Hurich 1996).

Interpretations and combined models of the NSDP84 reflection data and gravity data have been published by Holliger (1987), Holliger & Klemperer (1989) and Fichler & Hospers (1990); and Zervos (1987) modelled the gravity field along six regional seismic profiles taken from Ziegler (1982) and Glennie (1984). Hospers & Ediriweera (1988, 1991) published a map of depth to the crystalline basement, based on an integrated analysis of magnetic, gravity and seismic data.

Most papers agree in Moho depths ranging between 9.5 and 10.5 s twt identified as the base of lower-crustal reflectivity. However, no modern deep seismic refraction data have been available to constrain the depth conversion and confirm the published crustal thickness and depth to Moho maps. Furthermore, poor definition of the top of the crystalline basement makes it difficult to construct basement thickness maps that reflect the cumulative crustal thinning in response to several post-Caledonian rift phases.

Several models have been proposed to explain the crustal thinning and basin formation in the northern North Sea, and the same regional deep seismic lines have been interpreted according to both the simple and pure shear models. The symmetrical pure shear model (McKenzie 1978) was favoured for the Viking Graben by Giltner (1987) and Badley *et al.* (1988). Alternatively, the asymmetrical simple shear model (Wernicke

From: NØTTVEDT, A. *et al.* (eds) *Dynamics of the Norwegian Margin*. Geological Society, London, Special Publications, **167**, 15–40. 1-86239-056-8/00/$15.00

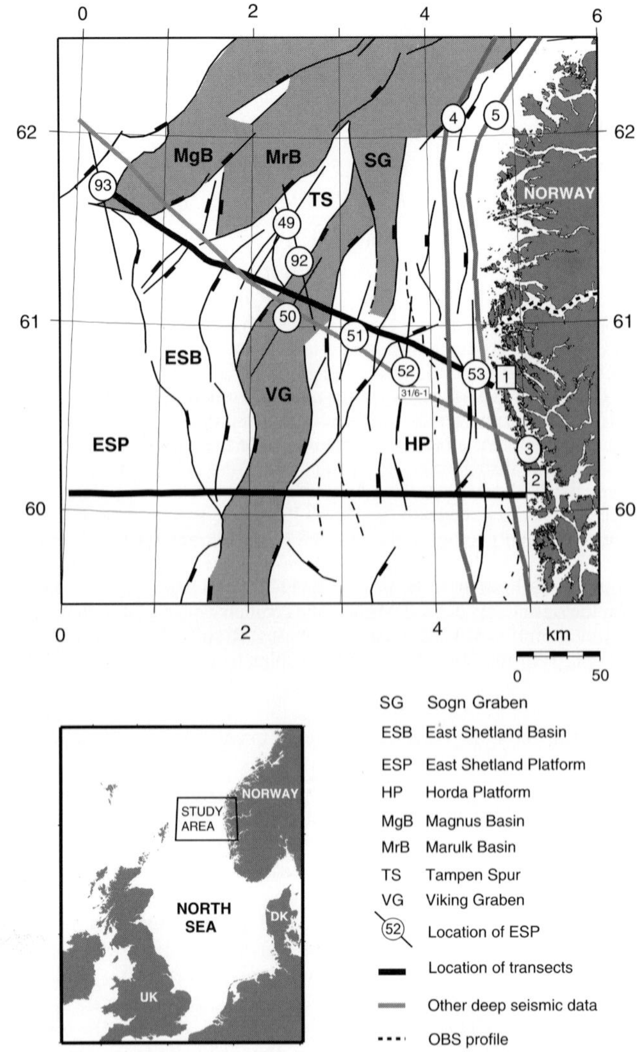

Fig. 1. Main structural elements within the study area. Location of the two regional transects and other deep seismic reflection and refraction profiles.

1985) has been used to explain the asymmetry of the graben structure (Beach 1986; Beach *et al.* 1987; Scott & Rosendahl 1989). Tectonic models combining simple and pure shear, characterized by structural asymmetry of the brittle upper crust and symmetrical thinning of the lower ductile crust, have gathered increasing support (Kusznir *et al.* 1991, 1995; Kusznir & Ziegler 1992). Odinsen *et al.* (1999*a*) have also favoured such a combined model, involving separation of the crust at different detachment levels caused by rheology and pre-existing zones of weakness.

The aim here is to present an integrated geophysical study of the crustal structure across the northern North Sea. The deep crustal structure will be related to the onshore geology of western Norway whereas the geometry of crustal thinning will be related to the post-Caledonian evolution including several rift phases. For this purpose, two regional transects have been constructed based on high-quality conventional seismic reflection data and reprocessed deep seismic reflection profiles (NSDP 84-1 and -2; Fig. 1). The crustal configuration is further constrained by integration of deep seismic refraction, gravity and magnetic data. Furthermore, seismic mapping of the study area including some 15 000 km of high-quality conventional seismic data has been carried out to establish a regional structural and stratigraphic framework.

Geological framework

The North Sea rift system, post-dating the Caledonian orogenic extensional collapse, was affected by two lithospheric extension events in Permian–?earliest Triassic time (in the literature often referred to as the Permo-Triassic event) and late Mid-Jurassic to earliest Cretaceous time (here referred to as Late Jurassic), respectively. The rift axis for the Permo-Triassic rift is thought to lie beneath the present Horda Platform whereas the Late Jurassic rift was centred beneath the present Viking Graben (Fig. 1). Each rift phase was followed by a thermal cooling stage, characterized by regional subsidence in the basin areas (Ziegler 1982; Gabrielsen 1986; Giltner 1987; Badley *et al.* 1988; Gabrielsen *et al.* 1990). The northern North Sea rift system is bounded by the East Shetland Platform in the west and the Øygarden Fault Zone in the east (Fig. 1). Structures within this area are characterized by large rotated fault blocks with sedimentary basins in asymmetric half-grabens associated with lithospheric extension and thinning of the crust.

Devonian strata, deposited in response to the extensional collapse of the Caledonides, are present in onshore western Norway (Hossack 1984; Steel *et al.* 1985; Norton 1986, 1987; Seranne & Seguret 1987; Osmundsen 1995), Shetland (Allen 1981; Flinn 1985; Seranne 1992) and eastern Greenland (Hartz & Andresen 1995). Although Devonian sediments in the northern North Sea have been reached in only a few wells, there are reasons to believe that Devonian sediments are present regionally in the deeper parts of the pre-Triassic half-grabens beneath the Horda Platform, Viking Graben and East Shetland Basin. Recent studies of seismic reflection data on the East Shetland Platform (Holloway *et al.* 1991; Platt 1995) indicate large sedimentary basins thought to contain Upper Palaeozoic (Devonian–Carboniferous) rocks. The presence of Upper Palaeozoic rocks, of both Devonian and Lower Permian age (Rotliegendes), has also been confirmed by drilling on the East Shetland Platform (Johnson *et al.* 1993; Duncan & Buxton 1995). Upper Palaeozoic rocks have not yet been encountered in the eastern part of the northern North Sea basin. However, the seismic data on the Horda Platform indicate that as much as 3–4 km of strata exist in the deepest parts of the half-grabens, which are not tested by drilling. These sediments probably date back into Permian and older times.

Permo-Triassic rifting is best documented in areas marginal to the Viking Graben, e.g. the Horda Platform and the Magnus Basin–Unst Basin region, because these were less influenced by Jurassic deformation. Here, seismic data display Triassic and older sediments, which show wedge-shaped geometries indicating that these areas were tectonically active in the early rift phase. On the Horda Platform, seismic data calibrated against well data (e.g. 31/6-1; Fig. 1) suggest that this phase terminated in early Triassic times. The initiation of this phase is not dated in the study area but is generally assumed to have been during Permian time (Færseth *et al.* 1995; Odinsen *et al.* 1999*a*). Palaeomagnetic dating of the Nordfjord–Sogn detachment, bounding the Devonian basins of Western Norway, indicates a tectonic event at around 260 Ma (Torsvik *et al.* 1992). Dolerite dykes in the coastal Sunnfjord region and immediately west of the Devonian basins are also of Permian age (250–270 Ma; Torsvik *et al.* 1997). Following the rift structures below the Horda Platform southward, tilted fault blocks are observed just below the Upper Permian Zechstein salt (Ditcha 1998). The north–south trending structures that dominated the early rift phase were inherited from earlier deformation of Precambrian basement (Færseth *et al.* 1995).

The Triassic to Middle Jurassic succession reflects a pattern of repeated outbuilding of clastic wedges from the Norwegian and East Shetland hinterlands within a generally evolving post-rift basin (Lervik *et al.* 1989; Steel & Ryseth 1990; Steel 1993). A broadly similar geometry of the megasequences in both continental Triassic and marine Jurassic successions was related by Steel (1993) to subsidence-rate variations. Differential subsidence across faults throughout Triassic time has also been reported (Steel & Ryseth 1990). The Øygarden Fault Zone, forming the eastern margin of the Permo-Triassic basin, was active throughout most of the time interval.

The latest tectonic event of major importance in the northern North Sea is the Late Jurassic stretching event, characterized by the northeast–southwest trending Viking Graben. The initiation of the Jurassic rift event in the northern North Sea is dated to late Mid-Jurassic time by most workers, with a climax in Late Jurassic time (Ziegler 1982; Beach *et al.* 1987; Badley

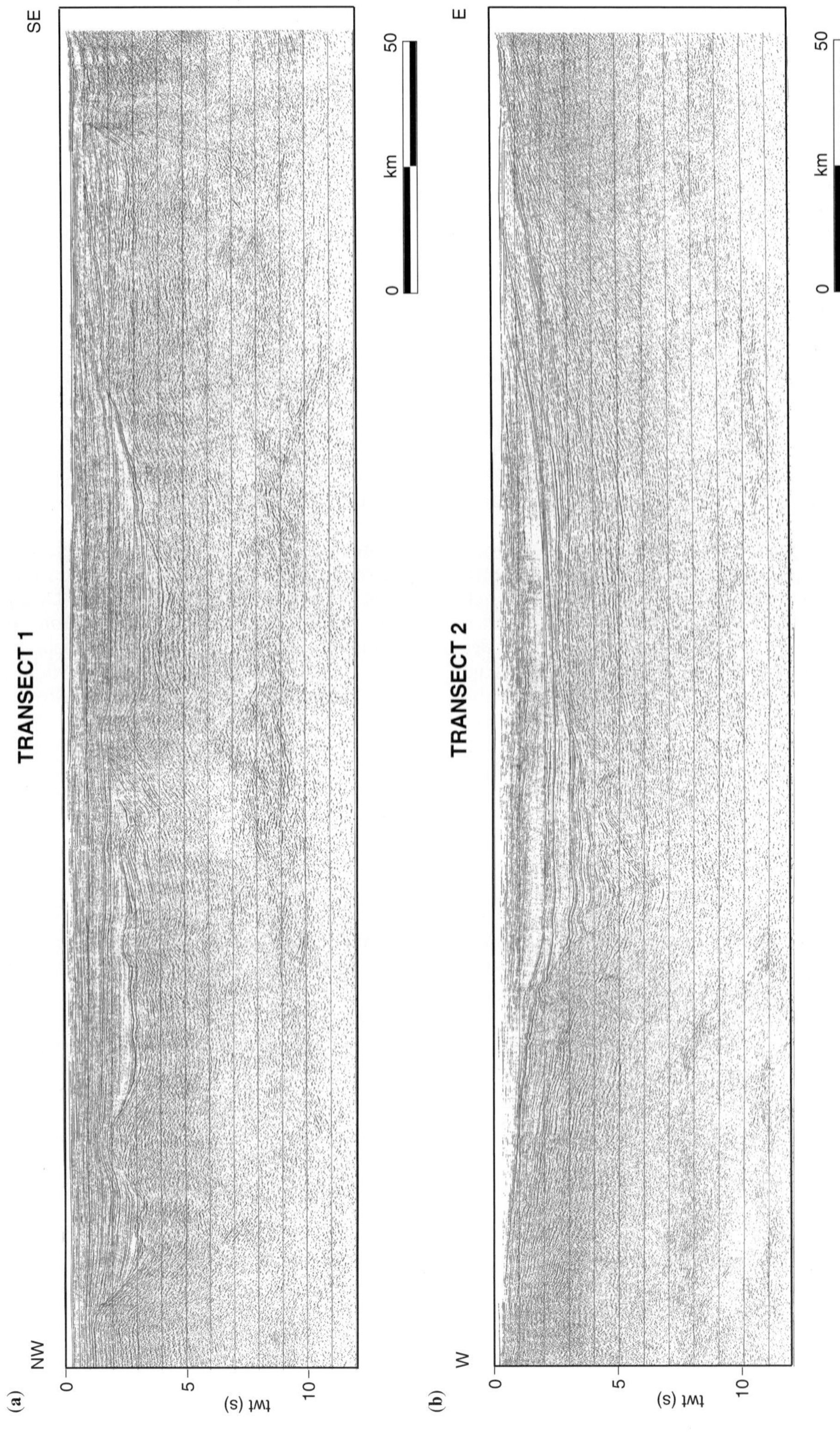

Fig. 2. Deep seismic reflection profiles across the northern North Sea. (**a**) NSDP 84-1 (Transect 1); (**b**) NSDP 84-2 (Transect 2). Location of profiles is shown in Fig. 1.

et al. 1988; Gabrielsen *et al.* 1990). The initial stages of Jurassic rifting influenced a broader area than did the final stages of graben formation (Gabrielsen *et al.* 1990). Continued extension was concentrated on fewer faults along the graben margins. Simultaneously, the internal graben relief became more pronounced, and the rift system developed a mature graben topography with platforms, sub-platforms, platform marginal highs and a graben feature with complex centre of subsidence along its axis (Gabrielsen *et al.* 1990; Nøttvedt *et al.* 1995).

During Early Cretaceous time continued differential subsidence was accompanied by the gradual infilling of the graben relief with deep-water shales and minor pelagic carbonates. Regional subsidence combined with a general rise in eustatic sea level resulted in the progressive overstepping of basin edges and a gradual decrease in clastic influx into the North Sea basin. The high Late Cretaceous eustatic sea level provided conditions suitable for widespread deposition of chalk. A thick Upper Cretaceous–Danian chalk succession covers most of the central and southern North Sea, whereas terrigenous mudstones dominate in the northern North Sea. The northward change from chalk to mudstone in the central Viking Graben is gradational, both laterally and vertically. Even at the northern limits of the Viking Graben, there are a few metres of Maastrichtian chalk (Hancock 1990).

The Cenozoic basin fill of the northern North Sea reflects a complex pattern of regional vertical movements controlling basin geometry and provenance of the sediments (Jordt *et al.* 1995, 1999). Paleogene sediments were mainly derived from areas in the west and north, whereas southern Norway became an important source area from Oligocene and throughout Late Cenozoic times. Depocentre locations and sequence geometries indicate that differential vertical movements of the basin floor were closely related to boundaries of underlying Mesozoic and Palaeozoic structures in the North Sea Basin (Jordt *et al.* 1999).

Data analysis

For a better understanding of the crustal structure and its implications for basin formation and evolution, we have integrated all available geophysical data. The upper crust is best displayed on high-quality conventional seismic reflection data (0–7 s twt) whereas the middle and lower crust is best studied on deep seismic reflection data (0–15 s twt) (Figs 2 and 3). Deep seismic refraction data (expanded spread profiles, ESPs) provide crustal velocities that are used in the depth conversion. Gravity and magnetic data constrain the deep basin geometry and crustal structure. Our integrated approach also includes correlations between the offshore and onshore geology.

The regional profiles in Fig. 4 were constructed by combining traditional interpretation of the main sequence boundaries with the pattern of crustal reflectivity on the deep seismic lines (Fig. 3). Transect 1 is oriented NW–SE and crosses the Magnus Basin, East Shetland Basin, Viking Graben and Horda Platform, and Transect 2 runs E–W across the East Shetland Platform, Viking Graben and Horda Platform (Figs 1 and 4).

Conventional seismic reflection data (0–7 s twt)

Despite the strong focus on the two crustal transects, we have interpreted a relatively dense grid of conventional regional seismic lines covering the Norwegian sector, but only a few lines have been available on the UK side. Key lines for the regional transects have been the reprocessed versions of SG8043-403 (Transect 1) and SG8043-101 (Transect 2), which show improved resolution of the deep basin stratigraphy after reprocessing. The seismic stratigraphy along the transects has been calibrated against selected wells, which, in combination, cover the stratigraphic interval from Triassic to Quaternary time.

The shallow part of the profiles (Fig. 4) is characterized by a relatively flat-lying Cenozoic and Cretaceous post-rift sequence overlying faulted Mesozoic and Upper Palaeozoic strata. The axis of the Viking Graben displays the thickest depocentres of post-rift sequences. Jurassic strata are also thickest in the Viking Graben, particularly on Transect 2 (Fig. 4), and thin at the basin flanks. A Permo-Triassic basin is best visualized on the Horda Platform. Here, growth sequences towards the major faults indicate pre-Jurassic fault activity giving rise to a system of half-grabens. The most pronounced difference in structural style between the two transects is the change in polarity of the tilted fault blocks beneath the eastern part of the Horda Platform (Fig. 4). The polarity shift is probably connected to transfer faults running in an east–west direction. A strong reflector tied to basement drilled in well 31/6-1 (Fig. 1) can be traced for some distance on the central Horda Platform, but is difficult to interpret regionally.

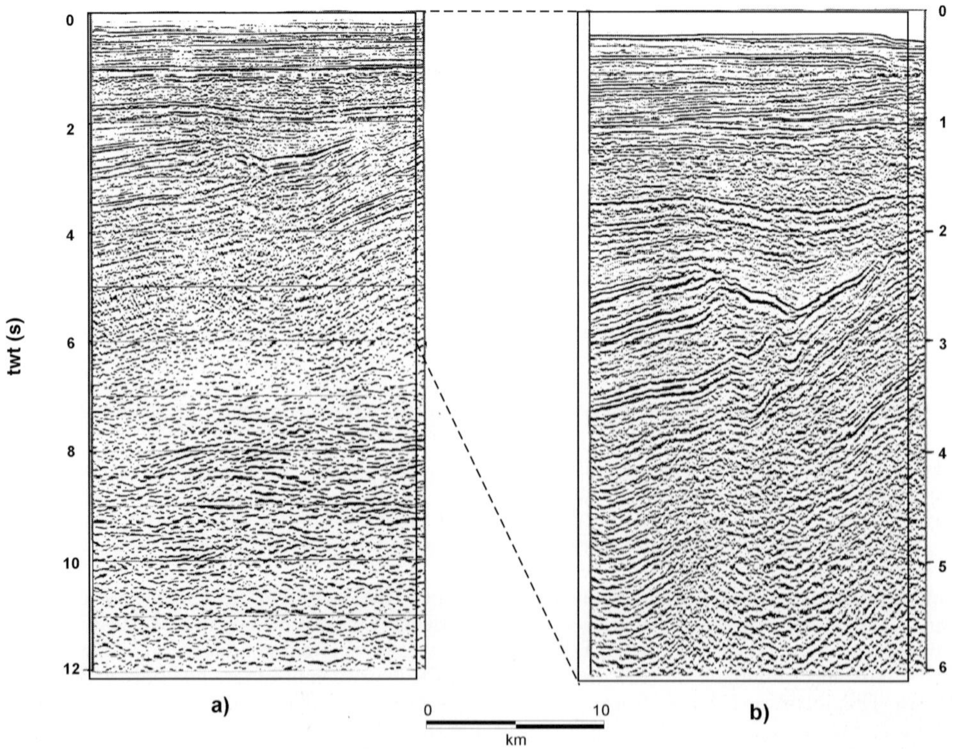

Fig. 3. Seismic data examples from the same part of Transect 1. (**a**) Deep seismic reflection line NSDP 84-1; (**b**) conventional seismic reflection line NVGTI-92-102.

Deep seismic reflection data (0–15 s twt)

The deep seismic reflection lines (NSDP84, Figs 1 and 2) were acquired and processed in 1984–1985 by GECO on behalf of BIRPS and several oil companies. They were further reprocessed in 1991 by BIRPS (Klemperer & Hobbs 1991). The dataset was also reprocessed by Norsk Hydro in 1994 for the IBS project to enhance the lower-crustal reflectivity and Moho definition. Important elements in the post-stack reprocessing were correct scaling and balancing of sections to preserve dynamic range, and proper migration. Reprocessing improved definition of lower-crustal structure, and the Moho can now be observed as a contrast in reflective character along most of the deep seismic lines (Figs 2 and 3). Migration, using the velocity fields established along the crustal transects, has provided more correct geometries and fault definition. However, the data are still contaminated by long-period multiples.

The general outline of the transects is similar to that of many extended areas (e.g. Blundell 1990), with a reflective sequence of sediments near the surface, a transparent middle crust, and a reflective lower crust (Figs 2 and 4). The upper reflective crust has already been described based on the conventional seismic reflection data. Here, we will focus on the middle and lower crust.

Although the middle crust is characterized by poor reflectivity, several important observations can be made on the reprocessed deep seismic data when these are combined with conventional data. Some graben-margin master faults have been traced through the transparent middle crust to the lower reflective crust, where faults seem to merge at *c.* 18 km depth (Fig. 4) (Odinsen *et al.* 1999*a*). The Hutton Fault on Transect 1 is mainly a Permo-Triassic feature whereas the 'Western Margin Fault' of the Viking Graben on Transect 2 is mainly a Late Jurassic feature (Fig. 4).

The large east-dipping fault bounding the horst structure at the central Horda Platform on Transect 2 seems to correlate with a zone of east-dipping mid-crustal reflections below the Øygarden Fault Zone (Figs 2 and 4). These dipping reflections might be linked to Caledonian structures beneath western Norway. East-dipping

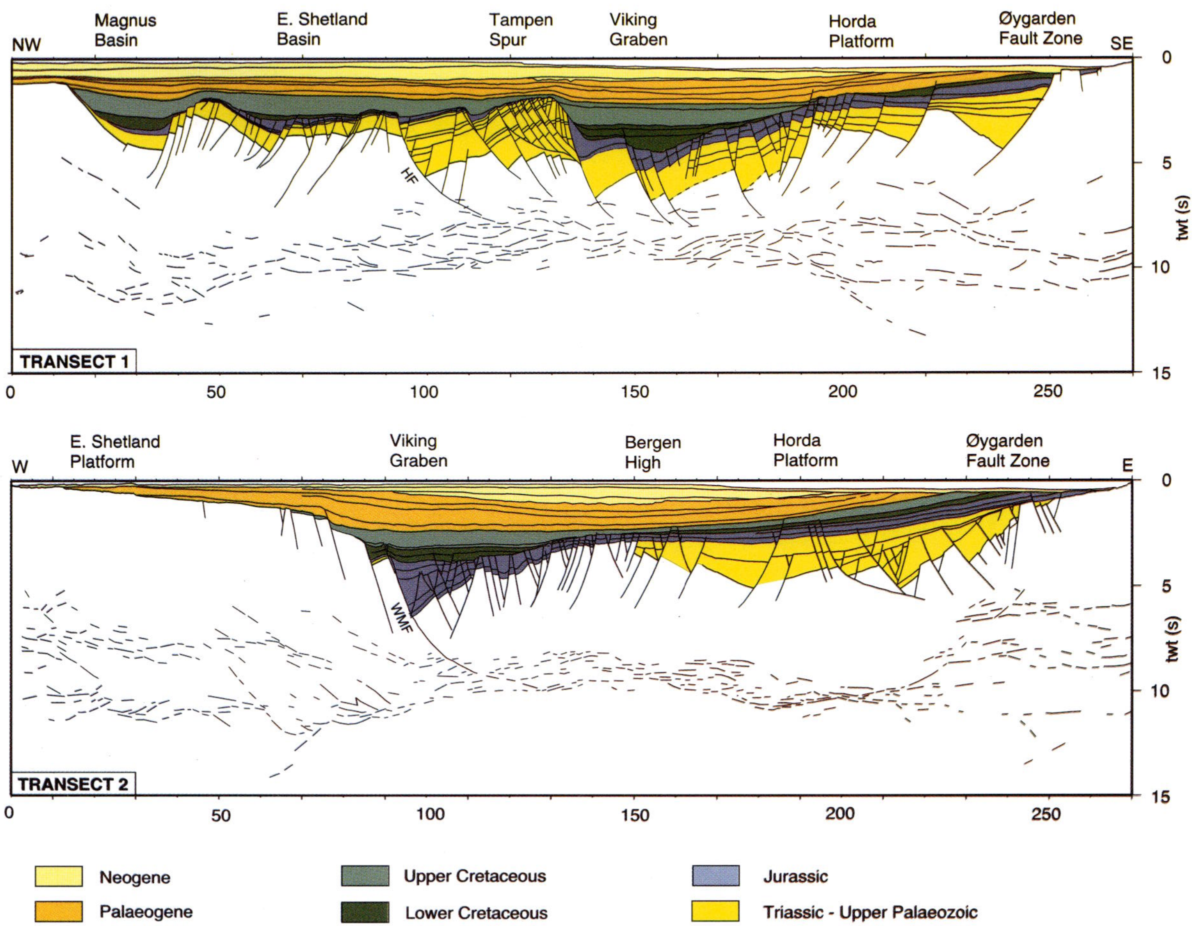

Fig. 4. Line drawing of the two regional crustal transects. The crustal reflectivity pattern is interpreted from the reprocessed deep seismic reflection profiles in Fig. 2, whereas the basin stratigraphy is based on conventional seismic reflection data. HF, Hutton Fault; WMF, 'Western Margin Fault'. Location of profiles is shown in Fig. 1.

mid-crustal reflections are also observed beneath the Magnus Basin on Transect 1 (Fig. 1).

The reflective lower crust is seen as an undulating transition zone between the overlying transparent middle and upper crystalline crust and the underlying upper mantle. The thickness of this reflective band normally varies between 3 and 6 km but locally increases to 10 km under the basin margins. The reflectivity is strongest under the basin flanks, close to the Norwegian coast, and becomes weaker under the Viking Graben, which is also the most extended area. The Moho is well displayed, either as a single reflector or as the base of the lower-crustal reflectivity, and shallows below the most strongly extended part of the rift system (Figs 2 and 4).

Separation of the lower crust into an upper and lower reflective zone can be seen in both transects beneath the Horda Platform, although the lower-crustal reflections are weaker on Transect 2 giving rise to two Moho candidates in this area. A similar geometry was observed by Beach (1986) on a nearby deep seismic reflection profile (profile 3 in Fig. 1). We will return to a discussion of the body bounded by the two Moho candidates after integration with velocity and density information from analyses of other geophysical data.

Reflections below the Moho are recorded in both transects (Figs 2 and 4). These features, dipping away from the graben axis, are located close to the transition between the thinned crust and the thicker crust beneath the platform areas. This pattern of intra-mantle reflections has also been described by previous work (Beach 1986; Beach *et al.* 1987; Klemperer 1988; Holliger & Klemperer 1989; Brun & Tron 1993). The interpretation of these structures offers several solutions, but the coincidence between the dipping reflectors and the dip gradient of the Moho may suggest that the reflections represent faults or shear zones within the mantle.

Deep seismic refraction data

The deep seismic refraction data (ESPs) were collected by Elf Aquitaine Norge in co-operation with the Universities of Oslo and Bergen. During the research programme, from 1983 to 1986, 103 ESPs and 173 heat-flow measurements were acquired from the Norwegian continental margin, the northern North Sea, the Jan Mayen Ridge and in the Barents Sea.

The ESPs were acquired with a two-ship technique (Stoffa & Buhl 1979) such that the shot and receiver were moved apart at equal intervals maintaining the same midpoint. The two ships, with one firing its airguns (6000 in.^{3}) while the other was recording with its multichannel streamer (2.4 km long), started from initial positions at the end points of the ESP, usually between 70 and 100 km apart, steaming to the opposite end points and crossing the midpoint at a separation of about 1 km. In this manner, two ESPs are collected in one experiment, one at closing ranges and another at expanding ranges. Asymmetry when comparing the left and right side of the ESP reflects two-dimensional effects related to dip–topography.

The data were processed by Institut Français du Petrole (IFP) in co-operation with Elf Aquitaine. The traces with common source–receiver offsets were sorted and stacked to one single trace. The data were then transformed from the distance–travel time (x–t) domain to the intercept time–ray parameter (tau–p) domain, according to the slant-stack procedure of Stoffa *et al.* (1981). One advantage of the ESPs common midpoint geometry is that dipping interfaces have minor effect on determination of interval velocities (Diebold & Stoffa 1981), in contrast to sonobuoy data for which common receiver geometry yields a velocity that might be significantly affected by dipping layers (Olafsson *et al.* 1992).

The seven northern North Sea ESPs (Fig. 1), from 1983–1984, have not been published before, although Pinet (1989) referred to the IFP–Elf interpretation. Both distance–travel time (x–t, Fig. 5a) and intercept time–ray parameter (tau–p, Fig. 6a) domain presentations were available for all profiles. To extract the maximum velocity–depth information, a data reduction procedure similar to that of Eldholm & Mutter (1986) and Jackson *et al.* (1990) was used with three different methods applied in the x–t as well as one in the tau-p domain. The travel time computations were carried out using in-house computer software developed by Kitterød (1986) and Hegna (1989). The interpretation procedure involved the following steps:

- All seismic arrivals identified as refracted events were first analysed with a slope–intercept method (Ewing 1963). The assumptions of this reduction technique are plane layers and a constant velocity within each. If these constraints are met, the data will consist of straight-line segments where head or critically refracted waves are produced.
- Next, the seismic arrivals were examined as diving waves and a flat-Earth Herglotz–Wiechert inversion was performed (Aki & Richards 1980). The assumption here is a continuous velocity increase with depth;

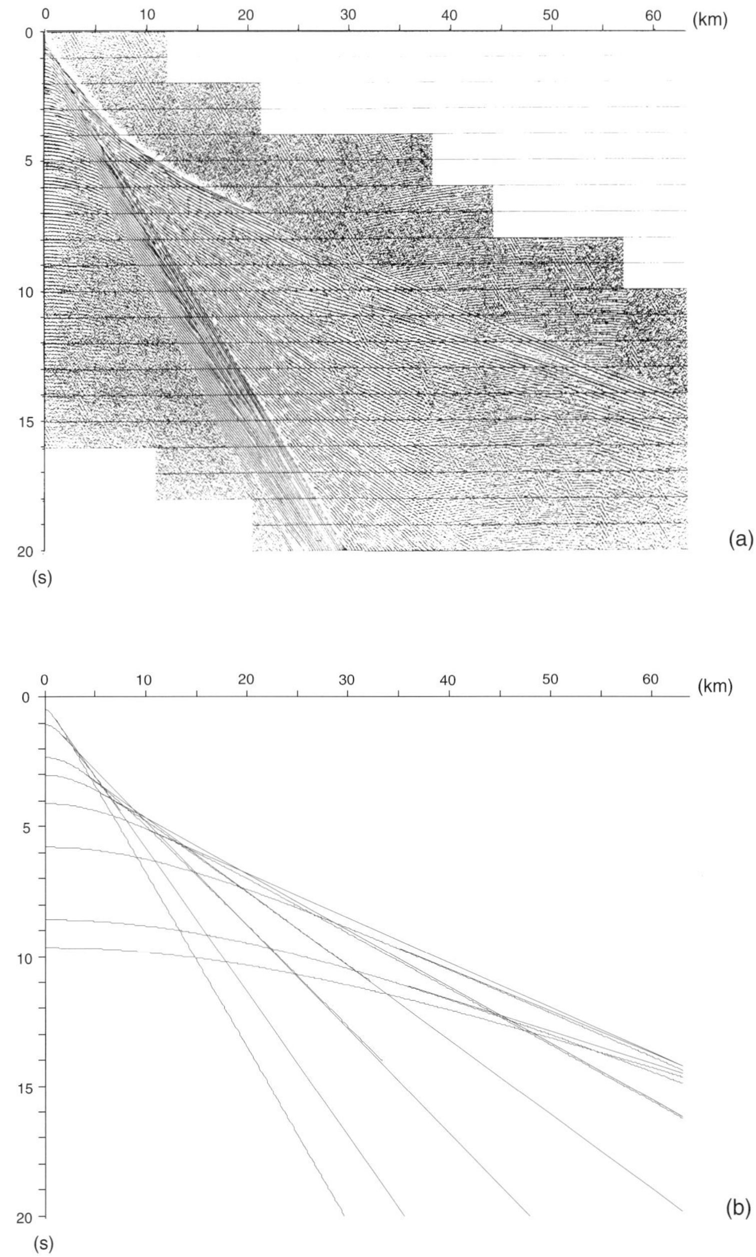

Fig. 5. ESP 51 in *x–t* domain. (**a**) Data example; (**b**) synthetic ray-tracing model. (See Fig. 1 for location of ESP 51.)

thus, ray paths were curved so that on the time v. distance plot the seismic arrivals form a curve also.

- A wide-angle reflection program based on the t^2–x^2 procedure of LePichon *et al.* (1968) was used for determining interval velocities. The near-vertical deep seismic reflection data were used to identify wide-angle reflections in the ESP data.
- Finally, a tau-sum interpretation (Diebold & Stoffa 1981) was carried out on all ESPs using the primary tau–*p* curve as input. This is equivalent to the standard first-arrival slope–intercept method except that it is valid

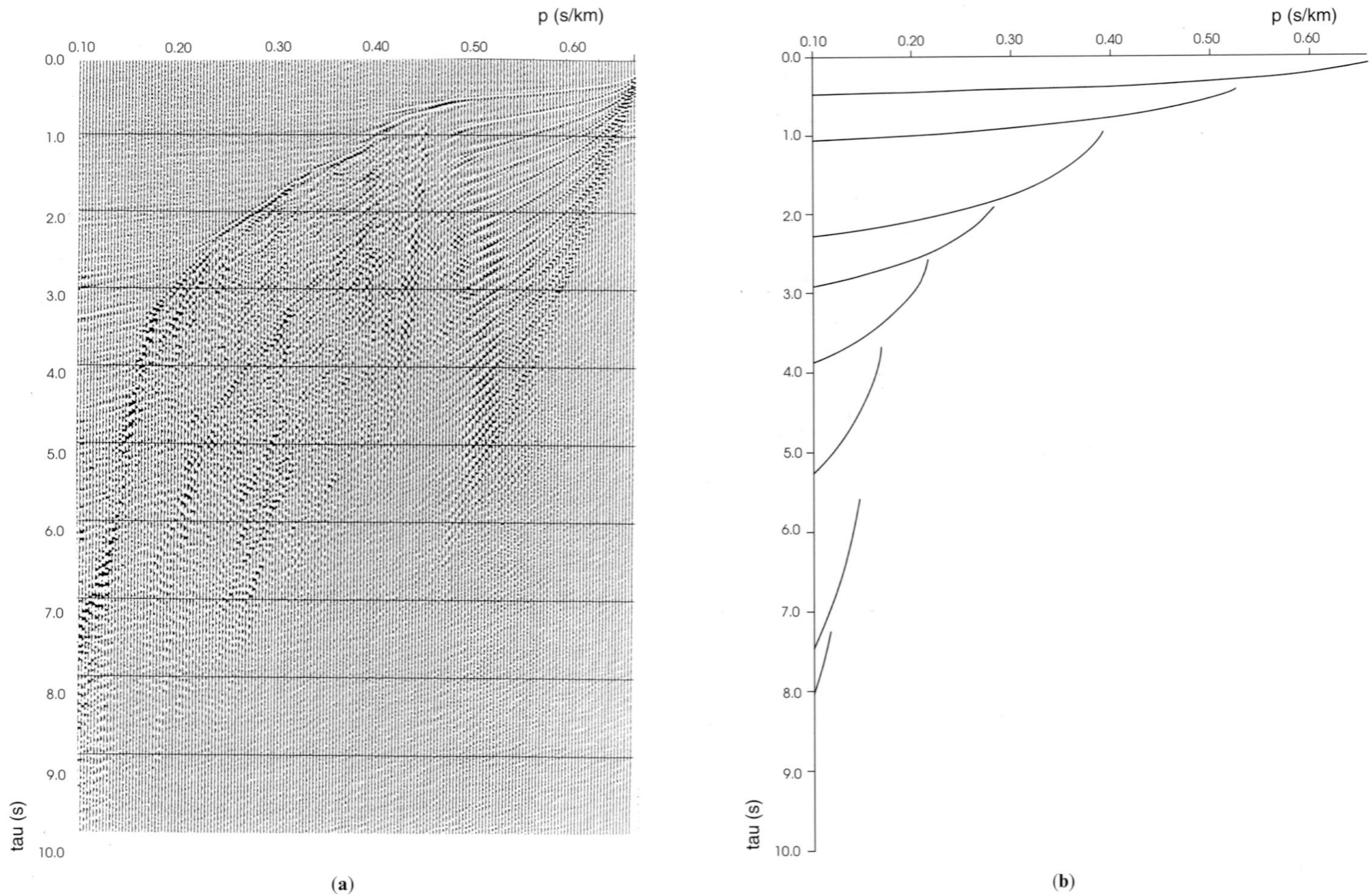

Fig. 6. ESP 51 in tau–p domain. (**a**) Data example; (**b**) synthetic ray-tracing model. (See Fig. 1 for location of ESP 51.)

for all critical and post-critical arrivals without distinction between reflections and refractions.

The velocity–depth curves derived from the various techniques are similar (Fig. 7). However, the slope–intercept and t^2–x^2 solutions provide a step-shaped rougher curve than the others. Computed (synthetic) travel time curves (Figs 5b and 6b) were produced by ray tracing to examine the accuracy of the inversions. After several iterations we arrived at a best fit final velocity–depth function (Fig. 7). The interpretation procedure is summarized in Fig. 8 and the final velocity–depth curves are plotted on a depth-converted Transect 1 in Fig. 9. The results of the analyses of all ESPs have been presented by Christiansson *et al.* (1999*a*), who also discussed the vertical and lateral velocity distribution along Transect 1 in more detail. The reduction of the ESPs in combination with deep seismic reflection data has improved our knowledge of the crustal velocity distribution across the northern North Sea, especially for the mid- and lower-crustal levels. Wide-angle reflection hyperbolas from the lower-crustal interfaces and possibly the Moho added more constraints to the deep crustal configuration.

The velocity field used in the depth conversion of the crustal transects was constructed from the ESP results and additional velocity data such as information from wells and stacking velocities. Typical velocities within the sedimentary sequences are: Tertiary 2.0–2.5 km s^{-1}; Cretaceous 3.0–4.1 km s^{-1}; Jurassic 3.3–4.6 km s^{-1}; pre-Jurassic 4.0–5.8 km s^{-1}. The uppermost

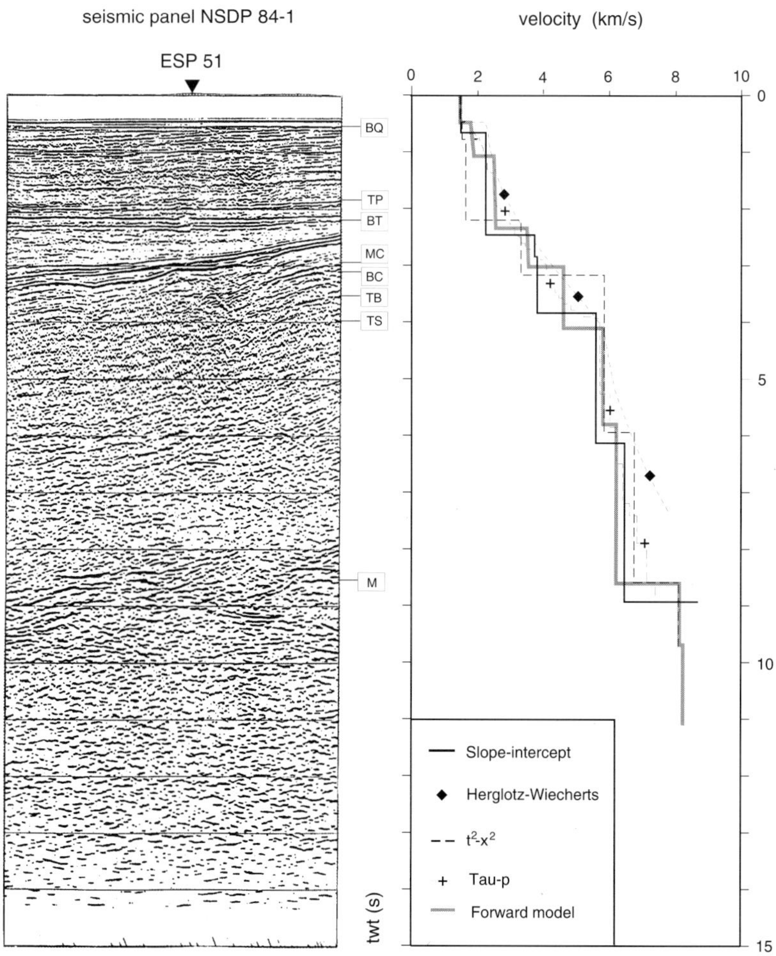

Fig. 7. Velocity–depth curves from the analysis of ESP 51. The velocity–depth curves are plotted against two-way time for direct comparison with the deep seismic panel.

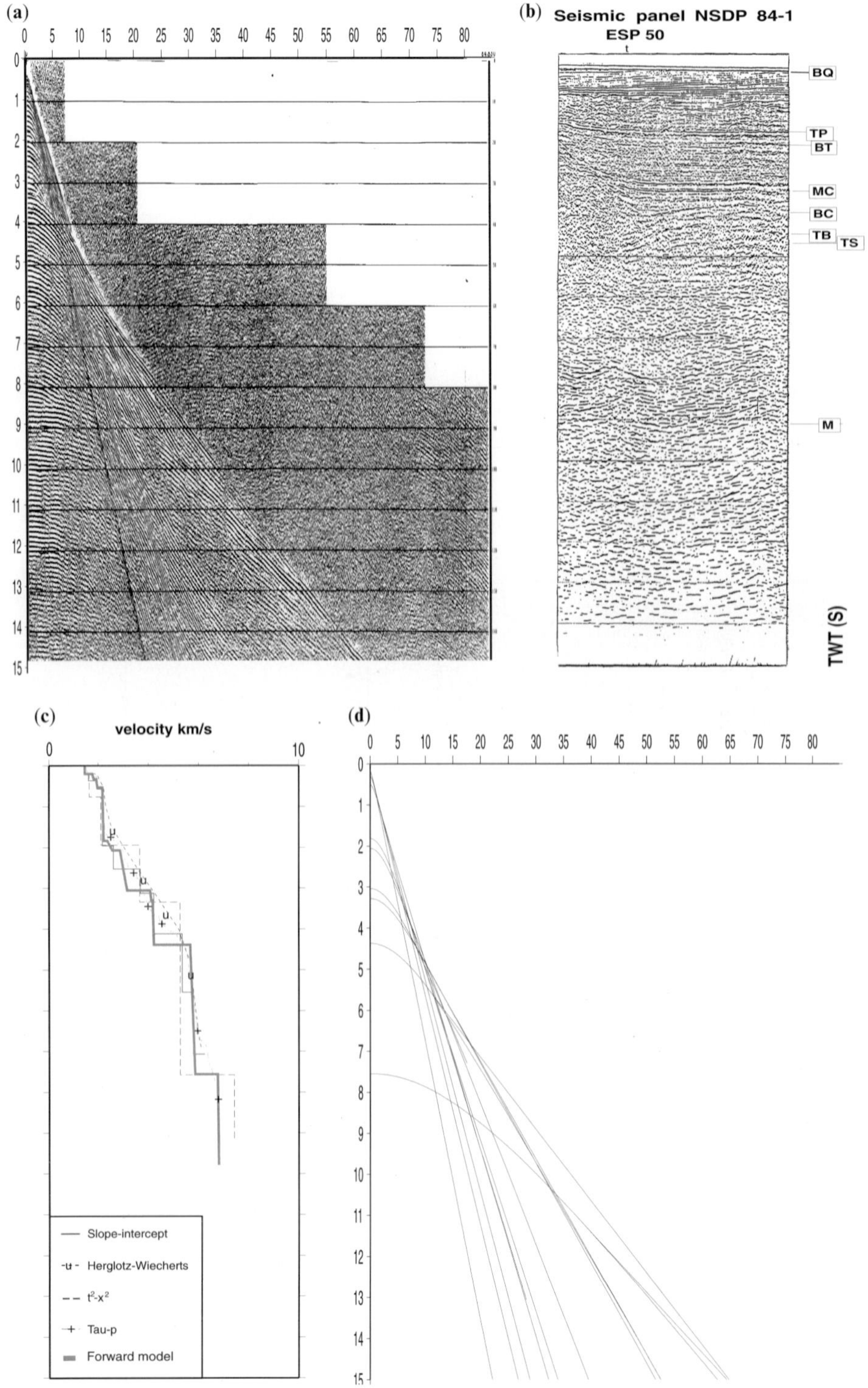

Fig. 8. Interpretation of deep seismic refraction and wide-angle reflection data. (**a**) ESP 50 data example; (**b**) deep seismic reflection data close to ESP 50; (**c**) velocity–depth curves from the analysis of ESP 50; (**d**) synthetic ray-tracing model. (See Fig. 1 for location of ESP 50.)

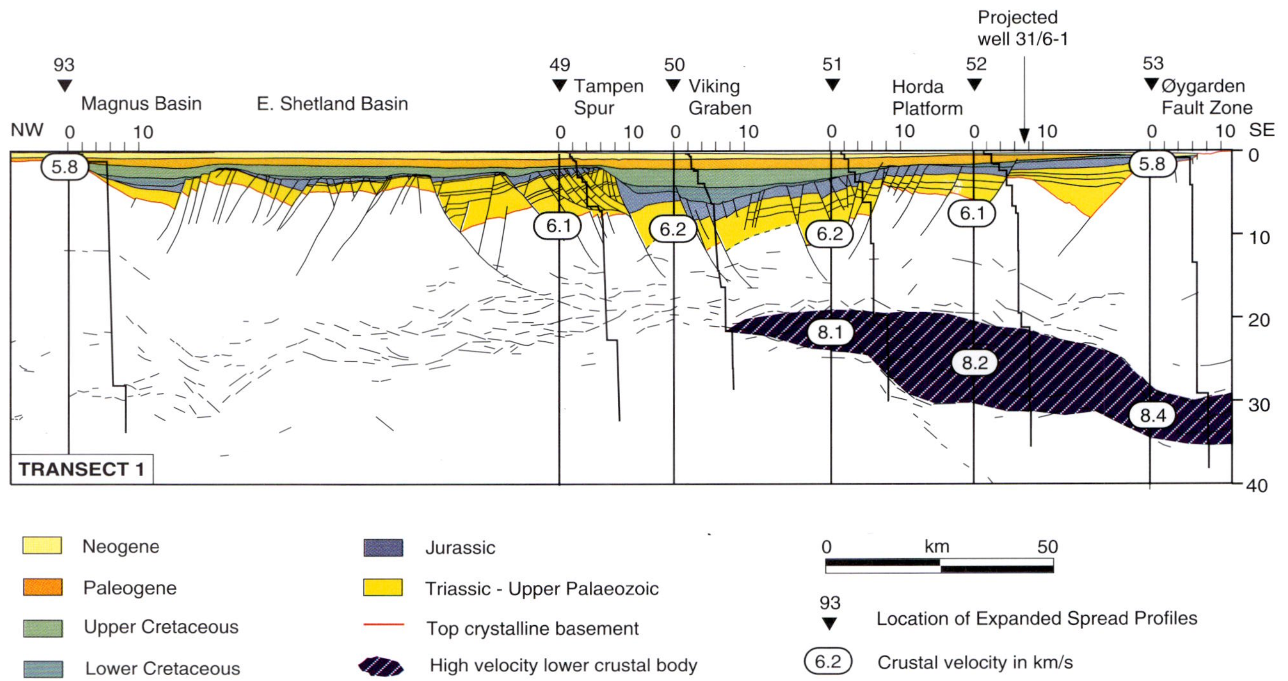

Fig. 9. The final velocity–depth solution for all ESPs plotted on a depth converted Transect 1. Velocities in the range 5.8–6.1 km s^{-1} are interpreted as typical for the upper part of the crystalline basement. The high-velocity lower-crustal body (8.1–8.4 km s^{-1}) detected in ESPs 51, 52 and 53 beneath the Horda Platform is outlined.

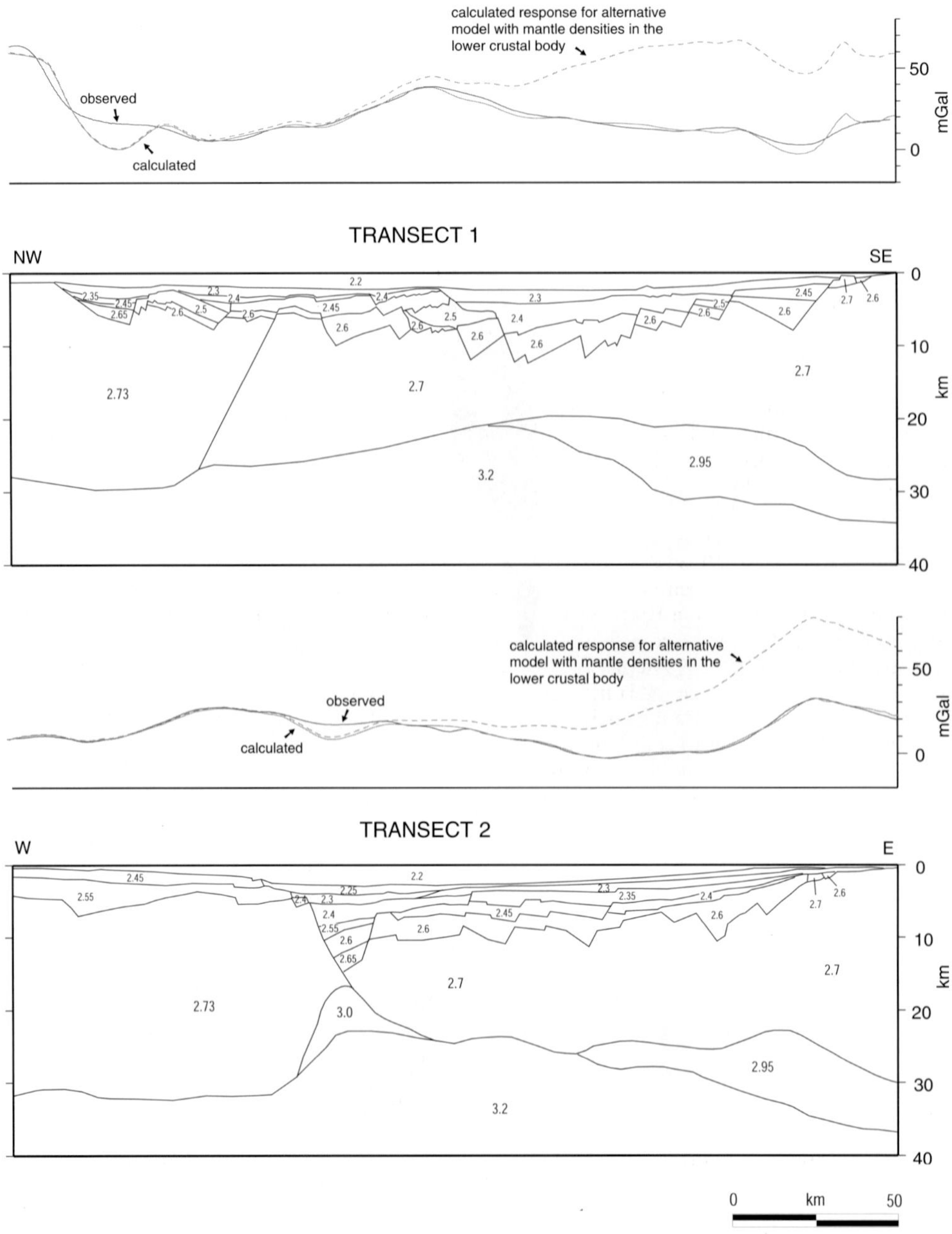

Fig. 10. Crustal-scale gravity models of Transects 1 and 2. The geometries and densities are constrained by the deep seismic reflection and refraction data.

crystalline crust shows velocities ranging from 5.8 km s^{-1} in areas with shallow basement, e.g. east of the Øygarden Fault Zone, to 6.1–6.2 km s^{-1} in more central areas (Fig. 9), which is in accordance with typical basement velocities from more global studies (Christensen & Mooney 1995). The top of the crystalline crust is not always easy to interpret on the seismic reflection data, especially in the deeper parts of the Viking Graben, which are infilled with sediments of 12 km thickness at the most. A typical velocity for the upper transparent crystalline crust is 6.4–6.5 km s^{-1}. Higher values within the middle–lower crust are observed under the eastern part of the Horda Platform in ESP 53, where wide-angle reflections give a P-wave velocity of 6.8 km s^{-1} for the depth interval between 13 and 29 km. A similar velocity was reported by Deemer & Hurich (1991) based on wide-aperture (WA) recording at land stations during acquisition of the deep seismic reflection line ILP-11 (profile 5 in Fig. 1). Lower-crustal velocities of 6.6–7.0 km s^{-1} were also derived from an extensive ocean-bottom seismograph (OBS) refraction experiment conducted on a 150 km profile along the Sognefjord (Fig. 1) (Iwazaki *et al.* 1994).

The ESPs located on the Horda Platform show strong wide-angle reflections corresponding to the two Moho candidates in the deep seismic reflection data, confining a body with interval velocities ranging between 8.1 and 8.4 km s^{-1} (Fig. 9). This high-velocity, lens-shaped body is *c.* 100 km wide and 5–10 km thick, with the base of the body located in the southeastern part of Transect 1 at depths of 30–35 km. First arrivals from this body have not been obtained because of limitations in offset of the ESPs. In the WA data, however, first arrivals from this body are recorded at offsets >130–140 km giving a velocity of 8.1–8.2 km s^{-1} (Deemer & Hurich 1991).

Velocities of 8+ km s^{-1} are normally assigned to mantle rocks, pointing to the shallow Moho candidate in the deep seismic reflection data as the base of the crust. However, as will be discussed below, this is not necessarily the case.

Gravity and magnetic data

In addition to the seismic data we have used gravity and magnetic data to constrain the crustal structure along Transects 1 and 2. Both regional and residual anomalies have been used in the gravity and magnetic modelling, and estimates of depth to magnetic basement show generally good agreement with the seismic data. Based on the seismic constraints, a crustal-scale polygon model for each transect was constructed. Densities were assigned to all polygons (Fig. 10) according to a standard velocity–density relation (Barton & Wood 1984; Christensen & Mooney 1995).

The observed Bouguer gravity anomalies along the two transects (Fig. 10) reflect the large-scale structures associated with crustal thinning. The regional field shows a positive gravity anomaly at the centre of the Viking Graben, an effect caused by the elevation of the Moho and the denser upper mantle. The shallow basement areas west of the Viking Graben and east of the Øygarden Fault Zone are also shown as gravity highs.

The gravity response of the crustal-scale models along Transects 1 and 2 was calculated and compared with the observed regional anomalies (Fig. 10). The gravity model of Transect 1 shows clearly that best fit with the observed gravity anomaly is achieved when the base of the crust is placed at the base of the high-velocity body, giving it a bulk density of 2.95 g cm^{-3}. The calculated gravity response shows a deviation of *c.* 40 mgal if we assign mantle densities to the high-velocity body. Similar results were obtained for Transect 2 (Fig. 10). Thus, it is necessary to explain the composition and origin of a rock body characterized by mantle velocities but densities intermediate between normal crust and mantle, before we can decide where to put the crust–mantle boundary beneath the Horda Platform. We will return to this below, when all results from the data analysis are integrated.

The density contrast between crust and mantle ranges from 0.47 to 0.50 g cm^{-3} along both transects, except for the lower-crustal high-velocity body, which has a crust–mantle density contrast of 0.25 g cm^{-3}. We also had to introduce a slightly higher basement density (2.73 g cm^{-3}) in the western parts of the two transects to obtain the best fit between the observed and calculated gravity anomalies. Under the central part of Transect 2 where thinning is most prominent, both gravity and magnetic data indicate a dense and magnetic body at lower-crustal level with a density of 3.0 g cm^{-3}. This high-density body may represent magmatic rocks intruding and underplating the crust beneath the southern Viking Graben in Jurassic time (Dixon *et al.* 1981; Underhill & Partington 1993).

Residual gravity and magnetic anomalies also indicate a Palaeozoic basin on the East Shetland Platform, which is given a density of 2.45 g cm^{-3} for the upper part and 2.55 g cm^{-3} for the lower part. Recent results from seismic reflection data on the East Shetland Platform just south of Transect 2 show a deep sedimentary basin of possible Palaeozoic age (Holloway *et al.* 1991; Platt 1995). Gravity modelling and magnetic depth estimates support the interpretation of this Palaeozoic basin, which has also been confirmed by drilling (Johnson *et al.* 1993; Duncan & Buxton 1995).

Crustal structure

Based on the integrated analysis of all geophysical data we have constructed the crustal transects in Fig. 11. In this paper we want to

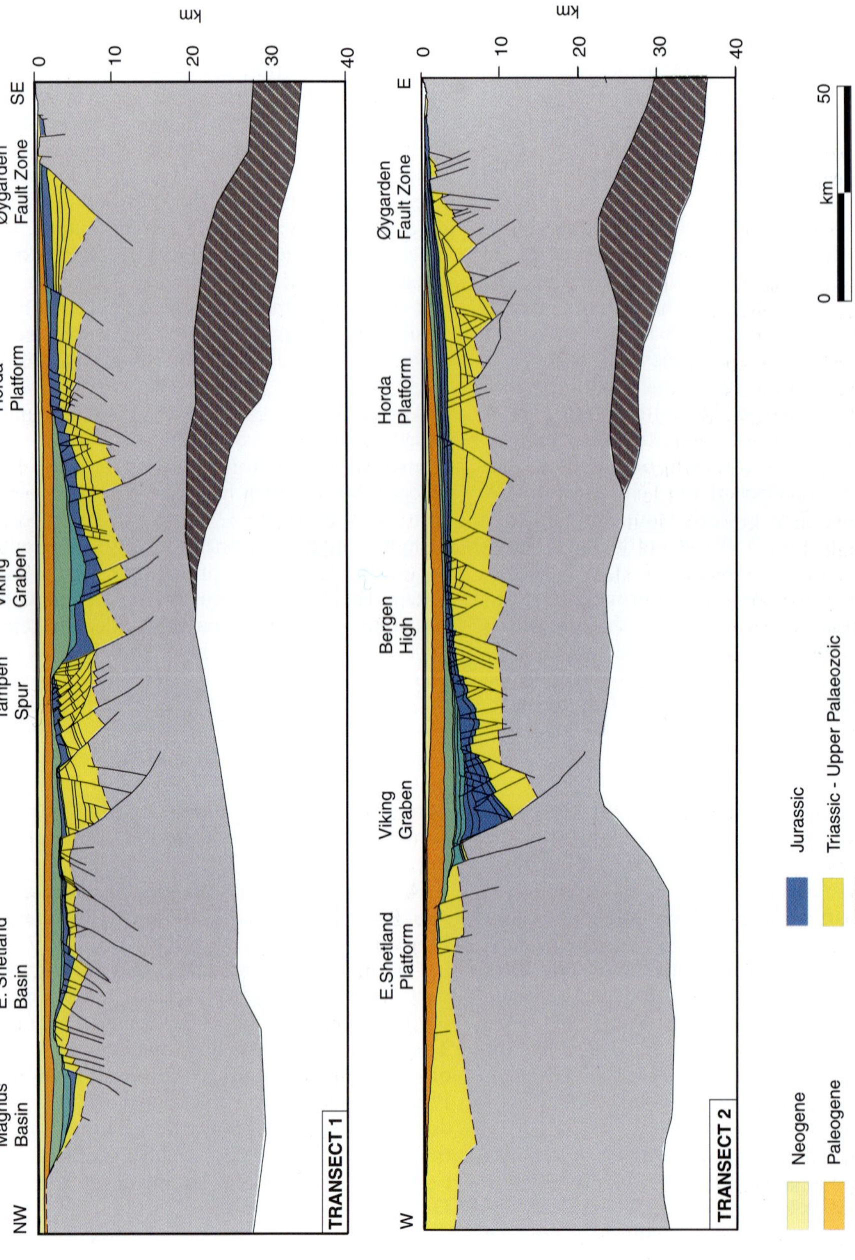

Fig. 11. Final crustal models for Transects 1 and 2, based on integration of all available geophysical and geological data.

focus on the geometry of crustal thinning and relate that to several phases of post-Caledonian rifting. However, before we can do that we must decide where the Moho is located beneath the Horda Platform and where the top of the crystalline basement is located beneath the Viking Graben.

Moho

The Moho discontinuity is defined by a rather abrupt increase of the P-wave velocity from about 7 km s^{-1} or less to typically in excess of 8.0 km s^{-1}, and is expected to be associated with a density contrast of about 0.5 g cm^{-3} (Steinhart 1967). Both geologists and geophysicists typically regard the seismically defined Moho as the base of the crust. However, the Moho does not necessarily correspond to the petrological boundary between crust and mantle (Griffin & O'Reilly 1987; Mengel & Kern 1992). Where crustal rocks are transformed into eclogites at the base of orogenically thickened crust, the position of the seismic Moho (defined as the first-order velocity discontinuity) is not identical to the petrological Moho (defined as the boundary between non-peridotitic crustal rocks and olivine-dominated mantle rocks). As a consequence, refraction seismic field studies may not detect eclogites as crustal rocks. The two contrasting bases for Moho definition have led to misunderstanding concerning the nature of the crust–mantle transition.

In the deep crust below the Horda Platform we are probably facing this problem. ESP data along Transect 1 demonstrate velocities of more than 8 km s^{-1} for the body, which is bounded by the two Moho candidates in the deep seismic reflection data (Fig. 9), suggesting that the body is not a part of the crust but represents mantle rocks. Gravity modelling, on the other hand, indicates a density of 2.95 g cm^{-3} for this body, intermediate between normal densities for crystalline basement and mantle. Introducing mantle densities in the high-velocity body gives rise to significant discrepancy between the observed and calculated gravity anomalies (Fig. 10). Furthermore, results from tectonostratigraphic modelling are not compatible with the shallow Moho candidate (Odinsen *et al.* 1999*b*).

This high-velocity, high-density body is interpreted to represent a deep crustal root of partially eclogitized rocks that are related to the onshore Western Gneiss Region following the conceptual model of Andersen *et al.* (1991) for the Late Caledonian extensional collapse in western Norway. In their model the crustal thickness was doubled during the Caledonian collision between the Laurentian and Baltic plates in Mid-Silurian time, followed by decoupling of the thermal boundary layer and extensional collapse of the orogen. This model permits the presence of eclogites within the lower crust as an effect of pro- and retrograde metamorphism during the subduction and eduction of the lower crust. These rocks were brought to the surface in the Western Gneiss Region in response to rapid uplift and exhumation. They are characterized by high velocities (up to 8.5 km s^{-1}) and densities (up to 3.6 g cm^{-3}). A lower crust where large segments underwent partial eclogitization as a result of fluid infiltration and deformation may explain the increased velocities observed toward the base of the crust.

Partial eclogitization leads to a mixture of granulite and eclogite as in the case of the eclogite breccias (Austrheim 1990) and creates rock bodies with intermediate densities between the protolith and the product. Eclogites found at the surface show variable degrees of back-reaction varying from nil to complete on the scale of an outcrop or even of a thin section. The variable degree of back-reaction is a question of fluid availability, rock permeability and kinetics rather than an indication of different uplift paths (Austrheim 1994). Detailed mapping of onshore western Norway (Austrheim 1987, 1990; Austrheim & Mørk 1988; Boundy *et al.* 1992) reveals that eclogites were formed along anastomosing shear zones up to 150 m thick and continuous over several kilometres. In an area of more than 50 km^2 the amount of eclogites formed was estimated as between 30 and 40% (Austrheim 1987). In a smaller area of 1 km^2, eclogites account for more than 50% of the exposed rocks (Austrheim & Mørk 1988).

A predominance of basic rocks is found in the Bergen Arcs whereas the Western Gneiss Region is much more heterogeneous with a large amount of quartzo-feldspathic rocks. The gabbro–eclogite relations studied so far have been limited to rather small bodies of a few hundred metres thickness. In extrapolating these observations to a lower crust of predominantly gabbroic composition, however, one would expect to find large volumes of incompletely reacted rocks. Internal shear zones or original fracture zones in such massifs should play an important role in metamorphic reactions. Thus the physical properties of such a crust would, to a large extent, be governed by the distribution of original zones of weakness and the generation of fractures and shear zones. The field evidence presented by Austrheim *et al.* (1997) demonstrates the importance of fluids in promoting reactions and hence controlling petrophysical properties. This fluid-controlled eclogitization can result in a partially eclogitized crust, with a mixture of rock types and metamorphic facies. These observations support indications of compositional heterogeneity in the lower crust suggested by the geophysical data.

The seismic velocity of a partly eclogitized crust is difficult to evaluate. Austrheim & Mørk (1988) assumed that the velocity would be a linear function of the volumetric ratio of the two rock types. However, Austrheim *et al.* (1997) suggested that it is possible that an abrupt change in velocity may occur at a certain percentage of eclogitization, which depends on seismic wavelength and the spatial distribution of eclogites and granulites with depth. If a marked increase in velocity occurs at low degree of eclogitization, the result will be a crust with a high velocity and relatively low density. This argument fits well with our geophysical observations and modelling results of Transects 1 and 2.

Brittan & Warner (1996) examined the importance of heterogeneity in biasing wide-angle seismic velocity measurements. Their synthetic seismic modelling indicates that velocities measured from wide-angle experiments may depend significantly upon the geometric arrangement and scale of the heterogeneities. Heterogeneities whose correlation length is larger than the dominant wavelength of the investigating wave will produce the greatest bias in seismic velocities; the seismic wave will always tend towards the fastest possible path. Thus, at large offsets, the inferred velocity will be larger than the true average velocity of crust traversed (Brittan & Warner 1996). Different geometric arrangements of the same heterogeneous composition can lead to a wide range of velocities measured by the surface observer. These results are important to consider when inferring material compositions from seismic velocities measured by wide-angle experiments.

We thus interpret the base of the crust to be at or near the base of the deepest reflections at 30–35 km depth, in contrast to most of the earlier published interpretations of deep seismic data in the northern North Sea (Beach 1986; Hurich & Kristoffersen 1988; Deemer & Hurich 1991; Færseth *et al.* 1995). On Transect 2 no ESPs are available, so the high-velocity body from Transect 1 cannot be detected by refraction and wide-angle reflection seismic methods. Instead, deep seismic reflection data (Figs 2 and 4) and gravity modelling (Fig. 10) provide the necessary constraints for the base of the crust (Fig. 11).

Hynes & Snyder (1991) also discussed the potential for crustal rocks below the Moho in the Scottish Caledonides. The rocks between the Moho at about 30 km and a strong reflector (W; McGeary & Warner 1985) at about 50 km have been interpreted as crustal, but with mantle velocities and densities. Hynes & Snyder (1991) investigated the mineralogy and seismic characteristics of such rocks. Their calculations indicated that continental crust of uniformly mafic composition could have seismic properties indistinguishable from those of mantle rocks. If, on the other hand, the continental crust contained a significant proportion of more felsic rocks (as in the Western Gneiss Region) it would not exhibit the high seismic velocities commonly attributed to the upper mantle.

Our results, supported by the work of Brittan & Warner (1996) and Austrheim *et al.* (1997), show that a partially eclogitized lower crust best fits the observed seismic and gravity responses. The non-typical velocity–density relation within the lower-crustal body may be explained by large-scale anisotropy related to the distribution of eclogites. The amount of eclogites and to what extent the lower-crustal rocks are more mafic in composition are difficult to constrain, making downward extrapolations from the surface distribution and composition of eclogites uncertain.

To illustrate the offshore–onshore correlation of the crustal units containing eclogites, we have constructed two regional transects across the northern North Sea and the Møre Basin continuing 200 km to the southeast onto the Norwegian mainland (Fig. 12). The North Sea transect (X–X′) is based on Transect 1 and onshore geology modified from Fossen (1992). Its onshore continuation crosses the Bergen Arcs, the allochthonous nappes, intersects the Hardangerfjord Shear Zone and finally ends at the Baltic Shield (Fig. 12). The crustal structure onshore is based on work by Sellevoll & Warrick (1971), who shot two seismic profiles across southwestern Norway. They found that the Moho discontinuity was at depths of 28–30 km beneath the coastline increasing eastward to 36–38 km beneath central parts of southern Norway. The depth to the Moho was determined from both Moho reflections (using t^2–x^2) and by travel times for the refracted wave, which gave a mantle velocity of 8.05 km s^{-1}.

Profile X–X′ (Fig. 12) shows how the crust gradually thins towards the Viking Graben, the depth to the base of the crust being reduced from *c.* 40 km to *c.* 20 km. It is important to note that the 8+ km s^{-1} velocities are confined to the lower-crustal body, indicating a different composition and/or amount of eclogites compared with the outcropping geology onshore.

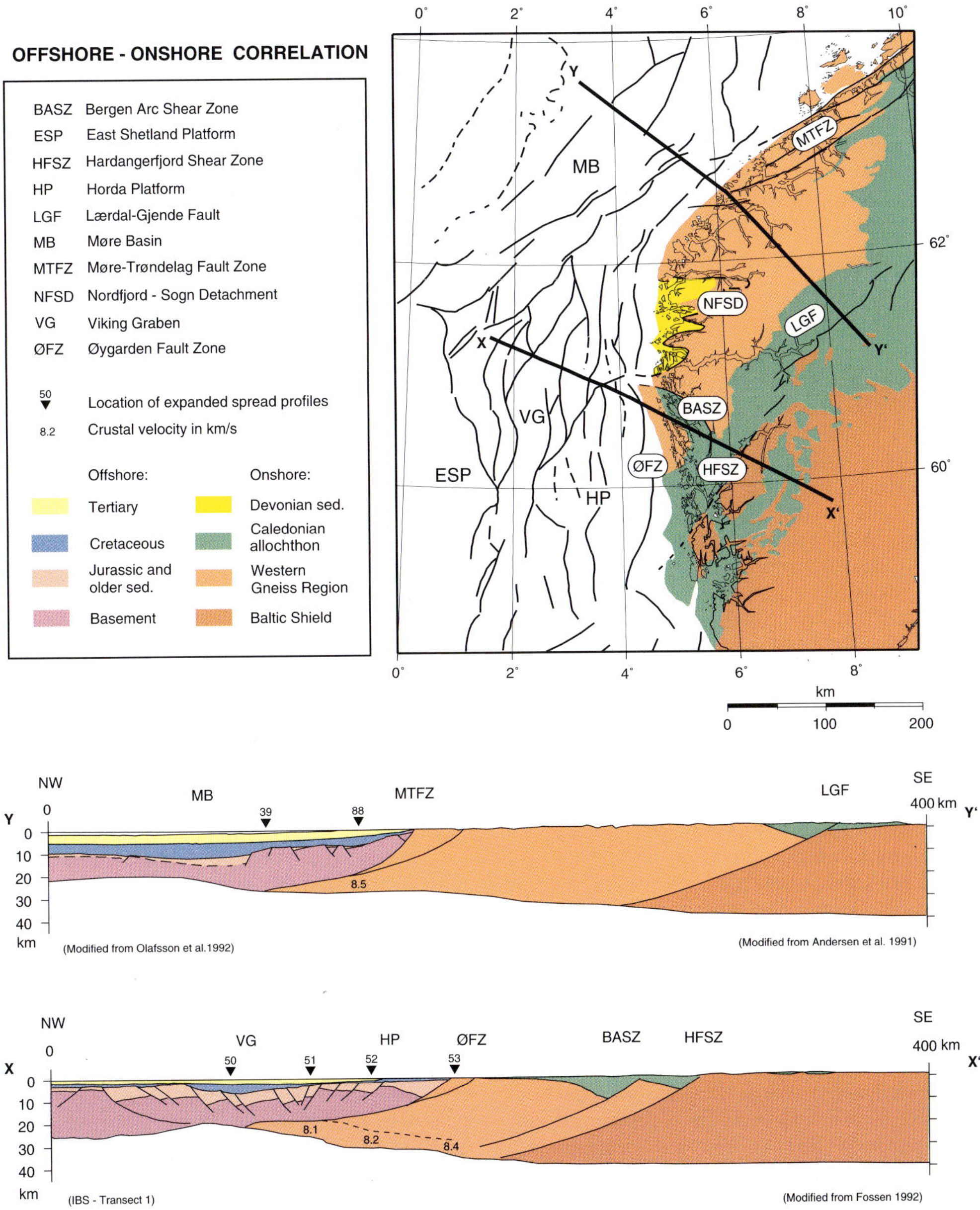

Fig. 12. Offshore–onshore correlation of major crustal features. Transect 1 is extended 200 km across the Caledonides of Western Norway onto the Baltic Shield (X–X′). A similar profile (Y–Y′) is constructed to the north across the Møre–Trøndelag Fault Zone. (Note the relationship between the lower-crustal high-velocity body offshore and the Western Gneiss Region exposed onshore.) References are given in figure.

The Møre transect (Y–Y′) is based on an offshore crustal-scale transect from Olafsson *et al.* (1992), with the onshore continuation modified from Andersen *et al.* (1991). Profile Y–Y′ (Fig. 12) crosses the Møre Basin and intersects the southern part of the Møre–Trøndelag Fault Zone at an almost perpendicular angle. It continues into the Norwegian mainland passing the Western Gneiss Region north of the Devonian basins, the allochthonous nappes, the Lærdal–Gjende Fault and ends at the Baltic Shield in the southeast. A high-velocity body ($8.5\,km\,s^{-1}$) was detected by Olafsson *et al.* (1992) beneath the eastern flank of the Møre Basin in a similar geological setting to our northern North Sea transect (Fig. 12). However, the density of this body has not been tested by gravity modelling. With only one ESP it is difficult to constrain the shape of this body.

Basement

The top of the crystalline crust is identified in parts of the study area as a strong seismic reflector, which is correlated with exploration well 31/6-1 drilled into crystalline basement on the Horda Platform (Fig. 1). However, it is not possible to trace this reflector over large distances. Wells penetrating crystalline basement are also known from the northeastern part of the northern North Sea (e.g. in blocks 35/3 and 35/9) and on the Shetland Platform, where about 30 UK wells have reached crystalline basement rocks (Johnson *et al.* 1993).

Further constraints on the top crystalline basement interpretation are based on high-quality filtered gravity and magnetic data. A fairly good match between magnetic basement depth estimates and interpretation from seismic data is achieved, except in the centre of the Viking Graben, where the interpretation is less constrained because of poorer seismic resolution. Here, the gravity and magnetic filtered residual anomalies in combination with the $6+\,km\,s^{-1}$ velocities in the ESP data are used in our interpretation of top crystalline basement.

The transects in Fig. 11 reveal large thicknesses of Upper Palaeozoic rocks. Magnetic depth estimates along the eastern part of Transect 2 coincide with the deepest reflector interpreted on line SG8043-101, indicating a considerable thickness of sedimentary strata (Devonian–?Carboniferous) beneath the wedge of syn-rift (Permian) origin preserved in the tilted half-grabens. The basin beneath the Tertiary cover of the East Shetland Platform in Transect 2 probably contains Upper Palaeozoic rocks. Devonian rocks were penetrated in UK well 3/29-2 located on Transect 2 on the terrace separating the East Shetland Platform and the Viking Graben (Fig. 11); and in several wells on the Platform (Johnson *et al.* 1993). Holloway *et al.* (1991) published a seismic line covering the East Shetland Platform just south of Transect 2 reflecting a deep (at least 6 km) Upper Palaeozoic basin. Unfortunately, we cannot differentiate between Devonian and Permian rocks in Transect 1 as was possible in Transect 2 (Fig. 11).

Crustal thinning and rifting

Basement and Moho relief along the transects (Fig. 11) allows us to study the geometry of crustal thinning. The basement thickness reflects cumulative crustal thinning in response to several post-Caledonian rift phases, assuming constant crustal thickness before the onset of crustal stretching and thinning. Stretching estimates based on crustal thinning have been compared with estimates based on other techniques such as subsidence analysis (Christiansson 2000; Fjeldskaar *et al.* 1999) or forward tectonostratigraphic modelling (Odinsen *et al.* 1999*b*).

The crustal thickness varies from *c.* 30 km and 36 km below the platforms on the graben flanks to 21–24 km below the axis of the Jurassic Viking Graben (Fig. 11). The transects show maximum basin depth where crystalline basement is thinnest and Moho shallowest, as predicted by the McKenzie model. Furthermore, the syn- and post-rift basin fill are superimposed, suggesting that the crustal thinning and lithosphere temperature field perturbations were laterally coincident.

A basement relief of >10 km is indicated along Transect 1 (Fig. 11). The crust is thickest in the east, close to the Norwegian coast, and becomes considerably thinner west of the Øygarden Fault Zone. The Horda Platform was relatively little affected by the Late Jurassic rifting, so most of this thinning must be related to Permian and earlier (Devonian?) stretching. Thus, the Permian rifting affected a wider area in the northern North Sea than did the Jurassic event. Transect 1 clearly shows that the Permian stretching must have been extensive, affecting an area from the Øygarden Fault Complex to the Hutton Fault in the East Shetland Basin (Figs 4 and 11). This is supported by correlation across pre-Jurassic faults with more than 3 km of vertical displacement in some cases, and a basin much wider than the present Viking Graben underlain by a thinned crystalline crust. However, when estimating the width of the Permian rift we must correct for the crustal stretching related to the younger (Late Jurassic) event.

Crustal thinning along Transect 2 follows the same pattern as for Transect 1, with maximum thinning of the crust under the present Viking Graben (Fig. 11). However, the pre-Jurassic basin geometry is here uncertain. There is hardly room for both a thick Permo-Triassic and Devonian–?Carboniferous sequence beneath the deeply buried Jurassic strata. A thick Devonian sequence is probably present in view of its existence both on the East Shetland Platform in the west and the Horda Platform in the east. If the deep basin fill consists mainly of Devonian–?Carboniferous strata, the main part of the Permian rift basin is located further east, on the Horda Platform. However, Permian movements on the 'Western Margin Fault' (Figs 4 and 11) cannot be ruled out.

Another pronounced difference in crustal structure between the two transects (Fig. 11) is the shift in polarity of some of the Late

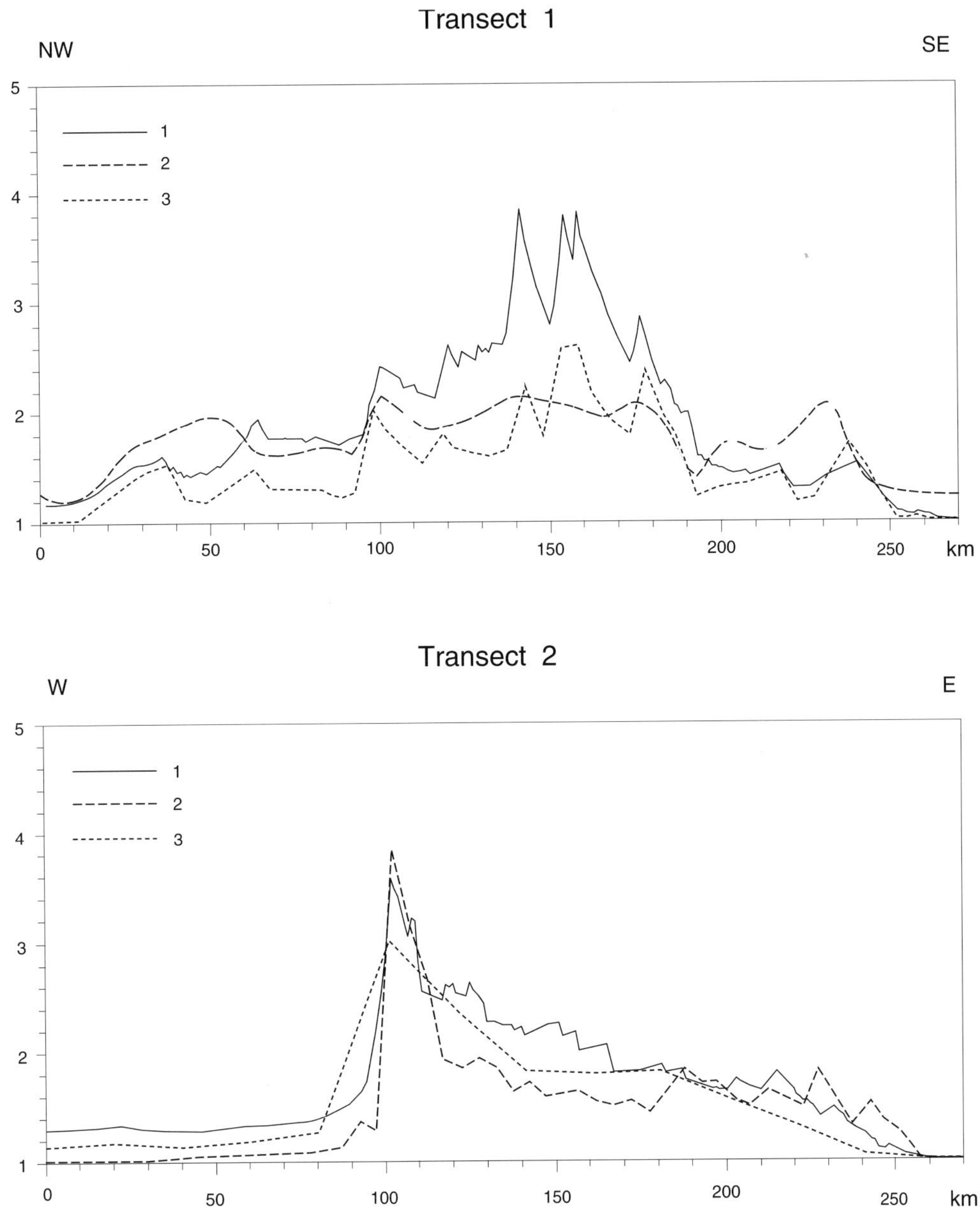

Fig. 13. β curves for crustal Transects 1 and 2. 1, From crustal thinning assuming an initial crustal thickness of 36 km; 2, from reverse modelling (Transect 1 Fjeldskaar *et al.* 1999; Transect 2, Christiansson 2000); 3, from forward modelling (Odinsen *et al.* 1999*b*).

Palaeozoic rift structures beneath the eastern Horda Platform. The polarity shift is probably connected to transfer faults running in east–west direction. Furthermore, Late Jurassic rifting appears to be more focused on Transect 2 compared with Transect 1 which gave rise to a very thick sequence of Upper Jurassic strata in the Viking Graben around 60°N (Figs 4 and 11). Most fault geometries seem to be planar, although some faults show listric geometries at depth. The Western Margin Fault, confining the Viking Graben to the west on Transect 2 show similarities in curvature and dip to the Hutton Fault and its deeper continuation on Transect 1 (Figs 4 and 11).

β-curves derived from crustal thinning have been calculated for the two transects assuming a constant initial crustal thickness of 36 km before Permian rifting (Fig. 13). There are, however, many uncertainties related to these curves. The uncertainties in identification of the Moho and top crystalline basement have been discussed in previous chapters. The Moho is identified along most of the transects but uncertainties exist with respect to velocities used in the depth conversion. Top crystalline basement is uncertain particularly beneath the Viking Graben. We do not have control on the distribution of Devonian strata all along the transects. Therefore we used the thickness of the crystalline part of the crust in our estimates. Strictly speaking this is not correct because the Devonian strata should have been considered as part of the crust involved in Permian and younger stretching. Furthermore, the assumption of a uniform crust of constant thickness is questionable. Lateral thickness variations were probably present following the extensional collapse of the Caledonides. We used the present-day thickness of 36 km east of the Øygarden Fault Zone as the initial crustal thickness assuming that Permian and younger stretching did not affect this area. This is not quite true, as we know that structures onshore were reactivated in Permian and Jurassic times (Torsvik *et al.* 1992).

All these uncertainties must be kept in mind when analysing the β curves derived from crustal thinning. Despite the uncertainties, the curves reflect the cumulative effect of several post-Caledonian rift events. In Fig. 13 we compare our β values with those obtained by forward modelling (Odinsen *et al.* 1999*b*) and reverse modelling (Christiansson 2000; Fjeldskaar *et al.* 1999) of both transects. The curves are similar in shape but the modelled values are systematically lower than those estimated from crustal thinning. A careful analysis of the various β curves is beyond the scope of this paper. For details and discussions we therefore refer to the modelling papers cited above.

Summary and conclusions

Two regional transects across the northern North Sea have been constructed based on an integrated analysis of deep seismic reflection and refraction data, as well as gravity and magnetic data. The shallow parts are based on high-quality conventional seismic reflection data calibrated against a large number of exploration wells. The new and partly reprocessed seismic data, combined with the other geophysical data, make possible a better documentation of the crustal configuration, such as the pre-Jurassic sediment distribution, basement and Moho relief, and deep fault geometry.

A lower-crustal body characterized by an 8+ km s^{-1} velocity and an average bulk density of 2.95 g cm^{-3} is present beneath the Horda Platform. This body probably represent a deep crustal root of partially eclogitized rocks, which formed in response to Caledonian collision and crustal thickening followed by extensional collapse of the orogen. The non-typical velocity–density relation within the lower-crustal body may be explained by heterogeneities related to the distribution of eclogites, which cause a bias in the seismic velocities measured from wide-angle data. The crust–mantle boundary is located at the base of this body and does not coincide with the seismically defined Moho at the top.

The geometry of crustal thinning reflects the cumulative effect of several post-Caledonian rift phases. Stretching estimates based on crustal thinning have been compared with estimates based on other techniques such as subsidence analysis and forward tectonostratigraphic modelling.

The interpretation presented here would not have been possible without the integration of various geological and geophysical data from both offshore and onshore areas.

This work was funded by the Commission of the European Union and the Norwegian Research Council in the framework of the DGXII – Joule Programme, sub-programme: Energy from fossil sources: Hydrocarbons, Integrated Basin Studies – The Dynamics of the Norwegian Margin. Norsk Hydro ASA, Saga Petroleum a.s. and Statoil a.s. provided data for this study. We are especially indebted to J. E. Lie, who reprocessed the NSDP lines. The paper also benefited from discussions with T. Andersen, P. T. Osmundsen, H. Austrheim, S. Planke and J. Skogseid. The close co-operation within the IBS project, particularly with R. Gabrielsen, A. Nøttvedt and T. Odinsen, is highly appreciated. We would like to thank N. Chamote-Rooke, Y. Kristoffersen and C. Hurich for thorough and constructive reviews of the manuscript. G. Farrow and S. Thompson improved the English.

References

AKI, K. & RICHARDS, P. G. 1980. *Quantitative Seismology, Theory and Methods, volume 1*. Freeman, San Francisco.

ALLEN, P. A. 1981. Devonian lake margin environments and processes, SE Shetland, Scotland. *Journal of the Geological Society, London*, **138**, 1–14.

ANDERSEN, T. B., JAMTVEIT, B., DEWEY, J. F. & SWENSSON, E. 1991. Subduction and eduction of continental crust: major mechanism during continent–continent collision and orogenic extensional collapse, a model based on the south Norwegian Caledonides. *Terra Nova*, **3**, 303–310.

AUSTRHEIM, H. 1987. Eclogitization of lower crustal granulites by fluids migration through shear zones. *Earth and Planetary Science Letters*, **81**, 221–232.

——1990. The granulite–eclogite facies transition: a comparison of experimental work and a natural occurrence in the Bergen Arcs, western Norway. *Lithos*, **25**, 163–169.

——1994. Eclogitization of the deep crust in continent collision zones. *Comptes Rendus de l'Académie des Sciences, Paris*, **319**, 761–774.

—— & MØRK, M. B. E. 1988. The lower continental crust of the Caledonian mountain chain: evidence from former deep crustal sections in western Norway. *In*: KRISTOFFERSEN, Y. (ed.) *Progress in Studies of the Lithosphere in Norway*. Norges Geologiske Undersøkelse Special Publication, **3**, 102–113.

——, ERAMBERT, M. & ENGVIK, A. K. 1997. Processing of crust in the root of the Caledonian continental collision zone: the role of eclogitization. *Tectonophysics*, **273**, 129–154.

BADLEY, M. E., PRICE, J. D., RAMBECH DAHL, C. & AGDESTEIN, T. 1988. The structural evolution of the northern North Viking Graben and its bearing upon extensional modes of graben formation. *Journal of the Geological Society, London*, **145**, 455–472.

BARTON, P. & WOOD, R. 1984. Tectonic evolution of the North Sea basin: crustal stretching and subsidence. *Geophysical Journal of the Royal Astronomical Society*, **79**, 987–1022.

BEACH, A. 1985. Some comments on sedimentary basin development in the northern North Sea. *Scottish Journal of Geology*, **21**, 493–512.

——1986. A deep seismic reflection profile across the northern North Sea. *Nature*, **323**, 53–55.

——, BIRD, T. & GIBBS, A. 1987. Extensional tectonics and crustal structure: deep seismic reflection data from the northern North Sea Viking Graben. *In*: COWARD, M. P., DEWEY, J. F. & HANCOCK, P. L. (eds) *Continental Extensional Tectonics*. Geological Society, London, Special Publications, **28**, 467–476.

BLUNDELL, D. J. 1990. Seismic images of continental lithosphere. *Journal of the Geological Society, London*, **147**, 895–913.

BOUNDY, T. M., FOUNTAIN, D. M. & AUSTRHEIM, H. 1992. Structural development and petrofabrics of eclogite facies shear zones, Bergen Arcs, western Norway: implication for deep crustal deformation processes. *Journal of Metamorphic Geology*, **10**, 127–146.

BRITTAN, J. & WARNER, M. 1996. Wide-angle seismic velocities in heterogeneous crust. *Geophysical Journal International*, **129**, 269–280.

BRUN, J. P. & TRON, V. 1993. Development of the North Viking Graben: inferences from laboratory modelling. *Sedimentary Geology*, **86**, 31–51.

CHRISTENSEN, N. I. & MOONEY, W. D. 1995. Seismic velocity structure and composition of the continental crust: a global view. *Journal of Geophysical Research*, **100**, 9761–9788.

CHRISTIANSSON, P. 2000. *Crustal structure and basin evolution of the Northern North Sea.* Dr Scient Thesis, University of Oslo.

——, BERGE, A. M. & FALEIDE, J. I. 1999*a*. *Analysis of Expanded Spread Profiles in the northern North Sea.* Department of Geology, University of Oslo, Internal Publication, **68**.

DEEMER, S. J. & HURICH, C. A. 1991. Comparison of coincident high-resolution wide-aperture and CDP profiling along the southwest coast of Norway. *In*: *Continental Lithosphere: Deep Seismic Reflections.* American Geophysical Union, Geodynamic Series, **22**, 435–442.

DIEBOLD, J. B. & STOFFA, P. L. 1981. The travel equation, tau–*p* mapping and the inversion of common midpoint data. *Geophysics*, **46**, 238–254.

DITCHA, E. 1998. *Pre-Tertiary evolution of the Stord Basin, northern North Sea.* Cand. scient. thesis, University of Oslo.

DIXON, J. E., FITTON, J. G. & FROST, R. T. C. 1981. The tectonic significance of post-Carboniferous igneous activity in the North Sea Basin. *In*: *Petroleum Geology of the Continental Shelf.* Heyden, London, 121–123.

DUNCAN, W. I. & BUXTON, N. W. K. 1995. New evidence for evaporitic Middle Devonian lacustrine sediments with hydrocarbon source potential on the East Shetland Platform, North Sea. *Journal of the Geological Society, London*, **152**, 251–258.

ELDHOLM, O. & MUTTER, J. C. 1986. Basin structure on the Norwegian margin from analysis of digitally recorded sonobuoys. *Journal of Geophysical Research*, **91**, 3763–3783.

EWING, J. 1963. Elementary theory of seismic reflection and refraction measurements. *In*: HILL, M. N. (ed.) *The Sea.* Wiley–Interscience, New York, 3–19.

FICHLER, C. & HOSPERS, J. 1990. Deep crustal structure of the northern North Viking Graben: results from deep reflection seismic and gravity data. *Tectonophysics*, **178**, 241–254.

FJELDSKAAR, W., JOHANSEN, H., TER VOORDE, M., CHRISTIANSSON, P. & FALEIDE, J. I. 1999. Modelling of crustal thinning and the effects of possible intra plate stress in the northern Viking Graben. *In*: *Basin Modelling.* AAPG Memoir.

FLINN, D. 1985. The Caledonides of Shetland. *In*: GEE, D. G. & STURT, B. A. (eds) *The Caledonide Orogen – Scandinavia and Related Areas*. Wiley, Chichester.

FOSSEN, H. 1992. The role of extensional tectonics in the Caledonides of south Norway. *Journal of Structural Geology*, **14**, 1033–1046.

FÆRSETH, R. B., GABRIELSEN, R. H. & HURICH, C. A. 1995. Influence of basement in structuring of the North Sea Basin, offshore southwest Norway. *Norsk Geologisk Tidsskrift*, **75**, 105–119.

GABRIELSEN, R. H. 1986. Structural elements in graben systems and their influence on hydrocarbon trap types. *In*: SPENCER, A. M. *et al.* (eds) *Habitat of Hydrocarbons on the Norwegian Continental Shelf*. Graham and Trotman, London, 55–60.

——, FÆRSETH, R., STEEL, R. J., IDIL, S. & KLØVJAN, O. S. 1990. Architectural styles of basinfill in the northern Viking Graben. *In*: BLUNDELL, D. J. & GIBBS, A. D. (eds) *Tectonic Evolution of the North Sea Rifts*. Clarendon, Oxford, 158–179.

GIBBS, A. 1987*a*. Linked tectonics of the northern North Sea basin. *In*: BEAUMONT, C. & TANKARD, A. J. (eds) *Sedimentary Basins and Basin Formation Mechanisms*. Canadian Society of Petroleum Geologists, Memoir, **12**, 163–171.

——1987*b*. Deep seismic profiles in the northern North Sea. *In*: BROOKS, J. & GLENNIE, K. (eds) *Petroleum Geology of North West Europe*. Graham and Trotman, London, 1025–1028.

GILTNER, J. P. 1987. Application of extensional models to the northern Viking Graben. *Norsk Geologisk Tidsskrift*, **67**, 339–352.

GLENNIE, K. W. 1984. The structural framework and the pre-Permian history of the North Sea Area. *In*: GLENNIE, K. W. (ed.) *Introduction to the Petroleum Geology of the North Sea*, Blackwell Scientific, Oxford, 17–39.

GRIFFIN, W. L. & O'REILLY, S. Y. 1987. Is the continental Moho the crust–mantle boundary? *Geology*, **15**, 241–244.

HANCOCK, J. M. 1990. Cretaceous. *In*: GLENNIE, K. W. (ed.) *Introduction to the Petroleum Geology of the North Sea*. Blackwell Scientific, Oxford, 61–73.

HARRISON, M. G. 1987. *An interpretation of deep seismic reflection profiles in the northern North Sea: implications for extension models*. MSc thesis, University of London, Royal Holloway and Bedford New College.

HARTZ, E. & ANDRESEN, A. 1995. Caledonian sole thrust of East Greenland: a crustal scale Devonian extensional detachment? *Geology*, **23**, 637–640.

HEGNA, S. 1989. *Kinematiske metoder for bestemmelse av hastigheter fra marin-seismiske data med anvendelse i østlige deler av Loppahøyden*. Cand. scient. thesis, University of Oslo.

HOLLIGER, K. 1987. *Crustal structure of the North Sea*. Diploma thesis, ETH Zurich.

—— & KLEMPERER, S. 1989. A comparison of the Moho interpreted from gravity data and from deep seismic reflection data in the northern North Sea. *Geophysical Journal*, **97**, 247–258.

HOLLOWAY, S., REAY, D. M., DONATO, J. A. & BEDDOE-STEPHENS, B. 1991. Distribution of granite and possible Devonian sediments in part of the East Shetland Platform, North Sea. *Journal of the Geological Society, London*, **148**, 635–638.

HOSPERS, J. & EDIRIWEERA, K. W. 1988. Mapping of the top of the crystalline continental crust in the Viking Graben area, North Sea. *In*: KRISTOFFERSEN, Y. (ed.) *Progress in Studies of the Lithospheres in Norway*. Norges Geologiske Undersøkelse, Special Publication, **3**, 21–28.

—— & ——1991. Depth and configuration of the crystalline basement in the Viking Graben area, northern North Sea. *Journal of the Geological Society, London*, **148**, 261–265.

HOSSACK, J. R. 1984. The geometry of listric growth faults in the Devonian basins of Sunnfjord, west Norway. *Journal of the Geological Society, London*, **141**, 629–638.

HURICH, C. A. 1996. Kinematic evolution of the lower plate during intracontinental subduction: an example from the Scandinavian Caledonides. *Tectonics*, **15**, 1248–1263.

—— & KRISTOFFERSEN, Y. 1988. Deep structure of the Caledonide orogen in southern Norway: new evidence from marine seismic reflection profiling. *In*: KRISTOFFERSEN, Y. (ed.) *Progress in Studies of the Lithosphere in Norway*. Norges Geologiske Undersøkelse, Special Publication, **3**, 96–101.

HYNES, A. & SNYDER, D. B. 1991. Deep-crustal mineral assemblages and potential for crustal rocks below the Moho in the Scottish Caledonides. *Geophysical Journal International*, **123**, 323–339.

IWAZAKI, T., SELLEVOLL, M. A., KANAZAWA, T., VEGGELAND, T. & SHIMAMURAH, H. 1994. Seismic refraction crustal study along the Sognefjord, south-west Norway, employing ocean-bottom seismometers. *Geophysical Journal International*, **119**, 791–808.

JACKSON, H. R., FALEIDE, J. I. & ELDHOLM, O. 1990. Crustal structure of the sheared southwestern Barents Sea continental margin. *Marine Geology*, **93**, 119–146.

JOHNSON, H., RICHARDS, P. C., LONG, D. & GRAHAM, C. C. 1993. *United Kingdom Offshore Regional Report: the Geology of the Northern North Sea*. British Geological Survey, HMSO, London.

JORDT, H., FALEIDE, J. I., BJØRLYKKE, K. & IBRAHIM, M. T. 1995. Cenozoic sequence stratigraphy of the Central and Northern North Sea basin. Tectonic development, sediment distribution and provenance areas. *Marine and Petroleum Geology*, **12**, 845–879.

——, ——, THYBERG, B. I. & NØTTVEDT, A. 1999. Cenozoic evolution of the central and northern North Sea with focus on differential vertical movements of the basin floor and surrounding clastic source areas. *This volume*.

KITTERØD, N.-O. 1986. *Hastighets variasjoner med dypet, tolkningsprosedyrer for ESP og sonarbøye data*. Cand. scient. thesis, University of Oslo.

KLEMPERER, S. L. 1988. Crustal thinning and nature of extension in the northern North Sea from deep seismic reflection profiling. *Tectonics*, **7**, 803–821.

—— & HOBBS, R. 1991. *Deep Seismic Reflection Profiles around the British Isles, the BIRPS Atlas*. Cambridge University Press, Cambridge.

—— & HURICH, C. A. 1990. Lithospheric structure of the North Sea from deep seismic reflection profiling. *In*: BLUNDELL, D. J. & GIBBS, A. (eds) *Tectonic Evolution of the North Sea Rifts*. Oxford University Press, Oxford, 37–63

—— & WHITE, N. 1989. Coaxial stretching or lithospheric simple shear in the North Sea? Evidence from deep seismic profiling and subsidence. *In*: TANKARD, A. J. & BALKWILL, H. (eds) *Extensional Tectonics and Stratigraphy of the North Atlantic Margins*. American Association of Petroleum Geologists, Memoir, **46**, 511–522.

KUSZNIR, N. J. & MATTHEWS, D. H. 1988. Deep seismic reflections and the deformational mechanism of the continental lithosphere. *Journal of Petrology*, Special Lithosphere Issue, 66–87.

—— & ZIEGLER, P. A. 1992. The mechanism of continental extension and sedimentary basin formation: a simple shear/pure shear flexural cantilever model. *Tectonophysics*, **215**, 117–131.

——, MARSDEN, G. & EGAN, S. S. 1991. A flexural cantilever simple shear/pure shear model of continental lithosphere extension: application to the Jeanne d'Arc basin and Viking Graben. *In*: ROBERTS, A. M., YIELDING, G. & FREEMAN, B. (eds) *The Geometry of Normal Faults*. Geological Society, London, Special Publications, **56**, 41–60.

——, ROBERTS, A. M. & MORLEY, C. 1995. Forward and reverse modelling of rift basin formation. *In*: LAMBIASE, J. (ed.) *Hydrocarbon Habitat in Rift Basins*. Geological Society, London, Special Publications, **80**, 33–56.

LEPICHON, X., EWING, J. & HOUTZ, R. E. 1968. Deep-sea sediment velocity determination made while reflection profiling. *Journal of Geophysical Research*, **73**, 2597–2614.

LERVIK, K. S., SPENCER, A. M. & WARRINGTON, G. 1989. Outline of Triassic stratigraphy and structure in central and northern North Sea. *In*: COLLINSON, J. D. (ed.) *Correlation in Hydrocarbon Exploration*. Graham and Trotman, London, 173–189.

MCGEARY, S. & WARNER, M. 1985. Seismic profiling the continental lithosphere. *Nature*, **317**, 795–797.

MCKENZIE, D. P. 1978. Some remarks on the development of sedimentary basins. *Earth and Planetary Science Letters*, **40**, 25–40.

MENGEL, K. & KERN, H. 1992. Evolution of the petrological and seismic Moho – implications for the continental crust–mantle boundary. *Terra Nova*, **4**, 109–116.

NORTON, M. 1986. Late Caledonian extension in western Norway: a response to extreme crustal thickening. *Tectonics*, **5**, 195–204.

——1987. The Nordfjord–Sogn Detachment, W. Norway. *Norsk Geologisk Tidsskrift*, **67**, 93–106.

NØTTVEDT, A., GABRIELSEN, R. H. & STEEL, R. J. 1995. Tectonostratigraphy and sedimentary architecture of rift basins; with reference to the northern North Sea. *Marine and Petroleum Geology*, **12**, 881–901.

ODINSEN, T., CHRISTIANSSON, P., GABRIELSEN, R. H., FALEIDE, J. I. & BERGE, A. M. 1999*a*. The geometries and deep structure of the northern North Sea rift system. *This volume*.

——, REEMST, P., VAN DER BEEK, P., FALEIDE, J. I. & GABRIELSEN, R. H. 1999*b*. Permo-Triassic and Jurassic extension in the northern North Sea: results from tectonostratigraphic forward modelling. *This volume*.

OLAFSSON, I., SUNDVOR, E., ELDHOLM, O. & GRUE, K. 1992. Møre Margin: crustal structures from analysis of expanded spread profiles. *Marine Geophysical Research*, **14**, 137–162.

OSMUNDSEN, P. T. 1995. *Late-orogenic structural geology and Devonian basin formation in western Norway: a study from the hanging wall of the Nordfjord–Sogn detachment in the Sunnfjord region*. Dr. scient. thesis, University of Oslo.

PINET, B. 1989. Deep seismic profiling and sedimentary basins. *Bulletin de la Société Géologique de France*, **8**, 749–766.

PLATT, N. H. 1995. Structures and tectonics of the northern North Sea: new insights from deep penetration regional seismic data. *In:* LAMBIASE, J. J. (ed.) *Hydrocarbon Habitat in Rift Basins*. Geological Society, London, Special Publications, **80**, 103–113.

RESTON, T. J. 1990. Shear in the lower crust during extension, not so pure and simple. *Tectonophysics*, **173**, 175–183.

SCOTT, D. L. & ROSENDAHL, B. R. 1989. North Viking Graben: an East African perspective. *AAPG Bulletin*, **73**, 155–165.

SELLEVOLL, M. & WARRICK, R. E. 1971. A refraction study of the crustal structure in southern Norway. *Bulletin of the Seismological Society of America*, **61**, 457–471.

SERANNE, M. 1992. Devonian extensional tectonics versus Carboniferous inversion in the northern Orcadian basin. *Journal of the Geological Society, London*, **149**, 27–37.

—— & SEGURET, M. 1987. The Devonian basins of western Norway: tectonics and kinematics of an extending crust. *In*: COWARD, M. P., DEWEY, J. F. & HANCOCK, P. L. (eds) *Continental Extensional Tectonics*. Geological Society, London, Special Publications, **28**, 537–548.

STEEL, R. J. 1993. Triassic–Jurassic megasequence stratigraphy in the northern North Sea: rift to post-rift evolution. *In*: PARKER, J. R. (ed.) *Petroleum Geology of Northwest Europe. Proceedings of the 4th Conference*. Geological Society, London, 299–315.

—— & RYSETH, A. 1990. The Triassic–early Jurassic succession in the northern North Sea: megasequence stratigraphy and intra-Triassic tectonics. *In*: HARDMAN, R. F. P. & BROOKS, J. (eds) *Tectonic Events Responsible for Britain's Oil and Gas Reserves*. Geological Society, London, Special Publications, **55**, 139–168.

——, SIEDLECKA, A. & ROBERTS, D. 1985. The Old Red Sandstone basins of Norway and their deformation: a review. *In*: GEE, D. G. & STURT, B. A. (eds) *The Caledonide Orogen – Scandinavia and Related Areas*. Wiley, New York, 293–315.

STEINHART, J. S. 1967. Mohorovicic discontinuity. *In*: RUNCORN, K. (ed.) *International Dictionary of Geophysics, Vol. 2*. Pergamon, Oxford, 991–994.

STOFFA, P. L. & BUHL, P. 1979. Two ship multichannel seismic experiments for deep crustal studies; expanded spread and constant offset profiles. *Journal of Geophysical Research*, **84**, 7645–7660.

——, ——, DIEBOLD, J. B. & WENTZEL, F. 1981. Direct mapping of seismic data in the domain of intercept time and ray parameter – a plane wave decomposition. *Geophysics*, **45**, 255–267.

TORSVIK, T. H., ANDERSEN, T. A., EIDE, E. A. & WALDERHAUG, H. J. 1997. The age and tectonic significance of dolerite dykes in western Norway. *Journal of the Geological Society, London*, **154**, 961–973.

——, STURT, B. A., SWENSSON, E., ANDERSEN, T. B. & DEWEY, J. F. 1992. Paleomagnetic dating of fault rocks: evidence for Permian and Mesozoic movements and brittle deformation along the Dalsfjord Fault, west Norway. *Geophysical Journal International*, **109**, 565–580.

UNDERHILL, J. R. & PARTINGTON, M. A. 1993. Jurassic thermal doming and deflation in the North Sea: implications of the sequence stratigraphic - evidence. *In*: PARKER, J. R. (ed.) *Petroleum Geology of Northwest Europe. Proceeding of the 4th Conference*. Geological Society, London, 337–345.

WERNICKE, B. 1985. Uniform-sense normal simple shear of the continental lithosphere. *Canadian Journal of Earth Science*, **22**, 108–125.

WHITE, N. 1989. Nature of lithospheric extension in the North Sea. *Geology*, **17**, 111–114.

——1990. Does the uniform stretching model work in the North Sea? *In*: BLUNDELL, D. J. & GIBBS, A. D. (eds) *Tectonic Evolution of the North Sea Rifts*. Oxford University Press, Oxford, 217–239.

—— & MCKENZIE, D. P. 1988. Formation of the 'steer's head' geometry of sedimentary basins by differential stretching of the crust and mantle. *Geology*, **16**, 250–253.

ZERVOS, F. A. 1987. A compilation and regional interpretation of the northern North Sea gravity map. *In*: COWARD, M. P., DEWEY, J. F. & HANCOCK, P. L. (eds) *Continental Extensional Tectonics*. Geological Society, London, Special Publications, **28**, 477–493.

ZIEGLER, P. A. 1982. *Geological Atlas of Western and Central Europe*. Shell International, The Hague.

The geometries and deep structure of the northern North Sea rift system

TORE ODINSEN,[1,4] PETER CHRISTIANSSON,[2,5] ROY H. GABRIELSEN,[1] JAN INGE FALEIDE[2] & ANKER M. BERGE[3]

[1] *Department of Geology, University of Bergen, Allégaten 41, N-5007 Bergen, Norway*
[2] *Department of Geology, University of Oslo, P.O. Box 1047 Blindern, N-0316 Oslo, Norway*
[3] *Norsk Hydro Research Centre, N-5020 Bergen, Norway*
[4] *Present address: Statoil, N-5020 Bergen, Norway*
[5] *Present address: Norsk Hydro ASA, P.O Box 200, N-1321 Stabekk, Norway*

Abstract: The enormous quantity of commercial reflection seismic lines across the North Sea Basin have made the area one of the most thoroughly studied continental settings in the world. Further insight in the deep architecture of the crust is provided by *c.* 10 000 km deep reflection seismic data. Unfortunately, these unique databases have rarely been combined systematically to constrain possible tectonic models for the area. This paper is built on a full integration of high-quality commercial lines (7 s twt) and the deep (15 s twt) NSDP84-1 and -2 lines. The deep lines have been post-stack reprocessed and depth-converted. A number of deep wells have provided stratigraphic control along the lines. The overall reflective pattern in the lines divides the crust in three, with a reflective upper and lower crust separated by a less reflective middle crust. The lateral changes in reflectivity matches the observed variation in crustal thickness, where the thinnest crust coincides with the Viking Graben area with a total crustal thickness of 21–24 km, increasing to 30–36 km in the platform areas. The lower crust is seen as an undulating 4–10 km thick band with shallow dipping reflections, with a Moho that consists of reflections with variable lateral thickness and amplitude, rather than one single strong reflection. The structural analysis shows that the crust is cut by a number of large normal faults with varying geometries. It is assumed that some of these major faults are long-lived features rooted in old basement grains. The most spectacular normal faults developed during the Permo-early Triassic extensional phase, but were often reactivated during the Jurassic extensional phase, and with continued minor fault movement into the Cretaceous thermal cooling period. Integration of commercial and deep reflection seismic sections shows that three detachment levels are present within the crust. These levels, which control changes in fault geometries, are believed to represent lateral rheological interfaces combined with or intersected by long-lived zones of weaknesses. The uppermost level is represented by supra-basement low-angle normal faults controlled by gravity and/or lithological changes during extension. An intra-basement (middle crust) level between 5 and 7 s (twt) coincides with decreasing dip of the larger basin bounding faults. The lower crust is the deepest detachment level, which probably exerts control on the geometric changes of the upper-mantle shear zones and the largest crustal normal faults.

The structural framework for Mesozoic to Tertiary basin development in the northern North Sea (Fig. 1) has become increasingly well documented during the last decade (McKenzie 1978; Sclater *et al.* 1986; Beach *et al.* 1987; Giltner 1987; Badley *et al.* 1988; Gabrielsen *et al.* 1990; Roberts *et al.* 1990). Discrete phases of crustal extension, with fault block rotation and sub-basin development, are separated by thermal cooling and broad basinal subsidence from Permian to Cretaceous time. However, the detailed nature, significance and areal extent of the faulting and its relation to differential subsidence and sedimentation are still under debate. In particular, the pre-Triassic evolution is debatable, as the thick sedimentary layers observed within the deep half-grabens on reflection seismic lines are not penetrated by wells. Data from the southwestern part of Norway and sequence stratigraphy studies indicate that rifting started in Permian time (Færseth 1978; Torsvik *et al.* 1992; Steel 1993). It is also acknowledged that Devonian

From: Nøttvedt, A. *et al.* (eds) *Dynamics of the Norwegian Margin*. Geological Society, London, Special Publications, **167**, 41–57. 1-86239-056-8/00/$15.00

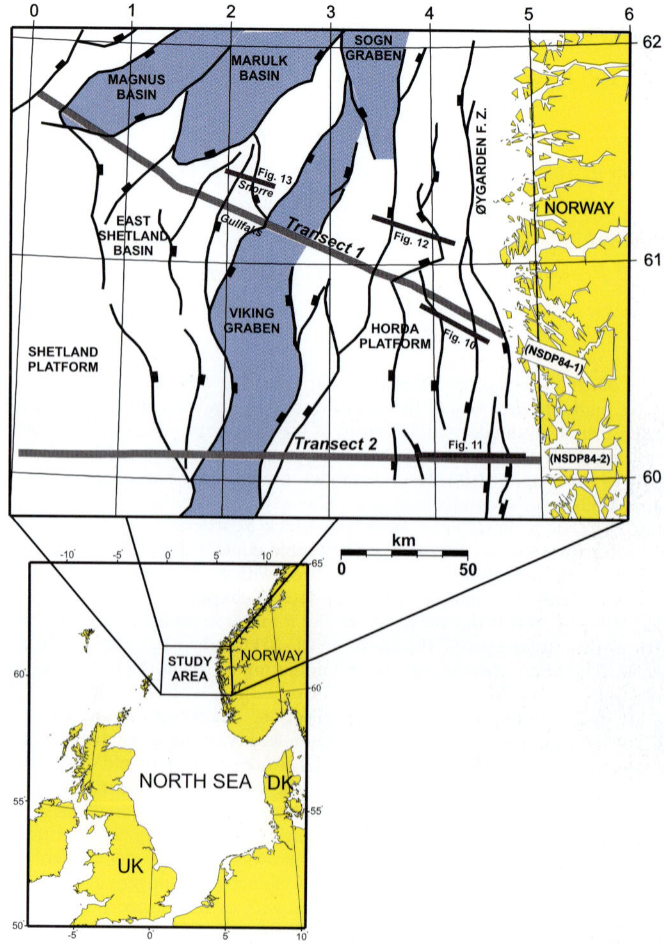

Fig. 1. Key map and main structural elements, northern North Sea (reference level is base Cretaceous). Areas that are primarily affected by mid-Jurassic–early Cretaceous stretching in grey. Locations of transects 1 and 2 and semi-regional commercial lines used in this paper are displayed.

(–Carboniferous?) sediments, which are well known from the UK sector, may cover basement in the deeper parts of the northern North Sea, as suggested by Platt (1995), Færseth *et al.* (1995*a*) and Færseth (1996).

For a more complete understanding of the evolution and interplay between the different extensional phases, basin geometry and subsidence history, it is essential to analyse this pre-Jurassic evolution, which also encompasses Caledonian and post-Caledonian structures. Compaction of Permo-Triassic basin fill and residual Permo-Triassic thermal anomalies may enhance Triassic and Jurassic subsidence and may therefore cause overestimation of stretching estimates derived from subsidence analysis assuming only one rift phase (in mid-Jurassic–early Cretaceous time). To succeed in this evaluation, it is essential to perform a structural and geometrical analysis of sufficient detail and completeness to provide a base for modelling. The present work is an attempt to provide this database, which has been used in the modelling of two crustal transects. The modelling results are presented separately (Odinsen *et al.* 2000).

To achieve this aim, we have combined and correlated the deep reflection seismic lines NSDP84-1 and -2 (15 s twt) (Figs 2a and 3a) with conventional (7 s twt) reflection seismic lines to study the structure at different crustal levels. This analysis of the deep crustal structure also allows us to recognize the pre-Jurassic structure

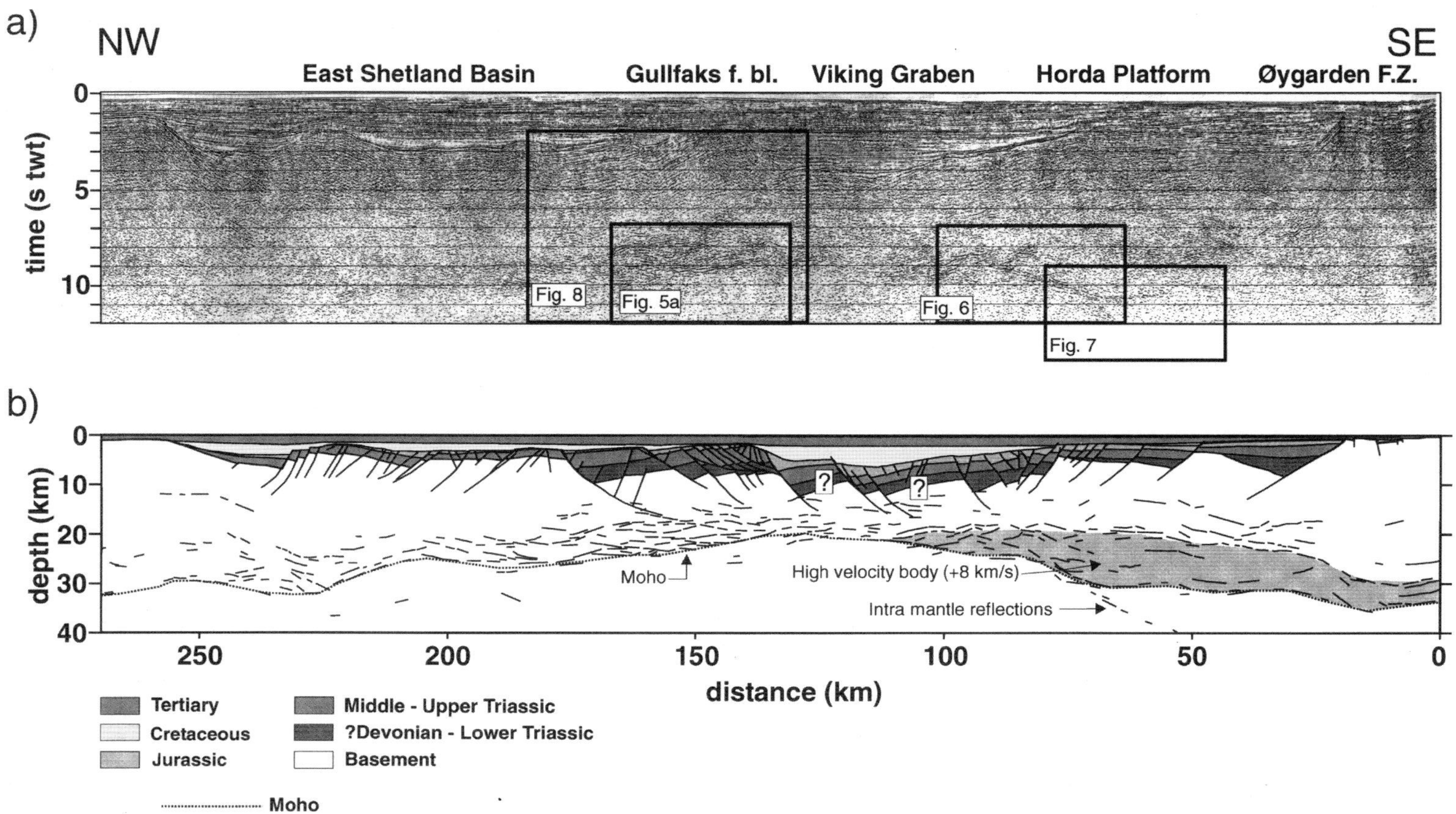

Fig. 2. (**a**) Time-section of crustal transect 1. Boxes show the locations of figures referred to in the text. (**b**) Depth-converted crustal transect 1. (For location, see Fig. 1.) No vertical exaggeration.

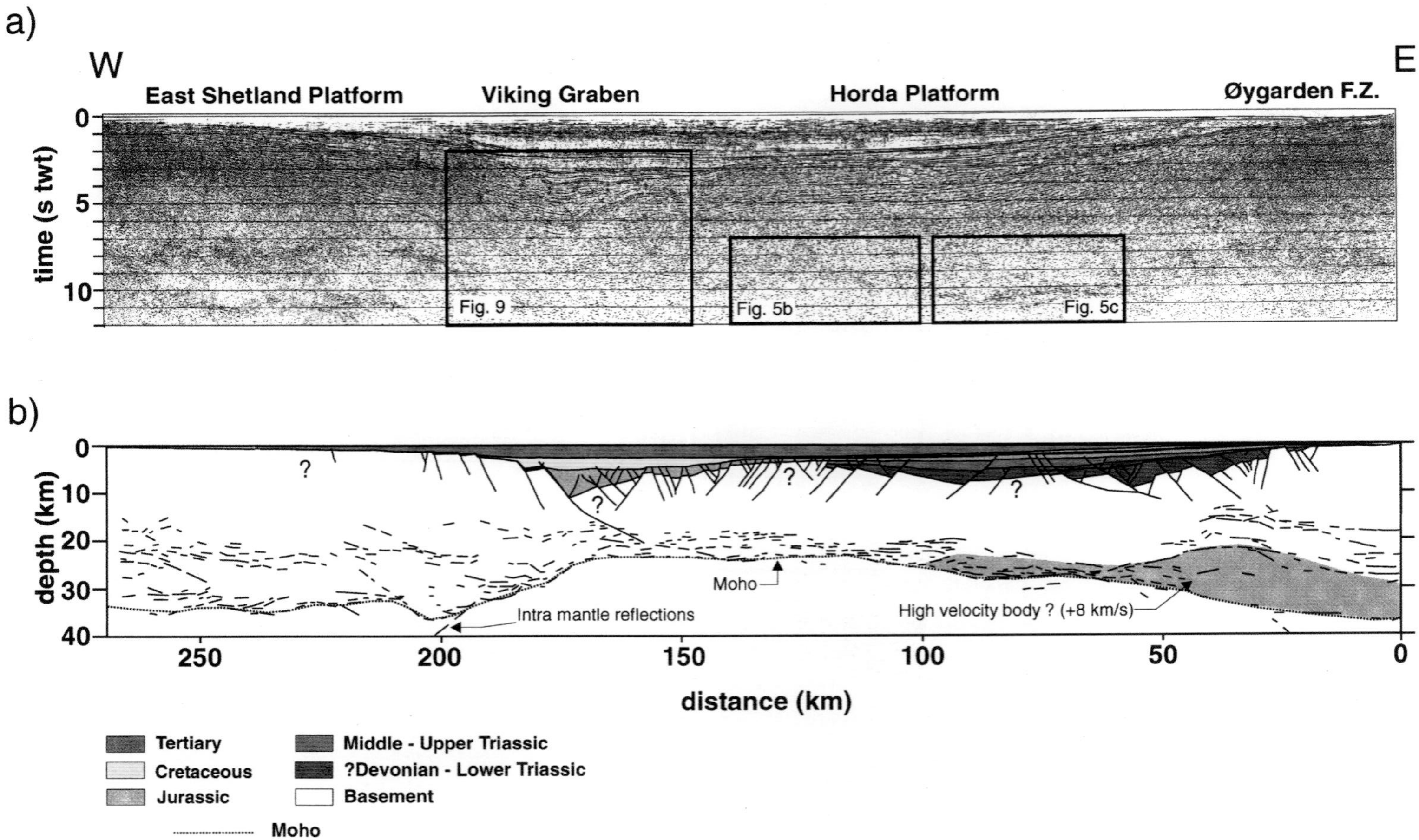

Fig. 3. (**a**) Time-section of crustal transect 2. Boxes show the locations of figures referred to in the text. (**b**) Depth-converted crustal transect 2. (For location, see Fig. 1.) No vertical exaggeration.

and particularly the architecture of the lower crust. Lines NSDP84-1 and -2 (hereafter referred to as transect 1 and transect 2 respectively) have been post-stack reprocessed and depth-converted (Figs 2b and 3b). Intersecting wells have been used to establish stratigraphic control and have provided velocity data in the upper 3–4 km of the crust, whereas analyses of expanded spread profiles (ESP) have given velocities at deeper crustal levels (Christansson *et al.* 2000). In the first stage of the analysis the geometry of the upper, middle and lower crust, as well as the reflection Moho, have been determined for transects 1 and 2. The second stage of the analysis used the observations derived from commercial reflection seismic lines.

Extensional models

Following the publication of deep seismic reflection data acquired in 1983 and 1984 (Beach 1986; Beach *et al.* 1987; Gibbs 1987; Klemperer 1988), three principal models (Fig. 4) for the deep structure of the North Sea Basin have coexisted. Both the symmetrical pure shear model, favoured for the Viking Graben by Giltner (1987) and Badley *et al.* (1988), and the inhomogeneous shear model, proposed by Klemperer (1988) for this area, may be seen as developments of the McKenzie (1978) model. The asymmetrical simple shear model, congruent with the model of Wernicke (1985), was proposed for the Viking Graben by Beach (1986), Beach *et al.* (1987), Gabrielsen (1989) and Scott & Rosendahl (1989). Finally, different types of delamination models (Lister *et al.* 1986), which allow for differential depth-dependent extension (Royden & Keen 1980), have been proposed for different stages of the development of the northern North Sea by Coward (1986), Fossen *et al.* (2000) and Ter Voorde *et al.* (2000).

Extensional models are generally based on the interpreted geometries of major crustal faults. However, such models also have to consider rheology and the general layering of the continental crust (Dunbar & Sawyer 1989). Such layering results in large jumps in yield strength at layer interfaces, and thereby a vertical alteration of stronger and weaker intra-crustal zones. In a simple two-layered mineralogical model, where felsic rocks dominate the upper crust and mafic rocks dominate the lower crust, one would expect normal faults to detach at the weaker interface between these two layers. However, as shown by Kusznir & Park (1987), depending

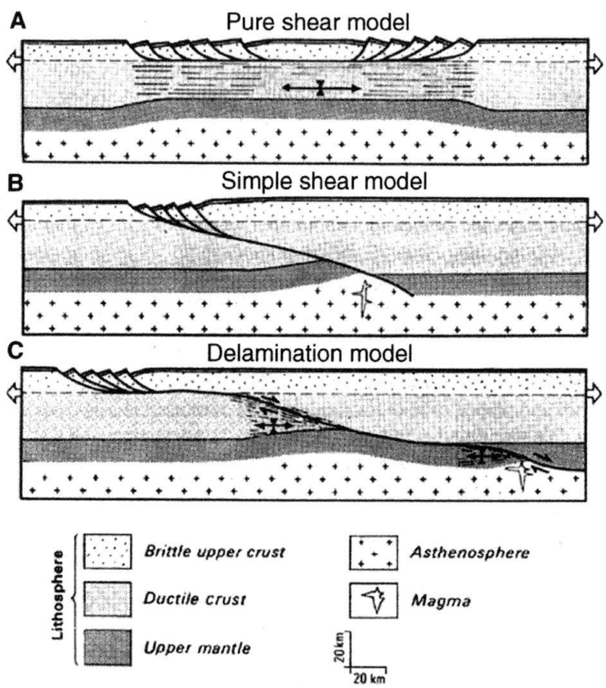

Fig. 4. Three models for continental extension (after Lister *et al.* 1986, fig. 1).

on the crustal composition, crustal thickness, geothermal gradient and strain rate, extensional faults may also penetrate through the lower crust. In a setting such as the northern North Sea, which has undergone several phases of tectonic activity, one would also expect that pre-existing weaknesses within the crust, such as faults or intrusions, play a major role during fault evolution (Dunbar & Sawyer 1989).

Thus, the upper crust is deformed by brittle localized faulting, whereas the lower crust extends by distributed ductile deformation controlled by non-Newtonian power-law creep (Kusznir & Park 1987). Evidence from exposed lower-crustal rocks and recent seismic reflection studies show that ductile deformation is very heterogeneous and localized (Blundell *et al.* 1989; Blundell 1990; Vilcotte *et al.* 1993). This scale-relationship implies that one has to differentiate between the bulk ductile (pure shear) and localized modes of deformation when describing extensional mechanisms of the lower crust.

Geological setting

The present crustal architecture in the northern North Sea is generally accepted to be a result of Permian–early Triassic and mid-Jurassic–early Cretaceous stretching, separated by thermal subsidence (Eynon 1981; Gabrielsen *et al.* 1990; Færseth *et al.* 1995*a*; Roberts *et al.* 1995; Færseth 1996). It is also anticipated that both Precambrian and Caledonian structures, as well as extensional collapse of the Caledonides, have influenced later extension and crustal reconfiguration in the area (Frost 1987; Hurich & Kristoffersen 1988; Klemperer & Hurich 1990; Færseth *et al.* 1995*a*).

Although the details of timing, significance and lateral extent of the Permo-early Triassic stretching are still a matter of debate (Giltner 1987; Gabrielsen *et al.* 1990; White 1990; Færseth *et al.* 1995a; Roberts *et al.* 1995), recent data suggest that this rift phase was more significant than the mid-Jurassic–early Cretaceous extensional event. Large tilted fault blocks with throws of the order of several kilometres formed a 150 km wide N–S oriented basin in late Palaeozoic time. During the thermal subsidence that followed the rifting, faulting occurred on both margins (Steel & Ryseth 1990) as a consequence of interaction of lateral variations in thermal subsidence, sediment loading, compaction and flexure (Gabrielsen 1986; Badley *et al.* 1988). Several studies have emphasized that thermal subsidence was still continuing when the mid-Jurassic rifting started. Hence, this must be taken into account in stretching calculations from subsidence analysis of the Cretaceous post-rift sequence (Giltner 1987; Gabrielsen *et al.* 1990; Roberts *et al.* 1995).

The Jurassic extension in the northern North Sea is well constrained. Rotation of major fault blocks shows that rifting was initiated in late Bajocian–early Bathonian time, and terminated in early Ryazanian time (Ziegler 1982; Leeder 1983; Badley *et al.* 1988; Rattey & Hayward 1993; Færseth *et al.* 1995*b*). The Permo-early Triassic master faults were partly reactivated in the Jurassic rifting, influencing the general structural pattern of the entire basin, and promoting segmentation and subsidence with opposing polarities in some areas (Færseth 1996). Later, during the rift climax in latest Jurassic time, fault activity was concentrated at fewer faults along the graben margin. As a result, the internal graben relief became more pronounced as the system developed a mature graben topography with platforms, sub-platforms, platform marginal highs and a graben feature *sensu stricto*, with a complex centre of subsidence along its axis. During the thermal cooling that followed the rifting, an early to mid-Cretaceous rapidly subsiding basin developed, accompanied by minor fault movement along some of the master faults (Gabrielsen 1986). Late Cretaceous to Tertiary modest subsidence affecting the northern North Sea is believed to have been affiliated with opening of the North Atlantic in early Tertiary times.

General crustal architecture, northern North Sea

In both deep reflection seismic transects (Figs 2 and 3), the geometry of the upper reflective crust is dominated by the Cenozoic, relatively flat-lying and unfaulted post-rift sequence, which

Fig. 5. Reflectivity pattern in the lower crust from reprocessed NSDP84-1 and 2. (See Figs 2a and 3a for location.) (**a**) Portion below Gullfaks fault block (transect 1) showing a highly reflective lower crust with reflector sets arranged as lenticular zones and a discontinuous Moho reflection (arrows). (**b**) Portion below the Horda Platform (transect 2) showing a diffuse lower crust reflectivity pattern and no well-defined Moho (arrows). (**c**) Portion below the Horda Platform (transect 2) showing a discrete Moho reflection (arrows) and a thin reflective lower crust above.

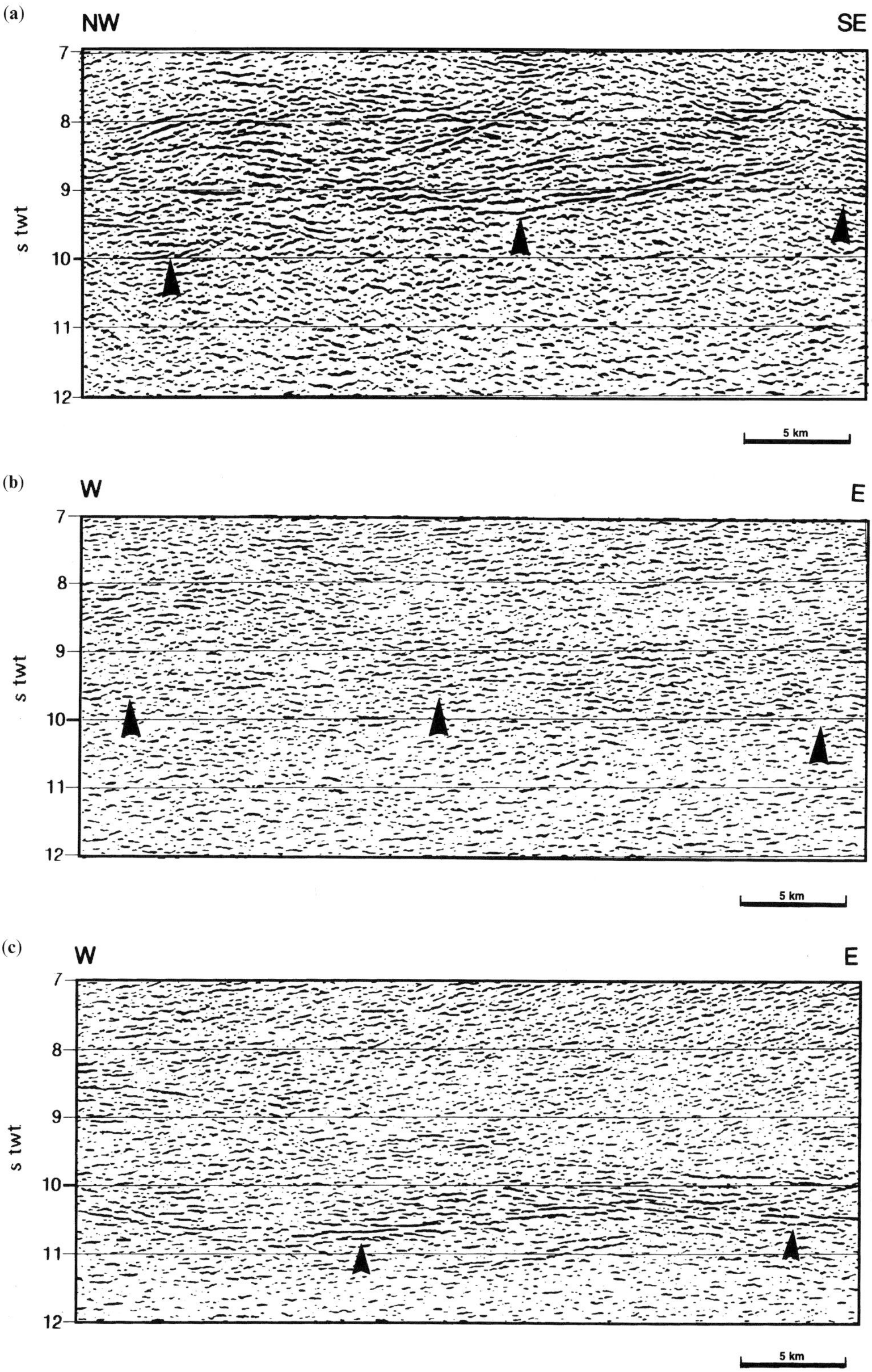
(a)
NW
SE
7
8
9
10
11
12
s twt
5 km
(b)
W
E
7
8
9
10
11
12
s twt
5 km
(c)
W
E
7
8
9
10
11
12
s twt
5 km

unconformably overlies the heavily faulted basement and its Permian to late Jurassic sedimentary cover (Jordt *et al.* 1995). Both transects reveal asymmetrical large-scale extensional geometries, with the deepest parts of the basin situated in the westernmost hanging wall of the inner western master fault of the Viking Graben. The present depth-converted data suggest that the fault geometries in the upper reflective crust vary with position within the basin and that more than one level of detachment may be identified. It is particularly noted that a local detachment at *c.* 6–7 km (4–5 s twt) is observed in the reflection seismic data. The existence of this supra-basement detachment on the western shoulder of the graben is also confirmed by detailed studies of Gullfaks fault block (Fossen 1989; Fossen *et al.* 2000). For the crustal-scale structures, many master faults have planar or slightly curviplanar geometries in their upper parts, but seem to flatten at a depth of 12–15 km.

It should also be noted that there are differences between the two transects. The major easterly throw at basement level is spread across five major faults in the more northerly transect 1 (Fig. 2b), whereas this throw is concentrated on only one master fault in the southerly transect 2 (Fig. 3b). Both transects clearly illustrate pre-Jurassic fault movements as well as fault polarity shift (transect 2), confirming the existence of pre-Jurassic development of separate N–S trending graben segments (Gabrielsen *et al.* 1990; Færseth *et al.* 1995*a*).

Although the middle crust is characterized by poor reflectivity, structural observations of major significance can be made in the reprocessed deep seismic reflection data, when these are combined with and correlated with commercial reflection seismic lines. Thus, it has been possible to trace some master faults into the more transparent middle crust and, in a few cases, to the top of the much more reflective lower crust. Although less clearly displayed, similar geometries are also indicated below the central part of the mid-Jurassic–early Cretaceous Viking Graben. The few strong reflections that are seen in the middle crust are primarily low-angled. More steeply dipping reflections have been mapped in the coast-parallel deep reflection seismic lines by Færseth *et al.* (1995*a*), who related these to the pre-late Mesozoic structuring.

The lower reflective crust is seen as an undulating reflective 4–10 km thick band with primarily gently dipping reflections, both in sections transverse to the strike of the N–S oriented Permo-Triassic and Jurassic basin axis (Figs 2 and 3), and along its strike in the Horda Platform. The top of the lower crust shallows from more than 20 km below the platform margins to 17 km beneath the present Viking Graben axis. This reflective zone is less pronounced below the graben axis. It is also noted that the reflections in the lower crust are not always very prominent, and can also appear as a diffuse pattern. At a larger scale, the reflective lower crust is undulating in a long-wavelength–low-amplitude mode, in which the shallowest part reflects the most thinned crust. In detail, this undulating mode is repeated by reflector sets, often arranged as lenticular zones. This is especially pronounced where the lower crust is most reflective below the Gullfaks area (Fig. 5a). The general expression of these zones is similar to the anastomosing shear bands described by Blundell *et al.* (1989) and Blundell (1990), who suggested that extension in the upper crust is accommodated by normal faulting whereas ductile shear dominates within the lower crust.

The reflective Moho is generally well defined in both transects. In the areas where the lower crust is more transparent, the Moho is seen only as a diffuse pattern of reflections (Fig. 5b). Our data show that the Moho consists of a set of reflectors with variable lateral thickness and amplitude, rather than one single strong reflector. Thus, the Moho reflection is displayed partly with abrupt cut-offs of the reflective band (Fig. 5a), and partly as a discrete reflection (Fig. 5c).

Below the Horda Platform the reflections of the lower crust split into two separate bands (Fig. 6). A dedicated velocity study has revealed that the body delineated by these two reflections is characterized by velocities typical for the upper mantle (8+ km s^{-1}) (Christiansson *et al.* 2000). However, results from forward modelling of the basin development (Odinsen *et al.* 2000) and gravity modelling (Christiansson *et al.* 2000), as well as the thickness and distribution of the associated sedimentary accumulation indicate that the base of the crust would be located at the base of the lower reflective band. Acknowledging these observations, the entire crustal thickness is interpreted to vary from *c.* 30 km and 36 km below the platforms on the graben flanks to 21–24 km below the axis of the base of the Jurassic sequence in the Viking Graben.

Beneath the reflective Moho intra-mantle reflectors are recorded in both transects (Fig. 7). These features, dipping 30–45° away from the graben axis, are focused at the transition between the basin and the platform areas, and follow the general pattern for intra-mantle reflections

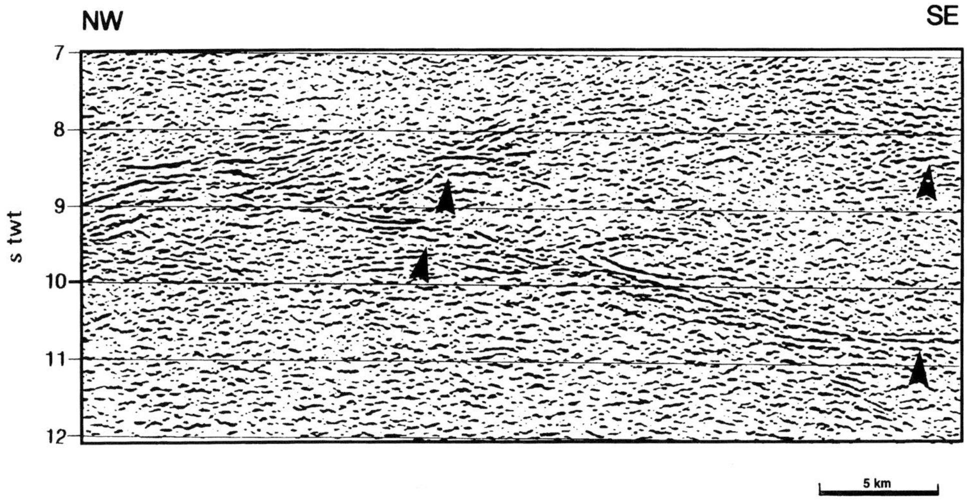

Fig. 6. Portion of the lower crust below the Horda Platform (see Fig. 2a for location) showing how the reflections split into two separate bands (arrows) enveloping a more transparent zone. (For discussion of this pattern, see text.)

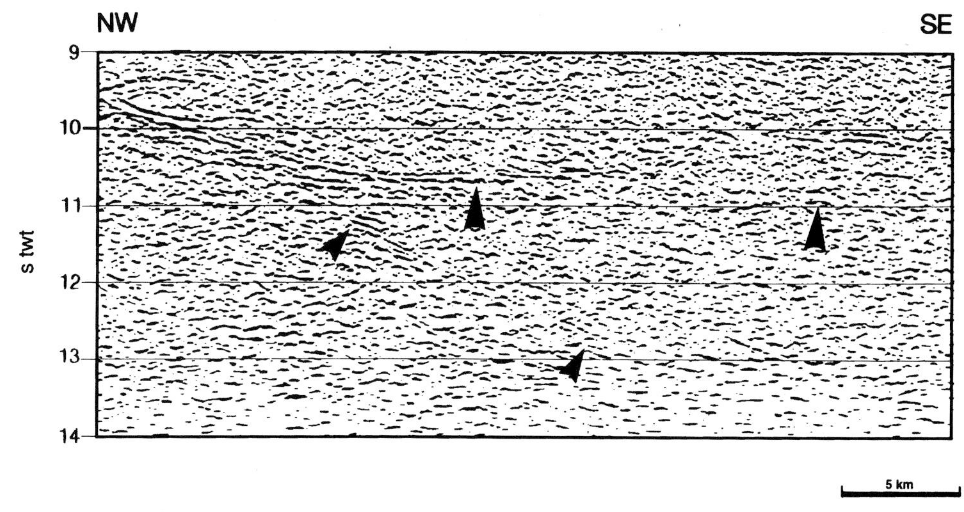

Fig. 7. Portion of the lower crust and upper mantle (see Fig. 2a for location) showing intra-mantle reflections (the two lower arrows).

described by Beach (1986) and Klemperer (1988). The interpretation of the mantle structures offers several alternative solutions, but the coincidence between the dipping reflections and the dip gradient of the Moho may suggest that the reflections represent mantle faults or shear zones, similar to those observed for offshore Britain in the BIRPS deep seismic data (McGeary *et al.* 1987; Blundell 1990; Reston 1990).

Large structural features in the northern North Sea

Although the relationship between the formation of the reflective patterns and the tectonic events in the northern North Sea is equivocal, a generally held assumption is that repeated crustal stretching may have modified the reflective architecture substantially (Klemperer 1988;

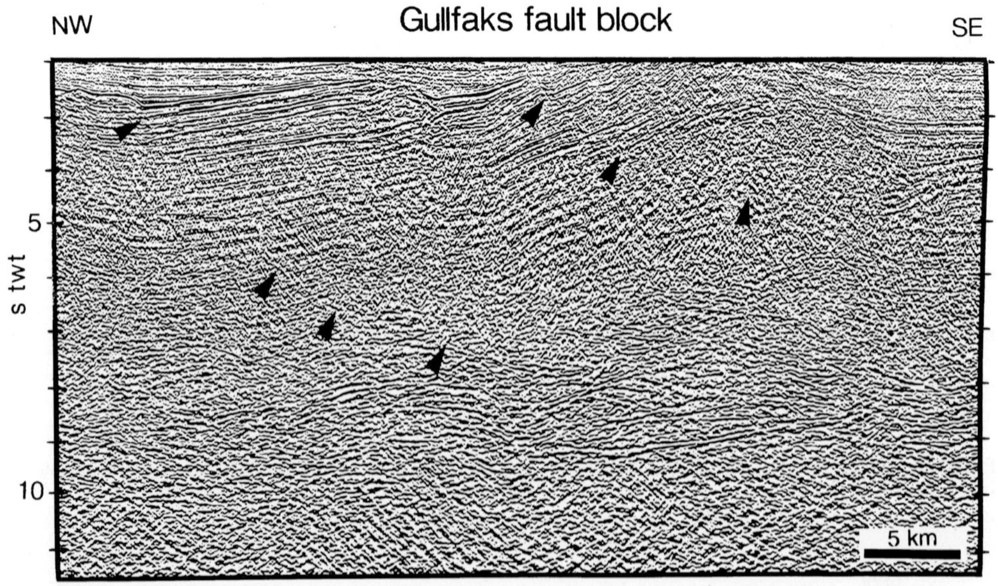

Fig. 8. Portion of transect 1 (see Fig. 2a for location) showing the fault-plane reflection of the principal east-dipping fault in the rift system (set of arrows to the left). Parts of the fault-plane reflection of the supra-crustal detachment below the Gullfaks fault block (D1) are also seen (set of arrows to the upper right).

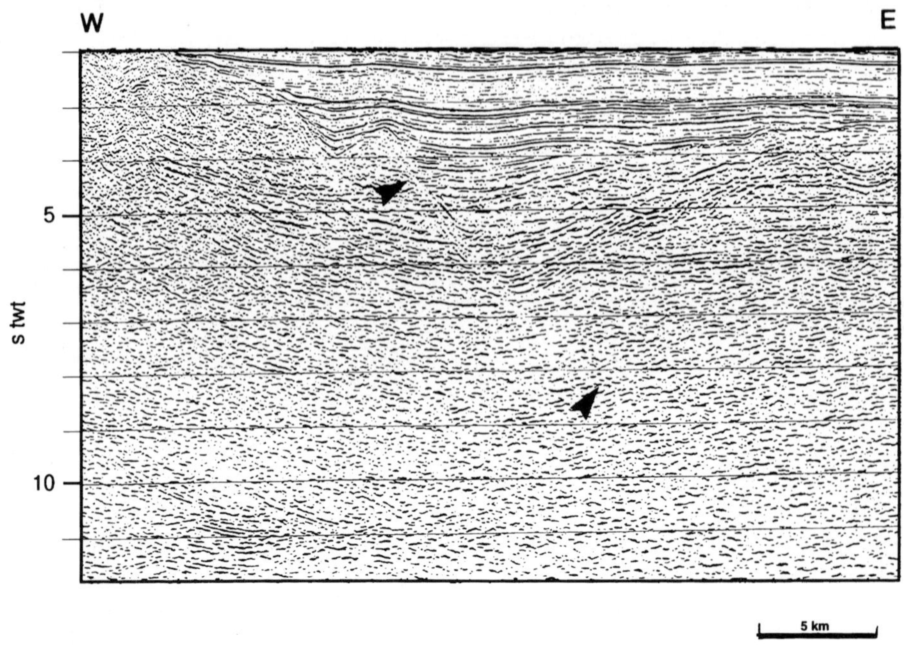

Fig. 9. Portion of transect 2 (see Fig. 3b for location) showing the fault-plane reflection (arrows) of the principal east-dipping fault in the rift system.

Blundell 1990; Klemperer & Hurich 1990). It is also accepted that the reflection Moho corresponds to the refraction Moho, which is the base of the crust in the area (Barton 1986).

One of the key factors in understanding how the lower crust deforms in response to lithospheric stretching, and thereby in constraining possible tectonic models for the area, is to study the reflection from large normal faults from the upper crust and its changes with depths. In previous studies, no deep lines crossing the North Sea Basin show a single reflector identified as a normal fault, which can be tracked from the upper crust down to or below Moho (Klemperer & White 1989; Blundell 1990). Our approach was to look specifically at the deepest penetrating normal faults and to analyse their fault plane reflection, both on the commercial lines and on the deep reprocessed lines.

The geometry of the principal east-dipping fault in the rift system below Gullfaks is illustrated in Fig. 8. Its western upper 3.5–7 km (3–5 s twt) is interpreted in transect 1 using cut-offs of the horizons as seen in the hanging wall and footwall. From 7 km depth (5 s twt) the fault is seen as an almost continuous reflection down to *c.* 18 km (8 s twt). Its upper part is characterized by a sub-planar, smooth fault plane dipping *c.* 55° (measured in depth-converted sections with no vertical exaggeration). Its dip gradually decreases, defining a sub-listric geometry when entering the lower crust. When the dip of the fault reflection becomes less than 10° it can no longer be distinguished from other reflections within the lower crust. A similar east-vergent master fault of the Mesozoic Viking Graben is observed in transect 2. But in contrast to what is seen in transect 1, the master fault in transect 2 can only be tracked continuously from the upper crust into the middle crust at *c.* 8 s (twt) (Fig. 9). Assuming the same geometry for the two faults, the detachment-depth in transect 2 may also be *c.* 20 km. It should here be noted that the deep stratigraphy and basement level in transect 2 is more uncertain than in transect 1.

The interference between the east-dipping master fault in transect 1 and the reflective band of the lower crust has been a subject of several papers (Beach *et al.* 1987; Klemperer 1988; Pinet 1989; Brun & Tron 1993). In this study, the intra lower-crustal reflections are sometimes found to be aligned with the continuation of the east-dipping master fault (Fig. 8). Although these lower-crustal reflections cannot be confirmed to represent a continuation of the fault in the form of shear zones down to the Moho, there is the possibility than the entire crust may have been offset by the master fault.

West-dipping faults dominate the eastern platform of the study area, especially to the north (transect 1; Fig. 2b). The most prominent structure of this set is the Øygarden Fault Zone (Fig. 10), which is characterized by multiple reactivation, and which defines the limit of the rift system *sensu stricto* (Gabrielsen 1986; Gabrielsen *et al.* 1995; Nøttvedt *et al.* 1995). The fault

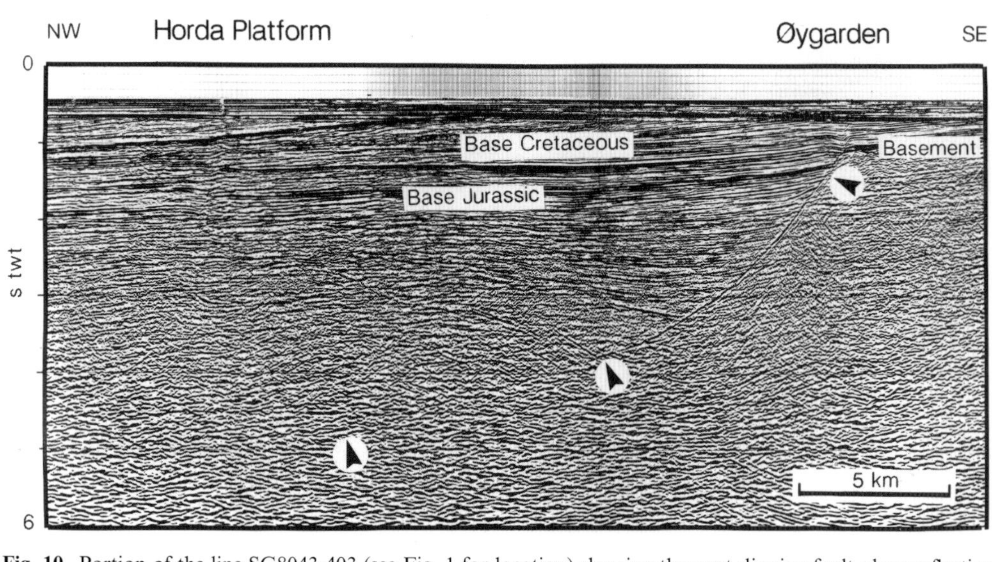

Fig. 10. Portion of the line SG8043-403 (see Fig. 1 for location) showing the west-dipping fault-plane reflection (indicated with arrows) of the Øygarden Fault.

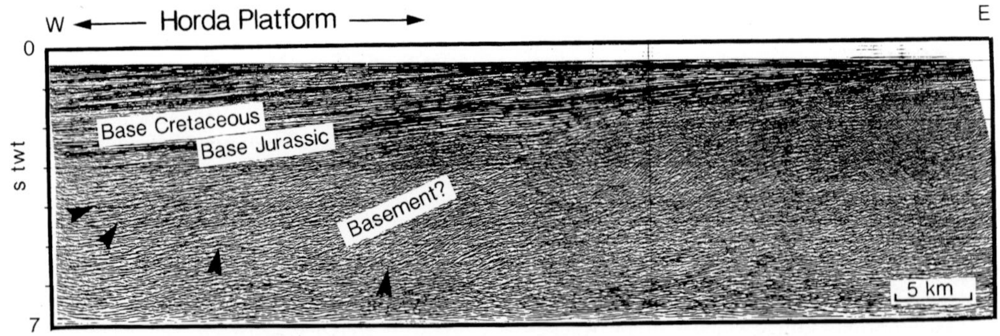

Fig. 11. Portion of the line SG8043-101 in the Horda Platform (see Fig. 1 for location) showing the low-angle east-dipping fault-plane reflection (indicated with arrows).

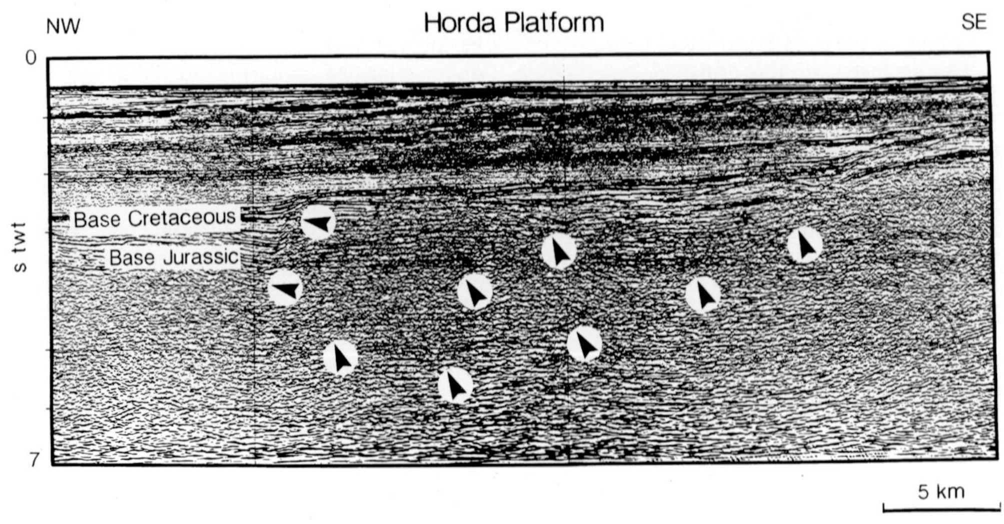

Fig. 12. Portion of the line NVGTI-92-105 in the Horda Platform (see Fig. 1 for location) showing three west-dipping faults with different dips. Parts of the fault-plane reflections are visible (indicated with arrows).

is clearly seen as an almost continuous reflection from 2 to 8 s (twt) depth (Fig. 10). The upper part of the fault, which cuts through the sedimentary column, typically dips at 40–50°. In the basement the dip decreases to 20–30°. The detachment level of this fault is seen to be near the top of the lower crust. It is noted that discontinuous westerly dipping reflections are seen below the Øygarden Fault Complex in both transects at mid-crustal levels (Figs 2 and 3). Also, a set of east-dipping extensional faults are seen in the Horda Platform area in transect 2. In general, the dip of the faults in the Horda Platform varies, going from low-angle, curviplanar to steep planar (Figs 11 and 12).

On the western graben margin, the faults are curviplanar and listric with detachments situated shallower than the major crustal faults, so that the shallower faults are transected by, or merge with, the upper part of the larger faults. This is particularly well documented below the Gullfaks and Snorre fields (Fossen *et al.* 2000) (Figs 8 and 13).

Constraints on extensional models for the northern North Sea

Tectonic models are generally based on the interpreted geometries of major normal faults. The different models proposed for the northern North Sea have therefore led to a debate on the geometric relations and the nature of rifts in general: planar rotational (Yielding *et al.* 1991),

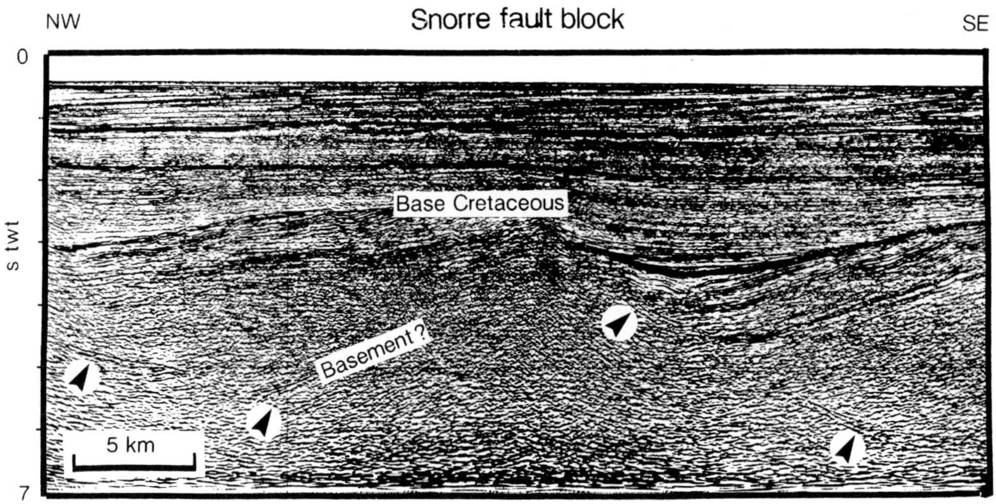

Fig. 13. Portion of line NVGTI-92-105 across the Snorre fault block (see Fig. 1 for location) showing the fault-plane reflection of two east-dipping intra-basement faults with decreasing dip at depths (indicated with arrows).

listric (Beach *et al.* 1987; Gibbs 1987) and combinations thereof (Gabrielsen 1986; Speck-snijder 1987; Gabrielsen *et al.* 1995) have been proposed. Any discussion and application of such models has to take into account that the area has experienced repeated extension and crustal reconfiguration, as is seen in the present basin configuration, which shows considerable differences between the two transects. Beneath the Horda Platform in transect 1, most of the Permo-early Triassic master faults dip towards the mid-Jurassic–early Cretaceous Viking Graben axis. In transect 2, it is seen that the eastern part of the Horda Platform is influenced by master faults that dip away from the graben axis, whereas the western part is characterized by a series of faults with shifting polarities. It is assumed that this contrast in polarity reflects the primary Permo-early Triassic basin configuration (Færseth *et al.* 1995*a*; Færseth 1996). This has been confirmed by restorations recently published by Færseth (1996). The width of the Permo-early Triassic basin was *c.* 150 km (transects 1 and 2). This corresponds to the location of the area of thinned crust as identified in transect 1 and where modelled stretching estimates display more extension in the Permo-early Triassic than in the mid-Jurassic–early Cretaceous (Roberts *et al.* 1995; Odinsen *et al.* 2000).

In spite of some differences in lower-crustal reflectivity as seen in transects 1 and 2 beneath the platform areas, the general impression is that the signature of the lower crust is similar between the two transects. Thus, transect 1 is characterized by eastward crustal thickening immediately below the outer limit of the Horda Platform, whereas a similar thickening is found further east in transect 2. The two transects show similarities both in the width of the area with Moho shallowing, and the depths to the Moho below the central part of the Viking Graben.

The intra-mantle reflections observed below the Horda Platform, as seen particularly in transect 1, have a dip angle of 30–45°. Such features are frequently recorded within the upper mantle in other extended areas, one example being offshore Britain (McGeary *et al.* 1987; Blundell 1990; Reston 1990). Although it has been proposed that they may represent magma, concentration of fluids, or diffraction from the lower crust, the current assumption is that they are localized mantle faults (Blundell 1990; Klemperer & Hurich 1990; Reston 1993). In transect 1 it is noted that these mantle reflections coincide with shift in the lower-crustal reflectivity, which is at the transition between the thinner crust below the Viking Graben and the thicker crust below the Horda Platform (Fig. 2). This, as well as the observed geometry of the feature, makes it reasonable to suggest a mantle fault. The coincidence between the location of the mantle fault, the lateral thinning of the crust and the high-velocity ($8+ \mathrm{km\,s^{-1}}$) body below the Horda Platform (transect 1) may have significance. Although this must be subjected to further study, this observation may indicate that the high-velocity body is a long-lived (Caledonian?) feature (Christiansson *et al.* 2000).

Figure 14 summarizes the main structural observations from the seismic reflection lines in the area. This conceptual model is based mainly on transect 1 but also has relevance to transect 2. The model shows that, although the most of the faults are linear to curvilinear, listric and more complex geometries are also observed. The latter geometries may be a result of pre-existing basement structures influencing renewed faulting. In the Horda Platform area, for example, it is indicated that listric faults are linked to or situated above steep planar normal faults (Badley *et al.* 1988; Gabrielsen 1986). It is acknowledged that the shifting polarity of the Permo-early Triassic (or older?) basin system beneath the Horda Platform imparts geometric problems that have not yet been solved (Figs 2 and 3). The resolution in the present data is not sufficient to solve the problem of cross-cutting relations between the major variably dipping faults.

Three levels of detachments (D1, D2 and D3) are present within the crust, as shown in Fig. 14. The supra-basement detachment (D1) is found within the sedimentary column, exemplified within the Gullfaks fault block (Fig. 8), but is also observed elsewhere in the area (Gabrielsen *et al.* 1995). According to Fossen *et al.* (2000), D1 is situated directly above basement, and is probably linked to the larger Gullfaks fault composing a ramp–flat–ramp geometry at depth.

The intra-basement detachment level (D2) is frequently seen in the commercial lines within the middle crystalline crust. The dip relations and the fact that these features can be easily followed along strike suggest that they are not side-reflections or noise from processing. They also generate deeper multiples in the commercial lines. The D2 level is similar on both sides of the Viking Graben.

As stated above, continuation between the master faults of the upper crust and structures in the lower crust–upper mantle is viable. Therefore the possibility of an easterly dipping simple shear zone exists. This is supported by recent studies showing that changes between coupled and decoupled modes of deformation are in best agreement with the observed pattern

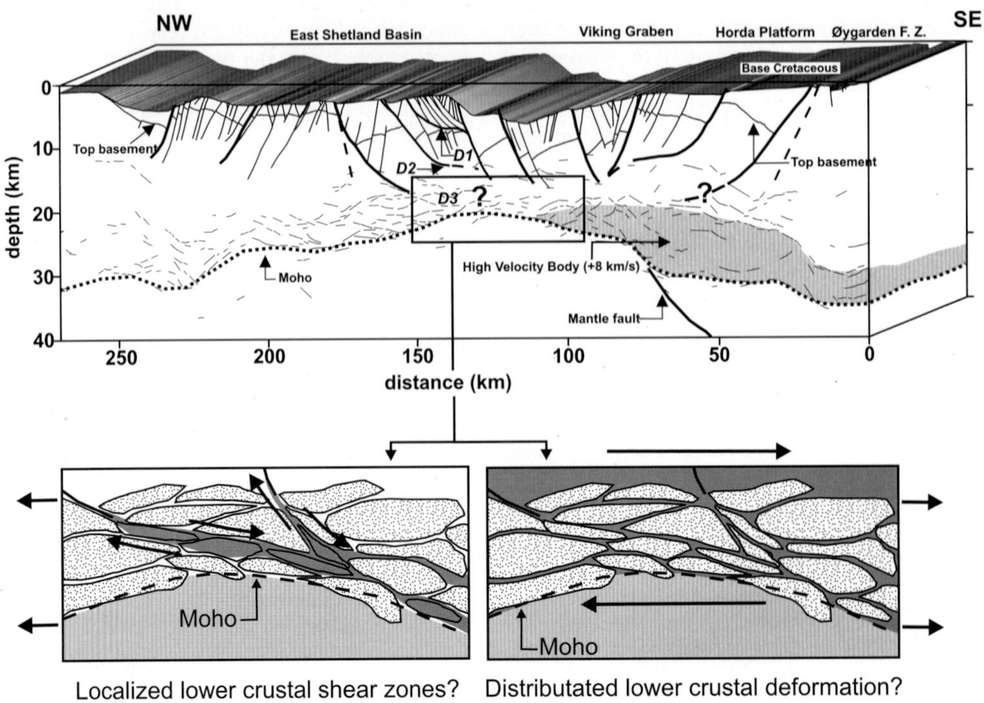

Fig. 14. Conceptual model with special reference to transect 1 summarizes the main structural observations from the deep and commercial lines. (Note the three detachment levels (D1, D2 and D3) and the possible link between the mantle fault and the principal east-dipping normal fault through the lower crust.) Two possible scenarios are illustrated for the deformation of the lower crust during movement of the largest normal fault. Also displayed are steep planar faults in the footwall of the rift system. (See text for further information.)

of subsidence (Ter Voorde *et al.* 2000). However, at least two types of contacts between the upper-crustal fault and that of the upper mantle can be depicted. In either case, the intra lower crust will then be the deepest detachment level (D3) as shown in Fig. 14.

The principal flattening levels D2 and D3 can easily be explained as a consequence of changes in rheology at depth, where flattening occurs within the weaker intra-crustal layers. In a simple layered model based on vertically changing mineralogical composition of the crust, one should expect that all large normal faults would detach at the weaker interface, for example, within the middle crust. In reality, any crust is much more heterogeneous than this, especially a continental crust that has undergone repeated heating and cooling. One also has to take into account the long-lived zones of weaknesses, which probably have an impact on the fault geometries. It is therefore not surprising that some larger listric faults seem to detach within the middle crust whereas others enter the lower crust. It is also noticed (Fig. 14) that some of the intra-mantle reflections are situated in the continuation of large-scale crustal faults. If this means that these are crustal-scale shear zones, this implies that (upper) crustal and mantle faults may be coupled through the lower crust. This can best be explained by a lower crust composed of shear lenses (Blundell 1990) comparable with lower metamorphic panels as displayed in the Basin-and-Range province. In this model the lower crust deforms by bulk pure shear (Allmendinger *et al.* 1987; Hamilton 1987).

Summary

The present crustal architecture of the northern North Sea is the result of a long-lasting tectonic history of crustal reconfiguration. Of the several tectonic phases that have affected the area, the Permo-Triassic and especially the mid-Jurassic to early Cretaceous extension are best known. The Permo-early Triassic normal faults are seen to be frequently reactivated during the Jurassic–early Cretaceous stretching. It is also observed that some of the large faults have been active during the thermal cooling periods that followed the two stretching phases.

The conceptual model shows that both supra- and intra-basement detachment levels are present within the upper (D1), middle (D2) and lower (D3) crust, respectively. The D1 level may partly be controlled by gravity and/or lithology during the Jurassic–early Cretaceous phase (Fossen *et al.* 2000). D2 and D3 may be caused by rheological interfaces within the crust, but are probably also influenced by pre-existing localized zones of weaknesses, such as faults and shear zones. Hence, the fault geometries support that decoupling occurred at more that one level, and that the principal decoupling surface can be traced into the lower crust, and possibly through the lower crust and into the upper mantle.

The deformation of the upper and middle crust is localized along a number of normal faults, which favours bulk simple shear. The geometries of the faults are variable, with steep planar, listric and composite geometries being observed. The much weaker lower crust is characterized by bulk pure shear, accommodated by anastomosing ductile shear zones separated by elongated rock bodies. An east-dipping and low-angle band of reflectors is recorded within the predominantly transparent upper mantle. Although other possibilities are evaluated, these features are believed to represent faults or shear zones.

This work was funded by the Commission of the European Union and the Norwegian Research Council in the framework of Integrated Basin Studies – The Dynamics of the Norwegian Margin. Norsk Hydro a.s., Saga Petroleum a.s. and Statoil a.s. provided data and research efforts for this study. We are especially indebted to J. E. Lie, who reprocessed the NSDP lines. Reviews and discussions by H. Fossen, as well as linguistic corrections by H. Brekke and J. McCrachen, are very much appreciated. We would like to thank H. Hjelmeland and an anonymous reviewer for thorough and very constructive criticism. The authors have honoured their suggestions on all major points.

References

ALLMENDINGER, R. W., HAUGE, T. A., HAUSER, E. C., POTTER, C. J. & OLIVER, J. 1987. Tectonic heredity and the layered lower crust in the Basin and Range Province, western United States. *In*: COWARD, M. P., DEWEY, J. F. & HANCOCK, P. L. (eds) *Continental Extensional Tectonics*. Geological Society, London, Special Publications, **28**, 233–246.

BADLEY, M. E., PRICE, J. D., RAMBECH-DAHL, C. & AGDESTEIN, T. 1988. The structural evolution of the northern Viking Graben and its bearing upon extensional modes of basin formation. *Journal of the Geological Society, London*, **145**, 455–472.

BARTON, P. J. 1986. Comparison of deep reflection and refraction structures in the North Sea. *In*: BARAZANGI, M. & BROWN, L. (eds) *Reflection Seismology: a Global Perspective*. American Geophysical Union Series, **13**, 297–300.

BEACH, A. 1986. Some comments on sedimentary basin development in the northern North Sea. *Scottish Journal of Geology*, **22**, 1–20.

——, BIRD, T. & GIBBS, A. 1987. Extensional tectonics and crustal structure: deep seismic reflection data from the northern North Sea Viking graben. *In*: COWARD, M. P., DEWEY, J. F. & HANCOCK, P. L. (eds) *Continental Extensional Tectonics*. Geological Society, London, Special Publications, **28**, 467–476.

BLUNDELL, D. J. 1990. Seismic images of continental lithosphere. *Journal of the Geological Society, London*, **147**, 895–913.

——, RESTON, T. J. & STEIN, A. M. 1989. Deep crustal structural controls on sedimentary basin geometry. American Geophysical Union Geophysical Monograph Series, **48**, 57–64.

BRUN, J.-P. & TRON, V. 1993. Development of the North Viking Graben: inference from laboratory modelling. *Sedimentary Geology*, **86**, 31–51.

CHRISTIANSSON, P., FALEIDE, J. I. & BERGE, A. M. 2000. Crustal structure in the North Sea; an integrated geophysical study. *This volume*.

COWARD, M. P. 1986. Heterogeneous stretching, simple shear and basin development. *Earth and Planetary Science Letters*, **80**, 325–336.

DUNBAR, J. A. & SAWYER, D. A. 1989. How pre-existing weaknesses control the style of continental breakup. *Journal of Geophysical Research*, **94**, 7278–7292.

EYNON, G. 1981. Basin development and sedimentation in the Middle Jurassic of the northern North Sea. *In*: ILLING, L. V. & HOBSON, G. D. (eds) *The Petroleum Geology of the Continental Shelf of North West Europe*. Heyden, London, 196–204.

FOSSEN, H. 1989. Indication of transpressional tectonics in the Gullfaks Oil Field, northern North Sea. *Marine and Petroleum Geology*, **6**, 22–30.

——, ODINSEN, T., FÆRSETH, R. B. & GABRIELSEN, R. H. 2000. Detachments and low-angle faults in the northern North Sea rift system. *This volume*.

FROST, R. E. 1987. The evolution of the Viking Graben tilted fault block structures: a compressional origin. *In*: BROOKS, J. & GLENNIE, K. (eds) *Petroleum Geology of North West Europe*. Graham and Trotman, London, 1009–1024.

FÆRSETH, R. B. 1978. Mantle-derived lherzolite nodules and megacrysts from Permo-Triassic dykes, Sunnhordland, western Norway. *Lithos*, **11**, 23–35.

——1996. Interaction of Permo-Triassic and Jurassic extensional fault-blocks during the development of the northern North Sea. *Journal of the Geological Society, London*, **153**, 931–944.

——, GABRIELSEN, R. H. & HURICH, C. A. 1995*a*. Influence of basement in structuring of the North Sea Basin offshore southwest Norway. *Norsk Geologisk Tidsskrift*, **75**, 105–119.

——, SJØBLOM, T. S., STEEL, R. J., LILJEDAHL, T., SAUAR, B. E. & TJELLAND, T. 1995*b*. Tectonic controls on Bathonian–Volgian syn-rift succession on the Visund Fault Block, northern North Sea. *In*: STEEL, R. J., FELT, V., JOHANNESSEN, E. & MATHIEU, C. (eds) *Sequence Stratigraphy on the Northwest European Margin*. Norwegian Petroleum Society, Special Publication, **5**, 325–346.

GABRIELSEN, R. H. 1986. Strucural elements in graben systems and their influence on hydrocarbon trap types. *In*: SPENCER, A. M. *et al.* (eds) *Habitat of Hydrocarbons on the Norwegian Continental Shelf*. Graham and Trotman, London, 55–60.

——1989. Reactivation of faults on the Norwegian continental shelf and its implications for earthquake occurrence. *In*: GREGERSEN, S. & BASHAM, P. (eds) *Causes and Effects of Earthquakes at Passive Margins and in Areas with Post-glacial Rebound on both Sides of the North Atlantic*. Elsevier, Amsterdam, 69–92.

——, FÆRSETH, R. B., STEEL, R. J., IDIL, S. & KLØVJAN, O. S. 1990. Architectural styles of basin fill in the northern Viking Graben. *In*: BLUNDELL, D. J. & GIBBS, A. D. (eds) *Tectonic Evolution of the North Sea Rifts*. Clarendon, Oxford, 158–179.

——, STEEL, R. J. & NØTTVEDT, A. 1995. Subtle traps in extensional terranes: a model with reference to the North Sea. *Petroleum Geoscience*, **1**, 223–235.

GIBBS, A. D. 1987. Deep seismic profiles in the northern North Sea. *In*: BROOKS, J. & GLENNIE, K. (eds) *Petroleum Geology of North West Europe*. Graham and Trotman, London, 1025–1028.

GILTNER, J. P. 1987. Application of extensional models to the northern Viking Graben. *Norsk Geologisk Tidsskrift*, **67**, 339–352.

HAMILTON, W. 1987. Crustal extension in the Basin and Range Province, southwestern United States. *In*: COWARD, M. P., DEWEY, J. F. & HANCOCK, P. L. (eds) *Continental Extensional Tectonics*. Geological Society, London, Special Publications, **28**, 155–176.

HURICH, C. A. & KRISTOFFERSEN, Y. 1988. Deep structure of the Caledonide Orogen in southern Norway: new evidence from marine seismic profiling. *In*: KRISTOFFERSEN, Y. (ed.) *Progress in Studies of the Lithosphere in Norway*. Norges Geologiske Undersøkelse, Special Publication, **3**, 96–101.

JORDT, H., FALEIDE, J. I., BJØRLYKKE, K. & IBRAHIM, M. T. 1995. Cenozoic sequence stratigraphy of the central and northern North Sea Basin: tectonic development, sediment distribution and provenance areas. *Marine and Petroleum Geology*, **12**, 845–879.

KLEMPERER, S. L. 1988. Crustal thinning and nature of extension in the northern North Sea from deep seismic reflection profiling. *Tectonics*, **7**, 803–821.

—— & HURICH, C. A. 1990. Lithospheric structure of the North Sea from deep seismic profiling. *In*: BLUNDELL, D. J. & GIBBS, A. D. (eds) *Tectonic Evolution of the North Sea Rifts*. Clarendon, Oxford, 37–63.

—— & WHITE, N. J. 1989. Coaxial stretching or lithospheric simple shear in the North Sea? Evidence from deep seismic profiling and subsidence. *In*: TANKARD, A. J. & BALKWILL, H. R. (eds) *Extensional Tectonics and Stratigraphy of the North Atlantic Margins*. American Association of Petroleum Geology Memoir, **46**, 511–522.

KUSZNIR, N. J. & PARK, R. G. 1987. The extensional strength of the continental lithosphere: its dependence on geothermal gradient, and crustal composition and thickness. *In*: COWARD, M. P., DEWEY, J. F. & HANCOCK, P. L. (eds) *Continental*

Extensional Tectonics. Geological Society, London, Special Publication, **28**, 35–52.

LEEDER, M. R. 1983. Lithospheric stretching and North Sea Jurassic clastic source lands. *Nature*, **305**, 510–514.

LISTER, G. S., ETHERIDGE, M. A. & SYMONDS, P. A. 1986. Detachment faulting and the evolution of passive continental margins. *Geology*, **14**, 246–250.

MCGEARY, S., CHEADLE, M. J., WARNER, M. R. & BLUNDELL, D. J. 1987. Crustal structure of the continental shelf around Britain derived from BIRPS deep seismic profiling. *In*: BROOKS, J. & GLENNIE, K. (eds) *Petroleum Geology of North West Europe*. Graham and Trotman, London, 33–41.

MCKENZIE, D. 1978. Some remarks on the development of sedimentary basins. *Earth and Planetary Science Letters*, **40**, 25–32.

NØTTVEDT, A., GABRIELSEN, R. H. & STEEL, R. J. 1995. Tectonostratigraphy and sedimentary architecture of rift basins; with reference to the northern North Sea. *Marine and Petroleum Geology*, **12**, 845–879.

ODINSEN, T., REEMST, P., VAN DER BEEK, P., FALEIDE, J. I. & GABRIELSEN, R. H. 2000. Permo-Triassic and Jurassic extension in the northern North Sea: results from tectonostratigraphic forward modelling. *This volume*.

PINET, B. 1989. Deep seismic profiling and sedimentary basins. *Bulletin de la Societé Géologique de France*, **8**, 749–766.

PLATT, N. H. 1995. Structure and tectonics of the northern North Sea: new insights from deep penetration regional seismic data. *In*: LAMBIASE, J. J. (ed.) *Hydrocarbon Habitat in Rift Basins*. Geological Society, London, Special Publications, **80**, 103–113.

RATTEY, R. P. & HAYWARD, A. B. 1993. Sequence stratigraphy of a failed rift system: the Middle Jurassic to Early Cretaceous basin evolution of the Central and Northern North Sea. *In*: PARKER, J. R. (ed.) *Petroleum Geology of Northwest Europe: Proceedings of the 4th Conference*. Geological Society, London, 215–249.

RESTON, T. J. 1990. The lower crust and the extension of the continental lithosphere: kinematic analysis of BIRPS deep seismic data. *Tectonics*, **9**, 1235–1248.

——1993. Evidence for extensional shear zones in the mantle, offshore Britain, and their implications for the extension of the continental lithosphere. *Tectonics*, **12**, 492–506.

ROBERTS, A. M., YIELDING, G. & BADLEY, M. E. 1990. A kinematic model for the orthogonal opening of the Late Jurassic North Sea Rift System, Denmark–Mid Norway. *In*: BLUNDELL, D. J. & GIBBS, A. D. (eds) *Tectonic Evolution of the North Sea Rifts*. Clarendon, Oxford, 180–199.

——, ——, KUSZNIR, N. J., WALKER, I. & DORN-LOPEZ, D. 1995. Quantitative analysis of Triassic extension in the northern North Sea. *Journal of the Geological Society, London*, **152**, 15–26.

ROYDEN, L. & Keen, C. E. 1980. Rifting processes and thermal evolution of the continental margin of eastern Canada determined from subsidence curves. *Earth and Planetary Science Letters*, **51**, 343–361.

SCLATER, J. G., HELLIGER, S. J. & SHOREY, M. 1986. *An analysis of the importance of extension in accounting for the post-Carboniferous subsidence of the North Sea basin*. University of Texas Institute for Geophysics, Internal Report.

SCOTT, D. L. & ROSENDAHL, B. R. 1989. North Viking Graben: an east African perspective. *AAPG Bulletin*, **73**, 155–165.

SPEKSNIJDER, A. 1987. The structural configuration of Cormorant Block IV in context of the northern Viking Graben structural framework. *Geologie en Mijnbouw*, **65**, 357–379.

STEEL, R. & RYSETH, A. 1990. The Triassic–early Jurassic succession in the northern North Sea: megasequence stratigraphy and intra-Triassic tectonics. *In*: HARDMAN, R. F. P. & BROOKS, J. (eds) *Tectonic Events Responsible for Britain's Oil and Gas Reserves*. Geological Society, London, Special Publications, **55**, 139–168.

——1993. Triassic–Jurassic megasequence stratigraphy in the Northern North Sea: rift to post-rift evolution. *In*: PARKER, J. R. (ed.) *Petroleum Geology of Northwest Europe: Proceedings of the 4th Conference*. Geological Society, London, 299–316.

TER VOORDE, M., FÆRSETH, R. B., GABRIELSEN, R. H. & CLOETINGH, S. A. P. L. 2000. Repeated lithospheric extension in the northern Viking Graben: a coupled or a decoupled rheology? *This volume*.

TORSVIK, T. H., STURT, B. A., SWENSSON, E., ANDERSEN, T. B. & DEWEY, J. F. 1992. Palaeomagnetic dating of fault rocks: evidence for Permian and Mesozoic movements and brittle deformation along the Dalsfjord Fault, west Norway. *Geophysical Journal International*, **109**, 565–580.

VILCOTTE, J. P., MELOSH, J., SASSI, W. & RANALLI, G. 1993. Lithosphere rheology and sedimentary basins. *Tectonophysics*, **226**, 89–95.

WERNICKE, B. 1985. Uniform-sense normal simple-shear of the continental lithosphere. *Canadian Journal of Earth Sciences*, **22**, 108–125.

WHITE, N. J. 1990. Does the uniform stretching model work in the North Sea? *In*: BLUNDELL, D. J. & GIBBS, A. D. (eds) *Tectonic Evolution of the North Sea Rifts*. Clarendon, Oxford, 217–240.

YIELDING, G., BADLEY, M. E. & FREEMAN, B. 1991. Seismic reflections from normal faults in the northern North Sea. *In*: ROBERTS, A. M., YIELDING, G. & FREEMAN, B. (eds) *The Geometry of Normal Faults*. Geological Society, London, Special Publications, **56**, 79–89.

ZIEGLER, P. A. 1982. *Geological Atlas of Western and Central Europe*. Shell Internationale, The Hague.

Repeated lithosphere extension in the northern Viking Graben: a coupled or a decoupled rheology?

M. TER VOORDE,[1] R. B. FÆRSETH,[2,3]
R. H. GABRIELSEN[3] & S. A. P. L. CLOETINGH[1]

[1] *Institute of Earth Sciences, De Boelelaan 1085, 1081 HV Amsterdam, the Netherlands*
[2] *Norsk Hydro, Exploration, P.O. Box 200, N-1321 Stabekk, Norway*
[3] *Department of Geology, University of Bergen, Allégaten 41, N-5007 Bergen, Norway*

Abstract: The Late Permian–Early Triassic and Jurassic–Cretaceous rift events that influenced the structure of the northern Viking Graben are examined. Here, a 2D forward numerical model including faults is applied on seismic line NVGT88-04, which crosses the basin in the E–W direction. We concentrate on (1) the amount of Jurassic basin subsidence that can be explained by thermal contracting resulting from the Permo-Triassic event, (2) the spatial connection between the two rift events and (3) the apparent mode of flexure (coupled v. decoupled) of the lithosphere. Modelling results indicate that post-rift subsidence as a result of Permo-Triassic rifting has practically ceased at the onset of Jurassic rifting. The position of the rift axis is shown to migrate with time, from the Horda Platform in Permo-Triassic time to the present Viking Graben centre in Jurassic–Cretaceous time. The major difference between the consequences of a coupled versus a decoupled mode of flexure is that the latter causes a larger amount of fault-block rotation. On this basis, we suggest a decoupled mode of flexure for the Permo-Triassic rift phase, a coupled mode of flexure for the Early Jurassic rift phase, and a decoupled mode of flexure again for the Late Jurassic period.

The Viking Graben is part of the series of linked half-grabens that compose the North Sea sedimentary basin (Beach *et al.* 1987; Badley *et al.* 1988; Færseth *et al.* 1995*a*; Christiansson *et al.* 2000; Odinsen *et al.* 2000; Fig. 1). This basin was formed during several extensional events following Caledonian collision. The first event occurred in Devonian time (e.g. Hossack 1984; McClay *et al.* 1986; Andersen & Jamtveit 1990; Fossen & Rykkelid 1992) and affected an area that extends far beyond the later margins of the rift system. This was followed by the pronounced Permo-Triassic(?) and Jurassic–Cretaceous rift phases. As well control is sparse for the pre-Triassic sequences, the age of the older of these events is still a matter of debate (see Table 1). However, a Late Permian–Early Triassic age is suggested by the occurrence of Permian sediments in the southern Viking Graben (Lervik *et al.* 1989) and in the Unst Basin (Johns & Andrews 1985; Platt 1995), by palaeomagnetic dating of fault rocks in the southwestern part of Norway (Torsvik *et al.* 1992), and by the occurrence of Permian dykes in the southwest Norwegian coastal area (Færseth *et al.* 1976; Færseth 1978; Furnes *et al.* 1982). Additional minor Late Triassic events were recognized by Morton *et al.* (1987) and Gabrielsen *et al.* (1990), and a phase of increased subsidence occurred in late Cretaceous time.

In spite of almost 30 years of research in the Viking Graben area, several problems remain to be solved:

- Because of overprint of later structuring, the Permo-Triassic rift stage was identified only with difficulty in seismic sections available for the early studies of the northern North Sea. Because of this and the limited well control, the relative magnitudes of the Permo-Triassic and Jurassic–Cretaceous events in the Viking Graben have been disputed. Some workers (e.g. White 1990; Lippard & Liu 1992) have argued that Jurassic rifting was dominant. However, White (1990) came to this conclusion by focusing on the East Shetland terrace area in the west, whereas major Triassic rifting seems to have occurred in the east, below the Horda Platform. Lippard & Liu (1992) inferred non-uniform thinning ($\beta_{crust} < \beta_{mantle}$) in the Jurassic Horda Platform, to obtain the best fit between observed and calculated basin subsidence. Others (e.g. Giltner 1987; Marsden *et al.* 1990) derived from numerical modelling that

From: NØTTVEDT, A. *et al.* (eds) *Dynamics of the Norwegian Margin*. Geological Society, London, Special Publications, **167**, 59–81. 1-86239-056-8/00/$15.00

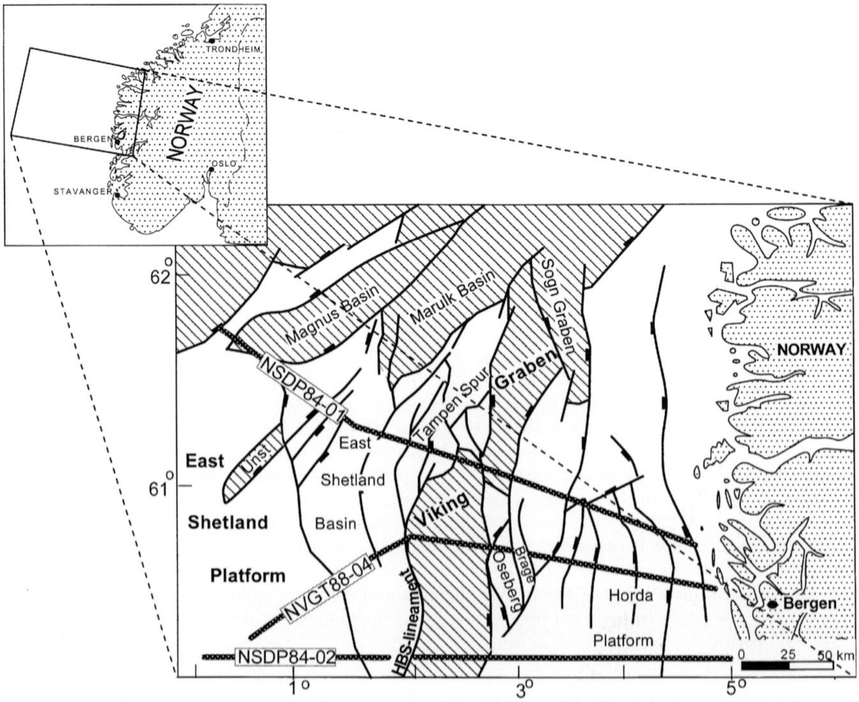

Fig. 1. Map of Viking Graben and its surroundings, showing the location of the seismic lines NVGT88-04, NSDP84-01 and NSDP84-02.

the major part of the extension occurred during the Permo-Triassic phase, and that the basin subsidence of the second phase is for a large part due to thermal contraction resulting from the first event. Giltner (1987) inferred this from a 1D, uniform extension model, including effects of a finite rift phase, whereas Marsden *et al.* (1990) used a 2D depth-dependent extension model, assuming instantaneous stretching. Although the importance of the Permo-Triassic event is now firmly established, by new, deep-penetration regional seismic lines (Platt 1995), improved interpretations of deep seismic reflection data (Christiansson *et al.* 2000) and basin modelling studies (Roberts *et al.* 1993, 1995; Odinsen *et al.* 2000), the amount of excess heat 'inherited' by the Jurassic event is still not well established.

- Another matter of debate is the position of the Triassic rift axis. It is often stated (e.g. Lippard & Liu 1992; Yielding *et al.* 1992) that the Triassic rift axis coincided with that of the Jurassic–Cretaceous basin. Badley *et al.* (1984, 1988) suggested that most of the major faults that were active during the latter episode were reactivated basement-involved faults of the first rift episode, and that greatest stretching and subsidence occurred above the Permian rift axis. However, they also observed occasional basement faults in places unaffected by the Permian rift stage. On the other hand, Færseth (1996) argued that the thickness distribution of the sediment pile suggests that the two observed rifting stages may have little spatial connection. The effects of the Permo-Triassic rift phase can be recognized to the east below the Horda Platform (e.g. Steel & Ryseth 1990; Sneider *et al.* 1995), whereas no unequivocal evidence for the rift is found below the Jurassic rifting axis. Therefore, Færseth (1996) proposed that the Permo-Triassic extension maximum may have been located in the eastern part of the basin, with the main rift situated beneath the present Horda Platform. This is supported by observations of other dominantly Triassic basins in the eastern North Sea, e.g. the Egersund Basin, the Horn Graben and the Åsta Graben (Lervik *et al.* 1989), which are also found east of the Jurassic rift axis.
- Several numerical basin modelling studies have been carried out on the northern Viking Graben (e.g. Giltner 1987; Marsden *et al.*

Table 1. *Dating of the major rift events, as proposed or used by various workers*

Reference	First rift phase	Second rift phase
Giltner (1987)	248–213 Ma*	169–98 Ma*
Gabrielsen *et al.* (1990)	?Late Palaeozoic–Scythian	Bathonian–Ryazanian
Marsden *et al.* (1990) (modelled as instantaneous events:	250–213 Ma 230 Ma	Bathonian–Ryazanian 170 Ma)
White (1990)	Triassic	160–100 Ma
Lippard & Liu (1992)	260–235 Ma† or 225–200 Ma†	156–131 Ma†
Roberts *et al.* (1995) (modelled as instantaneous events:	Early Triassic 250 Ma	Mid-Jurassic–Earliest Cretaceous 170 Ma)
Odinsen *et al.* (2000)	Kungurian–Anisian (261–236 Ma‡)	Callovian–Ryazanian (165–141 Ma‡)
This study	261–236 Ma‡	165–141 Ma‡

* Time scale of Harland *et al.* (1982).
† Time scale of Haq *et al.* (1987).
‡ Time scale of Harland *et al.* (1990).

1990; Roberts *et al.* 1995; Odinsen *et al.* 2000). In most of these studies an estimate of the lithosphere rigidity, often expressed as an 'effective elastic thickness' (EET), was made to calculate the flexural response on the mass redistribution as a result of the extension. The EET values estimated by different workers vary considerably, from 1.5 km (Roberts *et al.* 1995) to ±45 km (Odinsen *et al.* 2000). This discrepancy is often attributed to the possibility that the crust is decoupled from the mantle, because of vertical variations in the strength of the lithosphere. However, little attention has been paid so far to the influence of this strength distribution on the extension mechanism.

The purpose of this paper is to examine the nature of the various episodes of rifting in the Viking Graben, thereby focusing on the following questions:

- How much post-rift subsidence was induced by the excess heat, remnant from the Permo-Triassic stretching event? And how did this eventually influence the later development of the basin?
- Is there any spatial connection between the rift events? Where was the region of maximum Permo-Triassic extension, and did this influence the position of the Jurassic–Cretaceous basin?
- Is the lithosphere likely to be in a 'decoupled' state, and what does this mean for the flexural response on extension?

Whereas in early numerical modelling studies in the Viking Graben (e.g. Badley *et al.* 1988; White 1990) statements about the older rift phase were derived from studies on the amount of extension and subsidence of the latter, Roberts *et al.* (1995) focused particularly on the pre-Jurassic event, using a combined flexural backstripping and forward modelling method. Following these workers, we investigate the Permo-Triassic rift phase as well as the younger ones. We concentrate on the northern part of the Viking Graben (60°20′–61°N), using a depth-converted version of seismic line NVGT88-04 (Figs 1 and 2). A 2D, non-uniform numerical model, including faults and incorporating the effect of a finite period of extension, is used. The model allows for both coupled and decoupled behaviour (Ter Voorde *et al.* 1998). As the effect of compaction during burial is not included in the model, we first calculate the decompacted thicknesses of the sediment packages. These thicknesses are used to constrain the modelling.

Geological structure

Seismic line NVGT88-04 is situated between the deep seismic reflection lines NSDP84-01 and -2 published previously (Beach *et al.* 1987; Gibbs 1987) and recently reprocessed and reinterpreted (Christiansson *et al.* 2000; Odinsen *et al.* 2000; Fig. 3). The three lines together provide an outline of the cross-sectional structure of the northern Viking Graben. Heavily faulted Mesozoic synrift sediments are unconformably overlain by a Cenozoic, almost unfaulted post-rift sequence. Pre-Jurassic sediments can clearly

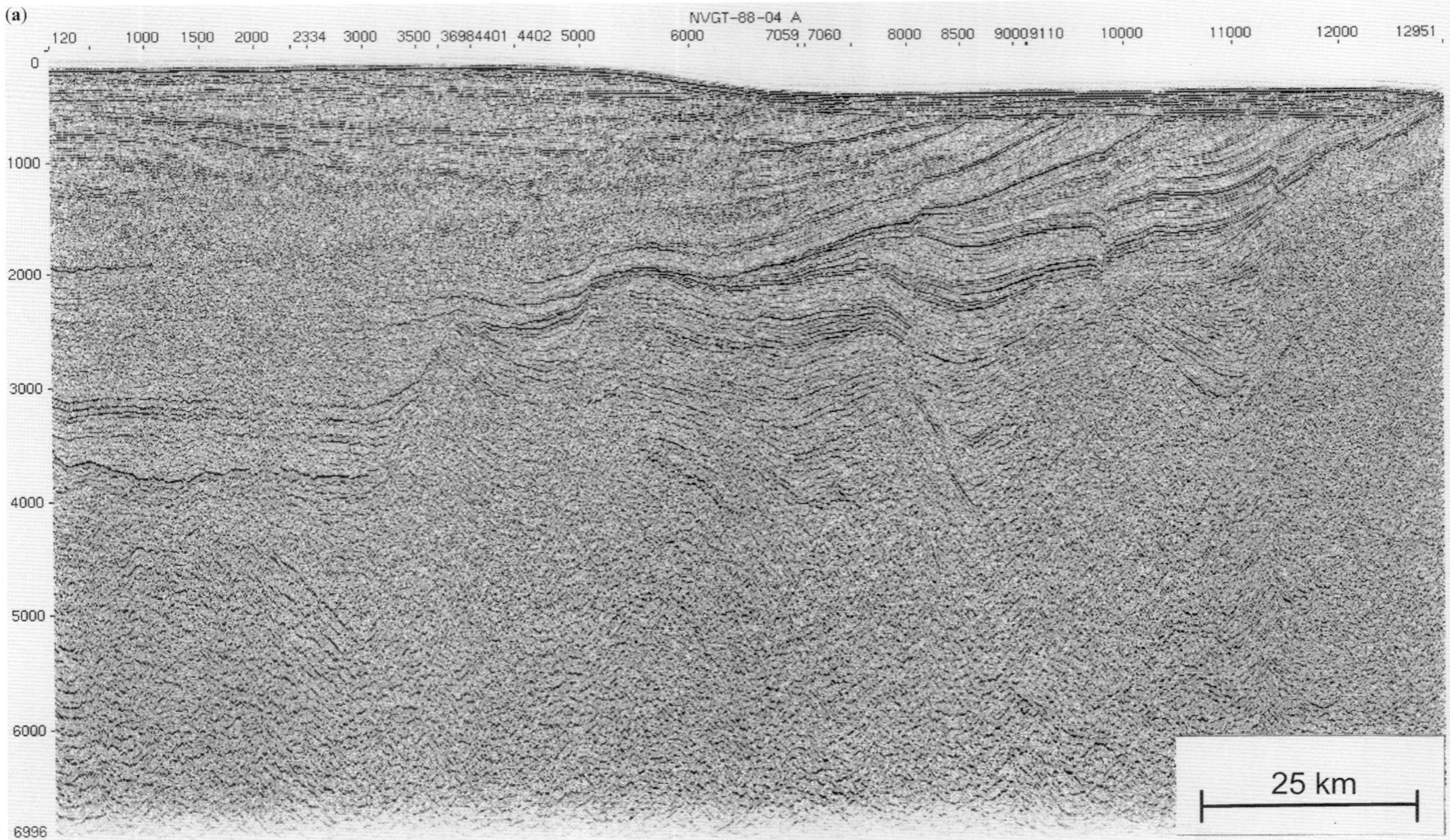

Fig. 2. (**a**) Seismic line NVGT88-04. Position of the line is indicated in Fig. 1. Vertical axis: two-way travel time (ms). (**b**) Depth-converted interpretation. ØFZ, Øygarden Fault Zone; HBS, Hild–Brent–Statfjord Fault; Os, Oseberg Fault; Br, Brage Fault; BrE, Brage East Fault; Øy, Øygarden Fault.

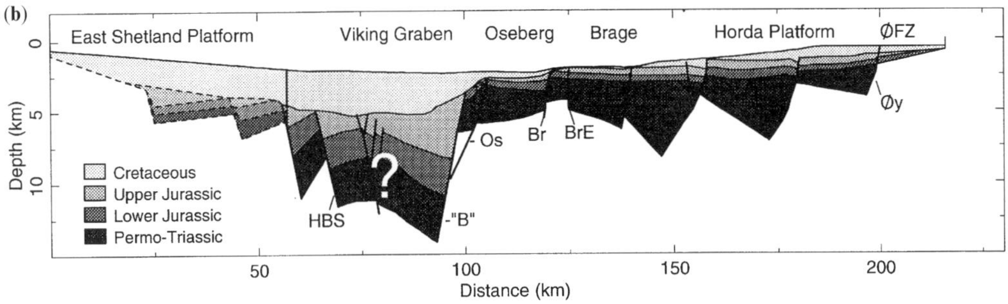

Fig. 2. (*continued*)

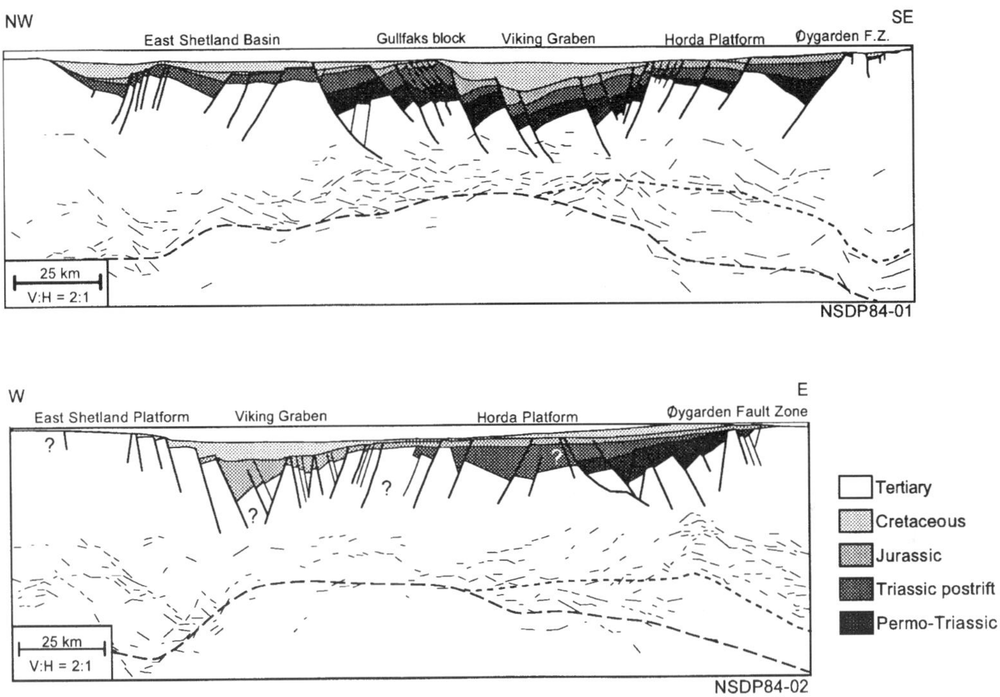

Fig. 3. Depth-converted interpretation of lines NSDP84-01 and NSDP84-02. After Christiansson *et al.* (2000). Position of the lines is indicated in Fig. 1.

be observed below the Horda Platform, and were recently also recognized further to the west (Platt 1995; Faleide *et al.* 2000). The Permo-Triassic event extended over a wider area than the Jurassic rift (e.g. Gabrielsen *et al.* 1990; Færseth *et al.* 1995*a*; Roberts *et al.* 1995; Odinsen *et al.* 2000).

Decompaction of the sediments

From line NVGT88-04, the compacted thicknesses of the sediments for four seperated time intervals (pre-Jurassic, early Jurassic, late Jurassic and Cretaceous) were extracted, and the decompacted thicknesses were calculated (Baldwin & Butler 1985; Dykstra 1987) (Fig. 4). As the Bajocian top of the Brent group is clearly visible on the seismic profiles, we choose this to be the boundary between the early Jurassic and the late Jurassic. For the compaction correction we used the porosity–depth relation

$$\phi = \phi_0 \exp(-cz) \qquad (1)$$

where ϕ is porosity and z is depth. We choose $\phi_0 = 0.56$ and $c = 0.39\,\mathrm{km}^{-1}$, as suggested by

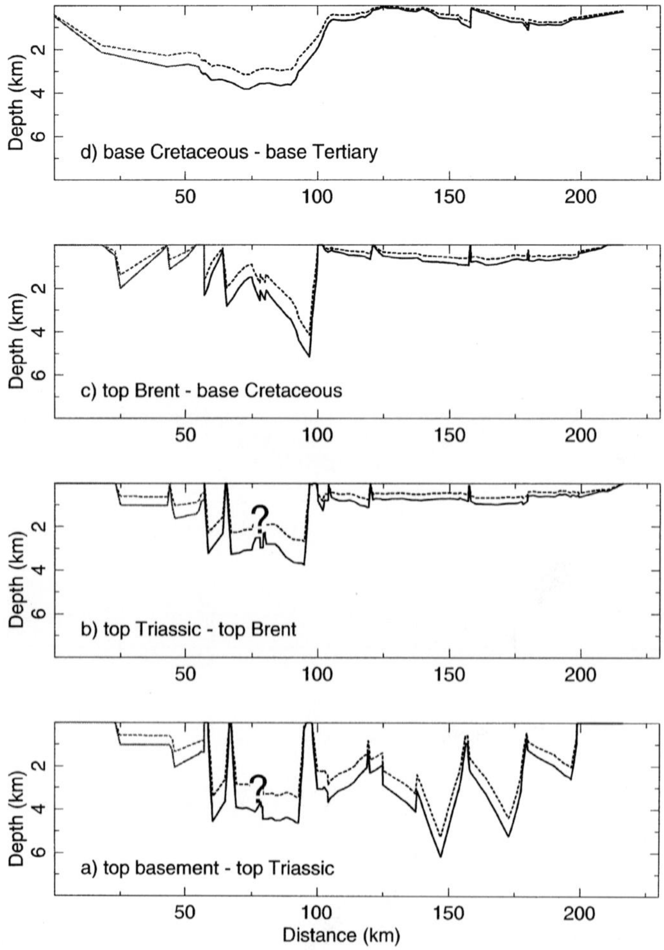

Fig. 4. Sediment thickness as measured along profile NVGT88-04, subdivided into four depth intervals. Dashed lines, before decompaction; continuous lines, after decompaction.

Sclater & Christie (1980) for an averaged sand–shale lithology in the North Sea. Subsequently, the reconstructed profiles for each time interval were obtained by correcting for the horizontal fault components at each level (Fig. 5). If the sedimentation rate was high enough to avoid the basin to become underfilled, the resulting profiles represent the basin subsidence during the given time interval. This was not always the case during the history of the study area. In Cretaceous and early Tertiary time, for example, the water depth was probably more than 500 m and may even have increased to a maximum of 1000 m in the central part of the basin (Nelson & Lamy 1987; Bertram & Milton 1988). As water depths are not included in the numerical model, we regard basin subsidence derived for such time intervals as underestimated.

Basin subsidence, subdivided into four time-intervals

Interval top basement–top Triassic (Fig. 5a)

The most striking features in this profile are the large, wedge-shaped basins at the easternmost part of the profile, beneath the present Horda Platform. These wedges are thinning to the west and show the largest subsidence of the Permo-Triassic basin system in the study area. Decompacted sediment thicknesses up to 6.2 km are derived. To the west, beneath the present rift axis, a sediment pile with an approximately even thickness of 4.5 km is found. However, because of the relatively poor quality of the seismic image beneath the Brent Group at this depth, this estimate should be regarded as uncertain.

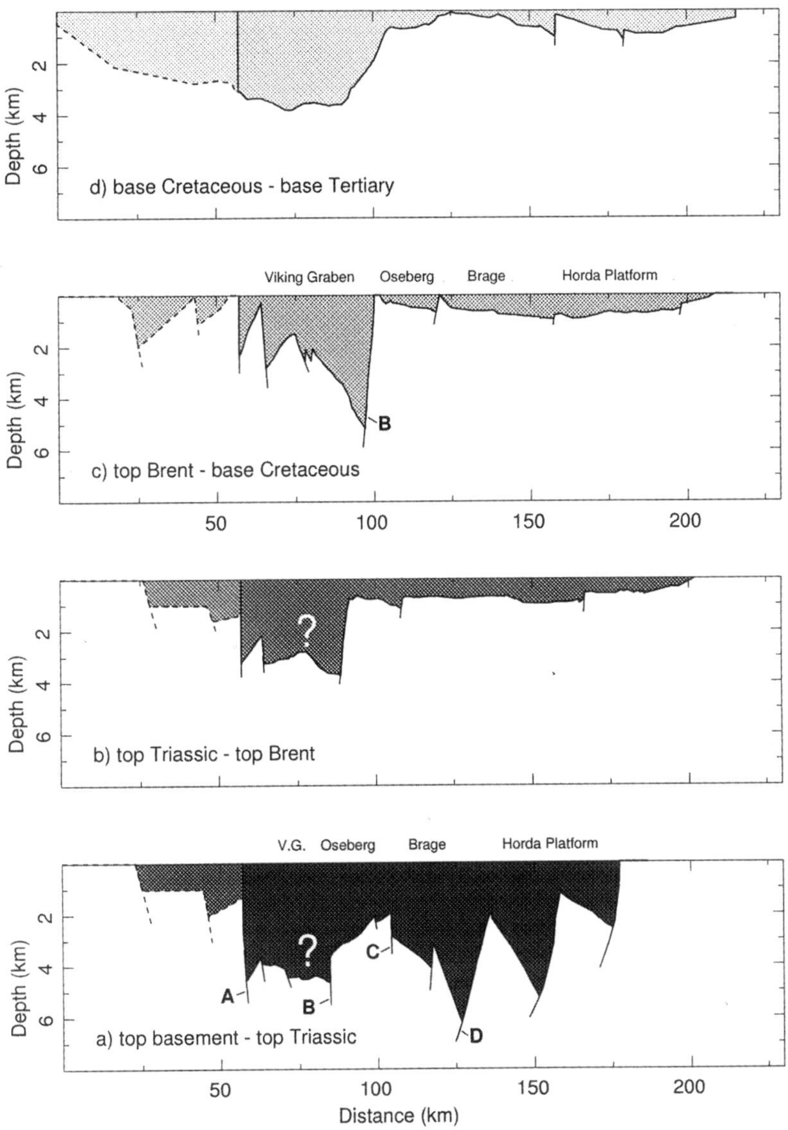

Fig. 5. Decompacted sediment thickness along profile NVGT88-04, subdivided into four intervals and corrected for subsequent faulting. Faults A, B, C and D are described in the text.

We interpreted four major, westerly dipping faults beneath the Horda Platform, whereas the fault pattern west of fault 'C' (the Brage East Fault) seems to be dominated by easterly dipping faults. This is emphasized by the sedimentary wedges slightly increasing in thickness towards the west above the Oseberg structure.

Interval top Triassic–top Brent (Fig. 5b)

At this time, subsidence was shifted towards the present rift axis, reaching a maximum basin depth of 3.7 km in a basin of less than 35 km width. The position of the present rift axis also corresponds to the site where the fault polarity shifted from easterly dipping faults in the west to westerly dipping faults in the east. This implies that this site has moved from the Brage area in Permo-Triassic time to the present rift axis in Early Jurassic time in this part of the basin. However, variations along strike in the relation between the Permo-Triassic and Jurassic centre of extension exist, possibly influenced by (deactivated?) transfer faults (Færseth 1996).

It should be noted that considerable faulting is derived for this period. This implies that the early Jurassic extensional episode may be more important than might be expected based on lines farther to the south, where this sequence is seen as a general progressive thickening to the west with only very limited fault control.

Interval top Brent–base Cretaceous (Fig. 5c)

Although the position of rifting in this period coincided with the Early Jurassic basin centre, activity along east-dipping faults in the west became more pronounced, reflecting initiation of the Shetland Platform. Fault rotation both east and west of the rift axis appears to have led to footwall uplift and erosion. The maximum basin depth was 5.2 km, measured in the strongly westerly thinning sediment wedge bounded by fault 'B'. The shallowest point of this wedge lies ± 17 km west of the deepest point, at a depth of 2.1 km.

Interval base Cretaceous–base Tertiary (Fig. 5d)

Except for a slight offset along two west-dipping faults at the Horda Platform, the effect of faulting is hardly noticeable in the Cretaceous sequence. Instead, subsidence in a wider area is observed. Maximum sediment thickness, along the rift axis, is measured to be 3.8 km. Possibly, thermal subsidence and sediment loading, combined with a general rise in sea level, resulted in gradual burial of Jurassic fault blocks during this period.

Although we acknowledge the fact that the top of the basement is difficult to constrain from line NVGT88-04, and is therefore not undisputed, a thick Triassic sediment pile was also suggested by Christiansson *et al.* (2000), who reinterpreted the deep structure of the northern Viking Graben from deep reflection seismic lines and gravimetric and magnetic data (Fig. 3). Bearing this in mind, some preliminary conclusions can be drawn. The Permo-Triassic rift phase affected a wider area than did the subsequent Jurassic–Cretaceous event. This is consistent with results of Færseth (1996). Also, Odinsen *et al.* (2000), using the interpretation of Christiansson *et al.* (2000), derived a regional stretching distribution in the Permo-Triassic compared with the more localized thinning in Jurassic time. The area of strongest subsidence associated with the Permo-Triassic event is positioned to the east of the location of major Jurassic extension. This is also consistent with a shift of the rift axis to the west after Triassic times, which is reflected in the reversal of fault polarity in time west of the Oseberg area.

The fault polarities derived from line NVGT88-04 are not necessarily representative for the whole Viking Graben area, as the fault polarity shifts along strike (Gabrielsen *et al.* 1990; Færseth *et al.* 1995*a*; Færseth 1996). However, a shift of the fault polarity in time is supported by what we regard as the most likely interpretations of lines further to the south (e.g. NVGT88-02). In this area, east-dipping faults within the Viking Graben seem to represent master faults in the Permo-Triassic period as well as in the early stage of Jurassic rifting, but are in places cross-cut by younger west-dipping faults during the late stage of Jurassic rifting (Færseth 1996).

Lithosphere rheology

One of the critical factors in a flexural basin model is the presumed lithosphere rheology. In most models, the lithosphere is assumed to react on a load in the same way as a thin elastic plate floating on a viscous fluid. The thickness of this (imaginary) plate, the so-called 'effective elastic thickness' (EET) determines the flexural response of the lithosphere to loading: a small EET yields a large-amplitude, short-wavelength response, whereas a large EET leads to small-amplitude, long-wavelength flexure. The EET can thus be estimated by considering it as a factor of the 'response function' describing the lithosphere deformation caused by loading. This approach has been followed for the Viking Graben by various workers, resulting in suprisingly large variations in estimates of the EET. Odinsen *et al.* (2000) derived an EET determined by the 450°C isotherm, which in their model corresponds to a depth of *c.* 45 km, using a kinematic numerical model developed by Kooi (1991). On the other hand, Kusznir *et al.* (1991), Roberts *et al.* (1995) and Ter Voorde *et al.* (1997), using flexural models for footwall uplift, arrive at EET values between 1.5 and 6 km in the same area. We propose that this discrepancy might be associated with the assumption of a 'coupled' versus a 'decoupled' lithosphere rheology, the latter of which is characterized by a very weak and ductile lower crust (i.e. Kusznir *et al.* 1991; Ter Voorde *et al.* 1998).

The continental plate may contain dipping or sub-horizontal weak ductile zones, which cannot support significant bending stresses. It is argued

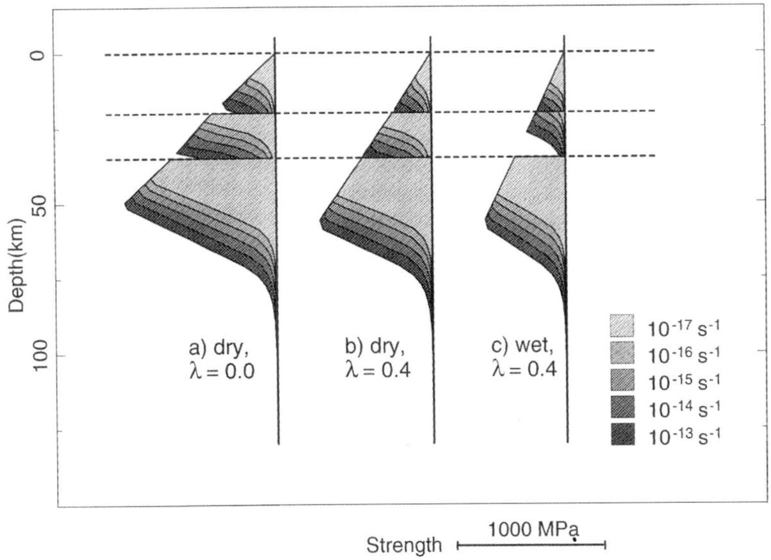

Fig. 6. Strength profiles for a typical continental lithosphere in an extensional regime, assuming a 20 km thick upper crust, a 15 km thick lower crust, and a total lithosphere thickness of 125 km, for various lithosphere rheologies. (**a**) and (**b**) assume a 'dry' rheology, the upper crust consisting of dry quartzite and the lower crust of diabase, and (**c**) assumes a 'wet' rheology, the upper crust consisting of wet quartzite and the lower crust of diorite. A zero pore pressure was assumed in (**a**), and a hydrostatic pore pressure in (**b**) and (**c**).

(e.g. Le Pichon & Chamot-Rooke 1991; Burov & Diament 1995) that the appearance of such zones in the lower crust permits mechanical decoupling of the upper crust from the mantle. A redistribution of loads in the upper crust, for example as a result of extensional faulting, would then be compensated for by elastic deformation in the upper crust exclusively, whereas the lower crust would adjust by ductile flow. This can explain the very low (1.5–6 km) effective elastic thicknesses obtained in modelling studies where the basin topography was used as the main constraining factor.

Figure 6 shows strength profiles for the tensional regime in a typical continental crust before extension. The profiles are constructed from the commonly used, laboratory-derived strength equations (e.g. Goetze & Evans 1979; Brace & Kohlstedt 1980; Carter & Tsenn 1987; see Appendix 1). The depth–strength curves demonstrate clearly the effect of the strain rate on the crustal configuration. If strain rates are higher than 10^{-13} s^{-1} the shallowest weak ductile layer occurs at the base of the lower crust, or perhaps even deeper. For lower strain rates, an additional ductile zone appears at the base of the upper crust.

A fully decoupled mode of flexure is likely to be an exceptional case, because the rate of lower crustal flow is generally not high enough to compensate for the entire upper-crustal deformation (Buck 1988; Ter Voorde *et al.* 1998). Therefore, part of the compensation probably mostly occurs by deep, very low viscosity mantle material causing the lithosphere to be in a 'partly decoupled' mode (see Ter Voorde *et al.* 1998). However, to make the effects of the assumption of coupled v. decoupled behaviour on the modelling results as clear as possible, we will consider only the end-members of the decoupled–coupled spectrum in this study.

The numerical model

We use a forward, 2D difference model, describing the lithospheric deformation resulting from extension, and including the effect of faulting in the upper crust. A finite syn-rift phase is considered (Waltham 1989, 1990). This is of major importance when calculating the post-rift thermal subsidence, as stressed, for example, by Ter Voorde & Cloetingh (1996). A presentation of the model is given in Fig. 7. Following Kusznir *et al.* (1991), the model is divided into an upper layer of faulting with vertical shear in the deforming hanging wall, and a lower layer of more distributed deformation. In both layers the

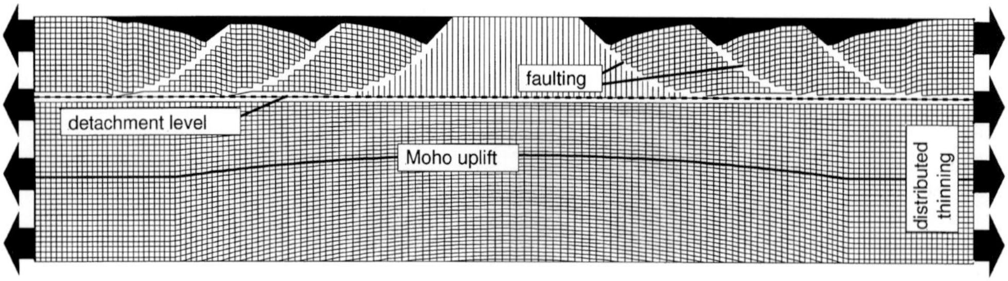

Fig. 7. Schematic representation of the model. Extension is achieved by distributed deformation (i.e. deformation along faults) in the upper crust, and by distributed thinning in the lower lithosphere.

condition of volume conservation is satisfied. Extension rates can be varied per time interval and per fault. The faults are assumed to flatten into the detachment that forms the boundary between the two layers. The depth of this detachment is an important factor, as it has the same function in the model as a 'necking depth' (e.g. Braun & Beaumont 1989; Weissel & Karner 1989; Kooi *et al.* 1992). The necking depth is defined as the level that remains horizontal in the absence of isostatic forces, and is a decisive factor for the existence and amount of footwall uplift (e.g. Ter Voorde & Cloetingh 1996).

Temperatures are calculated from the heat transfer equation:

$$\frac{\partial T}{\partial t} = \nabla(\kappa \nabla T) - \nu \nabla T + \frac{F}{\rho C} \qquad (2)$$

where T is temperature (°C), t is time (s), κ is thermal diffusivity ($m^2 s^{-1}$), v is velocity, F is heat production ($W m^{-3}$), ρ is density ($kg m^{-3}$) and C is specific heat ($J\,°C^{-1}\,kg^{-1}$).

The equation is solved using a finite difference method on a rectangular grid. Thermal properties on each grid point are obtained by interpolation from the second, moving grid representing the extending basin. Heat transfer during as well as after the deformation is calculated.

Flexure is calculated from the thin-plate approximation (e.g. Bodine *et al.* 1981):

$$\frac{\partial^2}{\partial x^2}\left[D(x)\frac{\partial^2 w}{\partial x^2}\right] + (\rho_{\text{comp1}} - \rho_{\text{comp2}})gw(x) = q(x) \qquad (3)$$

where w is deflection (m), D is rigidity (N m), ρ_{comp1} is density of the material underneath the plate ($kg m^{-3}$), ρ_{comp2} is density of the material above the plate ($kg m^{-3}$), g is gravitational acceleration ($m s^{-2}$) and q is vertical load ($N m^{-2}$).

The vertical load is calculated by vertical integration of the density contrasts caused by deformation of the crust, temperature changes and the mass of deposited sediments, relative to that of the undeformed crust:

$$q(x) = \int \Delta\rho(x,z)[1 - \alpha T(x,z)]g\,dz \qquad (4)$$

where α is thermal expansion coefficient (K^{-1}).

For the coupled mode of flexure, we assume the deflection w to be constant in each vertical column through the whole lithosphere, signifying that the crust and mantle are forged together by the lower crust. Hence, the integration interval for calculating the vertical load q comprises the whole lithosphere. ρ_{comp1} in equation (3) is the asthenosphere density, and ρ_{comp2} the sediment density. For the decoupled mode of flexure, the deflection w is assumed to be different for the crust and mantle. For the upper crust, q is calculated by integrating from the surface to the lower crust, ρ_{comp1} is taken as the lower-crustal density, ρ_{comp2} as the sediment density, and the rigidity is assumed to be low (EET $\leq$ 6 km). This is different from earlier studies using low EET values (e.g. Kusznir *et al.* 1991; Roberts *et al.* 1995), where the influence of decoupling on the compensation materials is not taken into account.

The thermal features of the model have been described in more detail by Ter Voorde & Bertotti (1994), the structural features by Ter Voorde & Cloetingh (1996), and the effects of decoupling have been discussed by Ter Voorde *et al.* (1998).

Model parameters

The model parameters we used in this study are summarized in Table 2. The Moho depth is assumed to be 35 km, based on the crustal thickness observed below the platform areas. The detachment depth is assumed to be 18 km,

Table 2. *Parameters used for the modelling*

Parameter	Value	Deviation used for sensitivity test
EET (coupled)	450°C isotherm	see Fig. 8
EET (decoupled)	1.5–6 km	see Fig. 8
Density sediments	1900 kg m^{-3}	±100 kg m^{-3}
Density upper crust	2700 kg m^{-3}	±50 kg m^{-3}
Density lower crust	2950 kg m^{-3}	±50 kg m^{-3}
Density asthenosphere	3300 kg m^{-3}	–
Moho depth	35 km	±2 km
Detachment depth	18 km	±5 km
Young's modulus	7×10^{10} N m	–
Poisson ratio	0.25	–

based on observations of Fossen *et al.* (1998). They described the flattening of normal faults at a maximum depth between 18 and 20 km, in the Tampen Spur area on the western flank of the Viking Graben.

Although Fossen *et al.* (1999) suggested that these flattened faults are inherited mechanical weak zones in the basement, related to Devonian extensional structures or Caledonian thrusts, the detachments could also be linked to ductile shear zones in the middle and lower crust and thus be explained by temperature- and pressure-controlled changes in the lithosphere rheology (e.g. Fossen & Gabrielsen 1996). The different detachments may have different causes, and may also have been initiated at different stages during rifting. Then, truly low-angle structures might also develop by flattening of originally steeper fault structures, during repeated phases of stretching, and perhaps act as detachments in a later stage (Fossen *et al.* 2000). Although the flattening of the faults with depth may have no unequivocal explanation, it may indicate the existence and position of ductile zones in the lower crust. In the case of the Viking Graben, the maximum observed depth of flattening should then also be interpreted as the depth of mechanical decoupling.

Model sensitivity and constraints

The overall state of flexure depends on the strength of the layer(s), the density and amount of sediments, the depth of the Moho, and the depth of necking during lithospheric extension. To obtain an impression of the model sensitivity to these variables for the Viking Graben configuration, we tested the effects of varying the EET, the densities and the Moho depth, using parameters and deviations given in Table 2 and the same fault configuration as in Fig. 5. Because, in our model, a change in necking level implies a change in fault shapes, effects of varying this level on the resulting flexural response are difficult to test independently. However, it is possible to calculate which change in sediment density would be equivalent to a change of 1 km in the level of necking (see Appendix 2), and we used this analogy to estimate the model sensitivity for the level of necking.

The deviations in densities and Moho depth cause maximum variations in the resulting basin subsidence of 60 m for the coupled case and 185 m for the decoupled case. A reasonable variation of 5 km in necking depth can be compared with a variation in sediment density of -148 kg m^{-3} (in the case of a deeper necking level) or 262 kg m^{-3} (in the case of a shallower necking level) for the coupled case (Appendix 2, equation (6)), leading to a maximum variation in basin subsidence of 100 m. For the decoupled case the level of necking has no influence (Appendix 2, equation (6)).

In Fig. 8 the effects of changes in EET values are shown for the simulation of the Permo-Triassic time interval (i.e. Fig. 5a). The figure shows clearly that the choice of the EET is of limited influence for the coupled scenario, but might be important in the case of decoupling. In the modelling, we used an EET defined by the 450°C isotherm for coupled flexure, and the EET giving the best fit for decoupled flexure.

Finally, effects of reasonable errors in the estimated fault shapes and deformation mechanism may be of the order of 1 km (White *et al.* 1986; White & Yielding 1991; Withjack & Peterson 1993).

In previous studies on the Viking Graben (Marsden *et al.* 1990; Roberts *et al.* 1995; Odinsen *et al.* 2000), modelling results and (seismic) data normally fit within 1000 m. Based on this, on the estimated sensitivity of the model, and on uncertainties in the seismic data, for example as a result of possible errors in chosen velocity models, we regard a fit of the model with the data within 1000 m as a good approximation.

Modelling results

The numerical model was used to simulate the sequential basin configurations shown in Fig. 5, in order to relate the amount of basin subsidence to the amount of extension along various faults, and to the strength and mode of flexure of the lithosphere. In the model, sediments are assumed to fill the basin up to the surface. This

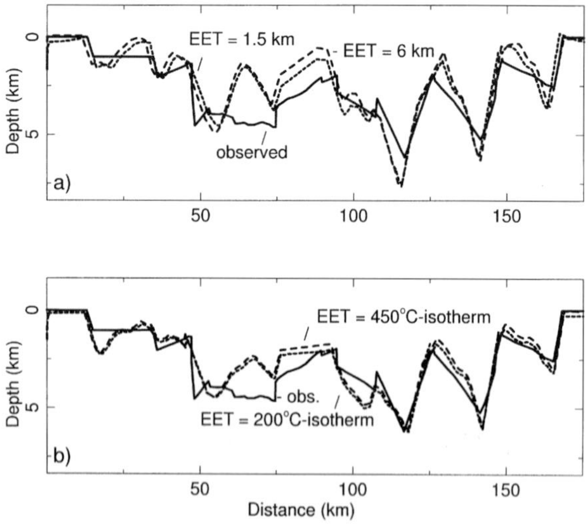

Fig. 8. Modelling results for interval top basement–top Triassic, assuming different values for the EET. (**a**) Decoupled rheology; (**b**) coupled rheology.

is a realistic approach, as long as the basin is not underfilled at the end of the modelled episode. The modelling parameters we used are summarized in Table 2.

Interval top basement–top Triassic (Fig. 9)

Triassic sediments are supposed to have been deposited in lacustrine environments, dominated by alluvial fans along the graben margins and with finer fluvial or lacustrine sediments in the 'lows' (e.g. Steel & Ryseth 1990). At the end of Triassic time the northern Viking Graben was approximately at sea level. The sediment thickness shown in Fig. 5a might thus be interpreted as the total basin subsidence during this time interval. We focused especially on the flanks of the graben, because of the poor seismic resolution in the basin centre. Consequently, the much better fit of the model beneath the platform areas than beneath the rift axis (Fig. 9a) reflects the higher confidence of the observed sediment thickness in that area.

To obtain the best modelling result, a fault configuration was required with a change in fault polarity on the flanks of the Viking Graben and of the Brage area (Fig. 9a). The thinning of the mapped sequence in the Oseberg area towards the east is then explained by extension along the east-dipping fault 'A', bounding the Viking Graben in the west, whereas the Viking Graben centre subsided further along the west-dipping fault 'B', 25 km farther to the east. The major extension occurred on the Horda Platform, provided that the top Triassic interpreted in the graben centre is correct.

Two scenarios were modelled for this time interval, one assuming a coupled rheology with an EET determined by the 450°C isotherm (corresponding to an initial depth of 43 km in the model), the other assuming a decoupled rheology with an EET of 1.5 km. A reasonable fit between the modelled basin configuration and the one interpreted from the seismic line could be obtained from both assumptions (Fig. 9b), but with different values for the amount of extension. In the case of a coupled configuration, the lithosphere is too strong to show a significant flexural response on the extension, and the basement subsidence is almost entirely due to crustal movements along the faults. However, if we adopt a decoupled mode of extension, the model shows a syn-rift uplift of the area, leading to a small amount of subaerial footwall erosion, followed by thermal subsidence (Fig. 9b and c).

The modelled uplift and subsidence patterns for the decoupled scenario agree with the findings of Roberts *et al.* (1995), and those of the coupled scenario with findings of Odinsen *et al.* (2000). Roberts *et al.* (1995) modelled the pre-Jurassic rifting stage in the Horda Platform region with a flexural cantilever model, using an EET of 1.5 km, and found that a stage of footwall uplift and erosion preceded a stage of overall thermal subsidence. On the other hand, the results of Odinsen *et al.* (2000), who used the

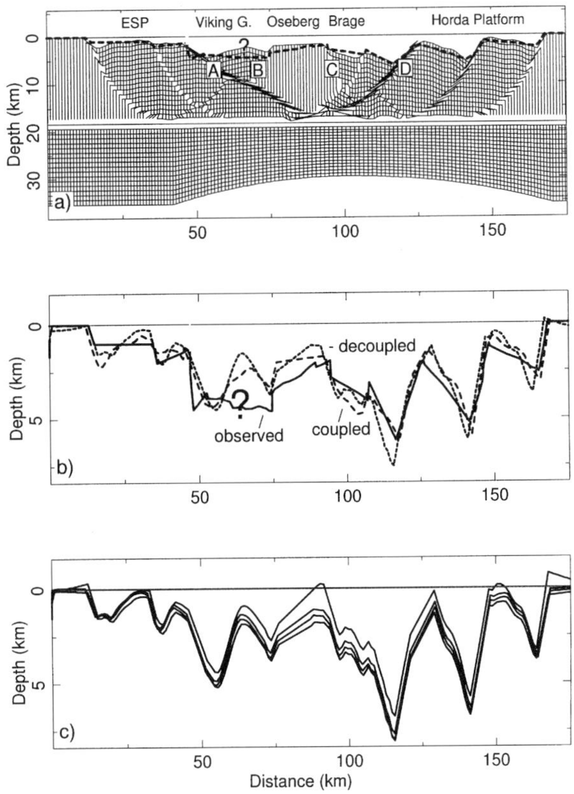

Fig. 9. Modelling results for interval top basement–top Triassic. (**a**) Modelled fault configuration; dashed line indicates basin depth as derived from profile NVGT88-04. (**b**) Long dashes, modelling result for coupled rheology; short dashes, modelling result for decoupled rheology; continuous line, basin depth as derived from profile NVGT88-04. (**c**) Modelled basin depth for decoupled rheology, at (from top to bottom) 0, 30, 74 and ∞ Ma after the end of rifting.

coupled approach, do not indicate such uplift. According to Roberts *et al.* (1995), the seismic expression of the better imaged Triassic fault blocks beneath the Horda Platform gives the impression of bevelling at the fault-block crests. This can be noticed in interpretations of Beach *et al.* (1987, fig. 2), Roberts *et al.* (1993, fig. 6a) and Odinsen *et al.* (2000, figs 2 and 5), and supports the decoupled scenario as the most likely one.

The total amounts of extension used in our model are 23 km for the coupled scenario and 34.5 km for the decoupled scenario, signifying an average β value of 1.18 for the coupled case, and 1.29 for the decoupled case (measured over the entire profile). Major extension in Permo-Triassic time took place on the Horda Platform, the total modelled amount of extension being 14 or 22% west of the Brage fault, and 22 or 38% east of it, for the coupled or decoupled case, respectively.

Roberts *et al.* (1995) reported a β value of 1.34 in the Horda Platform area, along the same section as in the present study, whereas Odinsen *et al.* (2000) calculated an average β factor of 1.27 for their northern transect, and 1.19 for their southern transect. They derived β factors for the Horda Platform of 1.33 and 1.39, but they did not indicate exactly over which width this was measured.

Figure 9c shows the syn-rift uplift and thermal subsidence with time for the decoupled scenario. Indicated are the top of the basement at the end of rifting, 30 Ma later (i.e. latest Triassic), 74 Ma later (i.e. when the second major rift phase is assumed to start) and after total thermal relaxation. In contrast to earlier proposals (e.g. Giltner 1987; Marsden *et al.* 1990; Roberts *et al.* 1995), but in accordance with Odinsen *et al.* (2000), the present model suggests that thermal subsidence as a result of Permo-Triassic rifting cannot have had much influence on the Jurassic event.

(It should be noted that thermal subsidence for the coupled case would have even less effect.) Researchers who stated that an inherited thermal anomaly influenced Jurassic rifting used 1D thermal calculations (e.g. Giltner 1987), instantaneous Triassic rifting (Marsden *et al.* 1990; Roberts *et al.* 1995), and/or derived their statements from studies of the Jurassic event alone (Giltner 1987).

Interval top Triassic–top Brent Group (Fig. 10)

The top of the Brent Group horizon is diachronous on a regional scale, and is situated in the Bathonian (Middle Jurassic) sequence in the northern Viking Graben. The lower Jurassic sequence consists predominantly of marine shales, whereas the Middle Jurassic sequence was deposited in non-marine to paralic environments (Fjellanger *et al.* 1996; Ravnås *et al.* 1999). The Tarbert Formation of the uppermost Brent Group consists of shallow marine sandstones. As an approximation, the profile shown in Fig. 5b might thus be regarded to represent the total basin subsidence during early Jurassic time, until mid-Bathonian time.

This time interval could be simulated with a much less complicated fault configuration than that of the Permo-Triassic phase. The position of the rift axis was in the Viking Graben centre, coinciding with the site of major subsidence as well as the position of shift in fault polarity along the profile (Fig. 10a). However, the subsidence in the basin centre might be overestimated, because, again, the seismic data at this depth in the basin centre do not allow for a decisive solution. Compared with the earlier rift phase, a change in fault dip was imposed between the Oseberg and Brage area (fault ‘C’). Figure 10 shows the modelling results for the coupled and the decoupled case. For the decoupled case, an EET of 6 km was used. The coupled scenario gives results that are more consistent with the observed basin configuration, as hardly any tilting of the fault blocks is observed. For this reason, thermal subsidence related to this phase is assumed to be negligible. To model the observed amount of subsidence, modelled offsets along the faults had to be set up to 2 km larger than the observed offsets, suggesting an overestimation of the amount of extension, and the existence of an additional source of basin subsidence, different from fault-related extension.

As shown by the modelling results of the top basement–top Triassic interval, only a small part of the additional subsidence (i.e. less then 350 m) can be explained by remnant thermal subsidence caused by Permo-Triassic rifting, which was not included in the modelling results. Another explanation for basin subsidence might be the existence of a ‘proto-rift stage’ (Gabrielsen 1986; Nøttvedt *et al.* 1995). Nøttvedt *et al.* (1995)

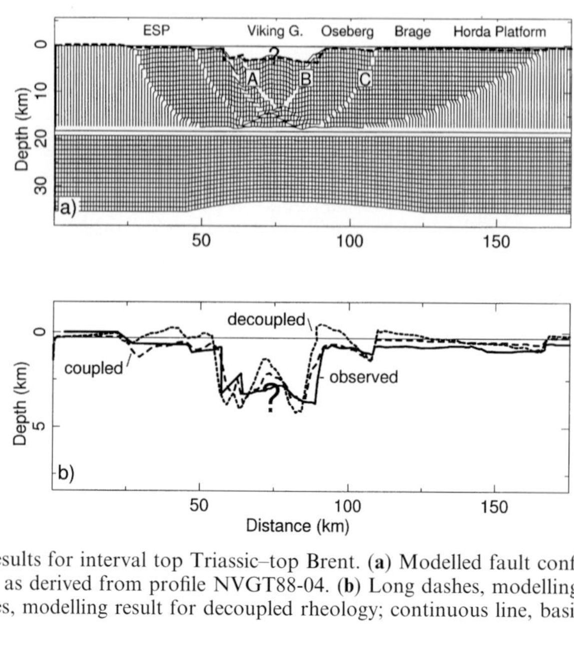

Fig. 10. Modelling results for interval top Triassic–top Brent. (**a**) Modelled fault configuration; dashed line indicates basin depth as derived from profile NVGT88-04. (**b**) Long dashes, modelling result for coupled rheology; short dashes, modelling result for decoupled rheology; continuous line, basin depth as derived from profile NVGT88-04.

proposed an idealized, three-stage model for rift evolution, involving a proto-rift stage, a syn-rift stage and a post-rift stage. The proto-rift stage is characterized by thermally induced domal uplift, or by deposition in a wide, slowly subsiding basin with only minor fault activity. The syn-rift stage describes the phase of active stretching and block rotation, and the post-rift stage consists of asymptotically decreasing subsidence, caused by thermal contraction. Interference between post-rift subsidence related to the Permo-Triassic rift phase and proto-rift subsidence related to the Jurassic rift phase (Steel 1993; Nøttvedt *et al.* 1995) might form an alternative source for the extra subsidence observed.

The total amount of extension in the modelled scenario (Fig. 10a) is 7.3 km, yielding an average β value of 1.05. The modelled β reaches a value of 1.16 in the basin centre. Apparently, a minor extensional event has occurred in earliest Jurassic time, as argued earlier by Røe & Steel (1985) and Gabrielsen *et al.* (1990). The magnitude of this event cannot be derived from the modelling results, because of the large uncertainty in the basin depth.

This early Jurassic extension, combined with the additional subsidence, caused the depositional environment to convert from continental to marine, implying that the creation of new accommodation space outpaced sediment supply. Subsequently, as sedimentation continued, the depositional environment passed to non-marine again, and in latest Bajocian–earliest Bathonian time the first rotational movements of the second major rift phase commenced (Færseth *et al.* 1995*b*; Ravnås & Bondevik 1999).

Interval top Brent Group–base Cretaceous (Fig. 11)

The Upper Jurassic sediments in the Viking Graben consist mostly of marine shales. The shales of the Bathonian–Oxfordian Heather

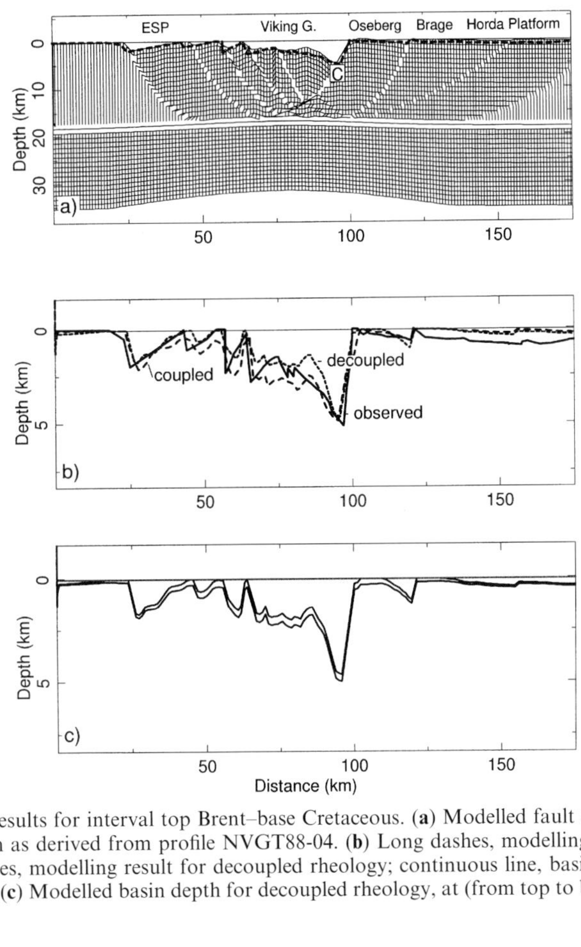

Fig. 11. Modelling results for interval top Brent–base Cretaceous. (**a**) Modelled fault configuration; dashed line indicates basin depth as derived from profile NVGT88-04. (**b**) Long dashes, modelling result for coupled rheology; short dashes, modelling result for decoupled rheology; continuous line, basin depth as derived from profile NVGT88-04. (**c**) Modelled basin depth for decoupled rheology, at (from top to bottom) 0 and ∞ Ma after the end of rifting.

formation are overlain by the organic-rich Draupne Formation (equivalent to the Kimmeridge Clay Formation), and both are interfingered by sandstone units (e.g. Ravnås *et al.* 1999). Based on environmental interpretations, the water depth at the end of this episode (Ryazanian) is estimated to be 200 m on the margins to 300 m in the graben centre (Badley *et al.* 1988), although this might be a very conservative estimate (e.g. Marsden *et al.* 1990; Yielding *et al.* 1992).

The modelling results are shown in Fig. 11. The EET used for the decoupled case was 6 km. Broadly the same fault configuration could be used as for the early Jurassic phase, but with a few more active faults. The basin now became asymmetric, deepening to the east in this part of the graben. However, this varies along strike (see, e.g. Odinsen *et al.* 2000). The footwall uplift in the west is best explained by using the decoupled approach, as shown in Fig. 11b. This approach yields an average β factor over the whole transect of only 1.07 (and a maximum of 1.2). Together with the early Jurassic phase, this makes a Jurassic β of 1.12. This is slightly lower than the values found for Jurassic stretching by Odinsen *et al.* (2000), who obtained 1.15 for the northern transect and 1.19 for the southern transect, or by Roberts *et al.* (1993), who reported a β of 1.3 in the basin centre, of 1.15 in the East Shetland Basin and western Horda Platform and of 1.05 in the eastern Horda Platform. That the β value we derived is relatively low can be related to the fact that our model assumes a zero water depth, which is an underestimation for this time interval. Figure 11c displays the basin configuration immediately after stretching as well as after thermal subsidence. From this figure, it is evident that the thermal anomaly caused by the extension alone cannot be responsible for the subsequent Cretaceous subsidence. This can partly be explained by the fact that the basin was underfilled at the end of this rift phase, which was not included in the modelling. The modelled amount of extension, and thus the thermal anomaly, is therefore probably an underestimation. During Cretaceous time, much of the sedimentation was accommodated by the infilling of previous rift bathymetry, causing additional subsidence because of sediment loading. This effect is further exaggerated by the compaction of earlier deposited sediments (Yielding *et al.* 1992), which is not taken into account in the present model.

Modelled Moho configuration

To compare the modelled Moho depth with the observations, we superimposed the Moho deformation of the three time intevals on each other, the result of which is shown in Fig. 12. The 'observed Moho' in Fig. 12 is derived by interpolation between the Moho reflectors observed in seismic lines NSDP84-01 and -2 (e.g. Christiansson *et al.* 2000; Odinsen *et al.* 2000), although we acknowledge that the Moho is not unequivocally imaged east of the Viking Graben (see Fig. 3). Starting from a constant initial value of 35 km, the modelled Moho depth is too large compared with the depth derived from the

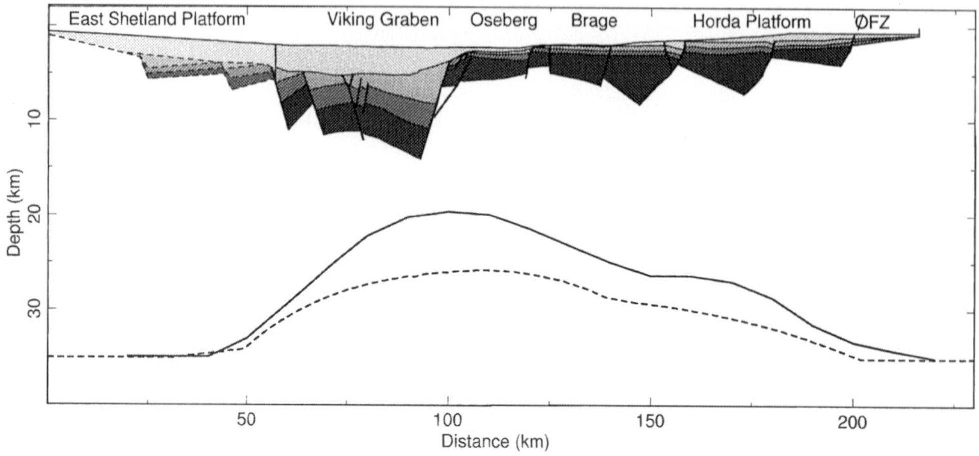

Fig. 12. Moho uplift as a consequence of rifting. Dashed line, modelled Moho depth resulting from three stages of rifting; continuous line, Moho depth along line NVGT88-04, as derived from interpolation between interpretations of line NSDP84-01 and NSDP84-02.

observations. The maximum difference, below the Viking Graben axis, is 6.5 km. This is in agreement with findings of Odinsen *et al.* (2000), who suggested that the Permian pre-rift Moho depth might have varied from ±35 km on the flanks to less than 30 km in the basin areas. To obtain the best fit for the subsidence data, Lippard & Liu (1992) needed to assume an initial crustal thickness varying from 28 km in the centre to 30–32 km on the flanks of the Graben. This pre-Permian variation in Moho depth is likely to be caused by Devonian crustal thinning, imposed on the thickened crust of the Caledonian Orogeny (Andersen & Jamtveit 1990; Fossen & Rykkelid 1992).

Mode of flexure in the Viking Graben

End-members: the coupled versus the decoupled mode

An important result of this modelling study is that it is hard to discriminate between the

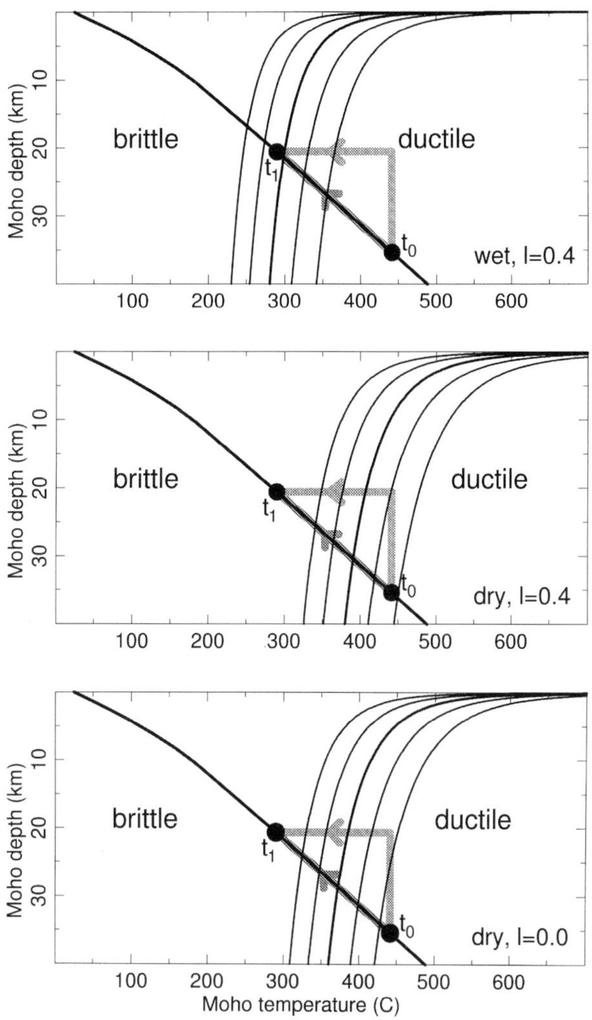

Fig. 13. Mode diagrams for the lithosphere, indicating the possibility of a decoupled rheology in Moho depth–Moho temperature space (Spadini & Podladchikov 1996), for three different lithosphere rheologies. Boundaries between the ductile and brittle lower crust are given for strain rates of (from left to right) 10^{-17}, 10^{-16}, 10^{-15}, 10^{-14} and 10^{-13} s^{-1}. Black dots indicate the assumed pre-rift (t_0) and post-rift (t_1) Moho depth (at the position of major Moho uplift) in the Viking Graben, as derived from Fig. 11. The light grey arrows indicate the change in the state of the lithosphere in cases of instantaneous rifting and very slow rifting.

coupled and the decoupled mode of flexure on the basis of the fit between modelling results and seismic data alone (e.g. Fig. 9). This might be an explanation for the large variations between EET values used for previous modelling studies of the Viking Graben. Nevertheless, we showed that the presence or absence of footwall uplift is a good indicator for the mode of flexure: the suspected presence of eroded fault blocks in the Permo-Triassic phase and of footwall uplift in the late Jurassic phase cannot be explained with the coupled approach (even though we used a rather deep necking level), whereas the absence of footwall uplift in early Jurassic time is impossible to reconstruct by decoupled flexure.

Although caution should be taken when interpreting the results (we modelled only the end-members of the coupled–decoupled spectrum, whereas the 'partly decoupled' mode is most likely to occur), the best results were obtained by assuming a decoupled mode of flexural response in the Permo-Triassic phase, a coupled mode of flexure in early Jurassic time, and a change back to decoupled flexure in late Jurassic time. The choice of the mode of flexure is important, as it can have a strong influence on the derived amount of extension. For the Permo-Triassic phase, for example, we derived a β factor of 1.18 for the coupled case and of 1.29 for the decoupled case.

A prerequisite for decoupled behaviour is the existence of a ductile lower crust. Figure 13 shows which Moho depth and temperature will cause the manifestation of a ductile lower crust according to the rheological laws. The curves were derived by calculating the depth at which brittle deformation gives way to ductile deformation (Spadini & Podladchikov 1996), i.e.

$$\sigma_b \geq \sigma_d$$

where σ_b is the brittle strength and σ_d is the ductile strength needed to cause deformation at a given strain rate (see Appendix 1). The parameters used are given in Table 3. The Moho depths and temperatures on the left side are associated with an entire brittle crust, whereas the conditions on the right side of the curves yield a ductile lower crust. If we assume the bold line to be the pre-rift geotherm in the northern Viking Graben, and the black dot at t_0 to indicate the state of the pre-rift Moho, we can conclude from the figure that the lower crust is at least partly ductile, and that the first condition for a decoupled mode of flexure is fulfilled. However, the uplift of mantle material combined with syn-rift cooling will eventually result in a shift from the decoupled to the coupled state of the lithosphere (Spadini & Podladchikov 1996), which agrees with our modelling results for the Permo-Triassic and the Early Jurassic phases.

The change back to the decoupled state in Bathonian time cannot be explained in this way, or be related to the change in extension (and thus strain) rate. In fact, at lower strain rates, decoupled behaviour is more likely to occur than at high strain rates (e.g. Brun & Tron 1993; Ter Voorde *et al.* 1998; see Figs 6 and 13). However, a thermal event could explain the weakening of the lithosphere and a shift to decoupled behaviour.

A Mid-Jurassic pre-rift thermal dome in the so-called 'triple junction' of the North Sea rift is

Table 3. *Parameters used for construction of strength profiles*

	Upper crust		Lower crust		Mantle olivine
	Quartzite (w)	Quartzite (d)	Diorite (wet)	Diabase (dry)	
E_p (kJ mol^{-1})	172.6	134	212	276	510
A_p (Pa^{-n} s^{-1})	1.26×10^{-13}	6.03×10^{-24}	1.26×10^{-16}	3.16×10^{-20}	7×10^{-14}
n	1.9	2.72	2.4	3.05	3.0
E_d (kJ mol^{-1})	–	–	–	–	535
A_d (s^{-1})	–	–	–	–	5.7×10^{11}
σ_D (GPa)	–	–	–	–	8.5
R (J (mol K)$^{-1}$)		8.314 J (mol K)$^{-1}$			
Thermal diffusivity sediments		0.75×10^{-6} m^2 s^{-1}			
Thermal diffusivity crust		1.0×10^{-6} m^2 s^{-1}			
Thickness heat producing layer		15 km			
Heat production		2.3×10^{-6} W m^{-3}			

Material constants for quartzite, diorite, diabase and olivine are adopted from Tsenn & Carter (1987).

well established (e.g. Ziegler & Van Hoorn 1989; Hendrie *et al.* 1993; Underhill & Partington 1993), evidenced by basin-wide subaerial erosion (the 'Mid-Cimmerian' unconformity), the extrusion of the Rattray volcanic rocks, and flooding events on basin margins. This thermal dome is interpreted to be the effect of a short-lived, anomalously hot asthenospheric pulse (Hendrie *et al.* 1993) or transient mantle plume (Underhill & Partington 1993). However, although the timing of this event is correct to explain the shift to decoupled behaviour in Bathonian time, the location might be too far away (i.e. ±250 km to the south) from the northern Viking Graben to have such a pronounced effect.

The partly decoupled mode of flexure

As mentioned above, in this study we only look at the end-members of the spectrum of decoupled–coupled modes of flexure. Nevertheless, the most probable mechanism to occur is the partly decoupled mode of flexure. In this case, the viscosity of the lower crust is too high to allow for an isostatic compensation level in the lower crust. Therefore, as a first approximation, the lithosphere will react in a coupled mode, and only short-wavelength lateral variation in the load is compensated by lower-crustal flow.

In general, modelled surface deflections obtained by assuming a partly decoupled lithosphere differ from the fully coupled mode in that an extra short-wavelength component can be observed. In contrast, the difference from the fully decoupled case is that the long-wavelength uplift component disappears (Ter Voorde *et al.* 1998). Translating this to Figs 9–11 raises the suggestion of partly decoupled flexure.

Conclusions

The evolution of the Viking Graben has been studied using a numerical model, constrained by seismic line NVGT88-04. Although the importance of the rift phase in Permo-Triassic time has been affirmed by the modelling results, the post-rift subsidence is shown to have practically ceased at the onset of Jurassic rifting. Hence, no significant amount of basin subsidence in the Jurassic phase can be ascribed to thermal contraction resulting from the Permo-Triassic event. The Jurassic subsidence should therefore be explained entirely by a new extensional event, possibly preceded by the 'proto-rift' stage as defined by Nøttvedt *et al.* (1995).

The position of the rift axis is shown to change with time. The Permo-Triassic rift axis was positioned beneath the present Horda Platform, whereas the Jurassic rift axis was positioned beneath the present Viking Graben centre. This result might be dependent on the position along strike (Færseth 1996). For example, Odinsen *et al.* (2000) found for line NDSP84-1 that the maximum β factors for both the Permo-Triassic and the Jurassic phases were in the present Viking Graben axis, whereas for the more southern line NSDP84-02 the maximum β factor in the Permo-Triassic phase was on the Horda Platform.

To our knowledge, this basin modelling study is the first one that attempts to discriminate between the coupled and the decoupled mode of flexure. The major difference between these two modes is the amount of fault-block rotation, which is larger when a decoupled rheology is assumed. The coupled and decoupled mode of flexure are sometimes hard to discriminate on the basis of basin geometries alone (e.g. Fig. 9), which may explain the large variation in EET values found in the literature. Nevertheless, the modelling results suggest a decoupled mode of flexure in the Permo-Triassic phase, a coupled mode of flexure in early Jurassic time, and a change back to decoupled flexure in Late Jurassic time. Subsequent research on this topic should focus on further constraints on the mode of flexure, such as the lower-crustal thickness and material, and the pre-rift geotherm.

We would like to thank R. Ravnås for bringing us together, and T. Odinsen for discussions about the development of the Viking Graben. Reviews by R. Boutilier and D. Waltham were greatly appreciated. This research was supported by the IBS (Integrated Basin Studies) project, part of the Joule II research programme funded by the Commission of European Communities (Contract JOU2-CT92-0110). This paper is Publication 981001 of the Netherlands Research School of Sedimentary Geology.

Appendix 1

The brittle yield strength is given by Byerlee's law (Byerlee 1978; Brace & Kohlstedt 1980):

$$\sigma_b = \alpha \rho g (1 - \lambda) z$$

where ρ is the density, g is the gravitational acceleration, z is the depth, $\alpha = 0.75$, $\lambda = 0$ for zero pore pressure, and $\lambda = 0.4$ for hydrostatic pore pressure. Ductile deformation is assumed to occur by power law creep (Kirby 1983):

$$\sigma_{\mathrm{d}} = A_{\mathrm{p}}^{-1/n}\dot{\epsilon}^{1/n}\exp\left(\frac{E_{\mathrm{p}}}{nRT}\right)$$

where $\dot{\epsilon}$ is strain rate, A, n and E_{p} are empirically derived material constants, R is the gas constant and T is the absolute temperature. For olivine with a strength exceeding 200 MPa, ductile deformation is described by Dorn creep (Goetze & Evans 1979; Tsenn & Carter 1987):

$$\sigma_{\mathrm{d}} = \sigma_{\mathrm{D}}\left\{1 - \left[\frac{-RT}{E_{\mathrm{d}}}\ln\left(\frac{\dot{\epsilon}}{A_{\mathrm{d}}}\right)\right]^{1/2}\right\}$$

Parameters are given in Table 3.

Appendix 2

The depth of necking is the level of zero vertical motion in the absence of isostatic forces. This level determines the ratio between thinning of the upper crust, where crustal material is replaced by sediments, and thinning of the lower lithosphere, where crustal material is replaced by mantle material. Therefore, the depth of necking is a decisive factor for the new load distribution caused by crustal extension. For a given amount of thinning, the change in load ΔP caused by a variation in necking depth ΔN is equal to the change in load ΔP caused by a variation in sediment density $\Delta\rho_{\mathrm{s}}$. $\Delta\rho_{\mathrm{s}}$ can be calculated as follows.

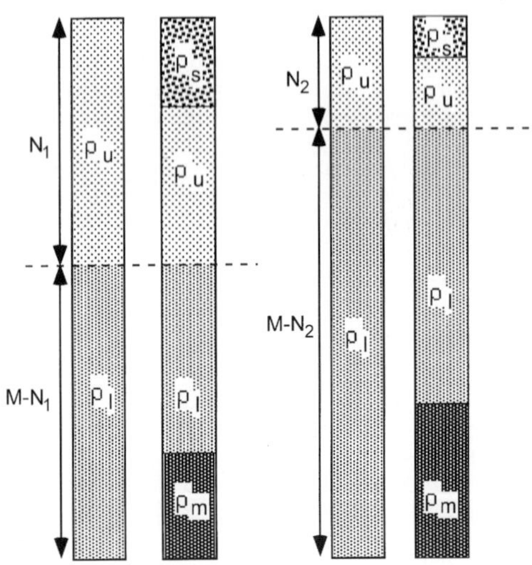

Fig. A1. Thinning of the crust around necking levels N_1 or N_2, assuming β is equal (i.e. the depression of the surface is different) for both cases.

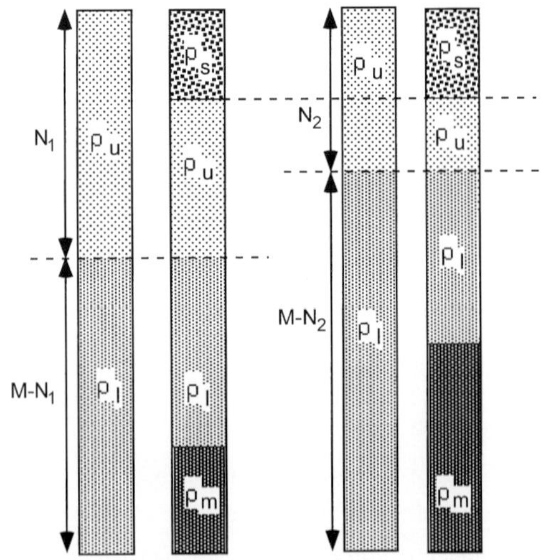

Fig. A2. Thinning of the crust around necking levels N_1 or N_2, assuming the depression of the surface is equal (i.e. β is different) for both cases.

Reference load P_1 caused by thinning around necking level N_1, using sediment density ρ_{s1} (see Fig. A1):

$$P_1 = gN_1\left(\frac{\beta-1}{\beta}\right)(\rho_{\mathrm{s1}} - \rho_{\mathrm{u}}) + g(M - N_1)\left(\frac{\beta-1}{\beta}\right)(\rho_{\mathrm{m}} - \rho_{\mathrm{l}}) \quad \text{(A.1)}$$

where g is gravitational acceleration (m s^{-2}), β is a stretching factor, ρ_{u} is density of the upper crust (kg m^{-3}), ρ_{l} is density of the lower crust (kg m^{-3}), ρ_{m} is density of the mantle (kg m^{-3}) and M is the Moho depth (km).

ΔP caused by thinning around N_2 instead of around N_1:

$$\Delta P = g(N_2 - N_1)\left(\frac{\beta-1}{\beta}\right)(\rho_{\mathrm{s1}} - \rho_{\mathrm{u}} - \rho_{\mathrm{m}} + \rho_{\mathrm{l}}) \quad \text{(A.2)}$$

ΔP caused by using sediment density ρ_2 instead of ρ_1:

$$\Delta P = g(\rho_{\mathrm{s2}} - \rho_{\mathrm{s1}})N_1\left(\frac{\beta-1}{\beta}\right) \quad \text{(A.3)}$$

This yields the following equivalence between $\Delta N (= N_1 - N_2)$ and $\Delta\rho (= \rho_1 - \rho_2)$:

$$\Delta\rho_{\mathrm{s}} = \left(\frac{\rho_{\mathrm{s1}} - \rho_{\mathrm{u}} - \rho_{\mathrm{m}} + \rho_{\mathrm{l}}}{N_1}\right)\Delta N \quad \text{(A.4)}$$

If we assume the depression of the surface as a result of crustal thinning to be constant (as it can be observed), a change in necking depth implies a change in the β factor (see Fig. A2). Then

$$N_1\left(\frac{\beta_1 - 1}{\beta_1}\right) = N_2\left(\frac{\beta_2 - 1}{\beta_2}\right) \quad (A.5)$$

and

$$\Delta\rho_s = M\left(\frac{\rho_m - \rho_1}{N_1 N_2}\right)\Delta N \quad (A.6)$$

which is the equation to be used for the case of the Viking Graben.

References

ANDERSEN, T. B. & JAMTVEIT, B. 1990. Uplift of deep crust during orogenic extensional collapse: a model based on field studies in the Sogn–Sunnfjord region of western Norway. *Tectonics*, **9**, 1097–1112.

BADLEY, M. E., EGEBERG, T. & NIPEN, O. 1984. Development of rift basins illustrated by the structural evolution of the Oseberg structure, Block 30/6, offshore Norway. *Journal of the Geological Society, London*, **141**, 639–649.

——, PRICE, J. D., RAMBECH DAHL, C. & AGDESTEIN, T. 1988. The structural evolution of the northern Viking Graben and its bearing upon extensional modes of graben formation. *Journal of the Geological Society, London*, **145**, 455–472.

BALDWIN, B. & BUTLER, C. O. 1985. Compaction curves. *AAPG Bulletin*, **69**, 622–626.

BEACH, A., BIRD, T. & GIBBS, A. D. 1987. Extensional tectonics and crustal structure: deep seismic reflection data from the northern North Sea Viking Graben. *In*: COWARD, M. P., DEWEY, J. F. & HANCOCK, P. L. (eds) *Continental Extensional Tectonics*. Geological Society, London, Special Publications, **28**, 467–476.

BERTRAM, G. T. & MILTON, N. J. 1988. Reconstructing basin evolution from sedimentary thickness: the importance of paleobathymetric control, with reference to the North Sea. *Basin Research*, **1**, 247–259.

BODINE, J. H., STECKLER, M. S. & WATTS, A. B. 1981. Observations of flexure and rheology of the oceanic lithosphere. *Journal of Geophysical Research*, **86**, 3695–3707.

BRACE, W. F. & KOHLSTEDT, D. L. 1980. Limits on lithospheric stress imposed by laboratory experiments. *Journal of Geophysical Research*, **85**, 6248–6252.

BRAUN, J. & BEAUMONT, C. 1989. A physical explanation of the relation between flank uplifts and the break–unconformity at rifted continental margins. *Geology*, **17**, 760–764.

BRUN, J. P & TRON, V. 1993. Development of the North Viking Graben: inferences from laboratory modelling. *Sedimentary Geology*, **86**, 31–51.

BUCK, W. R. 1988. Flexural rotation of normal faults. *Tectonics*, **7**, 959–973.

BUROV, E. & DIAMENT, M. 1995. The effective elastic thickness (T_e) of continental lithosphere: what does it really mean? *Journal of Geophysical Research*, **100**, 3905–3927.

BYERLEE, J. D. 1978. Friction of rocks. *Pure and Applied Geophysics*, **116**, 615–626.

CARTER, N. L. & TSENN, M. C. 1987. Flow properties of the continental lithosphere. *Tectonophysics*, **136**, 27–63.

CHRISTIANSSON, P., FALEIDE, I. J. & BERGE, A. M. 2000. Crustal structure in the northern North Sea: an integrated geophysical study. *This volume*.

DYKSTRA, 1987. Compaction correction for burial history curves: application to Lopatin's method for source rock maturation determination. *Geobyte*, Nov. 87, 16–23.

FALEIDE, J. I. *et al.* 2000. Deep basin geometries in the northern North Sea; constraints from gravity and magnetic data.

FJELLANGER, E., OLSEN, T. R. & RUBINO, J. L. 1996. Sequence stratigraphy and palaeogeography of the Middle Jurassic Brent and Vestland deltaic systems, Northern North Sea. *Norsk Geologisk Tidsskrift*, **76**, 75–106.

FOSSEN, H. & GABRIELSEN, R. H. 1996. Experimental modelling of extensional fault systems by use of plaster. *Journal of Structural Geology*, **18**, 673–687.

—— & RYKKELID, E. 1992. Post-collisional extension of the Caledonide orogen in Scandinavia: structural expressions and tectonic significance. *Geology*, **20**, 737–740.

——, ODINSEN, T., FÆRSETH, R. B. & GABRIELSEN, R. H. 2000. Detachments and low-angle faults in the northern North Sea rift system. *This volume*.

FURNES, H., MITCHELL, J. G., ROBINS, B., RYAN, P. D. & SKJERLIE, F. 1982. Petrography and geochemistry of peralkaline, ultrapotassic syenite dykes of Middle Permian age, Sunnfjord, West Norway. *Norsk Geologisk Tidsskrift*, **62**, 147–159.

FÆRSETH, R. B. 1978. Mantle-derived lherzolite xenoliths and megcrysts from Permo-Triassic dykes, Sun-nhordland, Western Norway. *Lithos*, **11**, 23–35.

——1996. Interaction of Permo-Triassic and Jurassic extensional fault-blocks during the development of the northern North Sea. *Journal of the Geological Society, London*, **153**, 931–944.

——, GABRIELSEN, R. H. & HURICH, C. A. 1995*a*. Influence of basement in structuring of the North Sea basin, offshore southwest Norway. *Norsk Geologisk Tidsskrift*, **75**, 105–119.

——, MCINTYRE, R. M. & NATERSTAD, J. 1976. Mesozoic alkaline dykes in the Sunnhordland region, western Norway: ages, geochemistry and regional significance. *Lithos*, **9**, 332–345.

——, SJØBLOM, T. S., STEEL, R. J., LILJEDAHL, T., SAUAR, B. E. & TJELLAND, T. 1995*b*. Tectonic controls on Bathonian–Volgian syn-rift successions on the Visund Fault Block, northern North Sea. *In*: STEEL, R. J., FELT, V., JOHANNESSEN, E. & MATHIEU, C. (eds) *Sequence Stratigraphy on the Northwest European Margin*. Norwegian Petroleum Society, Special Publication, **5**, 325–346.

GABRIELSEN, R. H. 1986. Structural elements in graben systems and their influence on hydrocarbon trap types. *In*: SPENCER, A. M. *et al.* (eds) *Habitat of Hydrocarbons on the Norwegian Continental Shelf*. Graham and Trotman, London, 55–60.

——, FÆRSETH, R. B., STEEL, R. J., IDIL, S. & KLØVJAN, O. S. 1990. Architectural styles of basin fill in the northern Viking Graben. *In*: BLUNDELL, D. J. & GIBBS, A. D. (eds) *Tectonic Evolution of the North Sea Rifts*. Clarendon, Oxford, 158–179.

GIBBS, A. 1987. Deep seismic profiles in the Northern North Sea. *In*: BROOKS, J. & GLENNIE, K. W. (eds) *Petroleum Geology of North-West Europe*. Graham and Trotman, London, 1925–1928.

GILTNER, J. P. 1987. Applications of extensional models to the Northern Viking Graben. *Norsk Geologisk Tidsskrift*, **67**, 339–352.

GOETZE, C. & EVANS, B. 1979. Stress and temperature in the bending lithosphere as constrained by experimental rock mechanics. *Geophysical Journal of the Royal Astronomical Society*, **59**, 463–478.

HAQ, B. U., HARDENBOL, J. & VAIL, P. R. 1987. Chronology of fluctuating sea levels since the Triassic (250 million years ago to present). *Science*, **235**, 1156–1166.

HARLAND, W. B., ARMSTRONG, R. L., COX, A. V., CRAIG, L. E., SMITH, A. G. & SMITH, D. G. 1990. *A Geologic Time Scale*. Cambridge University Press, Cambridge.

——, COX, A. V., LLEWELLYN, P. B., PICKTON, C. A., SMITH, A. G. & WALTERS, R. 1982. *A Geological Time Scale*. Cambridge University Press, Cambridge.

HENDRIE, D. B., KUSZNIR, N. J. & HUNTER, R. H. 1993. Jurassic extension estimates for the North Sea 'triple junction' from flexural backstripping: implications for decompression melting models. *Earth and Planetary Science Letters*, **116**, 113–127.

HOSSACK, J. R. 1984. The geometry of listric growth faults in the Devonian basins of Sunnfjord, West Norway. *Journal of the Geological Society, London*, **141**, 629–638.

JOHNS, C. R. & ANDREWS, I. J. 1985. The petroleum geology of the Unst Basin, North Sea. *Marine and Petroleum Geology*, **2**, 361–372.

KIRBY, S. H. 1983. Rheology of the lithosphere. *Review of Geophysics and Space Physics*, **21**, 1458–1487.

KOOI, H. 1991. *Tectonic modelling of extensional basins: the role of lithosphere flexure, intraplate stress and relative sea-level change*. PhD thesis, Vrije Universiteit, Amsterdam.

——, CLOETINGH, S. & BURRUS, J. 1992. Lithospheric necking and regional isostacy at extensional basins, 1. Subsidence and gravity modelling with an application to the Gulf of Lions margin (SE France). *Journal of Geophysical Research*, **97**, 17 553–17 571.

KUSZNIR, N. J., MARSDEN, G. & EGAN, S. S. 1991. A flexural cantilever simple-shear/pure-shear model of continental lithosphere extension: applications to the Jeanne d'Arc-Basin, Grand Banks and Viking Graben, North Sea. *In*: ROBERTS, A. M., YIELDING, G. & FREEMAN, B. (eds) *The Geometry of Normal Faults*. Geological Society, London, Special Publications, **56**, 41–60.

LE PICHON, X. & CHAMOT-ROOKE, N. 1991. Extension of continental crust. *In*: MÜLLER, D. W., MCKENZIE, J. A. & WEISSERT, H. (eds) *Controversies in Modern Geology: Evolution of Geological Theories in Sedimentology, Earth History and Tectonics*, Academic, London, 313–338.

LERVIK, K. S., SPENCER, A. M. & WARRINGTON, G. 1989. Outline of Triassic stratigraphy and structure in the central and northen North Sea. *In*: COLLINSON, J. (ed.) *Correlation in Hydrocarbon Exploration*. Graham and Trotman, London, 173–189.

LIPPARD, S. & LIU, G. 1992. Tectonic modelling of the northern North Sea using program RIFT. *In*: LARSEN, R. M., BREKKE, H., LARSEN, B. T. & TALLERAAS, E. (eds) *Structural and Tectonic Modelling and its Application to Petroleum Geology*. Norwegian Petroleum Society, Special Publication, **1**, 43–54.

MARSDEN, G., YIELDING, G., ROBERTS, A. M. & KUSZNIR, N. J. 1990. Application of a flexural cantilever simple-shear/pure-shear model of continental extension on the formation of the northern North Sea basin. *In*: BLUNDELL, D. J. & GIBBS, A. D. (eds) *Tectonic Evolution of the North Sea Rifts*. Clarendon, Oxford, 236–257.

MCCLAY, K. R., NORTON, M. G., CONEY, P. & DAVIS, G. H. 1986. Collapse of the Caledonian orogen and the Old Red Sandstone. *Nature*, **323**, 147–149.

MORTON, N., SMITH, R. M., GOLDEN, M. & JAMES, A. V. 1987. Comparative stratigraphic study of the Triassic–Jurassic sedimentation and basin evolution in the northern North Sea and northwest of the British Isles. *In*: BROOKS, J. & GLENNIE, K. W. (eds) *Petroleum Geology of North-West Europe*. Graham and Trotman, London, 697–710.

NELSON, P. H. H. & LAMY, J. M. 1987. The Møre/West Shetlands area: a review. *In*: BROOKS, J. & GLENNIE, K. W. (eds) *Petroleum Geology of North-West Europe*. Graham and Trotman, London, 775–784.

NØTTVEDT, A., GABRIELSEN, R. H. & STEEL, R. J. 1995. Tectonostratigraphy and sedimentary architecture of rift basins, with reference to the northern North Sea. *Marine and Petroleum Geology*, **12**, 881–901.

ODINSEN, T., REEMST, P., VAN DER BEEK, P., FALEIDE, J. I. & GABRIELSEN, R. H. 2000. Permo-Triassic and Jurassic extension in the northern North Sea: results from tectonostratigraphic forward modelling. *This volume*.

PLATT, N. H. 1995. Structure and tectonics of the northern North Sea: new insights from deep penetration regional seismic data. *In*: LAMBIASE, J. J. (ed.) *Hydrocarbon Habitat in Rift Basins*. Geological Society, London, Special Publications, **80**, 103–113.

RAVNÅS, R. & BONDEVIK, K. 1999. Architecture and controls on Bathonian–Kimmeridgian shallow-marine syn-rift wedges of the Oseberg–Brage area, northern North Sea. *Basin Research*, in press.

——, ——, HELLAND-HANSEN, W., LØMO, L., RYSETH, R. & STEEL, R. J. 1999. Sedimentation history as an indicator of rift initiation and development: the late Bajocian–Bathonian evolution of the Oseberg–Brage area, northern North Sea. *Norsk Geologisk Tidsskrift*, in press.

ROBERTS, A., YIELDING, G., KUSZNIR, N., WALKER, I. & DORN-LOPEZ, D. 1993. Mesozoic extension in the North Sea: constraints from flexural backstripping, forward modelling and fault populations. *In*: PARKER, J. R. (ed.) *Petroleum Geology of Northwest Europe: Proceedings of the 4th Conference*. Geological Society, London, 1123–1136.

——, ——, ——, —— & ——1995. Quantitative analysis of Triassic extension in the northern Viking Graben. *Journal of the Geological Society, London*, **152**, 15–26.

RØE, S. L. & STEEL, R. J. 1985. Sedimentation, sealevel rise and tectonics at the Triassic–Jurassic boundary (Statfjord Formation), Tampen Spur, northern North Sea. *Journal of Petroleum Geology*, **8**, 163–186.

SCLATER, J. G. & CHRISTIE, P. A. F. 1980. Continental stretching: an explanation of the post-mid-Cretaceous subsidence of the North Sea basin. *Journal of Geophysical Research*, **85**, 3711–3739.

SNEIDER, J. S., DE CLARENS, P. & VAIL, P. R. 1995. Sequence stratigraphy of the Middle to Upper Jurassic, Viking Graben, North Sea. *In*: STEEL, R. J., FELT, V., JOHANNESSEN, E. & MATHIEU, C. (eds) *Sequence Stratigraphy on the Northwest European Margin*. Norwegian Petroleum Society, Special Publication, **5**, 167–197.

SPADINI, G. & PODLADCHIKOV, Y. 1996. Spacing of consecutive normal faulting in the lithosphere: a dynamical model for rift axis migration (Tyrrhenian Sea). *Earth and Planetary Science Letters*, **144**, 21–34.

STEEL, R. 1993. Triassic–Jurassic megasequence stratigraphy in the Northern North Sea: rift to post-rift evolution. *In*: PARKER, J. R. (ed.) *Petroleum Geology of Northwest Europe. Proceedings of the 4th Conference*. Geological Society, London, 299–315.

—— & RYSETH, A. 1990. The Triassic–early Jurassic succession in the northern North Sea: megasequence stratigraphy and intra-Triassic tectonics. *In*: HARDMAN, R. P. F. & BROOKS, J. (eds) *Tectonic Events Responsible for Britain's Oil and Gas Reserves*. Geological Society, London, Special Publications, **55**, 139–168.

TER VOORDE, M. & BERTOTTI, G. 1994. Thermal effects of normal faulting during rifted basin formation. 1: A finite difference model. *Tectonophysics*, **240**, 133–144.

—— & CLOETINGH, S. 1996. Numerical modelling of extension in faulted crust : effects of localized and regional deformation on basin stratigraphy. *In*: BUCHANAN, P. G. & NIEUWLAND, D. A. (eds) *Modern Developments in Structural Interpretation, Validation and Modelling*. Geological Society, London, Special Publications, **99**, 283–296.

——, RAVNÅS, R., FÆRSETH, R. & CLOETINGH, S. 1997. Tectonic modelling of the middle Jurassic synrift stratigraphy in the Oseberg–Brage area, northern Viking Graben. *Basin Research*, **9**, 133–150.

——, VAN BALEN, R. T., BERTOTTI, G. & CLOETINGH, S. A. P. L. 1998. The influence of a stratified rheology on the flexural response of the lithosphere to (un)loading by extensional faulting. *Geophysical Journal International*, **134**, 721–735.

TORSVIK, T. H., STURT, B. A., SWENSSON, E., ANDERSEN, T. B. & DEWEY, J. F. 1992. Palaeomagnetic dating of fault rocks: evidence for Permian and Mesozoic movements and brittle deformation along the extensional Dalsfjord Fault, western Norway. *Geophysical Journal International*, **109**, 565–580.

TSENN, M. C. & CARTER, N. L. 1987. Upper limits of power law creep of rocks. *Tectonophysics*, **136**, 1–26.

UNDERHILL, J. R. & PARTINGTON, M. A. 1993. Jurassic thermal doming and deflation in the North Sea: implications of the sequence stratigraphic eveidence. *In*: PARKER, J. R. (ed.) *Petroleum Geology of Northwest Europe: Proceedings of the 4th Conference*. Geological Society, London, 337–345.

WALTHAM, D. 1989. Finite difference modelling of hanging wall deformation. *Journal of Structural Geology*, **11**, 433–437.

——1990. Finite difference modelling of sandbox analogues, compaction and detachment free deformation. *Journal of Structural Geology*, **12**, 375–381.

WEISSEL, J. K. & KARNER, G. D. 1989. Flexural uplift of rift flanks due to mechanical unloading of the lithosphere during extension. *Journal of Geophysical Research*, **94**, 13 919–13 950.

WHITE, N. J. 1990. Does the uniform stretching model work in the North Sea? *In*: BLUNDELL, D. J. & GIBBS, A. D. (eds) *Tectonic Evolution of the North Sea Rifts*. Clarendon, Oxford, 217–240.

—— & YIELDING, G. 1991. Calculating normal fault geometries at depth: theory and examples. *In*: ROBERTS, A. M., YIELDING, G. & FREEMAN, B. (eds) *The Geometry of Normal Faults*. Geological Society, London, Special Publications, **56**, 251–260.

——, JACKSON, J. A. & MCKENZIE, D. P. 1986. The relationship between the geometry of normal faults and that of the sedimentary layers in their hanging walls. *Journal of Structural Geology*, **9**, 789–795.

WITHJACK, M. O. & PETERSON, E. T. 1993. Prediction of normal-fault geometries – a sensitivity study. *AAPG Bulletin*, **11**, 1860–1873.

YIELDING, G., BADLEY, M. E. & ROBERTS, A. 1992. The structural evolution of the Brent Province. *In*: MORTON, A. C., HASELDINE, R. S., GILES, M. R. & BROWN, S. (eds) *Geology of the Brent Group*. Geological Society, London, Special Publications, **61**, 27–40.

ZIEGLER, P. & VAN HOORN, B. 1989. Evolution of North Sea rift systems. *In*: TANKARD, A. J. & BALKWILL, H. R. (eds) *Extensional Tectonics and Stratigraphy of the North Atlantic Margins*. American Association of Petroleum Geologists, Memoir, **46**, 471–500.

Permo-Triassic and Jurassic extension in the northern North Sea: results from tectonostratigraphic forward modelling

TORE ODINSEN,[1,4] PAUL REEMST,[2,5] PETER VAN DER BEEK,[2,6] JAN INGE FALEIDE[3] & ROY H. GABRIELSEN[1]

[1] *Department of Geology, University of Bergen, Allégaten 41, N-5007 Bergen, Norway*
[2] *Faculty of Earth Sciences, Vrije Universiteit, De Boelelaan, 1081 HV Amsterdam, The Netherlands*
[3] *Department of Geology, University of Oslo, P.O. Box 1047 Blindern, N-0316 Oslo, Norway*
[4] *Present address: Statoil, N-5020 Bergen, Norway*
[5] *Present address: Geologica AS, P.O. Box 8034, N-4029 Stavanger, Norway*
[6] *Present address: Université Joseph Fourier, 15 rue Maurice Gignoux, 38031 Grenoble Cedex, France*

Abstract: We have undertaken 2D forward modelling across the northern North Sea, based on reprocessed, interpreted and depth-converted deep reflection seismic lines NSDP84-1 and -2 (15 s twt) and refraction data. Two separate stretching phases, Permo-Triassic and Jurassic, are recognized. The cumulative stretching is consistent with the observed crustal structure and the overall basin configuration, as reproduced by forward modelling. Good agreement between observed and modelled top basement level, and crustal thickness below the platform areas are particularly emphasized. Crustal-scale modelling indicates that crustal thickness varied across the northern North Sea at the onset of the Permo-Triassic rifting, from *c.* 35 km in the platform areas to less than 30 km in the interior of the basin. This may be ascribed to Devonian(–Carboniferous?) crustal stretching. Thinning of the crust has progressively been narrowed, from post-Caledonian extensional collapse, to less regional Permo-Triassic basins, and finally development of the Viking Graben area in the Jurassic–early Cretaceous time. Most of the Permo-Triassic stretching occurred between the Øygarden Fault Zone to the east and the Shetland Platform (southern transect) and the Hutton Fault alignment to the west. The width of the Permo-Triassic basin was *c.* 120–125 km, with calculated β_{mean} between 1.38 and 1.40. Permo-Triassic β_{mean} estimates across the present Horda Platform vary between 1.33 and 1.39. The Jurassic β_{mean} estimates for the same area vary between 1.08 and 1.13. Across the Viking Graben, Permo-Triassic β_{mean} varies between 1.28 (southern transect) and 1.41 (northern transect). This is lower than estimates for the Jurassic β_{mean}, which amounts to 1.53 and 1.42. Permo-Triassic and Jurassic β_{mean} estimates across the East Shetland Basin are 1.29 and 1.11, respectively. Lithospheric thermal evolution reflects the general differences between Permo-Triassic and Jurassic stretching, with a much wider thermal perturbation during the former and a focusing and lateral migration towards the east of the peak thermal elevation during the latter. There are still uncertainties related to the degree of (de)coupling between the upper crust and upper mantle during the Permo-Triassic and the Jurassic rift phases. These uncertainties are related to the interplay between age, strain rate, crustal rheology, crustal thickness and long-lived zones of weaknesses.

The northern North Sea has been extensively described in the literature as an extensional sedimentary basin, which formed during repeated lithospheric stretching. As a result of hydrocarbon exploration of the area, much attention has focused on rotated fault blocks that evolved during the mid-Jurassic to early Cretaceous rift phase. A previous rift phase is best recognized in reflection seismic lines which cross the Horda Platform, as deeply buried, rotated large fault blocks. Because of the lack of well data from syn-rift sediments associated with these structures, the age of this earlier event is still a matter of debate, and both Permian (Eynon 1981;

From: Nøttvedt, A. *et al.* (eds) *Dynamics of the Norwegian Margin*. Geological Society, London, Special Publications, **167**, 83–103. 1-86239-056-8/00/$15.00

Badley *et al.* 1988; Gabrielsen *et al.* 1990; Færseth *et al.* 1995) and Triassic (Beach *et al.* 1987; Giltner 1987; Roberts *et al.* 1995) ages have been proposed.

Several studies have also focused on the crustal and lithospheric configuration across the northern North Sea, as interpreted from a deep seismic reflection survey NSDP84 and other deep lines (Beach 1986; Beach *et al.* 1987; Klemperer 1988; Klemperer & Hurich 1990; Brun & Tron 1993). Limitations in the data quality have permitted several interpretations of the crustal and lithospheric configuration. It has been suggested that lithospheric stretching was governed by symmetrical pure shear (Giltner 1987; Badley *et al.* 1988), decoupled symmetrical shear (Klemperer 1988), decoupled asymmetrical shear (Reston 1990), or asymmetrical simple shear (Beach 1986; Beach *et al.* 1987; Gabrielsen 1989).

Modelling of basin formation in the North Sea was initiated by the work of McKenzie (1978), who used the area as one of his type examples. His widely recognized model for lithospheric stretching has been followed by a number of related studies. Early works were concentrated on the Central Graben area, where 1D subsidence analysis was carried out (Sclater & Christie 1980; Wood 1981; Wood & Barton 1983). More recently, access to conventional seismic reflection data and well data moved attention towards the northern North Sea (Fig. 1). Such data facilitated calculation of

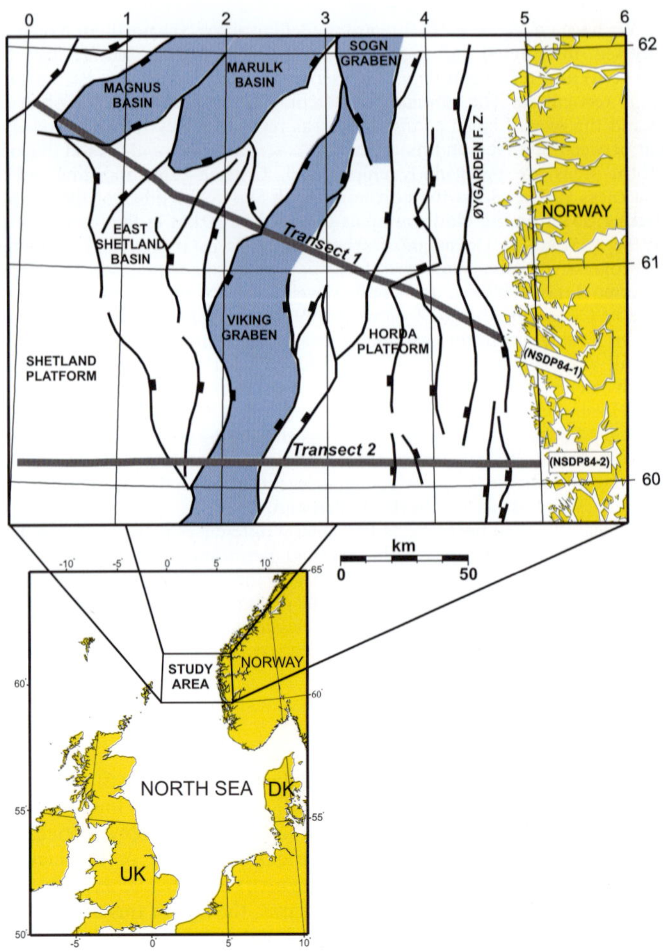

Fig. 1. Key map and main structural elements, northern North Sea (reference level is base Cretaceous). Areas that are primarily affected by Jurassic–early Cretaceous stretching are shown in grey. Locations of transects 1 and 2 are shown.

stretching estimates from subsidence and summation of fault heave (Giltner 1987; Badley *et al.* 1988). Some workers also have focused on stretching estimates looking at crustal thinning derived from deep seismic reflection lines (Beach *et al.* 1987; Klemperer 1988). The deep lines have not previously been subjected to more advanced quantitative studies such as forward or reverse modelling.

The present study is based on the frequently used deep (15 s twt) seismic reflection lines NSDP84-1 and -2 (transect 1 and 2 of the present study), which have been post-stack reprocessed and redisplayed (Figs 2a and 3a) and depth converted (Figs 2b and 3b). Detailed stratigraphic and structural architecture have been interpreted from high-quality conventional seismic lines (7 s twt). To obtain optimal velocity information, velocities from wells were used for the uppermost 3–4 s of the section, whereas refraction velocities from expanded spread profiles (ESPs) were applied to the deeper parts of the transects. The deep basin geometry is further constrained by gravity and magnetic data (Christiansson *et al.* 2000). We studied the thermal and tectonic evolution in the northern Viking Graben by forward modelling, integrating Permo-Triassic and mid-Jurassic–early Cretaceous lithospheric stretching. Three main topics were the specific subject of our study of transects 1 and 2: to estimate and compare Permo-Triassic and Jurassic stretching; to evaluate the modelled crustal configuration with respect to the stratigraphy, and the basement and Moho reliefs; to evaluate lithospheric thermal evolution.

The study extends previous studies in that we fully integrate commercial and deep seismic lines to better constrain the structural architecture and basement topography. Our modelling approach allows us to track the thermal and mechanical evolution of the lithosphere, as well as basin stratigraphy, through multiple finite-rate rifting phases, thus allowing meaningful predictions of stratigraphy to be made. We emphasize that this paper is part of continuing work in the North Sea Basin, and the reader will therefore benefit from additional reading of Christiansson *et al.* (2000), Fossen *et al.* (1999), Odinsen *et al.* (2000) and Ter Voorde *et al.* (2000). The work of Ter Voorde *et al.* (2000) is especially relevant as they take some of the present modelling results further into a new modelling tool.

Seismic data

Transect 1 has a length of 270 km and runs NW–SE. From SE, the transect crosses the Øygarden Fault Zone and Horda Platform. It continues across the interior of the Viking Graben and East Shetland Basin, onto the East Shetland Platform (Fig. 2).

Transect 2 is 270 km long and runs E–W (Fig. 3). At the eastern end of transect 2 it crosses the Øygarden Fault Zone and Horda Platform, before it passes across the Viking Graben, which is both deeper and narrower than in transect 1 further north. From the Viking Graben, transect 2 passes onto the Shetland Platform. Unfortunately, velocity data from ESPs were not available along transect 2. This forced us to calibrate the deep velocity profile with data from transect 1. Furthermore, the pre-Jurassic stratigraphy and depth to basement are not well constrained beneath the western part of the Horda Platform.

In both transects, the upper reflective crust is dominated by Cenozoic, relatively flat-lying and unfaulted post-rift sequences (Jordt *et al.* 1995), unconformably covering heavily faulted Mesozoic syn-rift sediments. Both transects reveal an asymmetrical pattern, with the deepest parts of the basin situated to the west, in the hanging wall of the inner western master fault of the Viking Graben.

The middle crust is characterized by poor reflectivity. This is in strong contrast to the undulating reflective lower crust. Although there are some differences in the reflective pattern of the lower crust in the two transects, the principal architecture is consistent. Also the width and area of Moho shallowing, and the depths of Moho, are similar in transects 1 and 2 (Odinsen *et al.* 2000). Reflections below Moho are recorded in both transects. These features dip away from the graben axis and are focused at the transition between strongly thinned crust and the thicker crust below the platform areas. They follow the general pattern for intra-mantle reflections as described by Klemperer (1988).

Modelling procedure and parameters

Early modelling studies of the northern North Sea (Giltner 1987; White 1990) were based on the McKenzie (1978) model. These models predict the first-order characteristics of the basin, but do not model fault-controlled syn-rift stratigraphy because they fail to incorporate the mechanics of rifting in a self-consistent manner (Kooi *et al.* 1992). Alternatively, shorter-wavelength models have been applied that explicitly take motion along rift border faults into account (Marsden *et al.* 1990; Roberts *et al.* 1993, 1995). The short-wave-length models require very low flexural

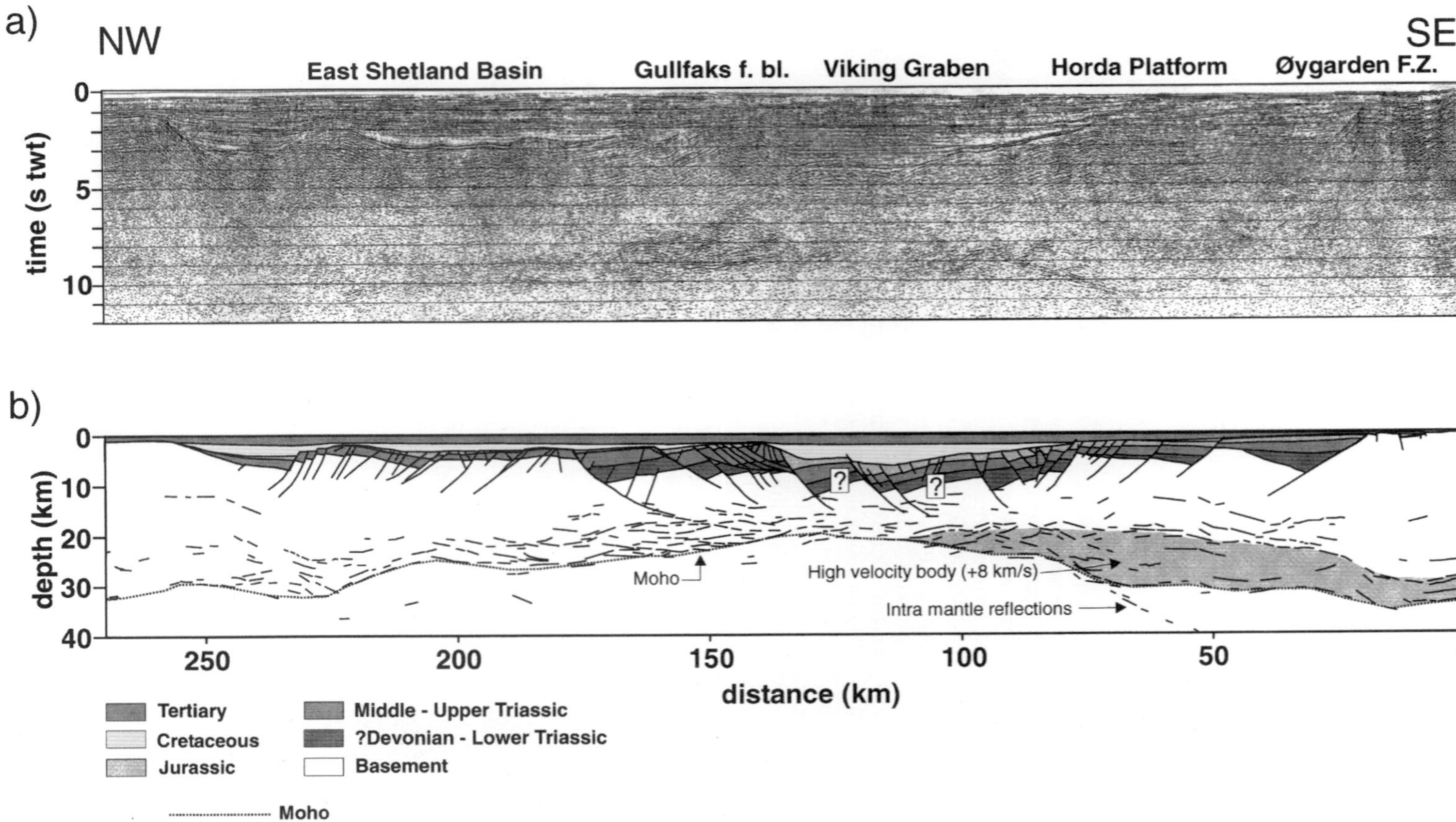

Fig. 2. (**a**) Post-stack reprocessed transect 1 in time. (For location, see Fig. 1.) (**b**) Depth-converted and interpreted transect 1. Question marks (?) refer to uncertainties of the Permo-Triassic sequence and basement level. No vertical exaggeration.

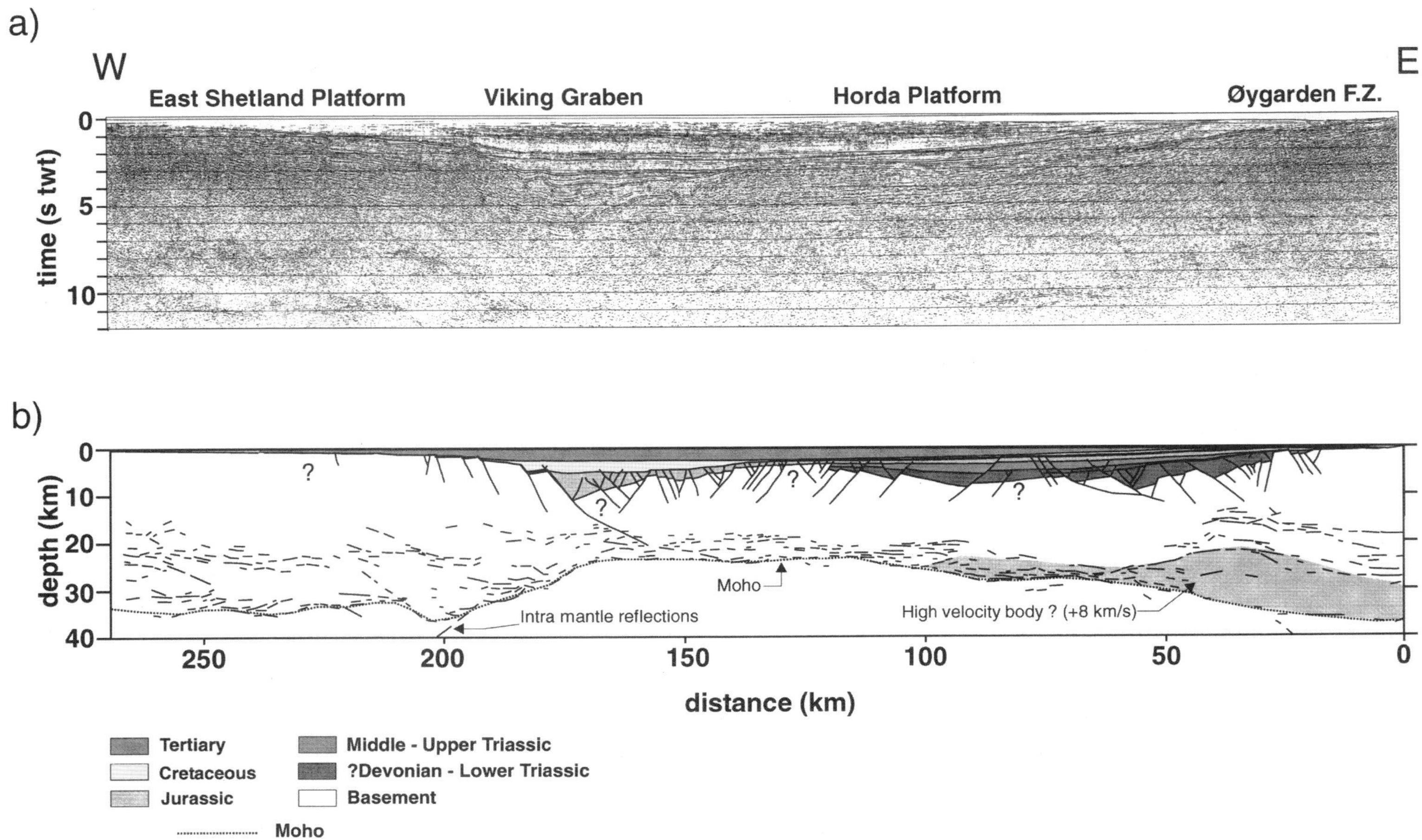

Fig. 3. (**a**) Post-stack reprocessed transect 2 in time. (For location, see Fig. 1.) (**b**) Depth-converted and interpreted transect 2. Question marks (?) refer to uncertainties of the westward continuation of the Permo-Triassic sequence and basement level. No vertical exaggeration.

rigidities of the lithosphere (effective elastic thickness (EET) 1.5–6 km) and are in apparent agreement with a gravity study of the central North Sea by Barton & Wood (1984), which suggested local isostatic compensation. The latter study, however, calculated the isostatic response of an initially unflexed and undeformed plate to post-rift loading by sediments and thermal contraction. As discussed by Kooi *et al.* (1992), such a treatment of isostatic compensation is internally inconsistent and is bound to yield very low elastic thicknesses. In addition, Barton & Wood (1984) did not take the gravity effect of sediment compaction into account, another factor that will lead to an underestimate of lithospheric strength (Cowie & Karner 1990). A more recent seismic and gravity study of the central North Sea by Holliger & Klemperer (1990) indicates significant departures from local isostasy over the basin, suggesting that the lithosphere has retained flexural rigidity. This conclusion is consistent with inferences from dynamic models of rifting (Braun & Beaumont 1989; Bassi *et al.* 1993; Burov & Diament 1995) that under moderate amounts of stretching, the lithosphere will retain significant strength. Ebinger *et al.* (1991) and Van der Beek (1997) have shown that rifting in the East African and Baikal rifts, which in many ways provide present-day analogues to rifting in the northern North Sea, is controlled by a strong lithosphere with EET of *c.* 30 km.

We use a 2D forward model (Kooi *et al.* 1992), which adopts a finite strength of the lithosphere during stretching. The kinematics of stretching and the flexural isostatic response are controlled by the depth of necking (Z_{neck}) (Fig. 4) (Braun & Beaumont 1989; Weissel & Karner 1989; Kooi *et al.* 1992). A surface depression with a depth (S) is then given by

$$S = [1 - (1/\beta)]Z_{neck}$$

where β is a variable stretching factor.

In this manner, fault-controlled syn-rift stratigraphy can be simulated without having to revert to extremely low flexural rigidities (Kooi *et al.* 1992; Spadini *et al.* 1995). The applied Z_{neck} is constrained by seismic data and trial and error, and considered robust. The model is run by dividing the crustal and subcrustal lithosphere, with a constant pre-rift lateral thickness (35 km), into a number of 5 km wide boxes. Each box is assigned a crustal and a subcrustal stretching factor (Royden & Keen 1980). The crustal and subcrustal factors are assumed to be equal in our model. It incorporates lateral heat flow and finite stretching (Cochran 1983). The

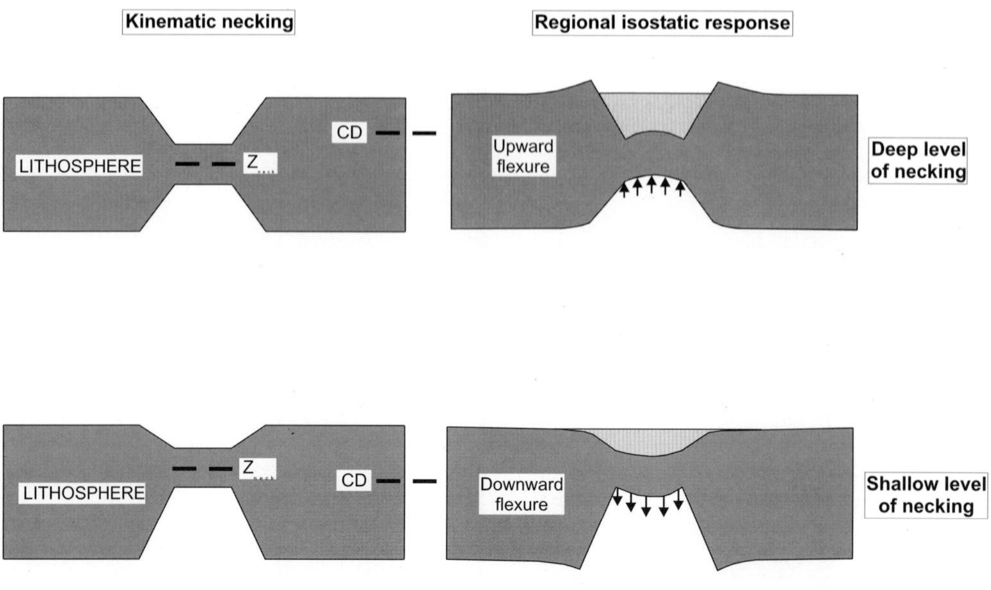

Fig. 4. Schematic illustration of the lithospheric necking (after Kooi *et al.* 1992; modified from Braun & Beaumont 1989). The level of necking (Z_{neck}) is defined as the level of no vertical motion in the absence of isostatic forces. A shallow level of necking (bottom) creates a surface depression that is shallower than the isostatically compensated basin depth (CD) and results in a downward state of flexure. Conversely, for a deep level of necking (top) an upward load acts on the lithosphere, resulting in flexural supported rift flanks.

thermal state and subsequent subsidence or uplift are calculated using a finite difference scheme. This basement subsidence is inverted to the corresponding stratigraphy by infill of sediments and calibrated by an estimated water-depth profile. The flexural strength of the lithosphere is controlled by the 450°C isotherm, which rises during thermal perturbation as a consequence of stretching. With this approach the flexural strength of the crust and mantle are always coupled. This is obviously a simplification of the elastic response as the model does not allow a further investigation of the interplay between strain rate, age, composition and thickness of the crust. As shown by Burov & Diament (1995) and Cloetingh & Burov (1996), these factors control the lithospheric strength, which may change totally before, during and after rifting. The subject has been further investigated by Ter Voorde *et al.* (2000), who tested both end-members (coupled or decoupled), and we will consider that study at the end of this paper.

Compaction of sediments is calculated using a standard porosity–depth relationship ($f(z) = 0.58e^{-0.41z}$) for all sediments (Sclater & Christie 1980). This is a simple approach where a sand/shale ratio of 40/60 reflects the 'average lithology' in the compaction scheme. Kyrkjebø *et al.* (2000) tested the effects of varying lithology on compaction by backstripping the Cretaceous and Cenozoic sequences along transect 2 assuming 100% of shale and 100% of sand, using the porosity–depth relation of Sclater & Christie (1980). The difference between shale and sand volumes (2D cross-sectional area) was of the order of 0–20%. Hence, the sensitivity analysis indicated that the uncertainties related to compaction–decompaction models do not undermine our regional conclusions.

The model also accounts for changes in sea level. We used a long-term sea-level curve based on Kominz (1984). For specified stages in basin evolution, tectonic subsidence and crustal structure can be examined. In addition, the forward model can predict a likely position of seismic horizons, which are otherwise difficult to track, for instance in the Viking Graben and below the western part of the Horda Platform in transect 2. The model parameters applied in the present study are shown in Table 1.

The timing of events responsible for the present structural framework in the northern North Sea has been debated in the literature. On the basis of published and our own data, two major stretching phases were used in the modelling, namely Artinskian to Ladinan (261–236 Ma), referred to as Permo-Triassic, and Bathonian to Berriasian (165–141 Ma), referred to as Jurassic (time scale of Harland *et al.* (1990)). It is acknowledged that the initiation of the first phase is especially controversial, as no wells have penetrated the thick pre-Triassic sediments observed within the deep half-grabens in the Norwegian part of the northern North Sea. However, age dating in the southwestern part of Norway indicate Permian ages (Færseth 1978; Torsvik *et al.* 1992), which is in accordance with seismic studies (Badley *et al.* 1988; Gabrielsen *et al.* 1990). Permo-Triassic rifting has also been suggested for the adjacent Unst Basin (Johns & Andrews 1985). We have chosen 261 Ma as initiation of Permian rifting but realize that this age is uncertain.

Precise dating of the initiation of the Jurassic rift phase has also been subject to disagreement (Gabrielsen *et al.* 1990). According to Helland-Hansen *et al.* (1992), increased subsidence rates and relative sea-level rise date back as far as late early Bajocian time, although significant sub-basin formation probably dates to Kimmeridgian time in many areas (Steel 1993). This make it reasonable to assume that most of the heat

Table 1. *Parameters used in the tectonostratigraphic modelling (after Bodine* et al. *1981; Steckler 1981; Kooi* et al. *1992)*

Symbol	Definition	Value
Z_{neck}	Depth of necking (regional istostasy)	18 km
T_e	Isotherm describing the effective elastic thickness (EET)	450°C
EET	Initial effective elastic thickness	44 km
L	Initial lithospheric thickness	130 km
C	Initial crustal thickness	35 km
α	Coefficient of thermal expansion	3.4×10^{-5} °C^{-1}
T_{asth}	Temperature of asthenosphere	1333°C
K	Thermal diffusivity	7.8×10^{-7} m^2 s^{-1}
m	Mantle density at surface conditions	3.33 g cm^{-3}
c	Crustal density at surface conditions	2.80 g cm^{-3}
c	Water density	1.03 g cm^{-3}

was introduced during Bajocian–Callovian time. Accepting the complexities in precise dating of the northern North Sea rift phases (Nøttvedt *et al.* 1995), it is important to note that reasonable changes of the ages we have used here will not change the stretching results or the crustal model.

Both phases are characterized by regional, E–W oriented extension associated with normal faulting and syn-rift sedimentation (e.g. Færseth *et al.* 1995), and followed by thermal cooling and sediment loading (e.g. Giltner 1987; Gabrielsen *et al.* 1990; White 1990; Roberts *et al.* 1995). Figure 5 summarizes the stratigraphy, relative

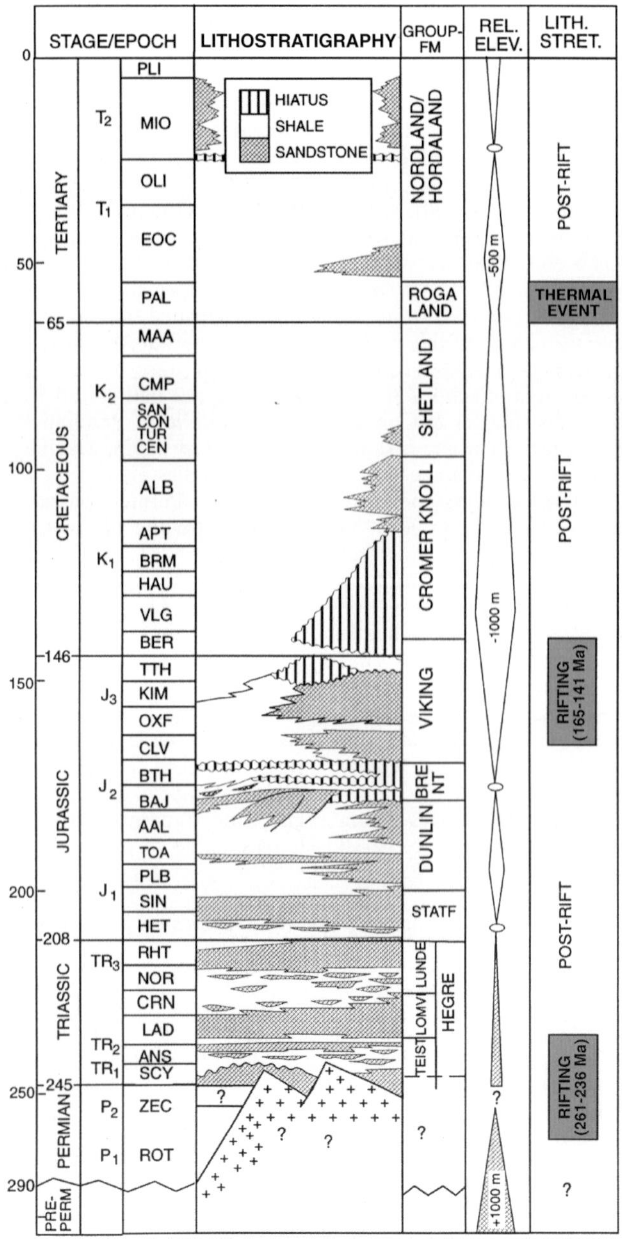

Fig. 5. Tectono-sedimentological events, northern North Sea (modified from Gabrielsen *et al.* 1990; Nøttvedt *et al.* 1995). Time scale from Harland *et al.* (1990).

Table 2. *List of ages assigned to the horizons*

Stratigraphy	Age (Ma)
Sea floor	0
Base Pliocene	5
Near top Oligocene	25
Inter lower Oligocene	31
Base Oligocene	35
Base Eocene	56
Base Tertiary	65
Inter Cenomanian	95
Inter lower Cretaceous (Ryazanian)	141
Base Cretaceous	144
Upper middle Jurassic	165
Lower Jurassic	200
Lower middle Triassic	236
Lower Triassic	244
Upper lower Permian	261

sea-level elevation and the tectonic events of the northern North Sea. Seismic horizons (Table 2) are calibrated to wells along the transects, and the ages of the horizons are correlated to the time scale of Harland *et al.* (1990).

Stratigraphic modelling and crustal thickness estimates

A clarification on our use of 'Viking Graben' may be needed in the following, as previous workers also have applied this term for the whole northern North Sea rift system. We make a distinction between the much more regional northern North Sea and the present *c.* 60 km wide Viking Graben, where most of the Jurassic rifting and the post-Jurassic subsidence occurred (Fig. 1). When the Permo-Triassic Viking Graben and the Jurassic Viking Graben are referred to, this is for comparison of stretching magnitudes in the same area, i.e. with respect to the present 60 km wide Viking Graben. This also goes for the Horda Platform and Shetland Basin and other areas.

Figures 6 and 7 show the modelled and observed stratigraphy, and the basement subsidence of transects 1 and 2. The Cretaceous and Tertiary stratigraphy is identically reproduced during modelling. Also, the Jurassic strata are generally in very good agreement, although some discrepancies occur in areas that have been uplifted and eroded. This is evident for the Gullfaks area and the western rim of the Horda Platform in transect 1 and towards the Øygarden Fault Zone in both transects. The modelled and observed Triassic strata are mostly in concert, although a misfit of 300–400 m occurs locally below the Gullfaks block in transect 1 and below the Horda Platform in transect 2. The good agreement at top basement level is emphasized, as previous definitions of this surface have been affected by considerable uncertainties (Klemperer 1988; Fichler & Hospers 1990; Marsden *et al.* 1990). A notion of the apparent thickness variation of the Jurassic and pre-Jurassic sediment distribution from the Gullfaks block to the Viking Graben (Fig. 6) is necessary. This portion of transect 1 has been backstripped by Fossen *et al.* (2000), and at Rogaland Research, and the results have been consistent when decompaction of the sediments is taken into account.

In areas where interpretation of the stratigraphy is uncertain or no interpretation is available, the model can suggest a first-order stratigraphy. This is evident in transect 2 within the deepest parts of the Viking Graben and the western Horda Platform, where the positions of the Triassic and pre-Triassic horizons are suggested by the forward model. Here, top basement level is further constrained by gravimetric and magnetic data (Christiansson *et al.* 2000).

Calculated tectonic and basement subsidence is demonstrated for different positions along the modelled transects (Figs 6a and 7a). The basement subsidence curves quantify the relative importance of the two rift phases, and show that the subsidence associated with the Permo-Triassic stretching was generally larger than for Jurassic stretching. It is only within the Viking Graben that Jurassic subsidence outpaced the Permo-Triassic event. The total basement subsidence reaches a maximum of 12–14 km within the deepest part of the Viking Graben. The calculated tectonic subsidence shows the flexural and thermal response, with its hanging-wall subsidence and footwall uplift in response to fault movement.

Exact estimates of the present-day crustal thickness are important during forward modelling. This constrains not only the final stratigraphic and crustal model, but also influences the parameters used during modelling. Figure 8 displays the modelled crustal structure of transects 1 and 2. The observed and modelled crustal thickness are in concert below most of the platform areas. This is important as these are the areas where the crustal thickness is best constrained. Below the basin areas in both transects, and below the western Horda Platform in transect 2 modelled crust is thicker than is observed (locally up to 9 km).

According to Christiansson *et al.* (2000), refraction data (ESP) close to transect 1 and gravity data, show that the base of the crust

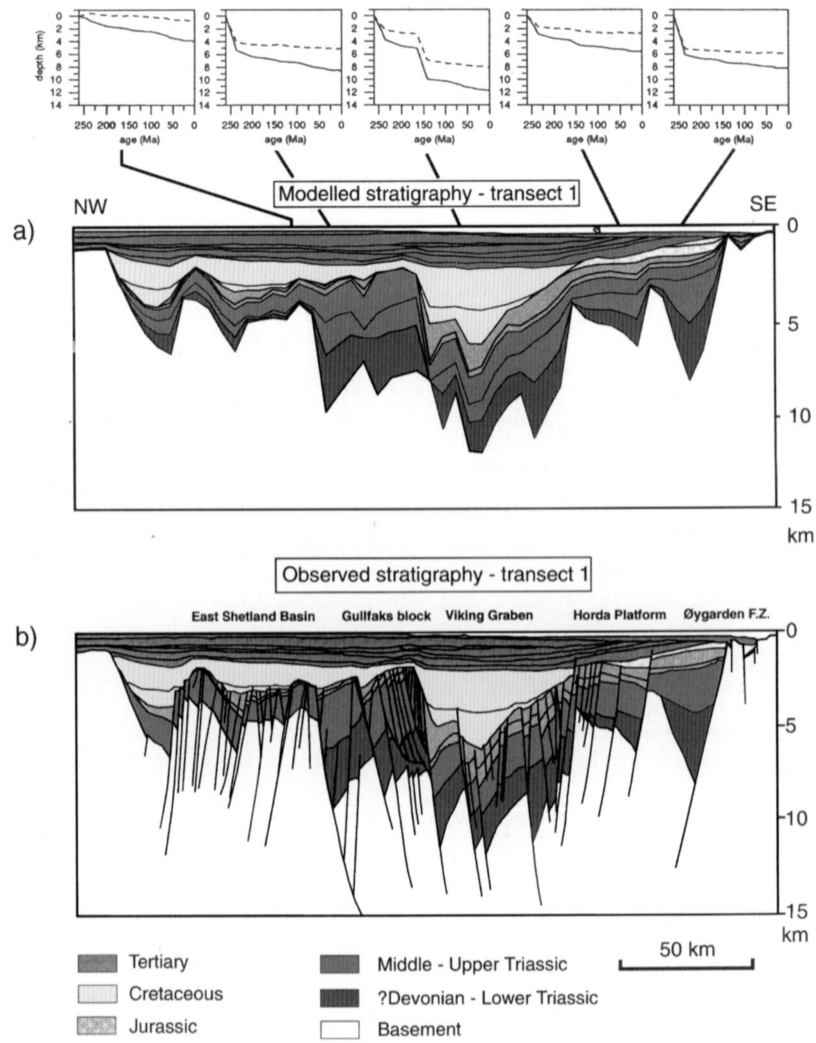

Fig. 6. (**a**) Modelled stratigraphy of transect 1. Boxes at top show calculated tectonic (dotted lines) and basement (continuous lines) subsidence along the modelled transect. (**b**) Observed stratigraphy of transect 1. (Note vertical exaggeration.)

coincides with the base of the reflective lower crust in both transects. A particular problem exists below the eastern part of the Viking Graben in transect 1 and below the Horda Platform in transect 2 where the Moho splits into two separate reflective bands (Figs 2 and 3) (Odinsen *et al.* 2000). In transect 1 the upper band deepens from 20 to 28 km, and the lower band deepens from 24 to 36 km to the SE. Here, the upper band delineates velocities of 6+ kms^{-1} above and 8+ $km\,s^{-1}$ below (Christiansson *et al.* 2000), which corresponds to the seismic definition of the Moho. However, this solution will place the Moho at the same or even shallower level below the northern Horda Platform than underneath the Viking Graben, implying a total crustal stretching of about the same order of magnitude from the Viking Graben to the Øygarden Fault Zone. However, both sediment distribution and subsidence curves contradict this. Thus, modelling results show that bulk stretching decreases eastwards from the Viking Graben, which would be compatible with an increase in the crustal thickness. Hence, when the bulk modelled stretching is inverted to Moho depth, one finds that the modelled Moho

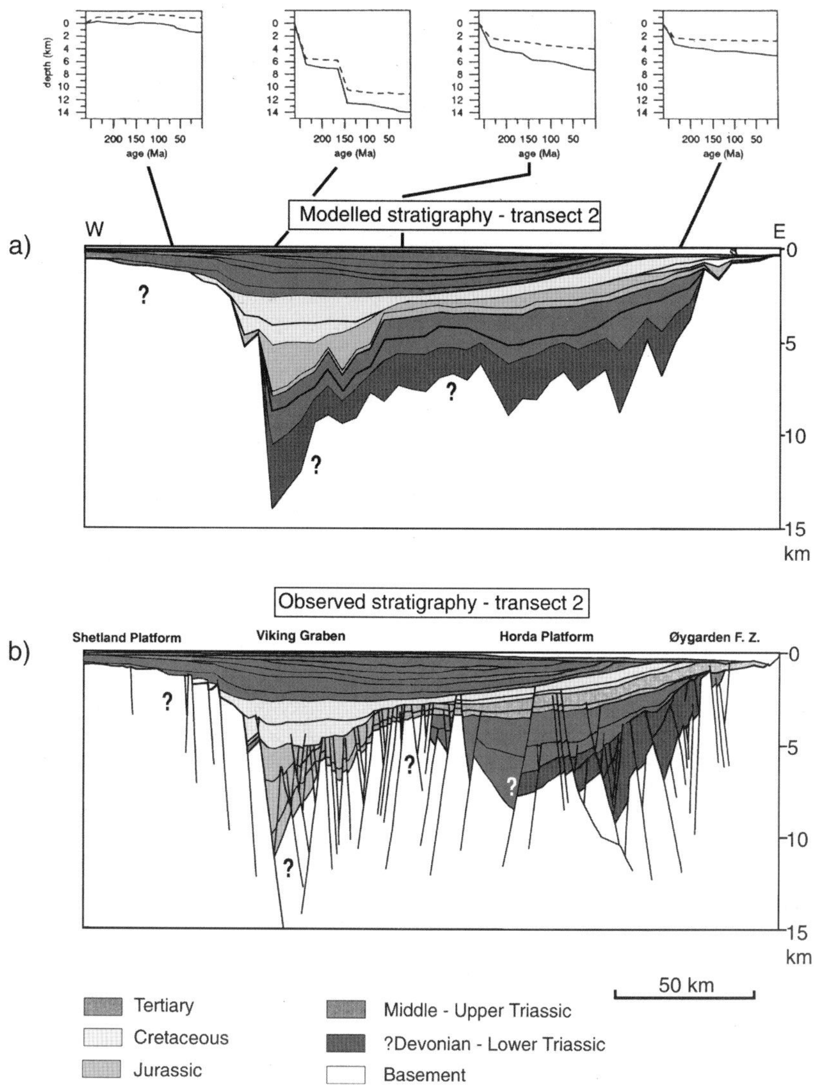

Fig. 7. (**a**) Modelled stratigraphy of transect 2. Proposed Permo-Triassic sequence emplaced from the Horda Platform to the Viking Graben. Boxes at top show calculated tectonic (dotted lines) and basement (continuous lines) subsidence along the modelled transect. (**b**) Observed stratigraphy of transect 2. (Note vertical exaggeration.)

corresponds to the lower of the two reflective bands (Fig. 8a). It is unclear how far south the 50 km wide and 5–10 km thick body continues, as no velocity data are available south of transect 1. However, it is noted that the same lower-crustal split can be interpreted from the crustal reflectivity pattern in transect 2 (Odinsen *et al.* 2000), and that the modelled Moho level (Fig. 8b) coincides with that of transect 1. Christiansson *et al.* (2000) obtained the same results from gravity modelling of the crustal transects. They also addressed the nature of this body in more detail.

Stretching distribution and thermal evolution

Estimating the stretching distribution in the northern North Sea has been the subject of many papers in recent years (Beach *et al.* 1987;

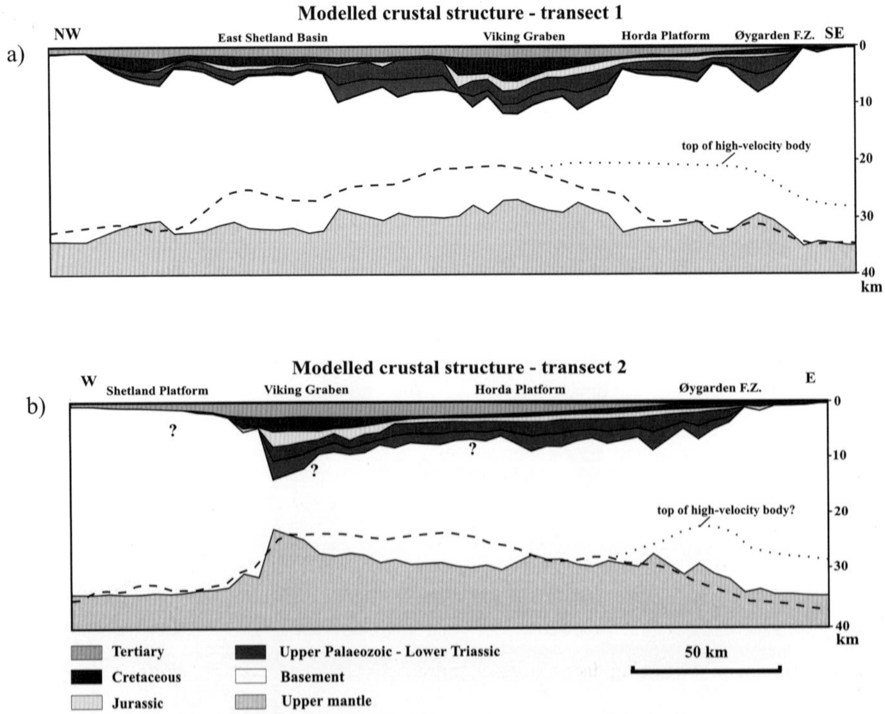

Fig. 8. (**a**) Modelled present-day crustal structure, and observed and modelled Moho depths for transect 1. Top of high-velocity body should be noted (from Christiansson *et al.* 2000). (**b**) Modelled present-day crustal structure, and observed and modelled Moho depths for transect 2. It should be noted that the modelled Moho level coincides with the base of the reflective lower crust (similar to transect 1).

Giltner 1987; Badley *et al.* 1988; Marsden *et al.* 1990; White 1990; Roberts *et al.* 1993, 1995; Ter Voorde *et al.* 2000), and have yielded a range of magnitudes for the specific areas. The large variations are dependent on the method used (summation of fault heaves, forward modelling, backstripping, measurements of crustal thinning), the cross-sectional area, and the interpretation for that specific area. This also indicates there may not be a unique β solution, as our understanding and simple measurement of stretching will never be able to reflect the true dynamics of an extending crust.

Recognizing the general limitations of stretching estimates, the present forward model compares the magnitude of the Permo-Triassic and Jurassic rift phases (and the coincidental thermal evolution) in one model run, without having to change method or approach during modelling. The β values are sampled each 5 km along the transects, and mirror the crustal stretching in a 'spiky' manner similar to Marsden *et al.* (1990, figs 2.15 and 2.16). However, unlike Marsden *et al.* (1990) and others (Roberts *et al.* 1993, 1995; Ter Vorde *et al.* 2000), we do not decouple the lower from the upper crust, although we likewise assume a ductile deformation mechanism in the lower crust. Hence, the present β profiles are not presented as smooth pure shear profiles, as they are meant to resemble the fault-controlled topography where laterally adjacent flexural loads are incorporated.

Late early Permian–early mid-Triassic stretching (Artinaskian (261 Ma)–Ladinian (236 Ma))

Figure 9 shows the modelled lateral stretching distribution along the two transects. The β distribution for Permo-Triassic rifting indicates that stretching was distributed in a broad basin, delineated to the east by the Øygarden Fault Zone and to the west by the Hutton Fault alignment in transect 1 and the Shetland Platform in transect 2. The basin is defined by a series of large tilted fault blocks, which are best imaged below the Horda Platform and the Tampen Spur area. The block-bounding normal faults have a

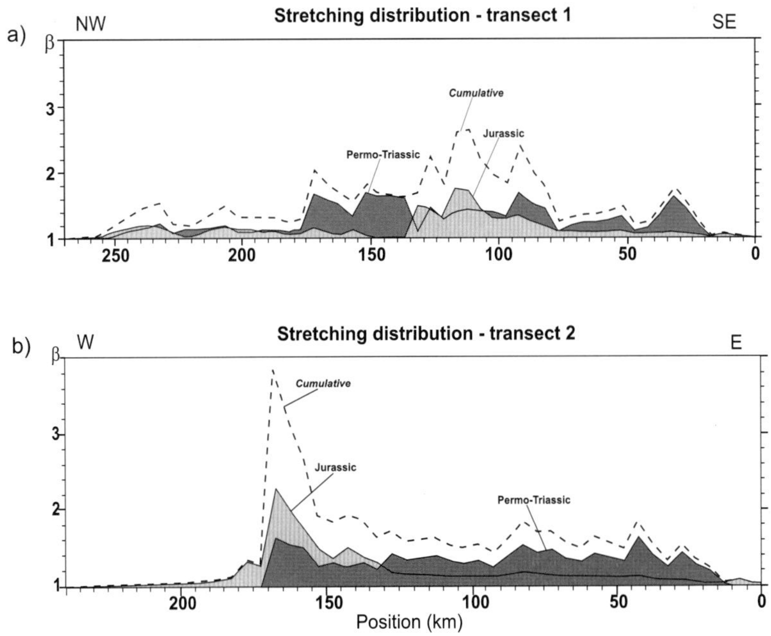

Fig. 9. (**a**) Modelled lateral stretching distribution of transect 1. (**b**) Modelled lateral stretching distribution of transect 2.

spacing of 15–20 km and throws up to 4–5 km (Færseth *et al.* 1995). The orientation of the faults was primarily the same as for the Jurassic faults, *c.* N–S below the Horda Platform and the Viking Graben, changing to NNE–SSW below the Shetland Basin.

The modelling procedure implies that an initial Permian crust of lateral uniform thickness existed. We believe that 35 km is probably a reasonable initial thickness if we consider the crust in adjacent areas that are little affected by the Permo-Triassic and Jurassic stretching. This is also in accordance with other studies (Sellevoll 1977; McClay *et al.* 1986; Kinck *et al.* 1991). Furthermore, tests have shown that changing the initial value from 35 km to 30 or 40 km will have some impact on the modelled stretching estimates, but will not alter the relative importance of the two rift phases. Table 3 gives an overview of the modelled mean stretching values for individual semi-regional areas along the two transects.

The modelled β_{mean} for the Permo-Triassic stretching is 1.27 (transect 1), and 1.19 (transect 2). If we consider only the interior of this basin β_{mean} becomes 1.4 in transect 1 (from the Øygarden Fault Zone to the Hutton alignment), and 1.38 in transect 2 (from the Øygarden Fault Zone to the East Shetland Platform). The width of this basin may have been *c.* 120–125 km when rifting ended in early mid-Triassic time. β_{mean} for the Horda Platform is 1.33 and 1.39 in transects 1 and 2 respectively. These values are very similar to those obtained by Roberts *et al.* (1995), who analysed an east–west oriented profile across the Horda Platform. Their profile was located south of transect 1. The β_{mean} values across the 'Permo-Triassic' Viking Graben are 1.41 in transect 1 and 1.28(?) in transect 2 (Table 3).

Table 3. *Modelled β_{mean} estimates for the Permo-Triassic and Jurassic rift phases*

Area	Permo-Triassic rifting mean β	Jurassic rifting mean β
Transect 1		
Horda Platform	1.33	1.08
Viking Graben	1.41	1.42
Western East Shetland Basin	1.14	1.13
East Shetland Basin	1.29	1.11
Whole transect	1.27	1.15
Transect 2		
Horda Platform	1.39	1.13
Viking Graben	1.28(?)	1.53
Shetland Platform	1.0(?)	1.03(?)
Whole transect	1.19	1.19

Permo-Triassic β_{mean} for the East Shetland Basin is 1.29. Here, the Hutton alignment had a similar status to the Øygarden Fault Zone on the east flank of the graben system. Modelling results indicate a basement subsidence of *c.* 7 km in the hanging wall of the Hutton alignment (Fig. 6a). The area northwest of the Hutton alignment is considerably less influenced by Permo-Triassic stretching, with a β_{mean} reduced to 1.14.

The modelling does not record any Permo-Triassic stretching history for the Shetland Platform. Here, Tertiary sediments unconformably overlie a thick sequence of Palaeozoic sediments, which again are locally draped by a very thin Permo-Triassic sequence. The thick pre-Permian sequence is supported by gravimetric and magnetic data (Christiansson *et al.* 2000) According to Platt (1995), these Palaeozoic sediments are most likely of Devonian(–Carboniferous?) age. The absence of Mesozoic strata on the Shetland Platform suggests that erosion has occurred, initiated by footwall uplift of the bounding fault system (Roberts & Yielding 1991). Our modelling analysis also shows that the Shetland Platform became elevated (and eroded) during rifting. It is uncertain how large a volume of sediment was eroded during long periods of elevation, and conversely how much sediment accumulated during the intervening periods of deposition.

Jurassic stretching (Bathonian (165 Ma)–Berriasian (141 Ma))

Overall Jurassic stretching amounts to approximately the same in transect 1 (β_{mean} is 1.15) and transect 2 (β_{mean} is 1.19). Although this is similar to the Permo-Triassic results for transect 2 it is much lower for transect 1 (Table 3).

The Jurassic β_{mean} across the Horda Platform amounts to 1.08 along transect 1 and 1.13 along transect 2. In other words, Jurassic stretching for the Horda Platform area was substantially less than the Permo-Triassic phase (1.33 and 1.39) for the same area. Estimated Jurassic β_{mean} across the Viking Graben is 1.42 in transect 1 and

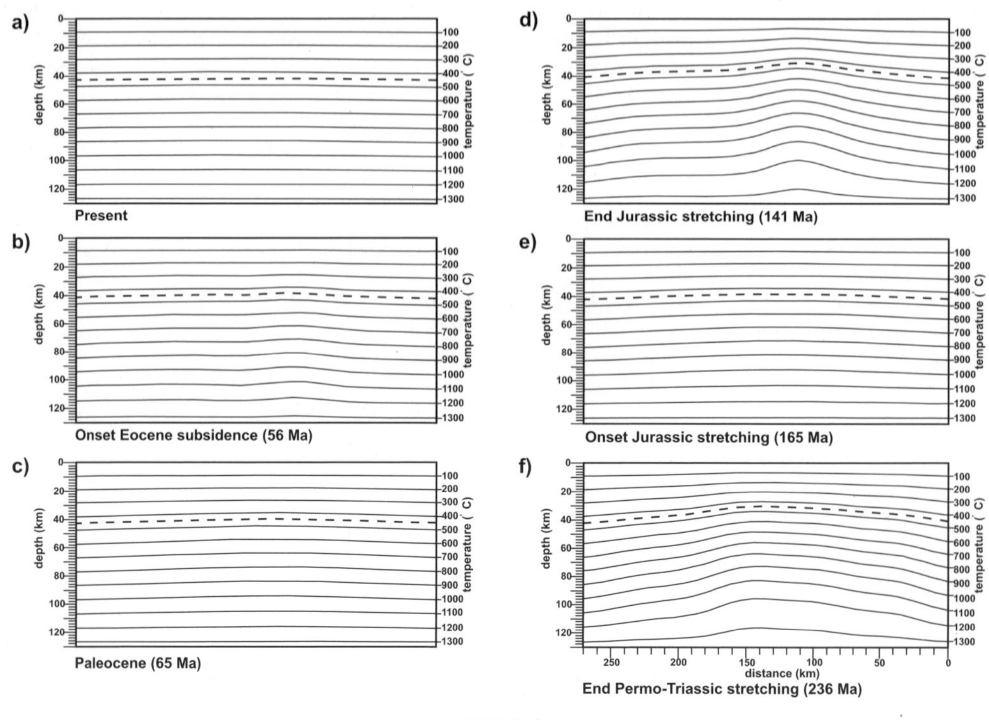

Fig. 10. Modelled thermal evolution of the lithosphere along transect 1 from the Triassic period to present. The base of the lithosphere is described by the 1300°C isotherm (contour interval is 100°C). Lateral EET value is displayed.

1.53 in transect 2. This is higher than the less constrained calculated Permo-Triassic values. β_{mean} in the East Shetland Basin is 1.11, which is much less than the Permo-Triassic stretching of 1.29. Stretching across the Shetland Platform (transect 2) is merely 1.03.

Modelled thermal evolution of transect 1

Modelled thermal evolution and the EET value in transect 1 are shown in Fig. 10. Maximum thermal perturbation at the end of the Permo-Triassic rift phase affected a much wider part of the lithosphere than Jurassic extension (Figs 10d and f). The lateral extent of the rifting is expected to be mirrored in the post-rift sequences in Figs 6 and 7; these figures demonstrate that the thermal subsidence that followed the first phase produced a broader and more uniformly distributed post-rift sequence than the latter phase. It is also noted that there is a shift towards the east and a focusing of the peak thermal elevation from the Permo-Triassic to the Jurassic rift phase.

The EET, which is controlled by the 450°C isotherm, started out with an initial value of 44 km. At the end of Permo-Triassic stretching, EET varied from 42 km to 30 km below the platform and basin areas, respectively (Fig. 10f). At the onset of the Jurassic stretching, EET is increased to 38 km below the basin areas (Fig. 10e). EET was reduced after the Jurassic stretching, to 40 km below the platform areas, and 30 km below the Viking Graben. These estimates are very close to those derived from the Baikal and East African rifts (EET *c.* 30 km) (Ebinger *et al.* 1991; Van der Beek 1997).

Several studies have emphasized that post Permo-Triassic thermal subsidence was not ended when Jurassic rifting started (Giltner 1987; Gabrielsen *et al.* 1990; Roberts *et al.* 1995), suggesting that the Jurassic northern North Sea was a hybrid basin affected simultaneously by thermal subsidence and renewed initiation of stretching. Lateral heat-flow modelling of the temperature structure for transect 1 supports this conclusion (Fig. 10e). The residual Permo-Triassic heat at the onset of the Jurassic rift phase indicates that there was a potential for minor thermal subsidence before thermal equilibrium.

The elevated thermal perturbation, and the reduced EET (from 40 km to 38 km), from Paleocene to Eocene time (Figs 10b and c), is a response to introduction of a minor stretching value of 1.02–1.03 during Paleocene time. This elevates the temperature field slightly and increases the subsidence in Paleocene and Eocene time, as recorded in the well data and seismic cross-sections. The additional post-Paleocene thermal cooling gave enough accommodation space to reproduce the observed stratigraphy without introducing unrealistically large water depths. Introducing this β value will obviously contribute to the thinning of the crust. Although this crustal thinning is speculative, it will not have any impact on the crustal-scale modelling that is addressed below. With respect to the modelled Permo-Triassic and Jurassic stretching results, the Paleocene values are within the limit of uncertainties on the regional scale that is studied here.

It is noted that recent studies seem to confirm that a mild thermal event occurred in the northern North Sea Basin in early Tertiary time (Hall & White 1994; Nadin & Kuznir 1995; Nadin *et al.* 1995), and various explanations for this departure from McKenzies' (1978) post-rift subsidence model have been offered. According to Bertram & Milton (1989) and Nadin & Kusznir (1995) no other mechanism is required for the formation of Tertiary accommodation space than post-Jurassic thermal subsidence 'buffered' by a transient Paleocene uplift event. On the other hand, Hall & White (1994) argued that a rapid increase in water-loaded subsidence occurred at the beginning of Tertiary time. This increase, up to 1 km in amplitude, is anomalous in that it is not predicted by the lithospheric stretching model, unless an additional phase of Tertiary stretching is invoked. There is, however, only limited evidence for sufficient Tertiary normal stretching. The present-day thermal status shows that the lithosphere has almost reached thermal equilibrium. Hence, the EET shows an almost uniform value of 42 km (Fig. 10a).

Comments on erosion, water depth and tectonic uplift

The problem of reproducing areas that have been elevated above sea level and eroded is present in all stratigraphic modelling. One way of dealing with this problem using the 2D forward modelling tool is to decrease the stretching values and thereby elevate the specific areas. This will reduce the stratigraphic misfit that appears in areas that has been uplifted and eroded. We have chosen to do this for the Gullfaks block and the western rim of the Horda Platform in transect 1 and close to the Øygarden Fault Zone in both transects. It is emphasized that this reduction of the Jurassic stretching values (<1.1) is local, and the net effect is negligible on the regional and semi-regional scales relevant in the discussion of the present analysis.

Trying to reproduce the Triassic sequence poses particular problems, which are partly reflected in the larger stratigraphic misfit (Figures 6 and 7). Because of the large Permo-Triassic stretching values, which lower the deep stratigraphy significantly, the problem here is not primarily elevation and erosion as described for the Jurassic sequence above. It is related to the model, which predicts a thicker Triassic post-rift sequence than is observed, as it simplifies the conditions for sediment preservation by assuming that base level is equal to sea level. If so, this would probably have produced a much thicker Triassic sedimentary coloumn than is observed. Where Triassic units are encountered in wells in the area, they are deposited into terminal flood basins, ephemeral lakes and alluvial plains (Eide 1989; Steel 1993; Nystuen & Fält 1995). The result is the formation of many bypass surfaces, along which sediments were transported downslope into the basin to the ultimate base level, thus giving rise to a much thinner continental Triassic succession than predicted by the model. As the model assumes marine sedimentation up to a specified water depth, we chose to compensate for these problems by introducing an enhanced Triassic water depth, which decreases the thickness of the lower and middle Triassic sediments (Fig. 11). It is important to note that we chose to reduce the stratigraphic misfit to avoid modelling problems for the post-Triassic sequence.

In general, water depth may be a critical factor in the subsidence and burial analysis because it assesses the amount of sediment underfill. An overestimation of palaeo-waterdepth may induce artificial uplift through time in the subsidence model, whereas underestimation artificially reduces subsidence rate. However, our introduction of enhanced palaeo-waterdepths actually provides a better stretching estimate, because it allows us to assess the subsidence relative to a non-zero base level.

When moving from a continental Triassic environment to a marine post-Triassic environment, the general water depth shallows to zero at the onset of Jurassic time, and increases again in late Jurassic to early Cretaceous time. In the deepest part of the Viking Graben water depths reach a maximum of 2000 m in early Cretaceous time. In the Shetland Basin and on the Horda Platform the depth is close to zero during early Cretaceous time. In previous studies, Nadin &

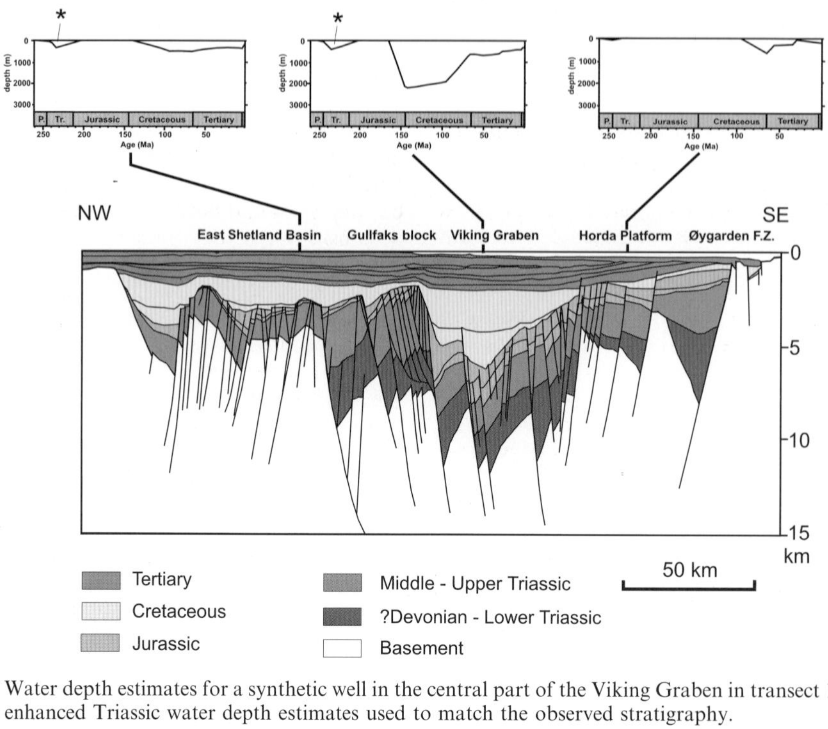

Fig. 11. Water depth estimates for a synthetic well in the central part of the Viking Graben in transect 1. Stars (*) indicate enhanced Triassic water depth estimates used to match the observed stratigraphy.

Kuznir (1995) predicted axial water depths in the Viking Graben of up to 1300 m for early Cretaceous time and 900 m for the late Cretaceous time, whereas Bertram & Milton (1989) and Barr (1991) suggested a Cretaceous water depth of the order of 1000 m for the northern North Sea Basin. At the onset of Tertiary time the water depths varied between 500 and 800 m across the basin (Fig. 11), which is in accordance with faunal trends (Gradstein *et al.* 1994). From this time the water depth shallows to the present values.

Uplift of the Norwegian mainland and adjacent areas in Tertiary time (Doré 1992) will have some impact on the stretching estimates, as these are calculated as a function of subsidence. This modelling problem is mostly limited to the Horda Platform. It is difficult to evaluate precisely how large this effect is, as the magnitude of uplift is affiliated with uncertainties. We observe that the western limit of the northern Horda Platform (transect 1) is situated much further east than seen to the south (transect 2). Therefore uplift of western Norway has also influenced the northern part of the Horda Platform more than the southern part. If we take into account the very few Jurassic faults seen along the southern Horda Platform, it is reasonable to expect that stretching during this phase was at least as great to the north. Acknowledging this, the importance of the Permo-Triassic stretching with respect to the Jurassic phase will still be the same.

Implications of crustal thickness estimates and the flexural strength of the lithosphere

The misfit between the observed and modelled Moho level below the basin areas in both transects, and below the western Horda Platform in transect 2 (Fig. 7), offers at least two explanations. First, one has to take into account that Permo-Triassic stretching may have been much larger in the basin areas than our results show. This is not likely, because the very high β values needed would reduce the crustal thickness to a few kilometres, which again would contradict the geophysical data (Christiansson *et al.* 2000).

The second and probably the most reasonable explanation deals with uncertainties of the initial crustal configuration. When the rifting was initiated in Permian time (set to 261 Ma), the crust has to be defined with a uniform lateral thickness in the model (we use 35 km). This is a simple approach where the influence of the unconstrained Devonian(–Carboniferous?) crustal stretching is not considered. According to Fossen & Rykkelid (1992), the Devonian extension of the thickened Caledonian North Sea crust was much more important than the Permo-Triassic and Jurassic extension. We limit the investigation on the Permian crustal configuration to changing the initial thickness during modelling. These tests show that reducing the initial crustal thickness from 35 km to 28–30 km gives a minimum misfit between the modelled and observed Moho depths below the basin areas, indicating that the Devonian(–Carboniferous?) stretching did not produce a crust with a uniform lateral thickness. It may have varied from *c.* 35 km in the platform areas to less than 30 km in the basin areas, if we accept that this crustal configuration lasted until the onset of the Permo-Triassic rifting.

It is acknowledged that Devonian(–Carboniferous?) sediments probably cover basement on the platform areas of the northern North Sea (McClay *et al.* 1986; Ziegler 1990; Færseth *et al.* 1995; Platt 1995), although their presence within the deeper parts of the basin is more uncertain. As the resolution of the present data does not allow a further investigation on this subject, we cannot fully integrate the pre-Permian tectonic evolution and sedimentary distribution in the modelling scheme at present. However, we realize that these factors may have some impact on the discussion of the Permo-Triassic stretching estimates and thermal configuration, as seen above for the modelled Moho level. The Devonian(–Carboniferous?) extension may, for example, have a similar thermal impact on the Permo-Triassic phase to that indicated for the Permo-Triassic phase on the Jurassic thermal evolution, depending on when the pre-Permian stretching ended. This is highly relevant on the Shetland Platform (transect 2), where a considerable sedimentary column of pre-Permian age covers the basement in half-grabens (Platt 1995). The age at which this topography was generated is generally unknown. We can, however, consider top basement as a time line (261 Ma) in the model. This is also a convenient approach for the Shetland Platform, where 'top basement' is defined as the top of this sequence. We believe this is a valid perspective, as the Permian and post-Permian stretching seems to be insignificant in this area.

With an isotherm of 450°C describing the EET, the flexural strength of the crust and mantle are always coupled. This approach is simple and has therefore been studied in more detail by Ter Voorde *et al.* (2000), who tested both coupled and decoupled EET along an E–W transect situated between our transect 1 and 2. Those workers suggested that the crust and mantle have been decoupled (EET 6 km) during the Permo-Triassic and Jurassic rift phases. The resulting stretching estimates may yield large differences, where a coupled flexural strength can give 50–60% less stretching than a fully decoupled one. However, the subject is complicated as coupled or decoupled flexural strength, and thereby stretching estimates, probably represent end-members, where a fully decoupled mode of flexure is likely to be an exceptional case (Buck 1988; Ter Voorde *et al.* 1998, 2000). Hence, one has to take into account that the degree, or mode of (de)coupling is controlled by a combination of age, crustal composition and weaknesses, crustal thickness, and strain rate (England 1983; Kuznir & Park 1987; Dunbar & Sawyer 1989; Bassi *et al.* 1993; Burov & Diament 1995; Cloetingh & Burov 1996). Future focus on the northern North Sea, following the present paper and the work of Ter Voorde *et al.* (2000), should therefore include studies on crustal rheology, and pre-Jurassic thickness and structural configuration, including an investigation on the interaction between the Devonian(–Carboniferous?) structural setting and the Permo-Triassic structure.

Conclusions

Three main topics were the specific subject of our forward modelling of transects 1 and 2.

- Estimation of the Permo-Triassic and Jurassic stretching shows that the first was generally much more important than the latter. It is only across the Viking Graben that the Jurassic stretching exceeds the older event. The Øygarden Fault Zone and the Hutton alignment appear to have been the principal structures of the Permo-Triassic basin.
- The modelled stratigraphy is generally similar to the observed one. Most of the discrepancies are limited to the Triassic sedimentary sequence. The modelled and observed Moho relief shows a minimum misfit below most of the platform areas. Here, the modelled Moho level coincides with the base of the reflective lower crust. The misfit increases below the basin areas in both transects, and below the western Horda Platform in transect 2. This misfit may be ascribed to a laterally varying crustal thickness at the onset of the Permo-Triassic lithospheric stretching, from *c.* 35 km in the platform areas to 28–30 km in the basin areas.
- The lithospheric thermal evolution shows that the maximum thermal perturbation at the end of the Permo-Triassic rift phase affected a much wider part of the lithosphere than did the Jurassic phase. This is also reflected in the post-rift sediment distribution following the two phases, where the first produces a much more uniform sedimentary sequence than the latter. There is also a focusing and migration towards the east of the peak thermal elevation from the Permo-Triassic to the Jurassic event. We assume that this focusing relates to changes of the crustal composition or rheology and structural framework from the Permo-Triassic to the Jurassic rift phase. It is also noted that the lithosphere did not fully reach a thermal equilibrium at the initiation of Jurassic stretching. The potential for enhanced subsidence in the Jurassic, however, seems to have been small.

We wish to thank A. M. Berge and P. Christiansson for co-operation and discussions, and J. E. Lie for reprocessing the NSDP lines. We are also grateful to W. Sassi, J. P. Nystuen, J. Skogseid and an anonymous reviewer for their constructive criticism and helpful comments, and to G. Sullivan for linguistic corrections. This work was funded by the Commission of the European Union and the Norwegian Research Council in the framework of Integrated Basin Studies – The Dynamics of the Norwegian Margin. Norsk Hydro a.s., Saga Petroleum a.s. and Statoil a.s. provided data and research efforts for this study.

References

BADLEY, M. E., PRICE, J. D., RAMBECH-DAHL, C. & AGDESTEIN, T. 1988. The structural evolution of the northern Viking Graben and its bearing upon extensional modes of basin formation. *Journal of the Geological Society, London*, **145**, 455–472.

BARR, D. 1991. Subsidence and sedimentation in semi-starved half-graben: a model based on North Sea data. *In*: ROBERTS, A. M., YIELDING, G. & FREEMAN, B. (eds) *The Geometry of Normal Faults*. Geological Society, London, Special Publications, **56**, 17–28.

BARTON, P. & WOOD, R. 1984. Tectonic evolution of the North Sea basin: crustal stretching and subsidence. *Geophysical Journal of the Royal Astronomical Society*, **79**, 987–1022.

BASSI, G., KEEN, C. E. & POTTER, P. 1993. Contrasting styles of rifting: models and examples from the eastern Canadian Margin. *Tectonics*, **12**, 639–655.

BEACH, A. 1986. Some comments on sedimentary basin development in the northern North Sea. *Scottish Journal of Geology*, **22**, 1–20.

——, BIRD, T. & GIBBS, A. 1987. Extensional tectonics and crustal structure: deep seismic reflection data from the northern North Sea Viking Graben. *In*: COWARD, M. P., DEWEY, J. F. & HANCOOK, P. L. (eds) *Continental Extensional Tectonics*. Geological Society, London, Special Publications, **28**, 467–476.

BERTRAM, G. T. & MILTON, N. J. 1989. Reconstructing basin evolution from sedimentary thickness; the importance of palaeobathymetric control, with reference to the North Sea. *Basin Research*. **1**, 247–257.

BODINE, J. H., STECKLER, M. S. & WATTS, A. B. 1981. Observations of flexure and the rheology of the oceanic lithosphere. *Journal of Geophysical Research*, **86**, 3695–3707.

BRAUN, J. & BEAUMONT, C. 1989. A physical explanation of the relation between flank uplift and the break-up unconformity at rifted continental margins. *Geology*, **17**, 760–764.

BRUN, J.-P. & TRON, V. 1993. Development of the North Viking Graben: inference from laboratory modelling. *Sedimentary Geology*, **86**, 31–51.

BUCK, W. R. 1988. Flexural rotation of normal faults. *Tectonics*, **7**, 959–973.

BUROV, E. & DIAMENT, M. 1995. The effective elastic thickness (T_e) of continental lithosphere: what does it really mean? *Journal of Geophysical Research*, **100**, 3905–3927.

CHRISTIANSSON, P., FALEIDE, J. I. & BERGE, A. M. 2000. Crustal structure in the northern North Sea; an integrated geophysical study. *This volume*.

CLOETINGH, S. & BUROV, E. B. 1996. Thermomechanical structure of European continental lithosphere: constraints from rheological profiles and EET estimates. *Geophysical Journal International*, **24**, 695–723.

COCHRAN, J. R. 1983. Effects of finite extension times on the development of sedimentary basins. *Earth and Planetary Science Letters*, **66**, 289–302.

COWIE, P. A. & KARNER, G. D. 1990. Gravity effect of sediment compaction: examples from the North Sea and Rhine Graben. *Earth and Planetary Science Letters*, **99**, 141–153.

DORÉ, A. G. 1992. The Base Tertiary surface of southern Norway and the northern North Sea. *Norsk Geologisk Tidsskrift*, **72**, 259–265.

DUNBAR, J. A. & SAWYER, D. A. 1989. How pre-existing weaknesses control the style of continental breakup. *Journal of Geophysical Research*, **94**, 7278–7292.

EBINGER, C. J., KARNER, G. D. & WEISSEL, J. K. 1991. Mechanical strength of extended continental lithosphere: constraints from the Western rift system, Africa. *Tectonics*, **10**, 1239–1256.

EIDE, F. 1989. Biostratigraphic correlation within the Triassic Lunde Formation in the Snorre Area. *In*: COLLINSON, J. D. (ed.) *Correlation in Hydrocarbon Exploration*. Graham and Trotman, London, 291–297.

ENGLAND, P. 1983. Constraints on extension of continental lithosphere. *Journal of Geophysical Research*, **88**, 1145–1152.

EYNON, G. 1981. Basin development and sedimentation in the Middle Jurassic of the northern North Sea. *In*: ILLING, L. V. & HOBSON, G. D. (eds) *Petroleum Geology of the Continental Shelf of North West Europe*. Heyden, London 196–204.

FICHLER, C. & HOSPERS, J. 1990. Deep crustal structure of the northern Viking Graben: results from deep reflection seismic and gravity data. *Tectonophysics*, **178**, 241–254.

FOSSEN, H. & RYKKELID, E. 1992. Post-collisional extension of the Caledonide orogen in Scandinavia: structural expressions and tectonic significance. *Geology*, **20**, 737–740.

——, ODINSEN, T., FÆRSETH, R. B. & GABRIELSEN, R. H. 2000. Detachments and low-angle faults in the northern North Sea rift system. *This volume*.

FÆRSETH, R. B. 1978. Mantle-derived lherzolite nodules and megacrysts from Permo-Triassic dykes, Sunnhordland, western Norway. *Lithos*, **11**, 23–35.

——, GABRIELSEN, R. H. & HURICH, C. A. 1995. Influence of basement in structuring of the North Sea Basin offshore southwest Norway. *Norsk Geologisk Tidsskrift*, **75**, 105–119.

GABRIELSEN, R. H. 1989. Reactivation of faults on the Norwegian continental shelf and its implications for earthquake occurrence. *In*: GREGERSEN, S. & BASHAM, P. (eds) *Causes and Effects of Earthquakes at Passive Margins and in Areas with Post-glacial Rebound on both Sides of the North Atlantic*. Elsevier, Amsterdam, 69–92.

——, FÆRSETH, R. B., STEEL, R. J., IDIL, S. & KLØVJAN, O. S. 1990. Architectural styles of basin fill in the northern Viking Graben. *In*: BLUNDELL, D. J. & GIBBS, A. D. (eds) *Tectonic Evolution of the North Sea Rifts*. Clarendon, Oxford, 158–179.

GILTNER, J. P. 1987. Application of extensional models to the northern Viking Graben. *Norsk Geologisk Tidsskrift*, **67**, 339–352.

GRADSTEIN, F. M., KAMINSKI, M. A., BERGGREN, W. A., KRISTIANSEN, I. L. & D'IORO, M. A. 1994. Cenozoic biostratigraphy of the North Sea and Labrador Shelf. *Micropaleontology*, **40**.

HALL, B. D. & WHITE, N. J. 1994. Origin of anomalous Tertary subsidence adjacent to North Atlantic continental margins. *Marine and Petroleum Geology*, **11**, 702–714.

HARLAND, W. B., ARMSTRONG, R. L., COX, A. V., CRAIG, L. E., SMITH, A. G. & SMITH, D. G. 1990. *A Geologic Time Scale 1990*. Cambridge University Press, Cambridge.

HELLAND-HANSEN, W., ASHTON, M., LOMO, L. & STEEL, R. J. 1992. Advance and retreat of the Brent delta: recent contributions to the depositional model. *In*: MORTON, A. C., HASZELDINE, R. S., GILES, M. R. & BROWN, S. (eds) *Geology of the Brent Group*. Geological Society, London, Special Publications, **61**, 109–127.

HOLLINGER, K. & KLEMPERER, S. L. 1990. Gravity and deep seismic reflection profiles across the North Sea rifts. *In*: BLUNDELL, D. J. & GIBBS, A. D. (eds) *Tectonic Evolution of the North Sea Rifts*. Clarendon, Oxford, 82–100.

JOHNS, C. & ANDREWS, I. J. 1985. The petroleum geology of the Unst Basin, North Sea. *Marine and Petroleum Geology*, **2**, 361–372.

JORDT, H., FALEIDE, J. I., BJØRLYKKE, K. & IBRAHIM, M. T. 1995. Cenozoic sequence stratigraphy of the central and northen North Sea Basin: tectonic development, sediment distribution and provenance areas. *Marine and Petroleum Geology*, **12**, 845–879.

KINCK, J. J., HUSEBYE, E. S. & LUND, C. E. 1991. The south Scandinavian crust: structural complexities from seismic reflection and refraction profiling. *Tectonophysics*, **189**, 117–133.

KLEMPERER, S. L. 1988. Crustal thinning and nature of extension in the northern North Sea from deep seismic reflection profiling. *Tectonics*, **7**, 803–821.

—— & HURICH, C. A. 1990. Lithospheric structure of the North Sea from deep seismic reflection profiling. *In*: BLUNDELL, D. J. & GIBBS, A. D. (eds) *Tectonic Evolution of the North Sea Rifts*. Clarendon, Oxford, 37–59.

KOMINZ, M. A. 1984. Ocean ridge volumes and sea-level change – and error analysis. *In*: SCHLEE, J. S. (ed.) *Interregional Unconformities and Hydrocarbon Accumulation*. American Association of Petroleum Geologists, Memoir, **36**, 109–127.

KOOI, H., CLOETINGH, S. & BURRUS, J. 1992. Lithospheric necking and regional istostacy at extensional basins 1. Subsidence and gravity modelling with an application to the Gulf of Lions margin (SE France). *Journal of Geophysical Research*, **97**, 17 553–17 571.

KUSZNIR, N. J. & PARK, R. G. 1987. The extensional strength of the continental lithosphere: its dependence on geothermal gradient, crustal composition and thickness. *In*: COWARD, M. P., DEWEY, J. F. & HANCOCK, P. L. (eds) *Continental Extensional Tectonics*. Geological Society, London, Special Publications, **28**, 35–52.

KYRKJEBØ, R., HAMBORG, M., FALEIDE, J. I., JORDT, H. & CHRISTIANSSON, P. 2000. Cenozoic tectonic subsidence from 2D depositional simulations of a regional transect in the northern North Sea basin. *This volume*.

MARSDEN, G., YIELDING, G., ROBERTS, A. M. & KUSZNIR, N. J. 1990. Application of a flexural cantilever simple-shear/pure-shear model of continental lithosphere extension to the formation of the northern North Sea Basin. *In*: BLUNDELL, D. J. & GIBBS, A. D. (eds) *Tectonic Evolution of the North Sea Rifts*. Clarendon, Oxford, 241–261.

MCCLAY, K. R., NORTON, M. G., CONEY, P. & DAVIS, G. H. 1986. Collapse of the Caledonian orogen and the Old Red Sandstone. *Nature*, **323**, 147–149.

MCKENZIE, D. 1978. Some remarks on the development of sedimentary basins. *Earth and Planetary Science Letters*, **40**, 25–32.

NADIN, P. A. & KUZNIR, N. J. 1995. Palaeocene uplift and subsidence in the northern North Sea Basin from 2D forward and reverse stratigraphic modelling. *Journal of the Geological Society, London*, **152**, 833–848.

——, —— & TOTH, J. 1995. Transient regional uplift in the Early Tertiary of the northern North Sea and the development of the Iceland Plume. *Journal of the Geological Society, London*, **152**, 953–958.

NYSTUEN, J. P. & FÄLT, L.-M. 1995. Upper Triassic–Lower Jurassic reservoir rocks in the Tampen Spur area, Norwegian North Sea. *In*: HANSLIEN, S. (ed.) *Petroleum Exploration and Exploitation in Norway*. Norwegian Petroleum Society Special Publication, **4**, 135–179.

NØTTVEDT, A., GABRIELSEN, R. H. & STEEL, R. J. 1995. Tectonostratigraphy and sedimentary architecture of rift basins with reference to the northern North Sea. *Marine and Petroleum Geology*, **12**, 881–901.

ODINSEN, T., CHRISTIANSSON, P., GABRIELSEN, R. H., FALEIDE, J. I. & BERGE, A. M. 2000. The geometries and deep structure of the northern North Sea rift system. *This volume*.

PLATT, N. H. 1995. Structure and tectonics of the northern North Sea: new insights from deep penetration regional seismic data. *In*: LAMBIASE, J. J. (ed.) *Hydrocarbon Habitat in Rift Basins*. Geological Society, London, Special Publications, **80**, 103–113.

RESTON, T. J. 1990. The lower crust and the extension of the continental lithosphere: kinematic analysis of BIRPS deep seismic data. *Tectonics*, **9**, 1235–1248.

ROBERTS, A. M. & YIELDING, G. 1991. Deformation around basin-margin faults in the North Sea/mid-Norway rift. *In*: ROBERTS, A. M., YIELDING, G. & FREEMAN, B. (eds) *The Geometry of Normal Faults*. Geological Society, London, Special Publications, **56**, 61–78.

——, ——, KUSZNIR, N. J., WALKER, I. & DORN-LOPEZ, D. 1993. Mesozoic extension in the North Sea: constraints from flexural backstripping, forward modelling and fault populations. *In*: PARKER, J. R. (ed.) *Petroleum Geology of Northwest Europe: Proceedings of the 4th Conference*. Geological Society, London, 1123–1136.

——, ——, ——, —— & ——1995. Quantitative analysis of Triassic extension in the northern North Sea. *Journal of the Geological Society, London*, **152**, 15–26.

ROYDEN, L. & KEEN, C. E. 1980. Rifting processes and thermal evolution of the continental margin of eastern Canada determined from subsidence curves. *Earth and Planetary Science Letters*, **51**, 343–361.

SCLATER, J. G. & CHRISTIE, P. A. F. 1980. Continental stretching: an explanation of the post mid-Cretaceous subsidence of the central North Sea Basin. *Journal of Geophysical Research*, **85**, 3711–3739.

SELLEVOLL, M. A. 1977. Moho beneath Fennoscandia and adjacent parts of the Norwegian Sea and the North Sea. *Tectonophysics*, **20**, 359–366.

SPADINI, G., CLOETINGH, S. & BERTOTTI, G. 1995. Thermo-mechanical modelling of the Tyrrhenian Sea: lithospheric necking and kinematics of rifting. *Tectonics*, **14**, 629–644.

STECKLER, M. S. 1981. *The thermal and mechanical evolution of Atlantic-type continental margins*. PhD thesis, Columbia University, Palisades, NY.

STEEL, R. J. 1993. Triassic–Jurassic megasequence stratigraphy in the Northern North Sea: rift to post-rift evolution. *In*: PARKER, J. R. (ed.) *Petroleum Geology of Northwest Europe: Proceedings of the 4th Conference*. Geological Society, London, 299–315.

TER VOORDE, M., FÆRSETH, R. B., GABRIELSEN, R. H. & CLOETINGH, S. A. P. L. 2000. Repeated lithosphere extension in the northern Viking Graben: a coupled or a decoupled rheology? *This volume.*

——, VAN BALEN, R. T., BERTOTTI, G. & CLOETINGH, S. A. P. L. 1998. The influence of a stratified rheology on the flexural response of the lithosphere to (un)loading by extensional faulting. *Geophysical Journal International*, **134**, 721–735.

TORSVIK, T. H., STURT, B. A., SWENSSON, E., ANDERSEN, T. B. & DEWEY, J. F. 1992. Palaeomagnetic dating of fault rocks: evidence for Permian and Mesozoic movements and brittle deformation along the Dalsfjord Fault, west Norway. *Geophysical Journal International*, **109**, 565–580.

VAN DER BEEK, P. 1997. Flank uplift and topography at the central Baikal rift (SE Siberia): a test of kinematic models for continental extension. *Tectonics*, **16**, 122–136.

WEISSEL, J. K. & KARNER, G. D. 1989. Flexural uplift of rift flanks due to mechanical unloading of the lithosphere during extension. *Journal of Geophysical Research*, **94**, 13 919–13 950.

WHITE, N. J. 1990. Does the uniform stretching model work in the North Sea? *In*: BLUNDELL, D. J. & GIBBS, A. D. (eds) *Tectonic Evolution of the North Sea Rifts.* Clarendon, Oxford, 217–240.

WOOD, R. 1981. The subsidence history of Conoco well 15/30-1, central North Sea. *Earth and Planetary Science Letters*, **54**, 306–312.

—— & BARTON, P. 1983. Crustal thinning and subsidence in the North Sea. Nature, **302**, 134–136.

ZIEGLER, P. A. 1990. *Geological Atlas of Western and Central Europe*, 2nd edn. Shell International, The Hague.

Detachments and low-angle faults in the northern North Sea rift system

HAAKON FOSSEN,[1] TORE ODINSEN,[1]
ROALD B. FÆRSETH[2] & ROY H. GABRIELSEN[1]

[1] *Department of Geology, University of Bergen, Allégaten 41, N-5007 Bergen, Norway*
[2] *Norsk Hydro, 1321 Stabekk, Norway*

Abstract: Several master faults in the North Sea basin tend to flatten to give low dips at depth, and in this sense form detachments in the rift system. Such low angle faults are identified in the western flank of the Viking Graben (Tampen Spur area), where they occur as both intra- and supra-basement detachments. Interference between detachments and steeper faults results in ramp–flat–ramp geometries. In the eastern part of the Gullfaks fault block, a supra-basement detachment is probably associated with anomalously high late Jurassic extension in the Gullfaks Field area. The low-angle Gullfaks detachment also helps explain the presence of sets of parallel east-dipping faults (domino systems), a common feature in the collapsed hanging wall to low-angle detachments. Similar detachments probably exist beneath the Gullfaks Sør block and SE of the Visund fault block. All of these are interpreted as late Jurassic collapse structures directly related to active late Jurassic extensional tectonics. Strong indications of intra-basement detachments are also found in the Tampen Spur area. These detachments are formed by major normal faults that flatten in the basement, as seen beneath the Visund fault block. This geometry may to some extent be related to fault rotation during repeated phases of extension in the Palaeozoic–Early Mesozoic period. However, abrupt flattening of some of the faults in the basement indicates that the master faults follow some of the many pre-existing mechanically weak zones in the basement, primarily low-angle Devonian extensional shear zones or Caledonian thrusts.

The most common hydrocarbon trap types in the North Sea are controlled or influenced by faults, as delineating structures of rotated fault blocks, as synsedimentary structures controlling the distribution of source and reservoir rocks, or as barriers to fluid flow (Hardman & Booth 1991). Accordingly, fault geometry is important for understanding many oilfields and gas fields in this area.

Experimental modelling, kinematic considerations and field examples all indicate that faults in extended regions may be planar, non-planar (e.g. listric), high angle or low angle, or a combination of these (e.g. Wernicke & Burchfiel 1982; Gibbs 1984; Gabrielsen 1986; McClay & Ellis 1987; Fossen & Gabrielsen 1996). However, there has been a tendency in the literature to prefer one of these types of fault geometries when making interpretations. Faults shown as steep, sub-planar features dominate the literature, possibly because downward-flattening or low-angle structures are more difficult to detect seismically. However, extensive interpretation of listric fault geometries was favoured by some in the 1980s (e.g. Beach 1984; Gibbs 1984), and the discovery of low-angle faults in extended terranes (Armstrong 1972) also led to consideration of such structures in the same time period (e.g. Wernicke & Burchfiel 1982; Lister *et al.* 1986). In general, a large-scale rift system can be expected to contain elements of all of these types of fault geometries. We believe that this also holds true for the northern North Sea rift system.

In this paper we present observations primarily based on a combination of commercial 2D and 3D seismic data and a reprocessed deep seismic line across the northern North Sea (NSDP84-1) (Fig. 1). We will focus on the Gullfaks–Visund–Snorre region and on the non-planar and low-angle fault geometries revealed by the seismic data. Downward-flattening faults and detachment faults have been previously described south of the present study area, where extensional detachments tend to form along the Zechstein salt (Clausen & Korstgård 1996; Thomas & Coward 1996). Downward-flattening faults are also interpreted at deeper as well as higher levels in the North Sea (Beach 1984; Gibbs 1984; Speksnijder 1987; Gabrielsen 1988; Graue 1992; Platt 1995). In a complementary work to the present contribution, Odinsen *et al.* (2000) suggest that multiple levels of detachment, as well as composite fault geometries, are common in the northern North Sea.

From: NØTTVEDT, A. *et al.* (eds) *Dynamics of the Norwegian Margin*. Geological Society, London, Special Publications, **167**, 105–131. 1-86239-056-8/00/$15.00

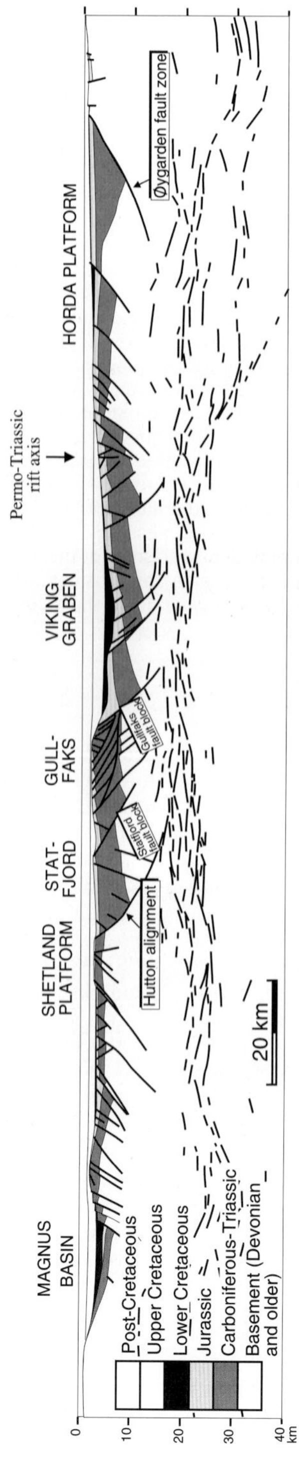

Fig. 1. Regional profile based on seismic line NSDP84-1. Modified from Christiansson *et al.* (1999).

We define the term detachment as: a low-angle (portion of a) fault or shear zone that separates rocks that show different amounts of extension and/or deformation styles. By low-angle fault we mean a fault with dip <30°. A listric fault is usually defined as a spoon-shaped fault in three dimensions, and appears as a curved, concave-upward fault on cross-sections. Faults with geometries different from both listric and planar faults are simply referred to as non-planar faults, with additional descriptive terms such as downward flattening. Below, we distinguish between supra-basement detachments occurring in the sedimentary sequence above the Caledonian basement, and intra-basement detachments, which partly or wholly are confined to basement rocks.

Regional geological development

The North Sea rift (Fig. 2) is a post-Caledonian graben system with a multiphase extensional history that started with Devonian extension of the thickened Caledonian crust. This extensional phase affected an area that extends far beyond the later (Permo-Triassic and Jurassic) margins of the North Sea rift system (Fossen & Rykkelid 1992; Fossen 1998). At least two important post-Devonian phases of rifting are recognized in the northern North Sea.

The first is a Permo-Triassic phase, which is not known in great detail but is thought to be of great importance, in terms of both extension and formation of a structural framework (Gabrielsen *et al.* 1990; Færseth *et al.* 1995*a*; Roberts *et al.* 1995). The Permo-Triassic extension appears to have affected a wider area than the Jurassic phase (Gabrielsen *et al.* 1990; Roberts *et al.* 1990, 1995; Færseth *et al.* 1995*a*; Odinsen *et al.* 2000*b*). The extensional faults defining the largest fault blocks in the North Sea rift are mostly of Permo-Triassic origin, although reactivated in Jurassic time. Thermal contraction and sediment loading prevailed throughout Triassic time, and around 2–4 km of Triassic sediments are deposited in the northern Viking Graben (e.g. Badley *et al.* 1988; Steel & Ryseth 1990). A major uplift (erosion) is recorded in the Lower–Middle Jurassic series of the central North Sea, where a major rift dome may have been located (e.g. Ziegler 1990). In the northern Viking Graben, doming-related regression led to the deposition of the Brent Group sandstones.

Although local fault activity is known for more or less the entire Jurassic time period, it is

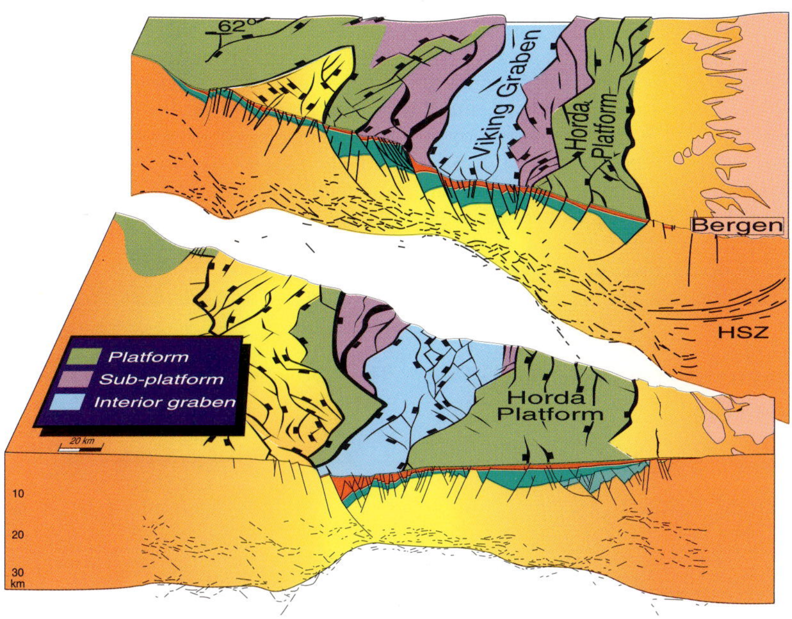

Fig. 2. Regional overview and internal subdivision of the northern North Sea rift system. Interpretations of lines NSDP84-1 and -2 are incorporated into the figure. It should be noted that colours on sections illustrate sedimentary sequences, whereas map areas are colour coded according to the topo-tectonic classification suggested by Gabrielsen (1986). HSZ, Hardangerfjord shear zone.

well established that most of the structural traps in the North Sea are the results of a late Mid- to Late Jurassic extension phase. The rate of crustal extension appears to have increased substantially from Mid- to Late Jurassic time. The Shetland Platform was formed, and differential subsidence was associated with the development and rotation of large fault blocks. Extensional structures related to this Jurassic phase are clearly imaged on seismic sections, and are therefore known in more detail than the Permo-Triassic ones. Stretching estimates vary from less than 15% in the platform/sub-platform areas to 40–50% in the interior part of the rift (Viking Graben) (Marsden *et al.* 1990; Roberts *et al.* 1993; Odinsen *et al.* 2000*b*). The fault blocks exhibit eroded crests, indicating that they were all at or near sea level during a late (Kimmeridgian) stage of the rifting history. Increased extension rate and accelerated fault movements resulted in increased water depth in late Kimmeridgian–Volgian time, and not until this time did the Viking Graben become the deepest structural–topographical element of the rift in the study area (Fig. 1) (Gabrielsen *et al.* 1990).

The Jurassic Viking Graben did not develop as a single, straight entity, but as a system of syn-rift units bounded by master faults. A system of graben segments formed, where the graben segments were linked through accommodation zones as described by Scott & Rosendahl (1989).

The rate of extension decreased at the Jurassic–Cretaceous transition, and was practically terminated by Ryazanian time. Thermal and sediment loading-related subsidence influenced the entire North Sea until Paleocene time. A general rise in sea level resulted in a progressive overstepping of the platform and burial of Jurassic fault blocks during Cretaceous time (e.g. Ziegler 1990), and the significant bathymetry at the end of the Jurassic period was infilled by marine Cretaceous shales.

General structure and low-angle faults of the northern North Sea

General structure of graben systems

Several structural features are common in extensional graben systems that have reached a certain stage of development (Johnson 1930; Robson 1971; Harding 1984). Several studies suggest that such features reflect the bulk geometry of the extending crust on a regional scale (Wernicke 1985; Coward 1986), and complex fault interaction at depth on the semi-regional and local scales (Wernicke & Burchfiel 1982).

On the basis of observations in the North Sea, Gabrielsen (1986) proposed a conceptual topotectonic model for extensional graben systems (Fig. 2). This model allows for non-rotating steep planar faults, rotating low-angle planar faults and listric faults to be developed at different stages of graben formation, and for such faults to be simultaneously active in different parts of the extending graben. Such relationships are also seen in analogue experimental models (Fossen & Gabrielsen 1996). In Gabrielsen's model, the extra-marginal fault system separates the area that has undergone stretching from its undeformed surrounding terrane. The extra-marginal faults, which may constitute a relatively small-scale horst-and-graben topography, are steep structures that were activated at the initial stage of the graben history. The platform, which commonly is separated from the extra-marginal fault system by a minor horst, represents a structural unit that is characterized by moderate fault activity and subsidence. Its width may vary considerably along the strike of the graben margin. On its graben-ward side, the platform is frequently bordered by a marginal platform high towards the heavily faulted sub-platform, which is characterized by an array of rotated fault blocks. The outer master fault system separates the margin from the interior graben.

Following Bosworth (1985) and Rosendahl (1987), we acknowledge that most graben systems are subdivided into separate units, the geometry of which reflects along-strike spatial interaction of the master fault systems and their interconnections at depth. Such relationships include transfer zones of different types, reflecting the degree of overlap and eventually shifting polarities between the fault segments.

Structure of the northern North Sea rift system

The North Sea basin displays most of the features described above, as illustrated in Fig. 2. Also, the interior graben of the late Jurassic–Cretaceous Viking Graben displays several centres of subsidence (Færseth 1983), indicating that separate graben units exist. A similar pattern of graben units with shifting polarities is recognized in the Permo-Triassic basin in the present Horda Platform (Scott & Rosendahl 1989; Gabrielsen *et al.* 1990; Færseth *et al.* 1995*a*). Finally, complex and composite fault geometries in the North Sea have been reported by several workers (Badley *et al.* 1984, 1988; Spekschnijder 1987). This has led to speculations about whether additional levels of detachment may have been active during its development (Gabrielsen 1988).

Utilizing reprocessed deep reflection data, high-quality commercial seismic lines and gravimetric and magnetic data, the deep structure of the northern Viking Graben has recently been reinterpreted by Christiansson *et al.* (2000) and Odinsen *et al.* (2000). These studies support the view that the area is truncated by a principal east-dipping crustal-scale fault, which subcrops along the eastern margin of the East Shetland Basin and flattens in the highly reflective lower crust beneath the western border of the (Jurassic) Viking Graben. Intra-mantle eastward-dipping reflections mapped beneath the Horda Platform (also reported by Klemperer (1988)) may represent the continuation of this master fault. A shallower (intra-Triassic) detachment was postulated for the Gullfaks Field (Fossen 1989; Koestler *et al.* 1992) and for the Cormorant Field (Speksnijder 1987), and has been supported in the later work by Odinsen *et al.* (2000). A similar detachment was observed below the Horda Platform, which is dominated by antithetical and more planar faults. The present study concentrates on analysis of these detachments, and elaborates upon their complex geometries.

A transect across the northern North Sea (Fig. 1) reveals the asymmetrical geometry of the rift. This part of the North Sea basin is at present bound by downward flattening, marginal faults, called the Øygarden Fault Zone and Hutton alignment, respectively. These faults, as well as most master faults within the rift, are of Permo-Triassic origin, but were reactivated in Jurassic time.

Between these marginal faults, the strata are rotated and generally dip toward the margins,

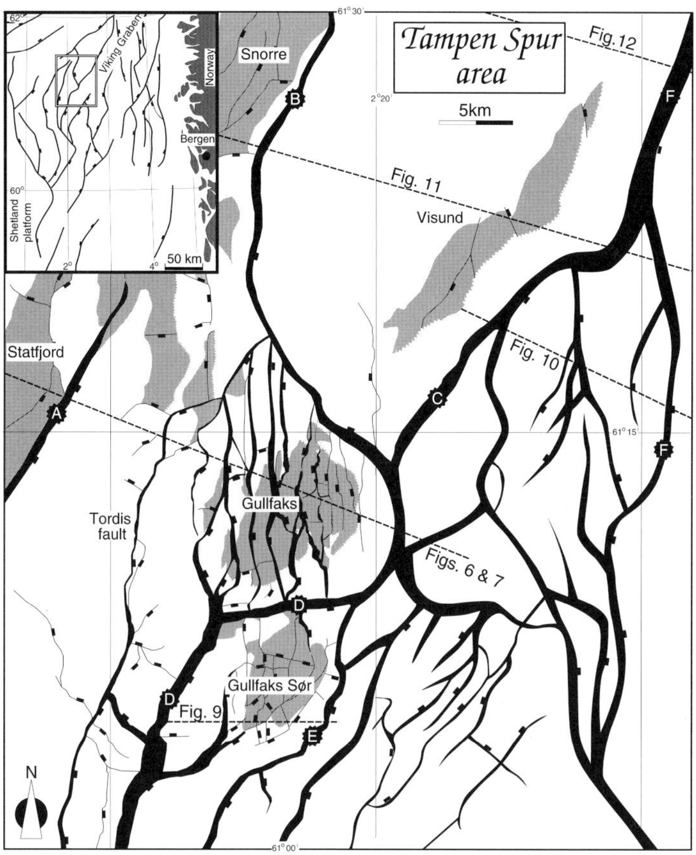

Fig. 3. Fault map of the Gullfaks–Visund–Snorre area, with profile locations. Faults at top Brent Group or base Cretaceous level (where the Brent Group is eroded).

away from the Permo-Triassic rift axis east of the Jurassic Viking Graben (Figs 1 and 2). Similarly, the main faults dip toward the Permo-Triassic rift axis, but somewhat more steeply on the eastern (Horda Platform) side. This difference in dip and the asymmetry of the system reflect that a larger portion of the extension accumulated on the western side of the Permo-Triassic rift axis than on the eastern side.

Low-angle faults are particularly related to the western margin of the Viking Graben proper (e.g. Swallow 1986; Harris & Fowler 1987; Cherry 1993; Færseth *et al.* 1995*b*; Platt 1995). Thus, the area west of the Hutton alignment represents the western footwall of the entire asymmetrical Jurassic graben system. However, low-angle faults of less regional significance also occur within the basin, and it may be useful to distinguish between marginal and intra-basin low-angle faults because of their different spatial and kinematic significance in rift systems.

The marginal faults to the northern North Sea rift exhibit steep upper parts (*c.* 55–60° for the Øygarden fault), and flatten downwards into the basement to locally form low-angle faults. It is possible that these faults are linked to low-angle or sub-horizontal ductile shear zones in the ductile middle to lower part of the crust, and in this sense are mechanically linked. The downward flattening of these marginal faults is reminiscent of that of simple extensional models where rigid footwalls require the marginal faults to be listric to develop sets of rotated (domino) fault blocks in their hanging walls (e.g. Wernicke & Burchfiel 1982, fig. 7). Rotation of the domino fault blocks is made possible by the non-planar geometry of the related marginal fault, and consequently, an abrupt change in dip is seen from the relatively horizontal beds in the footwall to rotated beds in the hanging wall. However, models that include (flexural–isostatic) footwall deformation (e.g. Kusznir *et al.* 1991) are not restricted by the same boundary conditions, and non-planar marginal faults are therefore not a necessity in rift systems.

The Tampen Spur area

Intra-basin low-angle faults or detachments are most common on the western side of the Viking Graben, particularly the Gullfaks–Visund–Snorre part of the Tampen Spur area (Fig. 3). The high density of oilfields and prospects has made the concentration of geological and geophysical data in this area remarkably high. Low-angle faults and detachments in that region will be the main focus of the rest of this paper.

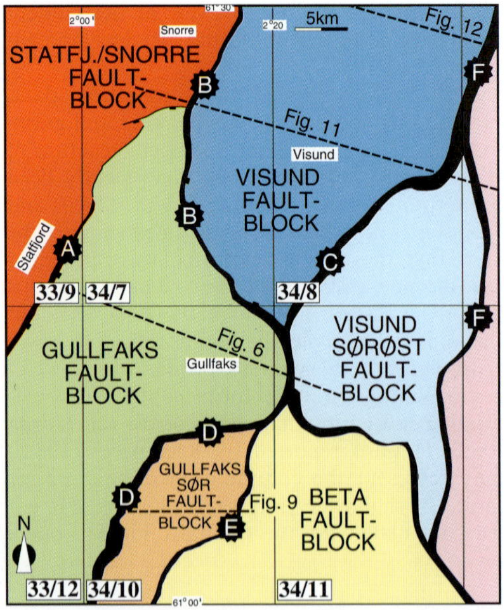

Fig. 4. Main fault blocks in the Gullfaks–Visund–Snorre area and the names applied in the text. A, the Statfjord fault; B, the Snorre fault; C, the Visund fault; D, the Gullfaks fault; E, the Gullfaks Sør fault; F, the Viking Graben boundary fault.

The most significant faults in the area are referred to as follows (Fig. 4): A, the Statfjord fault; B, the Snorre fault; C, the Visund fault; D, the Gullfaks fault; E, the Gullfaks Sør fault; F, the Viking Graben boundary fault. Similarly, the main fault blocks of the area (Fig. 4) are called the Statfjord, Gullfaks, Gullfaks Sør, Visund and Visund sørøst fault blocks. All these fault blocks are tilted so that the bedding is dipping gently to the west or northwest between the generally east- or southeast-dipping faults. We have found indications that several of these faults (A, B, C, D) have non-planar geometries, and that they form local basement-involved detachments in the area (Fig. 5). In addition, we see indications of a supra-basement detachment beneath the Gullfaks Field (Gullfaks detachment) and south and northeast of Gullfaks (Gullfaks Sør and Visund sørøst detachments; see below).

Supra-basement detachments

An example of supra-basement detachment is found beneath the Gullfaks oilfield (Petterson *et al.* 1990; Fossen & Hesthammer 1998), which occupies the eastern part of the Gullfaks fault

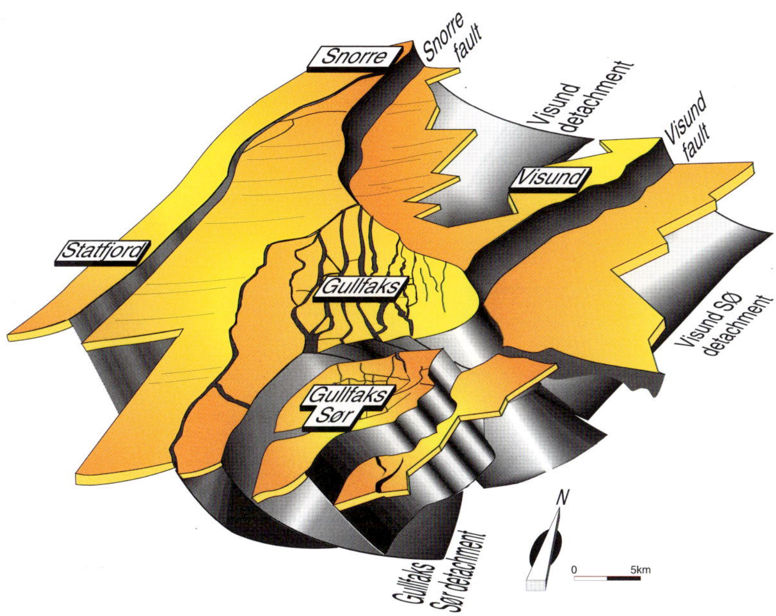

Fig. 5. Simplified 3D illustration of the Gullfaks–Visund–Snorre area. Layering at middle Jurassic level is indicated.

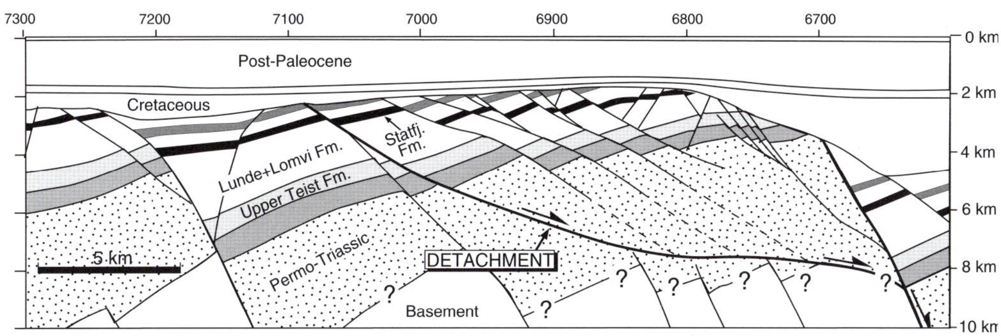

Fig. 6. Profile across the Gullfaks fault block, showing the presence of a low-angle detachment. Carefully interpreted 3D seismic data, well information and the deep seismic line shown in Fig. 7 form the basis for the interpretation. A standard 'lin-vel' depth conversion method with well data from the Gullfaks Field was applied and extended down to basement, which is assigned a constant velocity of 6.1 km s^{-1}. Parameters used for interval sea floor–base Cretaceous: $V_01 = 1790$ m s^{-1}, $k = 0.2653$; base Cretaceous–top Statfjord Fm: $V_02 = 800$ m s^{-1}, $k = 0.8523$; base Cretaceous–top Triassic: $V_02 = 1000$ m s^{-1}, $k = 0.8523$. (See Fig. 3 for location.)

block (Fig. 4). Master faults with kilometre-scale displacements separate the Gullfaks fault block from the Statfjord fault block to the west (fault A) and the Visund–Gullfaks Sør area to the east (faults D and B Fig. 4). Among these, faults D and B form a highly non-planar structure that wraps around the Gullfaks Field in a very characteristic manner. This fault structure splays both to the south and north (faults D, E, B and C), and delineates the Gullfaks Sør and Visund fault blocks, respectively (Fig. 4).

Detailed knowledge about the structural geology of the Gullfaks Field has been gained through systematic analysis of exploration and production data (Fossen & Hesthammer 1998). In general terms, the western part of the Gullfaks fault block is dominated by a classical domino fault system. The domino faults are dipping about 30° to the east, whereas the bedding within the blocks exhibits more gentle dips (generally 10–18°) to the west. This distinct and geometrically uniform domino system extends from the Tordis fault in the west (Fig. 3) to a horst complex in the easternmost part of the Gullfaks fault block (Fig. 6), spanning about 10–15 km in the E–W direction and slightly more in the N–S direction.

The extension in the domino area is considerably higher than in the rest of the Gullfaks fault block. A recent map-view restoration of the

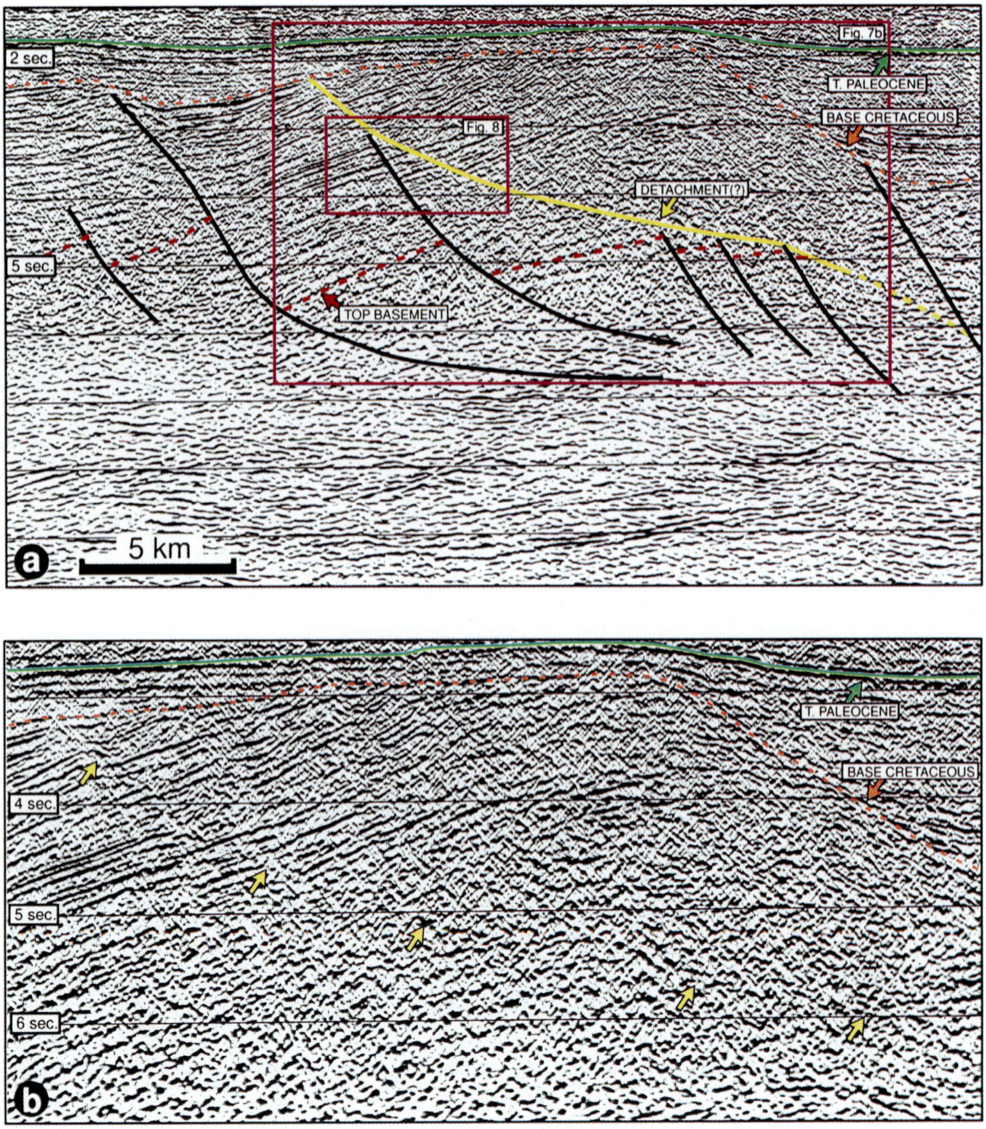

Fig. 7. (**a**) Part of the reprocessed deep seismic line NSDP-1 that traverses the Gullfaks fault block. The low-angle detachment is indicated together with some of the major basement-involving faults. The interpretation of faults within the basement is speculative and not well constrained. (See Fig. 3 for location.) (**b**) Close-up of the reflection interpreted as the Gullfaks detachment (arrows).

Gullfaks Field (Rouby *et al.* 1996) shows that the seismically resolvable Jurassic E–W extension across the field is of the order of 40–50% ($\beta = 1.4$–1.5). A similar estimate of the western part of the Gullfaks fault block gives only 10–15% extension ($\beta = 1.1$–1.15). Similar results can be obtained by summing the heaves of faults across E–W profile lines. There is good evidence that additional sub-seismic deformation is present in the domino system (Koestler *et al.* 1992; Fossen & Hesthammer 1998), possibly increasing the total extension to as much as 80% ($\beta = 1.8$).

The Gullfaks detachment

The presence of a low-angle, shallow detachment beneath the Gullfaks Field has been suggested by previous workers (Fossen 1989; Koestler *et al.* 1992), although no compelling evidence for its existence has been presented. Reprocessing of line NSDP84-1 has, however, revealed a gently E-dipping seismic event at about 3–5 s (Fig. 7). In the west the reflection appears to merge with a steeper fault. There are no similar reflectors shallower in the section, where the reflections are either sub-horizontal or westerly dipping. The reflection is therefore not a multiple of shallower reflections, and can be traced for a horizontal distance of more than 10 km. Considering the geometry, position and extent of the reflection, we interpret it as a low-angle fault (detachment) underlying the Gullfaks Field. A problem with this interpretation is that the possible detachment reflector cross-cuts a strong and continuous W-dipping signal in its western part (Fig. 8a). However, a simple restoration exercise (Fig. 8a–c) reveals that this continuous seismic signal can be explained as a coincidental alignment of two reflections from different stratigraphic levels on each side of the low-angle E-dipping reflection (detachment), a case that is familiar to most experienced seismic interpreters.

An interpreted geological section along this seismic line, based on both 2D and Gullfaks 3D seismic data and well information (Fig. 6), indicates that the detachment has a dip of 5–30°, and reaches a depth of about 7–8 km in the eastern part of the Gullfaks fault block. Gravity and magnetic data are consistent with the seismic interpretation of top basement beneath the Gullfaks Field as shown in Fig. 6 although a precise definition of top basement is difficult from any available dataset. According to our interpretation, the detachment soles out just above the basement–cover interface.

The contact relationship with the steeper and older (Permo-Triassic) master fault to the east (fault B–D) is not clear from the reflection seismic data. As no evidence is found for the detachment in the hanging wall to this fault, it is assumed that the detachment merges with fault B–D to form a ramp structure, similar to the Visund sørøst detachment (see below). In the west, the detachment appears to be connected to the Tordis fault, which therefore may have acted as the breakaway fault to the detached Gullfaks domino system.

The Gullfaks Sør detachment

Fault D (Fig. 4), which separates Gullfaks Sør from the Gullfaks fault block to the west, has a flattening-downward appearance on 3D seismic data (Fig. 9). Although the geometry is not well constrained in its eastern part, it appears that fault D flattens at depths of about 5.5–7 km to form a detachment at similar depth to the Gullfaks detachment. The entire Gullfaks Sør fault block rides on this detachment, which appears to join fault E east of Gullfaks Sør. Faults above the detachment are characterized by domino-style, sub-parallel E-dipping faults, similar to the style seen in the Gullfaks Field. The general interpretation of depth to basement in this area implies that the detachment is located within the Permo-Triassic sedimentary sequence. We therefore classify the Gullfaks Sør detachment as a supra-basement detachment.

The Visund sørøst detachment

Faults B and F formed the boundaries of the initial Visund mega fault block (combined area of the Visund and Visund sørøst fault blocks in Fig. 4). The upper part of fault F (Fig. 3) appears to be straight in cross-section. It is of Permo-Triassic origin (Færseth *et al.* 1995*b*), although the main throw is related to Jurassic extension, reaching a maximum of about 5 km at Middle Jurassic level.

The area at present bounded by faults C and F (Visund sørøst fault block in Fig. 4) was part of the initial Visund mega fault block, but was separated as a result of Volgian footwall collapse of the southeastern part of the mega-block (Færseth *et al.* 1995*b*). Fault C and some minor, east-dipping faults formed at this stage. The faults appear to converge and merge at depth (Fig. 10).

Depth conversion of fault C reveals a ramp–flat–ramp geometry with a present maximum

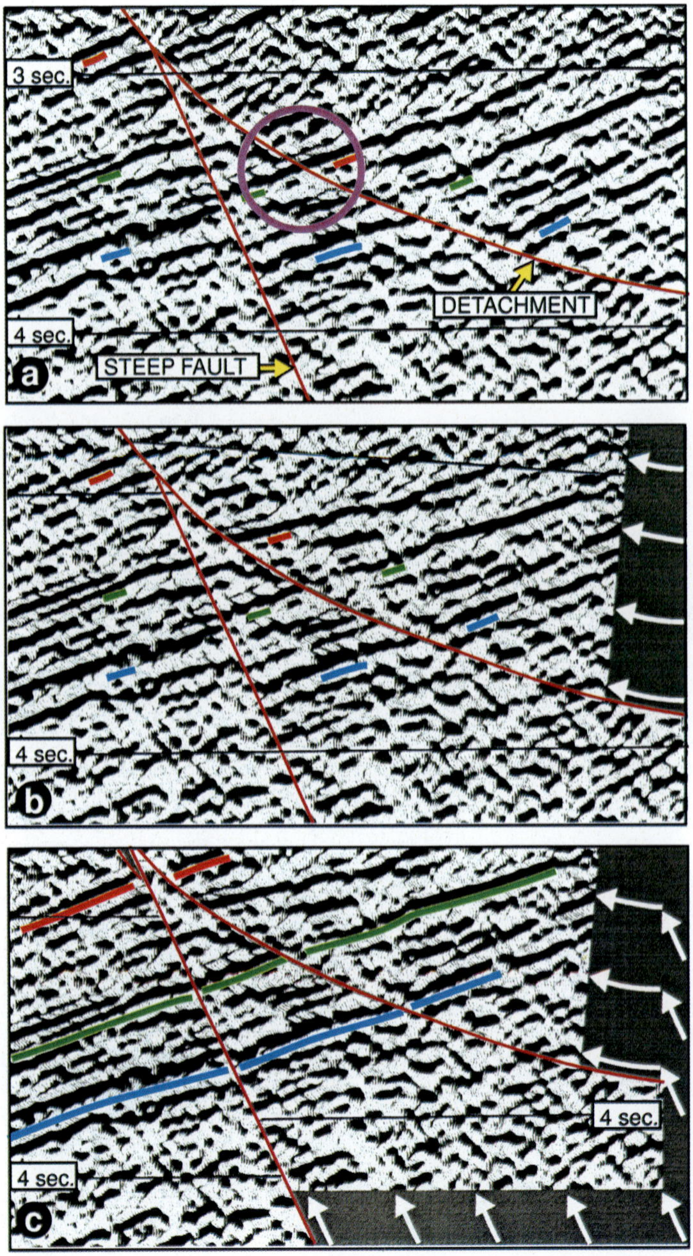

Fig. 8. Simple restoration of the portion of line NSDP84-1 where the detachment reflection merges with the steeper normal fault (see Fig. 7a for location). From (**a**) the detachment reflection appears to transect a continuous reflection (circled). The section was restored by rigid rotation of the hanging wall to the detachment (**b**) and subsequently by translation along the steep fault (**c**). In (**c**) the section is restored, after removal of an offset of a few hundred metres along the detachment. The rotation that was required is consistent with anti-clockwise rotation of the hanging wall of the detachment during deformation. The successful restoration shows that the continuous reflection may be the result of coincidental alignment of two different reflecting surfaces on each

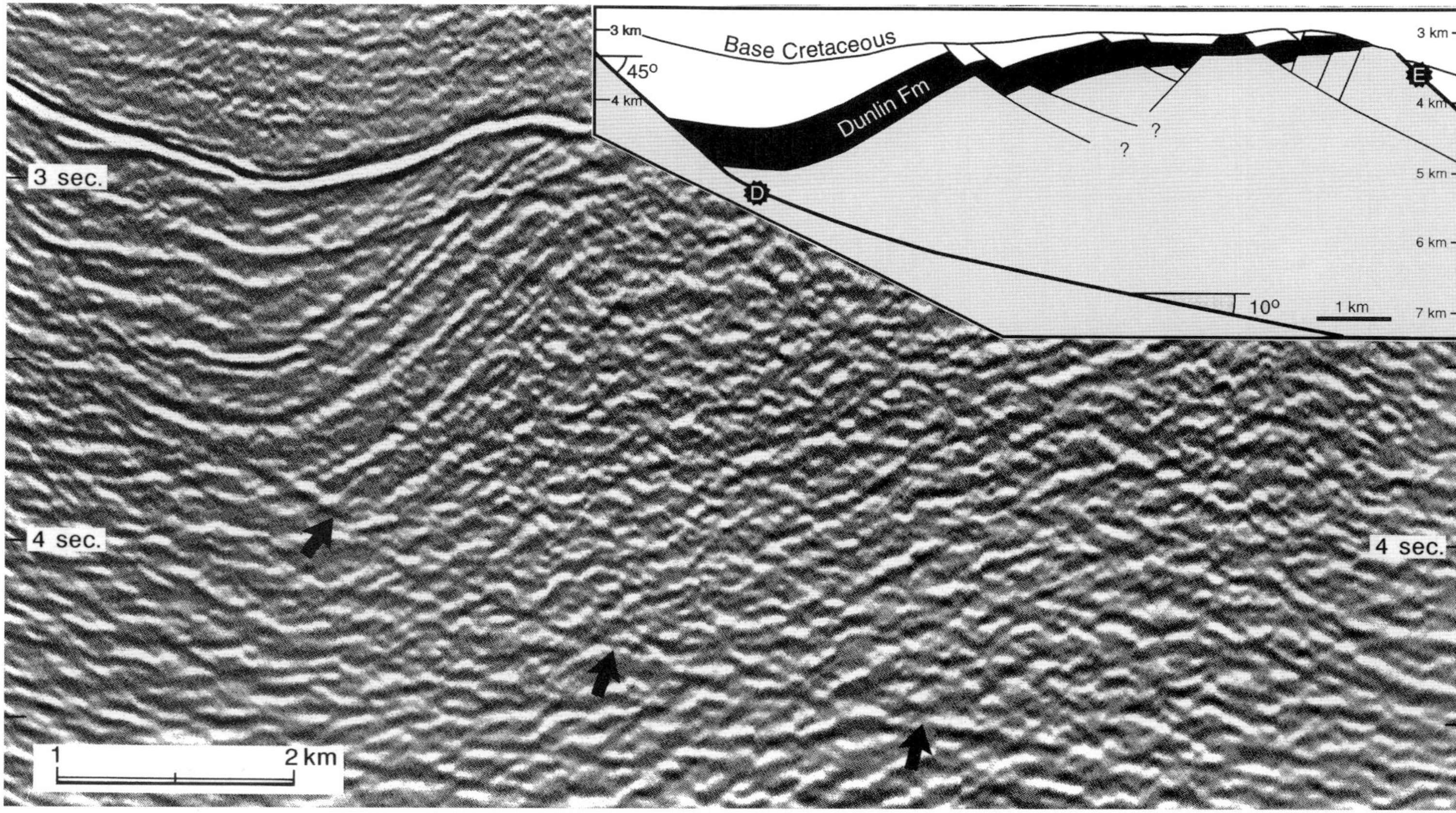

Fig. 9. Seismic profile through the Gullfaks Sør fault block, showing the downward-flattening geometry of fault D within the (Permo-?)Triassic sediments. Depth-converted version (no vertical exaggeration) is shown in the upper right corner. The low-angle part of fault D defines the Gullfaks Sør detachment.

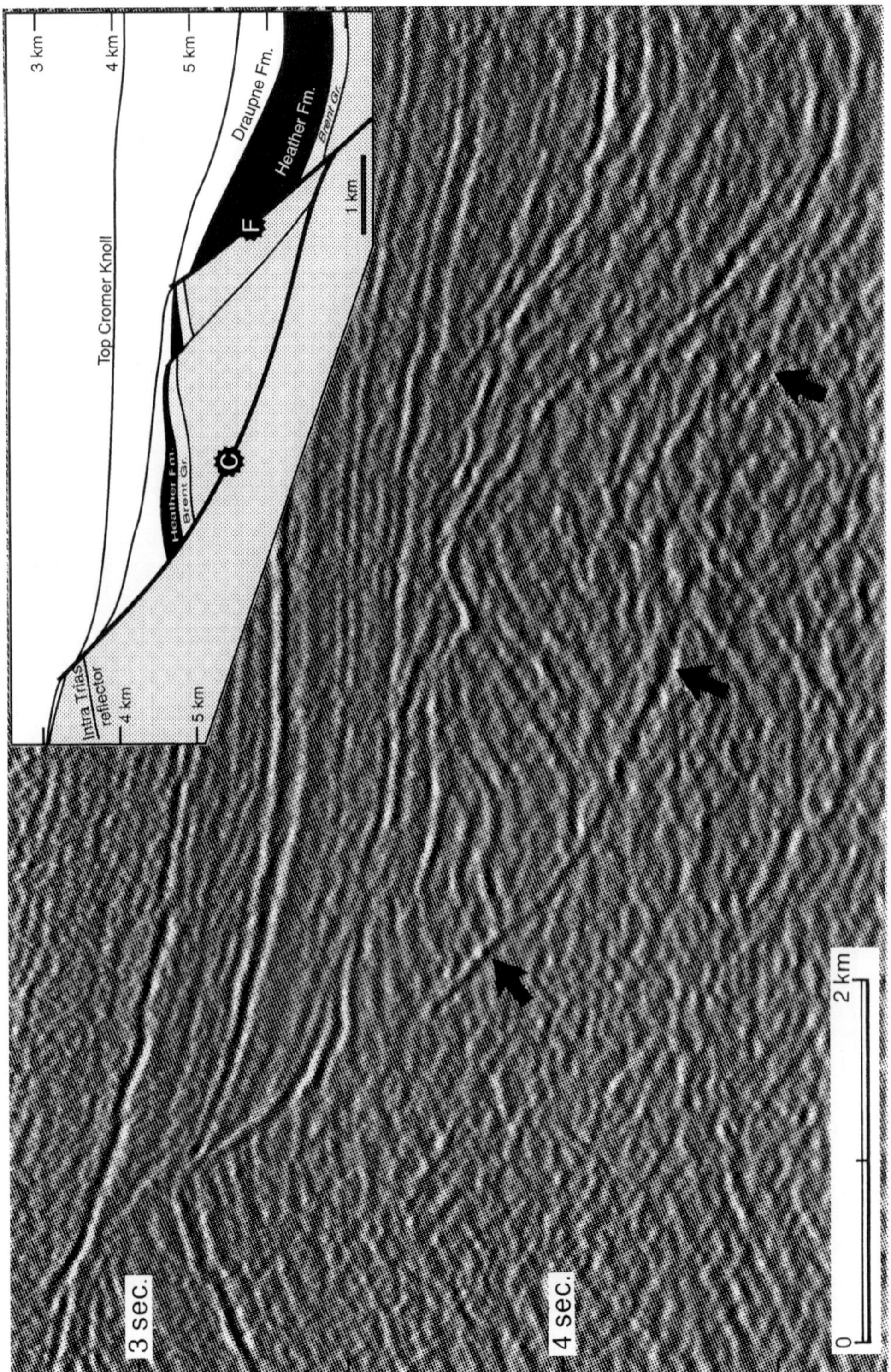

3 km
4 km
5 km
Draupne Fm.
Heather Fm.
Brent Gr.
Top Cromer Knoll
F
C
Heather Fm.
Brent Gr.
1 km
Intra Trias reflector
4 km
5 km
3 sec.
4 sec.
0
2 km

dip of 45–50°, decreasing towards 15° below the Visund sørøst fault block before merging with the steeper fault F (Fig. 10). The exact position of top basement is ambiguous from the seismic data, but from regional stratigraphic information it is likely that the fault detaches just above top basement. We therefore classify the fault as a supra-basement detachment similar to the Gullfaks detachment.

Detailed tectono-stratigraphic work by Færseth *et al.* (1995*b*) gives strong evidence that the Visund sørøst detachment (fault C) formed during latest Jurassic footwall collapse of the Visund fault block, which before this collapse was continuous with the Visund sørøst fault block.

Intra-basement detachments

The Visund detachment

The Snorre fault (B), which separates the Visund fault block from the Snorre Field, is well defined from fault-plane reflections and/or terminating reflectors in the hanging wall and footwall down to about 4 s on 2D lines (Figs 11 and 12) (see also Nelson & Lamy 1987). Below this depth the fault enters a zone of reduced reflectivity where the fault plane is not so clearly displayed. However, at lower depths (5–6.5 s) another reflector appears on the seismic line. This signal is much stronger than the surrounding reflections, and is gently east dipping (Fig. 12b) or locally sub-horizontal (Fig. 11b). It is clear from their strength and geometry that these reflections cannot be multiples from overlying events (they are overlain by a low-reflective zone below W-dipping reflections), and the fact that they generate unequivocal sea-bottom multiples (Figs 11b and c, and 12b and c) shows that they represent real physical structures in the crust. It is possible to correlate this signal from line to line, and a contoured map of the reflection is shown in Fig. 13. We suggest that the lower, strong reflection represents a low-angle fault within the basement, and that it is connected to the Snorre fault to define the Visund detachment. On some of the lines, a sub-horizontal reflection also appears to the west of the Snorre fault (e.g. Fig. 11), suggesting that the Visund detachment is part of a more extensive detachment system at about 6 s. Depth conversion of this detachment (see Fig. 15a, below) shows that the detachment remains sub-horizontal after depth conversion, and exists at a depth of about 14 km.

Gullfaks area

Interpretation of the deep seismic data across the Gullfaks Field, together with commercial 2D seismic lines, has resulted in the fault geometries shown in Fig. 14 (Odinsen *et al.* 2000*a*). The Statfjord fault (A) is here interpreted as a non-planar fault that becomes a low-angle detachment structure in the basement. A domino-style fault-block arrangement is tentatively interpreted to be situated above this detachment, similar to the Jurassic domino system above the overlying Gullfaks detachment. Hence, there are indications that two levels of detachments (a supra- and intra-basement detachment) may exist beneath the Gullfaks Field.

Evolution of the intra-basement detachments

Low-angle detachments of late to post-Caledonian age are known from both sides of the North Sea rift. The sub-horizontal décollement zone between Caledonian nappes and basement in southern Norway, the NW-dipping Hardangerfjord shear zone, and the Nordfjord–Sogn detachment are all well-exposed examples of low-angle structures accommodating substantial Devonian extension on the eastern side of the North Sea (Norton 1987; Séranne & Séguret 1987; Fossen 1992). Similarly, reactivation of Caledonian thrusts as low-angle extensional features is recognized in Scotland (McClay *et al.* 1986; White & Glasser 1987; Powell & Glendinning 1990). Some of these detachments continue under the North Sea and form zones of weakness in the crust. For mechanical reasons, these and similar low-angle structures of Caledonian or Devonian age were easily reactivated as detachments during the Permo-Triassic and Jurassic extension phases, especially in areas of high post-Devonian extension, and caused ramp–flat geometries of the type seen in Fig. 11. Reactivation of orogenic structures as low-angle extensional faults or shear zones is a common phenomenon in extended areas, such as in the western USA (Allmendinger *et al.* 1983; Coney & Harms 1984; Wernicke *et al.* 1987; Parrish *et al.* 1988; Constenius 1996). Assuming that this

Fig. 10. Seismic profile (from 3D data) across fault C (Visund sørøst detachment), which clearly appears as a downward-flattening fault on the seismic data. Depth-converted version (no vertical exaggeration) is shown in the upper right corner. (See Fig. 3 for location.)

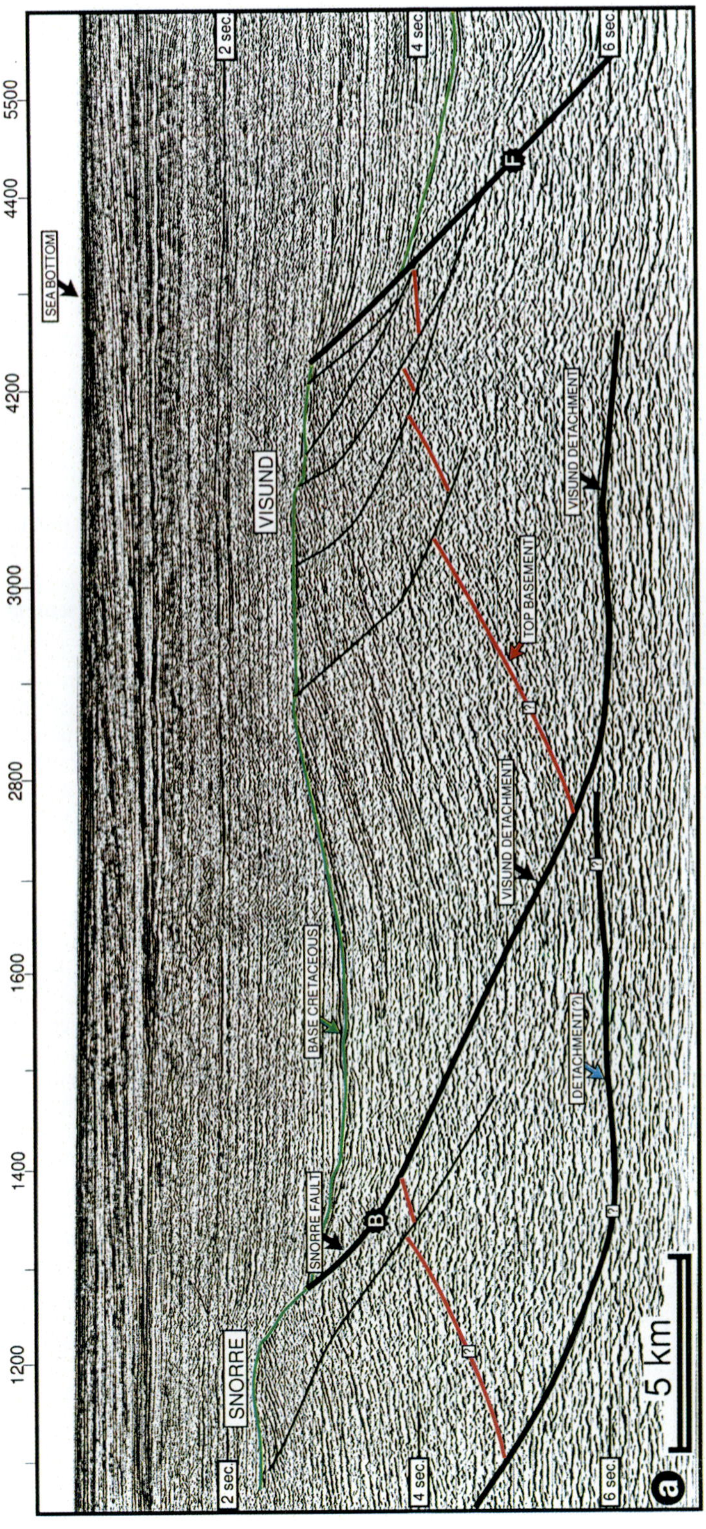
1200
1400
1600
2800
3000
4200
4400
5500
SEA BOTTOM
SNORRE
VISUND
BASE CRETACEOUS
SNORRE FAULT
B
F
TOP BASEMENT
VISUND DETACHMENT
VISUND DETACHMENT
DETACHMENT (?)
2 sec.
4 sec.
6 sec.
5 km
a

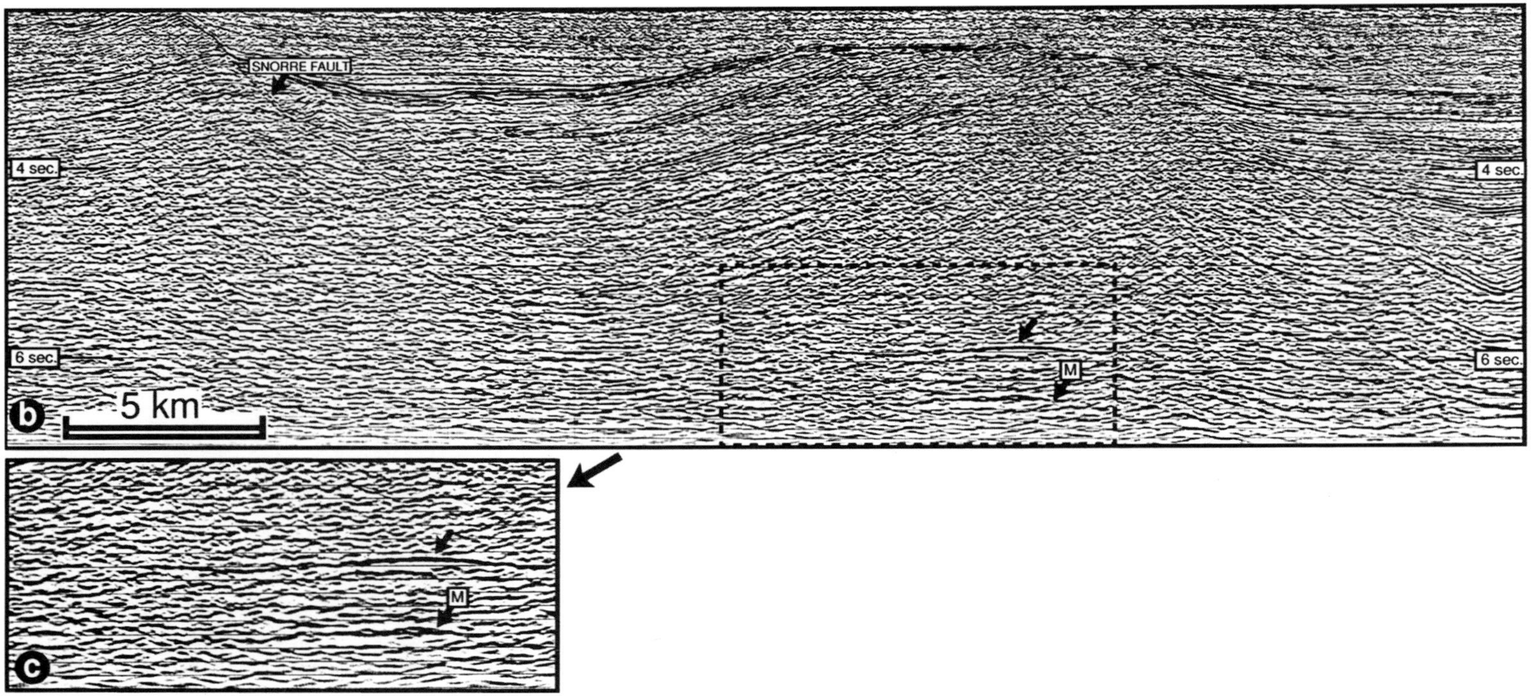

Fig. 11. (**a**) Seismic line NVGT-88-08 across the Visund fault block, showing how the Snorre fault flattens beneath the Visund fault block to form an intra-basement detachment (Visund detachment) at about 6 s. (**b**) Uninterpreted version of (**a**). The upper part of the Snorre fault is clearly visible as a downward-flattening reflection, and a sub-horizontal reflection is interpreted as a low-angle detachment. (**c**) Close-up of the strong sub-horizontal reflection and its multiple. (See Fig. 3 for location.)

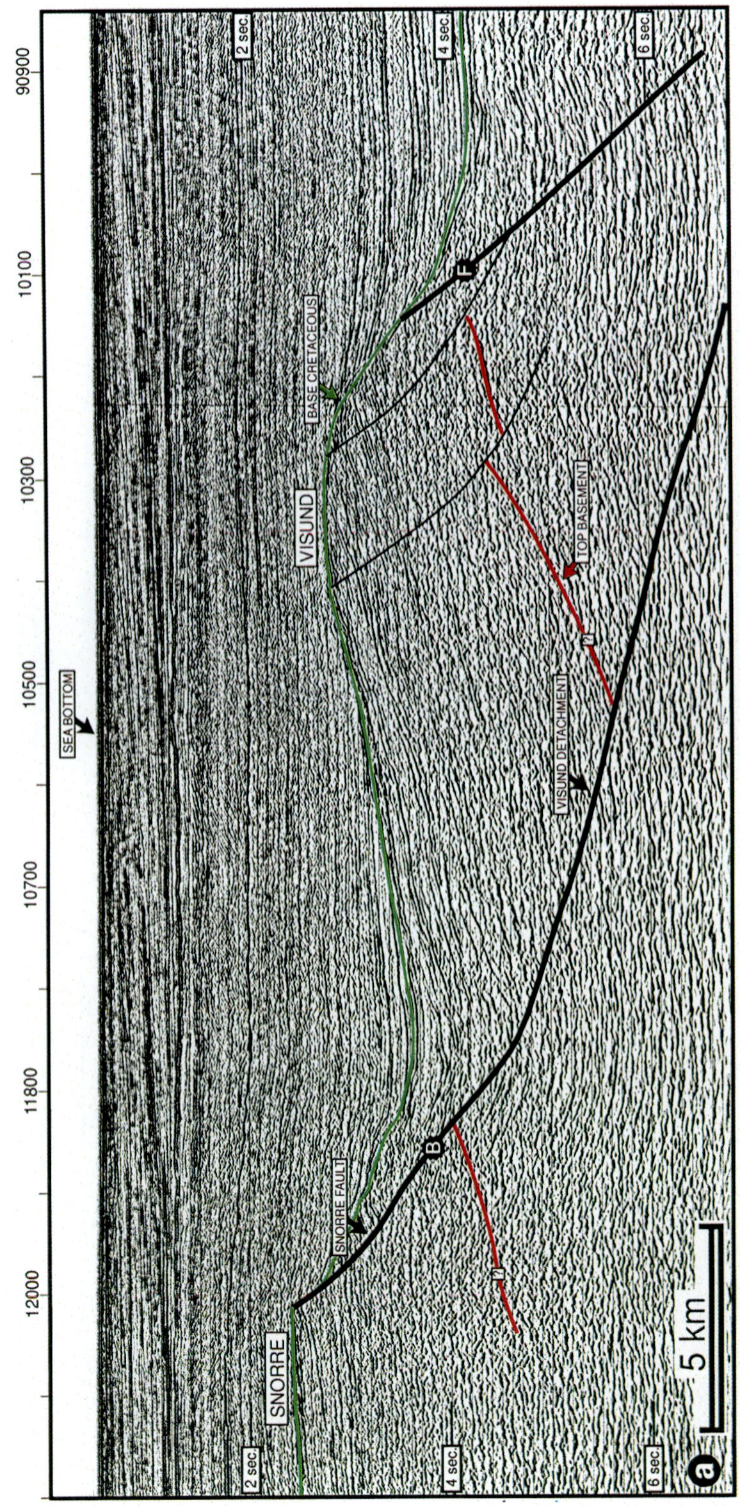
90900
10100
10300
10500
10700
11800
12000
2 sec.
4 sec.
6 sec.
SEA BOTTOM
BASE CRETACEOUS
VISUND
F
TOP BASEMENT
VISUND DETACHMENT
SNORRE FAULT
B
SNORRE
5 km
a

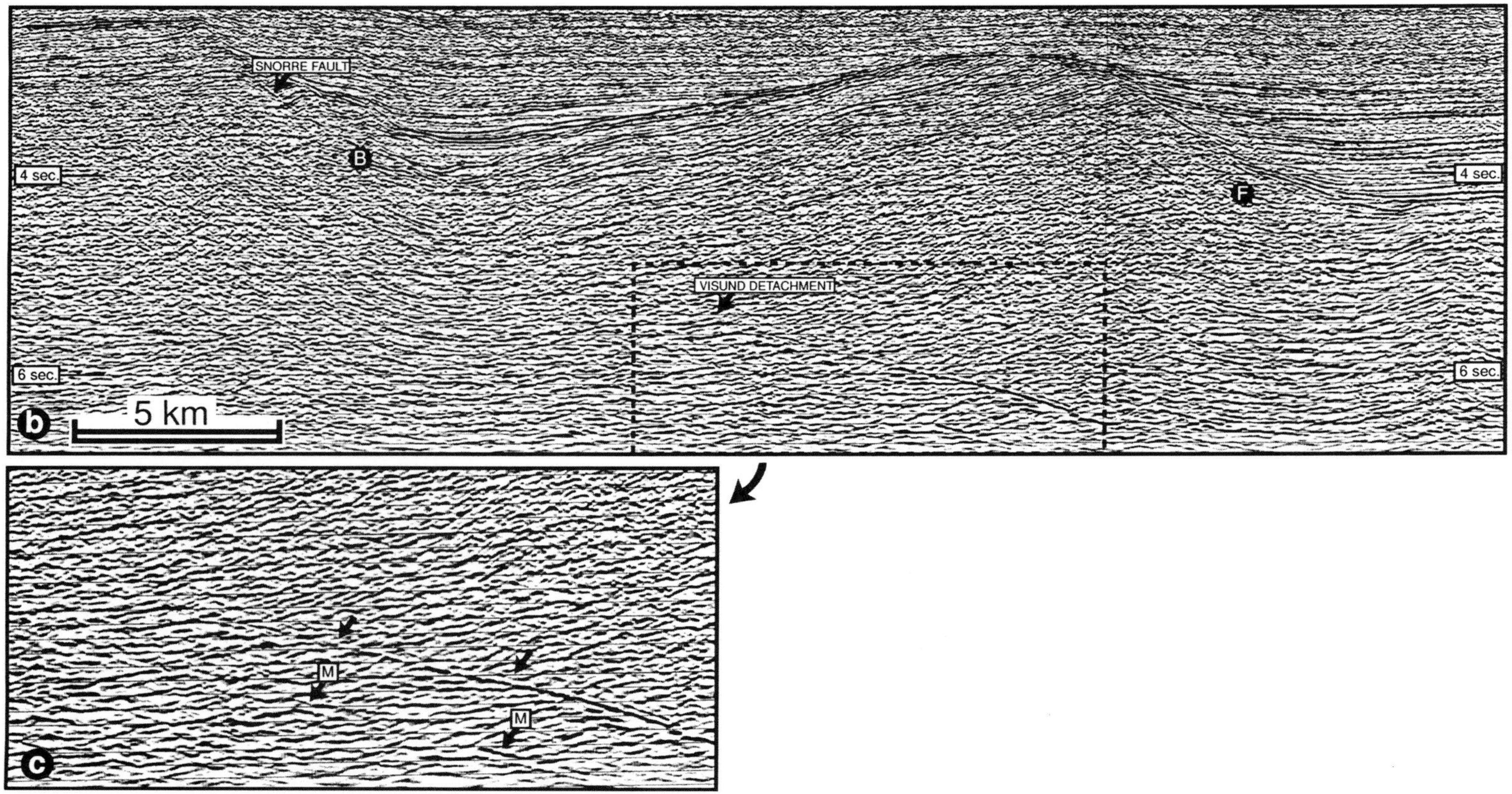

Fig. 12. (**a**) Seismic section NVGT-88-09 shows how the low-angle Visund detachment is somewhat steeper than in Fig. 11. (**b**) Uninterpreted version of (**a**). The Visund detachment is clearly visible as a multiple-generating reflection in the basement (**c**). (See Fig. 3 for location).

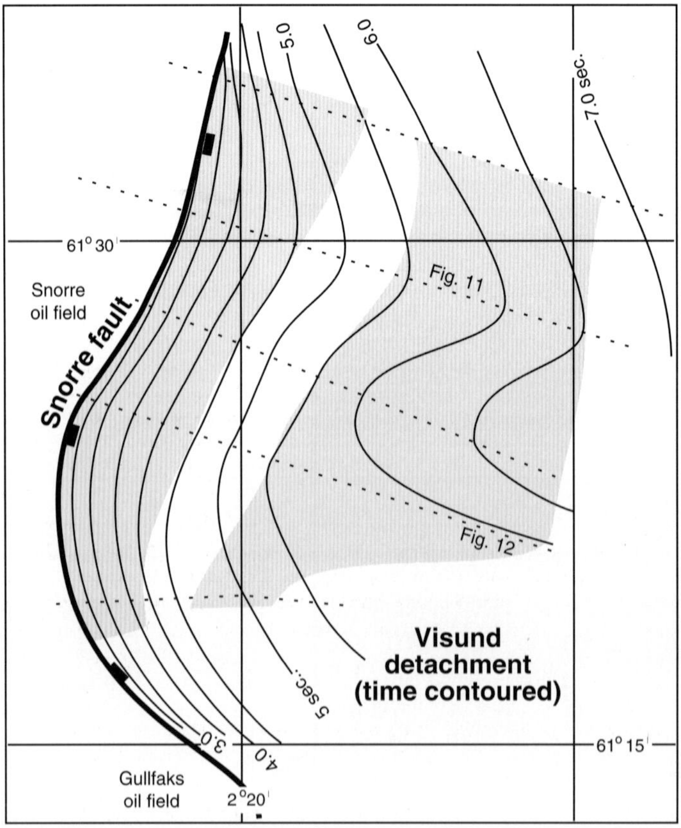

Fig. 13. Time-contoured map of the Visund detachment (Snorre fault) based on 2D seismic lines (stippled lines). Grey area indicates area where the fault-plane reflection is visible, or where terminating reflectors define the fault plane.

development also holds for the North Sea, the easterly dip of low-angle intra-basin faults or detachments in the Tampen area may indicate that the Caledonian suture zone is located east of the Tampen area, beneath the Viking Graben or close to the Permo-Triassic rift axis.

On the other hand, it is possible that the detachments were initially steeper faults that rotated to become low-angle structures during the Permo-Triassic and Jurassic extensional phases. A maximum estimate of the original (pre basin-fill) dip of the detachments is obtained from the dip of the top-basement surface in the footwall of the detachment faults. This estimate is valid only for rigid block rotation without significant internal, small-scale deformation. In the case of the Visund detachment, the top basement surface in the footwall is dipping about 25° to the west (Fig. 15a). Top basement restoration by predominantly rigid block rotations and translations would result in original dip of detachment of the same order (Fig. 15c). If additional small-scale deformation is allowed for (e.g. inclined shear, not shown in Fig. 15), original fault dips may be found to have been lower than 25°. Lowering of fault dips throughout the extensional history of the North Sea Rift is thus possible, but only to a limited extent (maximum 20–30°). Hence, faults that are currently dipping 30° to the east may have exhibited dips up to 50–60° in Permian time. It is possible that some of the faults developed into low-angle structures after repeated phases of extension, and acted as low-angle faults or detachments only during the Late Jurassic extension phase.

The above discussion relates to the model envisaged by Yielding *et al.* (1991). This model suggests that the downward shallowing of master faults in the North Sea is a result of multiple episodes of deformation. An early, Permo-Triassic phase of extension caused the formation and subsequent rotation of the faults in the

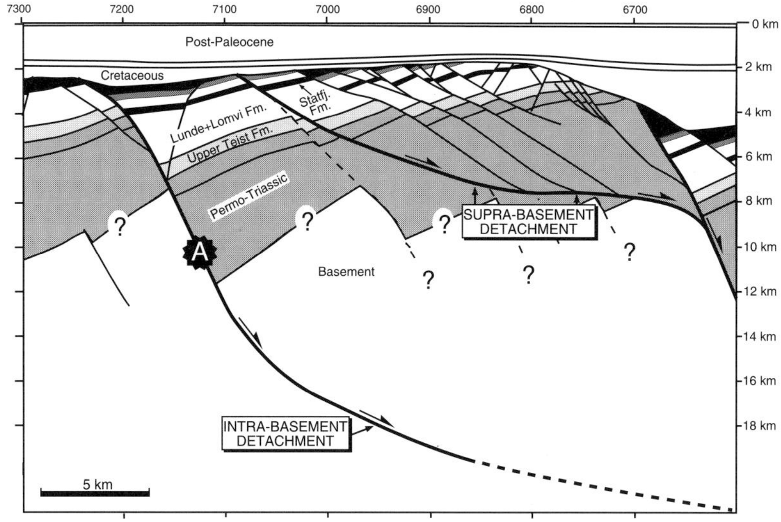

Fig. 14. Depth profile across the Gullfaks fault block, showing the interpretation of fault A as downward flattening, forming an intra-basement detachment underneath the Gullfaks supra-basement detachment. The interpretation on this line is based on a combination of line NSDP84-1 (Fig. 7), the Gullfaks 3D seismic data and well information, and conventional 2D seismic data from the area.

basement, whereas Jurassic extension caused the faults to grow into the Jurassic sequence with steeper dips. Faults with a lower (Permo-Triassic level), low-angle segment and an upper (Jurassic level), steeper segment would result from this model. However, examples such as Fig. 11, where the ramp–flat transition is located within the basement, demonstrate that the model of Yielding *et al.* (1991) cannot account for all of the fault geometries related to low-angle faults in the northern North Sea. We think that both the multiple-deformation model and the control of older (Devonian or Caledonian) shear zones in the anisotropic basement influenced the occurrence of low-angle fault movements in Mesozoic time.

Evolution of the supra-basement detachment

From the discussion above, we conclude that most of the intra-basement detachments have a history that goes back at least to the Permo-Triassic extension phase, and probably to the Devonian extensional and/or Caledonian contractional events. The supra-basement detachments are clearly of younger age, because they occur in rocks of (late) Triassic age. They are therefore likely to be partly or wholly related to the late Jurassic extension phase, as discussed below.

This is particularly clear for the Visund sørøst detachment, which has recently been shown to be a collapse structure that formed during late Jurassic rotation of the Visund fault block. This collapse occurred in the southeastern and highest portion of the original Visund mega-fault block. The initiation of the collapse can be dated to mid–late Kimmeridgian time, but the main activity is believed to have been in the Volgian period (Færseth *et al.* 1995*b*). These constraints show that supra-basement detachments formed at a relatively late stage in the late Jurassic extension phase and after the main faults were established.

The nearby Gullfaks Field is located on a bend in the Gullfaks fault block, which occurs between the southern tip of fault B and the northern termination of fault D. The characteristic 'Gullfaks bend' may be interpreted as an accommodation zone (family 3 case G′) between faults B and D according to the scheme of Rosendahl (1987) (Fig. 16). Throw variation data (Fig. 17) for faults B and D indicate a significant decrease in throw for both faults towards the apex of the 'Gullfaks bend'. This is consistent with a model in which these faults grew towards the Gullfaks field, where they finally intersected to form the Gullfaks bend accommodation zone. During this process, the Gullfaks Field area became an elevated part of the rotating Gullfaks fault block. Any elevated

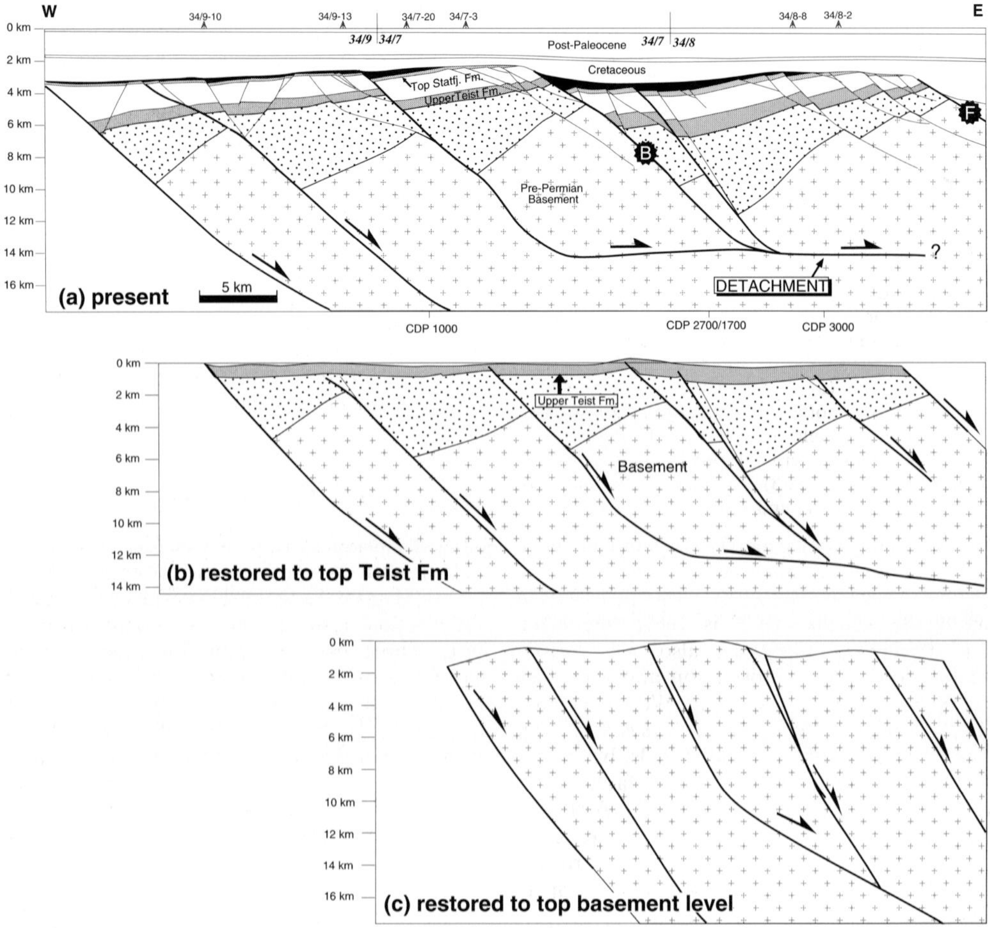

Fig. 15. (**a**) Depth-converted version of Fig. 11, showing that the detachment occurs at about 12–14 km depth. It should be noted that the change in dip of the master faults occurs well within the basement. The lin-vel depth conversion method based on well data from the Gullfaks Field is applied down to basement, where a constant velocity of 6.1 km s^{-1} is applied. (See text for discussion.) (**b**) Simple restoration to the Triassic top Teist Fm level. (**c**) Restoration of the top basement surface. (Note change in dip of the detachment fault.) The balancing performed here restores the section by making the marker horizon approximately horizontal. Rigid-body rotation and minor amounts of vertical shear are applied. Compaction-related effects are not considered. Vertical shear with no rigid body rotation gives a similar result.

part of a large fault block in a rift system is the potential object of extensional collapse, and it appears that the high portion of the Gullfaks fault block collapsed above the Gullfaks detachment. This process was probably dynamic rather than strictly sequential, so that the gradual uplift of the Gullfaks area led to formation of the Gullfaks detachment and repeated slip along the detachment as faults B and D grew.

The presence of a detachment beneath the Gullfaks Field explains several of the characteristics of the area. First, it provides a sound explanation for the presence of the domino system. All the main faults in the Gullfaks Field (i.e. in the upper plate of the Gullfaks detachment) dip to the east, antithetical to bedding, except for the marginal horst complex at its eastern edge (Fig. 6, sp. 6800). The formation of such uniformly dipping sets of faults is favoured if a dipping basement or detachment exists underneath the parallel fault system. In that case, the faults in the upper block may easily develop domino-style fault blocks with faults dipping synthetically to the detachment or basement. This has been demonstrated by extensional experiments performed with tilted

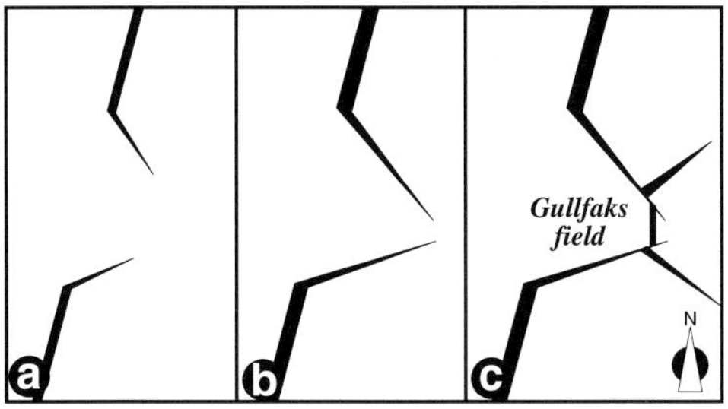

Fig. 16. Model for the development of the Gullfaks spur. The Snorre and Gullfaks Sør faults (faults D and B in Fig. 4) are thought to have defined half-grabens with a geometry similar to that described by Rosendahl (1987) (**a**). As they grew (**b**), the faults approached one another, and as they crossed (**c**), the Gullfaks spur was defined. It is possible that part of the complex fault pattern mapped to the east of the Gullfaks Field (see Fig. 3) is a result of simultaneous movement on faults D and B.

sandboxes (McClay & Ellis 1987; Vendeville *et al.* 1987) (Fig. 18), and is also characteristic for the upper-plate deformation above detachments in the western USA (e.g. Wernicke 1985; Lister & Davis 1989). Domino systems may also develop in untilted sand models, but rarely with the consistency and number of fault blocks as seen in the Gullfaks Field (Fig. 6). A similar argument can be made for the existence of a gently east-dipping detachment underneath Gullfaks Sør during development of the upper-block faults.

A Gullfaks detachment also helps explain the unusually large extensions in the Gullfaks Field (domino system) as compared with the rest of the Gullfaks fault block. Rouby *et al.* (1996) used a numerical map-view restoration method to show that the Late Jurassic E–W extension is much higher in the Gullfaks Field (*c.* 40%) than in the Gullfaks block in general (*c.* 15%). This observation fits well with a detachment model, where the upper plate can collapse and extend independently from the lower and western part of the block. In a similar way, the Gullfaks Sør block showed high extension (33%) compared with the area to the west, which can be explained by an underlying Gullfaks Sør detachment.

Insights from plaster experiments

The observations from the Gullfaks Field and physical modelling suggest that the geometries and interaction between master faults are crucial for the final result of the extensional deformation. Three basic relationships exist between supra-basement detachments and the steeper master fault associated with the basement scarp (ramp). In the first case (Fig. 19a), the detachment is older than the steep fault and is therefore offset by the latter. In the second example (Fig. 19b) the detachment is the younger structure, and accordingly cross-cuts and offsets the steep fault. Finally, if the two are active at the same time (Fig. 19c), they may join and merge into a single structure in the ramp region.

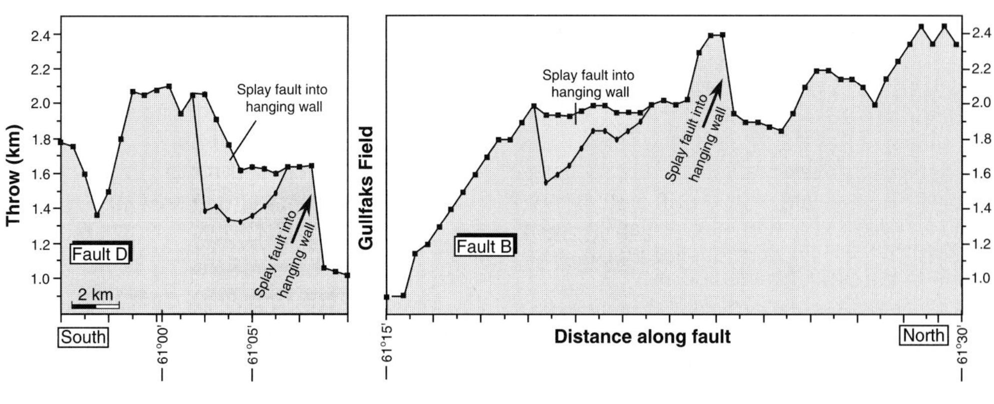

Fig. 17. Throw-variation diagrams for faults D and B (Fig. 4) at top Brent Group level. Fault D shows a maximum at about 61°00′, and a decrease in throw is recorded towards the Gullfaks Field. Similarly, fault B has maximum throw values north of the Gullfaks Field (north of 61°30′), and shows a decrease in throw to the south towards Gullfaks. These data indicate that faults D and B grew towards the Gullfaks Field, where they eventually met, as suggested in Fig. 16.

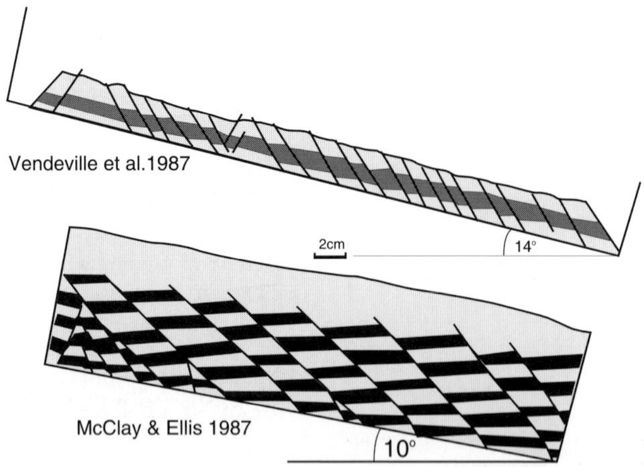

Fig. 18. Results of sandbox experiments above an inclined base. Redrawn from McClay & Ellis (1987) and Vendeville *et al.* (1987). In both cases a prominent domino-style fault pattern was produced, similar to the style seen above the Gullfaks supra-basement detachment.

However, if the displacement along the detachment is large, the upper part of the steep master fault is likely to become deactivated as it is transported towards the graben center. In this case, a new steep fault typically forms above the ramp in the footwall to the former fault (Fig. 19c).

It is possible to model supra-basement detachments and ramp–flat–ramp geometries experimentally with sand, clay or plaster. In such models, the temporal development of faulting and fault interaction can be studied in detail. Plaster experiments of the type described by Fossen & Gabrielsen (1996) have been found to be useful, particularly if wet plaster is extended together with a stronger, ductile basement of wet barite powder. In one of their experiments (run 1), a supra-basement detachment and fault-block geometries similar to the Gullfaks example were produced. During this run (Fig. 20), the plaster in the right-hand end of the deformation box moved relative to a barite basement ramp, indicating the presence of a fault or detachment between the two. Subsequently, a straight fault (fault 3 in Fig. 20) developed to the left and merged with the detachment, and a large, non-planar detachment structure formed (Fig. 20d). The resulting detachment does not cut the steep faults formed at an earlier point, but together these structures form a kinematically coherent system similar to Fig. 19c. Geometrically, this ramp bears similarities to the master fault east of the Gullfaks detachment (Fig. 6 except that the Gullfaks master fault continues downward).

In a different experiment, two steep master faults (1 and 2 in Fig. 21b) formed in a volume of wet plaster under plane strain extension. A new, listric fault (5 in Fig. 21c) grew from the free surface of the previously undeformed footwall, flattened out at about one-third of the total height of the model, and eventually cut across fault 1 to join fault 3 (Fig. 21d). Fault 1 was consequently deactivated and offset, and the resulting geometry and deformation history is an example of the case shown in Fig. 19b.

The two plaster experiments described here illustrate how two different situations can occur during a single plane strain extension history. For natural examples of supra-basement detachments, such as the Gullfaks and Gullfaks Sør detachments discussed above, the geometric relationships in the ramp zone (as illustrated in Fig. 19) should be examined. The apparent continuity of the relatively steep master fault that transects the basement and the sedimentary cover on the seismic data excludes the model shown in Fig. 19b. Although the available seismic data are not good enough to safely distinguish between Fig. 19a and c there is no indication of a continuation of the Gullfaks or Gullfaks Sør detachment on the downthrown side of the steeper master fault. Together with the similarities between the model shown in Fig. 20 and the Gullfaks Field (basement ramp, marginal horst), this supports the model shown in Fig. 19c, although the model shown in Fig. 19a was suggested by Koestler *et al.* (1992). As far as the Visund area is concerned, the age constraints discussed above (Færseth *et al.* 1995*b*) show that the Visund sørøst detachment is clearly younger than the steep master fault to the

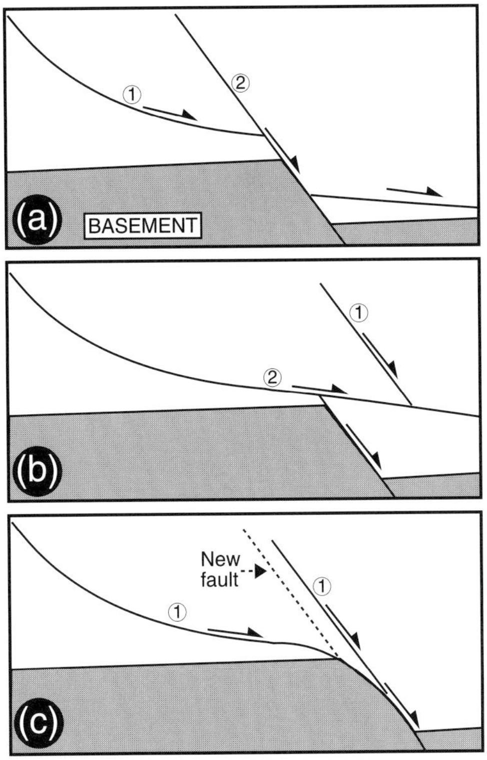

Fig. 19. The three possible relationships between detachments and steeper master faults above a basement escarpment. Numbers indicate relative timing. In (**a**) the steep fault post-dates the detachment, whereas in (**b**) the timing is opposite. In (**c**) the detachment and the steep fault are active simultaneously. If the displacement along the detachment is significant, the steep fault is transported across the ramp, and a new fault forms in its footwall (stippled). The result can be a complex fault zone above the ramp.

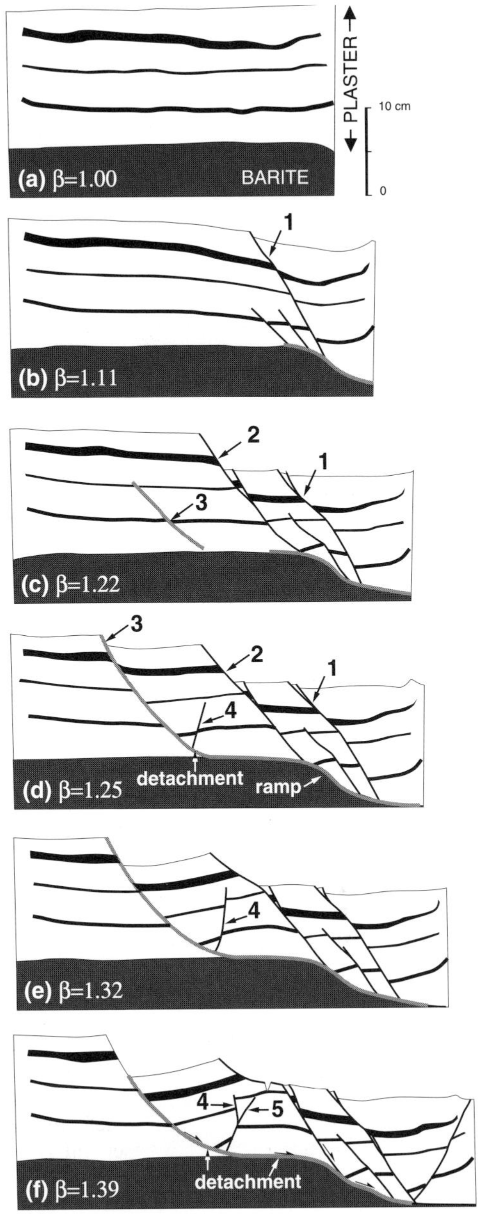

Fig. 20. Temporal development of plaster model, where the barite basement is stiffer than the overlying plaster, and therefore forms a ramp as the right-hand wall is pulled to the right. Fault 3 develops into a downward-flattening fault, which forms a detachment towards the end of the run. Modified from Fossen & Gabrielsen (1996).

east, but that they moved simultaneously in latest Jurassic time. Hence, the model shown in Fig. 19c also applies to this detachment system.

Conclusions

Some of the master faults on the western flank of the Viking Graben exhibit low dips in the basement, and acted as intra-basement detachments during the late Jurassic stretching phase. Some of these detachment faults might have had higher initial dips and rotated into less steep orientations through block rotations and internal deformation during pre-Jurassic rifting phases. In this model, faults that were originally intermediate or high angle rotated to form low-angle faults, which acted as detachments in Jurassic time. However, some faults are seen to flatten *within* the basement, indicating that they follow pre-existing E-dipping weak zones that

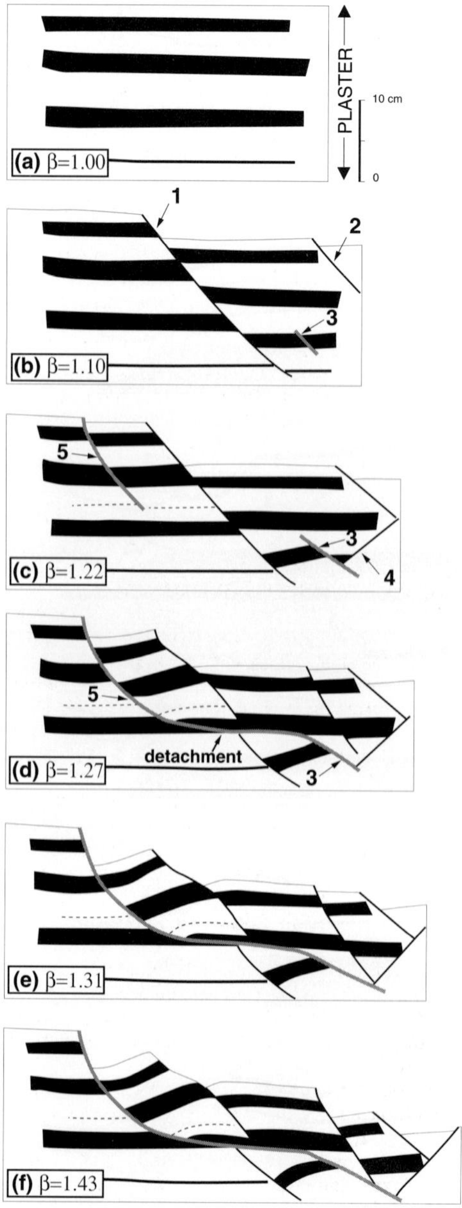

Fig. 21. Sequential development of plane strain extensional plaster model carried out at the 1992 TSGS meeting, Bergen. No basement is introduced, and a low-angle detachment develops and cuts earlier steep faults. (See Fossen & Gabrielsen (1996) for description of the experimental method.)

most likely are Devonian extensional shear zones and/or reactivated Caledonian thrusts.

Supra-basement detachment is identified in Triassic sediments beneath the Gullfaks Field, SE of the Visund fault block and underneath Gullfaks Sør. These detachments are therefore not reactivated Devonian extensional detachments or Caledonian thrusts. The development of a classical domino system on Gullfaks (consistent fault dip polarity) indicates that the detachment existed as an active, E-dipping slip surface at early stages of the late Jurassic development of the domino system. Support for this interpretation is found in experimental work, where such domino systems are found to develop more easily if the basal plate of the experiment is tilted. The Gullfaks detachment model also provides a sound explanation for the unusually high extension recorded in its hanging wall. Contemporaneous movement along the detachment and the master fault supports a model where the detachment merges with the master fault to form a single fault structure at depth (Fig. 19c), a model that certainly applies to the Visund sørøst detachment and probably also to the Gullfaks Sør detachment.

All of the three supra-basement detachments occur in the high parts of rotated, first-order fault blocks in the Triassic–Jurassic sequence of the Tampen area. This suggests that they are gravity-controlled collapse structures formed during extensional rotation of the first-order fault blocks.

Both supra- and intra-basement detachments may have a significant influence on the development of upper-crustal structures. A better mapping and understanding of these low-angle structures is important for improved understanding of the upper-crustal development of the North Sea rift system.

This paper has benefited from comments by J. Akselsen, N. Platt, M. Séranne, J. Walsh and D. Couturier.

References

Allmendinger, R. *et al.* 1983. Cenozoic and Mesozoic structure of the eastern Basin and Range province, Utah, from COCORP seismic reflection data. *Geology*, **11**, 532–536.

Armstrong, R. L. 1972. Low angle (denudation) faults, hinterland of the Sevier Orogenic Belt, Eastern Nevada and Western Utah. *Geological Society of America Bulletin*, **83**, 1729–1754.

Badley, M. E., Egeberg, T. & Nipen, O. 1984. Development of rift basins illustrated by the structural evolution of the Oseberg structure, Block 30/6, offshore Norway. *Journal of the Geological Society, London*, **141**, 639–649.

——, Price, J. D., Dahl, C. R. & Agdestein, T. 1988. The structural evolution of the northern Viking Graben and its bearing upon extensional modes of basin formation. *Journal of the Geological Society, London*, **145**, 455–472.

BEACH, A. 1984. Structural evolution of the Witch Ground Graben. *Journal of the Geological Society, London*, **141**, 621–628.

BOSWORTH, W. 1985. Geometry of propagating continental rifts. *Nature*, **316**, 625–627.

CHERRY, S. T. J. 1993. The interaction of structure and sedimentary process controlling deposition of the Upper Jurassic Brae Formation Conglomerate, Block 16/17, North Sea. *In*: PARKER, J. R. (ed.) *Petroleum Geology of Northwest Europe: Proceedings of the 4th Conference*. Geological Society, London, 387–400.

CHRISTIANSSON, P., FALEIDE, J. I. & BERGE, A. M. 1999. Crustal structure in the northern North Sea; an integrated geophysical study. *This volume*.

CLAUSEN, O. R. & KORSTGÅRD, J. A. 1996. Planar detachment faults in the southern Horn Graben, Danish North Sea. *Marine and Petroleum Geology*, **13**, 537–548.

CONEY, P. J. & HARMS, T. A. 1984. Cordilleran metamorphic core complexes: Cenozoic extensional relics of Mesozoic compression. *Geology*, **12**, 550–554.

CONSTENIUS, K. N. 1996. Late Paleogene extensional collapse of the Cordilleran foreland fold and thrust belt. *Geological Society of America Bulletin*, **108**, 20–39.

COWARD, M. P. 1986. Heterogeneous stretching, simple shear and basin development. *Earth and Planetary Science Letters*, **80**, 325–326.

FOSSEN, H. 1989. Indication of transpressional tectonics in the Gullfaks Oil Field, northern North Sea. *Marine and Petroleum Geology*, **6**, 22–30.

——1992. The role of extensional tectonics in the Caledonides of South Norway. *Journal of Structural Geology*, **14**, 1033–1046.

——1998. Advances in understanding the post-Caledonian structural evolution of the Bergen area, West Norway. *Norsk Geologisk Tidsskrift*, **78**, 33–46.

—— & GABRIELSEN, R. H. 1996. Experimental modeling of extensional fault systems. *Journal of Structural Geology*, **18**, 673–687.

—— & HESTHAMMER, J. 1998. Structural geology of the Gullfaks Field. *In:* COWARD, M. P., JOHNSON, H. & DALTABAN, T. S. (eds) *Structural Geology in Reservoir Characterization and Field Development*. Geological Society, London, Special Publications, **127**, 231–261.

—— & RYKKELID, E. 1992. Post-collisional extension of the Caledonide orogen in Scandinavia: structural expressions and tectonic significance. *Geology*, **20**, 737–740.

FÆRSETH, R. B. 1983. *Tectonic map showing: northern North Sea and west Norwegian mainland.* Norsk Hydro, a.s., Stabekk.

——, GABRIELSEN, R. H. & HURICH, C. A. 1995*a*. Infuence on basement in structuring of the North Sea basin, offshore southwest Norway. *Norsk Geologisk Tidsskrift*, **75**, 105–119.

——, SJØBLOM, T. S., STEEL, R. J., LILJEDAHL, T., SAUAR, B. E. & TJELLAND, T. 1995*b*. Tectonic controls on Bathonian–Volgian syn-rift successions on the Visund fault block, northern North Sea. *In*: STEEL, R. J. (ed.) *Sequence Stratigraphy on the Northwest European Margin*. Norwegian Petroleum Society, Special Publication, **5**, 325–346.

GABRIELSEN, R. H. 1986. Structural elements in graben systems and their influence on hydrocarbon trap types. *In*: SPENCER, A. M. *et al.* (eds) *Habitat of Hydrocarbons on the Norwegian Continental Shelf*. Graham and Trotman, London, 55–60.

——1988. Reactivation of faults on the Norwegian continental shelf and its implications for earthquake occurence. *In*: GREGERSEN, S. & BASHAM, P. (eds) *Causes and Effects of Earthquakes at Passive Margins and in Areas with Post-glacial Rebound on both Sides of the North Atlantic*. Elsevier, Amsterdam, 69–92.

——, FÆRSETH, R. B., STEEL, R. J., IDIL, S. & KLØVJAN, O. S. 1990. Architectural styles of basin fill in the northern Viking Graben. *In*: BLUNDELL, D. J. &. GIBBS, A. D. (eds) *Tectonic Evolution of the North Sea Rifts*. Clarendon, Oxford, 158–183.

GIBBS, A. D. 1984. Structural evolution of extensional basin margins. *Journal of the Geological Society, London*, **141**, 609–620.

GRAUE, K. 1992. Extensional tectonics in the northernmost North Sea: rifting, uplift, erosion and footwall collapse in Late Jurassic to Early Cretaceous times. *In*: SPENCER, A. M. (ed.) *Generation, Accumulation and Production of Europe's Hydrocarbons II*. Springer, Berlin, 23–34.

HARDING, T. P. 1984. Graben hydrocarbon occurences and structural styles. *AAPG Bulletin*, **68**, 333–362.

HARDMAN, R. F. P. & BOOTH, J. E. 1991. The significance of normal faults in the exploration and production of North Sea hydrocarbons. *In*: ROBERTS, A. M., YIELDING, G. & FREEMAN, B. (eds) *The Geometry of Normal Faults*. Geological Society, London, Special Publications, **56**, 1–13.

HARRIS, J. P. & FOWLER, R. M. 1987. Enhanced prospectivity of the mid–late Jurassic sediments of the south Viking Graben, northern North Sea. *In*: BROOKS, J. & GLENNIE, K. W. (eds) *Petroleum Geology of North West Europe*. Graham and Trotman, London, 879–898.

JOHNSON, D. 1930. Geomorphic aspects of rift valleys. *Proceedings 15th International Geological Congress*, **2**, 354–373.

KLEMPERER, S. L. 1988. Crustal thinning and nature of extension in the northern North Sea from deep seismic reflection profiling. *Tectonics*, **7**, 803–821.

KOESTLER, A. G., MILNES, A. G. & STORLI, A. 1992. Complex hanging-wall deformation above an extensional detachment – example: Gullfaks Field, northern North Sea. *In:* LARSEN, R. M., BREKKE, H., LARSEN, B. T. & TALLERAAS, E. (eds) *Tectonic Modelling and its Application to Petroleum Geology*. Norwegian Petroleum Society, Special Publication, **1**, 243–251.

KUSZNIR, N. J., MARSDEN, G. & EGAN, S. S. 1991. A flexural–cantilever simple-shear/pure-shear model of continental lithosphere extension: applications to the Jeanne d'Arc Basin, Grand Banks and Viking Graben, North Sea. *In*: ROBERTS, A. M.,

YIELDING, G. & FREEMAN, B. (eds) *The Geometry of Normal Faults*. Geological Society, London, Special Publications, **56**, 41–60.

LISTER, G. S. & DAVIS, G. A. 1989. The origin of metamorphic core complexes and detachment faults formed during Tertiary continental extension in the northern Colorado river region, USA *Journal of Structural Geology*, **11**, 65–94.

——, ETHERIDGE, M. A. & SYMONDS, P. A. 1986. Detachment faulting and the evolution of passive continental margins. *Geology*, **14**, 246–250.

MARSDEN, G., YIELDING, G., ROBERTS, A. M. & KUSZNIR, N. J. 1990. Application of a flexural cantilever simple-shear/pure shear model of continental lithosphere extension to the formation of the northern North Sea basin. *In*: BLUNDELL, D. J. & GIBBS, A. D. (eds) *Tectonic Evolution of the North Sea Rifts*. Clarendon Press, Oxford, 241–261.

MCCLAY, K. R. & ELLIS, P. G. 1987. Geometries of extensional fault systems developed in model experiments. *Geology*, **15**, 341–344.

——, NORTON, M. G., CONEY, P. & DAVIS, G. H. 1986. Collapse of the Caledonian orogen and the Old Red Sandstone. *Nature*, **323**, 147–149.

NELSON, P. H. H. & LAMY, J.-M. 1987. The Møre–West Shetlands area: a review. *In*: BROOKS, J. &. GLENNIE, K. (eds) *Petroleum Geology of North West Europe*. Graham and Trotman, London, 775–784.

NORTON, M. 1987. The Nordfjord–Sogn Detachment, W. Norway. *Norsk Geologisk Tidsskrift*, **67**, 93–106.

ODINSEN, T., CHRISTIANSSON, P., GABRIELSEN, R. H., FALEIDE, J. I. & BERGE, A. 2000*a*. The geometries and steep structures of the northern North Sea rift system. *Norsk Geologisk Tidsskrift*.

——, REEMST, P., VAN DER BEEK, P., FALEIDE, J. I. & GABRIELSEN, R. H. 2000*b*. Permo-Triassic and Jurassic extension in the northern North Sea: results from tectonostratigraphic forward modelling. *This volume*.

PARRISH, R. R., CARR, S. D. & PARKINSON, D. L. 1988. Eocene extensional tectonics and geochronology of the southern Omineca belt, British Columbia and Washington. *Tectonics*, **7**, 181–212.

PETTERSON, O., STORLI, A., LJOSLAND, E. & MASSIE, I. 1990. The Gullfaks Field: geology and reservoir development. *In*: *North Sea Oil and Gas Reservoirs – II*. Graham and Trotman, London, 67–90.

PLATT, N. H. 1995. Structure and tectonics of the northern North Sea: new insights from deep penetration regional seismic data. *In*: LAMBIASE, J. J. (ed.) *Hydrocarbon Habitat in Rift Basins*. Geological Society, London, Special Publications, **80**, 103–113.

POWELL, D. & GLENDINNING, N. R. W. 1990. Late Caledonian extensional reactivation of a ductile thrust in NW Scotland. *Journal of the Geological Society, London*, **147**, 979–987.

ROBERTS, A. M., YIELDING, G. & BADLEY, M. E. 1990. A kinematic model for the orthogonal opening of the late Jurassic North Sea rift system, Denmark–Mid Norway. *In*: BLUNDELL, D. J. & GIBBS, A. D. (eds) *Tectonic Evolution of the North Sea Rifts*. Clarendon, Oxford, 180–199.

——, ——, KUSZNIR, N. J., WALKER, I. & DORN-LOPEZ, D. 1993. Mesozoic extension in the North Sea: constraints from flexural backstripping, forward modelling and fault populations. *In*: PARKER, J. R. (ed.) *Petroleum Geology of Northwest Europe. Proceedings of the 4th Conference*. Geological Society, London, 1123–1136.

——, ——, ——, —— & ——1995. Quantitative analysis of Triassic extension in the northern Viking Graben. *Journal of the Geological Society, London*, **152**, 15–26.

ROBSON, D. A. 1971. The structure of the Gulf of Suez (Clysmic) rift with special reference to the eastern side. *Journal of the Geological Society, London*, **127**, 247–276.

ROSENDAHL, B. R. 1987. Architecture of continental rifts with special reference to East Africa. *Annual Review of Earth and Planetary Sciences*, **15**, 445–503.

ROUBY, D., FOSSEN, H. & COBBOLD, P. R. 1996. Extension, displacements and block rotations in the larger Gullfaks area, northern North Sea, as determined from plan view restoration. *AAPG Bulletin*, **80**, 875–890.

SCOTT, D. L. & ROSENDAHL, B. R. 1989. North Viking Graben: an East African perspective. *AAPG Bulletin*, **73**, 155–165.

SÉRANNE, M. & SÉGURET, M. 1987. The Devonian basins of western Norway: tectonics and kinematics of an extending crust. *In*: COWARD, M. P., DEWEY, J. F. & HANCOCK, P. L. (eds) *Continental Extensional Tectonics*. Geological Society, London, Special Publications, **28**, 537–548.

SPEKSNIJDER, A. 1987. The structural configuration of Cormorant Block IV in context of the northern Viking Graben structural framework. *Geologie en Mijnbouw*, **65**, 357–379.

STEEL, R. & RYSETH, A. 1990. The Triassic–Early Jurassic succession in the northern North Sea: megasequence stratigraphy and intra-Triassic tectonics. *In*: HARDMAN, R. F. P. & BROOKS, J. (eds) *Tectonic Events Responsible for Britain's Oil and Gas Reserves*. Geological Society, London, Special Publications, **55**, 139–168.

SWALLOW, J. L. 1986. The seismic expression of a low angle detachment (sole fault) from the Beryl Embayment, central Viking Graben. *Scottish Journal of Geology*, **22**, 315–324.

THOMAS, D. W. & COWARD, M. P. 1996. Mesozoic regional tectonics and South Viking Graben formation: evidence for localized thin-skinned detachments during rift development and inversion. *Marine and Petroleum Geology*, **13**, 149–177.

VENDEVILLE, B., COBBOLD, P. R., DAVY, P., BRUN, J. P. & CHOUKROUNE, P. 1987. Physical models of extensional tectonics at various scales. *In*: COWARD, M. P., DEWEY, J. F. & HANCOCK, P. L. (eds) *Continental Extensional Tectonics*. Geological Society, London, Special Publications, **28**, 95–107.

WERNICKE, B. 1985. Uniform-sense normal simple-shear of the continental lithosphere. *Canadian Journal of Earth Science*, **22**, 108–125.

—— & BURCHFIEL, B. C. 1982. Modes of extensional tectonics. *Journal of Structural Geology*, **4**, 105–115.

——, CHRISTIANSEN, R. L., ENGLAND, P. C. & SONDER, L. J. 1987. Tectonomagmatic evolution of Cenozoic extension in the North American Cordillera. *In*: COWARD, M. P., DEWEY, J. F. & HANCOCK, P. L. (eds) *Continental Extensional Tectonics*. Geological Society, London, Special Publications, **28**, 203–221.

WHITE, S. H. & GLASSER, J. 1987. The outer Hebrides Fault Zone: evidence for normal movements. *In*: PARK, R. G. & TARNEY, J. (eds) *Evolution of the Lewisian and Comparable Precambrian High Grade Terrains*. Geological Society, London, Special Publications, **27**, 175–183.

ZIEGLER, P. A. 1990. Tectonic and palaeogeographic development of the North Sea rift system. *In*: BLUNDELL, D. J. & GIBBS, A. D. (eds) *Tectonic Evolution of the North Sea Rifts*. Clarendon, Oxford, 1–36.

YIELDING, G., BADLEY, M. E. & FREEMAN, B. 1991. Seismic reflections from normal faults in the North Sea. *In*: ROBERTS, A. M., YIELDING, G. & FREEMAN, B. (eds) *The Geometry of Normal Faults*. Geological Society, London, Special Publications, **56**, 79–89.

Syn-rift sedimentary architectures in the Northern North Sea

R. RAVNÅS,[1] A. NØTTVEDT,[2] R. J. STEEL[3] & J. WINDELSTAD[4]

[1] *Norske Conoco A.S, P.O. Box 488, 4001 Stavanger, Norway*
[2] *Norsk Hydro Research Centre, Sandsliveien 90, Postboks 646, 5001 Bergen, Norway (Present address: Norsk Hydro Canada, 1190 111-5th Avenue SW, Calgary, Alberta T2P 3YZ, Canada)*
[3] *University of Wyoming, Department of Geology & Geophysics, Laramie, Wyoming 82071-3355, USA*
[4] *Statoil, LTEK, P.O. Box 300, 4001 Stavanger, Norway*

Abstract: From Permian to Jurassic times the northern Viking Graben and adjacent platform areas experienced multiple rifting, the Permian–early Triassic and middle–late Jurassic rift episodes, separated by an intervening middle Triassic–middle Jurassic inter-rift period dominated by relative tectonic quiescence. The associated syn- and inter-rift strata show large variations in sedimentary architecture as a result of temporal and spatial variations in tectonic deformation and subsidence, sediment supply, climate and accommodation creation.

The Permian–early Triassic syn-rift succession is believed to consist predominantly of non-marine, arid to semiarid, aeolian, sabkha, alluvial and lacustrine strata, probably interbedded with marine strata on the Horda Platform and in the Viking Graben.

The middle Triassic–middle Jurassic experienced several subsidence stages which, together with climatic variations, exerted a major control on the periodic outbuilding and retreat of rift marginal, alluvial and shallow marine clastic wedges. Evidence for fault block rotation suggests that the subsidence was caused partly by minor extensional stages. As such, the middle Triassic–middle Jurassic does not fit the type of development assumed to be typical for either post- or pre-rift basins. Hence, the notation *inter-rift* is assigned to this period and the associated succession.

The middle–late Jurassic rift episode was characterized by multiple rift phases separated by intervening stages of relative tectonic quiescence. The syn-rift infill is mixed non-marine and marine and consists of fluvial through shallow marine and shelfal deposits to deeper marine sediment gravity flow and (hemi-)pelagic strata. At the larger scale, related to the entire middle–late Jurassic rift episode, the syn-rift infill in general shows a two-fold sandstone–mudstone lithology motif, typical of underfilled rift basins. At the intermediate scale, related to single rift phases, threefold sandstone–mudstone–sandstone, twofold sandstone–mudstone and single mudstone lithology motifs are present, typical of sediment overfilled/sediment balanced, sediment underfilled and sediment starved rift basins, respectively. The spatial and temporal variations in the syn-rift infill reflect relative distance to the rift basin hinterland areas (which had a large sediment yield potential) and overall increased tectonic subsidence and enhanced rift topography as the rift basin evolved.

This suggest that the tectonostratigraphic evolution of the northern North Sea rift basin can be viewed at several scales: at the largest scale the rift basin evolved through multiple rift episodes, which commonly had a duration of several tens of Ma. The rift episodes are separated by inter-rift periods. Rift episodes are subdivided into intervals representing distinct rift phases. These rift phases were separeted by tectonic relatively quieter intervals, here referred to as tectonic quiescence stages. Inter-rift periods are subdivided into prolonged tectonic quiescence intervals separated by short-lived rift stages or minor rift phases. Distinct rift phases and inter-rift tectonic quiescence intervals commonly represent periods of a few to 10+ Ma, and correspond to second-order sequences or 'megasequences'. At the smaller scale, syn-rift successions can be subdivided into packages related to distinct rotational tilt event or faulting events (deformation spans), representing hundreds of ka to few Ma and corresponding to third-order sequences. Solitary, large-magnitude faulting events (deformation clines) are likely to exert a major control on high frequency base- or sea-level fluctuations and thus on the development of higher-order sequences. However, such a control is difficult to prove and can probably only be recognized in sub-basins with abundant wells and a dense well spacing.

From: NØTTVEDT, A. *et al.* (eds) *Dynamics of the Norwegian Margin*. Geological Society, London, Special Publications, **167**, 133–177. 1-86239-056-8/00/$15.00

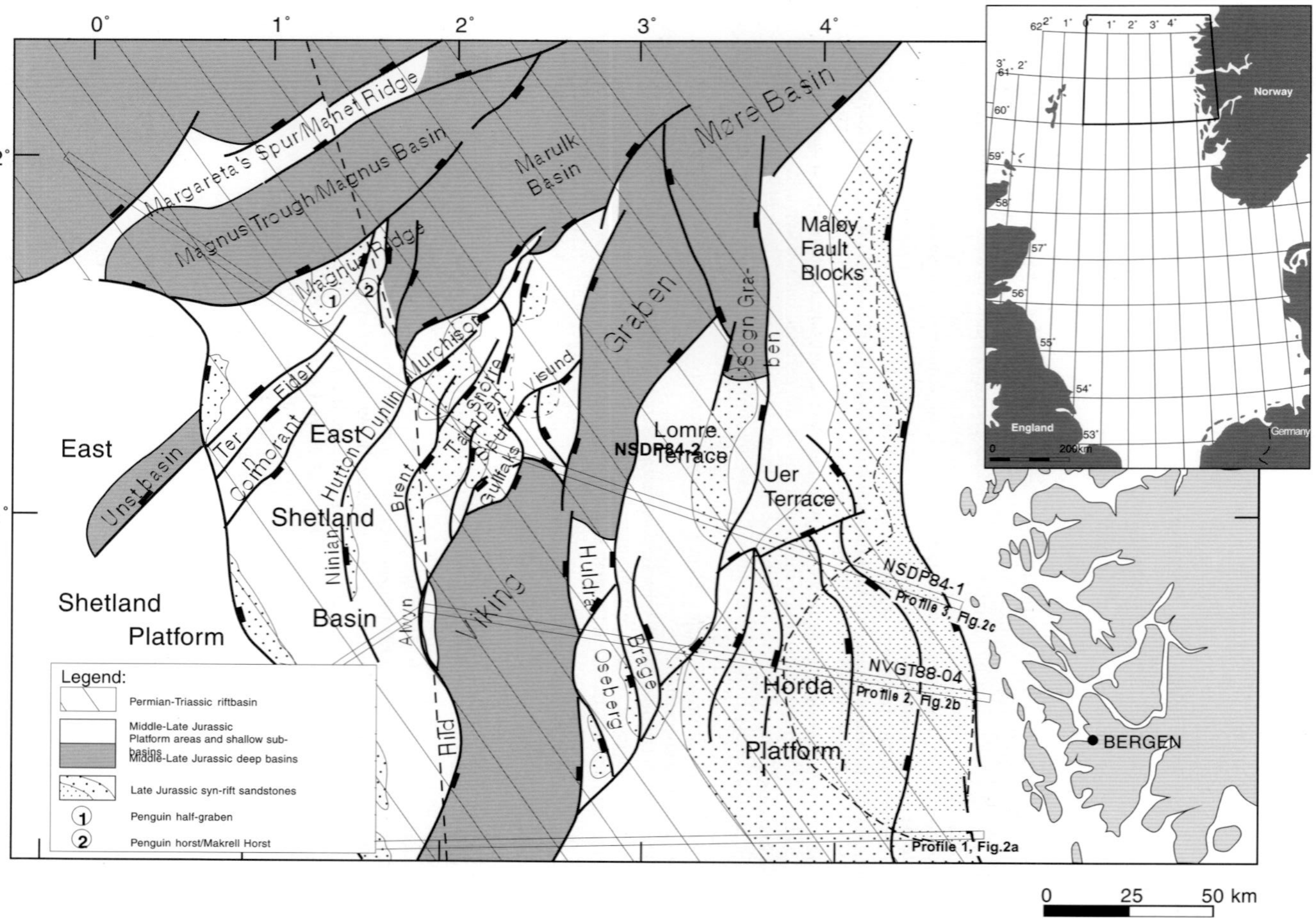

Fig. 1. Map of northern North Sea with location of seismic sections in Fig. 2 and study areas. Note different areal extent of the Permian–early Triassic and the middle Jurassic–early Cretaceous rift basins.

During recent years, considerable effort has been put into recognizing the architectural signature diagnostic of syn-rift basin-fills in order to develop predictive models for syn-rift plays. Distinct stratigraphical signatures of contrasting types of rift basins, non-marine as well as marine, have been documented or conceptualized (e.g. Surlyk 1978, 1989; Alexander & Leeder 1986; Frostick & Reid 1987; Leeder & Gawthorpe 1987; Hamblin & Rust 1989; Morley 1989; Mack & Seager 1990; Schlische & Olsen 1990; Lambiase 1990; Prosser 1993). Although attempts have been made to erect unifying syn-rift infill models, there is growing consensus that a variety of syn-rift infill patterns may be present. Their signature depends on the interplay between creation of syn-rift accommodation, climate, type and volume of sediment supply, and relative base- or sea-level stand and their changes (Leeder & Gawthorpe 1987; Surlyk 1989; Prosser 1993; Leeder 1995; Ravnås & Steel 1998).

The present contribution describes a series of syn-rift infill patterns observed in Permian–Triassic and Jurassic sub-basins of the North Sea (Fig. 1). We have focused on half-graben type sub-basins in areas underlying and next to three northern North Sea regional seismic lines (Fig. 2). The structural interpretations of these lines are discussed in detail by Odinsen *et al.* (in press) and by ter Voorde *et al.* (in press). They illustrate different styles of rift basin architecture: a southern segment (Fig. 2a), which forms a solitary, wide half-graben with flanking, broad platforms; a central segment (Fig. 2b), which is represented by a full-graben with flanking fault blocks and terraces, and bordering platform areas; and a northern segment (Fig. 2c), which consists of a series of half-graben sub-basins.

The intention of the present contribution is threefold: first, the significant variability present in non-marine and, in particular, marine syn-rift infills of regionally subsiding rift terrains is described. Secondly, the main controlling factors on syn-rift sedimentary architecture for specific fault blocks are discussed. Emphasis is put on the effects that Permian–Triassic and Jurassic extensional tectonics had upon the distribution of lithofacies and stratigraphic surfaces such as hiatuses, unconformities and condensed sections on a half-graben scale. This discussion illustrate how specific syn-rift signatures, as recognized from wells and seismic data, can be used to date activity on extensional faults and to separate distinct tectonic events and discrete rift phases. Lastly, the detailed temporal and spatial evolution of the syn-rift sedimentary architecture as well as of the larger-scale infill trends in the basin-history are discussed. This discussion aim to demonstrate who a detailed interpretation of the syn-rift sedimentary architecture can be used to further decipher the structural evolution of the North Sea rift basin.

The database from the northern North Sea is wide ranging, and consist of 2D and 3D seismic coverage, and abundant well data. The studied sections have been cored extensively in all areas, which, in addition to detailed biostratigraphy, have allowed interpretation of sedimentary processes and depositional environments, as well as the recognition and dating of key stratigraphical surfaces. This has, in turn, allowed the detailed dating of the sequential evolution of the rift basin's depositonal and structural history, and resulted in recognition of a series of tectonic events from the Triassic throughout the Jurassic. More data exist on syn-rift strata and their sedimentary architectures in the upper Triassic–Jurassic, and especially from the middle–upper Jurassic, which thus has received a more thorough attention. A generalized litho-stratigraphical column of the Triassic–Jurassic section in the northernmost North Sea and associated tectonic events are presented in Fig. 3.

Geological setting

The northernmost North Sea rift basin, i.e. the Viking Graben and adjacent platform areas, were moulded by at least two episodes of lithospheric stretching, dated as ?late Permian–early Triassic and middle Jurassic–earliest Cretaceous (e.g. Eynon 1981; Badley *et al.* 1984, 1988; Beach *et al.* 1987; Giltner 1987; Lervik *et al.* 1989; Blundell & Gibbs 1990; Gabrielsen *et al.* 1990; Ziegler 1990*a*, *b*; Yielding *et al.* 1992). The Permian–early Triassic and middle Jurassic–early Cretaceous rift episodes were characterized by distinct, but different structural configurations (Fig. 1; Lervik *et al.* 1989; Færseth 1996). The tectonic extension in the northern Viking Graben during these stretching episodes is known to have varied both spatially and temporarily (e.g. Roberts *et al.* 1993*a*, 1995; Odinsen *et al.* in press), leading to considerable variability in architecture and composition of the syn-rift infill. Moreover, as a result of the multiphase extension and crustal thinning, isostatic rebound caused a gradual lowering of the surface relief from mainly continental during Permian–early Triassic rifting to essentially marine during late Jurassic–early Cretaceous rifting (Nøttvedt *et al.* 1995).

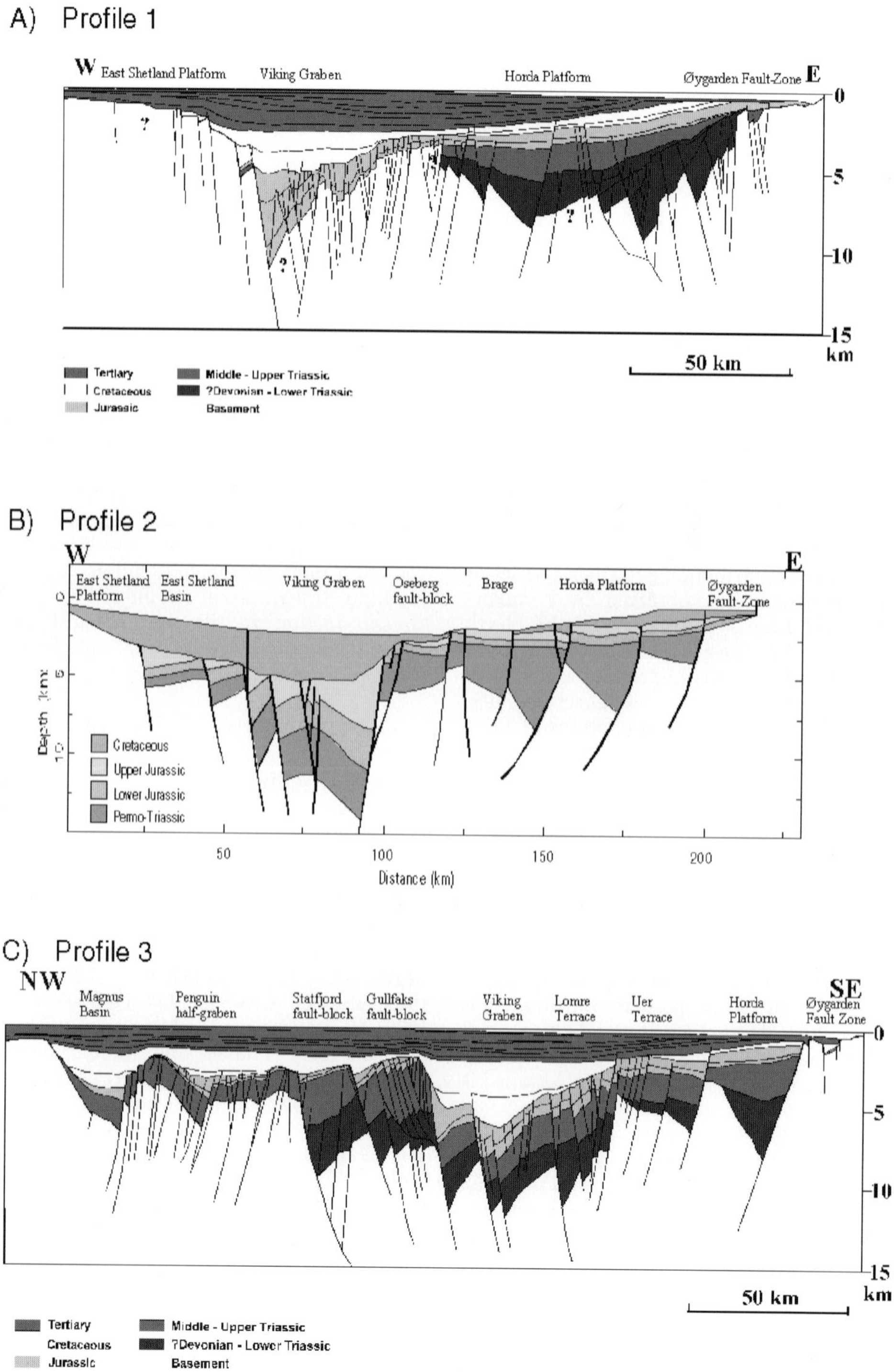

Fig. 2. Depth-converted interpretations of seismic lines (a) NSDP84-2, (b) NVGT88-04 and (c) NSDP84-1. Note difference in structural geometry from solitary wide half-graben in (a) through a full-graben in (b) to a series of half-graben sub-basins in (c). See Fig. 1 for location of seismic sections. Profiles 1 and 3 (a and c, respectively) are from Odinsen *et al.* (in press) whereas profile 2(b) is from Ter Voorde (in press).

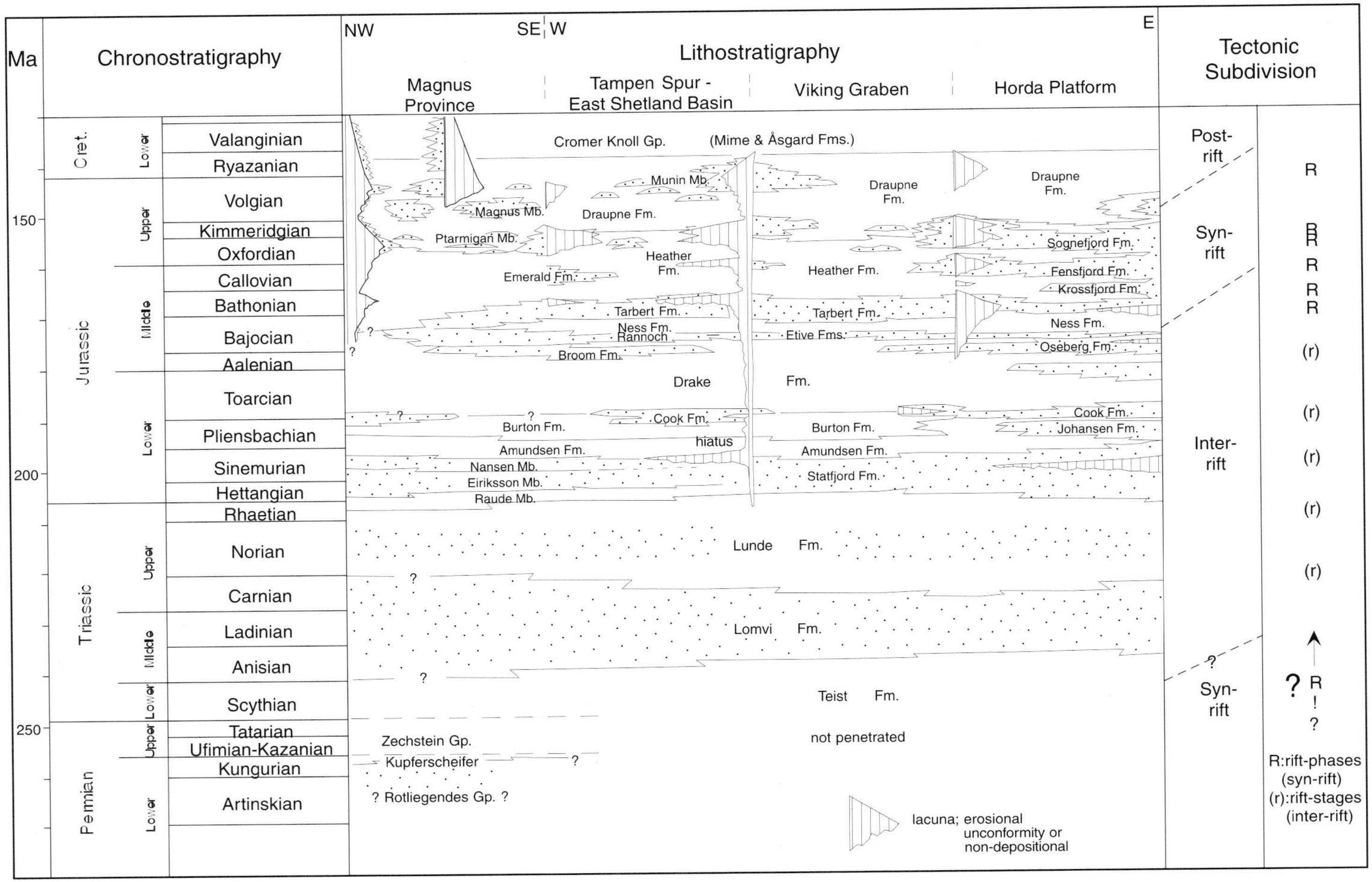

Fig. 3. Triassic–Jurassic lithostratigraphic nomenclature and tectonic subdivisions, i.e. rift episodes, inter-rift periods and rift phases, for the northernmost North sea. The lithostratigraphy is modified from Vollset & Dore (1984) and Richards *et al.* (1993).

The Permian–early Triassic rift basin extends westwards from the Norwegian coast and includes the major part of the East Shetland Platform and East Shetland Basin (Johns & Andrews 1985; Lervik *et al.* 1989; Roberts *et al.* 1995). It consists of a series of half-grabens with master-fault spacing of *c.* 15–35 km. (Badley *et al.* 1984, 1988; Lervik *et al.* 1989; Gabrielsen *et al.* 1990; Marsden *et al.* 1990). Wedge-shaped packages that are believed to contain mainly non-marine strata fill in the half-grabens (Lervik *et al.* 1989; Steel & Ryseth 1990; Steel 1993). The top of the syn-rift sequence has been argued to be within the Teist Formation and assigned a Scythian age (Fig. 3; Steel & Ryseth 1990). By analogy with East Greenland, the West Shetland basins, the southern North Sea and the Oslo Graben (e.g. Surlyk *et al.* 1984; Glennie 1990*a*, *b*; Booth *et al.* 1993; Olaussen *et al.* 1994), it seems reasonable to assume that the lower part of the syn-rift infill may date back to the late and possibly early Permian (see also Johns & Andrews 1985; Lervik *et al.* 1989; Gabrielsen *et al.* 1990; Steel 1993; Færseth *et al.* 1995a). This notion is supported by the presence of Permian dykes (Færseth *et al.* 1976; Færseth 1978; Furnes *et al.* 1982) and Permian extensional reactivation of faults of the Nordfjord–Sogn detachment onshore western Norway (Torsvik *et al.* 1992). Recent work by Christiansson *et al.* (2000) suggests that the Permian–early Triasic syn-rift succession is underlain by yet another sedimentary package, possibly early Permian and/or Carboniferous in age, and similar to what has been described below the East Shetland Platform (Platt 1995), and in the Oslo graben (e.g. Olaussen *et al.* 1994).

The overlying middle Triassic–middle Jurassic succession has generally been assigned a post-rift status, following Permian–early Triassic stretching. However, the middle Triassic–middle Jurassic basin development was not accompanied by simple, exponentially decreasing subsidence rate as a result of thermal relaxation (e.g. see McKenzie 1978). Instead, the interval was characterized by distinct phases of increased basinal subsidence, probably taken up by differential movement across rift marginal and certain intrabasinal lineaments (Steel & Ryseth 1990; Steel 1993). These variations in basinal subsidence rates and accompanying variations in sediment supply controlled the outbuilding and retreat of five large-scale, regressive-transgressive alluvial and shallow marine clastic wedges from the Norwegian and Scottish hinterlands. It is argued below, based on the sedimentary architecture of some of these clastic wedges, that gentle fault block tilting accompanied at least some of the intervening intervals with high basinal subsidence rates. Thus, minor extensional stages apparently occurred repeatedly also throughout the middle Triassic–middle Jurassic (Færseth & Ravnås 1998).

The Jurassic rift episode affected the Viking Graben from the Bathonian to the Ryazanian. Onset of block tilting was not synchronous throughout the basin, and is argued to have commenced as late as Kimmeridgian on the bordering Horda Platform (Badley *et al.* 1988). A diagnostic feature of the middle Jurassic–earliest Cretaceous evolution of the northern North Sea, is that the tilted fault block sub-basins appear to have developed through a series of distinct tectonic phases (Underhill 1991*a*, *b*; Stewart *et al.* 1992; Partington *et al.* 1993; Rattey & Hayward 1993; Færseth *et al.* 1995*b*; Færseth & Ravnås 1998). Dating and chronostratigraphical resolution is still too poor to determine whether there was a synchroneity or diachronism of tectonic maxima and minima, though there are indications of a slightly diachronous rifting acme. It is also worth noting that the intensity of the different rift phases is observed to vary spatially across the rift basin (see below). The middle Jurassic–early Cretaceous stretching resulted in characteristic wedge-shaped packages which only partially infill the fault block topography/bathymetry. In contrast to the Permian–early Triassic syn-rift basin-fill, the sediment infill of the middle Jurassic–earliest Cretaceous rift basins was essentially of marine nature. Deltaic and paralic sediments were deposited in rift marginal areas, peripheral to major intrabasinal highs and in some platform areas, whereas the more axial and deeper parts of the rift system received mainly (hemi-)pelagic clays and interbedded sediment gravity flow deposits (e.g. Færseth & Pederstad 1988; Steel 1993).

Permian–early Triassic syn-rift infills

In the northern North Sea, the Permian–early Triassic syn-rift succession has so far only been drilled on the Margareta's Spur and in the Unst Basin, where it consists of alluvial red beds (alluvial fan and fluvial sandstones, and alluvial plain mudstones) and lacustrine mudstones (Johns & Andrews 1985; Lervik *et al.* 1989). Basal Zechstein equivalent carbonates and evaporites are present on the Margareta's Spur, where they rest unconformably upon eroded Devonian strata (Lervik *et al.* 1989). On the Horda Platform, underneath the Viking Graben and in the East Shetland Basin syn-rift strata are postulated based on their wedge-shaped stratal geometries (e.g. Lervik *et al.* 1989; Gabrielsen *et al.* 1990; Færseth 1996). In these areas the Permian–early Triassic syn-rift succession is up to 2 km thick, infills N–S oriented half-grabens 15–35 km wide, and is believed to be mainly alluvial in nature. The uppermost part of the syn-rift succession has been drilled in well N31/2-4 (Steel & Ryseth 1990, their fig. 9), where it consists of a sand-prone alluvial package. The overlying mud-rich package is dated as Scythian-Anisian in the nearby N31/6-1 well (Lervik *et al.* 1989). Alluvial to lacustrine environments have also been inferred for the early Triassic syn-rift Teist Formation in the Beryl Basin, southern Viking Graben (Frostick *et al.* 1992), as well as for time-equivalent strata further south in the Norwegian–Danish Basin (e.g. Steel & Ryseth 1990). In the latter area there is a predominance of flood-basin and lacustrine deposits at increasing distance from the rift marginal hinterlands.

Extending the rift episode back into the late and possibly early Permian, and accepting the notion that the Permian–early Triassic rift-axis was centred underneath the Horda Platform (Færseth 1996; Odinsen *et al.* in press), imply that potential marine strata (Zechstein and Kupferschiefer equivalents) and possible arid or desert sediments (Rotliegendes equivalent) may form part of the Permian–early Triassic syn-rift succession. Possible desert deposits may include inland sabkha to aeolian and interdune deposits, similar to the Rotliegendes of the Moray Firth and Northern Permian Basin (e.g. Sørensen & Martinsen 1987). By analogue with East Greenland and the southern North Sea, potential marine deposits may include marine transgressive sandstones to offshore claystones and possible carbonates and evaporites. Early syn-rift desert sediments may be present across the entire Permian–early Triassic rift basin, whereas marine strata may have been deposited in the deeper half-grabens only, i.e. on the Horda Platform and possibly the northern Viking Graben. However, the northern North Sea is underlain by Caledonian crust that may have been thickened compared to the southern North Sea during Permian–early Triassic rifting, leaving a possibility that isostasy caused continental facies and environments to prevail in the northern North Sea relative to the encroaching marine waters from the north and south.

Hence, a wide variety of Permian–early Triassic non-marine syn-rift deposits may be present, ranging from arid to fluvial and lacustrine, possibly including marine strata in sub-basins close to and along the rift-axis. Spatial and temporal variations in infill architecture across the rift basin are likely to be present as a result of the interplay between sediment supply and extensional faulting, i.e. the creation of a tilted fault block topography and the half-graben's dependence on sediment supply from local or through flowing (sub-) regional drainage (Leeder & Gawthorpe 1987; Steel 1993). The rate and style of rifting during the Permian–early Triassic is uncertain. It is not known whether the Permian–early Triassic rift episode experienced pulsatory rifting, similar to the middle Jurassic–earliest Cretaceous rift episode (see below), or evolved through a single stretching interval, characterized by repeated fault events and associated topography rejuvenation *without* any prolonged intervening tectonic quiescence stages. Climatic variations, from a dry, desert-like environment in the Permian to more humid but still semiarid, fluvial and lacustrine conditions in the early Triassic, are also likely to have influenced on the sedimentary architecture of the rift basin infill.

Underlining the likely complexity of the Permian–early Triassic syn-rift infill, it is difficult to speculate further upon the detailed sedimentary architecture of the syn-rift infill without better well control.

Middle Triassic–middle Jurassic successions

Based on the documentation of a series of large-scale clastic wedges which built from both the east and west into the northern North Sea basin, Steel (1993) suggested that the middle Triassic–middle Jurassic was characterized by pronounced temporal variations in basinal subsidence and sediment supply. Relatively high basinal subsidence rates are postulated for the Late Carnian–Early Norian, Late Rhaetian, Middle Sinemurian, Early Toarcian, and possibly also the Late Pliensbachian (Steel & Ryseth 1990; Steel 1993). Moreover, local subsidence rates varied across a number of major fault zones, and it should be noted that some of the successions deposited during periods with high basinal subsidence rates are locally represented by wedge-shaped stratal packages (e.g. Gabrielsen *et al.* 1990; Steel & Ryseth 1990; Færseth & Ravnås 1998).

Late Carnian–early Norian and late Rhaetian clay-prone successions (lower-middle Lunde Formation and uppermost Lunde Formation–Raude Member of the Statfjord Formation), are characteristically thick downflank on large Permian–early Triassic fault blocks, whereas they thin progressively in an updip direction (e.g. Steel & Ryseth 1990, their fig. 19). On a basin-wide scale, the intervening sandprone units of the Lomvi, upper Lunde and Statfjord Formations have more tabular geometries (Steel & Ryseth 1990; Steel 1993), indicating basinwide more uniform subsidence rates. Locally also these sand-prone units may show distinct thickness variations across certain faults, e.g. as in the Gullfaks area (A. Dalland, pers. comm. 1997).

Increased basinal subsidence rates during the middle Sinemurian resulted in drowning of the Statfjord Formation alluvial plain and the establishment of marine conditions, the 'Dunlin sea', in the northern North Sea. It is noteworthy that the sediments representing the transition from alluvial-deltaic to marine conditions, the Nansen Member, are present mainly in downdip areas of Permian–early Triassic fault blocks. In updip reaches on the same fault blocks, e.g in the Tampen Spur area, marine mudstones of the Amundsen Formation is separated by a transgressive lag from, or rest directly upon alluvial strata of the Eiriksson Member (e.g. Nystuen *et al.* 1989, their fig. 14).

During the middle Triassic–early Jurassic, subsidence seems to have been accentuated along certain masterfaults, especially those forming the western boundary faults to the Viking Graben, the eastern Øygarden Fault Zone, and some intraplatformal faults on the Horda Platform (Gabrielsen *et al.* 1990; Steel & Ryseth 1990; Færseth & Ravnås 1998). Other faults, which grew to form masterfaults during the middle-late Jurassic rifting, probably were initiated during the late Triassic–early Jurassic (Færseth & Ravnås 1998).

Intermittent periods of uplift have been documented in the middle Triassic (Steel & Ryseth 1990) and in the late Triassic–early Jurassic (Røe & Steel 1985), although their precise dating is difficult due to the biostratigraphical barren nature of the successions. As most wells are situated on structural highs, the uplifts may possibly represent intermittent periods of footwall uplift related to rotational faulting (see also Hollander 1987; Roberts *et al.* 1987).

Periods of minor footwall uplift are recorded for the latest Pliensbachian–early Toarcian in the footwall of the Oseberg Fault (Livbjerg & Mjøs 1989), the Alwyn–Ninian–Hutton alignment (Johnson & Essautier 1987; Sawyer & Keegan 1996), and in the Statfjord–Gullfaks area along the western flank of the northern Viking Graben (Roberts *et al.* 1987; T. Dreyer, pers. comm. 1995). The uplift and ?submarine erosion resulted in the formation of partly footwall derived shallow marine sand ridges of the upper Cook Formation on the Oseberg Fault Block (Livbjerg & Mjøs 1989). This occurred concomitantly with the retreat of the rift marginal Cook Formation shoreline (Steel 1993) and drowning of downfaulted basinal areas to the west (Færseth & Ravnås 1998), and thus mirrors the Callovian–Kimmeridgian formation of shallow marine syn-rift wedges in the same area (see below; Ravnås & Bondevik 1997). Parts of the Cook Formation in the Gullfaks area may have a similar origin (T. Dreyer, pers. comm. 1995), although the unit in this area probably represents tidally influenced shorelines or tidal estuaries (Dreyer & Wiig 1995; Marjanac & Steel 1997). It is also noteworthy that some rotational faulting has been invoked to explain the sedimentary architecture of Sinemurian–Early Toarcian strata in the Beryl Embayment (Richards 1991).

Minor rotational faulting has further been postulated to occur locally in the Aalenian, e.g. in the East Shetland Basin (Roberts *et al.* 1987; Rattey & Hayward 1993; Nøttvedt *et al.* 1995) and on the Horda Platform, where it exerted a major control on the presence and distribution of the Aalenian Oseberg Formation (Graue *et al.* 1987; Muto & Steel 1997). This faulting event was possibly related to the middle Jurassic domal uplift centred beneath the tripple junction of the North sea rift basin (Ziegler 1990*a, b*; Underhill & Partington 1993), or to uplift and rejuvenation of the rift marginal hinterlands (Steel 1993).

The above review indicates that the middle Triassic–middle Jurassic period in the northern North Sea was characterized by a series of repeated subsidence events, which where at least in some areas associated with gentle rotational faulting. Deformation is generally located above the Permian–early Triassic rift structure, but apparently focused to areas beneath and adjacent to the late Jurassic–early Cretaceous rift-axis (Færseth & Ravnås 1998). On the bordering platformal areas, i.e. the Horda Platform and in the East Shetland Basin, the late Triassic–middle Jurassic succession is, in general, characterized by tabular to subtabular sediment geometries, resembling pre- or post-rift strata.

The mechanism behind the middle Triassic–middle Jurassic subsidence events is presently not clear. Two non-extensional modes of subsidence can be envisaged; differential compaction of the strongly wedge shaped Permian–early Triassic syn-rift half-graben infills must have led to differential subsidence and slip on many of the tilted block boundary faults during subsequent thermal relaxation, and the uneven nature of the thermal subsidence itself from a minimum at the basin margins to a maximum towards the basin axis is also likely to have been accommodated by several of the older fault systems. Compaction driven subsidence was most important immediately subsequent to the stretching phase. Later, and as an alternative mechanism, basin development was accompanied by renewed crustal stretching, albeit not of significant magnitude. The indications of differential subsidence and rotational faulting continuing into the early and middle Jurassic may support a component of repetitive, weak lithospheric stretching.

This suggests that the typical threefold subdivision into proto-, syn- and post-rift stages, commonly used in the analysis of rift basin evolution and infill history, needs further modifications when dealing with rift basins that experienced multiple rift episodes. Prolonged intervals with a duration of tens of Ma occurring between rift episodes, and characterized by more diffuse extension, may be better referred to as *inter-rift periods*.

Middle Jurassic–earliest Cretaceous marine syn-rift architectures

Viking Graben at 60°N (Profile 1)

The Viking Graben at 60°N as shown in Profile 1 (Fig. 2a), consists of a solitary half-graben, some 50 km wide. The half-graben is separated from bordering platform areas by a masterfault to the west and a flexured and faulted, gently sloping eastern margin. Some intrabasinal highs are present. Along the profile there are a few scattered 'basinal' wells (e.g. N30/7-7, N30/10-5, N30/10-6, N30/11-3, N30/11-4), which give some insight into the sedimentary architecture of the early, Bajocian–Bathonian syn-rift basinal infill. At present there are too little data to speculate on the sedimentary architecture of the remaining part of the syn-rift succession, i.e. the late Bathonian–Volgian Viking Group, along this profile.

The uppermost Ness and Tarbert Formations are late Bajocian–Bathonian in age and together they constitute three transgressive–regressive sequences, each some 150–200 m thick (Fig. 4). The uppermost part of the Tarbert Formation forms the transgressive segment of an additional sequence. The transgressive–regressive sequences constitute threefold, sandstone–mudstone–sandstone packages, and are believed to be a result of a pulsatory style of rotational faulting (Fig. 5). The sedimentary architecture of each transgressive–regressive sequence suggests that progradation occurred during tectonically dormant stages whereas backstepping of the shoreline took place during intervening stages with higher rates of rotational faulting. Four distinct rotational faulting events are recognized from the early Bajocian to the middle/late Bathonian, and are grouped together as the Bajocian–Bathonian rift maxima (Figs 3, 5). A similar relationship is found in the uppermost Ness and Tarbert Formations in the Hild and Oseberg areas to the north (see below; Ravnås *et al.* 1997).

Two trends of sediment partitioning within the sequences are worth noting. First, the transgressive segments are thick in downdip areas whereas they thin in an updip direction on the flexured half-graben margin. Towards the bordering platform they are represented by an unconformity. In contrast, the regressive segments appear to be of more even thickness in downdip and medial reaches of the flexured, hangingwall margin, whereas they also thin in an updip direction on the hangingwall. Secondly, the transgressive segments are commonly the thicker member of the lower sequences. This is gradually reversed upwards in the succession, where the regressive segments become thicker at the expense of the transgressive. It should also be noted that the transgressive segments usually are dominated by coarser sediment than the regressive. Although tidal influence is common throughout, it tends to be more prevalent in the transgressive segments.

In downdip areas, the regressive-transgressive turnaround is characterized by an aggradational shoreline and thick, heterolithic and often coal-bearing strata in more landward positions (e.g. well N30/10-6, Fig. 4). This is interpreted as a result of increased basinal subsidence rates and relative sea-level rise, which, in turn, are related to increased rates of rotational faulting. Initially, the high sediment supply kept pace with the increasing rates of relative sea-level rise, leading to a temporary stationary position of the palaeocoastline and an aggradational stacking pattern of paralic and shoreline lithosomes. Coal-bearing strata are therefore particularly common in the late progradational to early backstepping stage, before marine conditions were (re-) established along the axis of the rift basin.

Sediment gravity flow deposits are present in downdip areas in the upper sequences, implying that at this stage deeper marine conditions had been established along the axis of the early Viking Graben (e.g. well N30/7-7, Ravnås *et al.* 1997, their fig. 16). Thus, the retreat and eventual drowning of the Brent-delta was due to sediment supply not being able to keep pace with increasing basinal subsidence rates. Because of the assumed very low gradient of

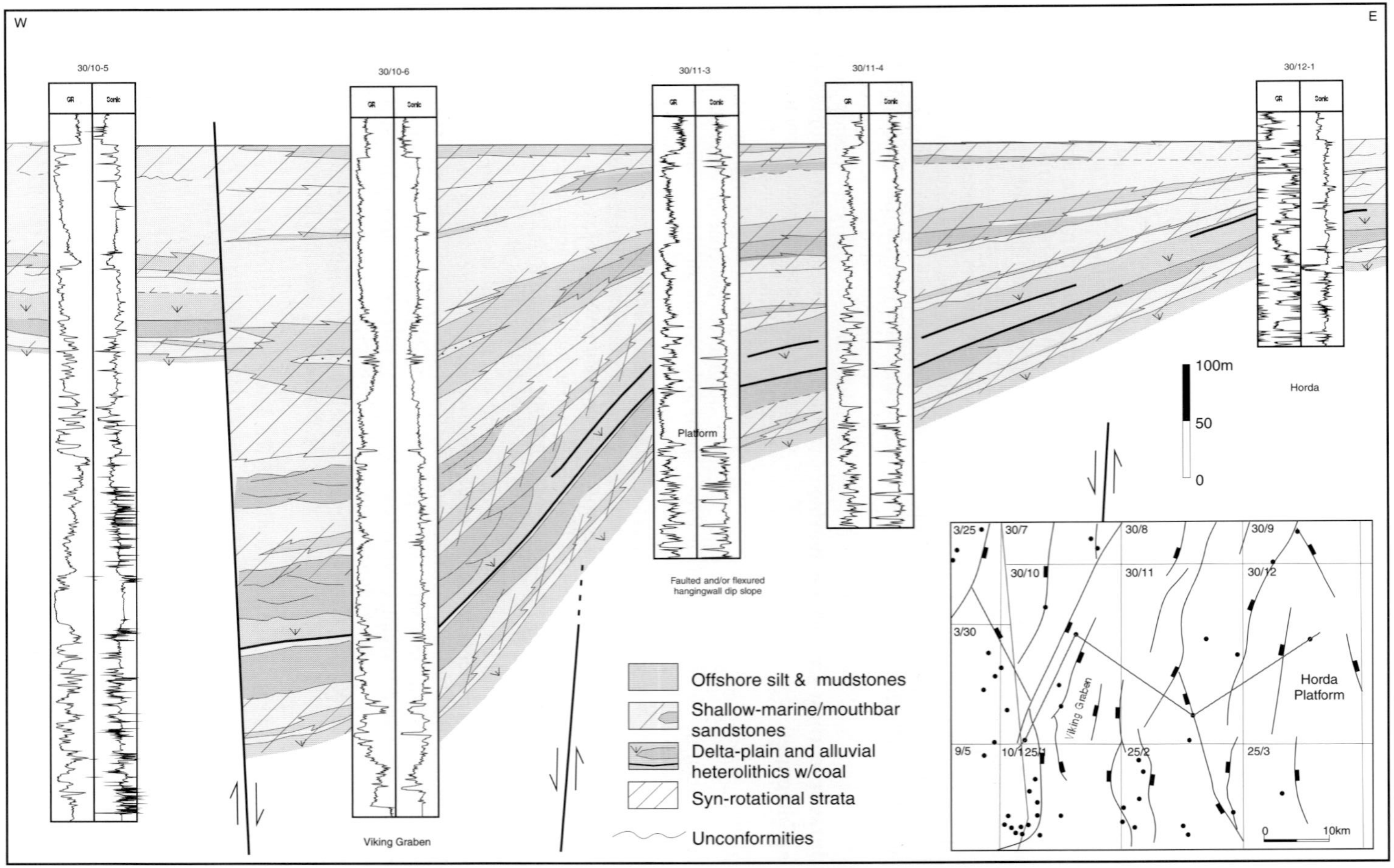

Fig. 4. E–W well correlation of the late Bajocian–Bathonian Upper Brent Group, central Viking Graben segment. Four intervals of higher tilt-rates are inferred from the sedimentary architecture. Syn-rotational strata form aggradational-to-backstepping packages, including the lower part of the progradational packages. Intervening tectonic quiescence strata are represented by the main body of the progradational packages.

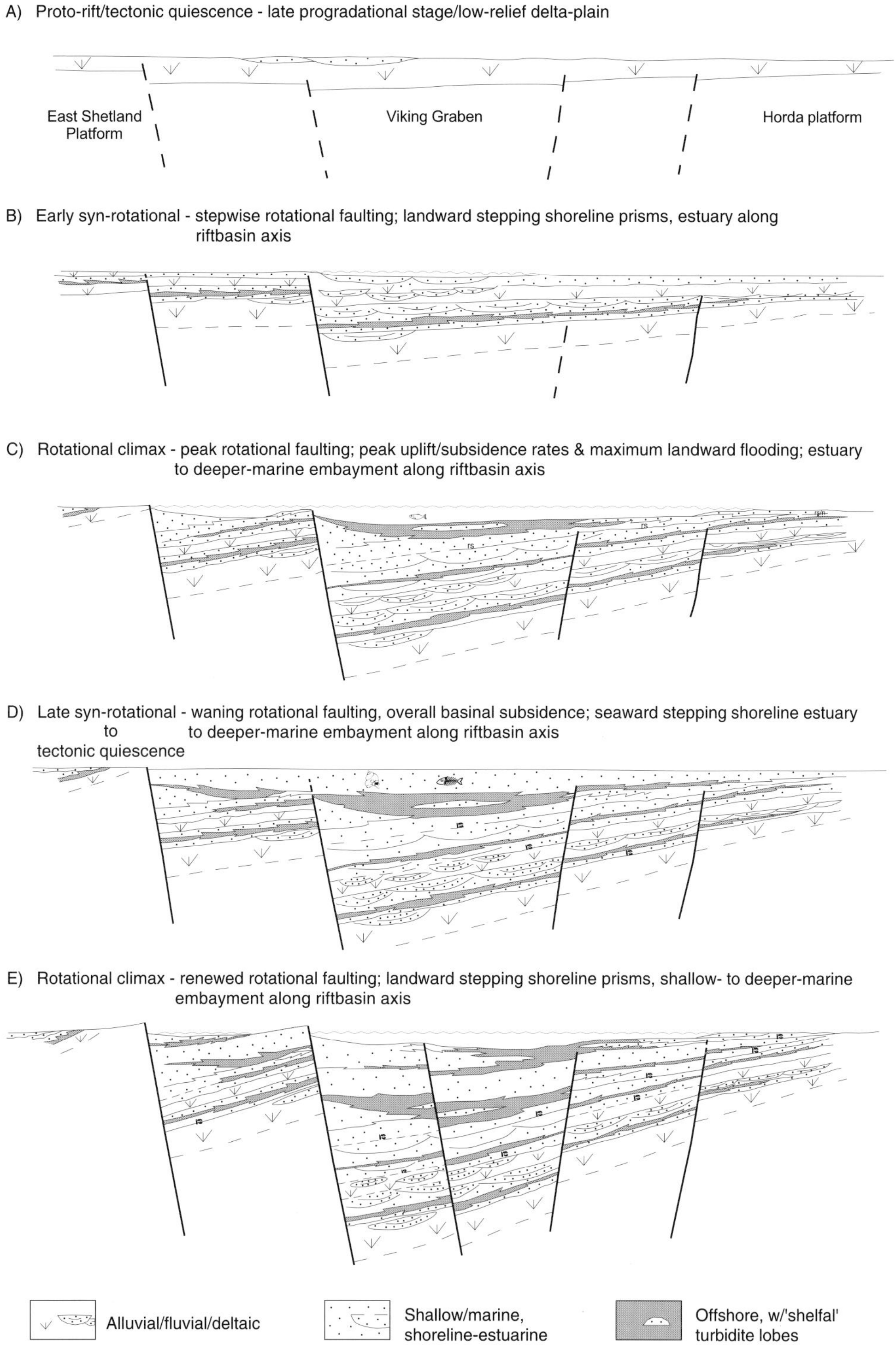

Fig. 5. Schematic late Bajocian–Bathonian infilling of the central Viking Graben segment (southern half of Norwegian Quadrant 30). The faulted right half of the profiles represent the hanging-wall dip slope in Fig. 4 whereas the deepest basin and terrace to the left represent positions of well 30/10-6 and 30/10-5, respectively. The inferred footwall further to the left is not shown in Fig. 4.

the Brent delta plain along a basinal N–S strike profile, single large earthquakes with associated slip on masterfaults (in the range of a few metres), potentially could have led to flooding events reaching several kilometres landward along the axis of the early rift basin. Repeated earthquake events in combination with eustasy may therefore have resulted in repeated, southwards encroaching flooding events across the Brent-delta and the formation of stacked higher-order sequences or parasequences. Each higher-order transgressive–regressive sequence may accordingly represent distinct faulting events, although such a linkage is difficult to confirm.

Alwyn–Hild area (Profile 2)

Profile 2 (Fig. 2b) crosses north of the Hild area, which forms a horst-and-graben faulted terrace, located at the terminal junction between two crescent shaped masterfaults. These faults constitute the western boundary fault system of the northern Viking Graben (Fig. 1). During the

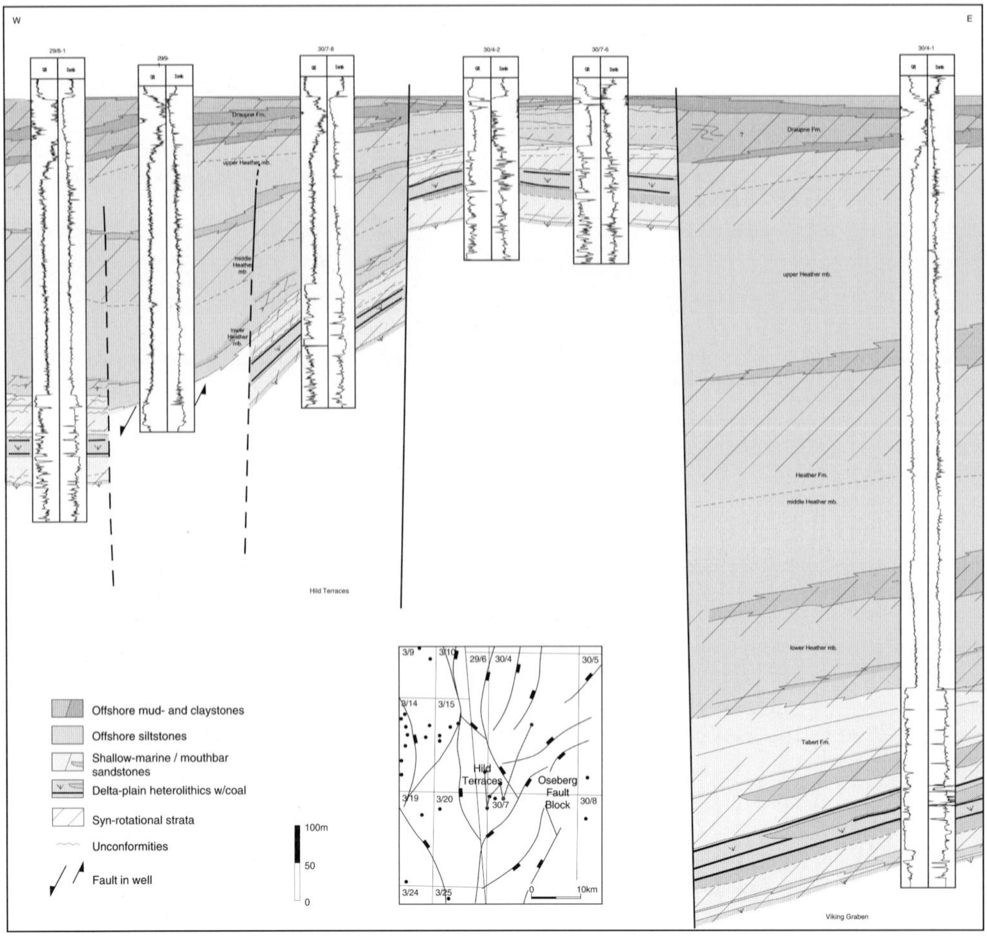

Fig. 6. E–W well correlation of the late Bajocian–Bathonian Tarbert Formation and the Bathonian–Ryazanian Viking Group in the Hild area. Intervals of higher tilt-rates in the Tarbert Formation typically are represented by a threefold aggradational-to-backstepping-to-intial progradational package, whereas tectonic quiescence strata constitute the reminder of the progradational package, i.e. similar to what is inferred for the Tarbert Formation in Fig. 4. (For a detailed correlation of the Tarbert Formation stacking pattern, the reader is referred to Ravnås & Steel (1998), their Fig. 9). Syn-rotational strata in the mud-prone Heather Formation are overall backstepping, whereas intervening tectonic quiescence strata are overall forestepping-to-aggradational. Syn-rotational strata in the Draupne Formation form a forestepping-to-backstepping wedge, underlain and overlain by tectonic quiescence, condensed claystones.

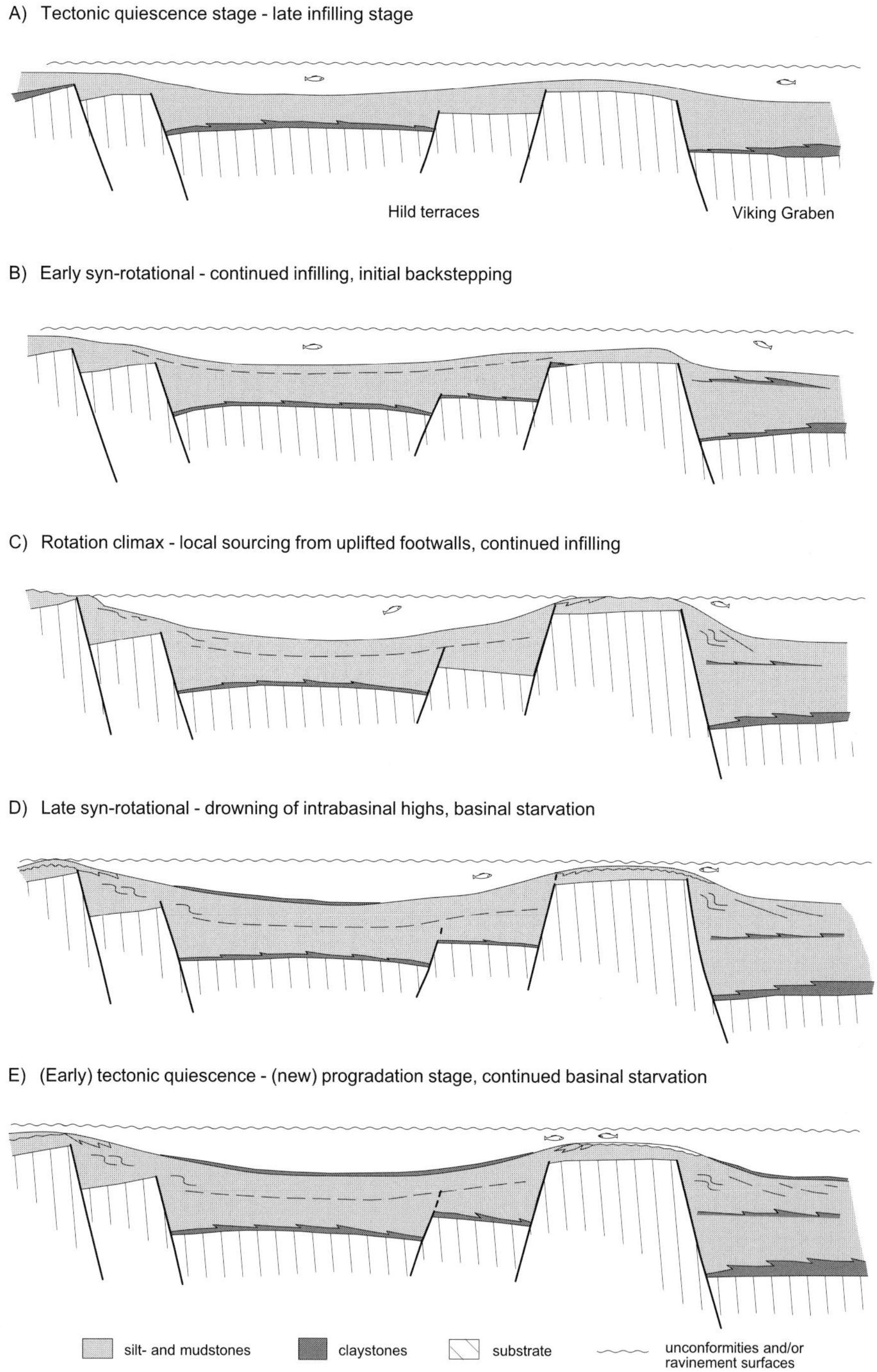

Fig. 7. Schematic structural developement and infilling during one of the Bathonian–Kimmeridgian (Heather Formation) rift phases in the Hild area. The fault block to the left represent the footwall to the Hild terraces, i.e. the southern continuation of the Alwyn structure. The central graben and horst topography represents the Hild terraces, whereas the deeper basin to the right represents the Viking Graben proper.

early (Bajocian–Bathonian) rift maxima the Hild area formed the western, downfaulted margin of the asymmetrical early Viking Graben (Færseth & Ravnås 1998), and the area experienced fault-related subsidence which caused slight tilting of several sub-basins. Already at this early stage, differential subsidence across segments of the initially broader Viking Graben caused west-dipping palaeoslopes with open marine connections in downdip areas and coal-bearing coastal plain environments in an updip and landward, i.e. southward, direction.

The Tarbert Formation (late Bajocian–Bathonian) is unusually thick in this area (Fig. 6), and four transgressive–regressive sequences, possibly reflecting the same number of differential subsidence events, can be mapped (see also Ravnås & Steel 1997, their fig. 9). There were significant focus and sediment volume partitioning between the coastal plain and the marine shoreface areas at this stage, though the tilting was never sufficient to produce unconformities over crestal reaches. Successive sequences show an irregular, but overall landward-stepping trend prior to the Heather transgression. Sediment volumes were partitioned in the sequences such that the transgressive segments, which include most of the coal-bearing back-barrier and coastal plain deposits, thicken landwards, whereas the shoreface-dominated regressive segments thicken seawards. Any sequence therefore consists of two oppositely thickening components.

From the Late Bathonian, major faulting shifted to the masterfaults to the east of the Hild area, i.e. the present western boundary faults of the Viking Graben. This caused the Hild area to become a terrace between the Viking Graben proper and the structurally shallower half-grabens of the East Shetland Basin to the west (Fig. 6). As such, the Hild area experienced a dramatic change in structural setting from being part of a wider basinal area during the Bajocian–Bathonian rift maxima to a position in the footwall of the rift basin's western boundary fault system during the subsequent phases. Accordingly, the Callovian–?Volgian rift phases are recorded by a series of unconformity strands, each forming a separate syn-rift unconformity (Færseth & Ravnås 1998), and overlying clay-prone or condensed sections. The unconformities may at least locally be of submarine nature (Fig. 7).

During the Callovian–?Volgian, minor tilting of the terrace area in addition to the formation of an intraterrace horst-and-graben topography, resulted in continued deposition in intraterrace structural lows. However, the syn-rotational deposits were derived by erosion of older, dominantly fine-grained middle–late Jurassic syn-rift sediments. The late Bathonian–Kimmeridgian Heather Formation is therefore dominated by a relatively homogenous mud-prone succession (Figs 6, 7). Possible candidates for coarse-grained syn-rotational deposits are located close to and abutting against the western and eastern terrace-bounding masterfaults. Despite the mudprone nature of the succession, three syn-rift wedges are recognized which appear to mirror their time-equivalent strata on the Oseberg Fault Block on the opposing eastern flank of the Viking Graben (see below). The well data, however, are scant and give little detailed information on the three-dimensional sedimentary architecture of the syn-rift wedges.

The Kimmeridgian–early Ryazanian Draupne Formation is too thin to be subdivided on seismic data. Again, the available seismic and well data suggest that the tectono-sedimentary regime prevalent during deposition of the Draupne Formation in the Hild area mirrored that of the Draupne Formation on the Oseberg Fault Block. Seismic data suggest that coarser material is present in the hangingwall of the larger faults, forming ?isolated fault-scarp aprons. Introduction of coarser material into the deeper marine realm is correlated with periods of increased tectonic activity. This was accompanied by pronounced fault block rotation, uplift, sediment instability and gravitational resedimentation. The latter was particularly common down the

Fig. 8. E–W well correlation of the late Bajocian–Bathonian Tarbert Formation and the Bathonian–Ryazanian Viking Group in the Oseberg–Brage area. Intervals of higher tilt-rates in the Tarbert Formation typically are represented by wedge-shaped, threefold aggradational-backstepping-intial progradational package, whereas tectonic quiescence strata constitute the main body of the progradational package, i.e. similar to what is inferred for the Tarbert Formation in Fig. 4. Syn-rotational strata in the Heather Formation are represented by forestepping-to-aggradational-to-backstepping sand-prone, fault-scarp derived succession downdip on the hangingwall, mud-prone basinal successions, and forestepping-to-backstepping sand-prone footwall successions. Intervening tectonic quiescence strata show a progradational-to-aggradational stacking pattern. Facies variations are observed between sub-basins, and between the syn-rotational and tectonic quiescence strata, reflecting spatial and temporal variations in basin physiography during both Tarbert and Heather Formations deposition (Ravnås & Bondevik 1997; Ravnås *et al.* 1997). Syn-rotational strata in the Draupne Formation form a forestepping-to-backstepping wedge, underlain and overlain by tectonic quiescence, condensed claystones, similar to what is inferred for the Hild area (Fig. 6).

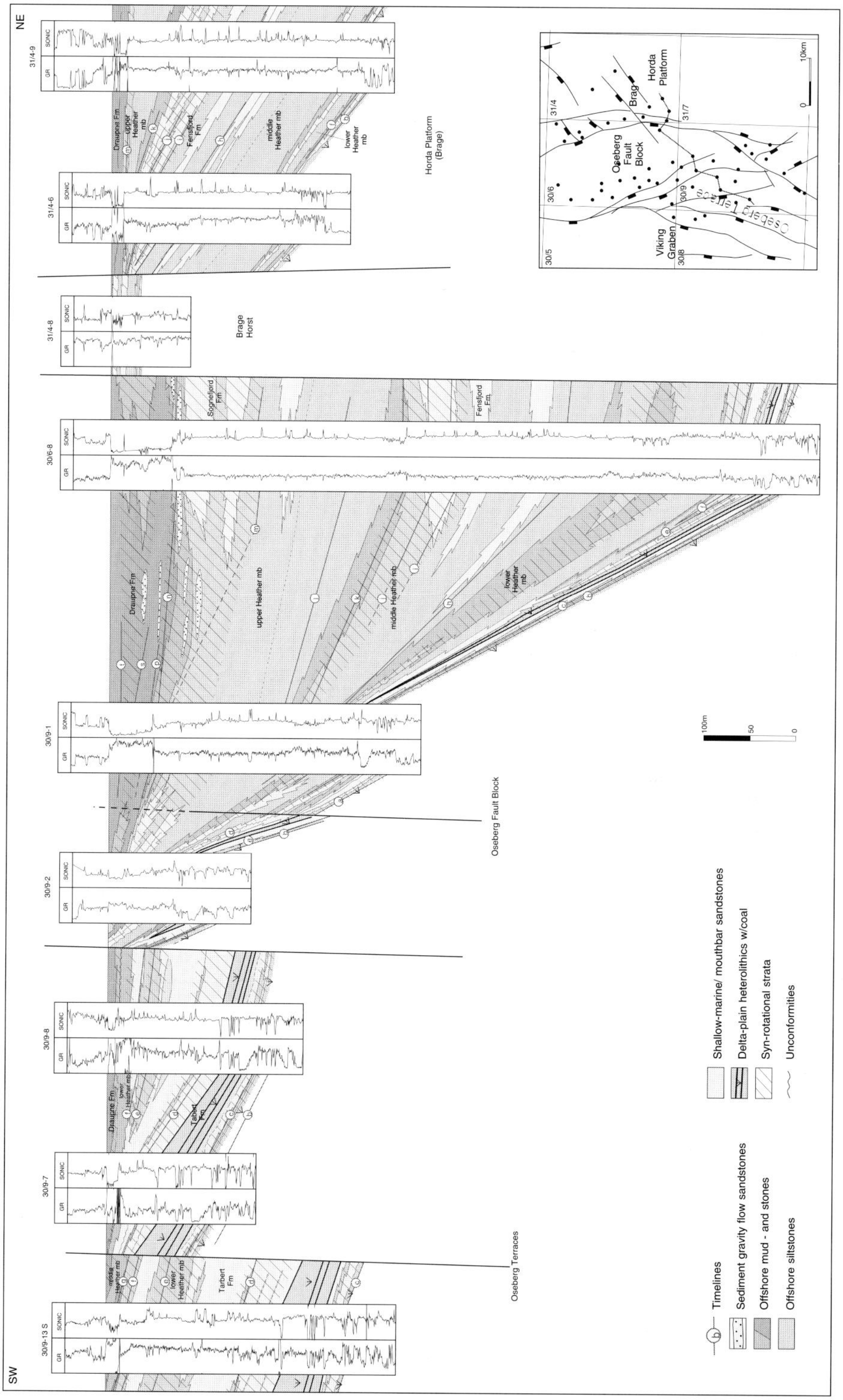
SW
NE
30/9-13 S
30/9-7
30/9-8
30/9-2
30/9-1
30/6-8
31/4-8
31/4-6
31/4-9
GR
SONIC
Oseberg Terraces
Oseberg Fault Block
Brage Horst
Horda Platform (Brage)
Draupne Fm
upper Heather mb
middle Heather mb
lower Heather mb
Tarbert Fm
Fensfjord Fm
Sognefjord Fm
Timelines
Sediment gravity flow sandstones
Offshore mud - and stones
Offshore siltstones
Shallow-marine/ mouthbar sandstones
Delta-plain heterolithics w/coal
Syn-rotational strata
Unconformities
100m
50
0
Oseberg Fault Block
Brage
Horda Platform
Viking Graben
Oseberg Terraces
30/5
30/6
30/8
30/9
31/4
31/7
10km
0

sub-basin-bounding masterfaults. In contrast, intervening periods of relative tectonic quiescence were dominated by draping of hemipelagic and pelagic claystones, commonly with a high source-rock potential, across the entire terrace area.

Oseberg–Brage area (Profile 2)

In the Oseberg and Brage areas (Fig. 2b), which form the eastern flank of the northern Viking Graben and the western margin of the Horda Platform respectively, middle Jurassic stretching commenced in the Bajocian–Bathonian (Helland-Hansen *et al.* 1992; Ravnås *et al.* 1997). During the early (Bajocian–Bathonian) rifting, the area formed the faulted eastern margin of the asymmetrical (westward sloping) early Viking Graben. From the late Bathonian to the Kimmeridgian, the Oseberg Fault Block together with the smaller fault blocks along its western flank, formed an up to 25 km wide, N–S elongate, eastward-tilted mega fault block. Collapse and segmentation of the Oseberg mega fault block occurred in the Kimmeridgian–Volgian, and resulted in the present structural configuration of the area (Færseth & Ravnås 1998). This interval was characterized by increased subsidence along the NE–SW trending structural grain.

The overall, large-scale trend of the middle-late Jurassic syn-rift infill is one of fining-upwards, from the paralic and shallow-marine sandstones of the Tarbert Formation, through the shallow-marine to shelfal sand and mudstones of the Heather Formation, to the deeper marine claystones of the Draupne Formation (Fig. 8). This large-scale fining-upward succession records the overall progressive deepening of the marine rift basin as the rift evolved, from a mixed non-marine and shallow marine basin in the Bajocian–Bathonian into a shallow marine basin in the Callovian–Kimmeridgian, and then into a deeper marine basin in the Kimmeridgian–Volgian. Successive rift phases in the Bajocian–Oxfordian/Kimmeridgian resulted in the deposition of a series of superimposed shallow marine, wedge-shaped packages. Each wedge-shaped package correlates with a distinct rift phase (Ravnås & Bondevik 1997; Ravnås *et al.* 1997; Færseth & Ravnås 1998). Sediment was supplied from rift marginal sources, first from the axial Brent-delta and later from the transverse Troll-delta, during periods of relative tectonic quiescence. During the intervening periods with high rates of rotational faulting, i.e. the syn-rotational stages, sediment was supplied predominantly from the uplifted updip areas of the fault blocks or rift interior sources.

The intervals corresponding to high extension rates in the Bajocian–Bathonian are characterized by a threefold lithosome model, consisting of basal and capping sandstones and intervening mudstones (Ravnås *et al.* 1997). Three intervals of high extension rates are recognized, the first in the latest Early Bajocian (uppermost Ness and lowermost Tarbert Formations), the next two in the late Bajocian (intra Tarbert Formation) and the early–middle Bathonian (Tarbert–Heather transition), respectively. Although these units are relatively thin on the Oseberg Fault Block, thicker successions (<200 m) are recognized in the bordering terrace area and in the basinal Viking Graben wells (Fig. 8; see also Ravnås *et al.* 1997, their fig. 15). The sedimentary architecture and inferred tectono-stratigraphical development of the uppermost Ness and Tarbert Formation transgressive–regressive sequences mirror that inferred for the time-equivalent strata in the central segment of the Viking Graben at 60°N (see above). Furthermore, Ravnås *et al.* (1997) argued, based on the stacking pattern of the transgressive–regressive sequences, that the late Bajocian–early Bathonian rift-maxima was characterized by a series of smaller-scale rotational tilt events. These shorter-term rotational tilt events are here referred to as deformation spans. Ravnås *et al.* (1997) also suggested that the threefold lithosome model established for mixed non-marine and shallow marine syn-rift successions applies at several scales; both the longer term late Bajocian–early Bathonian rift maxima as well as the shorter term rift events or deformation spans.

The intervals corresponding to high extension rates in the middle/?late Bathonian–Kimmeridgian shallow marine syn-rift successions are characterized by a twofold, sandstone-claystone lithosome model (Figs 8, 9; Ravnås & Bondevik 1997). Again, three intervals of high extension rate are recognized. These occurred in the middle–late Bathonian (lower Heather member, upper part), late Callovian–early Oxfordian (middle Heather member, upper part) and the late Oxfordian–Kimmeridgian (upper Heather member, upper part). The lack of sandstone cappings to the syn-rotational intervals is attributed to the fact that the sediments deposited during high extension rates were mainly sourced from adjacent footwall highs. The peneplanation of the rift interior sources during the late syn-rotational and subsequent tectonic quiescence stages, together with the large distance to the

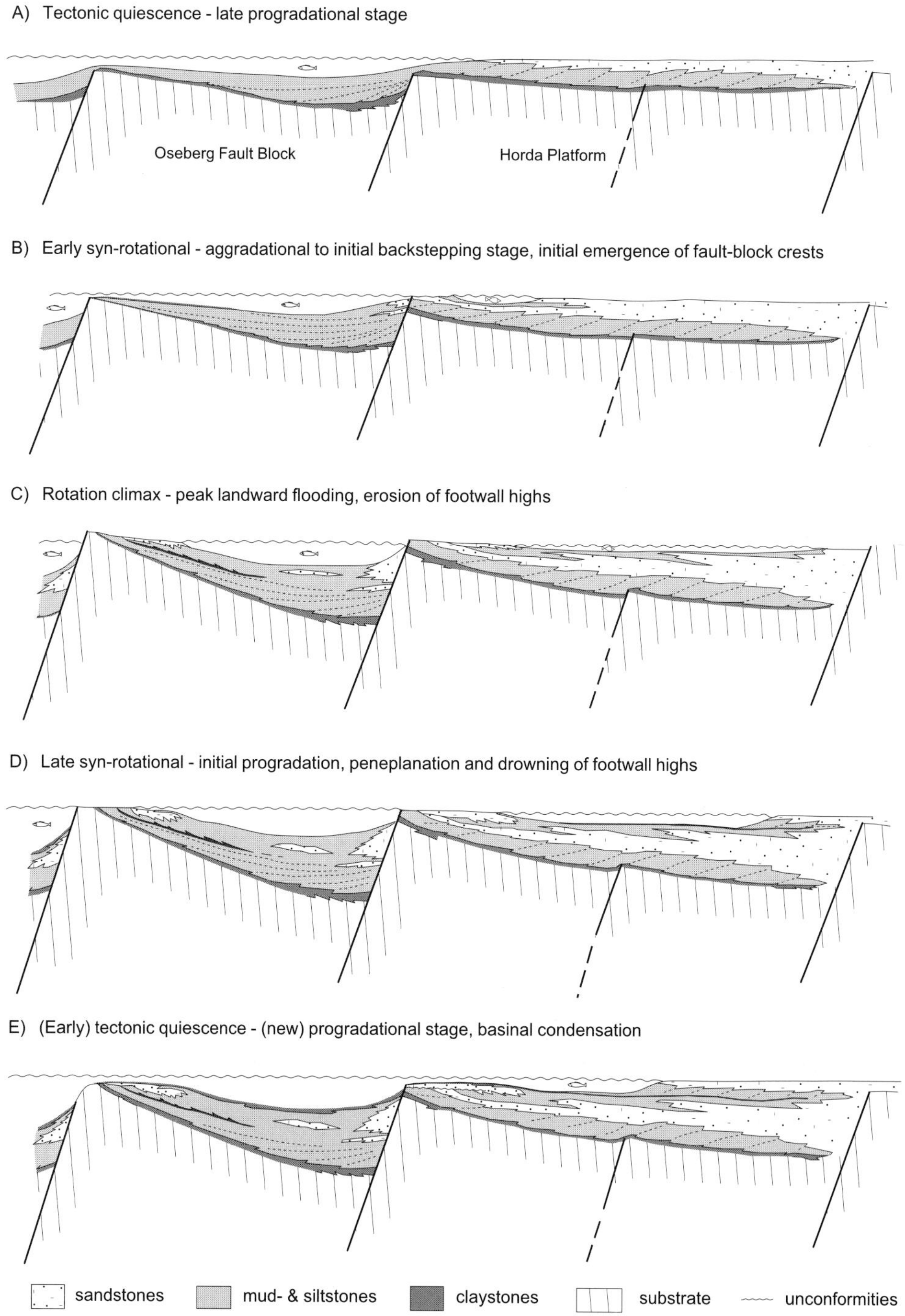

Fig. 9. Schematic structural developement and infilling during one of the Callovian–Kimmeridgian (Heather Formation) rift phases in the Oseberg–Brage area and on the Horda Platform. The half-graben to the right represent the Oseberg Fault Block, whereas the larger, mega fault block to the right represent the Horda Platform.

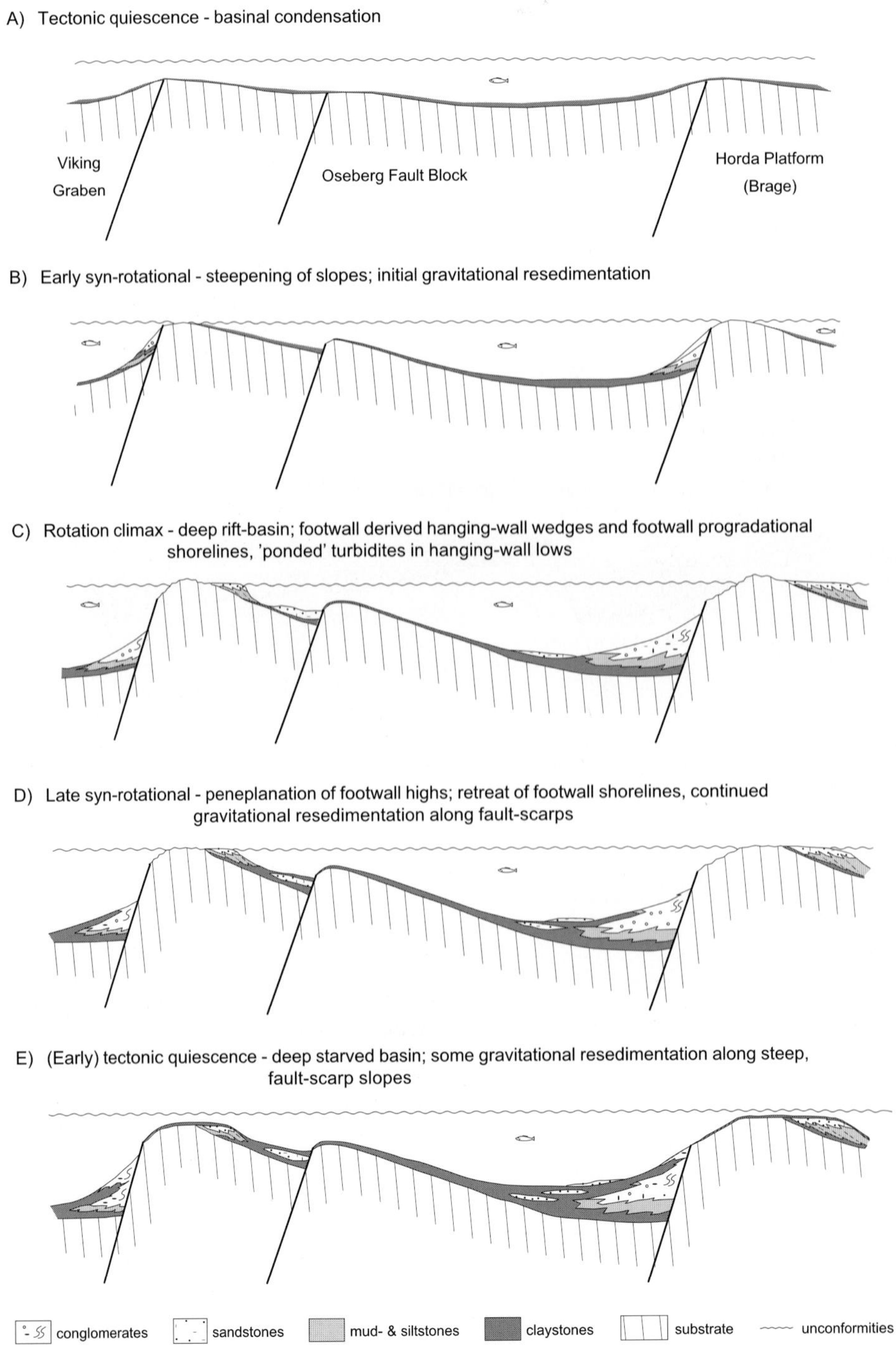

Fig. 10. Schematic structural development and infilling of during Kimmeridgian–Volgian (Draupne Formation) rifting in the Oseberg–Brage area. The Oseberg Fault-Block is presented in simplified, schematic manner as a medium-scale half-graben with a faulted hanging wall dip slope.

(then progradational) rift marginal shoreline, rendered the Oseberg area sediment starved well into the subsequent tectonic quiescence stage. During the Bathonian–Oxfordian rift phases, complete infilling of the half-graben could not occur before the rift marginal shoreline once again had approached the area (Fig. 9; Ravnås & Bondevik 1997). The syn-rotational strata in the Brage area and on the Oseberg Fault Block show a similar sediment architechture, only differing in sediment calibre of their infill (Fig. 8).

Intervals of higher extension rate in the deeper marine Kimmeridgian–Volgian syn-rift succession are also characterized by a twofold, ?conglomerate–sandstone–claystone lithosome model (Figs 8, 10). During this interval the sandprone strata and coarser lithologies are interpreted to have been derived by erosion from adjacent highs, and redeposited by gravitational resedimentation processes into the adjacent sub-basins (Ravnås & Steel 1997). The coarser syn-rotational material constitute stacked ?conglomerate–sandstone aprons along the half-graben-bounding masterfaults, and sand-prone turbidite successions in hangingwall structural lows. The latter units probably represent a series of ponded, sheet-like turbidites. The hangingwall fault-scarp aprons are assumed to be overall mud-prone, consisting of debris flow deposits and slumps/slides. The intervening tectonic quiescence intervals are characterized by suspension fall-out with a low influx of clastic sediment. The sediments deposited during the tectonic quiescence stages form half-graben-wide claystone drapes, which appear to have blanketed also most of the footwalls of the half-graben-bounding masterfaults (Ravnås & Steel 1997).

Claystones with a high source-rock potential predominantly formed during intervening tectonic quiescence intervals from the late Bathonian to the Volgian, especially during the Kimmeridgian–Volgian. In the latter period, displacement along N–S and NE–SW trending faults probably restricted water circulation considerably, resulting in the formation of rhomb-shaped sub-basins with a stratified water column and anaerobic bottom water conditions (Færseth & Ravnås 1998).

Horda Platform (Profile 2)

Prior to the late Oxfordian, the Horda Platform can be considered as a gently eastward tilted mega fault block, bounded by the Øygarden Fault Zone in the east and the Viking Graben and its flanking fault blocks and terraces to the west (Fig. 2b; Færseth 1996). Figure 11 shows an E–W well cross-section from the Troll field in the east to the Brage field in the west. This section illustrates the complex outbuilding and retreat of transverse rift marginal shallow marine clastic wedges during intervals with gentle syn-rift extension and fault block tilting.

Following a period of weak tilting and drowning in the Bathonian, the Fensfjord Formation shoreline advanced towards the Brage area during the early and middle Callovian (Hellem *et al.* 1986; Steel 1993; Stewart *et al.* 1995; Ravnås & Bondevik 1997). Renewed extension and increasing rates of basinal subsidence resulted in eastward retreat of the rift marginal shoreline, across the Horda Platform in the late Callovian (Fig. 9; Steel 1993). Concomitantly, there was uplift of the Brage area as a result of the combined effects of footwall uplift along the Brage fault and gentle clockwise rotation of the Horda Platform mega fault block (Færseth & Ravnås 1998). This resulted in the formation of locally derived shallow marine lithosomes, both on the Brage area and in the Oseberg sub-basin to the west (Ravnås & Bondevik 1997). Peak eastward progradation of the Brage hangingwall shoreline took place close to the rotational tilt climax, and was nearly coeval with maximum eastward retreat of the rift marginal shoreline on the Horda Platform. Syn-rotational sand-prone strata around the exposed footwall thus correlate with a clay-prone interval in a downdip and landward direction, i.e. towards the east (Figs 9, 11). During the following period with waning and low rates of extensional faulting, sediment supply from the lateral hinterlands again was sufficient to outpace basinal subsidence, and there was outbuilding of the rift marginal Sognefjord Formation shoreline (Steel 1993). The early stages of progradation of the Sognefjord Formation shoreline was contemporaneous with sourceward retreat of the locally derived shorelines in the Brage area and the eventual drowning and draping of the exposed footwall islands by marine condensed claystones.

These temporal and spatial variations in subsidence/uplift rates and sediment supply resulted in a marked diachroneity in the development of clay- and sand-prone intervals. In shallow marine half-grabens dominated by infilling from the transverse rift marginal shoreline such as the Callovian–Oxfordian Horda Platform, the rotational tilt climax was accompanied by maximum subsidence rates and peak landward flooding in downdip areas, whereas updip areas experienced uplift and commonly subaerial erosion. Although the basin as a whole

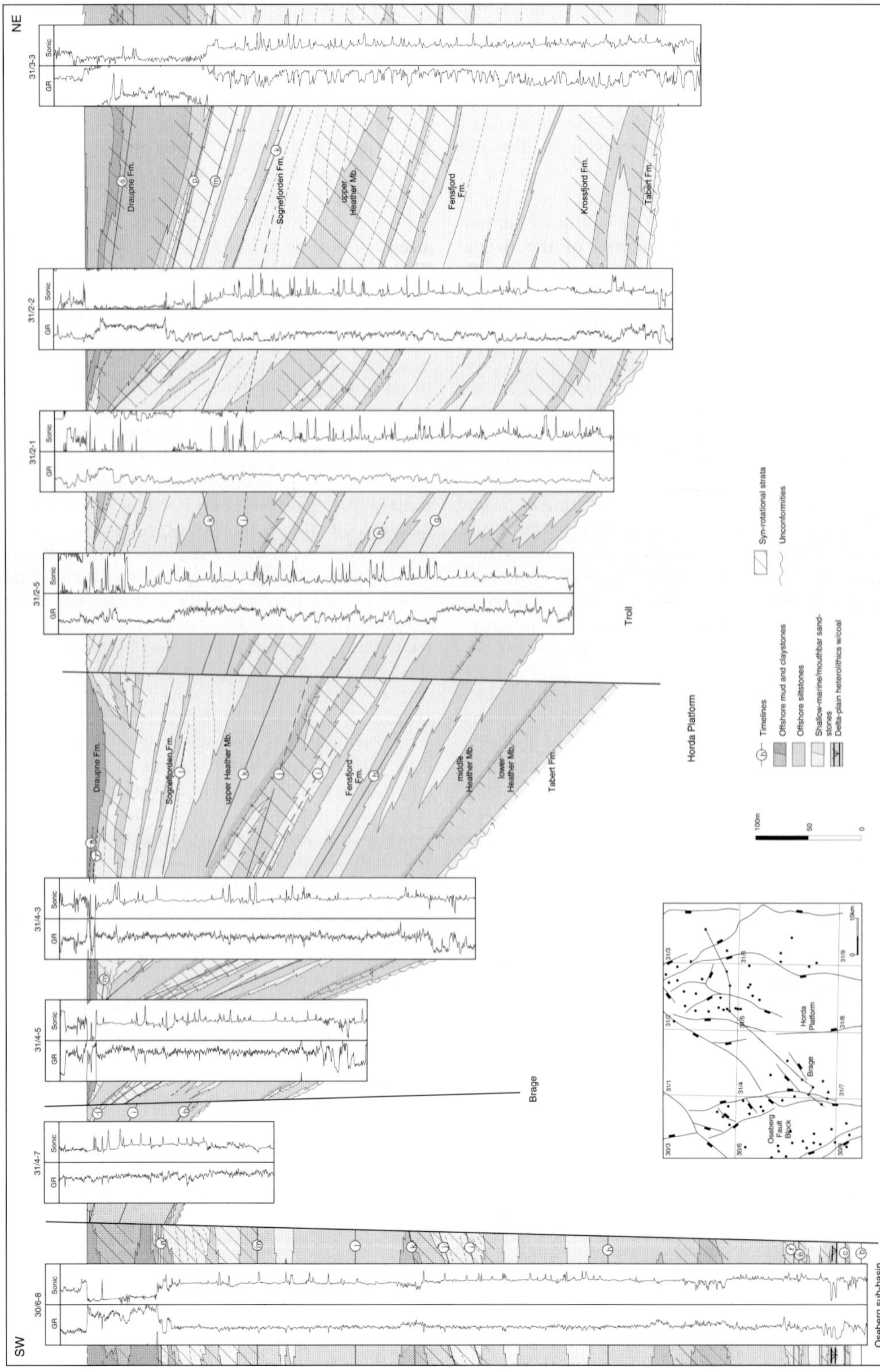
NE
SW
31/3-3
31/2-2
31/2-1
31/2-5
31/4-3
31/4-5
31/4-7
30/6-8
Sonic
GR
Draupne Fm.
Sognefjorden Fm.
upper Heather Mb.
Fensfjord Fm.
Krossfjord Fm.
Tabert Fm.
middle Heather Mb.
lower Heather Mb.
Troll
Horda Platform
Brage
Oseberg sub-basin
100m
50
0
Timelines
Offshore mud and claystones
Offshore siltstones
Shallow-marine/mouthbar sandstones
Delta-plain heterolithics w/coal
Syn-rotational strata
Unconformities
Oseberg Fault Block
Horda Platform
Brage
10km
30/3
30/6
30/9
31/1
31/4
31/7
31/2
31/5
31/8
31/3
31/6
31/9

is dominated by a threefold sandstone–mudstone–sandstone syn-rift succession (Figs 9, 11), there are unique syn-rift sedimentary signatures to each half-graben-specific setting: half-graben downdip or in basinal areas are characterized by a syn-rotational transgressive–regressive development (e.g. as seen in the Troll area wells), whereas fault block updip areas show progradation-aggradation-backstepping of locally derived sedimentary lobes (e.g. as documented for the evolution in the footwall of the Brage Fault, i.e. the Brage field, see also Ravnås & Bondevik 1997).

A similar scenario is also envisaged for the late Oxfordian–Kimmeridgian rift maxima. Higher rates of rotational faulting and basin floor subsidence resulted in the retreat of the rift marginal Sognefjord Formation shoreline (Fig. 11). The formation of late Oxfordian and younger unconformities in the footwall of intraplatformal faults (e.g. well N31/2-5, Fig. 11), suggest that the former Horda Platform mega fault block then had become disrupted into a series of smaller fault blocks. Accordingly, during the late Oxfordian–Kimmeridgian and subsequent rift phases there were several smaller intraplatformal source areas which complicated the more simple picture described for the late Callovian–early Oxfordian rift phase. However, it is anticipated that the easternmost, i.e. landward, half-grabens will show a similar syn-rotational infill trend to that outlined above. Increased basinal subsidence rates towards the late Jurassic rift-axis, suggest that half-grabens located successively towards the west were dominated by progressively deeper depositional environments. Thus, a change from mixed non-marine and shallow marine environments on the Horda Platform through predominantly shallow marine in the Brage area and on the Oseberg Fault Block, to deeper marine environments in the Viking Graben proper is expected.

Penguin half-graben (Profile 3)

The Penguin half-graben (Fig. 2c), located in the footwall of the End-of-the-World Fault, is a late Jurassic–early Cretaceous half-graben delineated by N–S striking middle-late Jurassic and NE–SW striking late Jurassic–early Cretaceous normal faults (Fig. 1). This sub-basin contains an example of deeper marine syn-rift infill, the late Oxfordian–Ryazanian Kimmeridge Clay Formation and its intercalated Ptarmigan and Magnus Sandstone Members (Fig. 12). Two intervals of high extension rates followed by periods of relative tectonic quiescence are recognized in the late Oxfordian–Ryazanian interval (Ravnås & Steel 1997). Each of these two rift phases span some 5–6 Ma. The Bathonian–Oxfordian shallow marine syn-rift strata, i.e. the Tarbert and Heather Formations, have not been studied in detail and will therefore not be described.

The older rift phase, dated as late Oxfordian–Kimmeridgian, was accommodated by movements mainly on N–S trending structures. The interval corresponding to high rates of extensional faulting is represented by a threefold, sandstone–mudstone–sandstone lithosome model, sometimes with a thin basal, coarsening-upward part (Fig. 13; Ravnås & Steel 1997). The basal coarsening-upward part reflects shedding of detritus from adjacent highs during the initial tilting stages (Fig. 13a, b). Deposition was from a wide range of sediment gravity flows, including slides/slumps, cohesive debris flows, non-cohesive sandy debris flows, and high- and low-density turbidity currents. The background mudstones represent hemipelagic and pelagic sedimentation. The overlying fining-upwards interval is a highly disorganized succession. The fining-upward signature is defined by an upward increase in mud content, whereas sandier intervals commonly are thinner and volumetrically less abundant. Sandy sediment gravity flow lobes (Ptarmigan Sandstone Member) stack in a random, but overall downdip shifting manner. These are particularly common during the interval of inferred maximum subsidence rate, i.e. the tilt climax (Fig. 13c). Sediment was delivered both from longitudinally dispersed flows as well as lateral sources. Fault-scarp successions, which represent submarine fans and

Fig. 11. E–W well correlation of the Bathonian–Ryazanian Viking Group on the Horda Platform, i.e. from the Troll to the Brage fields. In the Bathonian–Kimmeridgian Heather, Krossfjord, Fensfjord and Sognefjord Formations syn-rotational strata form threefold backstepping-to-forestepping packages in downdip positions, i.e. to the east, whereas forestepping-to-backstepping infill trends are developed in updip positions, i.e. to the west. In the upper syn-rotational sandstone package (Sognefjord–Draupne Formations) the upper forestepping part is less developed. In the Kimmeridgian–Volgian Draupne Formation potential syn-rotational strata show similar trends in updip and downdip positions as described for underlying syn-rotational packages, but are overall clay-prone. Syn-rotational forestepping-to-backstepping sandstones are only present locally on some fault-blocks, e.g. wells 31/2-5 (thin transgressive lag) and 31/2-1 (uppermost thin sandstones stringers in overall silty succession).

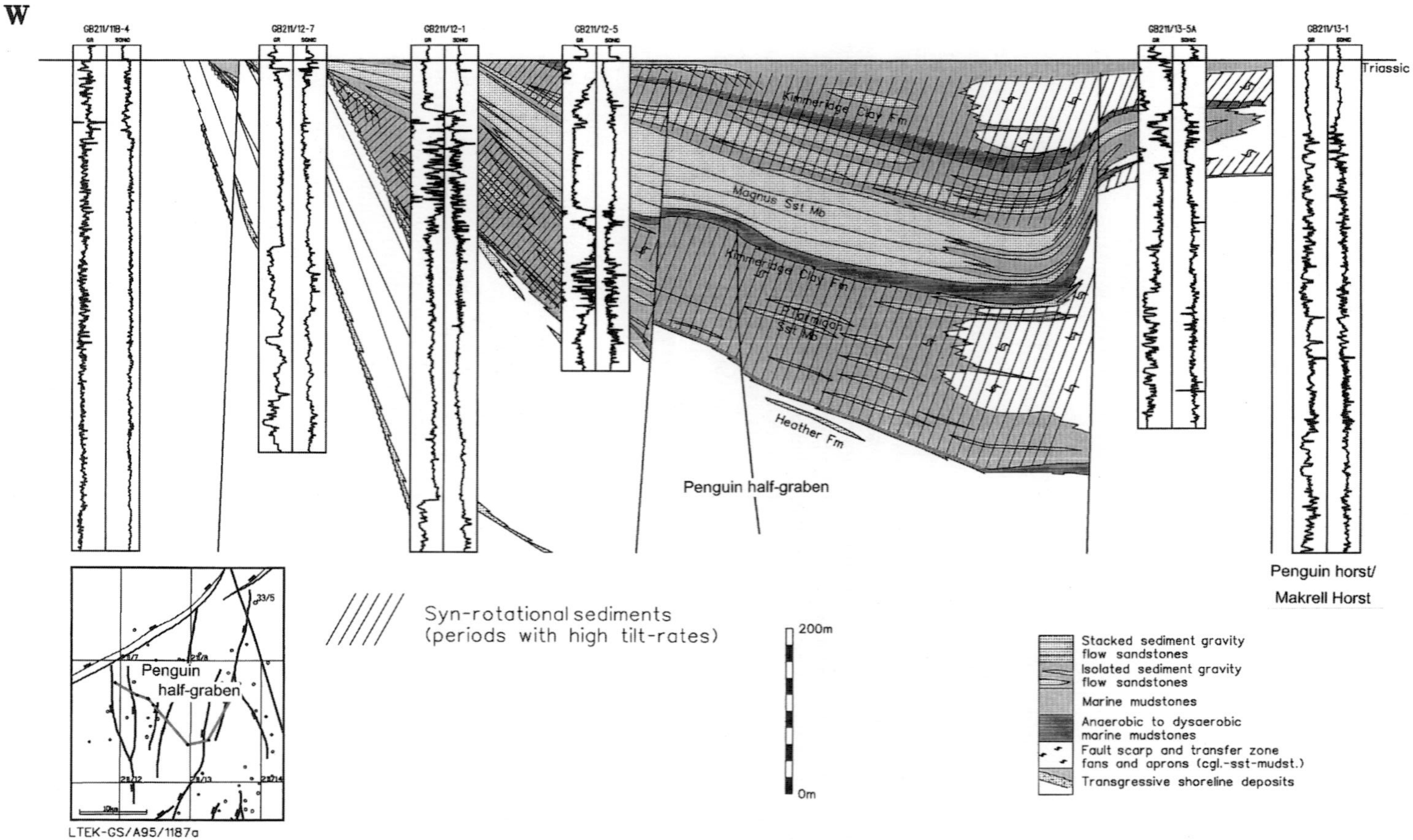

Fig. 12. E–W well correlation of the Kimmeridge Clay Formation and its intercalated Ptarmigan and Magnus Sandstone Members in the Penguin half-graben (modified from Ravnås & Steel 1997). Two syn-rotational packages are present: the lower form a threefold sandstone–mudstone–sandstone package with abundant sediment gravity flow sandstones interbedded with the middle mudstones. The upper syn-rotational packages shows a twofold sandstone–mudstone motif. Forestepping-to-aggradational-to-backstepping fault-scarp successions are present in both the lower and the upper syn-rotational package. Tectonic quiescene strata separating the two syn-rotational packages comprises a progradational-to-aggradational sandstone packagee, whereas the upper tectonic quiescence/early post-rift interval consists of condensed claystones.

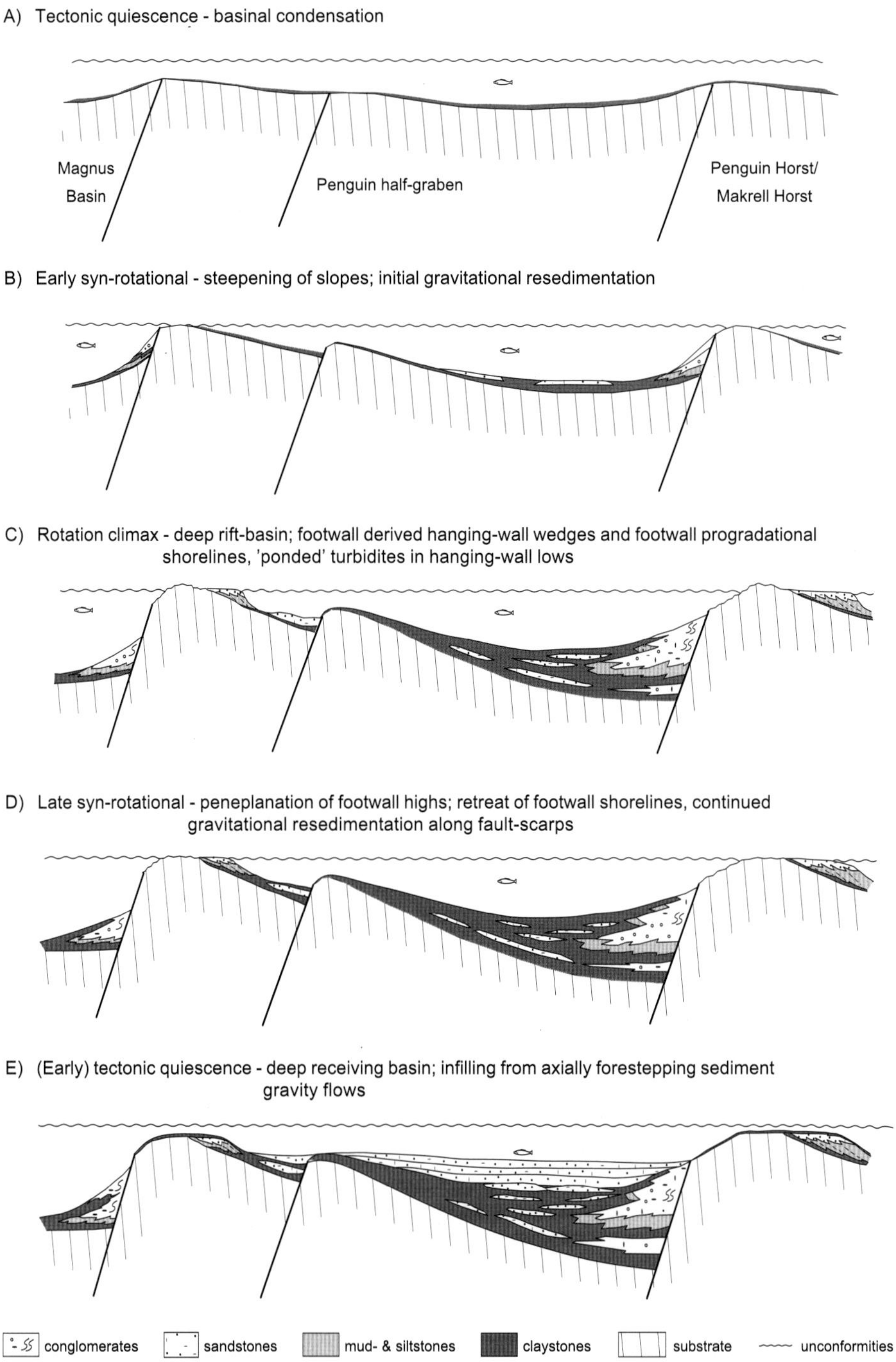

Fig. 13. Schematic structural development and infilling during Late Oxfordian–Kimmeridgian rifting and the subsequent Kimmeridgian–Early Volgian tectonic quiescence stage in the Penguin half-graben. The Penguin half-graben is presented in simplified, schematic manner as a medium-scale fault block with a faulted hanging wall dip slope.

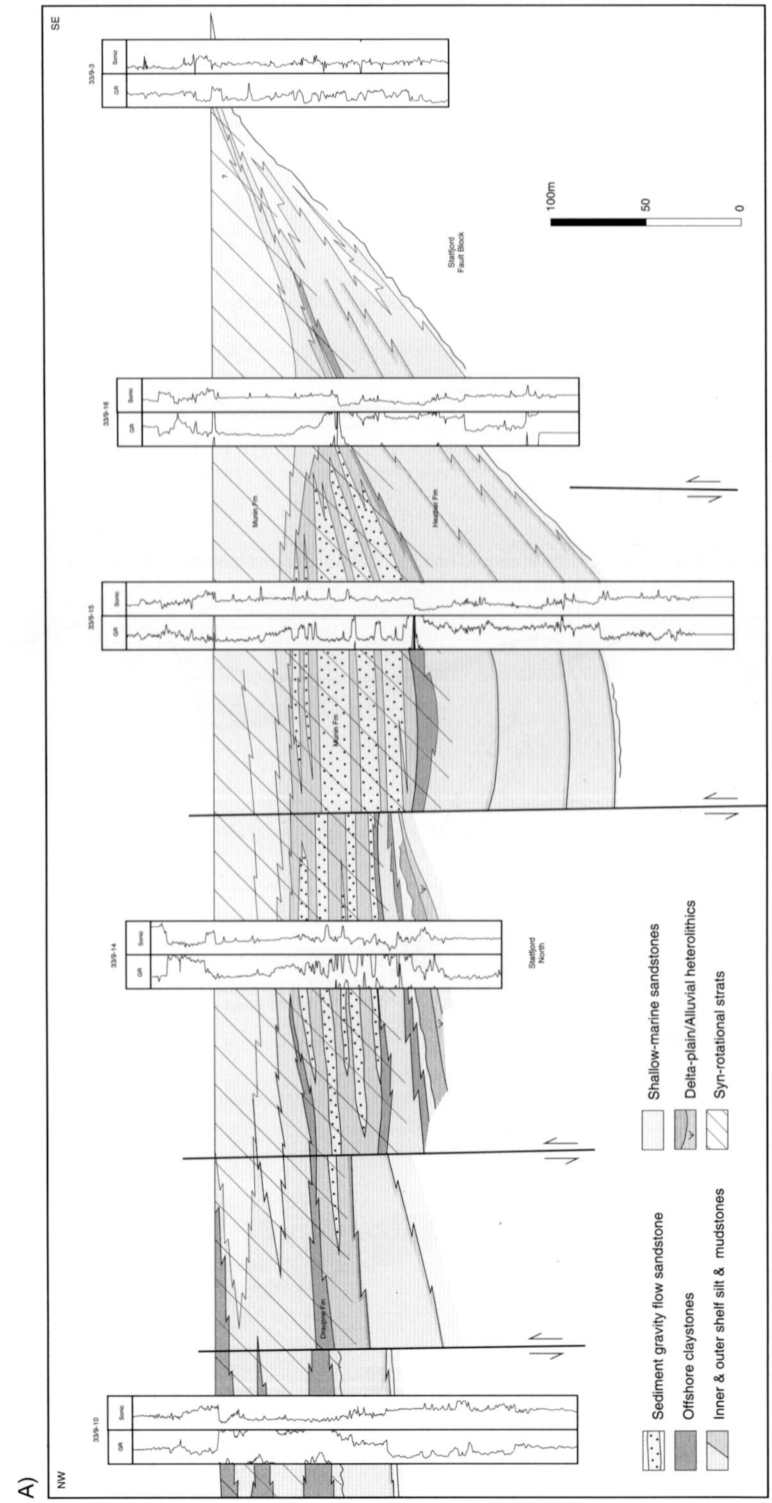
A)
NW
SE
33/9-10
33/9-14
33/9-15
33/9-16
33/9-3
GR
Sonic
Draupne Fm
Munin Fm
Heather Fm
Statfjord North
Statfjord Fault Block
100m
50
0
Sediment gravity flow sandstone
Offshore claystones
Inner & outer shelf silt & mudstones
Shallow-marine sandstones
Delta-plain/Alluvial heterolithics
Syn-rotational strats

B)

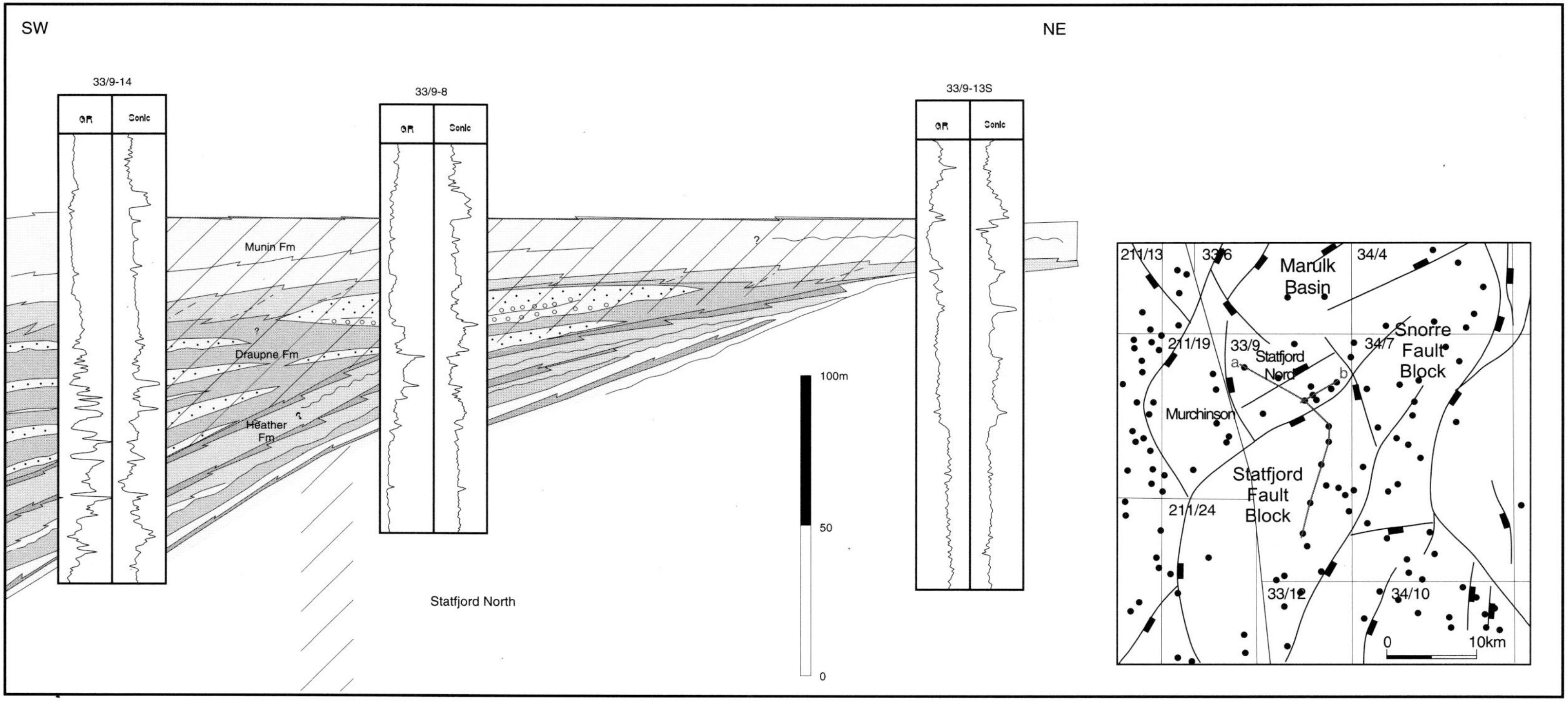

Fig. 14. (a) NW–SE and (b) SW–NE well correlations of the Draupne Formation and intercalated Munin Sandstone Member in the Statfjord and Statfjord North areas. Syn-rotational deeper-marine sandstones are present in half-graben sinks to the west and northwest where they are overlain by offshore mudstones, which, in turn, are overlain by shallow-marine sandstones. In more proximal positions the deeper-marine sandstones are lacking; only offshore mudstones and/or shallow-marine sandstone cappings are present. In the most proximal wells, i.e. towards the fault block crest, the interval is represented by an unconformity, locally with overlying transgressive lag deposits.

aprons, commonly have a coarsening-to-fining upwards signature (Fig. 12). Larger, thicker and commonly more sand-prone sedimentary lobes occur basinward of transfer zones. The fining-upward succession is capped by thick claystones (Fig. 13d) which are partly coeval with sandier sediment gravity flow deposits to the north, i.e. proximally with respect to the sub-basin's hinterland area. The sandstone cappings, i.e. the lower part of the Magnus Sandstone Member, have a sharp base, and record the income of confined, high-continuity, forestepping or progradational, sandy sediment gravity flow lobes. These had a northerly source, entered the Penguin half-graben axially, and define an overall progradational or forestepping stacking pattern; the older flows initially infilled basinal lows, whereas subsequent flows reached progressively higher up on the hangingwall and fault-scarp slopes and further out in the basin (Fig. 13e).

The younger interval characterized by high extension rates, tentatively dated as latest Early Volgian–middle/?late Volgian, was dominated by significant displacement on NE–SW trending faults, in addition to reactivation of the N–S trending structural grain (Ravnås & Steel 1997). This interval is represented by a two-fold, sandstone–claystone lithosome model. The basal fining-upward part is related to the sourceward retreat of the Magnus submarine fan/sheet system, and deposition of a backstepping turbidite package. Increased faulting along the NE–SW trending End-of-the-World Fault cut off the northerly sediment source, resulting in temporal basinal sediment starvation. Renewed input of coarser silicilastic material occurred as the half-graben updip area and Penguin Horst/Makrel Horst became exposed during the subsequent Middle Volgian tilt climax. This upper syn-rotational succession is terminated by late Volgian–Ryazanian claystones (Fig. 12). The claystones, forming a condensed section, were deposited during the late syn-rift and early post-rift stages (see development interpreted for the Draupne Formation in the Oseberg area, Fig. 10).

The tectonic quiescence strata differ considerably. A deep-marine sediment gravity flow sandstone package, the main body of the Magnus sandstone member, separates the two syn-rotational packlages, wheras the upper syn-rotational package is capped by a condensed claystone (compare Fig. 13 with Fig. 10). This reflect a closer distance to a sandier hinterland source with high sediment yield during the former tectonic quiescence stage. Changing structural regime during the last rift phase left the Penguin half-graben as a starved, intrabasinal sub-basin, sourced from small bordering highs (Ravnås & Steel 1997).

Statfjord and Statfjord North areas (Profile 3)

The Statfjord and Statfjord North areas constitute westward to northwestward tilted N–S to NE–SW trending half-grabens situated between the northern Viking Graben and the Marulk Basin (Fig. 1). The sub-basins are delineated by N–S and NE–SW trending masterfaults which downthrow to the east and southeast. The Statfjord and Statfjord North sub-basins contain examples of sand-prone Kimmeridgian–Volgian deeper-to-shallow marine syn-rift infill, the Draupne Formation and its intercalated Munin Sandstone Member. The early syn-rift Tarbert Formation has been discussed in some detail by Johannessen *et al.* (1995). The mud-prone Heather Formation is too thin to allow any detailed subdivision.

Downdip on the fault blocks, the Kimmeridgian–Volgian syn-rift strata form a threefold sandstone–mudstone–sandstone succession with basal sandy sediment gravity flow deposits, overlying offshore mudstones, and capping shallow marine sandstones. In the medial reaches of the hangingwall dip-slope, the succession is represented by a heterolithic interval with interbedded offshore mudstones and sediment gravity flow sandstones, overlain by shallow marine sandstones. In updip reaches only the shallow marine sandstones are present, commonly resting upon or showing an intra-sandstone unconformity, whereas the entire interval is represented by an unconformity and overlying reworked sandstones towards crestal areas of the fault blocks (Fig. 14).

The Kimmeridgian–Volgian succession has been discussed to some extent by Gradijan & Wiik (1987) and Solli (1995). Solli (1995) related each sandstone depositional event to distinct uplift phases of the footwalls along the half-graben masterfaults, especially the Inner and Outer Snorre Faults. An alternative explanation, which is favoured here, relates the basal sediment gravity flow deposits to the initial and early uplift stages. This early uplift resulted in sediment instability on the tilted slope, gravitational collapse and resedimentation of sediment formerly stored along the fault block updip areas by ensuing sediment gravity flows, leading to deposition of downdip or basinal sediment gravity flow lobes (Fig. 15a, b). Continued extensional

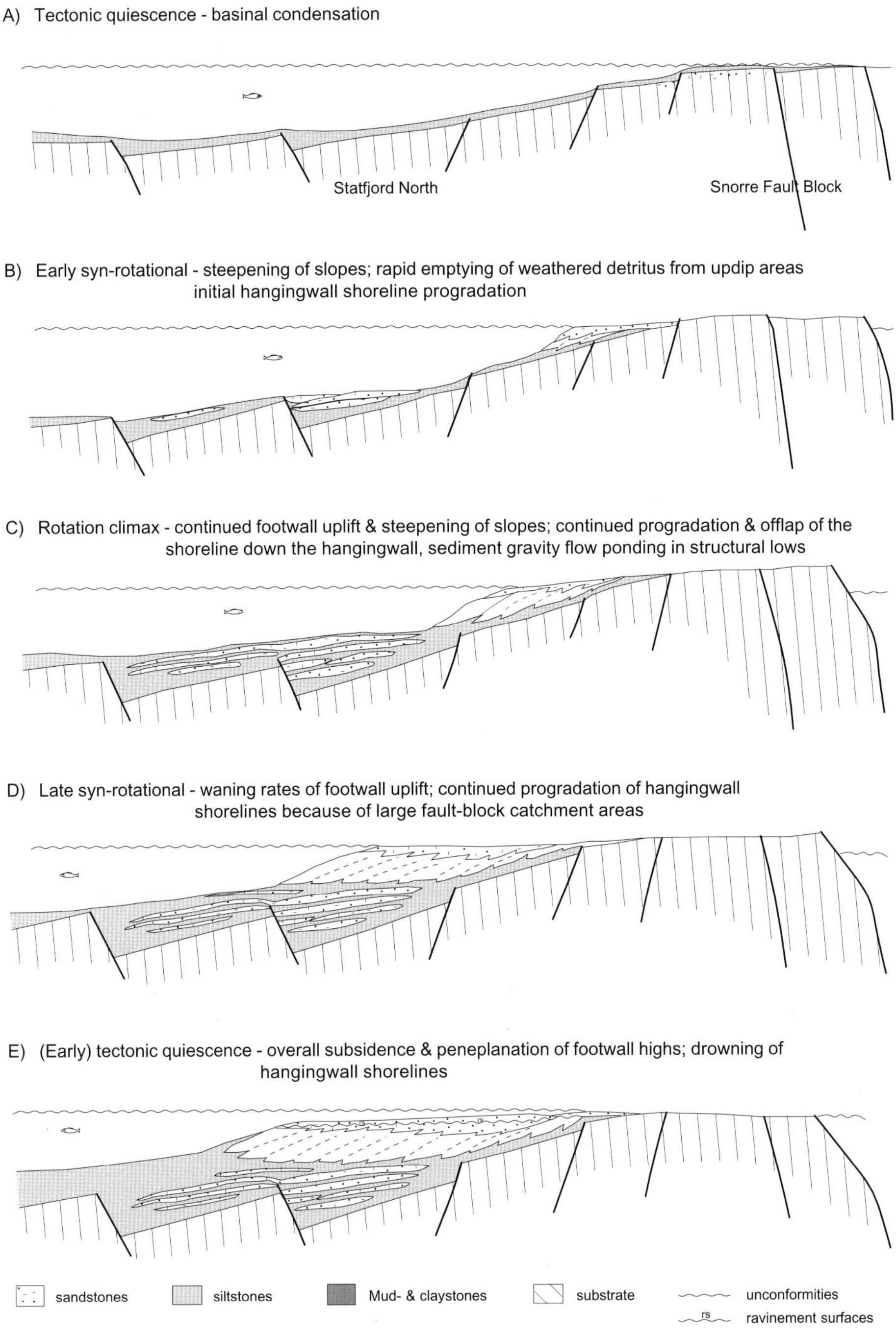

Fig. 15. Schematic structural development and infilling during Volgian rifting in the Statfjord–Statfjord North area. Only the upper part of the fault block dip-slope is shown. Sub-basins represent depositional sinks on the fault-block dip-slope itself, e.g. as on the Stattfjord North structure, or smaller sub-basins between two separate fault blocks, e.g. between the Statfjord fault block and the Statfjord North structure.

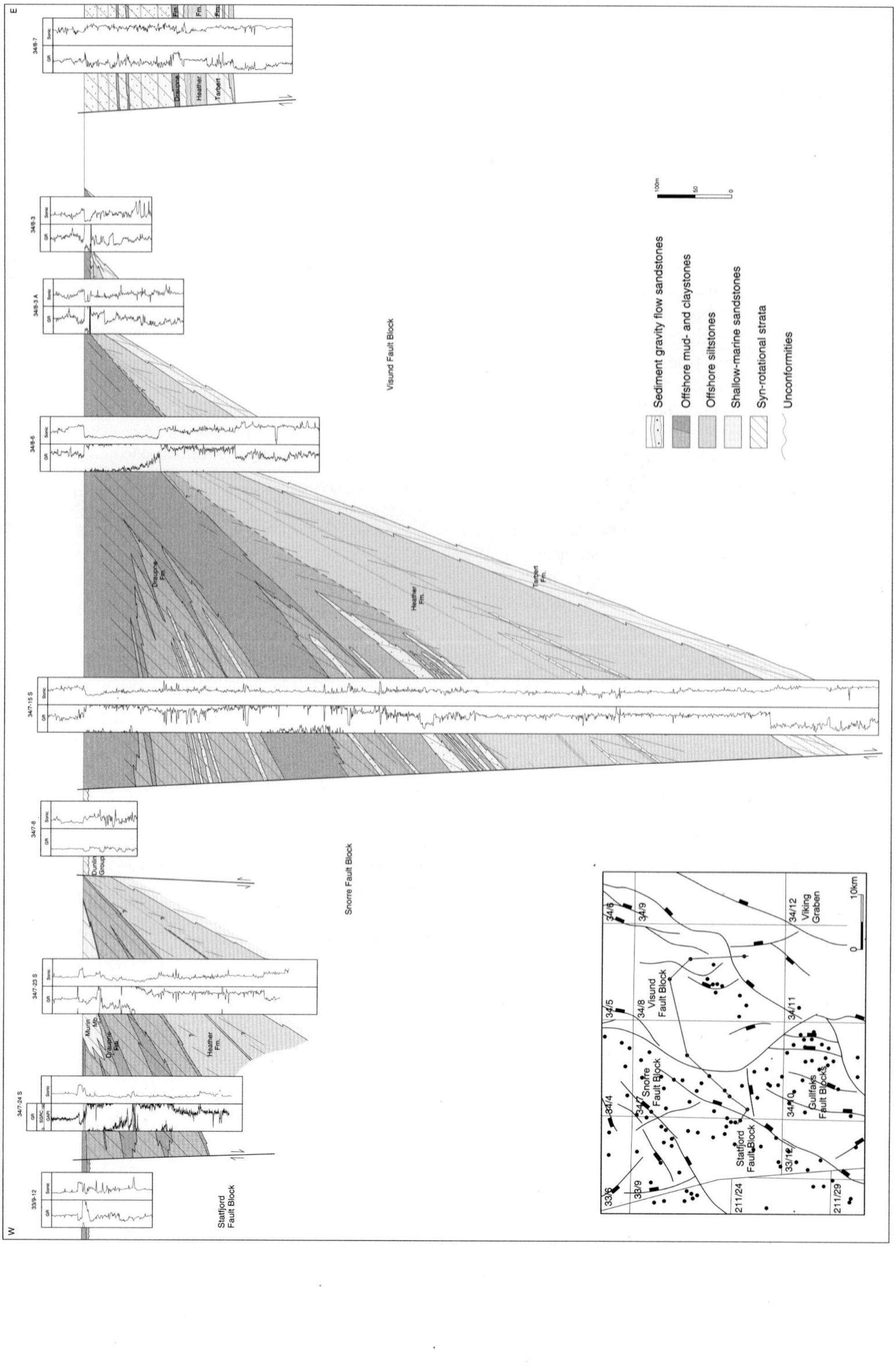
W
E
33/9-12
34/7-24 S
34/7-23 S
34/7-8
34/7-15 S
34/8-6
34/8-3 A
34/8-3
34/8-7
GR
Sonic
Statfjord Fault Block
Snorre Fault Block
Visund Fault Block
Dunlin Group
Munin Mb.
Draupne Fm.
Heather Fm.
Tarbert Fm.
Draupne
Heather
Tarbert
Sediment gravity flow sandstones
Offshore mud- and claystones
Offshore siltstones
Shallow-marine sandstones
Syn-rotational strata
Unconformities
100m
50
0
34/6
34/9
34/12
Viking Graben
34/5
34/8
Visund Fault Block
34/11
34/4
34/7
Snorre Fault Block
34/10
Gullfaks Fault Blocks
33/6
33/9
33/12
211/24
Statfjord Fault Block
211/29
0
10km

faulting resulted in further tilting of the fault blocks and continued uplift of their updip areas. The uplift resulted in widening of the footwall islands and created shallowing across the fault block in updip reaches. The widening of the footwall islands increased the fault blocks' sediment yield potential, whereas the progressive steepening may have increased erosion rates. These two processes in combination with the shoaling in updip reaches promoted the progradation of locally derived hangingwall fan-deltas and shorelines (Fig. 15c). This progradation is argued to have continued well into the late syn-rotational stage, resulting in deposition of the shallow marine sandstone capping of at least the updip reaches of the half-grabens. Sourceward retreat of the hangingwall shorelines was due either to diminishing sediment supply as the footwall islands became progressively deeper eroded, or because the shorelines had entered deeper waters downdip on the hangingwall and delivered their sediment directly into deeper waters. In either case, the capping of most of the fault block updip areas by thin Ryazanian offshore mudrocks suggest that by this time there was little subaerial relief and that the fault blocks by then had been peneplaned and drowned (Fig. 15d, e; cf. Roberts *et al.* 1993*b*). Accordingly, it is postulated that the threefold sandstone–mudstone–sandstone syn-rift succession represents one single rift phase.

The most diagnostic features of the syn-rotational hangingwall deposits are illustrated in the well cross-sections in Fig. 14. Note especially that the lower part of the succession defines a fining upward trend from deeper-marine sediment gravity flow deposits to offshore mudstones, which at least in updip reaches represents a period of upward shallowing. Seismic resolution is poor, yet there appears to be rapid lateral and basinal pinch-out of the turbidite lobes (Ghazi & Schmidt 1995), as also suggested by well data. Moreover, these early turbidite lobes were apparently ponded in structurally defined bathymetric lows. The coarsening upward part of the succession to the sandstone capping may record a variety of relative sea-level changes dependent on the structural position within the half-graben. Downdip of the fault block fulcrum the coarsening upward part of the succession reflects a normal regression (sensu Posamentier *et al.* 1992). The most basinal or downdip wells suggest that there were at least two punctuated progradational events. Although the flooding across the Munin shoreline sandstones may at least partly reflect eustatic sea-level fluctuations, in view of the tectonic setting they are tentatively correlated with distinct faulting events. Such faulting events will have resulted in punctuated tilting and gentle rejuvenation of the half-graben relief. Renewed progradation of the Munin hangingwall shoreline would then have occurred as a response although slightly delayed relative to the postulated faulting. Updip of the fault block fulcrum, the coarsening upward part of the succession reflects a forced regression, and no evidence of the punctuated shoreline progradations is evident. Further updip, transgressive shoreline sandstones or ravinement lags rest upon a deeply truncated substrate. In crestal settings it is only the late syn-rotational and subsequent post-rift transgressive stage that is recorded by deposits; most of the rotational stage is represented by an unconformity.

Snorre half-graben (Profile 3)

The westward tilted Snorre half-graben (Solli 1995) is situated at the southern extension of the southerly plunging Snorre Fault Block, between the Statfjord East and Visund Fault Blocks (Fig. 2c). Rotation along these major fault zones caused a rotation of the Snorre-H pre-rift strata to about 9 degrees by Ryazanian times (Nøttvedt *et al.* this volume). To the north, the Snorre and Statfjord Fault Blocks merge to form a complex horst structure that has suffered *c.* 1.5 km of composite uplift. The Snorre half graben is delineated and dissected by N–S and NE–SW trending faults, commonly with an E–SE dip, and contains a Bathonian to Ryazanian shallow-to-deeper marine syn-rift infill. The Heather Formation (Bathonian–early

Fig. 16. E–W well correlation of the Viking Group and Tarbert Formation from the Statfjord Fault Block across the Snorre and Visund Fault Blocks to the terraces southeast of the Visund Fault. Several syn-rift wedges are recognized in downdip basinal sinks and correlates with unconformity strands acroos footwall highs. The syn-rotational wedges are separated by tectonic quiescence packages. Syn rotational strata are sandy in the Tarbert Formation, mudprone in the Heather Formation, and becomes more sandy in the Draupne Formation. Syn-rotational shallow-marine sandstones are present in the Tarbert Formation. Syn-rotational shallow-marine sandstones are also present updip on the hanging-wall of individual fault blocks in the Draupne Formation wedges, whereas deep-marine gravity flow sandstones occupy the halfgraben depositional sinks.

Oxfordian) consists of silty mudstones with few sandstone interbeds. The Draupne Formation (late Oxfordian–Ryazanian) includes a lower, dark and pyritic claystone unit coarsening upwards into a micaceous sandstone unit in the upper part in medial hangingwall reaches and in the northern axial area. No wells have been drilled off the Statfjord East footwall, but from seismic interpretation and structural reconstruction it is suggested that the lower Draupne Formation may contain sediment gravity flow sandstones passing upwards into more fine-grained siltstones and claystones in the upper Draupne Formation.

The most diagnostic features of the hangingwall syn-rift infill of the Snorre Fault Block are illustrated in the well cross-section panel in Fig. 16. An early Bathonian tilt initiation has been advocated for the neighbouring Statfjord and Visund Fault Blocks (Roberts *et al.* 1993*b*; Færseth *et al.* 1995*a*), whereas a late Bajocian initiation of tilting has been suggested for the Gullfaks area to the south (Johannessen *et al.* 1995). This suggests that the sandstones of the Tarbert Formation may be part of the early syn-rift infill and that initial weak extension in the late Bajocian–Bathonian caused some thickening of the Tarbert Formation across the major fault blocks.

Accepting the interpretation that the late Bajocian–Bathonian Tarbert Formation represents the earliest rotational movements of the fault blocks, the syn-rift infill as a whole forms a threefold sandstone–mudstone–sandstone package in medial hangingwall and northern axial reaches of the plunging Snorre Fault Block.

Locally, on the hangingwall dip-slope, offshore marine mudstones of the Draupne Formation envelop sediment gravity flow sandstones. On a lithological basis, the data support a simple model of extension with a prolonged period of moderate rotational rates extending from the Callovian–Volgian. The model relates deposition of the upper Draupne sand unit to reduced rates of rotational faulting during the waning rift stage. An unconformable relationship between the Heather and Draupne Formations (see below), on the other hand, may favour an interpretation of at least two rift-maximas, one in the Callovian–early Oxfordian and one in the late Kimmeridgian–Volgian, separated by a period of tectonic quiescence in the late Oxfordian–early Kimmeridgian. To the south, in the more rapidly subsiding parts of the half-graben, a twofold lithosome model, a sandstone-mudstone succession, is developed. Sediment gravity flow sandstones are assumed to have been shed off the Statfjord East Fault Block in response to early uplift, corresponding to erosion of the Heather Formation and deposition of claystones in hangingwall areas. Towards the crestal areas of the Snorre Fault Block, the entire interval is represented by an unconformity. The unconformity cuts into the Brent Group in the south and deep into the lower Jurassic and upper Triassic, i.e. the Statfjord Formation, in the north. On the neighbouring Statfjord East Fault Block, the uppermost Draupne Formation is resting unconformably on eroded middle Jurassic Brent Group strata.

Slightly increased rates of extension in the late Bathonian–Callovian led to drowning of the Brent Group and deposition of the Heather Formation across a gently tilted, but submerged fault block topography (Fig. 17a). At this point, mudstones were widely deposited, and with the exception of the northern Tampen Spur, sediments were sourced from the rift marginal shoreline to the west and from the retreating Brent-delta to the south. Accelerated rates of rotation in the late Oxfordian led to exposure of footwall crests and segmentation of the Tampen Spur into distinct half-graben sub-basins (Fig. 17b). The relief across the emerging footwall highs was still too small to cause significant local erosion and redeposition, but because of the segmentation, the axial and transverse sediment supply no longer fed directly into the outer half-grabens, leading to starved basin conditions.

The Heather Formation appears to be unconformably overlain and slightly truncated by the overlying Draupne Formation in upflank positions on the Snorre hangingwall, as a result of renewed rotation and footwall uplift in the early Volgian (Fig. 17c). This uplift increased both the extent and relief of the emergent footwall islands, and thereby their sediment yield potential. In updip hangingwall areas, this led to a forced regression and significant local erosion of the Heather Formation. In the Statfjord East footwall, the Brent Group was extensively eroded. Because of the steep gradient of the footwall slope, sediments were unable to be stored in significant quantities along the footwall shoreline, and mixed sand and mud lithologies were therefore ponded as gravitational flow and turbidite deposits in structurally defined bathymetric lows.

Maximum uplift of the footwall islands were obtained in the late Volgian–Ryazanian (Fig. 17d), resulting in deep erosion into the Brent Group on the Snorre hangingwall dip-slope. Most of the eroded sand were stored along the hangingwall shorelines, but sediment instability and gravitational collapse caused some of the sand to be resedimented by ensuing gravity flows

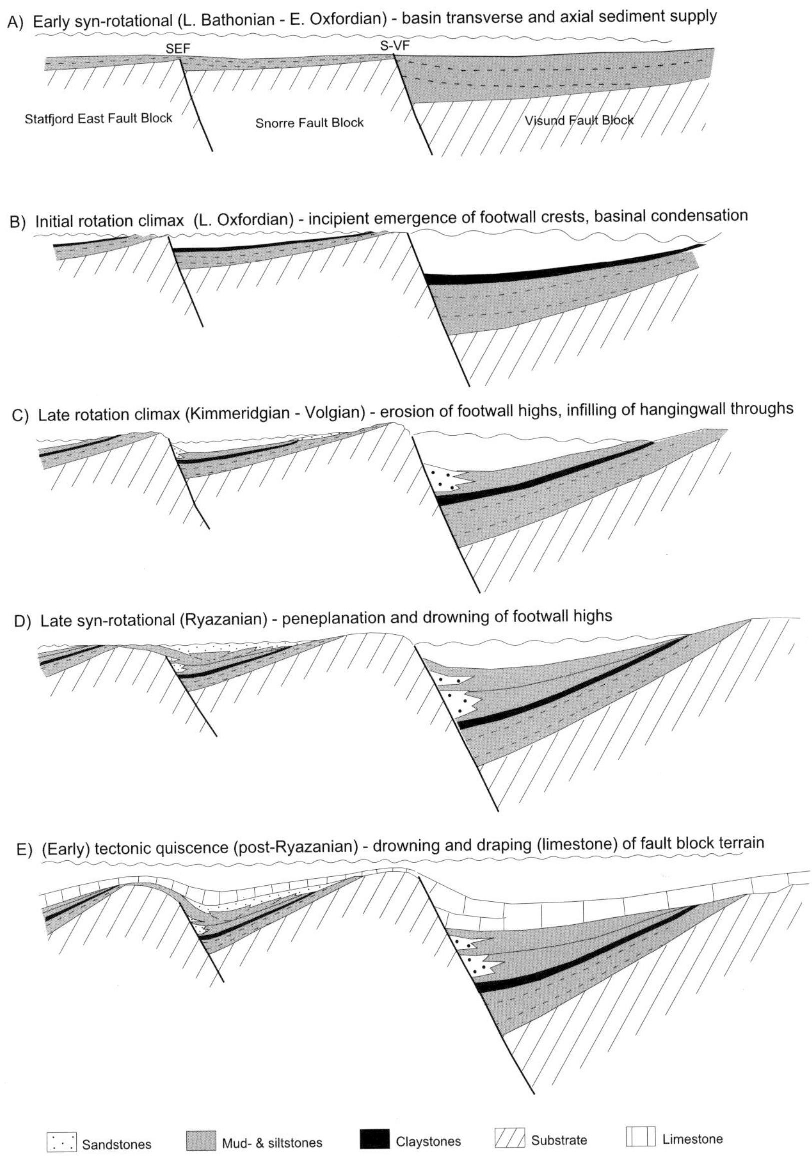

Fig. 17. Schematic tectonostratigraphical development of the Statfjord (left), Snorre (central) and Visund (right) Fault Blocks during the middle Jurassic-earliest Cretaceous, i.e. throughout the entire rift episode and the immediate post-rift stage. At the larger scale the evolution can be viewed as representing the early syn-rotational (a) through initial rotation climax (b) and late rotation climax (c), to the late syn-rotational (d) and early post-rift/tectonic quiescence stages (e). However, the longer-term evolution is likely to have been interrupted by a series of shorter-term rift phases, e.g. (a) trough (c), characterized by higher tectonic activity and increasing rates of faulting (see text for further discussion).

down the hangingwall dip-slope. Progradation and accumulation of the upper Draupne shallow marine sands across the Snorre H-area, is interpreted as a result of waning fault activity and decreasing rates of rotation (Fig. 17d, e).

Visund half-grabens (Profile 3)

The Visund fault block constitutes the easternmost structure in a series of N–S oriented, westerly tilted fault blocks between the northern

Viking Graben and the East Shetland Platform in the northernmost North Sea. The block is 23–26 km wide, is bounded in the west by the Inner Snorre Fault and in the east (against the Viking Graben proper) by the Visund Fault (Fig.1). Extension along these major basement-involved faults caused the pre-rift surface of the Visund half-graben to rotate by about ten degrees during the middle–late Jurassic syn-rift interval. This is steeper than normal in the area (Færseth *et al.* 1995*b*)

Syn-rift tilting of the Visund fault block occurred at a lower rate during the first *c.* 10 Ma interval (early Bathonian–middle Oxfordian) and more intensely for another 17 Ma (middle Oxfordian–early Ryazanian), causing both Heather and Draupne Formations to have clear wedge-shaped geometries (Fig. 17). The syn-rift infill is more than 1.5 km thick (decompacted) in the deepest part of the half-graben, but is absent over the crestal areas.

The pre-rift surface is estimated to have undergone a 2° tilt during deposition of the Bathonian–middle Oxfordian Heather Formation (Færseth *et al.* 1995b), during which time calcareous siltstones and mudstones formed an early syn-rift wedge of significant dimensions (Fig. 16). There appears to have been little sand deposition at this early stage and it is unlikely that footwall uplift created any substantial subaerial landscape. Although most of the present-day tilt on the fault block occurred later, this early syn-rift rotation was sufficient to create a modest syn-rift unconformity across the crestal areas of the block.

The middle Oxfordian–Ryazanian Draupne Formation, also developed as a clear wedge-shaped unit within the Visund half-graben (Fig. 16), accumulated during accelerated rotation of the fault block, resultant from a series of rift phases from middle Oxfordian (from a 2° tilt) to late Volgian–Ryazanian (to a 10° tilt; Færseth *et al.* 1995*b*). Structural complications on the block occurred by middle Kimmeridgian times when a *c.* 6° tilt had been attained. A sizable part of the crestal area collapsed at this time, forming the smaller, tilted fault block which now occurs on the SE-flank of the Visund structure (Figs 17a, b). The present-day crestal area therefore developed as an emergent footwall island during the final 4° of tilt on the structure (Færseth *et al.* 1995*b*).

Both the main Visund half-graben and the smaller flanking half-graben were infilled mainly from middle Kimmeridgian times. The main sub-basin was sediment-starved and filled largely by claystones, though there were at least two episodes of sandy turbidite infill. The smaller half-graben received much coarser sediment from accelerated (Volgian) footwall uplift and drainage from the new crestal area. A wedge-shaped unit with some 150 m of slumped material, and conglomeratic debris flows and turbiditic sandstones has been documented here, at a distance of some 2 km out from the main fault scarp (well 34/8-7, Fig. 16). As with other syn-rift infills, upward-fining (sometimes upward coarsening-to-fining) signatures, on a scale of 20–30 m are common in this conglomeratic succession, and are possibly symptomatic of sediment supply pulses, or of sediment supply passing certain threshold conditions in the drainage basin.

The overall succession in the smaller half-graben, as in the main half-graben succession, shows a general fining upward trend. Early sandy or gravelly mass flow deposition evolved through time to give a mudstone-prone succession, probably symptomatic of underfilling and sediment starvation in half-grabens at large distance from the rift hinterland regions (Ravnås & Steel 1998).

Lomre Terrace–Uer Terrace–Horda Platform (Profile 3)

This section describes deeper marine successions deposited during the late Oxfordian–Volgian rift on the Uer and Lomre terraces along the eastern flank of the northern Viking Graben (Fig. 1). The Uer and Lomre Terraces form two wide, segmented, gently eastward and westward tilted terraces, respectively. These structures are situated between the northern Viking Graben, Sogn Graben and the Horda Platform (Fig. 2c). The two terraces can be considered as forming a large-scale relay zone or transfer zone between the Horda Platform and the deep Sogn Graben to the north (Fig. 1). Two syn-rift wedges are recognized and these correspond to a late Oxfordian–Kimmeridgian and a Volgian rift phase, respectively (corresponding to the lower wedge between timeline 18/19-22 and the overlying wedge, respectively, in Fig. 18).

In basinal areas of the Lomre and Uer Terraces, several kilometres from the sub-basin's boundary faults, the syn-rotational succession of the lower wedge forms a fining upward succession of mudstones and intercalated, generally thin (commonly less than few tens of metres) sediment gravity flow sandstones (e.g 35/8 wells, Fig. 18). The basal part of the wedge is generally sand-prone, and consists of a fining-upwards succession of stacked, smaller-scale

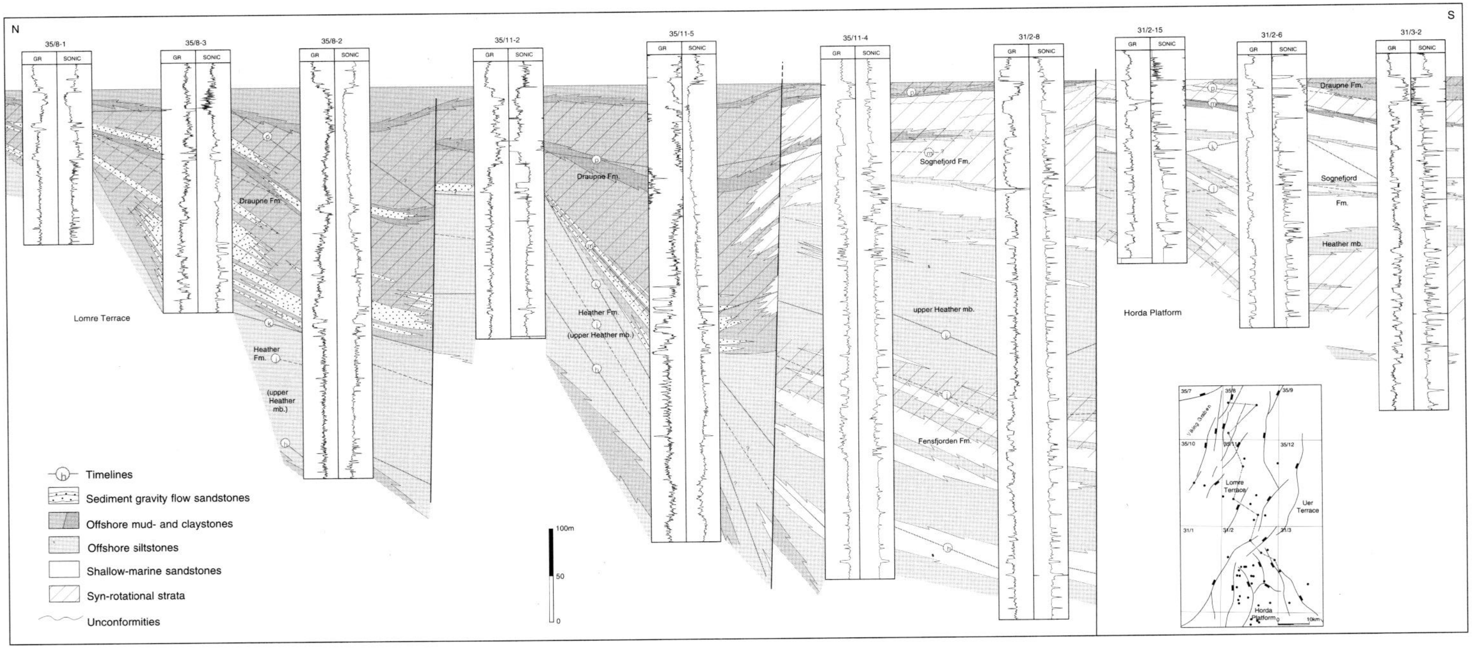

Fig. 18. NW–SE well correlation of the upper parts of the Viking Group (middle and upper Heather members, and the Fensfjord, Sognefjord and Draupne Formations) from the Sogn Graben/Lomre Terrace across the Uer Terrace to the Horda Platform. Two syn-rotational wedges are recognized in the upper Heather member, and the Sognefjord and Draupne Formations. The lower wedge is represented by an unconformity in the footwall, i.e. updip on the Horda Platform; forestepping-to-backstepping shallow-marine sandstones on the platform-bounding terraces; and backstepping or a fining-upward offshore/deeper-marine sediment gravity flow sandstone-hemipelagic mudstone package in basinal positions, i.e. to the northwest. The upper wedge is represented by an unconformity in footwall positions and on the terrace area; and a forestepping-to-aggradational-to-backstepping hemipelagite package in the sub-basins to northwest. Intervening and capping tectonic quiescence/early post-rift strata consist of condensed claystones, which locally in the basinal sinks envelop sediment gravity flow sandstones (e.g. well 35/8-2).

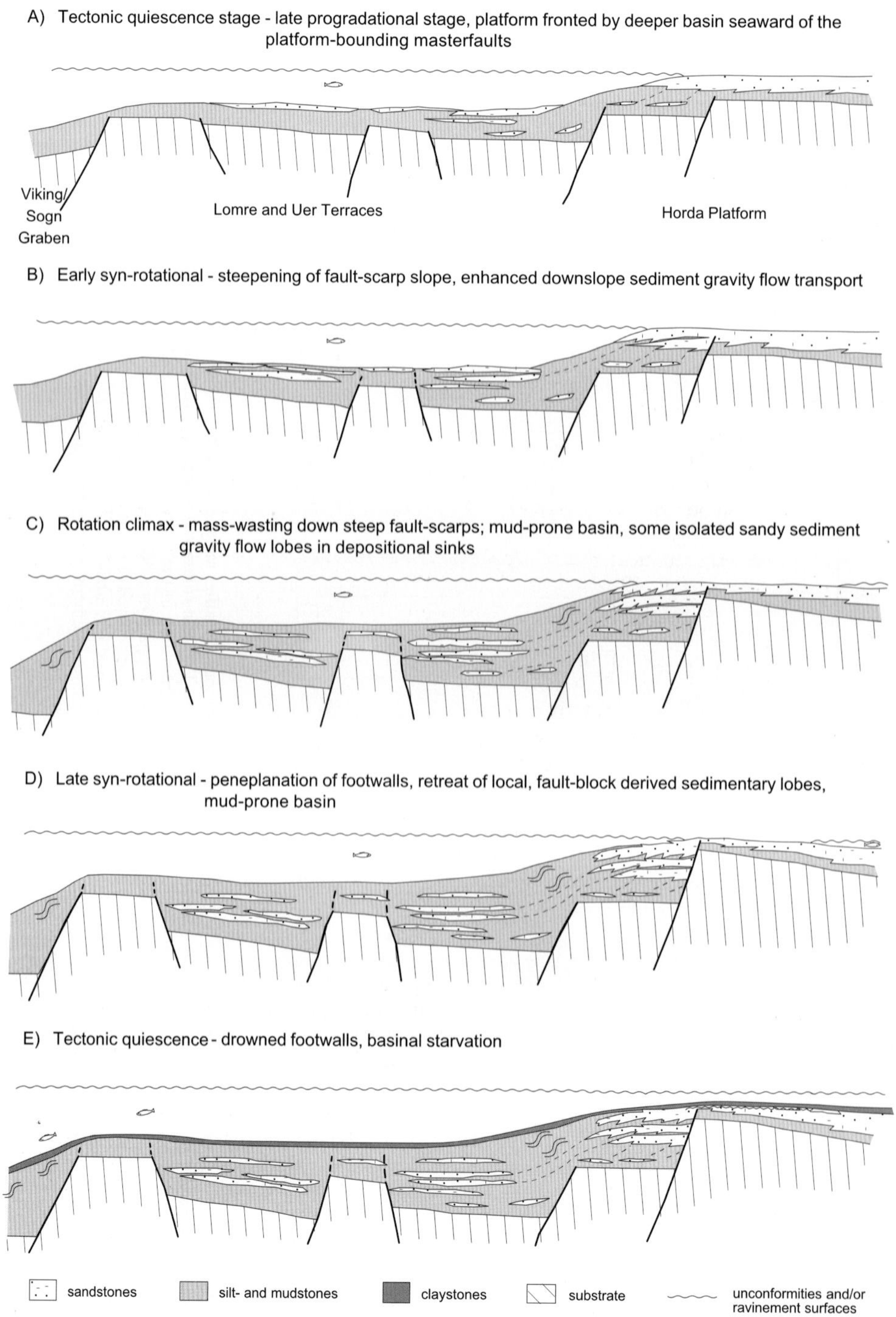

Fig. 19. Schematic structural development and infilling during Late Oxfordian–Volgian rifting along the northwestern edge of the Horda Platform and Lomre Terrace. The fault block to the right represents the updip edge of the Horda Platform, whereas the next, small fault block to the left represent the terraces bounding the Horda Platform to the northwest. The central horst and graben topography represents the larger Uer and Lomre Terraces, whereas the deeper basin to the left represents the northern Viking Graben or Sogn Graben.

fining-upwards sediment gravity flow-hemipelagite packages. These are interpreted to have been fed by a shallow marine system temporarily stationed at the seaward margin of the Horda Platform (Fig. 19a, b). The stationary position of the shallow marine feeder system was due to increasing subsidence rates along the Horda Platform's western boundary fault, which, in turn, positioned the palaeo-shelfbreak along the surface trace of this deep-seated basement fault.

In hangingwall positions close to the sub-basin-bounding masterfaults (e.g. well N35/11-5, Fig. 18), a coarsening-to-fining upward wedge is present overlying the basal sediment gravity flow-hemipelagite unit. This reflects the advance and retreat of a coarse submarine apron, and consists of interbedded sandstone, sometimes several tens of metres thick, and finer-grained mud- and siltstones. These wedges were predominantly derived by erosion of the adjacent footwall, though temporarily and locally the overall retreating rift marginal Sognefjord shoreline may have delivered sediment directly to the deeper marine wedges (Fig. 19c, d). The latter may be particularly common on the smaller, downfaulted fault blocks or terraces immediately basinward of the Horda Platform. In footwalls of the major basin-bounding faults the syn-rotational stage is represented by a footwall unconformity and overlying, thin transgressive deposits or ravinement lags only (e.g. well N31/2-15, Fig. 18).

The upper wedge is fine-grained throughout, although with a distinct coarsening-to-fining upward signature (e.g. well N35/11-5, Fig. 18). This is related to the renewed influx from adjacent highs during the rotational tilt climax. The syn-rotational stage is represented by an unconformity across the most pronounced footwall highs, similar to that observed in the underlying wedge. This unconformity may, in some places, have formed by submarine wave erosion and/or be related to gravitational collapse along the fault-scarps. Across smaller and less distinct highs the syn-rotational stage is represented by a condensed section. This suggests that these highs were submerged throughout the rotational tilt stage. Their starved syn-rotational sedimentary signature indicates that coarser sediment delivered by sediment gravity flows were ponded or bypassed in the adjacent lows.

As in other deeper marine sub-basins, the intervening tectonic quiescence stages are represented by clay-prone deposits with a high source-rock potential which blanket the entire sub-basins (Fig. 19e). The claystone blankets are commonly thin across the intrabasinal highs, whereas they thicken towards the intrabasinal structural lows. This is probably due to the addition of some fine-grained turbidites in such settings. Thicker, coarse, tectonic quiescence deposits are observed only locally (e.g. well N35/8-2, Fig. 18).

In the western, updip reaches of the Horda Platform, Cretaceous and locally Tertiary deposits rest unconformably upon truncated Late Jurassic strata. The updip areas therefore appear to have been subaerially exposed during the late syn-rift stage (possibly from the Late Oxfordian or Kimmeridgian) and well into the post-rift stage. The limited erosion of the updip reaches suggests that there was limited relief along the fault-scarp.

Northern North Seas rift basin fills: discussion and conclusions

Controls on syn-rift sedimentation

Cyclic development of syn-rift deposits may result from variations in either the rate of extensional faulting, sediment supply and/or eustatic fluctuations. In the basinal situations discussed above, varying sediment supply superimposed on a varying rate of rotational faulting is advocated as the major control on the resultant syn-rift sedimentary architecture. Syn-depositional temporal variations in the rate of rotational faulting is evidenced by a combination of two or more of the following factors: (1) wedge-shaped stratal packages showing internal stratal fanning alternating with packages with internal tabular geometries; (2) differential growth of time-equivalent stratal packages and/or evidence of differential change in palaeobathymetry across extensional faults; (3) change in drainage development and sediment supply directions from a predominantly regional rift marginal drainage or starved basinal conditions to local fault-block drainage systems and sediment supply; (4) evidence of downdip drowning/deepening concomittant with updip shallowing/emergence in individual half-grabens; (5) change in basin physiography from a larger open basin with similar or gradual changes in depositional environments across the entire rift-basin to large variations in basinal conditions between individual sub-basins. In additon, the rate of extensional faulting appear to have varied spatially. Spatial variations in extension rate resulted in variations in the onset of syn-rift conditions across the basin, as well as in variations in structural topography and accommodation creation between individual sub-basins. Also the background, basinwide subsidence/uplift rate may have varied temporarily and spatially, though this is more difficult to address and quantify. The background, basinwide uplift/subsidence nevertheless were likely highly influential on the resultant tectonoeustatic evolution of the rift basin (e.g. see Surlyk *et al.* 1993), and needs to be taken into account when discussing basinwide tectonic evolution, depositional history and/or correlation of events and sedimentary strata.

Sediment supply to individual sub-basins is likely to have changed as a response to changing structural styles and the evolving structurally formed topography/bathymetry (e.g. see Ravnås & Steel 1998). Also the regional or rift marginal drainage and sediment supply is likely to have changed during the rift cycle as a response to hinterlands tectonics. Changes in rift marginal drainage were, however, most likely more instrumental on the sediment supply to the rift marginal sub-basins, i.e. on the Horda Platform and Måløy Fault Blocks, and to the rift marginal sub-basins on the opposing UK margin. With respect to the middle–late Jurassic rift episode, the syn-rift infill were derived by erosion of poorly consolidated or semi-consolidated Triassic and early–middle Jurassic sedimentary successions. Drainage development and sedimentary response to tilting and uplift events are therefore likely to have been fairly rapid, at least for the North Sea rift interior sub-basins. With respect to the rift marginal depositional systems, these probably drained a basement terrain with overlying Paleozoic–Mesozoic strata. Hence, the sedimentary response of rift marginal drainage systems to hinterlands tectonics may also have been fairly rapid. Potential lags in response time due to erosional threshold needing to be exceeded, may have formed in areas were the hinterland consisted of crystalline basement terrains or was underlain by a metamorphosed substrate.

The main control on the sedimentary architecture of rift basin infill was thus sediment supply in relation to rate and style of accommodation creation. Creation of accommodation space was controlled by eustasy, tectonically induced subsidence/uplift, pre-rift waterdepth/elevation, basin geometry and basinfloor tilt rate. During the middle-late Jurassic, which was a non-glacial period, the rate and magnitude of eustatic sea-level fluctuation are likely to have been suppressed by tectonically induced base- and sea-level variations (Steel 1993; Ravnås & Steel 1998). Sediment flux was controlled by a wide range of interrelated factors, the most important being climate, tectonism, hinterland physiography, hinterland lithologies, hinterland and basinal drainage patterns, depositional patterns, especially whether point or linear sourced, basinal sediment dispersal patterns, and relative sea-level fluctuations. In rift basins consisting of a series of sub-basins such as the northernmost North Sea, extensional faulting causes considerable topography within what used to be a single, broader, low-relief basin, or enhances the previous topography. Sediments delivered from rift marginal sources was thus trapped in sub-basins adjacent to the hinterland. These sub-basins became sediment-overfilled or sediment-balanced (Nøttvedt *et al.* 1995; Ravnås & Steel 1998). Sub-basins at increasing distance from the hinterlands relied progressively on supply from rift interior sources. As these sources commonly were small compared to the adjacent half-graben sinks, sub-basins at increasing distance from the hinterlands received progressively less sediment, and thus became either sediment-underfilled or sediment-starved. The detailed syn-rift sedimentary architecture of these four basin infill types, i.e. the sediment-overfilled, sediment-balanced, sediment-underfilled and sediment-starved, is discussed by Ravnås & Steel (1998).

Milton (1993) demonstrated how the dip of the depositional foundation influence the resultant aspect ratio, i.e. the dip length : average thickness, of the syn-rift sedimentary geometries. Narrower facies belts with more rapid transtions between individual facies tracts should be expected on steeper dipping slopes compared to gentlier dipping slopes (Ravnås & Steel 1998). Such changes in depositional geometries are generally observed when comparing syn-rift infill related to earlier phases of the middle-late Jurassic rift episodes to those related to later phases. Syn-rift infill related to later phases generally have narrower and sometimes thicker facies belt, mainly due to the steep dips of the depositional foundation (up to 10°) due to the cummulative tilt resulting from recurrent rift phases.

Climate also exerted a major control on the resultant rift basin infill, as demonstrated by a change from arid to semiarid conditions in the late Permian–early Triassic, from a semi-arid to a more humid environment in the late Triassic–early Jurassic, and a possible return to a dryer climate in the Kimmeridgian–Volgian (e.g Clemmensen *et al.* 1980; Røe & Steel 1985; Wignall & Ruffell 1990; Wignall & Pickering 1993). Although such climatic changes may be diachronous across a large area, and significant in large, elongate rifts (such as the East African Rift–Red Sea–Gulf of Suez, and the Rio Grande Rift), they are likely to have occured relatively rapidly and caused less lateral variations within a restricted area such as the northern Viking Graben and adjacent platform areas.

Hierarchy of rift basin successions

Rift basin successions commonly contain infill segments related to the various stages of active rifting and are enveloped by sediments deposited prior to and subsequent to the rift episode proper, leading to a subdivision into pre-, proto-, syn- and post-rift strata (e.g. Nøttvedt *et al.* 1995). In cases where the rift basin was subjected to repeated stretching episodes, the sediments deposited during the intervening periods with relative tectonic quiescence, e.g. the middle Triassic-middle Jurassic of the Northern North Sea, differ in many respect from typical post- or pre-rift strata. We suggest therefore that such successions should be referred to as inter-rift.

As demonstrated by the middle-late Jurassic syn-rift successions of the northern North Sea, rift episodes can consist of higher-frequency tectonic events or rift phases. In such cases the syn-rift succession consists of a series of syn-rotational and intervening tectonic quiescence stratal packages. Whether such rift phases reflect true temporal variations in stretching rate recorded in basinwide near-synchronous or slightly diachronous rifting acmes, or can be attributed to initiation, propagation, migration, lateral shift or cessation of faulting along the sub-basin-bounding masterfaults, is yet not clear. At present, it is also difficult to suggest whether a pulsatory or a single-phase rift development is the more common.

In any case, the presence of repetitive rift episodes and the potential of repetitive rift phases occuring within a single rift episode, suggest that the rift basin successions can be ordered hierarchically (Fig. 20a, b). At the highest rank are the syn-rift strata representing the main rift episodes and the intervening inter-rift strata, and possible pre- and post-rift strata. These commonly represent a time interval of a few tens of million years, and are given a lowest or first order. Next are the syn-rotational strata and the intervening tectonic quiescence packages which relate to a single rift phase. In the middle-upper Jurassic of the northern North Sea, a combined syn-rotational and tectonic quiescence unit commonly represents some 4–6 Ma. In some cases it is possible to further subdivide the syn-rotational strata into higher-order packages related to distinct rotational tilt events, e.g. as established for the late Bajocian–early Bathonian Tarbert Formation transgressive–regressive sequences. These units commonly span some 1–2 Ma and possibly represent a cluster of faulting events. Such rotational tilt events correlate with discrete deformation spans. Theoretically, it should be possible to further subdivide syn-rotational successions into even higher-order units representing discrete, short-lived faulting events or deformation clines. At present, this is rather difficult, if not impossible, in the subsurface without good seismic database and/or abundant wells.

As with the syn-rift strata, the intervening inter-rift strata can sometimes be subdivided into packages which represents prolonged intervals of tectonic quiescence separated by units which correlate with discrete rift stages or minor rift phases (Fig. 20c). In such cases, syn-tectonic strata are expected to represent quite short time-intervals compared to the time span represented by the entire inter-rift package. Although the inter-rift tectonic stages may be insignificant with respect to the basinwide stretching factor, they appear to exert a major influence on the sedimentary architecture of the inter-rift succession (Steel 1993). As such, these events need to be taken into consideration if a proper understanding of the development and distribution of potential inter-rift reservoir and source rock intervals are to be achieved.

Sedimentary architecture of syn-rift packages related to the entire rift episode

At a large scale, a rift episode may be represented by a threefold, sandstone–mudstone–sandstone package, and this is typically developed in half-grabens where there is an abundant sediment supply, for example non-marine, lacustrine or fluvially dominated rifts (Fig. 21a; see also Hamblin & Rust 1989; Lambiase 1990), but has been advocated for marine rift basins as well (Fig. 21b, c; Prosser 1993; Leinfelder & Wilson in press). In the northern North Sea, marine syn-rift successions related to an entire rift episode, e.g. the middle-upper Jurassic, appear to be dominated by twofold, fining-upward, sandstone–mudstone packages (Ravnås *et al.* 1994; Nøttvedt *et al.* 1995). In this rift basin individual sub-basins seem to be completely infilled either well into the post-rift period or subsequent to continent-continent separation (in the case of abortive or successful rifting, respectively).

These large-scale threefold or twofold architectural lithosome models are in some cases

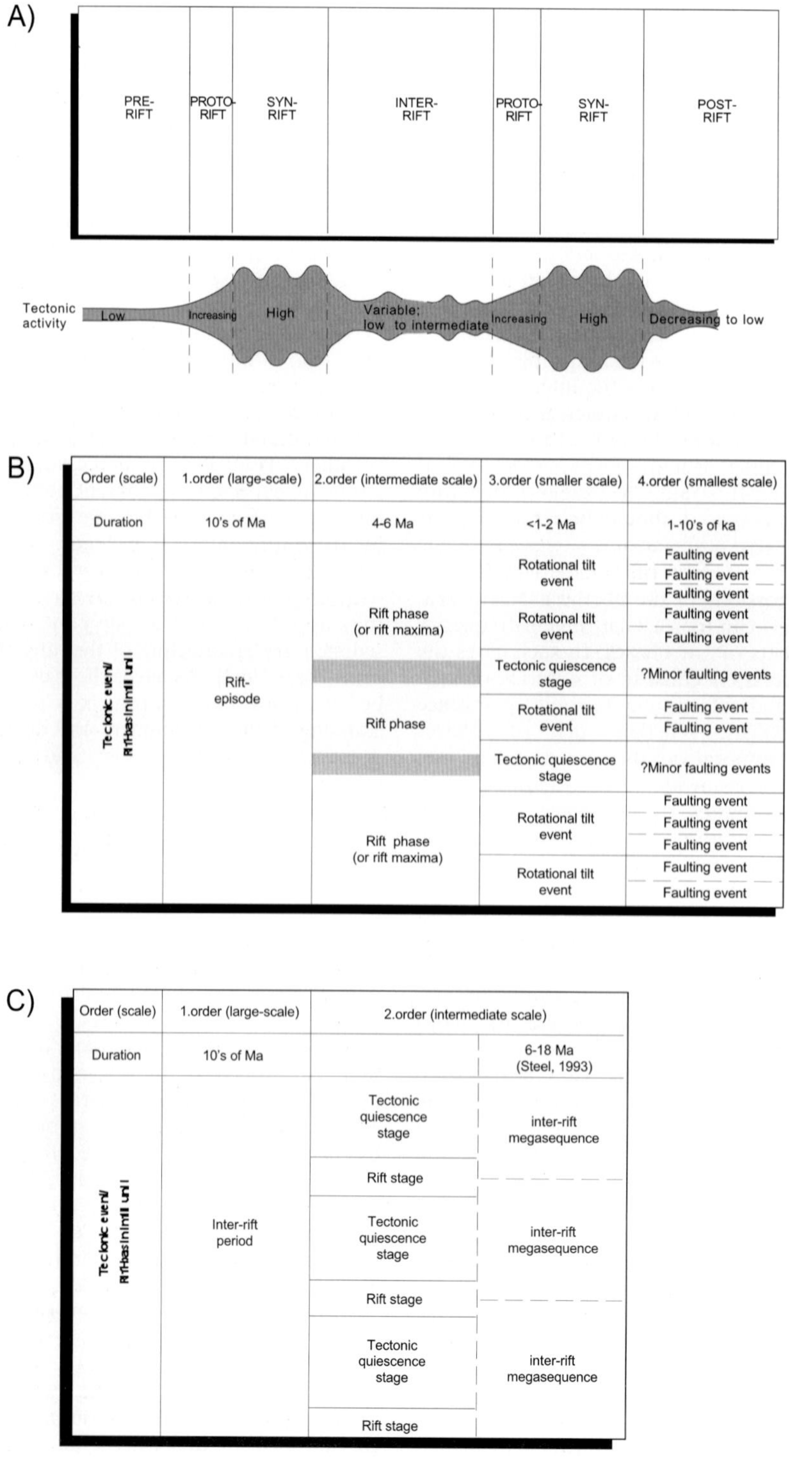

Fig. 20. Hierarchy of tectonic events during rift basin formation and development, and proposed nomenclature for rift basin sediment infill. (a) Relates to the entire rift basin evolution whereas (b) relates to a single, multiphase rift episode, and (c) relates to the inter-rift period.

paralleled by a vertical stacking of repetitive syn-rotational and tectonic quiescence packages. In such cases, the entire (or parts of the) syn-rift succession may display a cyclic development, the cyclicity reflecting the complex interplay between temporal variation in rates of extensional faulting, sediment supply and, for marine syn-rift successions, eustatic sea-level fluctuations. So far, syn-rift cyclicity due to temporal variation in the rate of extensional faulting, whether basin-wide or pertaining to individual half-grabens, has been demonstrated only for marine, mixed non-marine/marine, and lacustrine syn-rift successions (e.g. Sellwood & Netherwood 1984; Purser *et al.* 1990; Lambiase 1990; Underhill 1991*a*, *b*; Partington *et al.* 1993; Rattey & Hayward 1993; Færseth *et al.* 1995*a*). This documentation may be achieved if there is a reference level such as relative sea-level or local lake level, which is sensitive to and immediately responds to temporally changing tectonics in the basin. Non-marine rift basins may similarly have a multiphase rift-history (P. Theriault, pers. comm. 1995), though often there is a lack of stratigraphic evidence to suggest such a complex basin development.

Sedimentary architecture of syn-rotational packages related to a single rift phase

In the middle–upper Jurassic of the northern North Sea, syn-rotational strata related to a single rift phase show a wide range of possible sedimentary architectures as well. The subdivision of rift basins into a continuum of basin-types from sediment-overfilled through sediment-balanced and sediment-underfilled to sediment-starved, allows a distinct sequential development and hence specific sedimentary architecture to be erected for each basin type. The tectonic significance of stratigraphic surfaces and the variable sedimentary infill trends, from the early syn-rift through the rift climax and to the late syn-rift stage, related to a single rift phase for each of these four rift basin types, are discussed in more detail by Ravnås & Steel (1998). Notably, lithosome models for marine syn-rift successions related to a single rift phase, whether shallow- or deep-marine, may show threefold sandstone–mudstone–sandstone, two-fold sandstone–mudstone, or single mudstone) signatures. The main difference between the threefold and twofold marine syn-rift signatures relates to the extent to which the (changing) sediment supply is able to keep pace with varying basinal subsidence rates.

Shallow marine, threefold syn-rift signatures formed in sub-basins with low accommodation generation (low basinal subsidence rates) and/or high sediment supply. In the northern North Sea rift basin, such infill appears to be restricted to the early phase of the middle–late Jurassic stretching episode, and to rift marginal sub-basins during later phases. Examples of such infill types are the Bajocian–Bathonian Tarbert Formation in the Viking Graben, and the Bathonian–Kimmeridigian Heather, Krossfjord, Fensfjord and Sognefjord Formations on the Horda Platform (Figs 4 & 11).

Shallow marine twofold syn-rift signatures formed in sub-basins adjacent to source areas with limited sediment yield potential. Such sub-basins were commonly located at some distance from the regional hinterlands of the rift basin, or represent the transition from a mixed non-marine and shallow marine to a deeper marine rift basin as the rift evolved and subsided. The Bathonian–Kimmerdigian Heather Formation in the Viking Graben proper and in flanking half-grabens such as the Oseberg Fault Block (Fig. 8), are good examples of this infill type.

Deeper marine sub-basins with a threefold syn-rift stratigraphy had access to large hinterland areas, either the rift marginal hinterland or large rift interior sources. In such infill types, shallow marine lithosomes may or may not be present updip on larger fault blocks. This variant is rather unusual in the middle-upper Jurassic of the northern North Sea. The late Oxfordian–early Volgian of the Penguin half-graben (Fig. 12) constitutes the only presently recognized examples. The deeper-marine variant of the threefold lithosome model contrasts with its shallow marine counterpart by containing abundant sandstones in the strata corresponding to the rotational tilt climax, and by the common sharp basal contact of its capping of forestepping turbidites. In the combined shallow-to-'deeper'-marine variant exemplified by the Kimmeridgian–Volgian of the Stafjord and Snorre areas (Figs 14 & 16) there is a gradual transition from the offshore mudrocks to the shallow marine sandstone capping.

Deeper marine sub-basins with a twofold syn-rift signature, owe their lithosome character to the limited sediment yield potential of the adjacent source areas. This infill type is by far the most common of the deeper marine variants, and examples include the Early Volgian–Ryazanian in the Penguin half-graben (Fig. 12), and the Kimmeridgian–Ryazanian in the Viking Graben (Fig. 6), Oseberg Fault Block (Fig. 8), Visund Fault Block (Fig. 16) and Sogn Graben (Fig. 18).

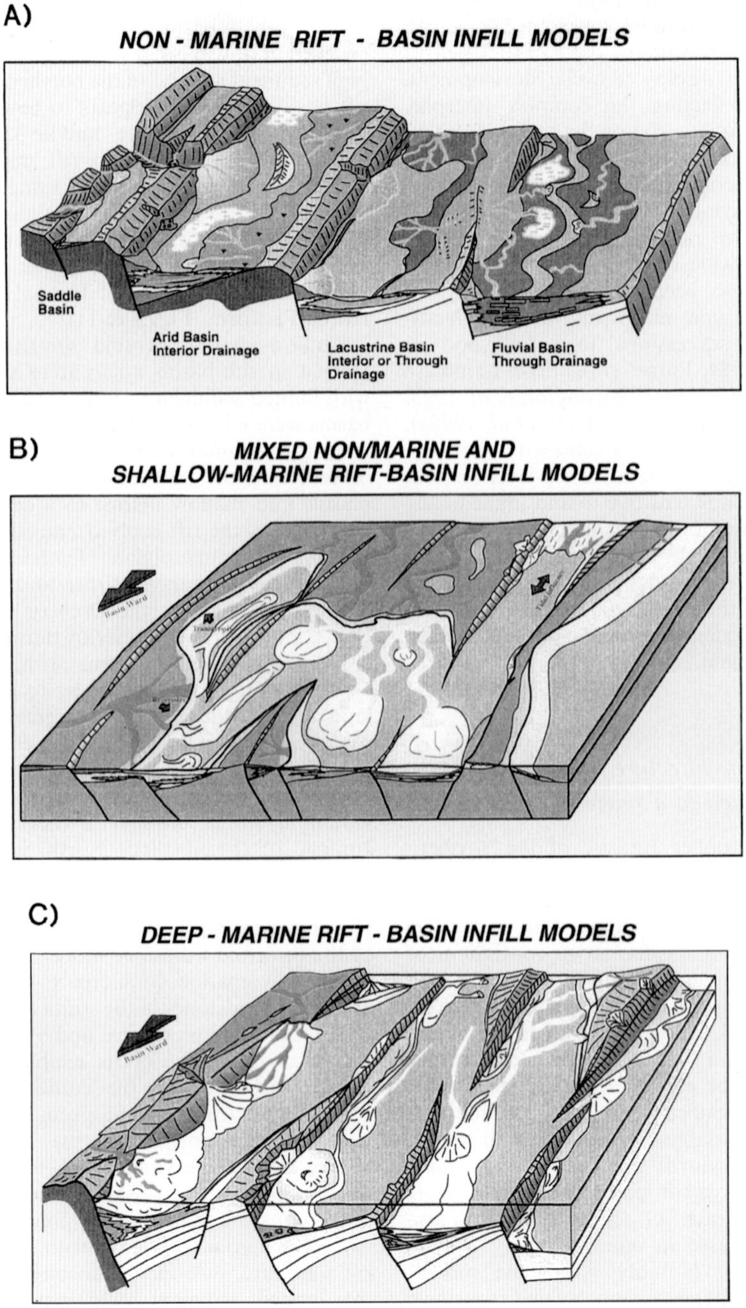

Fig. 21. Rift basin infill models: (a) Non-marine rift basins; (b) mixed non-marine-marine and shallow marine rift basins; and (c) deep-marine rift basins.

Concluding remarks

The syn-rift infill of the northernmost North Sea rift basin vary from predominantly non-marine (Fig. 21a) during the late Permian–early Triassic rift episode to dominantly marine (Fig. 21b, c) during the middle–late Jurassic rift episode. The middle Triassic–middle Jurassic non-marine to marine inter-rift succession show evidence that discrete rift stages or minor rift phases occurred intermittently also throughout the inter-rift period.

The Permian–early Triassic rift basin and the broader middle–late Triassic non-marine basin probably could not easily access marine waters. Data from the Permian–early Triassic syn-rift infill are too scant to suggest any evolutionary trends in time. Probable subsidence maxima in the middle–upper Triassic non-marine inter-rift succession correspond with parts of Teist, Middle Lunde and uppermost Lunde Formations lacustrine intervals with some brackish incursions (Steel 1993). This has resulted in an infill related to each inter-rift rift stage/rift phase that show a threefold architectural signatures typical for sediment overfilled and sediment balanced basins. Otherwise there appears to be no particular evolutionary trends in time associated with the syn-rift infill related to middle–late Triassic tectonic maxima. Potential source rocks include lacustrine claystones and coal-bearing successions.

The mixed non-marine and marine syn-rift infill related to early-middle Jurassic rift stages show a sedimentary architecture mimicing the infill related middle-late Jurassic rift phases, although with a subdued asymmetry of the syn-rift infill and only weakly developed footwall unconformities.

The middle–late Jurassic marine syn-rift successions of the northern North Sea rift basin reflect an evolutionary path from a mixed non-marine/marine to a fully marine environment. Characteristically the middle–late Jurassic rift episode developed through a series of rift phases separated by intervals of relative tectonic quiescence. Furthermore, the nature of the syn-rift infill varies considerably between the distinct rift phases. The transition from a threefold mixed non-marine and shallow marine syn-rift infill through a twofold shallow marine syn-rift infill to a twofold deep-marine syn-rift infill, reflects the evolution from a sediment-overfilled or sediment-balanced status to an increasingly sediment-underfilled and sediment-starved status. This evolution tends to have been paralleled by an increase in the pre-rotational elevation/waterdepth and the distance to the rift marginal hinterland areas, as well as to a decrease in the rift interior sediment yield potential.

In the lower part of the middle–late Jurassic syn-rift infill, both syn-rotational and tectonic quiescence sand-prone strata form reservoir intervals. As the rift evolved and subsided, shallow marine syn-rift reservoir intervals became confined to narrow belts along the rift basin's margin, i.e. to the east on the Horda Platform and west in the East Shetland Basin. Within the rift itself, significant shallow marine strata were deposited only along larger footwall islands. Deeper marine reservoir intervals are either syn-rotational in nature, typically developed as fault-scarp or relay ramp aprons and fans; or sub-basinwide, sheet-like sediment gravity flow packages, deposited during tectonic quiescence stages. Influx of syn-rift sandstones from rift interior sources usually represent distinct tectonic phases accompanied by pronounced footwall uplift or major reorganization of the tectonic regime which lead to the initiation of new masterfaults (e.g. Færseth *et al.* 1995*a*; Færseth & Ravnås 1998). Rift interior sand-prone tectonic quiescence packages should be expected only in sub-basins bordering large sediment sources (Fig. 21b, c; Ravnås & Steel 1997).

Syn-rift source rocks are more common in the latest Jurassic deeper marine successions. Although intervals with high source rock potential are present in the shallow marine to shelfal Heather Formation, it is the claystones of the Kimmeridge Clay and Draupne Formations that are the prime source rock in the northern North Sea rift basin. The source rock potential varies considerably also within these units: claystones deposited during tectonic quiescence intervals generally appear to have better source rock potential than their syn-rotational equivalents (Ravnås & Steel 1997). Secondly, claystone drapes across intrabasinal highs and fault block updip areas also seems to be richer than their downdip correlative. This may be attributed to an extreme starvation of siliciclastic material across such areas, and a higher export rate of organic matter to the bottom due to the relatively shallow palaeowater depths. The source rock potential of claystones deposited in downdip areas may be diluted by the higher frequency of intercalated silty and sandy turbidites. The formation of the Late Jurassic organic rich claystones in the northern North Sea was at least partly due to the formation of tectonically silled, rhomb-shaped sub-basins, the basin shape reflecting the interplay between the existing N–S and the subsequently formed NE–SW trending structural grains (Færseth & Ravnås 1998).

We would like to thank our fellow colleagues and research students in the Joule II: Integrated Basin Studies–Dynamics of the Norwegian Margin (IBS–DNM) project, and colleagues at respective companies and research institutions for stimulating co-operation and discussions. Reviews by Arne Dalland, Bjørn Tore Larsen, Snorre Olaussen and Sarah Prosser sharpened the final product and are appreciated. The drafting office at Statoil, Norsk Hydro and Norske Conoco prepared the illustrations. The work has been funded by the Research Council of Norway as part of the Joule II Research programme (CEC contract No. JOU2-CT92-0110).

References

ALEXANDER, J. & LEEDER, M. R. 1986. Active tectonic control on alluvial architecture. *In*: FLORES, R., ETHRIDGE, F. & HARVEY, M. (eds) *Recent Developments in Fluvial Sedimentology*. Society of Economic Paleontologists and Mineralogists Special Publications, **39**, 243–252.

BADLEY, M. E., EGEBERG, T. & NIPEN, O. 1984. Development of rift basins illustrated by the structural evolution of the Oseberg feature, block 30/6, offshore Norway. *Journal of the Geological Society, London*, **41**, 639–649.

——, PRICE, J. D., RAMBECH DAHL, C. & AGDESTEIN, T. 1988. The structural evolution of the northern Viking Graben and its bearing upon extensional modes of basin formation. *Journal of the Geological Society, London*, **145**, 455–472.

BEACH, A., BIRD, T. & GIBBS, A. D. 1987. Extensional tectonics and crustal structure: deep seismic reflection data from the northern North Sea Viking Graben. *In*: COWARD, M. P., DEWEY, J. F. & HANCOCK, P. L. (eds) *Continental Extensional Tectonics*. Geological Society, London, Special Publications, **28**, 467–476.

BLUNDELL, D. J. & GIBBS, A. D. 1990. *Tectonic Evolution of the North Sea Rifts*. Oxford University Press, Oxford.

BOOTH, J., SWIECICKI, T. & WILCOCKSON, P. 1993. The tectono-stratigraphy of the Solan Basin, west of Shetland. *In*: PARKER, J. R. (ed.) *Petroleum Geology of Northwest Europe: Proceedings of the 4th Conference*. Geological Society, London, 987–998.

CHRISTIANSSON, P., FALEIDE, J. I. & BERGE, A. M. 1999. Crustal structure in the North Sea – an integrated geophysical study. *This volume*.

CLEMMENSEN, L. B., JACOBSEN, V. W. & STEEL, R. J. 1980. Some aspects of Triassic sedimentation and basin development: East Greenland, North Scotland and North Sea. *In*: *The Sedimentation of the North Sea Reservoir Rocks*. Norwegian Petroleum Society, Geilo, paper XVII.

DREYER, T. & WIIG, M. 1995. Reservoir architecture of the Cook Formation on the Gullfaks field based on sequence stratigraphic concepts. *In*: STEEL, R. J., FELT, V. L., JOHANNESSEN, E. P. & MATHIEU, C. (eds) *Sequence Stratigraphy on the Northwest European Margin*. Norwegian Petroleum Society Special Publication, **5**, 109–142.

EYNON, G. 1981. Basin development and sedimentation in the Middle Jurassic of the Northern North Sea. *In*: ILLING, L. V. & HOBSON, G. D. (eds) *Petroleum Geology of the Continental Shelf of Northwest Europe*. Institute of Petroleum, Heyden, London, 765–774.

FROSTICK, L. E. & Reid, I. 1987. Tectonic controls of desert sediments in rift basins ancient and modern. *In*: FROSTICK, L. E. & REID, I. (eds) *Desert Sediments: Ancient and Modern*. Geological Society, London, Special Publications, **35**, 53–68.

——, LINSEY, T. K. & REID, I. 1992. Tectonic and climatic control of Triassic sedimentation in the Beryl Basin, northern North Sea. *Journal of the Geological Society, London*, **149**, 13–26.

FURNES, H., MITCHELL, J. G., ROBINS, B., RYAN, P. D. & SKJERLIE, F. 1982. Petrography and geochemistry of peralkaline, ultrapotassic syenite dykes of Middle Permian age, Sunnfjord, West Norway. *Norsk Geologisk Tidsskrift*, **62**, 147–159.

FÆRSETH, R. B. 1978. Mantle-derived lherzolite xenoliths and megacrysts from Permo-Triassic dykes, Sunnhordland, Western Norway. *Lithos*, **11**, 23–35.

——1996 Interference of Permo-Triassic and Jurassic stretching phases and associated megafault-blocks in the northern North Sea. *Journal of the Geological Society, London*, **153**, 931–944.

—— & PEDERSTAD, K. 1988. Regional sedimentology and petroleum geology of marine, late Bathonian–Valanginian sandstone in the North Sea. *Marine and Petroleum Geology*, **5**, 17–33.

—— & RAVNÅS, R. 1998. Evolution of the Oseberg Fault Block in the context of the northern North Sea structural framework. *Marine and Petroleum Geology*, **15**, 467–490.

——, MCINTYRE, R. M. & NATERSTAD, J. 1976. Mesozoic alkaline dykes in the Sunnhordland region, western Norway: ages, geochemistry and regional significance. *Lithos*, **9**, 332–345.

——, GABRIELSEN, R. H. & HURICH, C. H. 1995*a*. Influence of basement in structuring of the North Sea basin, offshore southwest Norway. *Norsk Geologisk Tidsskrift*, **75**, 105–119.

——, SJØBLOM, T. S., STEEL, R. J., LILJEDAHL, T., SAUAR, B. E. & TJELLAND, T. 1995*b*. Tectonic controls on Bathonian–Volgian syn-rift successions on the Visund Fault Block, Northern North Sea. *In*: STEEL, R. J., FELT, V. L., JOHANNESSEN, E. P. & MATHIEU, C. (eds) *Sequence Stratigraphy on the Northwest European Margin*. Norwegian Petroleum Society Special Publication, **5**, 325–346.

GABRIELSEN, R. H., FÆRSETH, R. B., STEEL, R. J., IDIL, S. & KLØVJAN, O. S. 1990. Architectural styles of basin fill in the northern Viking Graben. *In*: BLUNDELL, D. J. & GIBBS, A. D. (eds) *Tectonic Evolution of the North Sea Rifts*. Oxford University Press, Oxford, 158–179.

GHAZI, S. A. & SCHMIDT, W. J. 1995. Exploring stratigraphic types: an example from the Upper Jurassic in Block 33/9, the Norwegian North Sea. *In*: HANSLIEN, S. (ed.) *Petroleum Exploration and Exploitation in Norway*. Norwegian Petroleum Society Special Publication, **4** 181–197.

GILTNER, J. P. 1987. Application of extensional modes to the Northern Viking Graben. *Norsk Geologisk Tidsskrift*, **67**, 339–352.

GLENNIE, K. W. 1990*a*. Outline of North Sea history and structural framework. *In*: GLENNIE, K. W. (ed.) *Introduction to the Petroleum Geology of the North Sea*, 3rd edition, Blackwell, Oxford, 34–77.

——1990*b*. Lower Permian–Rotliegend. *In*: GLENNIE, K. W. (ed.) *Introduction to the Petroleum Geology of the North Sea*, 3rd edition, Blackwell, Oxford, 120–152.

GRADIJAN, S. J. & WIIK, M. 1987. Statfjord North. *In*: SPENCER, A. M. *et al.* (eds) *Geology of Norwegian Oil and Gas Fields*. Norwegian Petroleum Society, Graham & Trotman, 341–350.

GRAUE, E., HELLAND-HANSEN, W., JOHNSEN, J. R., LØMO, L., NØTTVEDT, A., RØNNING, K., RYSETH, A. & STEEL, R. J. 1987. Advance and retreat of Brent Delta System, Norwegian North Sea. *In*: BROOKS, J. & GLENNIE, K. (eds) *Petroleum Geology of North West Europe*. Graham & Trotman, London, 915–937.

HAMBLIN, A. P. & RUST, B. R. 1989. Tectono-sedimentary analyses of alternate-polarity half-graben basin-fill successions: Late Devonian–Early Carboniferous Horton Group, Cape Breton Island, Nova Scotia. *Basin Research*, **2**, 239–255.

HELLAND-HANSEN, W., ASHTON, M., LØMO, L. & STEEL, R. J. 1992. Advance and retreat of the Brent delta: recent contributions to the depositional model. *In*: MORTON, A. C., HASZELDINE, R. S., GILES, M. R. & BROWN, S. (eds) *Geology of the Brent Group*. Geological Society, London, Special Publication, **61**, 109–127.

HELLEM, T., KJEMPERUD, A. & ØVREBØ, K. O. 1986. The Troll Field: a geological/geophysical model established by the PL085 Group. *In*: SPENCER, A. M. *et al.* (eds) *Hydrocarbon Habitat on the Norwegian Continental Shelf*. Norwegian Petroleum Society, Graham & Trotman, 217–238.

HOLLANDER, N. B. 1987. Snorre. *In*: SPENCER, A. M. *et al.* (eds) *Geology of the Norwegian Oil and Gas Fields*. Graham & Trotman, London, 307–318.

JOHANNESSEN, E. P., MJØS, R., RENSHAW, D., DALLAND, A. & JACOBSEN, T. 1995. Northern limit of the 'Brent delta' at the Tampen Spur – a sequence stratigraphic approach for sandstone prediction. *In*: STEEL, R. J., FELT, V. L., JOHANNESSEN, E. P. & MATHIEU, C. (eds) *Sequence Stratigraphy on the Northwest European Margin*. Norwegian Petroleum Society Special Publication, **5**, 213–256.

JOHNS, C. & ANDREWS, I. J. 1985. The petroleum geology of the Unst Basin, North Sea. *Marine and Petroleum Geology*, **2**, 361–372.

JOHNSON, A. & EYSSAUTIER, M. 1987. Alwyn North Field and its regional geological context. *In*: BROOKS, J. & GLENNIE, K. (eds) *Petroleum Geology of North West Europe*. Graham & Trotman, London, 299–312.

LAMBIASE, J. J. 1990 A model for tectonic control of lacustrine stratigraphic sequences in continental rift basins. *In*: KATZ, B. J. (ed.) *Lacustrine Basin Exploration: Case Studies and Modern Analogs*. American Association of Petroleum Geologists Memoir, **50**, 265–276.

LEEDER, M. R. 1995. Continental rifts and proto-oceanic rift troughs. *In*: BUSBY-SPERA, C. J. & INGERSOLL, R. V. (eds) *Tectonics of Sedimentary Basin*. Blackwell, Oxford, 119–148.

—— & GAWTHORPE, R. L. 1987. Sedimentary models for extensional tilt block/half-graben basins. *In*: COWARD, M. P., DEWEY, J. F. & HANCOCK, P: L. (eds) *Continental Extensional Tectonics*. Geological Society, London, Special Publications, **28**, 139–152.

LEINFELDER, R. R. & WILSON, R. C. L. in press. Third order sequences in an Upper Jurassic rift-related second order sequence, central Lusitanian Basin, Portugal. *In*: *Sequence Stratigraphy of Western European Basins*. Society of Economical Paleontologists and Mineralogists Special Publication, **57**.

LERVIK, K. S., SPENCER, A. M. & WARRINGTON, G. 1989. Outline of Triassic stratigraphy and structure in the central and northern North Sea. *In*: COLLINSON, J. D. (ed.) *Correlation in Hydrocarbon Exploration*. Norwegian Petroleum Society, Graham & Trotman, London, 173–189.

LIVBJERG, F. & MJØS, R. 1989. The Cook Formation, an offshore sand ridge in the Oseberg area, northern North Sea. *In*: COLLINSON, J. D. (ed.) *Correlation in Hydrocarbon Exploration*. Norwegian Petroleum Society, Graham & Trotman, London, 299–312.

MACK, G. H. & SEAGER, W. R. 1990. Tectonic control on facies distribution of the Camp Rice and Palomas Formations (Pliocene–Pleistocene) on the southern Rio Grande Rift. *Geological Society of America Bulletin*, **102**, 45–53.

MARJANAC, T. & STEEL, R. J. 1997. Dunlin Group sequence stratigraphy in the northern North Sea: a model for Cook sandstone deposition. *AAPG Bulletin*, **81**, 276–292.

MARSDEN, G., YIELDING, G., ROBERTS, A. M. & KUSZNIR, N. J. 1990. Application of a flexural cantilever simple-shear/pure-shear model of continental lithosphere extension to the formation of the northern North Sea basin. *In*: BLUNDELL, D. J. & GIBBS, A. D. (eds) *Tectonic Evolution of the North Sea Rifts*. Oxford University Press, Oxford, 240–261.

MCKENZIE, D. P. 1978. Some remarks on the development of sedimentary basins. *Earth and Planetary Science Letters*, **40**, 25–32.

MILTON, N. J. 1993. Evolving depositional geometries in the North Sea Jurassic rift. *In*: PARKER, J. R. (ed.) *Petroleum Geology of Northwest Europe: Proceedings of the 4th Conference*. Geological Society, London, 425–442.

MORLEY, C. K. 1989. Extension, detachments, and sedimentation in continental rifts (with particular reference to East Africa). *Tectonics*, **8**, 1175–1192.

MUTO, T. & STEEL, R. J. 1997. The Middle Jurassic Oseberg delta, northern North Sea: a sedimentological and sequence stratigraphic interpretation. *AAPG Bulletin*, **81**, 1070–1086.

NØTTVEDT, A., GABRIELSEN, R. H. & STEEL, R. J. 1995. Tectonostratigraphy and sedimentary architecture of rift basins, with reference to the northern North Sea. *Marine and Petroleum Geology*, **12**, 881–901.

NYSTUEN, J. P., KNARUD, R., JORDE, K. & STANLEY, K. O. 1989. Correlation of Triassic to Lower Jurassic sequences, Snorre Field and adjacent areas, northern North Sea. *In*: COLLINSON, J. D.

(ed.) *Correlation in Hydrocarbon Exploration.* Norwegian Petroleum Society, Graham & Trotman, London, 273–289.

Odinsen, T., Reemst, P. & van der Beek, P. in press. Permo-Triassic and Jurassic extension in the northern North Sea: results from tectonostratigraphic forward modelling. *Norsk Geologisk Tidsskrift.*

Olaussen, S., Larsen, B. T. & Steel, R. J. 1994. The Carboniferous–Permian Oslo Rift: basin fill in relation to the tectonic development. *In*: Embry, A. M. (ed.) *Pangea: Global Environments and Resources.* Canadian Society of Petroleum Geologists Memoir, **17**, 175–197.

Partington, M. A., Mitchener, B. C., Milton, N. J. & Fraser, A. J. 1993. Genetic sequence stratigraphy for the North Sea Late Jurassic and Early Cretaceous: distribution and prediction of Kimmeridgian–Late Ryazanian reservoirs in the North Sea and adjacent areas. *In*: Parker, J. R. (ed.) *Petroleum Geology of Northwest Europe: Proceedings of the 4th Conference.* Geological Society, London, 347–370.

Platt, N. H. 1995. Structure and tectonics of the northern North Sea: new insights from deep penetration regional seismic data. *In*: Lambiase, J. J. (ed.) *Hydrocarbon Habitat in Rift Basins.* Geological Society, London, Special Publications, **80**, 103–113.

Posamentier, H. W., Allen, G. P., James, D. P. & Tesson, M. 1992. Forced regressions in a sequence stratigraphic framework: concepts, examples and exploration significance. *AAPG Bulletin*, **76**, 1687–1709.

Prosser, S. 1993. Rift-related linked depositional systems and their seismic expression. *In*: Williams, G. D. & Dobb, A. (eds) *Tectonics and Seismic Sequence Stratigraphy.* Geological Society, London, Special Publications, **71**, 35–66.

Purser, B. H., Philobbos, E. R. & Soliman, M. 1990. Sedimentation and rifting in the NW parts of the Red Sea: a review. *Bull. Soc. gèol. France*, **8**, 371–384.

Rattey, R. P. & Hayward, A. B. 1993. Sequence stratigraphy of a failed rift system: the Middle Jurassic to Early Cretaceous basin evolution of the Central and Northern North Sea. *In*: Parker, J. R. (ed.) *Petroleum Geology of Northwest Europe*: Proceedings of the 4th Conference. Geological Society, London, 215–249.

Ravnås, R. & Bondevik, K. 1997. Architecture and controls on Bathonian–Kimmeridgian shallow-marine syn-rift wedges of the Oseberg–Brage area, northern North Sea. *Basin Research*, **9**, 197–226.

—— & Steel, R. J. 1997. Contrasting styles of late Jurassic syn-rift turbidite sedimentation: a comparable study of the Magnus and Oseberg areas, northern North Sea. *Marine and Petroleum Geology*, **14**, 417–449.

—— & ——1998. Architecture of marine rift basin successions. *AAPG Bulletin*, **82**, 110–146.

——, —— & Theriault, P. 1994. Stratigraphic signature of some marine rift successions. *Abstract, EAPG 6th Conference and Technical Exhibition, Vienna, Austria 1994.*

——, Bondevik, K., Helland-Hansen, W., Lømo, L., Ryseth, A. & Steel, R. J. 1997. Sedimentation history as an indicator or rift initiation and development: the Late Bajocian–Bathonian evolution of the Oseberg–Brage area, northern North Sea. *Norsk Geologisk Tidsskrift*, **77**, 205–232.

Richards, P. C. 1991. Evolution of Lower Jurassic coastal plain and fan delta sediments in the Beryl embayment, North Sea. *Journal of the Geological Society, London*, **148**, 1037–1047.

——, Lott, G. K., Johnsson, H., Knox, R. W. O'B. & Riding, J. B. 1993. Jurassic of the Central and Northern North Sea. *In*: Knox, R. W. O'B. & Cordey, W. G. (eds) *Lithostratigraphic Nomenclature of the UK North Sea.* British Geological Survey, Nottingham.

Roberts, A. M., Yielding, G., Kusznir, N., Walker, I. & Dorn-Lopez, D. 1993*a*. Mesozoic extension in the North Sea: constraints from flexural backstripping, forward modelling and fault populations. *In*: Parker, J. R. (ed.) *Petroleum Geology of Northwest Europe: Proceedings of the 4th Conference.* Geological Society, London, 1123–1136.

——, —— & Badley, M. E. 1993*b*. Tectonic and bathymetric controls on stratigraphic sequences within evolving half-grabens. *In*: Williams, G. D. & Dobb, A. (eds) *Tectonics and Seismic Sequence Stratigraphy.* Geological Society, London, Special Publications, **71**, 87–121.

——, ——, Kusznir, N., Walker, I. & Dorn-Lopez, D. 1995. Quantitative analysis of Triassic extension in the northern Viking Graben. *Journal of the Geological Society, London*, **152**, 15–26.

Roberts, J. D., Mathieson, A. S. & Hampson, J. M. 1987. Statfjord. *In*: Spencer, A. M. *et al.* (eds) *Geology of the Norwegian Oil and Gas Fields.* Norwegian Petroleum Society, Graham & Trotman, London, 319–340.

Røe, S.-L. & Steel, R. J. 1985. Sedimentation, sea-level rise and tectonics at the Triassic–Jurassic boundary (Statfjord Formation), Tampen Spur, northern North Sea. *Journal of Petroleum Geology*, **8**, 163–186.

Sawyer, M. & Keegan, J. 1996. Use of palynofacies characterization in sand-dominated sequences, Brent Group, Ninian Field, UK North Sea. *Petroleum Geoscience*, **2**, 289–297.

Schlische, R. W. & Olsen, P. E. 1990. Quantitative filling models for continental extensional basins with applications to Early Mesozoic rifts of eastern America. *Journal of Geology*, **98**, 135–155.

Sellwood, B. W. & Netherwood, R. E. 1984. Facies evolution in the Gulf of Suez area: sedimentation history as an indicator of rift initiation and development. *Modern Geology*, **9**, 43–69.

Solli, T. 1995. Upper Jurassic play concept – an integrated study in Block 34/7, Norway. *First Break*, **13**, 21–30.

Steel, R. J. 1993. Triassic–Jurassic megasequence stratigraphy in the Northern North Sea: rift to post-rift evolutionl. *In*: PARKER, J. R. (ed.) *Petroleum Geology of Northwest Europe: Proceedings of the 4th Conference*. The Geological Society, London, 299–315.

—— & RYSETH, A. 1990. The Triassic–early Jurassic succession in the northern North Sea: megasequence stratigraphy and intra-Triassic tectonics. *In*: HARDMAN, R. F. P. & BROOKS, J. (eds) *Tectonic Events Responsible for Britain's Oil and Gas Reserves*. Geological Society, London, Special Publications, **55**, 139–168.

STEWART, D. J., SCHWANDER, M. & BOLLE, L. 1995. Jurassic depositional systems of the Horda Platform, Norwegian North Sea: practical consequences of applying sequence stratigraphic models. *In*: STEEL, R. J., FELT, V. L., JOHANNESSEN, E. P. & MATHIEU, C. (eds) *Sequence Stratigraphy on the Northwest European Margin*. Norwegian Petroleum Society Special Publication, **5**, 291–323.

STEWART, I. J., RATTEY, R. P & VANN, I. R. 1992. Structural style and the habitat of hydrocarbons in the North Sea. *In*: LARSEN, R. M., BREKKE, H., LARSEN, B. T. & TALLERAAS, E. (eds) *Structural and Tectonic Modelling and its Application to Petroleum Geology*. Norwegian Petroleum Society Special Publication, **1**, 197–220.

SURLYK, F. 1978. Submarine fan sedimentation along fault scarps on tilted fault blocks (Jurassic–Cretaceous boundary, East Greenland). Bulletin Grønlands Geologiske Undersøgelse, **128**, 1–108.

——1989. Mid-Mesozoic syn-rift turbidite systems: controls and predictions. *In*: COLLINSON, J. D. (ed.) *Correlation in Hydrocarbon Exploration*. Norwegian Petroleum Society, Graham & Trotman, London, 231–241.

——, PIASECKI, S., ROLLE, F., STEMMERIK, L., THOMSEN, E. & WRANG, P. 1984. The Permian basin of East Greenland. *In*: SPENCER, A. M. *et al.* (eds) *Petroleum Geology of the North European*. Norwegian Petroleum Society, Graham & Trotman, London, 303–315.

SURLYK, F., NOE-NYGAARD, N. & DAM, G. 1993. High and low resolution sequence stratigraphy in lithology prediction – examples from the Mesozoic around the northern North Atlantic. *In*: PARKER, R. J. (ed.) *Petroleum Geology of Northwest Europe: Proceedings of the 4th Conference*. Geological Society, London 199–214.

SØRENSEN, S. & MARTINSEN, M. 1987. A palaeogeographic reconstruction of the Rotliegendes deposits in the northeastern North Sea. *In*: BROOKS, J. & GLENNIE, K. (eds) *Petroleum Geology of North West Europe*. Graham & Trotman, London, 497–508.

TER VOORDE, M., FÆRSETH, R. B., GABRIELSEN, R. H. & CLOETINGH, S. A. P. L. (in press) Repeated lithosphere extension in the northern Viking Graben: a coupled or a decoupled rheology. *Norsk Geologisk Tidsskrift*.

TORSVIK, T. H., STURT, B. A., SWENSSON, E., ANDERSEN, T. B. & DEWEY, J. F. 1992. Palaeomagnetic dating of fault rocks: evidence for Permian and Mesozoic movements and brittle deformation along the extensional Dalsfjord Fault, western Norway. *International Geophysical Journal*, **109**, 565–580.

UNDERHILL, J. R. 1991*a*. Controls on Late Jurassic seismic sequences, Inner Moray Firth, UK North Sea: a critical test of a key segment of Exxon's original global cycle chart. *Basin Research*, **3**, 79–98.

——1991*b*. Implications of Mesozoic–Recent basin development in the western Inner Moray Firth Basin. *Marine and Petroleum Geology*, **8**, 359–369.

—— & PARTINGTON, M. A. 1993. Jurassic thermal doming and deflation in the North Sea: implications of the sequence stratigraphic evidence. *In*: PARKER, R. J. (ed.) *Petroleum Geology of Northwest Europe: Proceedings of the 4th Conference*. Geological Society, London, 337–345.

VOLLSET, J. & DORE, A. G. 1984. A revised Triassic and Jurassic lithostratigraphic nomenclature for the Norwegian North Sea. *Norwegian Petroelum directorate Bulletin*, **3**.

WIGNALL, P. B. & RUFFELL, A. H. 1990. The influence of a sudden climatic change on marine deposition in the Kimmeridgian of northwest Europe. *Journal of the Geological Society, London*, **147**, 365–371.

—— & PICKERING, K. T. 1993. Palaeoecology and sedimentology across a Jurassic fault scarp, NE Scotland. *Journal of the Geological Society, London*, **150**, 323–340.

YIELDING, G., BADLEY, M. E. & ROBERTS, A. M. 1992. The structural evolution of the Brent Province. *In*: MORTON, A. C., HASZELDINE, R. S., GILES, M. R. & BROWN, S. (eds) *Geology of the Brent Group*. Geological Society, London, Special Publication, **61**, 27–43.

ZIEGLER, P. A. 1990*a*. *Geological Atlas of Western and Central Europe (2nd edition)*. Shell International Petroleum Maatschappij, B.V., The Hague.

——1990*b*. Tectonic and palaeogeographic development of the North Sea rift system. *In*: BLUNDELL, D. J. & GIBBS, A. D. (eds) *Tectonic Evolution of the North Sea Rifts*. Oxford University Press, Oxford, 1–36.

Syn-rift evolution and resulting play models in the Snorre-H area, northern North Sea

ARVID NØTTVEDT,[1] ANKER M. BERGE,[2] NANCYE H. DAWERS,[3] ROALD B. FÆRSETH,[4] KJELL O. HÄGER,[2] GUNN MANGERUD[2] & CAI PUIGDEFABREGAS[5]

[1] *Norsk Hydro Research Centre, N-5020 Bergen, Norway*
Present address: Norsk Hydro Canada, 111–5th Avenue SW, Calgary, AB T2P 3Y6, Canada
[2] *Norsk Hydro Research Centre, N-5020 Bergen, Norway*
[3] *Department of Geology and Geophysics, University of Edinburgh, Edinburgh EH9 3JW, UK*
[4] *Norsk Hydro Exploration, 1321 Stabekk, Norway*
[5] *Norsk Hydro Research Centre, N-5020 Bergen, Norway*
Present address: Centre for Advanced Studies, Cami de Santa Barbara, E-17300 Blanes, Girona, Spain

Abstract: An upper Jurassic, wedge-shaped syn-rift succession, comprising the Heather and Draupne Formations, is present in the hangingwall trough of the Snorre Fault Block. The succession is bounded to the west by the Statfjord East Fault, whereas it onlaps the Snorre Fault Block to the east. It consists of a two-fold coarsening-upward sequence from shale to sandstone of shallow marine/shoreline origin.

Active fault block rotation and subsidence in the Snorre-H area commenced in the Mid-Bathonian and lasted through the Ryazanian. The Heather Formation was deposited during the early rift stage (Mid-Bathonian–Early Oxfordian; 3° cumulative tilt), whereas the Draupne Formation (Late Oxfordian–Ryazanian; 9° cumulative tilt) accumulated during the main and late rift stages. The lower part of the Heather Formation was likely deposited across a submerged tilted fault block terrain, with a predominant extra-basinal sediment supply. Deposition of the upper Heather Formation, however, was governed by gradually emerging footwall islands, albeit yet without significant local erosion. As a result of increased fault/block rotation during deposition of the Draupne Formation Shale, Sequences I–II (late Early Oxfordian–early Mid-Volgian) footwall islands became firmly established, providing a predominant local sediment source to the Snorre-H sub-basin. Clay and silt were supplied from erosion of the Heather Formation on the Snorre-H hangingwall, with subordinate input of sand from the Statfjord East footwall. Subsequent deposition of the Draupne Formation, Sequences III–V (late Mid-Volgian–Ryazanian), was governed by significant relief on the footwall islands, causing deep erosion into the Brent Group on the Snorre-H hangingwall dip-slope and leading to progradation of the Upper Draupne Sandstone shoreline complex across the Snorre-H area. Deposition of the Draupne Formation and temporal shoreline position were likely partly controlled by northwards fault-tip propagation of the Statfjord East Fault.

Various syn-rift play models and depositional reservoir facies are present within the Snorre-H hangingwall basin. They include dip-slope shallow marine/shoreline sands, basin floor gravity transported sands and likely footwall talus sands enveloped in organic rich shales of the Draupne Formation. The distribution of reservoir facies is intimately linked to exposure and erosion of the middle Jurassic Brent Group below the syn-rift unconformity.

In recent years, considerable effort has been put into developing predictive models for syn-rift plays, recognizing that such plays may hold significant hydrocarbon volumes. Syn-rift plays stem from a complex interplay between tectonics and sedimentation, leading to localized and subtle occurrence of reservoir facies. The signature of syn-rift plays depends on the interplay between creation of accommodation space, climate, sediment supply, and relative base- or

From: NØTTVEDT, A. *et al.* (eds) *Dynamics of the Norwegian Margin*. Geological Society, London, Special Publications, **167**, 179–218. 1-86239-056-8/00/$15.00

sea-level stand and their changes (Leeder & Gawthorpe 1987; Surlyk 1989; Prosser 1993; Gabrielsen *et al.* 1995; Nøttvedt *et al.* 1995; Ravnås & Steel 1997, 1998; Ravnås *et al.* 2000).

In the North Sea, oil has been proven in upper Jurassic syn-rift sandstone reservoirs of various kinds. The Piper (Maher 1981), Troll (Rønnevik & Johnsen 1984), Fulmar (Johnson *et al.* 1986), Claymore (Maher & Harker 1987), Clyde (Smith 1987) and Ula (Home 1987) discoveries comprise marginal to shallow marine sandstone reservoirs partly fringing emergent footwall islands, whereas submarine fan sandstones deposited in topographic lows constitute the reservoir in the Magnus (De'Ath & Scuyleman 1971), Brae (Turner *et al.* 1987) and Miller (McLure & Brown 1992) discoveries.

The present contribution describes the hangingwall dip-slope evolution of the Snorre Fault Block, the so-called Snorre-H area, in the Norwegian northern North Sea Block 34/7 (Fig. 1) (Dahl & Solli 1993; Solli 1995). We

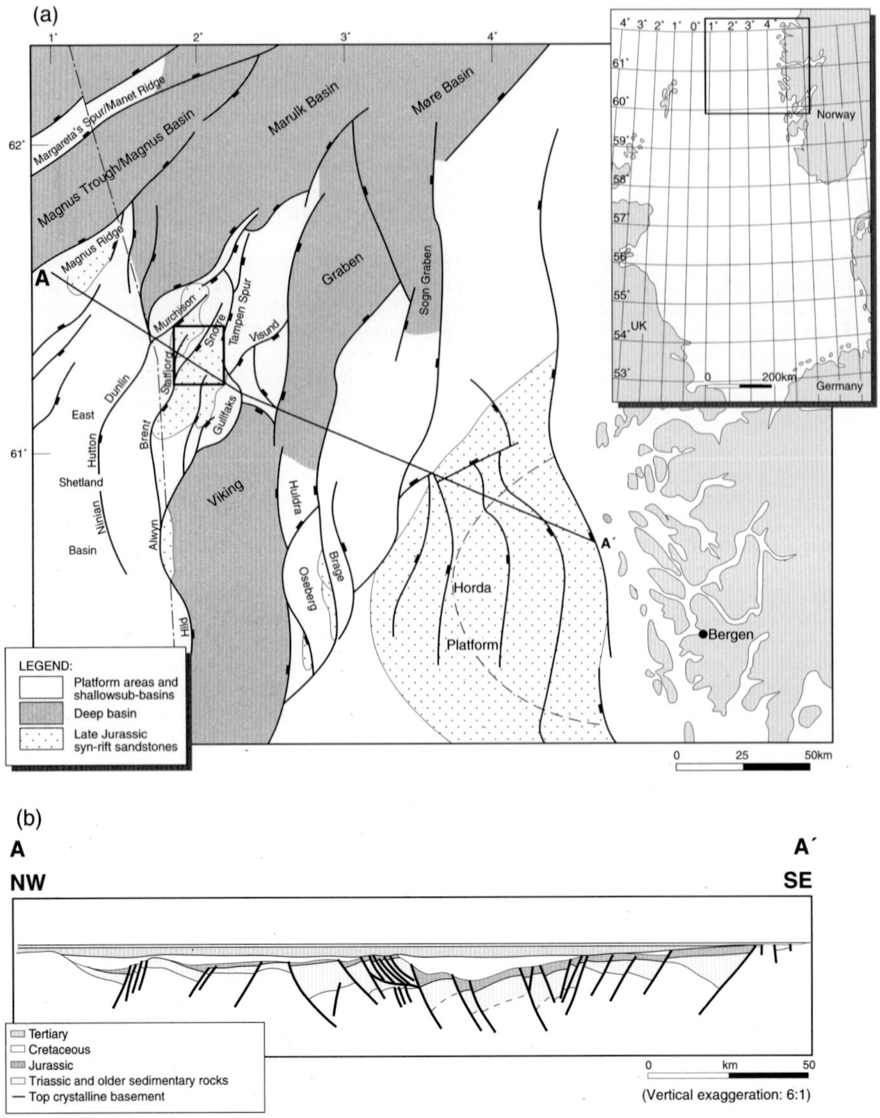

Fig 1. (**a**) Location map, showing the study area and array of westerly tilted fault blocks on the Tampen Spur to the west of the northern Viking Graben. (**b**) Geoseismic crustal section across the northern Viking Graben. See Fig. 1a for location.

attempt to present a dynamic model for the syn-rift evolution and to outline the various reservoir facies and play types associated with extension and rotation of the Snorre Fault Block and deposition of the upper Jurassic Draupne Formation in particular (Fig. 2). The focus is on the erosion and redistribution of sand related to footwall uplift and exposure of both the Snorre and neighbouring Statfjord East Fault Blocks.

In addition, we show how a systematic and integrated approach may significantly increase our understanding of syn-rift play potential and predictive ability of syn-rift facies and reservoir distribution. The study involves detailed structural and seismic stratigraphic interpretation of multiple 3D-seismic data, combined with sedimentological and biostratigraphical *in situ* and reworking analyses in a number of wells. Spatial and temporal linking of processes and mapping of mass budget between local drainage areas and associated depositional basins have been key elements in the process.

Database

The seismic interpretation of the southern part of the Snorre Fault Block is based on 3D surveys E86, SG9201, SG8431 and GE83 (Fig. 3). Regional information from merged and reprocessed survey CTM94 is also used. This survey combines E86, SG9201 and SG8431 with

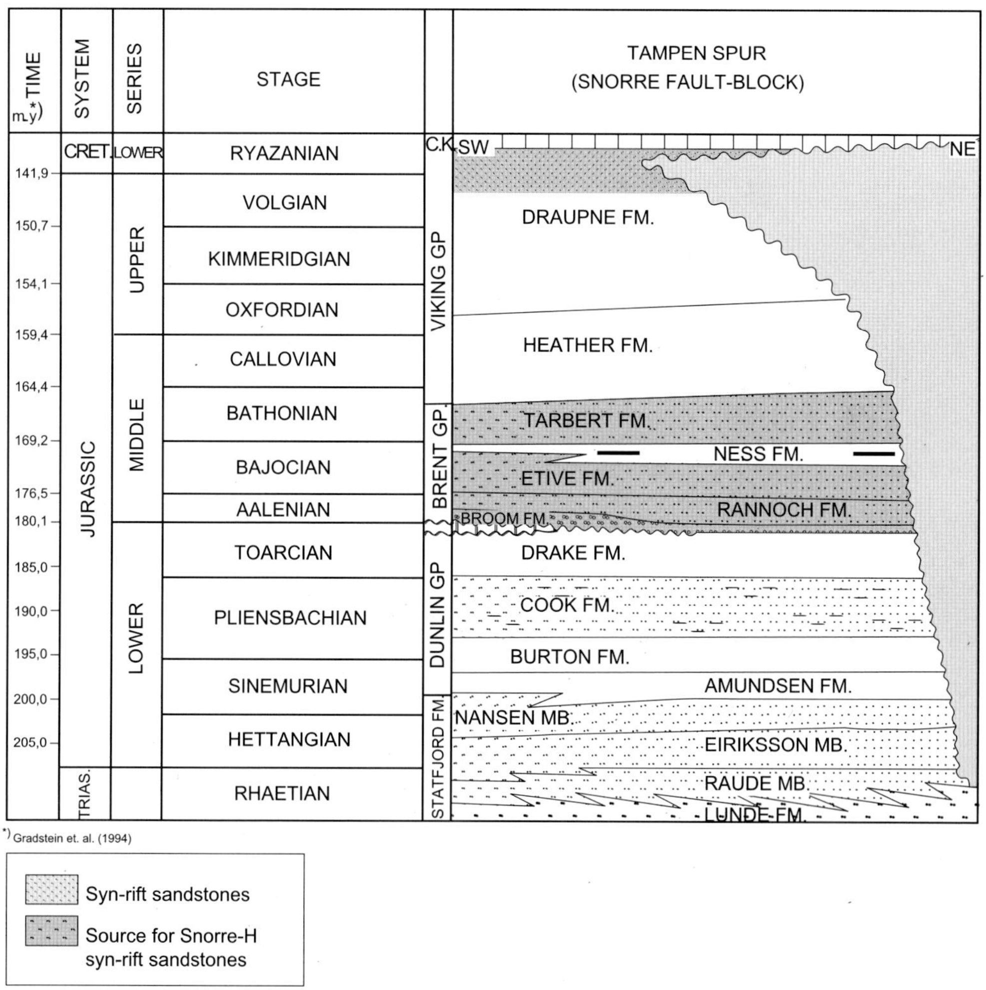

Fig. 2. Jurassic lithostratigraphy of the Tampen Spur area, following the nomenclature of Vollset & Dore (1984). Timescale is after Gradstein *et al.* (1994).

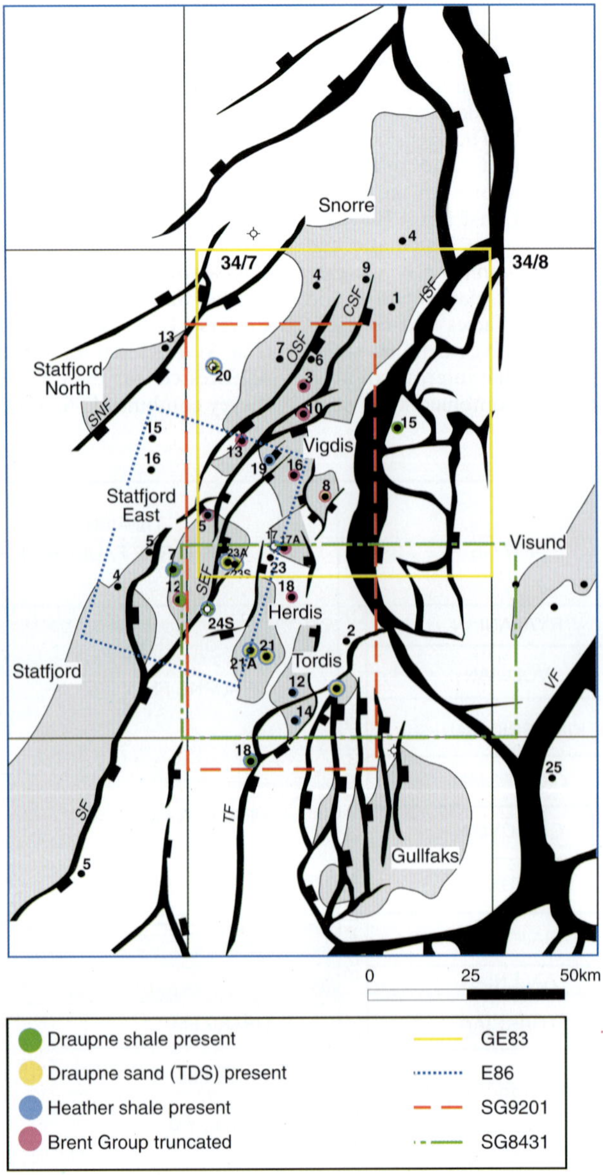

Fig. 3. Base Cretaceous structural map of the Snorre–Visund–Gullfaks area, showing location of the 3D seismic surveys E86, SG9201 and SG8431 and wells used in this study. Note annotated well symbols, showing which part of the syn-rift succession is present and if the Brent Group is truncated. SF, Statfjord Fault; SNF, Statfjord North Fault; SEF, Statfjord East Fault; OSF, Outer Snorre Fault; CSF, Central Snorre Fault; ISF, Inner Snorre Fault; TF, Tordis Fault;

ST9101, ST8511 and ST9207. The quality of the surveys varies and the quality within a single survey can vary over the area. The seismic data are assumed to be zero-phase, but differences in phase between surveys make it difficult to interpret the same reflectors in detail over the whole area. In the areas adjoining the structurally complex Gullfaks field the horizon definition is particularly poor. The density of interpretation of the upper Jurassic–lower Cretaceous syn-rift sequence varies between every second and every tenth inline. A detailed

Table 1. *Well database and main use of wells*

Well	Main use	Well	Main use
33/09-03	S	34/07-12	O
33/09-04	S	34/07-13	O
33/09–05	S	34/07-14	O
33/09-07	S	34/07-15S	B
33/09-09	S	34/07-16	O
33/09–12	S	34/07-17	B, S
33/09-15	C, B	34/07-17A	O
33/09-16	C, B	34/07-18	B
34/07-01	O	34/07-19	O
34/07-02	B	34/07-20	B
34/07-03	O	34/07-21	C, B, S
34/07-04	O	34/07-21A	C, B, S
34/07-05	S	34/07-22	B
34/07-06	S	34/07-23A	C, B, S
34/07-07	S	34/07-23S	C, B S
34/07-08	O	34/07-24S	B, S,
34/07-09	O	34/07-25	B
34/07-10	O	34/10-18	O

C, new core descriptions; B, biostratigraphical review or new analysis; S, used in seismic interpretation; O, used for general information.

interpretation of the Snorre-H basin has been done using the E86 survey, which is considered to be of good quality with reflectors and faults being well imaged.

Besides the discovery wells, the well database consists of wells on the high areas surrounding the Snorre half-graben, both on the Snorre footwall itself and on the Statfjord East Fault Block (Fig. 3). The wells used in the study are listed in Table 1 together with notes on how they have been used. A good tie between wells and seismic data has been obtained (Fig. 4).

Core descriptions have been made of the Snorre-H wells (Table 1), as well as biostratigraphical *in situ* zonation and reworking studies. As can be seen from the table, several of the

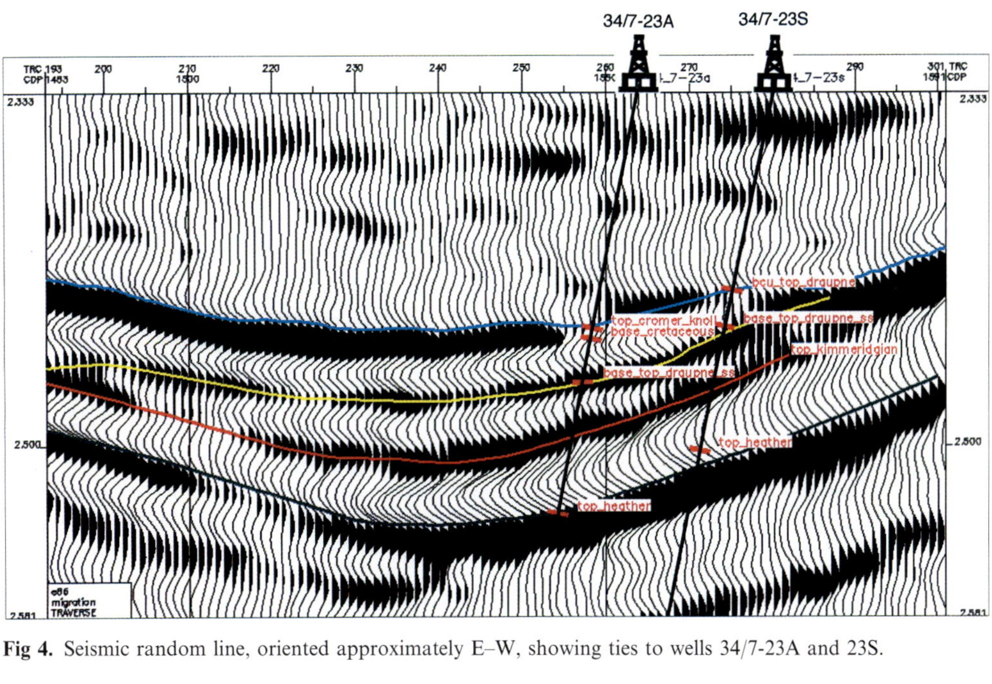

Fig 4. Seismic random line, oriented approximately E–W, showing ties to wells 34/7-23A and 23S.

analyses are quantitative and considered of good quality. The zonations are mainly based on new analyses carried out by Norsk Hydro/IKU.

Geological setting

The Snorre Fault Block (34/7) belongs to an array of westerly tilted, major fault blocks constituting the Tampen Spur area (Fig. 1). The Tampen Spur is bordered to the east by the northern North Sea Viking Graben and to the west by the Marulk and Magnus Basins.

The Viking Graben formed in response to Jurassic rifting, commencing in the late Middle Jurassic and in places continuing into the earliest Cretaceous. However, neither the onset nor the cessation of block tilting was synchronous throughout the basin. There is evidence for variable rates of extension and fault block rotation, with the characteristics of a pulsed syn-rift sedimentation (Partington *et al.* 1993; Rattey & Hayward 1993; Færseth *et al.* 1995, 1997; Ravnås & Bondevik 1997; Færseth & Ravnås 1998).

The early rift-stage coincides approximately with drowning of the axial Brent delta system and deposition of the lower part of the mud-prone, marine Heather Formation (Fig. 2). It may in places encompass deltaic sediments of the uppermost Brent Group (Graue *et al.* 1987; Badley *et al.* 1988; Helland-Hansen *et al.* 1992; Johannesen *et al.* 1995; Ravnås *et al.* 1997). A characteristic feature of the early rift-stage is that major fault blocks exhibit modest rotation, without developing major footwall islands.

The main rift stage is characterized by increased extension and accelerated fault block rotation, leading to the development of deep, hangingwall half-grabens and footwall islands with significant relief and to further segmentation of the rift structure. The clay-prone upper Heather and Draupne Formations (Fig. 2) were deposited locally in these sediment-starved sub-basins, as well as within the axial parts of the northern North Sea, where there was significant bathymetry. Erosion of the Heather Formation shales on the emergent footwall islands contributed to mud deposition, but where erosion was deep and sand prone strata like the Brent and Statfjord Formations became exposed, local deposition of sand took place.

The late rift stage is characterized by waning rates of fault block rotation, allowing filling and, in some places, progradation of upper Draupne Formation sandstones over the previously deposited shales of the Heather and lower Draupne Formations.

Some authors have invoked a significant strike-slip component within the North Sea Jurassic rift system (Gowers & Sæbøe 1985; Cartwright 1987; Beach *et al.* 1987; Larsen 1987). Elevated structures in the Tampen Spur area (Gullfaks, Visund Fault Blocks) have been interpreted as results of transpression (Beach 1985; Speksnijder 1987; Fossen 1989). However, it is now generally accepted that structuring results from extension and that most of the main Jurassic faults are close to dip-slip (Faerseth 1996; Rouby *et al.* 1996; Færseth *et al.* 1997).

The Marulk and Magnus Basins form essentially an extension of the Møre Basin, which is part of the North Atlantic rift structure. The timing of this structure differs slightly from that of the northern North Sea, with rifting continuing into the early Cretaceous, and the structural evolution of Block 34/7 must therefore be viewed in this context.

Major NE–SW striking faults in the Møre, Marulk and Magnus basins cut across N–S trending Jurassic structural elements of the Viking Graben. The northernmost part of the Tampen Spur (Block 34/4) has suffered *c.* 1.5 km of composite uplift (Yielding 1990) in the footwall of such crosscutting master faults. The block-bounding (first-order) master faults penetrate the brittle, upper crust (12–14 km) and presumably sole out at the brittle-ductile boundary (Fig. 1). Faults that define detachments above or within basement are particularly related to the eastern Tampen Spur (Færseth *et al.* 1995; Odinsen *et al.* 2000; Fossen *et al.* 2000).

The Snorre-H Structure

The Snorre-H structure (= Snorre-H Fault Block) is defined herein as the southern part of the Snorre Fault Block, comprising the uplifted Snorre-H footwall and neighbouring Snorre-H hangingwall sub-basin to the west. It is confined to the west by the NNE–SSW striking Statfjord and Statfjord East Faults and to the east by the Inner Snorre Fault (Fig. 3). To the north, the main Snorre Fault Block merges to some extent with the Statfjord Fault Block. The main Snorre Field is segmented by the Outer and Central Snorre Faults.

The Statfjord East Fault system is composed of a series of NE-striking, left-stepping, en-echelon fault segments that are linked by northerly-striking segments (Fig. 3) (Dawers *et al.* in press). The NE-striking en echelon segments are particularly evident at the top Brent level, where they define a series of sub-basins along the Statfjord East footwall.

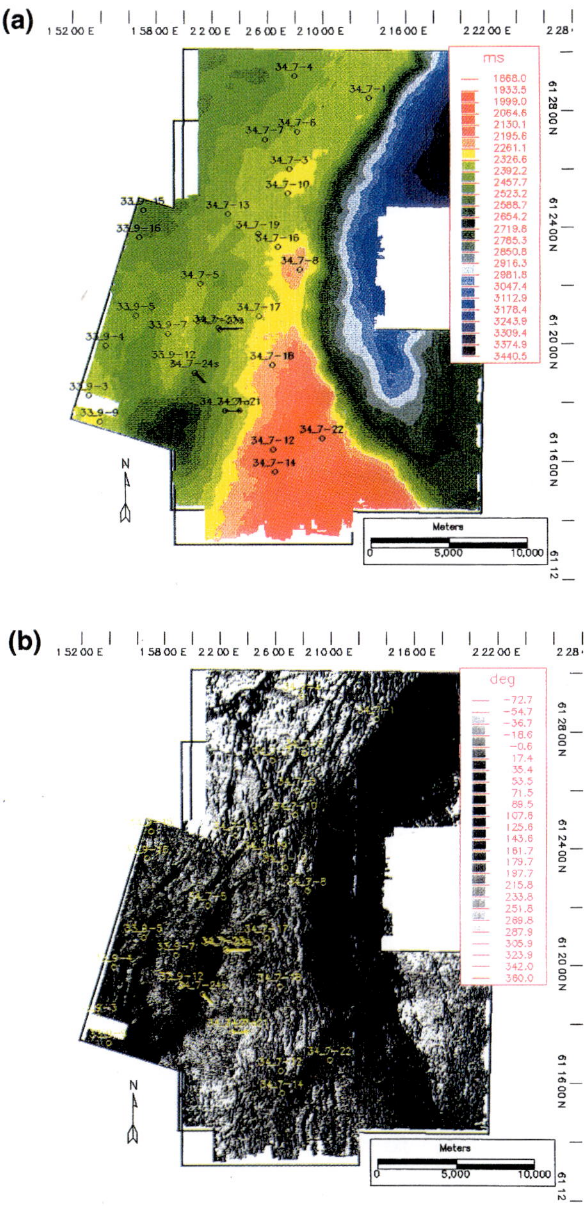

Fig. 5. (**a**) Time map of the base Cretaceous horizon. The horizon was interpreted on every second inline and every tenth crossline using surveys E86, SG9201, SG8431 and GE83. (**b**) Dip-azimuth map of the autotracked base Cretaceous map, with the grey scale simulating illumination from the northwest. The map was generated with light from the NW at an elevation of 30°.

As can be seen from the Base Cretaceous time map (Fig. 5a), the larger Gullfaks-Snorre footwall structure plunges North. The Snorre-H hangingwall sub-basin, on the other hand, plunges SW and becomes increasingly wider along strike. The Snorre-H structure can be seen to widen from less than 10 km north of wells 34/7-23 and -23S to more than 15 km south of wells 34/7-21 and -21A (Fig. 3). Close to well 34/7-24S it is about 12 km in width. The depth varies between 1722 and 3558 ms, with a maximum in the down-thrown side of the Statfjord

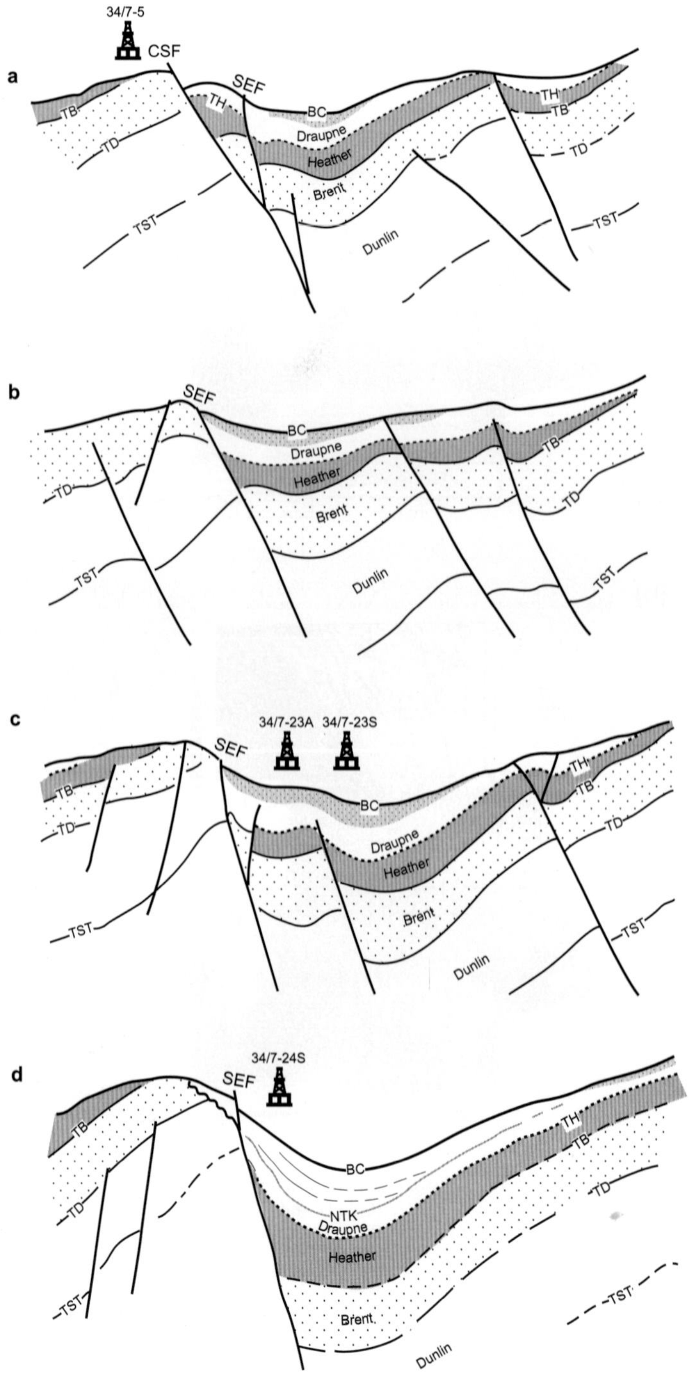
a
34/7-5
CSF
SEF
TB
TH
BC
Draupne
Heather
Brent
TD
TST
Dunlin
TH
TB
TD
TST
b
SEF
BC
Draupne
Heather
Brent
TD
TST
Dunlin
TB
TD
TST
c
34/7-23A
34/7-23S
SEF
TB
TD
TST
BC
Draupne
Heather
Brent
Dunlin
TH
TB
TD
TST
d
34/7-24S
SEF
TB
TD
TST
BC
NTK
Draupne
Heather
Brent
Dunlin
TH
TB
TD
TST

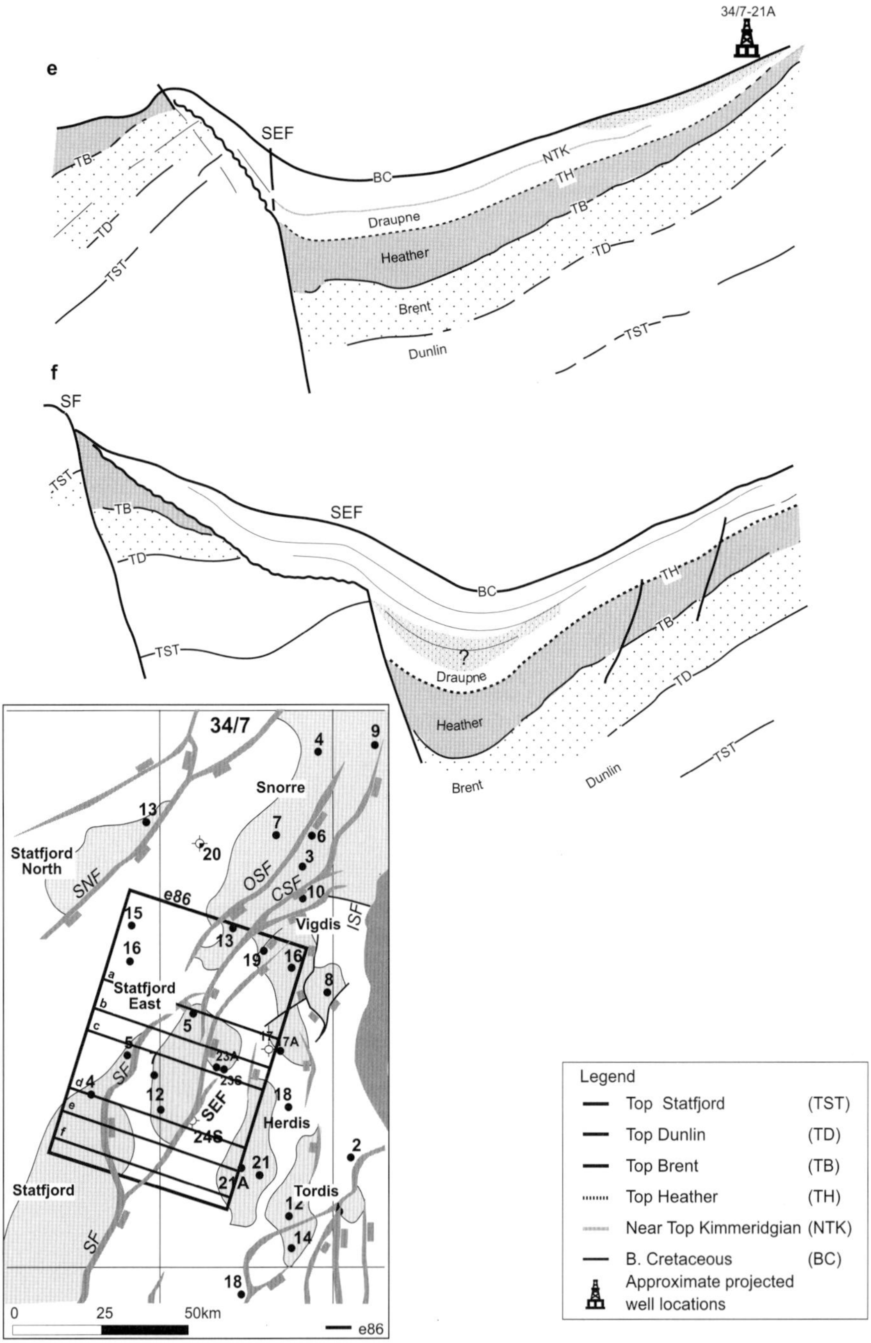

Fig. 6. Snorre Fault Block cross-section panel. See map insert for location of cross-sections. Fault abbreviations are the same as in Fig. 3.

Fault. The Snorre-H hangingwall sub-basin exhibits a current dip of 9–13° at pre-rift (top Brent Group, or close) level within the area under consideration. Maximum dips are recorded in the SSW of the area.

The internal fault pattern of the Snorre Fault Block is illustrated on the base Cretaceous dip-azimuth map (Fig. 5b). The smaller intra fault block (second-order) faults that result in the compartmentalization of the mega-blocks are generally basement-detached. Within the Snorre Fault Block such faults strike predominantly N–S, NE–SW and E–W.

A more detailed description of the structural characteristics and evolution of the Statfjord East Fault and associated Snorre-H sub-basin is given in Dawers *et al.* (1999).

The Snorre-H syn-rift succession

A wedge-shaped syn-rift succession, comprising the Heather and Draupne Formations, is present in the hangingwall trough of the Snorre-H Fault Block. The wedge ranges from Mid-Bathonian to Ryazanian in age. It is not clear if the syn-rift succession also includes the upper part of the Brent Group, as has been suggested on the Statfjord Fault Block to the south (Johannesen *et al.* 1995). A recent study of the Brent Group/Tarbert Formation in the Snorre area, however, suggests that intra-basinal faults controlled very small depocenters in Bajocian times, signifying the earliest pattern of fault initiation (Davies *et al.* 1998).

The Heather Formation (Mid-Bathonian–Early Oxfordian) consists predominantly of silty shales of offshore marine origin. The Draupne Formation (late Early Oxfordian–Ryazanian) consists of anoxic, black shales of offshore marine to slope origin (Draupne Shale), coarsening upwards into fine-grained sandstones of a prograding shoreline complex (Upper Draupne Sandstone), occasionally including an intercalated, assumed incised slope channel sandstone unit (Intra Draupne Sandstone). The Draupne Formation rests, partly unconformably, on the Heather Formation.

Syn-rift deposits are generally absent in crestal areas of the Snorre-H and Statfjord East Fault Blocks, but thin Draupne Formation shales of Late Volgian age cap parts of the present day crestal areas of the Snorre Fault Block. In cross-sectional view, the upper Jurassic syn-rift infill can be seen to butt against the NE–SW striking Statfjord East Fault to the north, whereas it blankets parts of the degraded footwall to the south (Fig. 6). The syn-rift succession is presently deepest buried in the south-west on the Snorre-H hangingwall and towards the east on the Visund Field (Figs 5 & 6). It thickens southward, reaching some 200 ms TWT south of well 34/7-24S.

Significant thickness variations occur in the upper Jurassic syn-rift infill succession across the en echelon segments of the Statfjord East Fault (Dawers *et al.* 1999). A series of fault-controlled depocenters developed during the Oxfordian through Kimmeridgian, involving a progressive younging of the syn-rift infill from south to north in the Snorre-H sub-basin.

A thin limestone marker bed (Cromer Knoll Group) of Hauterivian/Valanginian age drapes the Snorre Fault Block. It overlies the Draupne Formation in the distal hangingwall areas and rests directly on the syn-rift unconformity in updip footwall crestal areas.

Stratigraphic zonation

A fairly complete stratigraphic succession is recorded in the Snorre-H wells from the top Bajocian through Kimmeridgian, although several minor stratigraphic breaks seem to be present (Fig. 7). The types and number of events observed vary as a result of variable sample density within and between the wells, however, giving a variable stratigraphic resolution. The sampling density, on average, is 1–2 m in wells 34/7-21, 21A, 23A, 23S and 2–3 m in well 34/7-24S.

In well 34/7-21, zones PJ 5A2 and PJ 6C1 are not observed, suggesting minor stratigraphic breaks may be present in the Mid-Callovian and Mid-Oxfordian. Missing zones suggesting minor breaks are speculated also in the late Early–early Mid-Volgian. The late Mid-Volgian–Late Volgian and Late Ryazanian intervals are all considered present in the well, but the transition from Mid- to Late Volgian is not well defined. From comparisons to well 34/7-21A, deposition of the Upper Draupne Sandstone is interpreted to begin at this boundary. The sandstone is mainly Late Volgian in age, while the uppermost few meters are Ryazanian. An acme of *Botryococcus* is recorded in the upper part of the Upper Draupne Sandstone.

Well 34/7-21A is located only 600 m west of 34/7-21, and the stratigraphic succession is very similar. Some differences are seen in the late Early–Mid Volgian interval, however, where some zones are identified that were not seen in 34/7-21, and vice versa.

No zones have been identified in the Mid-Oxfordian in well 34/7-23A. Similarly, the Mid-Volgian is not conclusively identified. Therefore, a stratigraphic break may be present

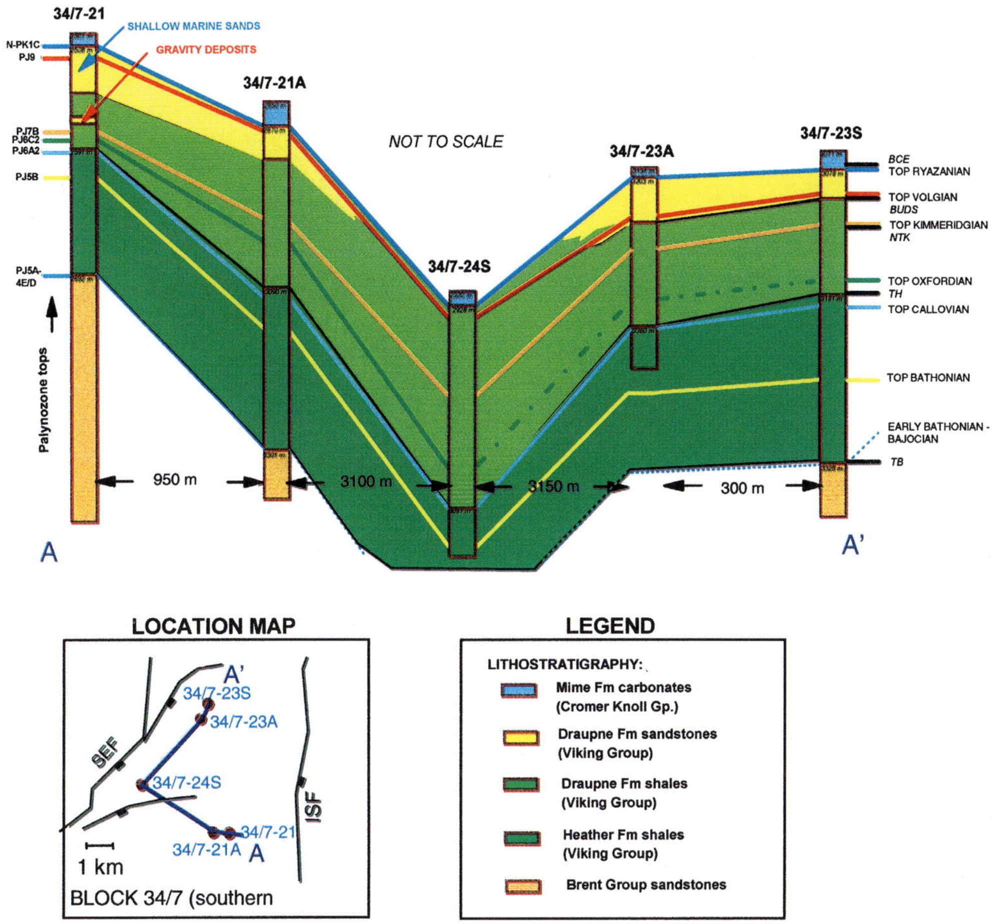

Fig. 7. Biostratigraphic zonation and well correlation, including chronostratigraphic and lithostratigraphic correlations. Palynozonations for the Snorre-H wells are indicated. Depths to formation tops are in mRKB MD.

in the succession, although, as the Mid-Volgian interval is not very thick, this may reflect a sampling density problem. Late Ryazanian taxa have been identified lower in the Upper Draupne Sandstone than in the 34/7-21 wells, suggesting that the main part of the sandstone in this well is Ryazanian in age. The lowermost part is thought to be of Late Volgian age. An acme of *Botryococcus* has been identified in the very top of the Draupne Shale, just before onset of sand deposition and another acme is recorded just beneath the top Kimmeridgian.

Well 34/7-23S is located about 250 m west of 34/7-23A and the succession is very similar. However, the Mid-Volgian is less well defined compared to well 34/7-23A. A *Botryococcus* acme is evident at the same relative position as in well 34/7-23A.

The 34/7-24S well has a succession very similar to wells 34/7-21 and -21A, but with no Upper Draupne Sandstone present. The uppermost part of the shale is Ryazanian in age and is underlain by Late Volgian strata. The Mid-Oxfordian and early Mid-Volgian are also here absent, or at least no indicative taxa have been recorded.

The acmes of *Botryococcus* described above have no age significance, but may indicate environmental stress, notably influx of fresh water into the basin or reworking of previously deposited fresh/brackish water sediments. The acmes observed near the base of the sands in 34/7-23S and -23A coincide with a cored section showing frequent coal fragments in the sediment (see below). This event is thought to be correlatable between the wells.

Biostratigraphical reworking

A considerable amount of biostratigraphical reworked material is present in the syn-rift succession, originating from erosion of local source uplands. Reworked taxa in the wells 34/7-21, -23S and -24S have been quantitatively represented and related to the formation where they normally occur *in situ* (Fig. 8), providing information on the history and timing of erosion and unroofing of upland source areas (Häger & Smelror 1997).

Typical Brent/Dunlin dinocyst flagellates (*Nannoceratopsis gracilis*, *Susadinium scrofoides*) first appear in zone PJ6A2/6B (Late Callovian/ Early Oxfordian), but in very limited numbers. 'Dunlin/Brent taxa' thereafter are present throughout the section. The distribution of taxa, according to the lithostratigraphical grouping, is relatively even until the Ryazanian, where 'Draupne taxa' dominate. No indications of reworking of strata older than Toarcian–Aalenian have been seen, i.e corresponding to Statfjord Formation or Triassic rocks, even though it has been documented that such deep erosion occurs further north and to the east on the Snorre Fault Block (Solli 1995).

A marked increase in reworking occurs in the Kimmeridgian and continues into Early Volgian. In the Middle Volgian there is an apparent decrease and lack of reworking. However, this probably relates mostly to the fact that the Middle Volgian interval has not been conclusively identified in any of the Snorre-H wells. It is considered unlikely that there was a significant reduction in reworking across this interval. In Late Volgian–Ryazanian times reworking was very intense and encompasses taxa from Draupne, Heather as well as Brent lithologies.

The timing and sequence of reworking vary between the wells, with the events coming in slightly earlier in well 34/7-21 compared to well 34/7-23S. The reworked taxa found in the Ryazanian sandstones in wells 34/7-23S and -23A are more numerous than the *in situ* forms and many of the species are characteristic of the Oxfordian–Volgian, while they do not occur (or are less common) in the Ryazanian. This indicates active erosion of previously deposited upper Jurassic sediments and redeposition within the prograding Upper Draupne Sandstone system. Since the main sedimentary end product in this case is sand, it is believed that erosion (cannibalization) of earlier deposited syn-rift sands in structurally higher positions may have occurred.

In some cases it can be shown that reworking of a depositional unit occurred a 'fairly short time' after deposition. In well 34/7-23S,

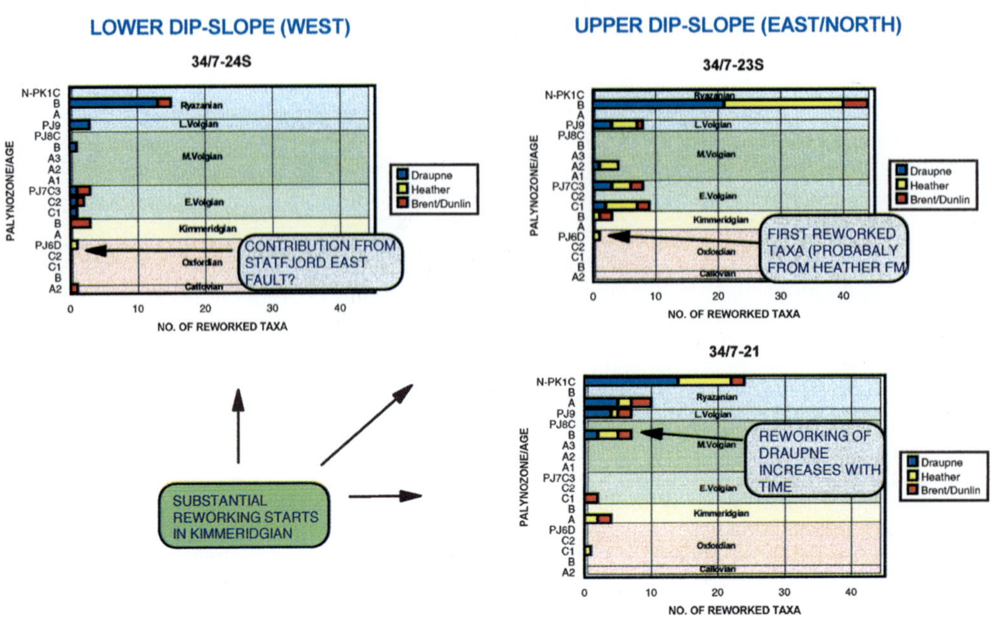

Fig. 8. Reworked taxa in selected Snorre-H wells related to formation of origin and structural position. The amount of reworked taxa is shown relative to the age of the strata where they were found and from which formation/group they originate (Draupne, Heather and Brent/Dunlin).

reworking of Late Kimmeridgian–Early Volgian sediments started already in the Mid-Volgian, as shown by the presence of *Perisseiasphaeridium pannosum*. In well 34/7-24S, reworking of Mid-Oxfordian–Early Kimmeridgian sediments occurred in the Early Volgian (*Ctenidodinium chondrum*, *Gonyaulacysta jurassica longicornis*) and of Kimmeridgian–Early Volgian sediments in late Mid-Volgian times (*Cribroperidinium longicorne*). These reworked taxa occur within Draupne Formation shales of uniform appearance, representing low-energy, offshore marine conditions. Shortly after, however, in Late Volgian times, sand progradation reached the area of wells 34/7-21, -21A, -23S, and -23A. It is worth noting, however, that a time lag between deposition and erosion/reworking of some few Ma, as noted above, is not a particularly 'short time' in terms of fault activity.

Because erosion on footwall scarps tends to cut into the entire elevated stratigraphy simultaneously, erosion is likely to have cut into Brent Group strata in the Statfjord East footwall at an early stage. On the adjoining Snorre-H hangingwall, erosion likely cut sequentially into the stratigraphy incorporated in fault block rotation. Hence, Heather Formation strata expectedly would be eroded for some time on the Snorre hangingwall prior to exhumation and erosion of the Brent Group. There is a weak indication of an inverted succession of reworked sediments, suggesting progressive unroofing. Overall, however, erosion into the Heather Formation and Brent Group probably was more or less simultaneous within the available time resolution. This is likely a result of the semi-enclosed basin physiography of the plunging Snorre Fault Block, providing both simple updip derived and longshore transported sediments and leading to a mixing of reworked taxa from source areas that were at different levels of erosion.

Description of the Draupne Formation

Facies characteristics

Nineteen cores from the wells 34/7-21, 34/-21A, 34/7-23A and 34/7-23S have been described and interpreted (Fig. 9).

The uppermost Heather Formation is cored in well 34/7-21, where it is represented by 4 m of strongly bioturbated silty shale with abundant glauconite and pyrite nodules. *Paleophycus* and *Skolithos* are the common ichnofacies. The top surface is well defined, intensely burrowed, and includes Belemnite remnants.

The lower part of the overlying Draupne Formation consists of thinly laminated, little bioturbated black shale, the Draupne Shale, with thin interbeds of carbonate shell-debris and pyrite laminae and nodules. The black shales pass gradually upwards into laminated dark grey shale with thin, graded or ripple cross-bedded sandy lenses or sandy laminae, often appearing as double layering. Outsize sandy beds of 1 to 2 cm are present in the lower part and become more frequent upsection. Contorted bedding structures and small-scale syn-sedimentary faulting have been observed. Coal debris is abundant in the upper part, whereas Belemnites are present throughout. *Muensteria* and *Asterosoma* are the common ichnofacies, but some *Planolites* burrowing has also been observed. The Draupne Shale is interpreted to reflect deposition on a tilted hangingwall dip-slope with general stagnant water circulation and anoxic water bottom conditions. The ripple bedded sandy lenses, together with small-scale slumping and resedimented clay and coal fragments, may indicate overall gradient steepening accompanied by coastline progradation and sand reaching the upper slope environment by storm processes. As shown by the reworking analysis, local erosion of the Heather Formation contributed to deposition of the Draupne Shale. The presence of sand layers and coal fragments in the upper part indicate erosion of the Brent Group.

A 3 m thick interbedded sandstone unit, the Intra Draupne Sandstone, has been observed in well 34/7-21. It is also recognized on well-logs in 34/7-21A, but in this well it is not cored. It consists of fine- to medium-grained, flat-bedded to cross-bedded, micaceous sandstone with abundant mud clasts and coal debris. Contorted bedding and frequent internal erosion surfaces testify to frequent failure. The Intra Draupne Sandstone is interpreted as a submarine channel deposit with ensuing gravity flow and traction currents, indicative of high rates of fault block rotation and gradient steepening. The black shale clasts were probably derived from the underlying Draupne Shale, possibly through channel wall failure. The micaceous sand and coal fragments, on the other hand, were likely sourced from erosion of the Brent Group. It is suggested that the sand was supplied from failure and remobilization of an incipient updip sandy shoreline or mouth bar system, likely due to oversteepening and instability resulting from progressive fault block rotation.

Thickness and frequency of sandy laminae in the Draupne Shale increase upwards into 10 to 40 cm thick graded units, which have a lower

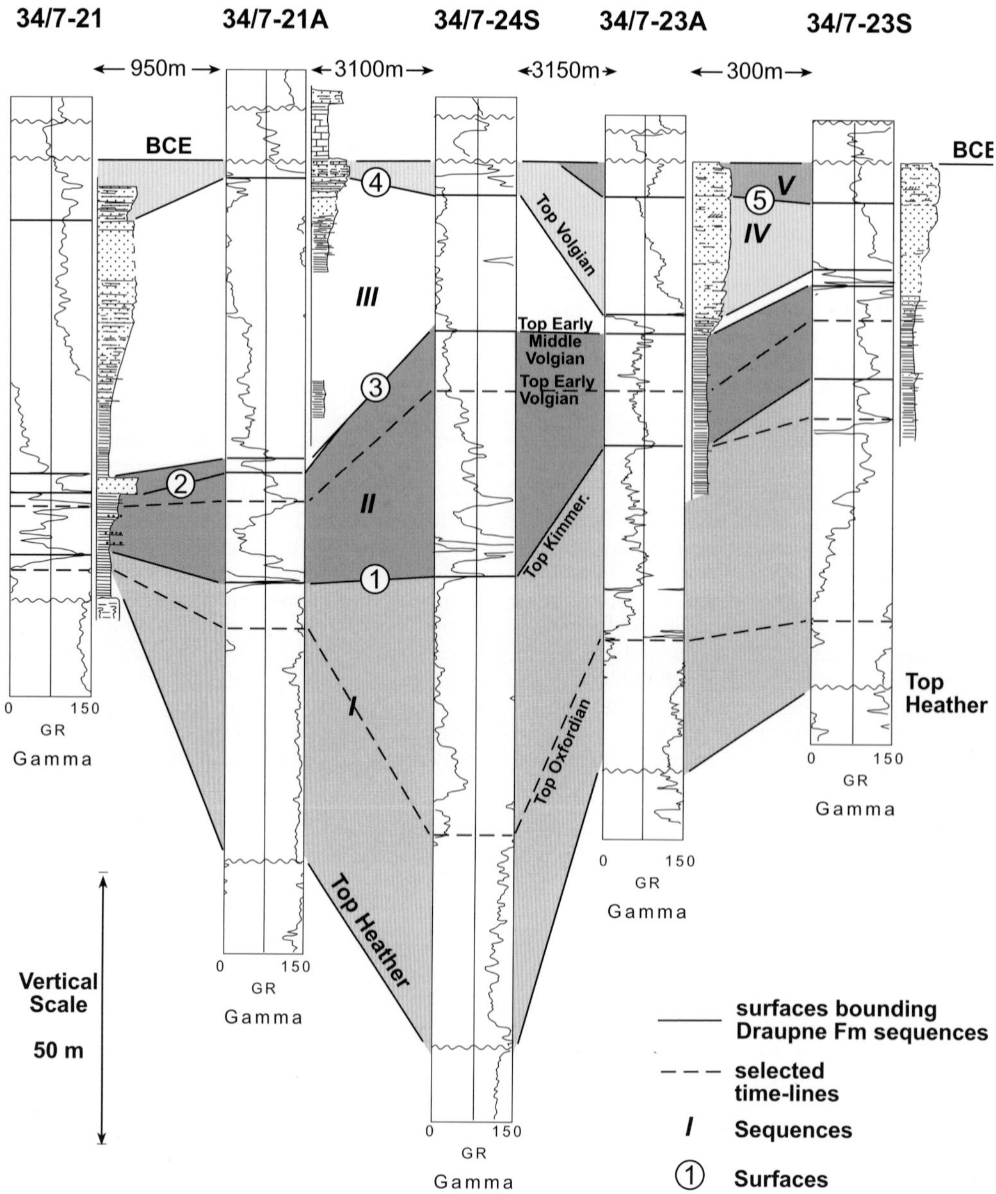

Fig. 9. Well correlation, including core descriptions, and sequence architecture of the Snorre-H Draupne Formation.

burrowed sandy part and an upper laminated shaly part. This heterolithic facies passes transitionally upwards into a tabular sandstone unit, the Upper Draupne Sandstone, which is up to 35 m thick. It consists of cm to dm thick units of micaceous, fine- to medium-grained, ripple- and cross-laminated sandstone. Bioturbation decreases upwards and horizontal lamination and cross-bedding, occasionally low-angle, characterise the uppermost part. *Paleophycus* and *Skolithos* burrowing occasionally overprint the stratification. A lower to upper shoreface, prograding shoreline environment is suggested. The lack of top set deposits inhibits a conclusive interpretation of the shoreline type, but the general character suggests a mixed fluvial- (protuberated) and wave-dominated (linear) shoreline progradation.

The uppermost 6 m of the Upper Draupne Sandstone in wells 34/7-23A and -23S consist of medium grained, parallel bedded sandstone with intercalated clay laminae. Faint cross-bedding has been observed and the sandstone is moderately bioturbated, mostly by *Skolithos* burrows. It is interpreted as a transgressive sandstone capping the Upper Draupne Sandstone in these wells.

In wells 34/7-21A, -23A and -23S, the top part of the Upper Draupne Sandstone is impregnated by diffuse carbonate growth passing gradually into carbonate nodules, related to the subsequent Cromer Knoll transgression. The lowermost part of the Cromer Knoll Group was cored in well 34/7-21A, where it consists of limestone nodules interstratified with pyrite rich greenish marls passing into interbedded nodular carbonates and red shales. Shell debris is very abundant and ammonites are occasionally present. The Cromer Knoll Group is interpreted as a deepening carbonate succession, terminating the overall shallowing upward nature of the Upper Draupne Sandstone. The red-coloured interbeds indicate oxygenated water conditions, possibly below the carbonate compensation depth.

Key bounding surfaces

Five main surfaces (S-1 to S-5) can be distinguished between the top of the Heather Formation and base of the Cromer Knoll Group. The four lower surfaces are tied to well 34/7-21, as it is the only well being cored all the way down to the top Heather Formation. The fifth surface is only represented in wells 34/7-23A and -23S. A correlation panel has been constructed between the wells 34/7-21, -21A, -23A, -23S and -24S (Table 2; Fig. 9).

Surface 1 (S-1) occurs at 2580.0 m in well 34/7-21, within the Draupne Shale, and corresponds to the change from a muddy section with frequent minor syn-sedimentary slumps and faults to a laminated muddy section that gradually intercalates millimetre thin sandy laminae. The surface gives a distinct gamma-ray signal. In well 34/7-21A, it is well below the deepest cored interval (core no. 5). The gamma ray signal seems to be correlatable to wells 34/23-A and -23S, although no lithological changes have been noticed. S-1 corresponds approximately to the top Kimmeridgian timeline and intra Draupne seismic event. It is characterized by an acme of *Botryococcus* (fresh-brackish water algae) in the 34/7-23S and -23A wells, suggesting increased freshwater run-off or reworking of older fresh-brackish water deposits at this time. S-1 is interpreted as a flooding or major basin deepening surface. The combination of increased run-off/reworking and accompanying basin deepening is believed to indicate a major tilt event or pulse of increased rates of fault block rotation.

Surface 2 (S-2) occurs at 2573.0 m in well 34/7-21. It corresponds to the base of the Intra Draupne Sandstone and appears to be related to

Table 2. *Key surface calibration*

Surface	Type	Biostrat calibration	34/7-21	34/7-21A	34/7-24S	34/7-23A	34/7-23S
S-5	FS	FO S. cf. palmula	NP	NP	NP	L?, SED	B?, L?, SED
S-4	FS	FO O. diluculum, FO B. pomum, ACME Pterospermella & Tasmanites	B, SED	B, L, SED	B	L, SED	L
S-3	FS	LO G. mutabilis, LO S. jurassica, ACME Botryococcus	B, L, SED	L	B, L	B?, L	B?, L
S-2	Base IDS		L, SED	L	Merges with S 3?	Merges with S 3?	Merges with S 3?
Near Base IDS	TL	LO O. patulum, FO D. spinosum	L	B, L	B	L	B, L
S-1	FS/TL	LO G. jurassica, FO C. panneum, LO P. pannosum	B, L, SED	L	B, L	B	B
Top Oxfordian	TL	LO S. crystallinum	B, L	L	B, L	B, L	B, L

(FS, flooding surface; TL, timeline; NP, not present; B, correlated on palynological event; L, correlated on log response; SED, correlated on sedimentary features; ?, calibration uncertain.

the channel morphology of this sand body. S-2 is characterized by a strong gamma-ray signal. It is still present in well 34/7-21A, but probably merges with surface S-3 in the other wells.

Surface 3 (S-3) occurs at 2566.0 m in well 34/7-21, immediately above the Intra Draupne Sandstone. It marks the base of the Upper Draupne Sandstone prograding shoreline sequence. S-3 has a distinct gamma-ray signal. It can be correlated with the base of the upper Draupne coarsening-upward sequence in well 34/7-23A and with the strongest gamma signal in the non-cored section of well 34/7-23S. As for S-1 an acme of *Botryococcus* is observed at this level in the 34/7-23S and -23A wells, suggesting increased freshwater run-off or reworking of older fresh-brackish water deposits at this time. S-3 is believed to represent a flooding or major basin deepening surface. The combination of increased run-off/reworking and accompanying basin deepening is believed to indicate a pulse of increased rates of fault block rotation.

Surface 4 (S-4) occurs at 2521.0 m in well 34/7-21 and corresponds to the top of the progradational Upper Draupne Sandstone shoreline sequence. The surface correlates to 2878.0 in well 34/7-21A, to 3244.0 m in well 34/7-23A and falls in the non cored section in well 34/7-23S, where it is indicated by a strong gamma peak. The correlation of S-4 is based on biostratigraphical dating, however, and is not well constrained in wells 34/7-23S and -23A. It coincides approximately with the boundary between Late Volgian and Ryazanian. S-4 is interpreted as a flooding or major basin deepening surface.

Surface 5 (S-5) occurs at 3211.8 m in well 34/7-23A, as a conspicuous shell bed of Belemnites. The facies description of core no.1 of well 34/7-23S suggests that the surface correlates to the lithology change at 3086.5 m. S-5 delineates the base of the uppermost part of the Upper Draupne Sandstone sequence in these wells, and is truncated by the unconformity underlying the Cromer Knoll Group. S-5 is believed to represent the final flooding or basin deepening surface in the syn-rift succession prior to lower Cretaceous transgression and drowning of the rift terrain.

Sequence architecture

The key surfaces described above can be used to define depositional sequences (Seq-I to Seq-V) within the Draupne Formation, with the top Heather Formation, S-1, S-3, S-4 and base Cromer Knoll Group as the main sequence-bounding surfaces. A sequence correlation between the wells is shown in Fig. 9 where the more obviously correlatable flooding surfaces have been used as sequence bounding surfaces.

Sequence I (Seq-I) is late Early Oxfordian to Kimmeridgian in age. It is bounded by the unconformity/hiatus at top Heather Formation and the flooding surface S-1, which coincides approximately with the intra Draupne seismic event. Seq-I has been cored only in well 34/7-21, where it consists of anoxic and laminated clays, typically characterized by frequent indications of minor slumping and syn-sedimentary faulting. It is only a few metres thick in well 34/7-21, but from well log correlation it appears to thicken in wells 34/7-23 and -24. The core descriptions point to starved conditions, likely related to basin compartmentalization and steepening depositional gradients.

Sequence II (Seq-II) is bounded by the S-1 and S-3 surfaces. It comprises the Early Volgian and the lower part of Mid-Volgian. Seq-II is cored in wells 34/7-21, -23A and -23S and coincides with the appearance of the first sand influx, first as millimetre thick laminae, and gradually thickening up to more frequent, centimetre-thick intercalations of ripple cross-laminated sandstones. The 3 m thick Intra Draupne Sandstone slope channel cored in well 34/7-21 is not seen in wells 34/7-23A and -23S, but an associated strong gamma peak seems to be recognizable throughout the area. This suggests that, whereas the sandstone unit itself is essentially discontinuous and localized, the time surface at which it occurs is of more lateral extent. The correlation is supported by the position of the acme *Botryococcus* in the last two wells. The base and top surfaces of the Intra Draupne Sandstone likely merge into one single surface locally, depending on the presence or absence of channel sand-bodies. Seq-II is interpreted as a shallowing upward slope succession, reflecting increased local sediment supply.

Sequence III (Seq-III) is Mid- to Late Volgian in age. It is bounded by the S-3 and S-4 flooding surfaces and includes most of the Upper Draupne Sandstone in wells 34/7-21 and -21A. It consists of a 45 m thick shoreline succession, coarsening upward from laminated mud to coarse-grained and bioturbated shoreface sandstones. In wells 34/7-23A and -23S, it is represented only by a 3 to 4 m thick, more fine-grained coarsening upward sequence. Seq-III reflects a gradual shallowing of the basin, likely as a result of reduced rates of fault block tilting.

Sequence IV (Seq-IV) is bounded by S-4 and S-5 in wells 23/7-23A and -23S, and by S-4 and the base of Cromer Knoll in wells 34/7-21 and -21A. It is Ryazanian in age. In wells 34/7-23A and -23S it is represented by an upward coarsening shoreline succession that is similar, in all aspects, to Seq-III in wells 34/7-21 and -21A. It is interpreted to onlap S-4, and in wells 34/7-21 and -21A it appears as a thin, transgressive sand capping the Upper Draupne Sandstone. Seq-IV is interpreted as a progradational increment in an overall back stepping shoreline system.

Sequence V (Seq-V) is bounded by S-5 and the base of the Cromer Knoll Group and is assumed latest Ryazanian in age. It is represented by about 6 m of transgressive sand constituting the uppermost part of the Upper Draupne Sandstone in wells 34/7-23A and -23S. Seq-V may represent the final retreat and drowning of the coastal progradation, but could as well be seen as the lateral equivalent to a thicker sandy sequence deposited even further updip in the basin.

Draupne Formation seismic character

Two well-defined seismic reflectors, the base Cretaceous and top Heather Formation, mark the top and base of the Draupne Formation and have been mapped out regionally (Figs 4 & 10). In addition, several intra-Draupne Formation reflectors have been mapped locally across parts of the Snorre-H area. Here, the Draupne Formation is seismically defined as consisting of a lower unit and an upper unit, separated by an intra-Draupne seismic event. The intra Draupne seismic event underlies the base of the Upper and Intra Draupne Sandstones (Figs 6 & 10) and corresponds approximately to the top Kimmeridgian timeline (bounding surface S-1). The intra Draupne seismic event, is observed to onlap the top Heather along the hangingwall dip-slope in the southern part of the Snorre-H basin. However, along the north-eastern edge of the sub-basin, near wells -23S and 23A, it appears to be truncated by the Base Cretaceous Event (BCE), possibly related to footwall uplift along a small fault near the wells. Alternatively, it may just be too thin to resolve.

The lower unit is relatively homogeneous in seismic character and corresponds to depositional sequence Seq-I, the shale dominated lower part of the Draupne Formation (= Draupne Shale). However, as the unit is not more than one cycle thick, its internal character and geometry are not well resolved. It thickens slightly westward, with several local depocenters developed along segments of the Statfjord East Fault.

The upper seismic unit is more heterogeneous in character and contains a number of spatially discontinuous internal reflectors. It corresponds to depositional sequences Seq-II through Seq-V, the layered mudstones, siltstones and sandstones of the upper Draupne Formation (incl. the Upper Draupne Sandstone), and it has a distinct wedge shaped geometry. To the south in the Snorre-H area (Fig. 6), the upper unit onlaps onto the lower unit. The lower part of the upper unit has a sub-parallel internal reflection pattern, whereas the upper part has a multiple lobe shaped character. It is described in more detail below.

In the southern part of the area, relatively continuous reflections overstep the degraded Statfjord East footwall (Fig. 10c). The completion log for well 33/9-12, located just north of the most intensively degraded footwall area, records a thin drape of Late Volgian Draupne shale.

Sandstones in the upper Draupne Formation

Sandstones in the upper Draupne Formation have been observed in many wells in the larger Statfjord and Snorre area (Fig. 3). It is difficult to map out these sands seismically, as they are thin and do not commonly set up reflections that can be followed regionally. However, a set of reflections in the upper part of the Draupne Formation, corresponding to the level of sandstone development, have been mapped locally across the Snorre-H area (Dawers *et al.* 1999).

Figure 10 shows several traverses through the Snorre-H wells. The base of the Upper Draupne Sandstone is seismically resolvable in wells 34/7-23S and -23A (Fig. 10a). The sandstone is interpreted as a lobe-like unit lying immediately below the BCE. In wells 34/7-21 and -21A, a reflector correlates with the position of the Intra Draupne Sandstone (Fig. 10b). This horizon can be mapped northward and southward from wells 34/7-21 and -21A, forming the base of a separate lobate unit, which is interpreted to contain both the Intra and Upper Draupne Sandstone units in this area. The seismic lobe in wells 34/7-23S and -23A overlies the seismic lobe in wells 34/7-21 and -21A, supporting the noted biostratigraphical younging of the Upper Draupne Sandstone from late Volgian in wells 34/7-21 and -21A to Ryazanian in wells 34/7-23S and -23A.

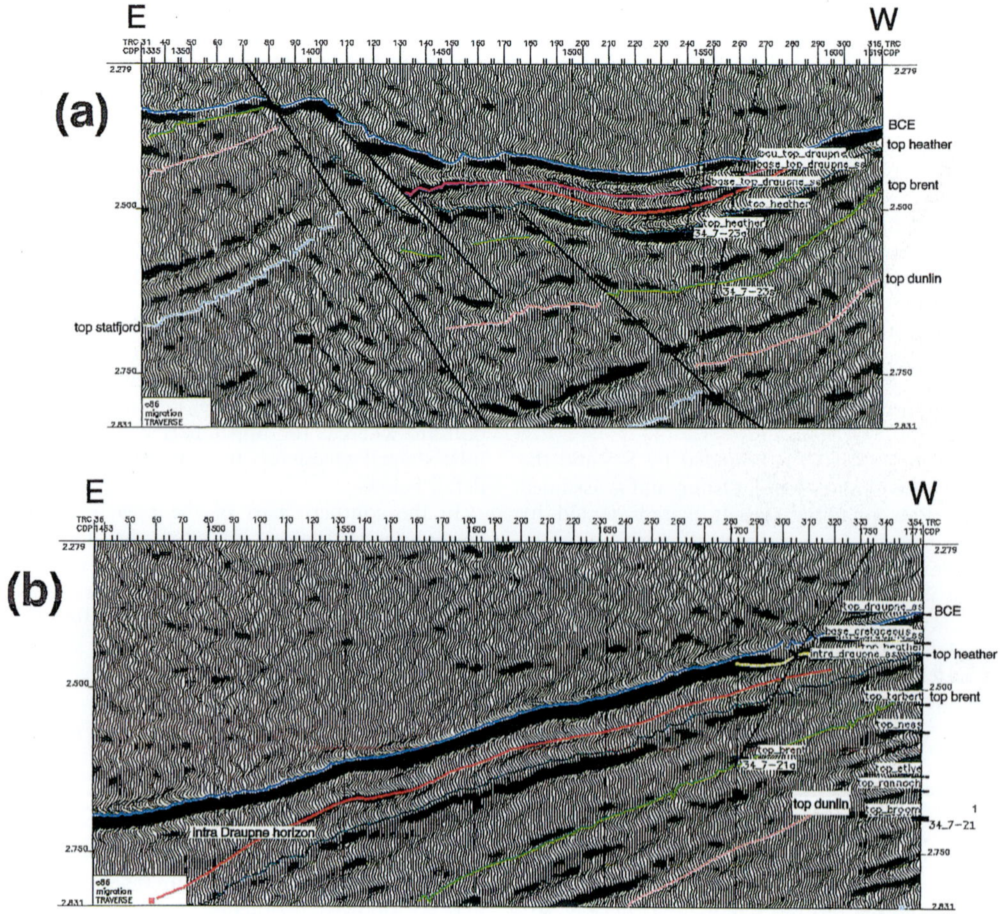

Fig. 10. Seismic sections through the Snorre-H area wells. See map insert for location of lines. (**a**) Section through wells 34/7-23A and -23S, showing the base of the Upper Draupne Sandstone in pink. (**b**) Traverse through wells 34/7-21A and -21, showing the base of the sand-prone section in yellow. (**c**) Inline 170 from survey E86 showing the Statfjord Fault (left) and Statfjord East Fault (right). Several internal reflectors overlie the top Kimmeridgian horizon, shown in red; the lower ones onlap top Kimmeridgian, whereas the uppermost ones appear to drape a degraded portion of the Statfjord East footwall which has been displaced by the Statfjord Fault. (**d**) Detail of part c showing the chaotic wedge above the top Kimmeridgian horizon. Note the subtle downlap of internal reflectors.

Both seismic lobes were mapped in seismic section on every second inline in the E86 survey. In areas where a base sandstone reflector was observed in seismic section, a weakening in the amplitude of the base Cretaceous reflector was also observed (Fig. 11a). This reduction of the otherwise strong impedance contrast associated with the passage from Cromer Knoll limestones to Draupne shales is thought to arise when sandstones are present in the uppermost part of the Draupne Formation, as a result of less velocity contrast between limestones and sandstones than between limestones and shales. Thus, there is a potential to discriminate the Draupne sandstones using the seismic attributes of the BCE. In order to investigate this possibility a number of different seismic attribute maps, using various measures of amplitude and various window widths, were made for the autotracked interpretation of the BCE reflector in the E86 data set. The results show that amplitude anomalies along the BCE correlate with the pattern of base sandstone reflectors mapped in the uppermost part of the

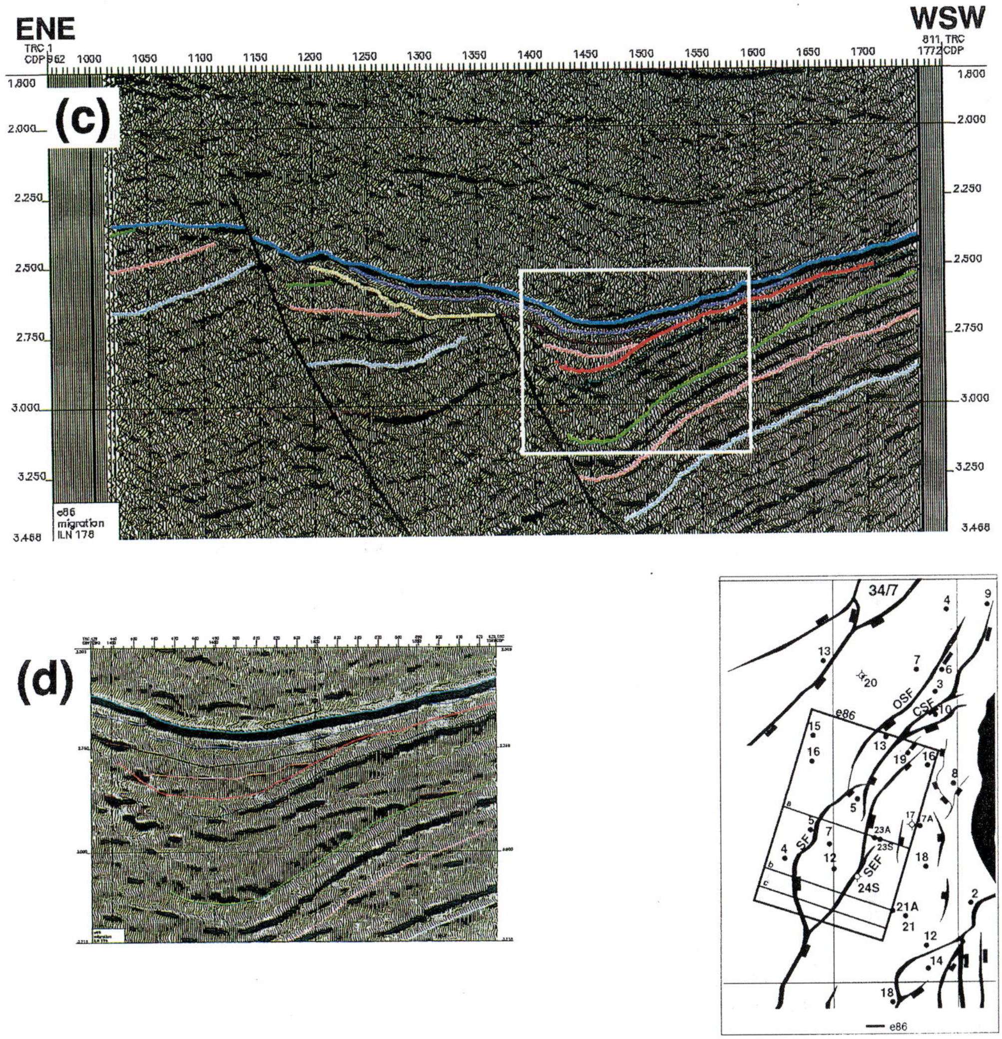

Draupne Formation. For example, Fig. 11b shows the spatial variation of the rms amplitude within a window from 4 ms above the BCE to 24 ms below. The yellow through red areas correspond to low amplitudes and the blue areas correspond to higher amplitudes. Note that the sandstone occurrence in the 34/7-21 wells and 34/7-23 wells corresponds to low amplitudes. The aerial extent of the interpretation of the base of the sandstone, which is resolvable in seismic sections (Figs 10a and 10b), is also shown on the attribute map (Fig. 11b). The interpretation of the mapped sandstone extent correlates very well with the areas of low amplitude observed in this attribute map.

Other significant Draupne Formation seismic anomalies

In the deepest part of the Snorre-H basin, several stacked reflectors are observed to onlap the top Kimmeridgian reflector described above (Fig. 10c). The lowermost of these reflectors form the top of a seismically chaotic wedge that occasionally shows internal reflections that downlap onto the top Kimmeridgian horizon (Fig. 10d). The extent of the wedge is shown in Fig. 12. The chaotic wedge appears as an area of low amplitudes relative to the surrounding region, suggesting that it may be lithologically distinct. The Statfjord East footwall immediately adjacent to this area has been degraded

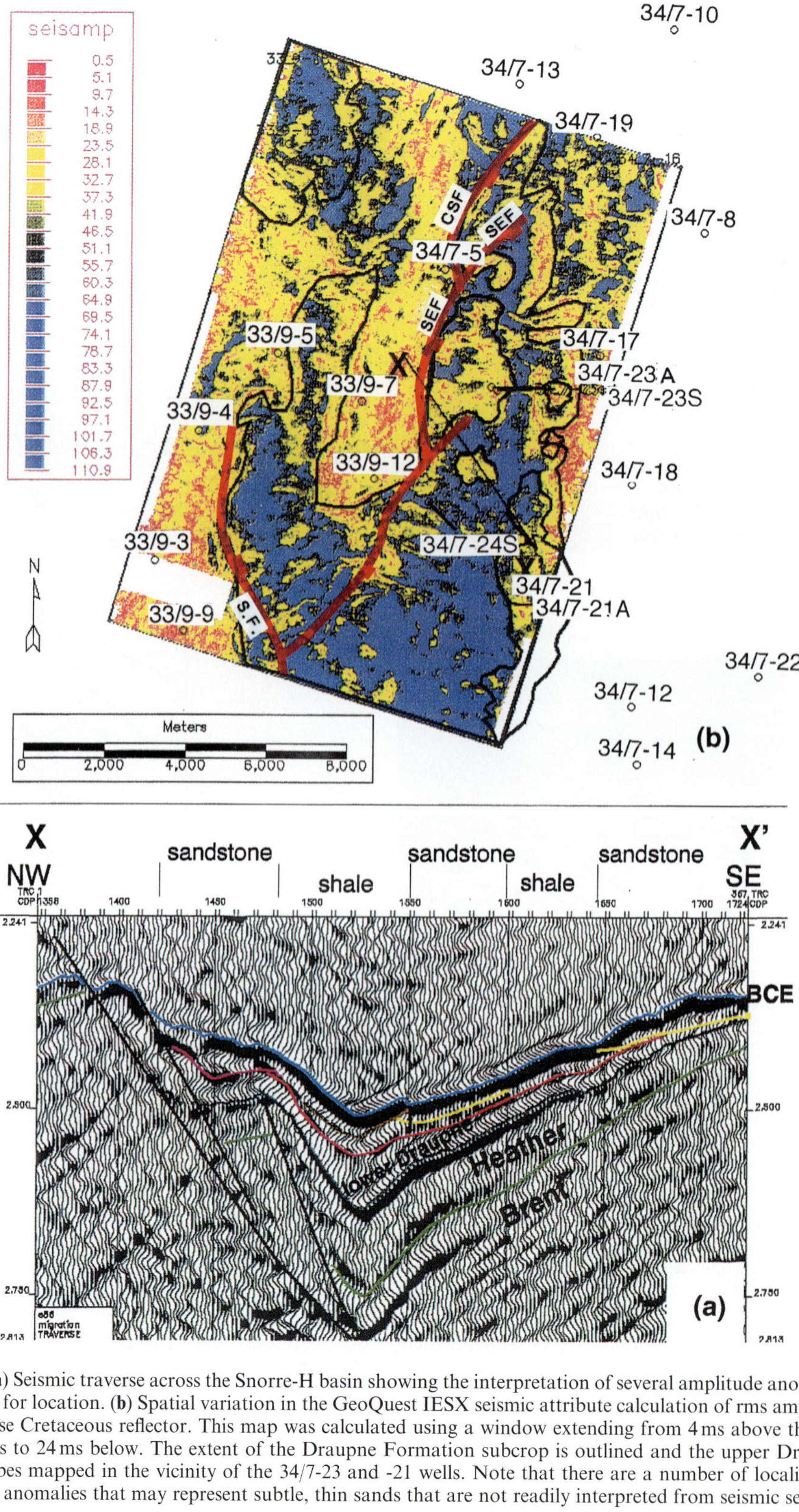

Fig. 11. (**a**) Seismic traverse across the Snorre-H basin showing the interpretation of several amplitude anomalies. See part b for location. (**b**) Spatial variation in the GeoQuest IESX seismic attribute calculation of rms amplitude for the base Cretaceous reflector. This map was calculated using a window extending from 4 ms above the base Cretaceous to 24 ms below. The extent of the Draupne Formation subcrop is outlined and the upper Draupne seismic lobes mapped in the vicinity of the 34/7-23 and -21 wells. Note that there are a number of localized amplitude anomalies that may represent subtle, thin sands that are not readily interpreted from seismic sections.

down through the Brent/Dunlin levels (Figs 10c, 13). This suggests that the chaotic wedge and deeper internal reflectors may be related to processes and products of footwall degradation, and may in fact be sandy. Similar low amplitudes are observed also in the vicinity of well 34/7-24S, on the other hand, which is entirely shaly. However, footwall degradation adjacent to this well does not involve substantial erosion of sand prone formations like the Brent Group, as seen to the south, next to the chaotic wedge.

East of the wedge, some elongated N–S trending anomalies can be seen (indicated by arrow in Fig. 12). When flattening on the base Draupne/top Kimmeridgian reflector, they seem to lie in the deepest part of the half-graben during this time. The interpretation of these features is not clear, but as they are oriented parallel to the column direction of the 3D survey, they may just represent geophysical noise. Alternatively, but less likely, they may represent axially transported gravity deposits.

Figure 13 summarizes the key elements of the seismic interpretation, including the distribution of the subtle Upper Draupne Sandstone, based on the seismic attribute maps and in relation to the main structural elements.

Discussion

In the following, the tectonostratigraphic evolution of the Snorre-H area is summarized and erosion and redistribution of sand in the syn-rift succession is discussed in relation to potential hydrocarbon play models.

Tilting of the Snorre Fault Block

In order to understand the tectonostratigraphic evolution of rotational fault blocks it is necessary to determine the tilt history. As a first approximation, tilt can be measured as the angle between a pre-rotational, assumed horizontal, datum and a given syn-rift surface in a seismic section (Fig. 14a). As a result of thinning and pinch-out of syn-rift packages onto the footwalls, however, post-depositional differential compaction will cause thicker basinal deposits

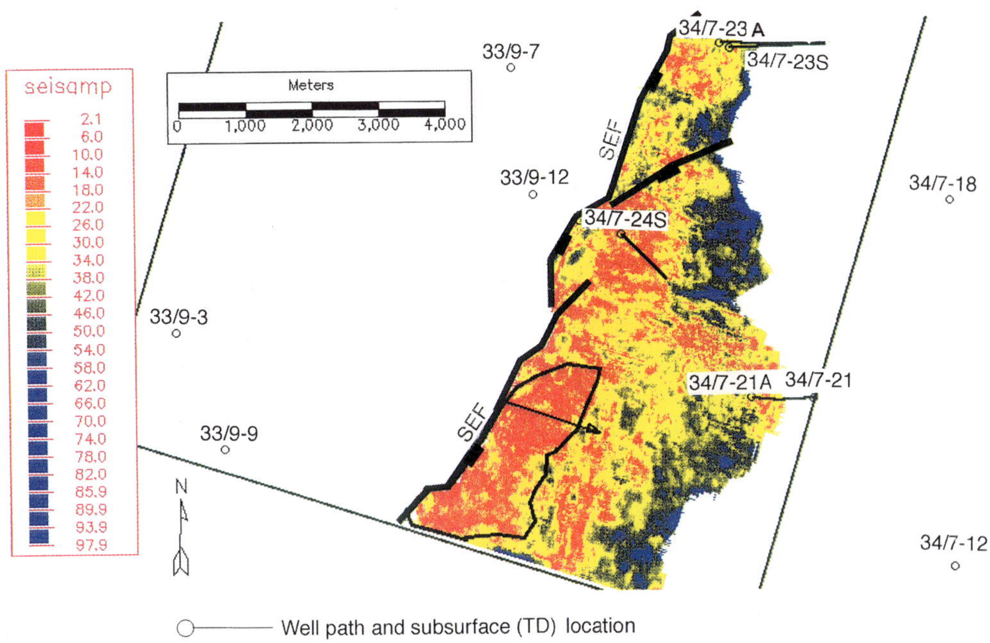

Fig. 12. Spatial variation in the rms amplitude for the intra Draupne top Kimmeridgian reflector. This map was calculated using a window extending from 8 ms below the reflector to 24 ms above. The outlined area represents the extent of the wedge-shaped unit lying on top of the Kimmeridgian horizon. Note the N–S trending amplitude anomalies to the south in the map view. They most likely represent geophysical noise, but could, alternatively, represent axially transported gravity deposits.

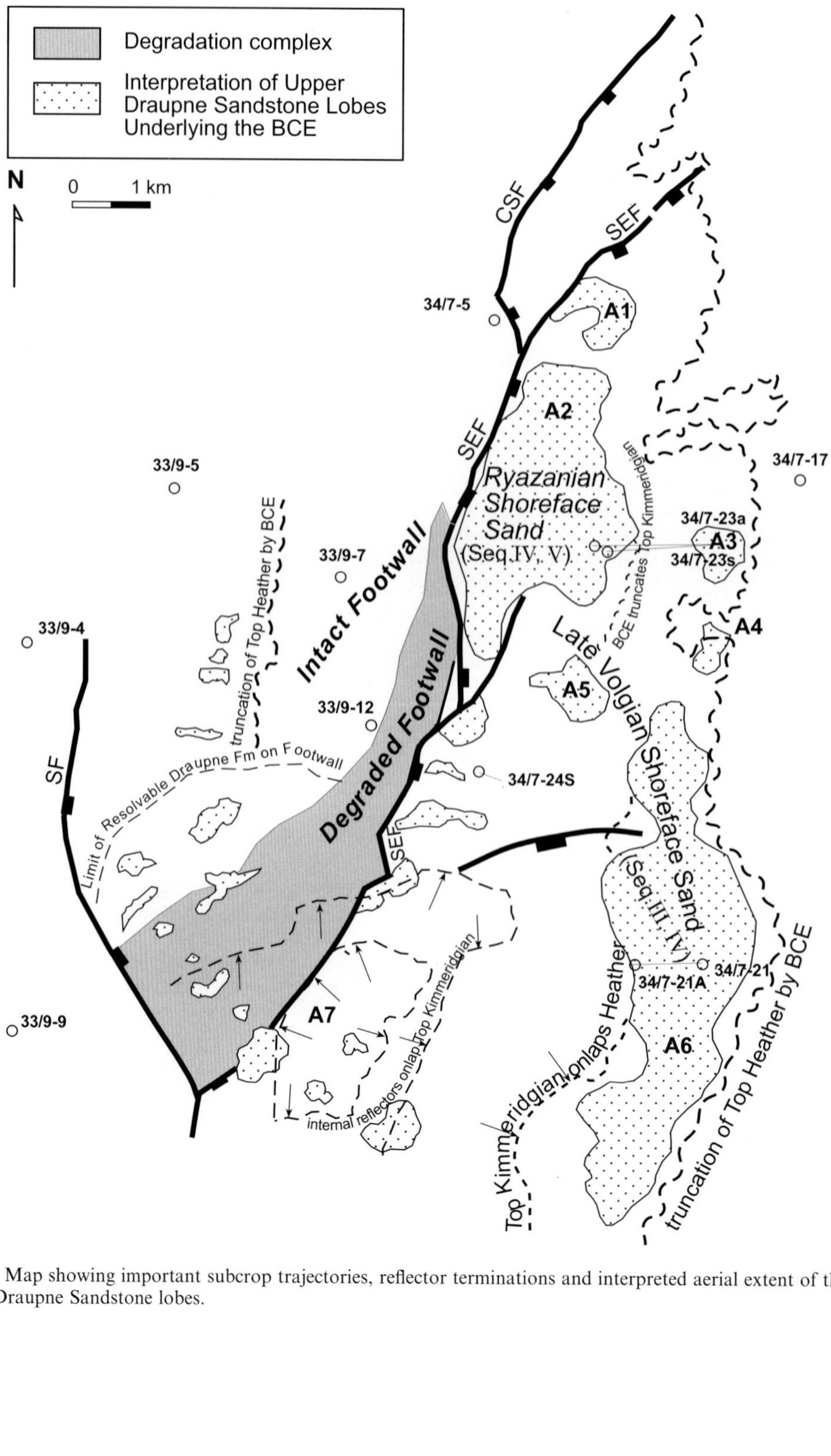

Fig. 13. Map showing important subcrop trajectories, reflector terminations and interpreted aerial extent of the Upper Draupne Sandstone lobes.

to compact more than thinner flank deposits, reducing the angle between top and base of the packages. Assuming a typical compaction coefficient between some 30–40° (Sclater & Christie 1980), the measured angle therefore needs to be corrected by a similar number (Fig. 14b). In addition, the assumed depositional gradient in the basin at the time of deposition needs to be added.

The widespread occurrence of coals in the Brent Group, especially in the Ness Formation, provides an excellent definition of palaeo-delta-top (i.e. sea-level) in the Snorre-H area. Apparent parallel reflectors from the top Brent Group downwards suggest it was deposited largely before onset of stretching and fault block rotation. To determine the tilt history of the Snorre Fault Block, the top Brent Group horizon is therefore used as datum and set horizontal. The amount of tilt varies from north to south along the Snorre-H structure, however, the tilt history discussed below has been estimated in the central Snorre-H area. The tilt measurements have been averaged over the width of the hangingwall sub-basin. Locally along the steepest parts of the hangingwall dip slope, much higher dips are attained (Fig. 6c–d).

The Brent Group in the central Snorre-H area is tilted relative to the base Draupne/top Heather (Early/Mid-Oxfordian) surface by some 2° towards the WSW. Decompaction of the Heather Formation by 30–40% (i.e. Sclater & Christie 1980) increases the tilt to about 2.5°. In addition, the top Heather depositional surface had a primary gradient that is set to 0.5–1°. If higher, the depositional surface would most likely become unstable and trigger significant gravity collapse, of which there is little evidence. This implies that the fault block tilt by end-Heather Formation (Mid-Oxfordian) times is likely to have been *c.* 3° (Fig. 14b, 15).

The top lower Draupne seismic surface (top Kimmeridgian), which can only be mapped in the southern part of the Snorre-H sub-basin, has less than 1° of additional tilt compared to the base Draupne surface. However, deposition of the laminated black shales of the lower Draupne seismic unit (= Draupne Shale) suggests a bathymetric deepening and increase in depositional gradient during deposition of this unit, tentatively set to 2°. Consequently, the fault block tilt by Late Kimmeridgian is estimated to *c.* 4–5°.

The tilt of the Brent Group increases to 5° relative to the top Draupne (Late Ryazanian) surface. Correcting for a 30–40% decompaction of the syn-rift infill increases the tilt to 6.5–7°. In well 34/7-24S, the gamma-ray log suggests a weakly coarsening upward sequence from black shales to silty shales, but with no evidence of sand deposition (Fig. 9), resulting from underfilling of the basin. As cores of the Draupne Formation shales in wells 34/7-21 and -21A to the south-east contain slump deposits, indicative of gravitational instability, the primary depositional gradient in the 34/7-24S area is set higher than for the Heather Formation, tentatively to 2°. This gives a top Draupne (Late Ryazanian) fault block tilt of *c.* 9°, as measured in the central Snorre-H area. This represents an incremental increase of 4° from Late Kimmeridgian.

The cumulative late Jurassic–early Cretaceous tilt amounts to 13° in the southern part of the Snorre-H sub-basin, suggesting that this part of the Snorre-H Fault Block experienced higher rates of rotation than did the central and northern area. A top Draupne depositional gradient in excess of 4° is needed in order to explain the present tilt in this area. Such a high gradient

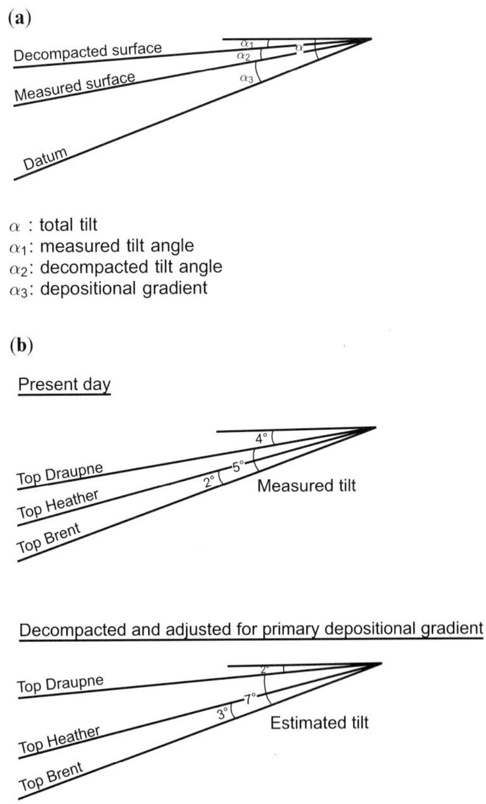

Fig. 14. (**a**) Methodology for estimating tilt angles in a syn-rift succession. (**b**) Measured and estimated (corrected) tilt angles for the Snorre-H hangingwall basin.

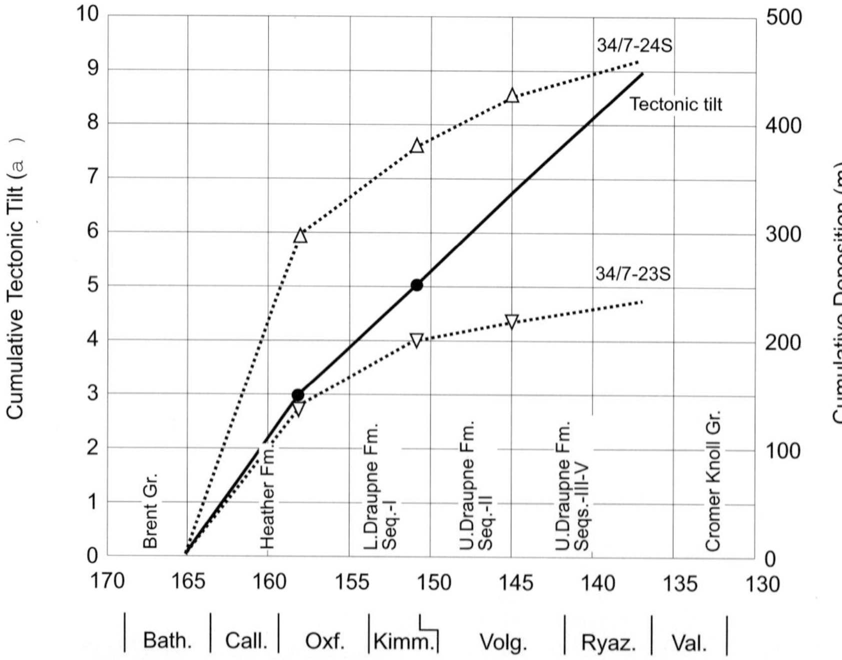

Fig. 15. Plot showing the Snorre-H Fault Block cumulative tectonic tilt and sediment thickness vs. geological time. The timescale is after Gradstein *et al.* (1994).

would normally trigger significant instability and gravity collapse, of which there is little evidence. The draping of the Draupne Formation by the Cromer Knoll limestones and lack of differential loading, however, may have prevented such collapse. Alternatively, it raises a question if there has been some additional, post-rift tilting of the Snorre-H Fault Block.

Snorre-H footwall uplift model

To determine the nature and history of the Snorre-H Fault Block footwall uplift, the so-called 'shoreline trajectory' during syn-rift evolution has been investigated (Fig. 16). The shoreline trajectory refers to the cross-sectional path of the shoreline as it migrates in response to changing relative sea-level (Helland-Hansen & Gjelberg 1994). The shoreline trajectory is synonymous with the temporal migration of the 'basement emergence point' of Roberts *et al.* (1993) in cross-sectional view. It gives a measure of footwall island width through time. The basement emergence point should not be confused with the structural fulcrum or lever point, which separates area of net uplift from area of net subsidence across a rotating fault block. Depending on the position of the fulcrum relative to the basement emergence point, the shoreline trajectory will vary and rotational fault blocks may experience footwall island expansion, stillstand or contraction subject to an identical tilt history (Ravnås & Steel 1998).

In the central Snorre-H area the BCE truncates the top Brent Group about halfway between the Statfjord East Fault and the Snorre-H footwall crest (6/6 km). The top Heather Formation is truncated by the BCE some 7 km westward of the Snorre-H footwall crest and 5 km eastward of the Statfjord East Fault (Fig. 13). As argued in the discussion on tectonostratigraphic evolution below, the lower part of the Heather Formation is believed primarily to have extended far to the east, possibly covering the Snorre-H footwall crest (Fig. 16). The shoreline position (= basement emergence point) by the Mid-Oxfordian (end Heather Formation) is not well understood, but we suggest it was located some distance (2–4 km) towards the west.

The top Kimmeridgian onlaps the Heather Formation approximately 1–2 km west of the BCE/top Heather truncation line (Fig. 13). However, the Kimmeridgian succession is clearly offshore/deep marine in origin and the

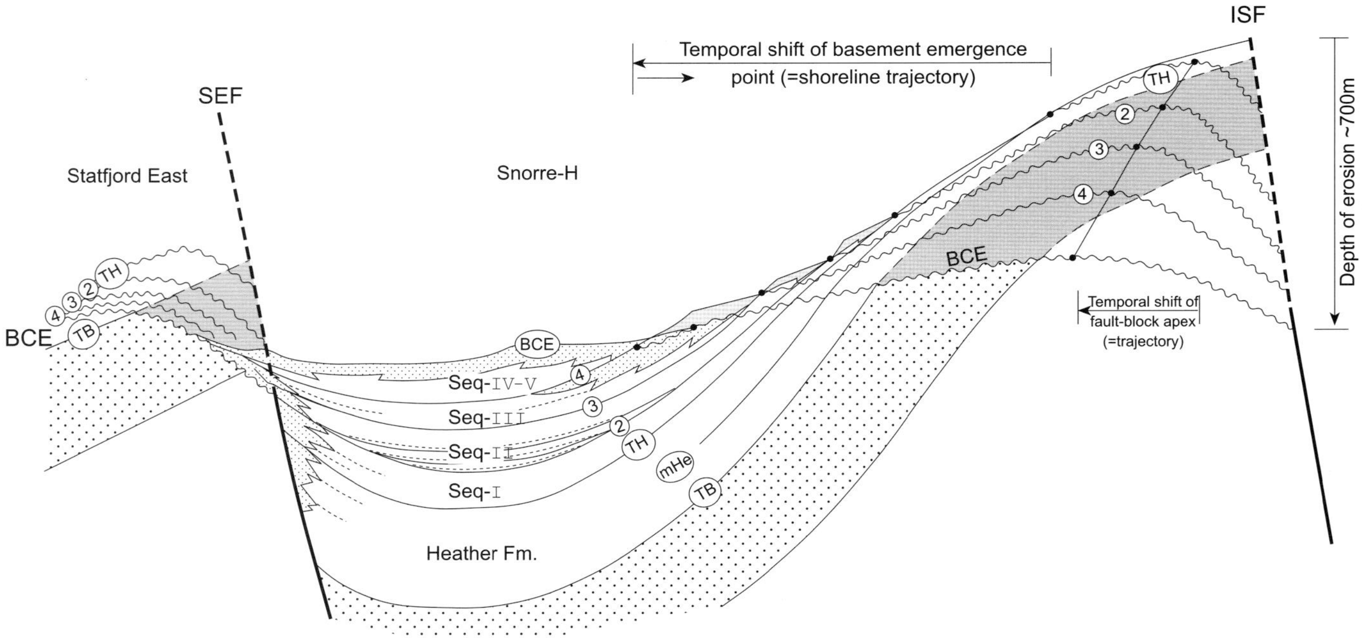

Fig. 16. Composite tectonostratigraphic model of the Snorre-H area, linking Snorre-H hangingwall depositional units to uplift and erosion of the Snorre-H and Statfjord East footwalls. Note that the model represents a combination of the tectonostratigraphic evolution in the southern part of the sub-basin, in the 34/7-21 wells area (Seqs-I–II), and the northern part of the sub-basin, near the 34/7-23 wells (Seq-III–V).

observed onlap must therefore be of submarine origin, possibly resulting from axial and transverse sediment supply to the Snorre-H half-graben from the Statfjord East footwall. The biostratigraphical data shows an overall compression of the Late Oxfordian–Kimmeridgian interval in the 34/7-21 well, indicating that the shoreline by the end of Kimmeridgian probably was located east of the present BCE/top Heather truncation line (Fig. 16).

Volgian strata further onlap and drape Kimmeridgian deposits, and Late Volgian prograding shoreline sandstones (Seq-III; Figs 9, 13) fringe the Snorre-H hangingwall roughly along the BCE/top Heather Formation truncation line, *c.* 1–2 km eastward of the top Kimmeridgian onlap line. The shoreline unit is overlain by a thin Ryazanian transgressive sandstone. This suggests initial rapid expansion of the footwall island during late Early Oxfordian–Early Volgian, followed by reduced expansion and stagnation/contraction during the Late Volgian–Ryazanian (Fig. 16).

To the north, towards the 34/7-23 wells, the stratigraphic development is slightly different. Here, the top Kimmeridgian is truncated by the BCE immediately west of the BCE/top Heather truncation line (Fig. 13). However, the top Kimmeridgian is also truncated by the BCE some further 1–2 km to the west, along an intra-basinal fault that was likely activated in the Volgian. Volgian–Ryazanian strata (Seqs-II–V; Fig. 9) are banked against the northern segments of the Statfjord East Fault and are largely bounded to the east by the intra-basinal fault. Patches of assumed Late Volgian/Ryazanian sandstones (Seqs-III–V, Fig. 9) are present also eastward of the intra-basinal fault (Fig. 6a–c), but these strata seem to have been later partly uplifted and eroded. In summary, the observations indicate a continuously expanding footwall island since the Oxfordian. Footwall contraction in this area was likely limited to the Late Ryazanian, in response to termination of rifting and thermal subsidence of the rift terrain.

The shoreline trajectory on the Snorre-H hangingwall shifted basinward at a low angle through most of the syn-rift stage (Fig. 16), representing an 'accretionary forced regression' in the classification by Helland-Hansen & Martinsen (1996). The accretionary forced regression typically results from abundant sediment supply (i.e. footwall erosion) superimposed on a falling relative sea-level (i.e. footwall expansion). The turnaround and reverse movement of the shoreline trajectory in the very late stages of rift evolution represents a 'non-accretionary transgression', resulting from a high rate of subsidence relative to rate of sediment supply.

The modified domino/variable rates of faulting model discussed by Roberts *et al.* (1993) predicts an increasing width of footwall islands as a result of successive fault block rotation, but not of the magnitude seen here. It is believed, therefore, that uplift and expansion of the Snorre-H footwall island became amplified by sub-regional uplift caused by the buoyantly rising main Snorre Fault Block and Tampen Spur to the north. Footwall island expansion also was amplified by an overall eustatic sea-level fall from the Late Kimmeridgian through Ryazanian (Hallam 1969; Haq *et al.* 1988; Surlyk 1989). A general sea-level rise from Bathonian to Early Kimmeridgian probably reduced the effect of footwall uplift and generation of subaerial relief during the early stages of rifting.

In the domino/variable rates of faulting model progressive increase in rates of faulting produces continuous offlap, as seen in the northern Snorre-H area. An offlap/onlap pattern, as seen in the central and southern Snorre-H area, results from increased, followed by reduced, rates of faulting. This variability in style of faulting is coherent with the noted northward propagation by segment linkage of the Statfjord East Fault (Dawers *et al.* 1999; Dawers & Underhill in press), leaving the southern portion of the Statfjord East Fault, though apparently quite active from the Bathonian through Kimmeridgian, tectonically inactive and drowned by Late Volgian times. At this time, fault movements in the Snorre-H area localized around the main Statfjord Fault further to the west.

The total throw on the Statfjord East Fault next to well 34/724S is roughly 1 km corresponding to a fault displacement rate of 0.05 mm/y over the Mid-Bathonian–Mid-Volgian interval. By comparison, measurements of recent tilting of fault blocks in Greece (Gawthorpe *et al.* 1994) show displacement rates up to 5 mm/y, or two orders of magnitude higher. This indicates that North Sea stretching was generally slow. Alternatively, rifting may have been considerably faster in shorter periods, corresponding to an overall pulsed mode of rifting.

Cumulative uplift and erosional yield

In clastic syn-rift terraines, usually local erosion follows as an immediate response to footwall uplift. Due to the unlithified nature of the sediments involved in rotation of the Snorre and Statfjord East Fault Blocks, such erosion expectedly would be very fast and efficient. As a simplification, the erosional yield can be considered proportional to the vertical relief created and the total area of uplift.

Based on the estimated cumulative tilt and temporal and spatial position of the shoreline/basement emergence point, theoretical footwall uplift and hangingwall subsidence across the central Snorre-H Fault Block have been calculated (Table 3). As the Snorre-H Fault Block dips become significantly reduced towards the footwall crest (Solli 1995; his Fig. 2), however, footwall uplift estimates based on average tilt angles in the Snorre-H sub-basin will be too high. Estimates of hangingwall subsidence, on the other hand, are considered more representative. In order to correct the uplift model, estimated uplift has been compared with the current depth of Late Jurassic erosion along the Snorre-H footwall crest. A cumulative tilt of 9° by the end of Ryazanian gives approximately 1100 m of cumulative uplift. Estimates by Solli (1995) suggest between 500–1000 m of erosion of this part of the Snorre Fault Block, likely close to 700 m due east of well 34/7-24S. This is some 400 m or 40%, less than the estimated uplift using 9° of tilt. As the Snorre-H footwall scarp erosional base was at sea-level the observed depth of erosion likely corresponds closely to the total uplift, suggesting that all estimated uplift numbers be corrected accordingly. Corrected uplift estimates suggest some 60 m of uplift by Mid-Oxfordian increasing to about 700 m by the end-Ryazanian (Table 3). Hangingwall subsidence increased less, from 400 m by Mid-Oxfordian to some 800 m by the end-Ryazanian, reflecting the rapidly expanding footwall island width. The Snorre-H dip variations have not been measured and integrated over the entire fault block width, but, as a sensitivity, a cumulative uplift of about 700 m corresponds to a Late Ryazanian tilt angle averaging 6°.

To test the uplift model, the sedimentation rate in the Snorre-H hangingwall basin has been investigated (Fig. 15). The amount of reworked bioflora in the syn-rift succession (Fig. 8) has been used as a measure of actual footwall relief generated and local erosional yield. The wells 34/7-23 and -24S in the Snorre-H sub-basin clearly have an exponential reduction in sedimentation rate through time, although the effect of variations in sediment supply cannot be easily separated from variations in generation of accommodation space. This observation may seem contradictory to the tilt evolution and footwall uplift model, which suggests overall increasing footwall uplift and generation of relief throughout the same time period. It is suggested, however, that the data be explained in terms of initial high rates of sedimentation from a predominant regional, extra-basinal sediment supply in the Bathonian–Oxfordian, during deposition of the Heather Formation. The scarcity of reworked taxa before the Late Oxfordian supports the notion of a pre-dominant extra-basinal sediment supply. A reduced rate of sedimentation from the Kimmeridgian onwards likely reflects a shutoff of the extra-basinal sediment source and transition to erosion and supply of sediment from local footwall highs. This interpretation is supported by the observed increasing number of reworked taxa in this part of the succession.

It should be noted that the observed reduction in reworked taxa in the Mid-Volgian (Fig. 9; transition Seqs-II/III) does not fit with a continuously expanding footwall island or uplift of the magnitude estimated above. This observation, however, is regarded erroneous and not representative for the Mid-Volgian uplift and erosional yield. The Late Volgian and Ryazanian Upper Draupne Sandstone lobes (Seqs-III–V) contain abundant reworked taxa, indicating significant local erosion in the Ryazanian. The reduction in sedimentation rate observed for the upper Draupne Formation is

Table 3. *Estimates of footwall uplift and hangingwall subsidence of the Snorre-H Fault Block, along an E–W transect near well 34/7-24S*

Time	Measured tilt-angle	Footwall island width	Cumulative hangingwall subsidence	Cumulative footwall uplift uncorrected	Footwall uplift corrected (40% reduction)
Mid-Oxfordian	3°	2–4 km	400–600 m	100–200 m	60–120 m
End Kimmeridgian	5°	7 km	450 m	600 m	350 m
End Ryazanian	9°	7 km	800 m	1100 m	700 m

In order to accommodate a decreasing fault-block tilt from the hangingwall across the footwall, initial footwall uplift estimates have been corrected by applying a fixed 40% reduction.Note that a fixed tilt angle of 5° has been applied for top Kimmeridgian, whereas the estimated tilt angle is between 4–5°.

believed to reflect bevelling of the footwall highs and reduced erosional yield during the waning rift stage.

Tectonostratigraphic evolution

In order to summarize the tectonostratigraphic evolution of the Snorre-H area, each of the main sequences described above has been discussed in terms of seismic character, structural context, depositional facies and palaeogeographic setting (Fig. 17).

Pre-Mid-Bathonian

The pre-Mid-Bathonian succession has not been included in this study. Nevertheless, during Late Bajocian–Early Bathonian (=Tarbert Formation) the Snorre-H area was likely governed by differential subsidence across the main block bounding faults (Fig. 17a). Our understanding of the stratigraphic development of the Tarbert Formation is hampered by the lack of well data, but emerging evidence from the Tampen Spur area (Johannesen *et al.* 1995; Davies *et al.* 1998) suggest that it has a weak wedge shaped geometry and was deposited in response to incipient rotational movements along the synrift master faults.

Mid-Bathonian–Early Oxfordian (Heather Fm)

Mid-Bathonian–Early Oxfordian subsidence (=Heather Formation) was concentrated along NNW–SSE striking master faults (Fig. 3). The seismic data does not allow a detailed interpretation of the Heather Formation internal reflectors and it is not possible to determine if rates of fault block rotation did vary over this interval. Overall rates of rotation were significant, however, reaching a tilt of *c.* 3° by the Early Oxfordian (Fig. 15). This was sufficient to create a wedge shaped geometry of the Heather Formation and a pronounced depocenter developed along the southern segment of the Statfjord East Fault (Dawers *et al.* 1999).

It is presently unclear to what extent the Heather Formation was deposited across the southern Snorre Fault Block during Mid-Bathonian–Early Oxfordian time (fig. 17b), as demonstrated on the Statfjord and Visund Fault Blocks (Roberts *et al.* 1993; Færseth *et al.* 1995). The scarcity of reworked bioflora before the Late Oxfordian suggests that footwall crestal islands were poorly developed at this stage. It should be noted also that no sand has been detected within the Heather Formation in the Snorre-H area, which may indicate that at least the lower part the Heather Formation was deposited across the footwall crestal areas of the southern part of the Snorre Fault Block.

By Mid-Oxfordian times the Snorre-H hangingwall shoreline and footwall morphological apex (=palaeodrainage divide) were located far to the east on the fault block (Fig. 16). The main sediment source for the Heather Formation was likely extra-basinal. It is suggested, however, that forced regression on the Snorre-H hangingwall dip-slope during deposition of the upper Heather Formation led to erosion and cannibalisation of the lower Heather Formation. Reflecting the infant stage of uplift, erosion of the Brent Group was confined to the footwall sides of the Snorre and Statfjord East Fault Blocks, as suggested by the presence of reworked Brent Group taxa in the Callovian succession in well 34/7-24S (Fig. 8).

Figure 18a shows a schematic palaeogeographic reconstruction of the larger Snorre area during Mid-Callovian times. As noted above, the Snorre-H and Statfjord East footwalls were likely submerged, whereas a source upland is believed to have existed to the north, as a result of progressive southward tilting of the Tampen Spur/Snorre Fault Block.

Late Early Oxfordian–Kimmeridgian (Draupne Fm; Seq I)

Continued rotation during the late Early Oxfordian–Kimmeridgian (=lower Draupne Formation; Seq-I) increased the tilt to some 5° (Fig. 15). Subsidence still was concentrated along the NNW–SSE striking master faults, leading to a gradual thickening of the lower Draupne Formation, Seq-I (=lower Draupne seismic unit; Draupne shale), towards the Statfjord East Fault, whereas it thins more abruptly onto the Snorre-H hangingwall. Small basinward dipping faults occur along the top of the lower Draupne Formation on the hangingwall dip-slope, but these are tentatively suggested to be post-depositional and related to subsequent tilting.

During deposition of the lower Draupne Formation, Seq I, footwall crestal islands had become firmly established, leading to increased rift-basin segmentation (Fig. 17c). Consequently, axial and transverse sediment supply did no longer feed directly into the outer half-grabens. This is supported by the apparent condensation or hiatus in the Mid-Oxfordian (Fig. 7), suggesting a period of widespread starvation at this time. The high gamma-ray log responses seen in the Kimmeridgian interval (Fig. 9) further indicate basin deepening and restricted circulation. As the entire interval is less than one cycle thick there is no seismic resolution within the time interval, but the seismic data indicates a

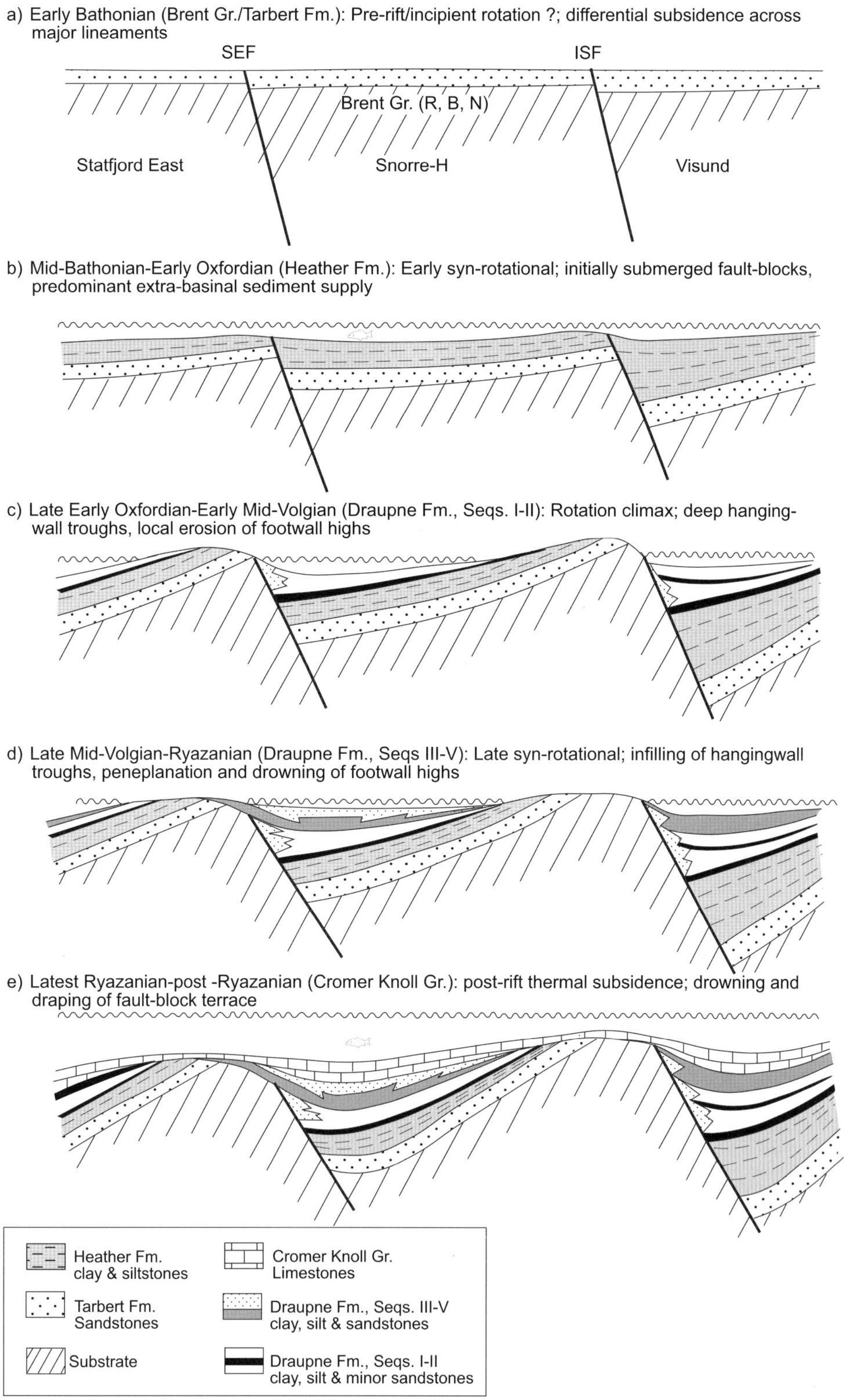

Fig. 17. Schematic illustration of the tectonostratigraphic evolution of the Snorre-H Fault Block.

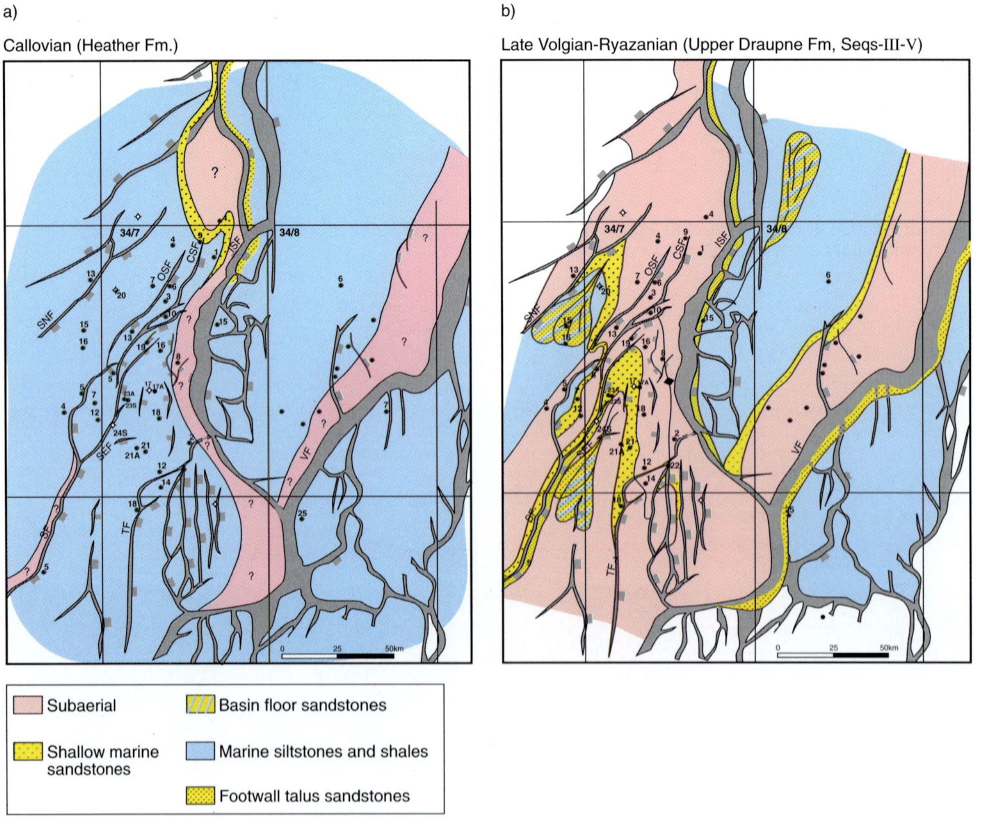

Fig. 18. Schematic illustration of the larger Snorre area palaeogeographic evolution. (**a**) Callovian. (**b**) Composite Late Volgian–Ryazanian.

weak onlap of the lower Draupne Formation and top Kimmeridgian surface onto the Heather Formation on the Snorre-H hangingwall (Fig. 13). There is also a pronounced updip thinning from wells 34/7-24S to 34/7-21A and -21 (Figs 6e, 9), and a slightly angular, unconformable relationship between the lower Draupne Formation and the underlying Heather Formation locally is observed in upflank positions on the Snorre hangingwall. This may indicate a pulse of rapid uplift following deposition of the Heather Formation, with subsequent intensification of footwall erosion and local sediment yield.

By the end of Kimmeridgian the Snorre-H hangingwall shoreline, as well as the footwall erosional apex, had been shifted westwards, in response to footwall uplift and expansion and accompanying fault-scarp degradation and retreat (Fig. 16). Deposition of the lower Draupne Formation, Seq-I, was sourced by significant local erosion of the Heather Formation and to some degree the Brent Group, particularly from the Statfjord East footwall. The presence of blue clay fragments (probably Heather clay) in the dark laminated mud is likely derived from neighbouring footwall islands, although an efficient drainage network had not yet been established. Alternatively, the clay particles may stem from failure and resedimentation on the destabilized submarine slope. By end-Kimmeridgian erosion on the hangingwall had cut into the Brent Group, suggesting that incipient shoreline sand systems may have been fringing the Snorre hangingwall at this time (Fig. 16). As a result of continued footwall uplift and expansion into the Volgian, however, they had zero preservation potential and later became cannibalized.

Early–early Mid-Volgian (Draupne Fm; Seq-II)
In the Early Volgian–early Mid-Volgian (= upper Draupne Formation; Seq-II), subsidence along the NNW–SSE striking master faults became accompanied by second-order

N–S, NE–SW and E–W faulting, leading to segmentation of the larger fault blocks (Figs 3 & 6). The seismic data does not provide sufficient resolution to properly map out the upper Draupne Formation, but, generally, thicker syn-rift packages appear to be present in the bathymetric lows (Fig. 6).

During deposition of the upper Draupne Formation, Seq-II, continued high rates of fault block rotation and associated uplift led to increased footwall island relief and erosion (Fig. 17c). At this time, rotational faulting along the Statfjord East Fault had propagated to the north of the central Snorre-H area. The Snorre-H sub-basin evolved into a northward closed and southward deepening embayment, separated from the Statfjord East sub-basin to the west. East of the Snorre Fault Block, the Visund sub-basin was closed to the south and not connected to the Snorre-H sub-basin. This suggests that each of the sub-basins by now was sourced predominantly from erosion of the adjacent footwall crestal islands. As the rate of sediment input was not sufficient to keep pace with the fault-controlled subsidence, this led to starved, under-filled conditions and increased bathymetry in the half-graben.

A pronounced onlapping relationship of the upper Draupne seismic unit (= upper Draupne Formation) onto the lower Draupne seismic unit and sub-parallel passive infill character of the upper Draupne seismic unit (Figs 6d, f, 10c, d) may indicate that there was a pulse of rapid rotation and uplift following deposition of the lower Draupne Formation. As a result of the onlapping relationship (Fig. 6d, f) thicker syn-rift packages are present in the bathymetric lows and a depositional hiatus may be speculated at the Kimmeridgian boundary between Seq-I and Seq-II in updip areas (Fig. 10c, d). The biostratigraphical data shows an overall compression of the Late Kimmeridgian–Early Volgian interval in the 34/7-21 well, but no apparent stratigraphic break has been observed (Figs 7 & 9). This may be a result of lack of stratigraphic resolution and deposition in the bathymetric low occurring over a relatively short time period accompanying the increased rotation and uplift. The seismic onlap of the upper Draupne Formation is observed in a position immediately adjacent to the Statfjord East Fault, suggesting that the deeper internal reflectors relate to processes of fault-scarp degradation.

By early mid-Volgian the Snorre-H hangingwall shoreline and footwall erosional apex had shifted further westward relative to the end Kimmeridgian, indicating continued forced regression as well as increasingly deeper erosion on the Snorre-H hangingwall during the Early Volgian (Fig. 16). Deposition of the upper Draupne Formation, Seq-II, was sourced predominantly by erosion of the Heather and Brent Formations and, to some degree, the lower Draupne Formation, as evidenced by the reworked taxa present (Fig. 8). The Intra Draupne Sandstone is believed to have been sourced from erosion of Kimmeridgian and Early Volgian basin stepping shoreline deposits, in response to the noted forced regression.

The seismic character, sedimentary facies and sequence architecture of the lower/upper Draupne Formation Seqs-I–II may suggest a pulsed mode of rifting and general rift climax during the Late Oxfordian–Mid-Volgian. It should be emphasized, however, that variations in fault displacement rates along a fault system may promote 'rift climax' stratal patterns near active fault segment centres contemporaneously with 'early/late rift' stratal patterns at fault tips and in overlap zones between unlinked faults (Dawers & Underhill in press). The neighbouring Visund Fault Block to the east also had maximum stretching and footwall uplift in the Volgian (Færseth *et al.* 1995), whereas the Brent/Statfjord Fault Block to the SW had maximum rates of rotation in the Oxfordian (Roberts *et al.* 1993; Nøttvedt *et al.* 1995). In a regional perspective, this may point to a spatial and temporal partitioning of fault displacement in the Snorre-H area.

Late Mid-Volgian–Ryazanian (Draupne Fm; Seqs III–IV)

During late Mid-Volgian–Ryazanian (= upper Draupne Formation, Seqs-III–V) the Snorre-H area experienced waning rates of fault block rotation and increasingly more uniform subsidence, leading to an estimated tilt of 9–13° by Late Ryazanian (Fig. 15). In response to continued intra-block deformation, however, the Snorre-H area became compartmentalized and subject to differential subsidence by N–S as well as E–W faults, and secondary half-graben structures developed (Fig. 6a–c). Footwall crestal uplift along some of the faults led to the emergence of minor, intra-basinal footwall islands.

Deposition of the upper Draupne Formation, Seq-III–V, was governed by reduced rates of fault block rotation and uplift, leading to significant reduction in new topographic relief generated. As a result of continued westward shift of the Snorre-H hangingwall shoreline and footwall erosional apex (Fig. 16), erosion now cut deep into the Brent Group on the hangingwall dip-slope, releasing substantial volumes of sand into the Snorre-H sub-basin. In combination with reduced rates of subsidence, this led to progradation and accumulation of the Upper Draupne Sandstone across the eastern and northern Snorre-H area (Fig. 13). The overall shallowing-upward nature of the upper Draupne Formation, Seqs-III–IV, as seen in the wells in the central Snorre-H sub-basin (Fig. 9), and the fact that it fills across the entire basin in the wells 34/7-23 and 23S area, suggests reduced fault related subsidence and generation of new accommodation space after deposition of Seq-II.

A major depocenter developed along the Statfjord East Fault to the south of well 34/7-24S. The depocenter represents largely (axial) infilling of pre-existing bathymetry, however, as the draping of the southern portion of the Statfjord East Fault by the upper Draupne seismic unit indicates it had become tectonically inactive by Late Volgian times. The upper Draupne seismic unit is overstepping the lower Draupne seismic unit (top Kimmeridgian surface) and onlapping also onto the southern hangingwall of the Snorre Fault Block, around wells 34/7-21 and 21A (Figs 6e, 13), suggesting overall transgression in this area. As the Draupne Formation overstepped the Statfjord East Fault Block, however, the supply of sand from the SW was significantly reduced.

The seismic as well as biostratigraphical data suggest that there is a younging of sandstone depositional units towards the north (Figs 9, 13). The sandstones are believed to consist of minor regressive increments that prograded into the basin during regional, but punctuated, transgression. The close association between sandstone occurrence and structural position led Dawers *et al.* (1999) to suggest that the distribution of thicker shoreline sandstones may be partly controlled by the northwards fault-tip propagation of the Statfjord East Fault. It is important to emphasise, however, that fault displacement during Late Volgian–Ryazanian were less compared to those obtained during the Late Oxfordian–Early Volgian.

A composite palaeogeographic reconstruction of the Late Volgian–Ryazanian is shown in Fig. 18b. Shoreface and beach cliff erosion together with fault-scarp retreat from gravity collapse probably contributed most to the denudation of uplifted fault block crests. As local upland areas were too small to allow larger river systems to develop fluvial incision most likely was less important. Sands were deposited in a prograding shoreline complex along the rim of the emergent fault blocks and filled across the entire embayment to the north.

The lack of evidences for significant rotational faulting during the Ryazanian suggests that subsidence at this time may have been (partly) related to thermal relaxation and the rift structure entering into a post-rift state of development. As a corollary, the neighbouring Visund Fault Block has the syn- to post-rift transition documented as latest Volgian (Færseth *et al.* 1995).

Latest Ryazanian–post-Ryazanian

From the latest Ryazanian onwards the Snorre-H sub-basin subsided in response to post-rift thermal relaxation and post-depositional compaction. Following deposition of the Draupne Formation, Hauterivian/Valanginian Cromer Knoll Group limestones draping the syn-rift succession and rift terrain suggest a relative sea-level rise, leading to progressive drowning of the inherent footwall relief and shutoff of the local as well as extra-basinal clastic supply to the half-grabens. The relative sea-level rise is believed to have been accentuated by an early Cretaceous eustatic sea-level rise (Haq *et al.* 1988, Surlyk 1989).

It should be noted that regional seismic lines indicate an onlapping relationship of lower Cretaceous strata onto the top Draupne surface. The lower Cretaceous succession is believed to represent primarily passive infill of pre-existing bathymetry. By the latest Ryazanian the Statfjord-Statfjord East and Snorre footwall islands were largely levelled by erosion, and the subaerial relief remaining was therefore insufficient to provide for any significant clastic input to the Snorre-H area.

Erosion and redistribution of sand

The base Cretaceous subcrop map outlines roughly areas of Late Jurassic erosion vs. deposition. The boundary between the two was very dynamic, however, and shifted laterally in response to progressive fault block rotation (cf. discussion above). By superimposing the apex lines of the base Cretaceous structural time map to delineate the palaeodrainage divides, the area can be subdivided into separate sediment source basins and corresponding depositional basins (Fig. 19a). By further combining the base Cretaceous and base Draupne subcrop maps, an estimate of erosion of previously deposited sandy formations that provided a source of sand to the Snorre-H sub-basin can be made (Fig. 19b). This approach is based on the assumption that there has been no tilting of the area since Late Ryazanian, however, which is probably untrue. Post-rift thermal subsidence commonly leads to rotation of the entire rift stratigraphy towards the rift axis (Nøttvedt *et al.* 1995) and the present N plunge of the Snorre footwall may yield velocity effects that may also shift the fault block apex. Any of these effects are considered minor, however, relative to the scale of the estimates presented here.

As shown in Fig. 19b, the Brent Group is partly to completely eroded both to the west, north and east of the Snorre-H area. As a result of the southerly structural plunge, the Tampen Spur has the largest uplift and deepest erosion to the north, exposing Middle Jurassic–Triassic lithologies in this area. Upper Jurassic syn-rift sands in the Snorre-H sub-basin were sourced almost exclusively from erosion of the Brent Group, however, as within the local drainage area erosion did not cut deep enough to encounter other sandy formations.

Sediments were dispersed into the Snorre-H sub-basin both from the north and east down the Snorre hangingwall dip-slope and to a minor extent from the Statfjord East fault scarp to the west. During infant stages of erosion, the footwall erosional apex (=palaeodrainage divide) was shifted closer towards the footwall crests than the present base Cretaceous structural apex, simply as a result of the backward retreating nature of fault scarp erosion. This means that a somewhat larger portion of the eroded material was likely transported down the hangingwalls compared to the footwalls of the fault blocks. In performing the Snorre-H sub-basin material balancing,

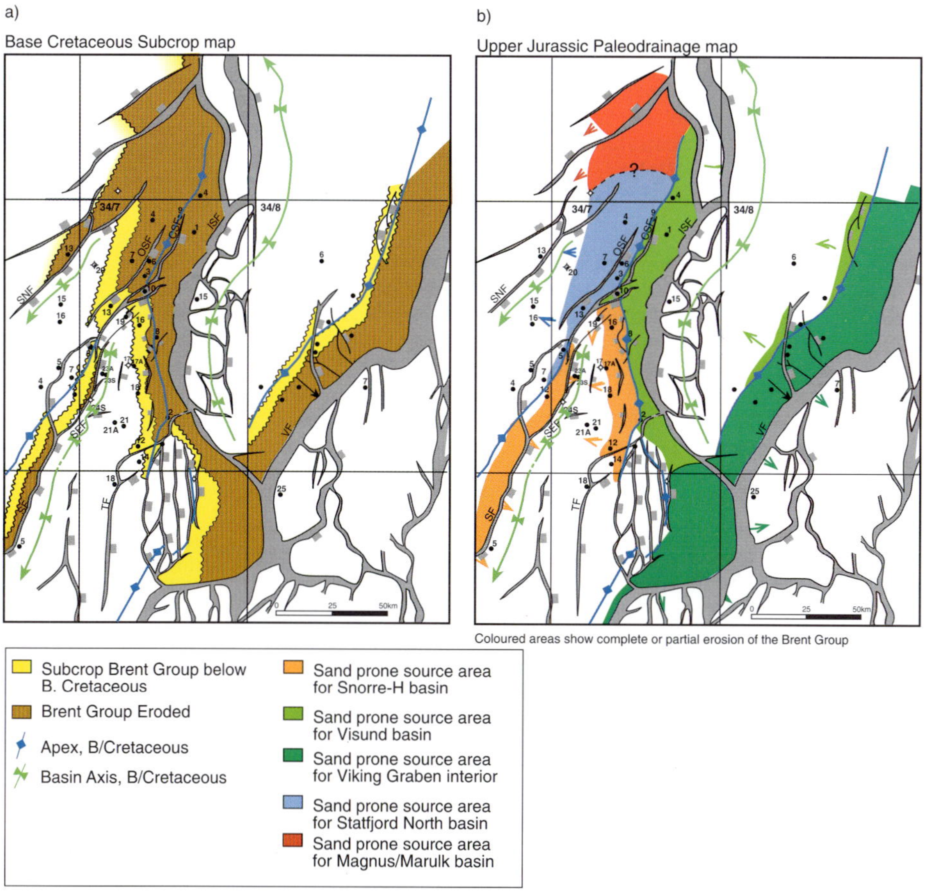

Fig. 19. Structural map of the base Cretaceous in the eastern Tampen Spur area. (**a**) Subcrop and areas of erosion of the Brent Group. Note also structural map apex and basin axis trajectories. (**b**) Upland sediment source areas, transport routes and depositional. The structural apex trajectories have been used as basis for constructing the palaeodrainage divides.

however, underestimation of material down the Snorre-H hangingwall is considered largely compensated for by overestimation of material down the Statfjord East footwall. In a regional perspective, large volumes of eroded material were shed also from the Snorre Fault Block into to the Visund sub-basin and into the Marulk basin to the NW.

Biostratigraphical reworking data suggests that the Brent Group within the established drainage domain became exposed to erosion from Late Oxfordian times. Accordingly, eroded Brent Group sediments may potentially be distributed within the complete Draupne Formation thickness interval. As it became overstepped and drowned by Late Volgian times, material eroded off the Statfjord East footwall implicitly must be distributed largely in the Mid-Oxfordian–Early Volgian succession (Fig. 16; Seqs-I–II). Effective hangingwall erosion and sediment transport did not commence until the Late Kimmeridgian (Fig. 8), however, suggesting that most of the Brent Group eroded off the Snorre hangingwall dip-slope was removed and dispersed towards the WSW during Volgian–Ryazanian, and may be concentrated within this time interval (Fig. 16; Seqs-III–V).

Mass budget

In the Snorre-H sub-basin sediment source area (= Snorre-H + Statfjord East footwall), the Brent Group is completely eroded across an area corresponding to roughly 10 km^2 (Fig. 19b). The thickness of the Brent Group in the Snorre-H sub-basin is *c.* 200 m giving an eroded Brent Group volume of 2 km^3. In addition, the Brent Group is partly eroded across approximately 40 km^2 of the sediment source area. Using an average thickness of 100 m or half the normal thickness of the Brent Group, the eroded volume corresponds to about 4 km^3. This gives a total volume of eroded Brent Group of 6 km^3. Applying a N/G value of 0.5, *c.* 3 km^3 of coarse clastic material (sand) was eroded and supplied from the Snorre-H sediment source area to the Snorre-H sub-basin (Table 4). The main Statfjord footwall has not been included in the above estimate, as it probably did not contribute significantly to the Snorre-H depositional basin sediment supply.

Volumes of mapped anomalies are calculated as follows:

(i) The aerial extent of the Snorre-H central and flanking seismic anomalies (shoreline/shallow marine sandstones), has been estimated to approximately 14 km^2 (Tables 3, 4 and Fig. 13; A1–A6). Using sand thickness numbers as given in Table 4 a total sand volume of 0.5 km^3 has been calculated.

(ii) The seismic anomaly to the south (basin floor sandstones) has a mapped areal extent of about 35 km^2 (Figs 12 and 13; A7). It may, however, continue somewhat south of the present survey area. By using the mapped area and a seismic thickness of 44 ms (62 m), a gross sand volume of 0.21 km^3 has been estimated.

The volumetric calculations show that the two mapped seismic anomalies added up may contain about 0.7 km^3 of sand, or about one quarter of the eroded volume. Doubling the sand thickness in all the mapped seismic anomalies raises the deposited sand volume only to 1.4 km^3, or a little less than half of the eroded volume. This suggests that a substantial volume of sand supplied from the Snorre-H drainage domain to the Snorre-H sub-basin has not yet been identified. Some of the excess sand may have been deposited as footwall talus sands off the Statfjord East footwall and some may be distributed in intra Draupne sand accumulations so far undetected. It is further speculated if some of the sand delivered to the Snorre-H sub-basin may have bypassed and became deposited in the half-graben to the south.

Syn-rift play models

Several different play models occur in association with late Jurassic extension and rotation of the Snorre Fault Block (Fig. 20). They are all intimately linked to erosion into the sand prone Brent Group and subsequent deposition of pinch-out sandstone reservoirs within Draupne Formation shales. The enveloping upper Jurassic Draupne Formation constitutes the main source rock in the area and burial depths are within the range of the oil window, allowing generation and migration of hydrocarbons into structural/stratigraphic pinch-out traps. Below, various reservoir facies models and distribution of syn-rift reservoirs in the Snorre-H area are discussed.

(1) Shallow marine/shoreline sandstones

Upper Draupne shallow marine/shoreline sandstones have been drilled in the hangingwall of the Snorre Fault Block, in the previously described wells 34/7-21 and -23 (Fig. 9). The sandstones fringe the Snorre-H sub-basin and their presence depend on local supply of sand from erosion into the Brent Group, mainly on the Snorre hangingwall. Good quality reservoir

Table 4. *Comparison of eroded sand in the sediment source area and volumes of sand deposited in the mapped seismic anomalies in the Snorre-H area*

Source area	Erosional area	Net sand (50%)	Eroded sand volume
Snorre-hw; BG completely eroded	10 km^2	100 m	1.0 km^3
Snorre-hw + Statfj.E-fw; BG ptly eroded	40 km^2	50 m	2.0 km^3
Total			3.0 km^3
Depositional basin	**Depositional area**	**Reservoir thickness (assuming all sand)**	**Deposited sand volume**
Snorre-H central area (A1)	0.51 km^2	30 m	0.01530 km^3
Snorre-H central area (A2)	5.10 km^2	30 m	0.15300 km^3
Snorre-H central area (A3*)	0.34 km^2	25/2 msec × 2.13 m/msec (=27 m)	0.00905 km^3
Snorre-H central area (A4*)	0.18 km^2	20/2 msec × 2.13 m/msec (=21 m)	0.00383 km^3
Snorre-H central area (A5**)	0.47 km^2	28/2 msec × 2.80 m/msec (=39 m)	0.01842 km^3
Snorre-H central area (A6)	7.21 km^2	40 m	0.28840 km^3
Snorre-H structural low (A7***)	3.46 km^2	44/2 msec × 2.80 m/msec (=62 m)	0.21314 km^3
Total			0.7 km^3

Snorre-hw, Snorre hangingwall; Statfjord E-fw, Statfjord East footwall; BG, Brent Group. (*) Used interval velocity of 2.13 m/msec from 34/7-23 wells. (**) Used interval velocity of 2.8 m/msec from 34/7-21 wells. (***) No well information near this area; used 2.8 m/msec interval velocity.

sandstones occur in a belt along the eastern margin of the Snorre half-graben, and in a more axial position to the north of well 34/7-24. However, such syn-rift sandstones have been penetrated mostly in updip positions and reservoir sandstones may thicken considerably down the hangingwall of the Snorre Fault Block.

The Upper Draupne Sandstone (Seqs-III–V) was deposited during the late rift stage, when the rate of fault block rotation was significantly reduced. This enabled sand to prograde across and fills the northerly embayment of the southerly plunging Snorre-H half-graben. Earlier syn-rift shoreline sandstones (Seq-I–II) were probably cannibalized completely during the forced regressive evolution of the Snorre-H hangingwall. If locally preserved, however, they likely have a different geometry. Due to relatively high depositional gradients established during rift climax, such shoreline sands probably aggraded more vertically than laterally.

(2) Basin floor gravity transported sandstones

It is suggested that a significant amount of sand must have bypassed the central Snorre-H subbasin to be deposited further to the south, likely by gravity flow processes. Basin floor gravity transported sandstones have not yet been proven within the Snorre-H half-graben, but the basin physiography and large volumes of eroded sand that cannot be accounted for make such a reservoir model very likely. Such sandstones may be detached from the shoreface sandstone sequences in the Snorre-H area and there may be a stratigraphic trap potential related to them.

A significant basin floor deposit (Seq-II) has been mapped in a topographic low in the southerly reaches of the Snorre-Gullfaks half-graben (Figs 6, 13 & 16). Degradation of the Statfjord East footwall likely contributed significantly to the supply of sand to this bathymetric deepening, however, and there is a risk that the sandstones are interbedded with shale and that the net/gross is reduced. On the other hand, sand derived from footwall erosion probably was stored intermediately in shoreline systems fringing the footwall crestal islands, subsequent to being released and transported to the basin floor setting (i.e. the Intra Draupne Sandstone, Seq-II). This increases the probability of better sorted and thicker packages of basin floor sandstone reservoir facies being developed.

Similar topographic low areas are present in the larger map outline, in the Statfjord East and northern Visund half-grabens. These are potential accumulation sites for basin floor gravity transported sand (Fig. 18b). Basin floor sandstones have been found in the Draupne Formation in the 33/9-15 and -16 wells.

(3) Footwall talus sandstones

Footwall talus sandstones have not been proven on the Snorre Fault Block. On the contrary, well 34/7-15S sits in a footwall position to the Snorre

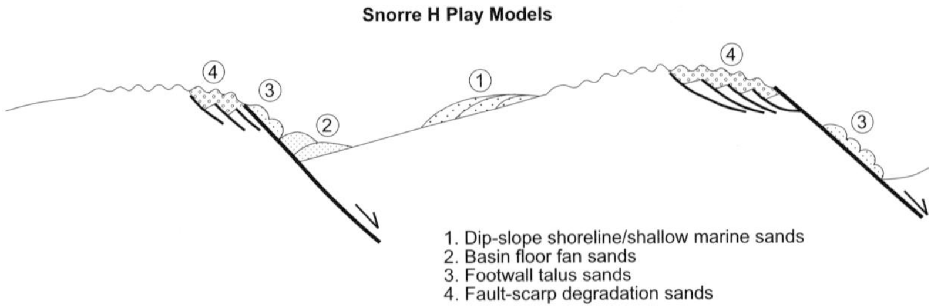

Fig. 20. Schematic illustration of various reservoir models across the Snorre Fault-block.

Fault Zone, and found nothing but shales and very minor sandstone stringers. On the Visund Fault Block, however, thick footwall talus sandstones were shed off the main fault scarp next to deep erosional incisions (Færseth *et al.* 1995). Such incisions cannot be seen in the footwalls of the Snorre nor the Statfjord East Fault Blocks, but this does not rule out the possibility that footwall talus sandstones may be locally present. In fact, the excess eroded sand noted in the volumetric calculations may support this reservoir model.

Because of the steep footwall gradients, sand is likely to have been supplied by rock fall and gravity transport associated with fault scarp retreat and would therefore be less sorted and clean compared to the easterly, dip-slope derived sandstones that were subject to shoreline and shallow marine reworking.

(4) Footwall degradation complexes

The present study has not looked in detail into potential fault-scarp degradation play models. The Snorre-H and Statfjord East footwalls have been degraded down through Brent and Dunlin Formations, however, and it is possible that the footwall surfaces may have been subject to slumping and disintegration to form small-scale deformation complexes (Fig. 20) similar to what has been shown for the Brent Field to the south (Coutts *et al.* 1996).

Towards a methodology for evaluating syn-rift plays

The present analysis of the Snorre-H area serves to illustrate a methodology for evaluating syn-rift plays. In evaluating such plays it is important to extract information on a very detailed level from all available data, with the objective to predict the distribution of reservoir and trap potential in space and time. In the Snorre-H example, good quality 3D seismic data and well data from footwall as well as hangingwall locations were available. More commonly, variable data types and quality will be available.

The analysis of the Snorre-H syn-rift play involved the following basic steps:

(1) Mapping in detail of key seismic surfaces and sequences
(2) Proper dating and sedimentological analysis of the syn-rift succession
(3) Volumetric mapping and mass budget estimates of sand

Time maps and fault maps of the base and top of the Snorre-H syn-rift succession were constructed from the seismic data. Careful mapping of the Statfjord East boundary fault system and the intra-fault block fault pattern provided key information on the Snorre-H hangingwall basin development and accommodation. By extracting various seismic attributes, particularly amplitude and reflection intensity, a correlation between observed sand occurrence in the wells and the seismic data was obtained. The map outline was extended to cover the Snorre-H and Statfjord East footwall crestal areas in addition to the Snorre-H hangingwall, in order to map out the shape, apex and topography of the syn-rift unconformity surface. This enabled a definition of sediment source area's and sediment transport routes into the Snorre-H sub-basin. Examination of the syn-rift seismic sequence geometries provided the basis for understanding the tilt history, which is essential to establish the rate and magnitude of footwall uplift and hangingwall subsidence through time.

High-resolution biostratigraphy allowed a precise dating of the Snorre-H syn-rift succession. Equally important, by a systematic analysis of the reworked bioflora, information on timing of erosion of upland sandy formations as source for the local syn-rift sand deposits was obtained.

Facies analysis and sequence stratigraphy of cored, as well as non-cored, well sections allowed the construction of a detailed depositional model for the syn-rift succession. Special attention was paid to the definition of key boundary surfaces and the establishment of a generic sequence architecture.

The final step in the evaluation process included volumetric mapping of the eroded loss of sand above the unconformity surface in the Snorre-H sediment source area relative to models for sand deposition within the Snorre-H syn-rift succession. By superimposing the information on structural gradients, basin morphology and timing of erosion, prediction of potential sand rich systems and definition of play models were enabled.

Conclusions

The upper Jurassic infill (Heather and Draupne Formations) of the Snorre-H area has been evaluated in terms of its potential for syn-rift plays. The petroleum geological relevance of the work includes the development of a model for the Snorre-H hangingwall basin, as well as the erection of a methodology for evaluating syn-rift plays.

It is concluded from the structural analysis that the development of the Snorre-H sub-basin was closely coupled to the development of the western, basin bounding Statfjord East Fault. The growth of the Statfjord East Fault through northward propagating segment linkage resulted in incremental, northward migration and back stepping of the upper Jurassic syn-rift infill along the axis of the Snorre hangingwall dip-slope.

The sedimentological facies analysis has linked the development of the syn-rift succession to the deformation history of the Snorre Fault Block. A phase of basin deepening and increasing sediment instability related to increased fault block rotation (early rift and rift climax) led to predominantly shale deposition with increasing evidences of sediment instability. This was followed by reduced rates of rotation (late rift), leading to the progradation of a shallow marine/ shoreline sandstone unit, the Upper Draupne Sandstone, capping the succession in updip areas. Suppressed generation of new relief finally led to levelling of the rift terrain and subsequent drowning.

From high-resolution biostratigraphy, the Upper Draupne Sandstone has been shown to be slightly diachronous (Late Volgian–Ryazanian) and younging northward. Detailed seismic sequence and attribute analyses have demonstrated that the sandstone unit is complex in nature and consists of distinct lobe shaped units that were deposited as incremental progradational sequences in an overall retrogradational setting.

A thorough study of the reworked bioflora suggests that local erosion of footwall highs started in the Late Callovian/Early Oxfordian, becoming more substantial in Kimmeridgian times. A significant increase in reworking has been identified from Kimmeridgian through Ryazanian times, including pre-rift (Brent Group) as well as syn-rift (Heather and Draupne Formations) taxa. It is interpreted to reflect expanding upland topography and increased local erosional yield following increased fault block rotation and footwall uplift.

By mapping out the syn-rift depositional basin and complementary drainage basin as a whole, it has been possible to demonstrate a dynamic relationship between upland erosion, sediment transport and basin deposition. Progressive, but stepwise, tilting of the Snorre Fault-block to some 9° (central Snorre-H area) in the Ryazanian led to deposition of a twofold (threefold if Tarbert Formation is included), coarsening upward (sandstone)-shale-sandstone syn-rift succession with variable reservoir potential. The Upper Draupne Sandstone has been proved by drilling across the Snorre-H area, but reservoir sandstone facies may thicken considerably down the hangingwall of the Snorre Fault Block. Additional reservoir potential is suggested in the upper Draupne Formation, related to deposition of basin floor sands to the south on the Snorre Fault-block.

The analysis of the Snorre-H area provides a powerful methodology for evaluating syn-rift plays. By attempting to quantify the parameters controlling syn-rift sedimentation into the Snorre-H sub-basin, predictive models for sandstone reservoir distribution and hydrocarbon play models have been established.

The work is based on results from a research project carried out between the Norsk Hydro Research Centre in Bergen and the University of Edinburgh. Morten Smelror, IKU, performed most of the new palynological analyses. Norsk Hydro funded the project and Ingrid Johnsrud is gratefully acknowledged for supporting the work. We also thank John Underhill for stimulating discussions and Graham yielding for a very constructive review of the manuscript. Finally, we like to thank Nina Anthun at the Norsk Hydro drafting department in Bergen, who prepared the illustrations.

Note added in revision. Since this paper was written, three new wells have been drilled in the Snorre-H area, Norwegian Block 34/7. These have proven sandstones in the Upper Draupne Formation, validating the current interpretation of the seismic attribute map and model for the Snorre-H tectonostratigraphic evolution.

References

BADLEY, M. E., PRICE, J. D., RAMBECH DAHL, C. & AGDESTEIN, T. 1988. The structural evolution of the northern Viking Graben and its bearing upon extensional modes of basin formation. *Journal of the Geological Society, London*, **145**, 455–472.

BEACH, A. 1985. Some comments on sedimentary basin development in the northern North Sea. *Scottish Journal of Geology*, **21**, 493–512.

——, BIRD, T. & GIBBS, A. D. 1987. Extensional tectonics and crustal structure: deep seismic reflection data from the northern North Sea Viking Graben. *In*: COWARD, M. P., DEWEY, J. F. & HANCOCK, P. L. (eds) *Continental Extensional Tectonics*. Geological Society, London, Special Publications, **28**, 467–476.

CARTWRIGHT, J. A. 1987. Transverse structural zones in continental rifts – an example from the Danish Sector of the North Sea. *In*: BROOKS, J. & GLENNIE, K. W. *Petroleum Geology of Northwest Europe*. Graham and Trotman, London, 441–452.

COUTTS, S. D., LARSSON, S. Y. & ROSMAN, R. 1996. Development of the slumped crestal area of the Brent Reservoir, Brent Field: an integrated approach. *Petroleum Geoscience*, **2**, 219–229.

DAHL, N. & SOLLI, T. 1993. The structural evolution of the Snorre field and surrounding areas. *In*: PARKER, J. R. (ed.) *Petroleum Geology of Northwest Europe: Proceedings of the 4th Conference*. Geological Society, London, 1159–1166.

DAVIES, S. J., UNDERHILL, J. R., DAWERS, N. H. & MCLEOD, A. E. 1998. The evolution of early syn-rift depositional systems. *In*: *Integrated Structural and Sequence Stratigraphic Analysis in Rift Settings*. American Association of Petroleum Geologists, Hedberg Research Conference, Abstracts, p. 33.

DAWERS, N. H. & UNDERHILL, J. R. 1999. The role of fault interaction and linkage in controlling syn-rift strati raphic sequences: Late Jurassic, Statfjord [illegible] area, northern North Sea. *American Association of Petroleum Geologists Bulletin*.

——, BERGE, A. M., HÄGER, K.-O., PUIGDEFABREGAS, C. & UNDERHILL, J. R. 1999. Controls on Late Jurassic, subtle sand distribution in the Tampen Spur area, Northern North Sea. *In*: FLEET, A. J. & BOLDY, S. A. R. (eds) *Petroleum Geology of Northwest Europe: Proceedings of the 5th Conference*. Geological Society, London, 827–838.

DE'ATH, N. G. & SCHUYLEMAN, S. F. 1981. The geology of the Magnus oil field. *In*: ILLING, L. V. & HOBSON, G. D. (eds) *Petroleum Geology of the Continental Shelf of North-West Europe*. Heyden, London, 342–351.

FOSSEN, H. 1989. Indication of transpressional tectonics in the Gullfaks oil field, northern North Sea. *Marine and Petroleum Geology*, **6**, 22–30.

——, ODINSEN, T., FÆRSETH, R. B. & GABRIELSEN, R. H. 2000. Detachments and low-angle faults in the northern North Sea rift system. *This volume*.

FÆRSETH, R. B. 1996. Interaction of Permo-Triassic and Jurassic extensional fault blocks during the development of the northern North Sea. *Journal of the Geological Society, London*, **153**, 931–944.

—— & RAVNÅS, R. 1998. Evolution of the Oseberg Fault Block in the context of the northern North Sea structural framework. *Marine and Petroleum Geology*, **15**, 467–490.

——, SJØBLOM, T. S., STEEL, R. J., LILJEDAHL, T., SAUAR, B. E. & TJELLAND, T. 1995. Tectonic controls on Bathonian–Volgian syn-rift successions on the Visund Fault-block, northern North Sea. *In*: STEEL, R. J., FELT, V. L., JOHANNESSEN, E. P. & MATHIEU, C. (eds) *Sequence Stratigraphy on the Northwest European Margin*. Norwegian Petroleum Society, Special Publication, **5**, 325–346.

——, KNUDSEN, B.-E., LILJEDAHL, T., MIDBØE, P. S. & SØDERSTRØM, B. 1997. Oblique rifting and sequential faulting in the Jurassic development of the northern North Sea. *Journal of Structural Geology*, **19**, 1285–1302.

GABRIELSEN, R. H, STEEL, R. J. & NØTTVEDT, A. 1995. Subtle traps in extensional terranes: a model with reference to the North Sea. *Petroleum Geoscience*, **1**, 223–235.

GAWTHORPE, R. L., FRAZER, A. J. & COLLIER, R. E. L. 1994. Sequence stratigraphy in active extensional basins: Implications for the interpretation of ancient basin fills. *Marine and Petroleum Geology*, **11**, 642–658.

GOWERS, M. B. & SÆBØE, A. 1985. On the structural evolution of the Central Trough in the Norwegian and Danish sectors of the North Sea. *Marine and Petroleum Geology*, **2**, 298–318.

GRADSTEIN, F. M., AGTERBERG, F. P., OGG, J. G., HARDENBOL, J., VAN VEEN, P., THIERRY, J. & HUANG, Z. 1994. A Mesozoic Time Scale. *Journal of Geophysical Research*, **99**, (B12), 24 051–24 074.

GRAUE, E., HELLAND-HANSEN, W., JOHNSEN, J. R., LØMO, L., NØTTVEDT, A., RØNNING, K., RYSETH, A. & STEEL, R. J. 1987. Advance and retreat of Brent Delta System, Norwegian North Sea. *In*: BROOKS, J. & GLENNIE, K. W. (eds) *Petroleum Geology of North West Europe*. Graham & Trotman, London, 915–937.

HALLAM, A. 1969. Tectonism and eustacy in the Jurassic. *Earth Science Reviews*, **5**, 45–68.

HAQ, B. U., HARDENBOHL, J. & VAIL, P. R. 1988. Mesozoic and Cenozoic chronostratigraphy and cycles of sea-level change. *In*: WILGUS, C. K., HASTINGS, B. S., ROSS, C. A., POSAMENTIER, H. W., VAN WAGONER, J. & KENDALL, G. ST. C. (eds) *Sea-Level Changes: An Integrated Approach*. Society of Economic Paleontologists and Mineralogists, Special Publication, **42**, 71–108.

HELLAND-HANSEN, W. & GJELBERG, J. G. 1994. Conceptual basis and variability in sequence stratigraphy: a different perspective. *Sedimentary Geology*, **92**, 31–52.

—— & MARTINSEN, O. J. 1996. Shoreline trajectories and sequences: description of variable depositional dip-scenarios. *Journal of Sedimentary Research*, **66**, 670–688.

——, ASTON, M., LØMO, L. & STEEL, R. J. 1992. Advance and retreat of the Brent delta: recent contributions to the depositional model. *In*: MORTON, A. C., HASZELDINE, R. S., GILES, M. R. & BROWN, S. (eds) *Geology of the Brent Group*. Geological Society, London, Special Publications, **61**, 109–127.

HOME, P. C. 1987. Ula. *In*: SPENCER, A. M. (ed.) *Geology of the Norwegian Oil and Gas Fields*. Graham and Trotman, London, 143–151.

HÄGER, K.-J. & SMELROR, M. 1997. Reworking of dinocysts as a means of obtaining useful geological information: A study of the Upper Jurassic Draupne Formation in the Tampen area, Norwegian North Sea. *American Association of Stratigraphic Palynologists*, Abstracts, 30th Annual Meeting, Woods Hole.

JOHANNESSEN, E. P., MJØS, R., RENSHAW, D., DALLAND, A. & JACOBSEN, T. 1995. Northern limit of the 'Brent delta' at the Tampen Spur – a sequence stratigraphic approach for sandstone prediction. *In*: STEEL, R. J., FELT, V., JOHANNESSEN, E. P. & MATHIEU, C. (eds) *Sequence Stratigraphy on the Northwest European Margin*. Norwegian Petroleum Society, Special Publication, **5**, 213–256.

JOHNSON, H. D., MACKAY, T. A. & STEWART, D. J. 1986. The Fulmar oil field: geological aspects of its discovery, appraisal and development. *Marine and Petroleum Geology*, **3**, 99–125.

LARSEN, V. B. 1987. A synthesis of tectonically related stratigraphy in the North Atlantic–Arctic region from Aalenian–Cenomanian time. *Norsk Geologisk Tidsskrift*, **7**, 281–294.

LEEDER, M. R. & GAWTHORPE, R. L. 1987. Sedimentary models for extensional tilt block/half-graben basins. *In*: COWARD, M. P., DEWEY, J. F. & HANCOCK, P. L. (eds) *Continental Extensional Tectonics*. Geological Society, London, Special Publication, **28**, 139–152.

MAHER, C. E. 1981. The Piper oil field. *In*: ILLING, L. V. & HOBSON, G. D. (eds) *Petroleum Geology of the Continental Shelf of North-West Europe*. Heyden, London, 358–370.

—— & HARKER, S. D. 1987. The Claymore oil field. *In*: BROOKS, J. & GLENNIE, K. W. (eds) *Petroleum Geology of North West Europe*. Graham & Trotman, London, 835–845.

MCLURE, N. M. & BROWN, A. A. 1992. Miller Field: a subtle Upper Jurassic submarine fan trap in the South Viking Graben, UK sector North Sea. *In*: HALBOUTY, M. T. (ed.) *Giant Oil and Gas Fields of the Decade*. American Association of Petroleum Geologists, Memoir, **54**.

NØTTVEDT, A., GABRIELSEN, R. H. & STEEL, R. J. 1995. Tectonostratigraphy and sedimentary architecture of rift basins, with reference to the northern North Sea. *Marine and Petroleum Geology*, **12**, 881–901.

ODINSEN, T., CHRISTIANSSON, P., GABRIELSEN, R. H., FALEIDE, J. I. & BERGE, A. M. 2000. The geometries and deep structure of the northern North Sea rift system. *This volume*.

PARTINGTON, M. A., MITCHENER, B. C., MILTON, N. J. & FRASER, A. J. 1993. Genetic sequence stratigraphy for the North Sea Late Jurassic and Early Cretaceous: distribution and prediction of Kimmeridgian–Late Ryazanian reservoirs in the North Sea and adjacent areas. *In*: PARKER, J. R. (ed.) *Petroleum Geology of Northwest Europe: Proceedings of the 4th Conference*. Geological Society, London, 347–370.

PROSSER, S. 1993. Rift-related linked depositional systems and their seismic expression. *In*: WILLIAMS, G. D. & DOBB, A. (eds) *Tectonics and Seismic Sequence Stratigraphy*. Geological Society, London, Special Publications, **71**, 35–66.

RATTEY, R. P. & HAYWARD, A. B. 1993. Sequence stratigraphy of a failed rift system: the Middle Jurassic to Early Cretaceous basin evolution of the Central and Northern North Sea. *In*: PARKER, J. R. (ed.) *Petroleum Geology of Northwest Europe: Proceedings of the 4th Conference*. Geological Society, London, 215–249.

RAVNÅS, R. & BONDEVIK, K. 1997. Architecture and controls on Bathonian–Kimmeridgian shallow-marine syn-rift wedges of the Oseberg-Brage area, northern North Sea. *Basin Research*, **9**, 197–226.

—— & STEEL, R. J. 1997. Contrasting styles of Late Jurassic syn-rift turbidite sedimentation: a comparative study of the Magnus and Oseberg areas, northern North Sea. *Marine and Petroleum Geology*, **14**, 417–449.

—— & ——1998. Architecture of marine rift basin successions. *AAPG Bulletin*, **82**, 110–146.

——, BONDEVIK, K., HELLAND-HANSEN, W., LØMO, L., RYSETH, A. & STEEL, R. J. 1997. Sedimentation history as an indicator of rift initiation and development: Late Bajocian–Bathonian evolution of the Oseberg–Brage area, northern North Sea. *Norsk Geologisk Tidsskrift*, **77**, 205–232.

——, NØTTVEDT, A., STEEL, R. J. & WINDELSTAD, J. 2000. Syn-rift sedimentary architectures in the northern North Sea. *This volume*.

ROBERTS, A. M., YIELDING, G. & BADLEY, M. E. 1993. Tectonic and bathymetric controls on stratigraphic sequences within evolving half-grabens. *In*: WILLIAMS, G. D. & DOBB, A. (eds) *Tectonics and Seismic Sequence Stratigraphy*. Geological Society, London, Special Publications, **71**, 87–121.

ROUBY, D., FOSSEN, H. & COBBOLD, P. R. 1996. Extension, displacement and block rotation in the Larger Gullfaks Area, northern North Sea. *AAPG Bulletin*, **80**, 875–890.

RØNNEVIK, H. & JOHNSON, S. 1984. Geology of the Greater Troll Field area. *Oil and Gas Journal*, **82**, 100–106.

SCLATER, J. G. & CHRISTIE, P. A. F. 1980. Continental stretching: an explanation of the post mid-Cretaceous subsidence of the Central North Sea Basin. *Journal of Geophysical Research*, **85**, 3711–3739.

SCHLISCHE, R. W. & OLSEN, P. E. 1990. Quantitative filling models for continental extensional basins with applications to Early Mesozoic rifts of eastern America. *Journal of Geology*, **98**, 135–155.

SMITH, R. L. 1987. The structural development of the Clyde Field. *In*: BROOKS, J. & GLENNIE, K. W. (eds) *Petroleum Geology of North West Europe*. Graham & Trotman, London, 523–531.

SOLLI, T. 1995. Upper Jurassic play concept – an integrated study in Block 34/7, Norway. *First Break*, **13**, 21–30.

SPEKSNIJDER, A. 1987. The structural configuration of Cormorant Block IV in context of the northern Viking Graben structural framework. *Geologie en Mijnbouw*, **65**, 357–379.

SURLYK, F. 1989. Mid-Mesozoic syn-rift turbidite systems: controls and predictions. *In*: COLLINSON, J. D. (ed.) *Correlation in Hydrocarbon Exploration*. Norwegian Petroleum Society, Graham & Trotman, London, 231–241.

TURNER, C. C., COHEN, J. M., CONNEL, E. R. & COOPER, D. M. 1987. A depositional model for the South Brae oilfield. *In*: BROOKS, J. & GLENNIE, K. W. (eds) *Petroleum Geology of North West Europe*. Graham & Trotman, London, 853–864.

VOLSET, J. & DORE, A. G. 1984. Revised Jurassic lithostratigraphy of the Norwegian North Sea, northern area. *Norwegian Petroleum Directorate, Bulletin*, **3**.

WHEATLEY, T. J., BIGGINS, D., BUCKINGHAM, J. & HOLLOWAY, N. H. 1987. The geology and exploration of the Transitional Shelf, an area to the west of the Viking Graben. *In*: BROOKS, J. & GLENNIE, K. W. (eds) *Petroleum Geology of North West Europe*. Graham & Trotman, London, 979–989.

YIELDING, G. 1990. Footwall uplift associated with Late Jurassic normal faulting in the northern North Sea. *Journal of the Geological Society, London*, **147**, 219–222.

Cenozoic evolution of the central and northern North Sea with focus on differential vertical movements of the basin floor and surrounding clastic source areas

HENRIK JORDT,[1,2] BRIT I. THYBERG[1] & ARVID NØTTVEDT[1,3]

[1] *Department of Geology, University of Oslo, P.O. Box 1047 Blindern, N-0316 Oslo, Norway*
[2] *Present address: Statoil, N-4035 Stavanger, Norway*
[3] *Present address: Norsk Hydro a.s., 1190, 111–5th Avenue SW, Calgary, Alberta, T2P 3Y6, Canada*

Abstract: Changes in outbuilding directions and shifts in sediment provenance are key factors for understanding the development of the Cenozoic succession in the central and northern North Sea. By establishing a seismic stratigraphic framework for the Cenozoic succession, and through utilization of velocity data and analyses of sediment compositions, these factors are investigated. Sequence geometries and reflection terminations are identified and related to basin-floor geometry and variations in erosional base level. Mineralogical and geochemical analyses of sediment samples from selected wells are compared with their seismic sequences. Paleogene clastic wedges rich in smectite were built from the East Shetland Platform and probably drained a source area west and northwest of the present North Sea. Outbuilding from southern Norway took place during Late Paleocene–Early Eocene and in Oligocene to mid-Miocene time. In mid-Miocene time outbuilding from the west stopped and subsequently sediments with a high content of feldspar, detrital chlorite and carbonate were deposited from easterly directions in the central and northern North Sea. The seismic velocity of the Cenozoic sediments increases down to a mid-Miocene unconformity, but the velocities of Oligocene and Eocene sediments below are lower and nearly unaffected by burial depth, resulting in a marked velocity inversion. Two tectonic subsidence events, late Eocene–early Oligocene and late Miocene–early Pliocene, have been identified subsequent to general eustatic sea-level lows. These subsidence events cannot be explained by thermal subsidence, and they may have crucial implications for the understanding of hydrocarbon generation, migration and trapping in the Tertiary and deeper stratigraphic levels.

The North Sea Basin is an extensional sedimentary basin on the northwest European continental shelf (Ziegler 1982, 1990; Glennie 1990) (Fig. 1). It was formed by lithospheric extension during Permo-Triassic and late Jurassic–early Cretaceous time. It is generally agreed that the Tertiary sediments in the North Sea constitute post-rift or thermal subsidence fill related to these earlier rifting episodes (Glennie 1990). Recently, however, Hall & White (1994) have interpreted anomalously high rates of Early Tertiary subsidence in the northern North Sea, and Jordt *et al.* (1995) have suggested that a complex pattern of differential vertical movement of the basin floor occurred. Several episodes of uplift along the margins of the North Sea have been suggested by several workers. A comprehensive bibliography of studies involving uplift of Norway has been given by Stuevold & Eldholm (1996). These studies indicate that uplifts concentrate in three periods: late Cretaceous–early Paleocene, late Eocene–early Oligocene, and late Pliocene–Pleistocene time.

The East Shetland Platform and the Mid North Sea High are prominent highs, west and south of the study area, that are separated from the Central and Viking Graben by major faults (Fig. 1). The Fjerritslev Fault Zone, which may be considered as a WNW continuation of the Sorgenfrei–Tornquist Zone, and the Øygarden Fault Zone, are zones of weakness along the Scandinavian Shield. Depocentres developed along these Mesozoic and older fault systems during late Paleocene, early Eocene and early Oligocene time (Fig. 2). Late Paleocene to early Oligocene, and early Miocene outbuilding developed depocentres along boundary faults in the Central and Viking Grabens.

From: Nøttvedt, A. *et al.* (eds) *Dynamics of the Norwegian Margin*. Geological Society, London, Special Publications, **167**, 219–243. 1-86239-056-8/00/$15.00

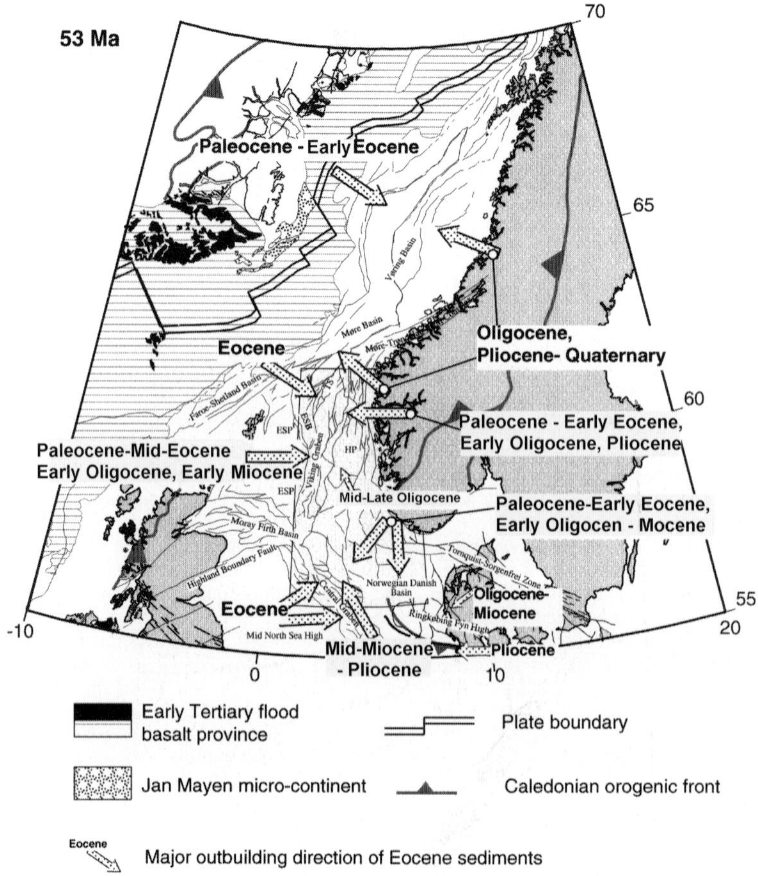

Fig. 1. Regional geological setting of the North Sea and North Atlantic continental rift zone. Position of Greenland restored to 53 Ma. Cenozoic outbuilding directions are indicated along the northwest European Atlantic margin. ESB, East Shetland Basin; ESP, East Shetland Platform; HP, Horda Platform; TS, Tampen Spur. (Major outbuilding directions are based on Spjeldnæs 1975; Stuevold *et al.* 1992; Cameron *et al.* 1993; Jordt *et al.* 1995).

Sediments were supplied from the west and northwest into the North Sea, and further north into the Vøring margin during Paleocene and Eocene times (Fig. 1). However, the outbuilding direction shifted from west to east during the Eocene–Oligocene transition in the central North Sea (Michelsen *et al.* 1995, 1998), and the sediment supply from the west stopped in the Møre and Vøring Basins (Fig. 1). Sediment supply from the east increased considerably and dominated during Miocene and Pliocene time. Abundant sediments, which may have come from the Scandinavian and Baltic Shields east of the basin, were probably transported into the basin by the so-called Baltic River system (Bijlsma 1981; Gibbard 1988; Cameron *et al.* 1993). A late Miocene or Pliocene system of NNW–SSE trending ridges and troughs in the UK sector may have formed on a pro-delta slope (Cartwright 1995), possibly reflecting the early evolution of a large Baltic delta. North–south trending incised valleys in the Middle and Upper Pleistocene succession in the Danish North Sea (Salomonsen 1995) suggest that this area was part of a deltaic system controlled by the Baltic River up to Pleistocene time.

This paper focuses on the Cenozoic succession in the northern Danish and the Norwegian North Sea basins. The main objective is to discuss the geological evolution of this area, with focus on differential vertical movements of the basin floor. Our intent is to show that seismic sequence geometries and outbuilding directions provide information about changes in the basin

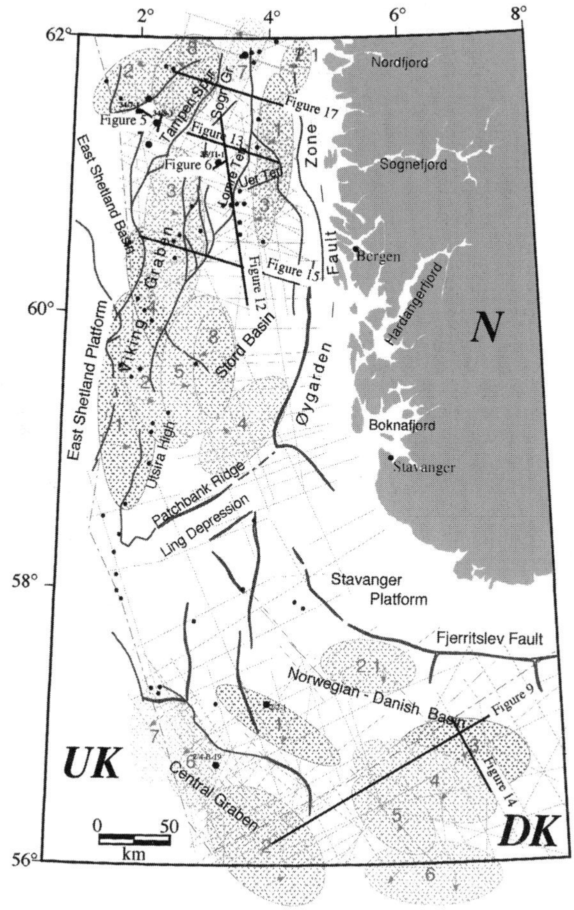

Fig. 2. Structure map of the study area showing the seismic data coverage and location of wells with logs and/or biostratigraphic data. Shaded ovals indicate positions of depocentres and numbers within these refer to CSS-sequence number shown in Fig. 3. Indicated seismic profiles are used as examples later in the text. Outbuilding directions and depocentre locations of CSS-1 to CSS-7 south of 57°N are based on Jordt (1995).

topography, and that these are related to underlying structures, to shifts in provenance area and to sediment accumulation rates.

Methods

In the study, we interpreted about 17 000 km of conventional multi-channel seismic reflection profiles (Fig. 2), and integrated mineralogical and geochemical analyses of cuttings from more than 40 wells (Thyberg *et al.* 2000). Velocity logs from the investigated wells were analysed and used to convert seismic travel times to depth. Published high-resolution biostratigraphic data were used to date the seismic sequences. Accumulation rates for the mapped seismic sequences were estimated and compared with sediment composition and stacking pattern. Results from the mineralogical and geochemical analyses of the sediments have been presented by Thyberg *et al.* (2000). The established seismic sequence framework, lithostratigraphic correlations, mapped outbuilding directions, and hiatuses indicated by seismic and biostratigraphic data are summarized in Figs 3 and 4.

Constraints on the seismic interpretation

Seismic mapping and interpretations are based on the identification and correlation of seismic unconformities that separate and bound packages that have relatively uniform seismic signatures. These seismic unconformities are marked by reflection terminations (i.e. onlap, downlap and toplap), by differences in the amplitude and frequency spectra of the adjoining sequences and/or by regionally continuous reflections.

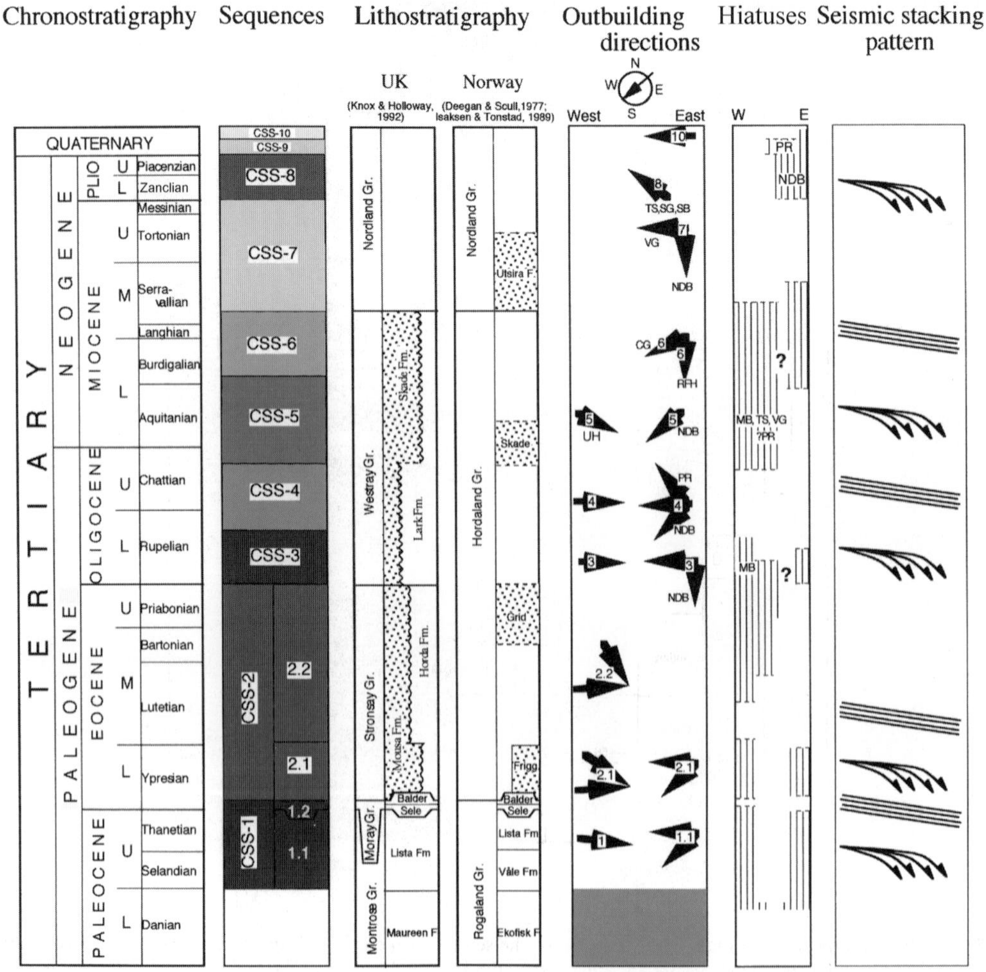

Fig. 3. Summary of main observations in the study area correlated with lithostratigraphy in the UK and Norwegian North Sea. Depocentre locations are indicated in Fig. 2. CG, Central Graben; ESP, East Shetland Platform; HP, Horda Platform; MB, Marulk Basin; NDB, Norwegian–Danish Basin; PR, Patchbank Ridge; TS, Tampen Spur; RFH, Ringkøbing Fyn High; SB, Stord Basin; SG, Sogn Graben; SN, southern Norway; UH, Utsira High; VG, Viking Graben.

Seismic mapping and interpretation were initially carried out independently of biostratigraphic and petrographic data. However, in cases where several seismic interpretations were possible, for example, because of noise, the interpretation was guided by available high-resolution biostratigraphy and completion log stratigraphic subdivision.

In this paper, we interpret seismic onlap to reflect a base-discordant relationship in which initially horizontal strata terminate or thin (below seismic resolution) progressively against an initially inclined surface, or in which initially inclined strata terminate or thin (below seismic resolution) progressively updip against a surface of greater initial inclination (see Mitchum 1977).

Progradational sequences interpreted from the seismic data outline the net outbuilding of sedimentary wedges, but do not necessarily reflect the direction of sediment transport. As an example, the main direction of sediment transport is parallel to the coast along non-deltaic coastlines, whereas it is more perpendicular or oblique to the coast where deltas are present (Galloway 1989). Both depositional settings may show net progradation, and result in similar facies associations (Galloway 1989). Implicitly, westward progradation observed on a seismic profile may be the result of dominating northward sediment transport.

It may also be deceptive to interpret current bathymetric conditions based on seismic geometries,

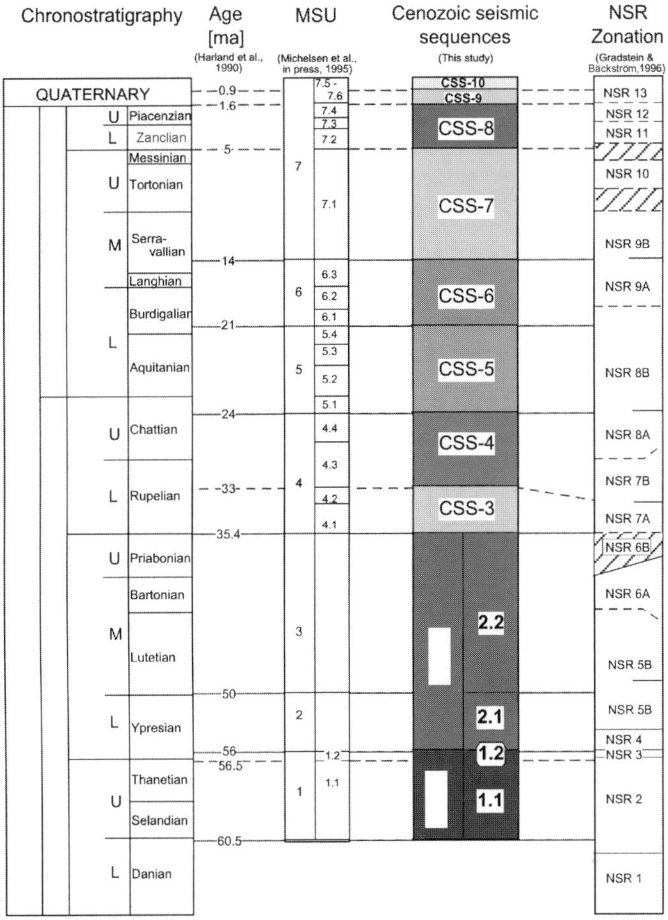

Fig. 4. Stratigraphic scheme showing correlation of the seismic sequences with the NSR zonation (Gradstein & Bäckström 1996) and the depositional sequences from the Danish North Sea (Michelsen *et al.* 1995, 1998). Dashed correlation lines indicate correlation without direct seismic tie to the Danish North Sea.

as the progradation of clastic sediments along shelf margins (Orton & Reading 1993) and in deep-sea settings (Reading & Richards 1994) may result in similar, but undifferentiable, depositional geometries. Recognition of onlap is of little help in the interpretation of depositional environment, as it is not possible to differentiate between coastal onlap and marine onlap on seismic data without having core data or palaeobathymetric information (Bertram & Milton 1988; Roberts *et al.* 1993).

Net accumulation rates

Absolute ages of the mapped sequences were used to calculate net accumulation rates (Table 1) using the following expression:

$$S = \frac{\mathrm{TWT} \times \mathrm{V}}{2D}$$

where S is net accumulation rate in m Ma^{-1}, TWT is thickness in s two-way travel time, V is interval velocity in m s^{-1} and D is the time duration for formation of the actual sequence. S is a minimum estimate of the accumulation rate as no account is taken of erosion after deposition or mechanical compaction.

Mineralogical analyses

The methods used in the mineralogical and geochemical analyses have been described in detail by Thyberg *et al.* (2000) and are only briefly summarized here. Cutting samples from the Cenozoic succession were analysed with X-ray fluorescence (XRF) and X-ray diffraction (XRD). The inorganic geochemical data were used to estimate whole-rock mineralogy, and in the assessment of main and trace elements.

The clay mineralogy was categorized as smectite (including expandable mixed-layer illite–smectite),

Table 1. *The calculated net accumulation rates are based on seismic velocities derived from well velocity logs (Fig. 10) and on observed seismic two-way travel times*

Sequence	Thickness (ms)		Duration (Ma)	Velocity (m/s)	Net accumulation rate (m/Ma)	
	Max	Min			Max	Min
CSS-8	800	100	3.4	2200	235	29
CSS-7	900	100	9	2000	100	11
CSS-6	500	50*	7	2000	71	7
CSS-5	500	50*	3	2000	166	16
CSS-4	400	100	9	2000	44	11
CSS-3	600	100	2.4	2000	250	41
CSS-2	800	100	20.6	2000	39	5
CSS-1	800	50	4.5	2200	195	12

The sequences were mechanically compacted and possibly eroded, therefore the accumulation rates represent minimum estimates. Minimum seismic two-way travel times are taken basinwards in the mapped sequences, and therefore probably reflect a sedimentary environment marked by starvation and limited erosion.
* Minimum thicknesses from the German North Sea (Michelsen *et al.* 1998).

illite, kaolinite and chlorite. The clay mineralogy at the time of deposition can be related to the sediment provenance and sedimentary facies. This includes not only the nature of the erosion products of the source rock but also the rates of erosion and the processes of rock weathering acting in different climates. Shales contain a great variety of non-clay minerals (e.g. detrital feldspar), which make up a significant fraction of these rocks. The feldspar content is a good indication of a rock's mineralogical maturity. The Al_2O_3/SiO_2 ratio is used as an indication of the clay/sand ratio because Al_2O_3 is commonly found in clays, feldspar and mica.

Seismic stratigraphy

The stratigraphic framework established is shown in Fig. 4. The following discussion on dating of the seismic sequences is based on the work by Jordt *et al.* (1995). However, the dating has been improved by new data illustrating the tie between the seismic data and well information. Biostratigraphic data from well 34/7-1, 34/8-1 and 35/11-1, published by Gradstein & Bäckström (1996) and Steurbaut *et al.* (1991), are included in Figs 5 and 6 to document the suggested ages of the mapped seismic sequences in the present study.

Age determination

Palaeontological ages

The CSS-1 sequence seems to correlate with the zones NSR-2 and NSR-3 (Gradstein & Bäckström 1996) in well 34/7-1, suggesting a Late Paleocene to earliest Eocene age (Fig. 5). The top of sequence CSS-1 has an Early Eocene age in well 35/11-1 (Fig. 6), where it is found below the nannoplankton zone NP11 (Steurbaut *et al.* 1991). This suggests an earliest Eocene age for the top of sequence CSS-1 in the northern North Sea, which is in agreement with the dating of this seismic boundary in the central North Sea (Michelsen *et al.* 1995, 1998). The top of sequence CSS-1 is found near the top of a narrow interval with low gamma-ray values and high sonic velocities, which corresponds to the top of the Balder tuffs (Fig. 5). The top of sequence CSS-1.1 correlates with the Paleocene–Eocene transition in well 35/11-1.

The seismic boundary at the top of the CSS-2 sequence correlates with the Eocene–Oligocene transition (Fig. 4). In well 34/7-1, 34/8-1 (Fig. 5) and 35/11-1 (Fig. 6), the top of sequence CSS-2 correlates with a major hiatus between Middle Eocene and Early Oligocene sediments. In the Danish wells Inez-1, F-1 and K-1, this seismic boundary is located below the NP22 zone (Michelsen *et al.* 1998). Steurbaut *et al.* (1991) indicated that zone NP22 is present in Oligocene sediments overlying the CSS-2 sequence in well 35/11-1. We correlate the top of sequence CSS-2 with the base of the Oligocene sediments and suggest that the top of sequence CSS-2 correlates with top Eocene. The top of sequence CSS-2.1 is found in a succession that has been assigned an Early, possibly Mid-Eocene age in well 35/11-1 (Fig. 6). We suggest that the CSS-2.1 sequence correlates with unit 2 of Michelsen *et al.* (1998) (Fig. 4), indicating an Early Eocene age.

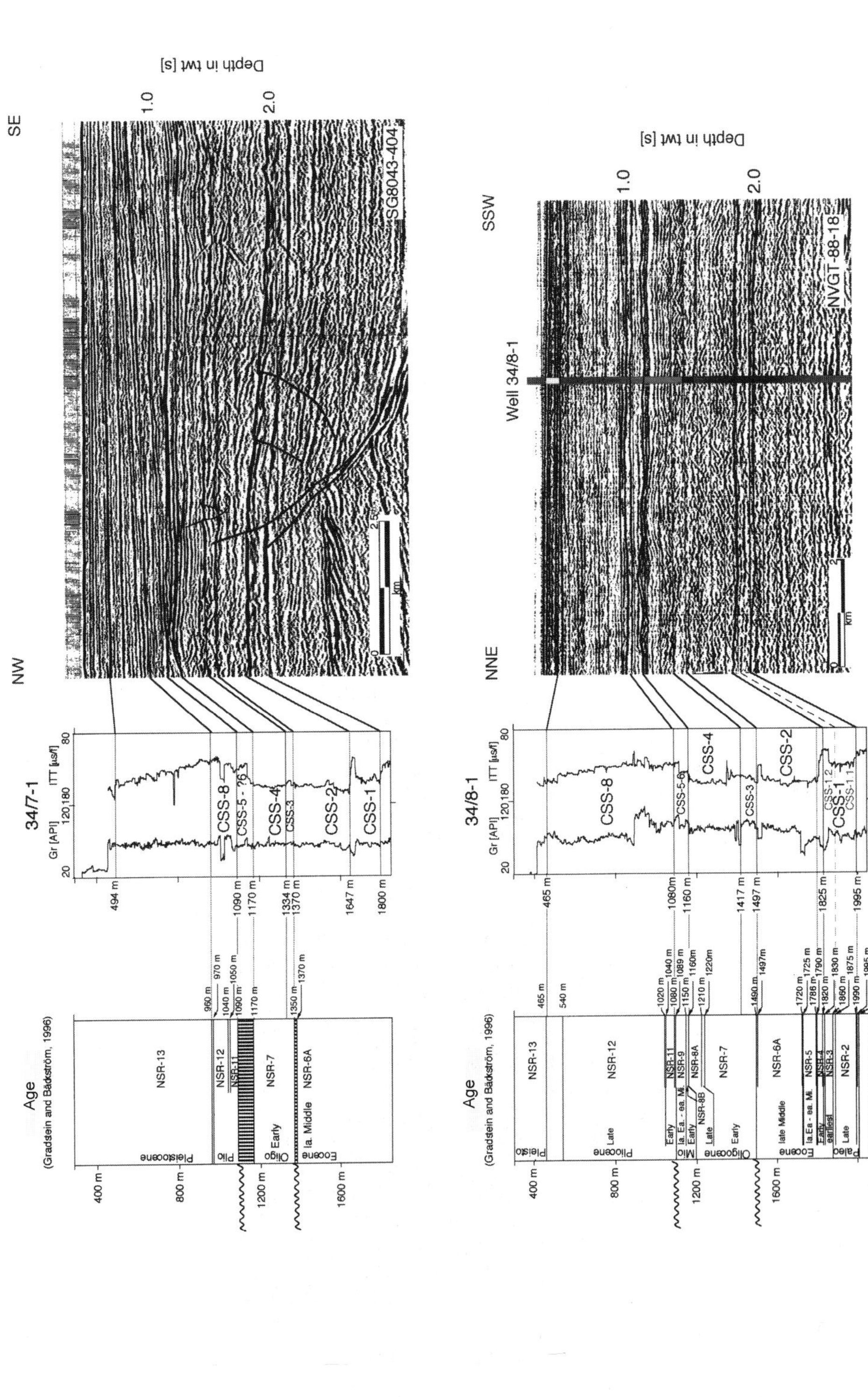

Fig. 5. Wells 34/7-1 and 34/8-1. Correlation of seismic data with biostratigraphy and log data from wells 34/7-1 and 34/8-1.

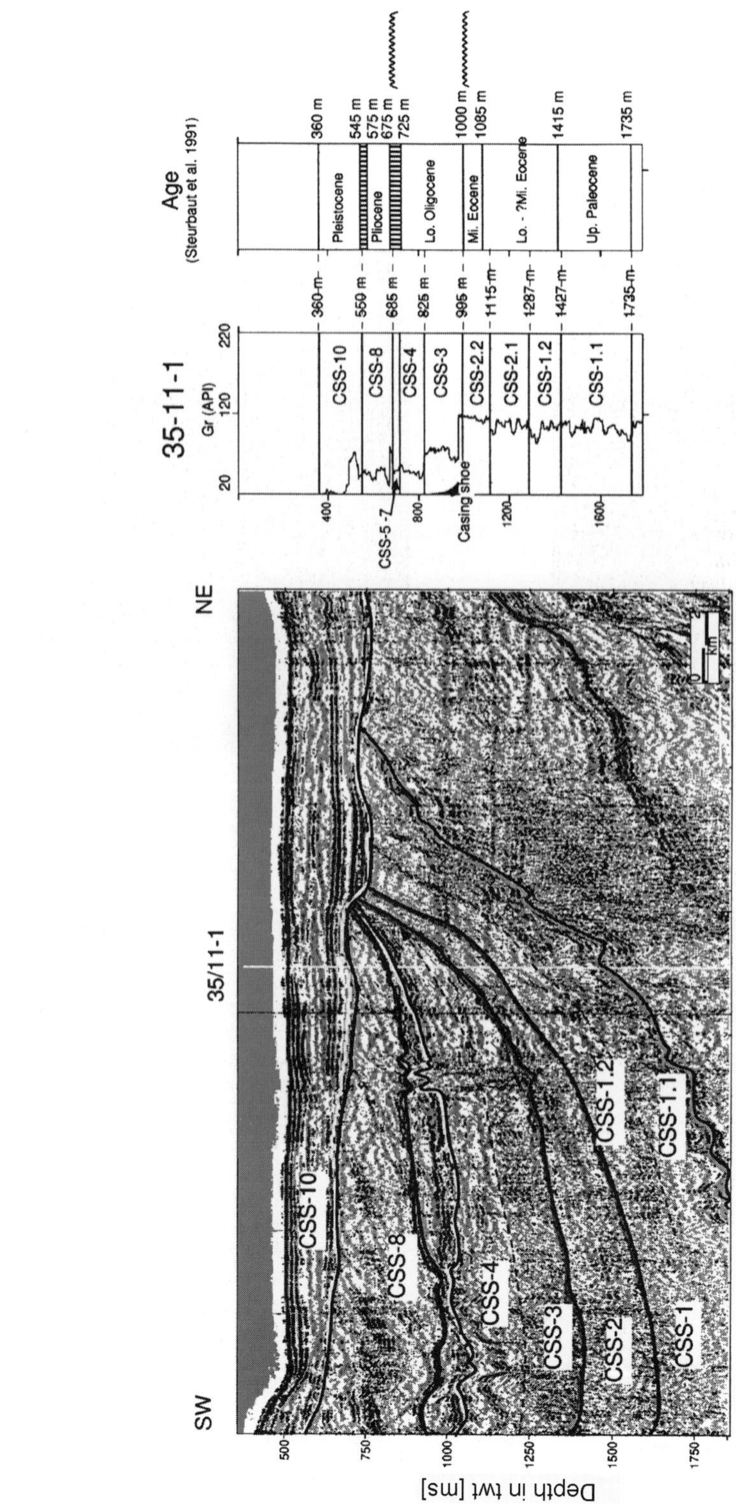

Fig. 6. Well 35/11-1. Correlation of seismic data with biostratigraphy and log data.

The top of sequence CSS-3 has no direct seismic tie to the Danish North Sea. In wells 34/7-1 and 34/8-1, the top of sequence CSS-3 is situated in stratigraphic unit NSR-7 (Gradstein & Bäckström 1996), suggesting an Early Oligocene age (Fig. 5). A similar age is supported by the biostratigraphy from well 35/11-1 (Fig. 6). An upward change in seismic signature from progradation to marked aggradation and onlap occurs against the top of sequence CSS-3 in the northern North Sea. A similar change in sequence geometry is observed across the top of sequence 4.2 in the Danish North Sea–central North Sea (Michelsen *et al.* 1998). On the basis of similarities in the seismic signature, and the ages indicated by the biostratigraphy, we believe that the top of sequence 4.2 of Michelsen *et al.* (1998) correlates with the top of sequence CSS-3 (Fig. 4), indicating a mid-Early Oligocene age.

The seismic boundary representing the top of sequence CSS-4 has been dated to latest Oligocene time by Michelsen *et al.* (1998). The top of sequence CSS-4 correlates with the transition between NSR-8A and NSR-8B in well 34/8-1 (Fig. 5), indicating a Late Oligocene or Early Miocene age. Although the ages indicated by the biostratigraphy are not quite in agreement, we suggest that the top of sequence CSS-4 is of latest Oligocene age.

The Miocene succession is divided into three seismic sequences, CSS-5, CSS-6 and CSS-7, and the boundaries separating these are dated in the central North Sea. The boundaries of the CSS-5 and CSS-6 sequences correlate seismically with the boundaries of Unit 5 and Unit 6 of Michelsen *et al.* (1998), respectively. This correlation suggests that the top of sequence CSS-5 is of late Early Miocene age and the top of sequence CSS-6 is of early Mid-Miocene age (Fig. 4). Miocene sediments are also present in the northern North Sea; however, the biostratigraphy is equivocal (Fig. 7). Eidvin & Riis (1992) indicated that Early Miocene sediments are overlain by late Mid- and Late Miocene strata in well 34/8-1. This suggests that sequences CSS-5 and CSS-7 are present and separated by a hiatus of late Early or early Mid-Miocene age. Dating of the same sediments in the same well (i.e. 34/8-1) by Gradstein & Bäckström (1996) also indicates that the Early Miocene sequence is present and in agreement with the biostratigraphy of Eidvin & Riis (1992). However, Gradstein & Bäckström (1996) dated the overlying sediments to late Early and early Mid-Miocene age, suggesting that sequences CSS-5 and CSS-6 are present and that the CSS-7 sequence is absent in well 34/8-1. The datings of Gradstein & Bäckström (1996) also indicate the presence of a major hiatus of Mid- and Late Miocene age in the northern North Sea. Steurbaut *et al.* (1991) have not identified upper Oligocene or Miocene sediments in well 35/11-1; however, as they have no datings between 675 m and 725 m the presence of sediments of this age is still possible (Fig. 6).

The CSS-8 sequence is correlated with Pliocene sediments. It is, however, uncertain whether a Lower Pliocene sequence is present in the northern North Sea. Age datings by Gradstein & Bäckström (1996) in wells 34/7-1 and 34/8-1 (Fig. 5) indicate that Lower Pliocene sediments are present. The datings of Eidvin & Riis (1992), however, are in disagreement with that interpretation, indicating the presence of only Upper Pliocene sediments in well 34/8-1 (Fig. 7). The top of sequence CSS-8 is correlated with the Pliocene–Pleistocene transition in wells 34/8-1 (Fig. 5) and 35/11-1 (Fig. 6). However, in well 34/7-1, the upper boundary of the CSS-8 sequence is of Pleistocene age. This age inconsistency is related to ambiguities in the seismic interpretation caused by noise from sea-floor multiples and by the westward change in toplap because of erosional truncation, and to toplap caused by progradation.

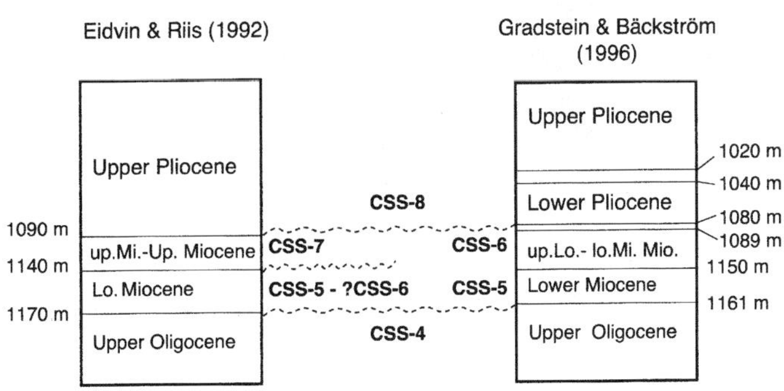

Fig. 7. Biostratigraphy from well 34/8-1.

Absolute ages

The time scale devised by Harland *et al.* (1990) was used to convert the suggested palaeontological ages of the seismic sequence boundaries to absolute ages (Fig. 4). The accuracy and precision of time resolution vary with the chronostratigraphic methods. The achievable resolution for integrated stratigraphy in Cenozoic marine successions varies between 5 ka and 1 Ma (Miall 1995). The duration of deposition of the Fur Formation (Spjeldnæs 1975), which forms the upper part of sequence CSS-1.2, is used here as an example of the uncertainty related to absolute age dating. Spjeldnæs (1975) summarized two opposite views about the sedimentation rate of the Fur Formation. If it is assumed that the lamination in the Fur Formation represents annual varves (Bonde 1973), then it may represent a period of only 60 ka. However, studies of magnetic reversals of the volcanic ash beds indicate a much longer time for deposition of this interval of about 3 Ma (Sharma 1969). Both views are compatible with the fact that there are more than 130 well-defined ash beds in the Fur Formation, and with the observations on sedimentation rates of similar recent sediments (Spjeldnæs 1975). The period of formation of most of the mapped sequences (i.e. CSS-1 to CSS-8) (Fig. 4) is beyond the uncertainty level (±1 Ma) related to the absolute age dating of marine Cenozoic successions. For example, the shortest period of formation is indicated for sequence CSS-3 of only 2.4 Ma (i.e. 35.4–33 Ma), and the longest is indicated for the CSS-2 sequence, nearly 21 Ma covering the time interval 56–35.4 Ma. The absolute ages of the mapped sequences are used in the discussion below.

Sequence description

Upper Paleocene–Eocene sequence

The CSS-1 sequence (Upper Paleocene–lowermost Eocene) is marked by prograding wedges that built out from southern Norway and from the East Shetland Platform (Figs 8 and 9). The depocentre is recognized as a major westward prograding wedge off Sognefjorden.

The sediment accumulation rates for sequence CSS-1 vary between 12 and 195 m Ma^{-1} (Table 1, Fig. 10), with maximum net accumulation in the depocentre. Sequence CSS-1 has a relatively uniform thickness of 200 ms TWT in the northwestern part of the study area; thus it attains a low accumulation rate along the Møre Basin. The upper sub-sequence of CSS-1 (CSS-1.2) is characterized by high concentrations of carbonate, smectitic clays and sometimes P_2O_5. The tuffaceous Balder Formation causes a marked increase in the smectite content uppermost in sequence CSS-1 (Fig. 11).

The CSS-2 sequences (Eocene) prograded from the East Shetland Platform and major depocentres formed in the Viking and the Central Grabens (Figs 8 and 9). CSS-2.1, the lower sub-sequence of sequence CSS-2, prograded from southern Norway and forms a minor depocentre in the Norwegian–Danish Basin (Fig. 8). Sequence CSS-2 as a whole shows a marked eastward thinning in the central North Sea, in the same area where Michelsen *et al.* (1998) indicated continuous sedimentation during Eocene time. The Eocene sequence also thins along the Øygarden Fault Zone further to the north (Fig. 8). However, biostratigraphical data indicate a hiatus between the Eocene and the Oligocene sequences in that area (Fig. 6).

The maximum net accumulation rate attains only 39 m Ma^{-1} for the Eocene sequence, which is very low compared with the other Tertiary sequences (Table 1). Sequence CSS-2 is dominated by clay, consisting mainly of smectite, and with a low quartz and feldspar content (Fig. 11). The high content of smectite in the CSS-2 sequence suggests a dominantly volcanic influenced source.

Although the seismic and mineralogical information gives a regionally consistent picture of a clay-dominated sequence, completion logs from wells in licence block 35/8 indicate locally a high sand content. These sand-rich Eocene sediments reflect a local thickening of the CSS-2 sequence northwest of the depocentre in the Viking Graben (Fig. 8), correlating with a mound at the base of sequence CSS-2 on the northern Horda Platform (Figs 12 and 13). The precise palaeontological age of these sand deposits is unknown, but their location at the base of sequence CSS-2 (Fig. 12) suggests an early Eocene age. The mound has a thickness of 300 ms TWT (3–400 m) and length of more than 25 km, thus it has a low topography. It is located on the crest of the downthrown block of the Lomre Terrace (Fig. 12), and west of the Øygarden Fault Zone (Fig. 9). Northward progradation of the CSS-2 sequence is reflected by the stacking pattern on the Horda Platform (Fig. 12). The base of sequence CSS-2 is at 1700 ms TWT depth north on the Lomre Terrace, but only at 1400 ms TWT depth on the Horda Platform to the south (Fig. 12).

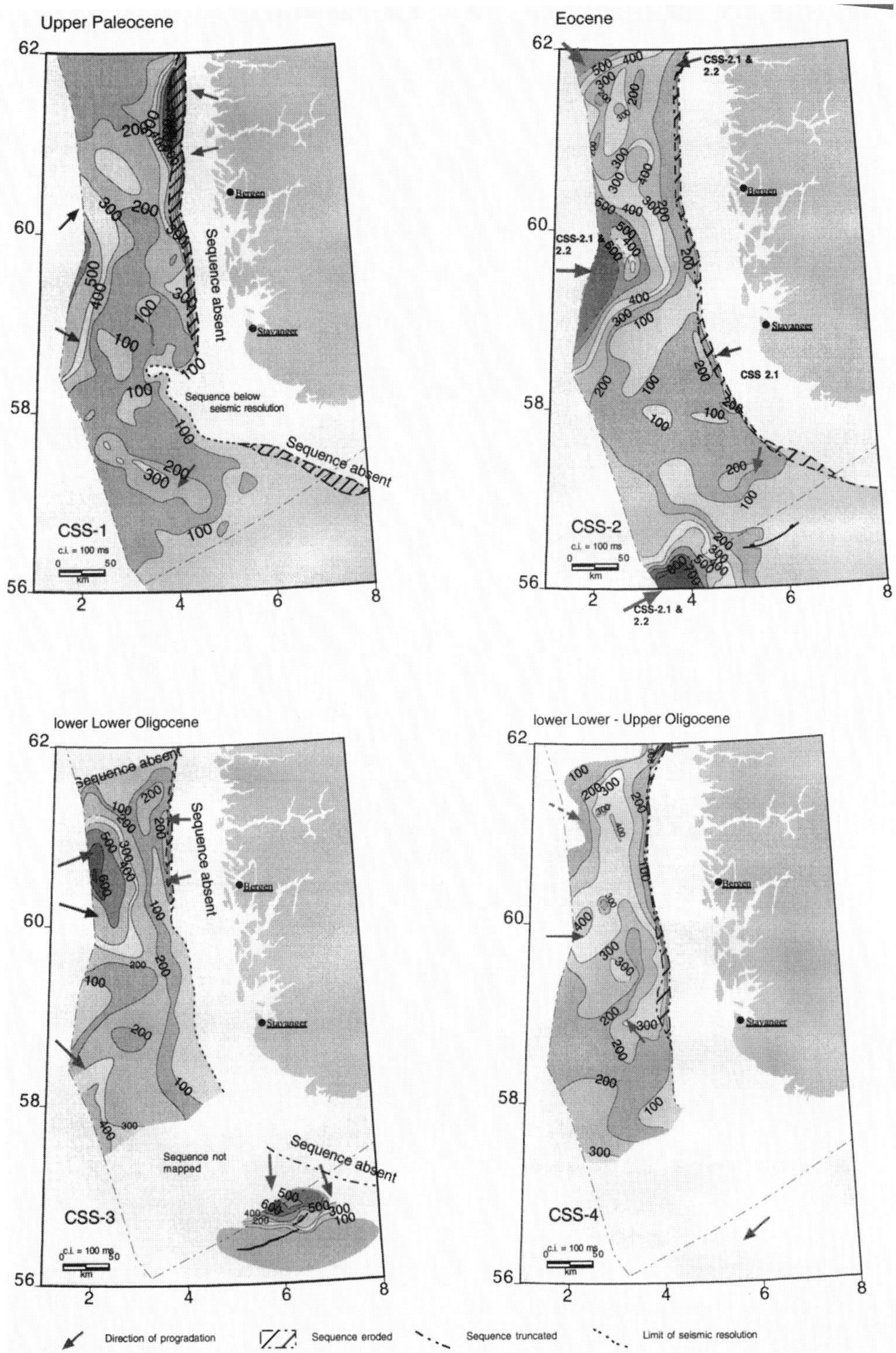

Fig. 8. Isopach maps of Upper Paleocene (CSS-1), Eocene (CSS-2) and Oligocene (CSS-3 and CSS-4) sequences. Outbuilding directions and contours of CSS-3 and CSS-4 in the Norwegian–Danish Basin are based on Jordt (1995).

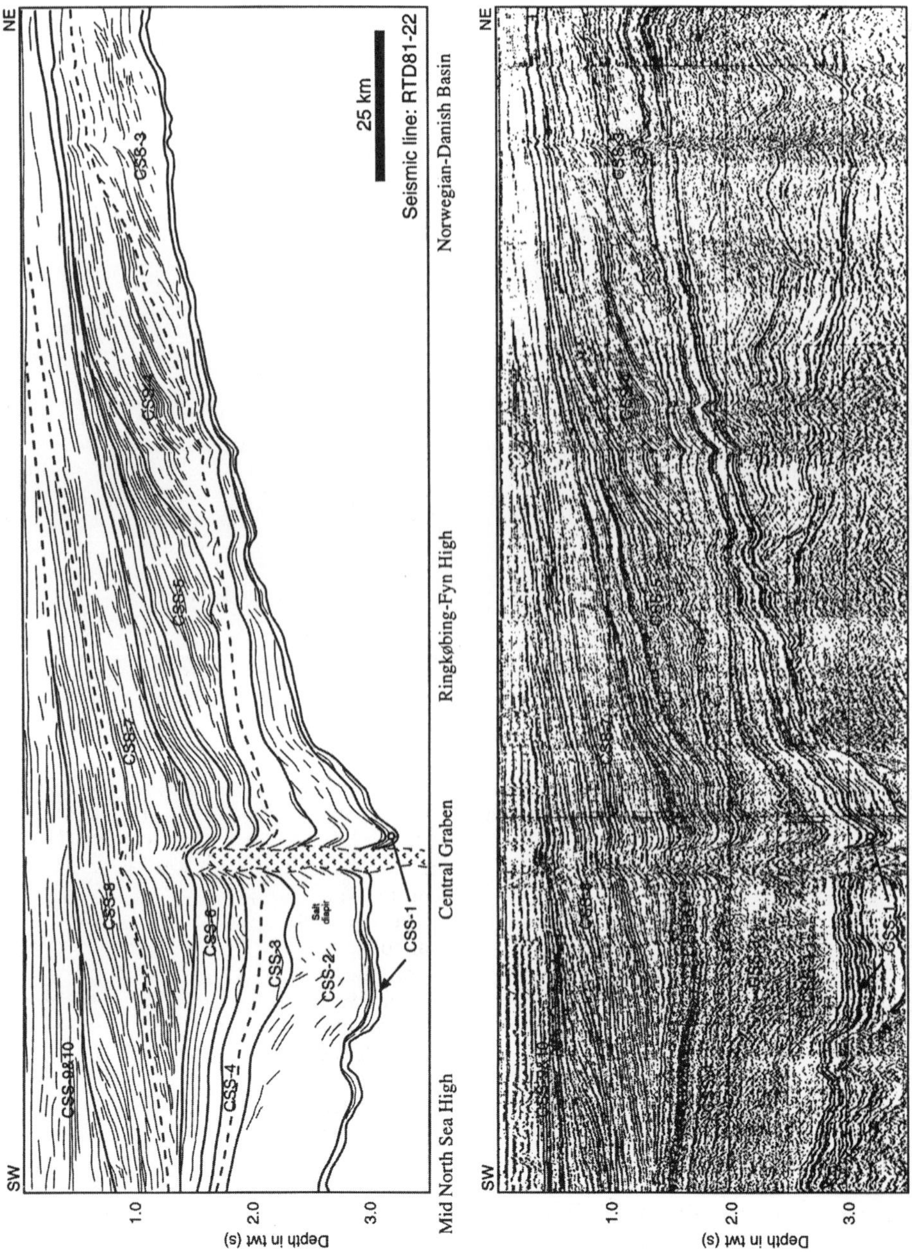

Fig. 9. NE–SW striking seismic section showing the mapped sequences in the Norwegian–Danish Basin. Line location is shown in Fig. 2.

Oligocene sequence

The CSS-3 sequence (lower Early Oligocene) was built from the East Shetland Platform and from the Øygarden Fault Zone towards the centre of the northern North Sea (Fig. 8), and towards the south (Fig. 14) into the Norwegian–Danish Basin. Sequence CSS-3 is thin (Fig. 5) or absent (Fig. 8) in the northern part of the study area. Hiatuses at the Eocene–Oligocene transition are indicated in the Tampen Spur area (Fig. 5) and along the Øygarden Fault Zone (Fig. 6).

The net accumulation rate for the CSS-3 sequence varies between 41 and 250 m Ma^{-1} (Table 1), and these rates are almost an order of magnitude higher than those of the underlying Eocene sequence both in the depocentre and in proximal positions. Such a large difference can probably be attributed to post-depositional erosion of the Eocene strata. The Oligocene CSS-3 depocentres in the Norwegian–Danish Basin and in the Viking Graben attain approximately the same maximum thickness of 5–600 ms TWT, indicating similar high rates of net accumulation along those basin margins.

The sediments of sequence CSS-3 contain less Co, Ni, Cu and Zn than the underlying Eocene sediments, whereas the smectite content appears higher than in the sediments immediately below (Fig. 11). These shifts in the sediment composition occur at the same depth as the suggested hiatus between the Eocene and Oligocene in well 34/7-1 (Fig. 5) and make the CSS-2–CSS-3 boundary easy to detect on the compositional data. A higher content of mica and illite in the lower Oligocene than in the underlying Eocene sediments occurs in the Norwegian–Danish Basin (Michelsen *et al.* 1998).

The CSS-4 sequence (upper Lower to Upper Oligocene) is characterized by a relatively uniform sediment distribution without marked depocentres (Fig. 8). Locally, in the northern North Sea, the thickness of sequence CSS-4 varies much because of remobilization of underlying smectite-rich clay of sequence CSS-3 (Fig. 15). The direction of progradation was from the northeast in the Norwegian–Danish Basin and mainly from the west in the northern North Sea. Progradational stacking is generally not pronounced in the CSS-4 sequence and internal reflection terminations are generally subtle. Therefore, sequence CSS-4 appears to be an aggradational sequence that onlaps against the underlying CSS-3 sequence and pinches out toward the basin margins (Figs 9 and 13). Sequence CSS-4 is possibly represented by a prograding wedge in the northwesternmost part of the study area, and by northward progradation with toplap on the Patchbank Ridge (Jordt *et al.* 1995).

The net accumulation rate for the CSS-4 sequence varies between 11 and 44 m Ma^{-1}, significantly lower than those of sequence CSS-3, but of similar magnitude to those of sequence CSS-2 (Table 1).

The content of smectite decreases whereas the feldspar content appears to increase upwards within sequence CSS-4 (Fig. 11). The content of Co, Ni, Cu and Zn is almost constant, but lower than in the Eocene sediments.

Miocene sequence

Development of the North Sea Basin during Miocene time (sequences CSS-5, CSS-6 and CSS-7) is characterized by a reduction in outbuilding from the west. The mapped Miocene depocentres are all located south of 60°N in the North Sea (Fig. 16). Farther north, the Miocene sequences thin to close to (or below) seismic resolution, and, consequently, it is not possible to tie their boundaries within the seismic grid.

The CSS-5 sequence was built out into the Viking Graben area from a basin margin to the west, and from the northeast in the Norwegian–Danish Basin. A marked basinward shift in onlap suggests that deposition was influenced by basin-floor topography, generated by the underlying Oligocene deposits (Fig. 9).

Sequence CSS-5 shows a net accumulation rate between 16 and 166 m Ma^{-1}, comparable with that of sequence CSS-1 (Table 1). It attains the highest net accumulation rate in the depocentre of the Viking Graben.

The overlying sequences, CSS-6 and CSS-7, thin northward and may be absent in large parts of the study area north of 60°N (Fig. 16). CSS-6 is a relatively thin, aggrading sequence that was built from the Norwegian–Danish Basin toward the south and southwest across the Ringkøbing–Fyn High and towards the Central Graben. Sequence CSS-7 was built from southern Norway toward the south and southwest, and downlaps toward the south in the Norwegian–Danish Basin (Jordt 1995). The CSS-7 sequence is composed of westward-dipping, sub-parallel reflections that onlap distally against the Mid North Sea High (Fig. 9). Sequence CSS-7 is truncated by the base Quaternary, as shown by a marked angular unconformity, reflecting post-Miocene uplift and erosion in the Norwegian–Danish Basin (Fig. 9).

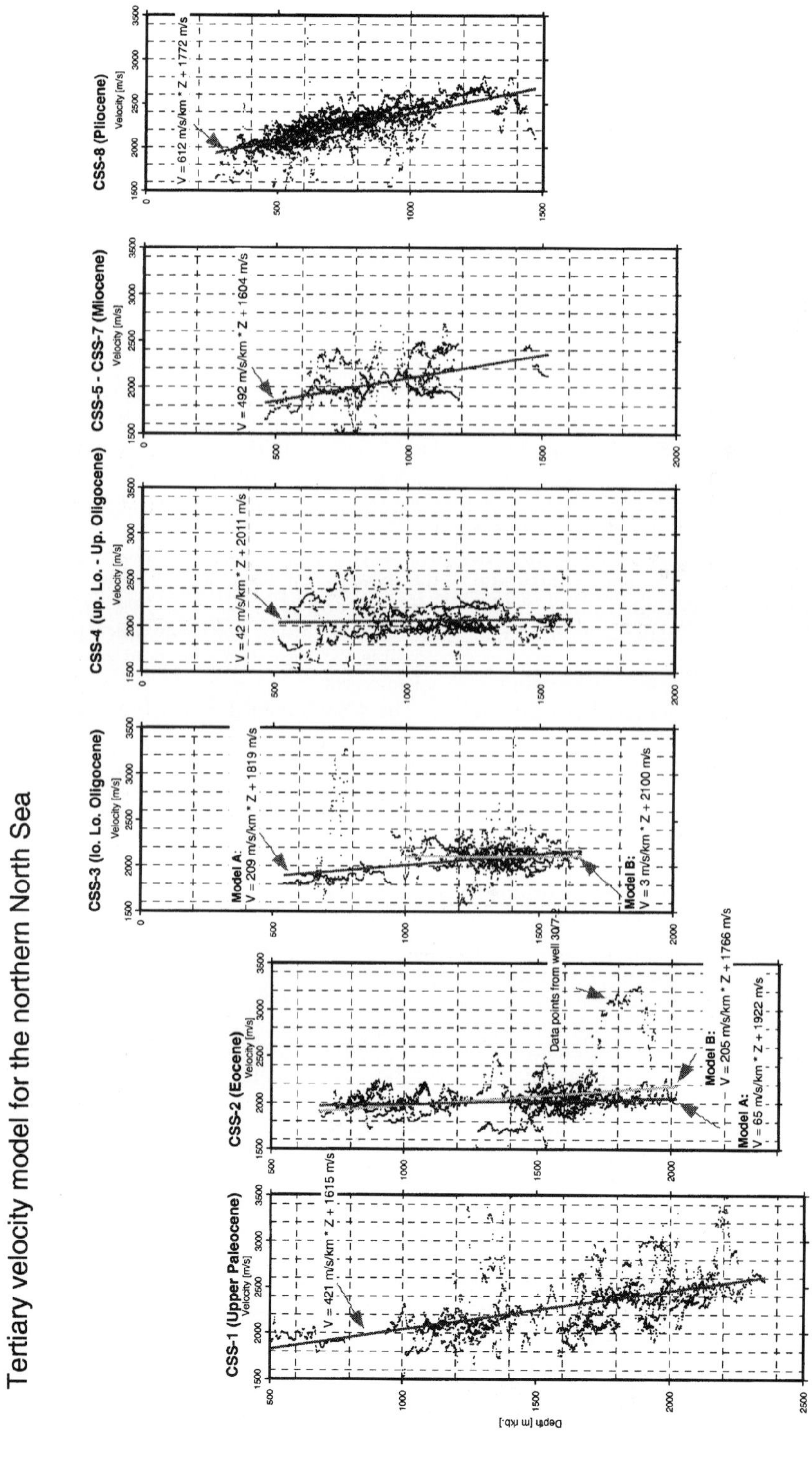

Fig. 10. Sonic log velocities from the northern North Sea. The velocity data are based on sonic logs from 40 wells between 60 and 62°N.

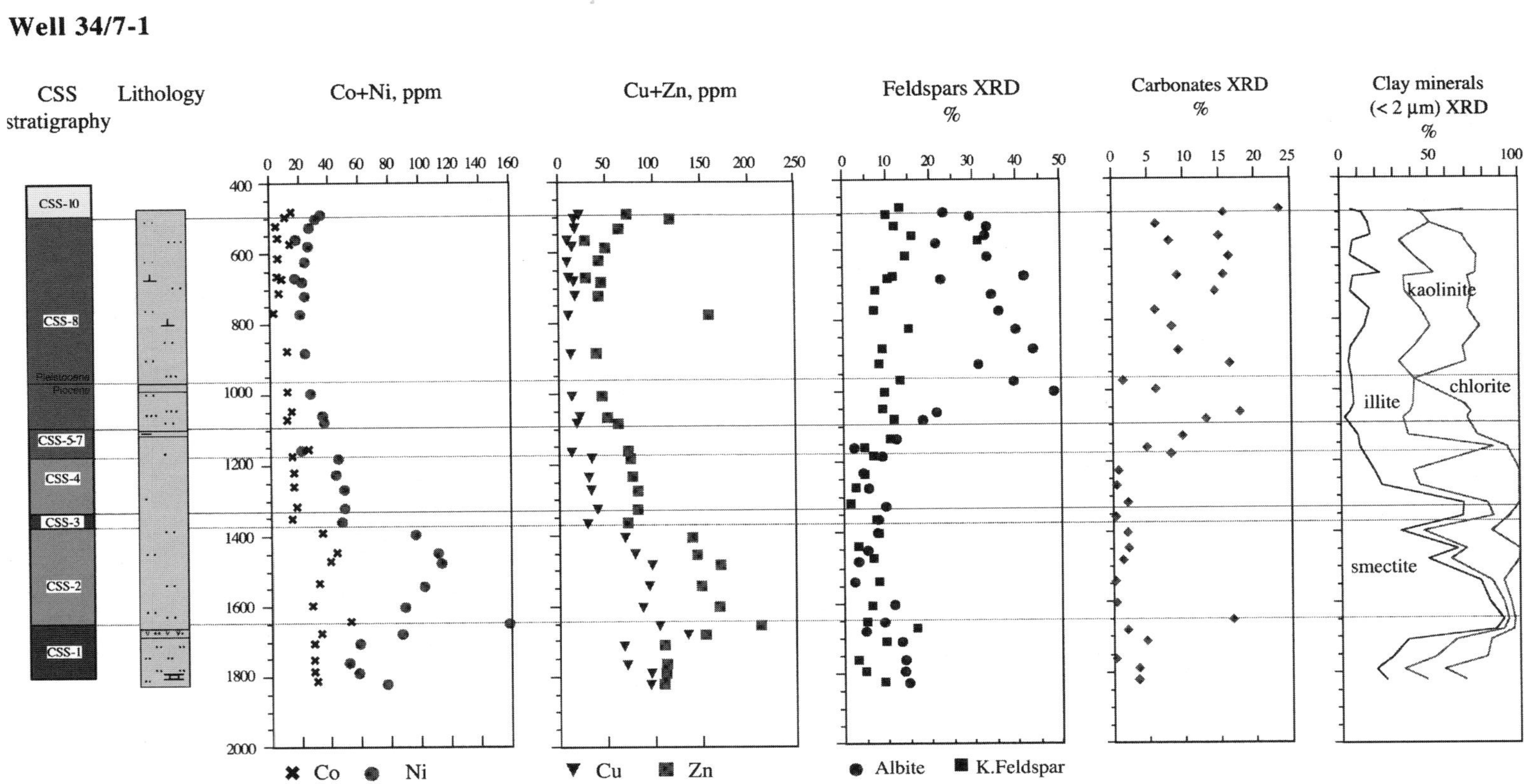

Fig. 11. Mineralogical and geochemical data from well 34/7-1 based on the established seismic framework and biostratigraphy (from Thyberg *et al.* 1999).

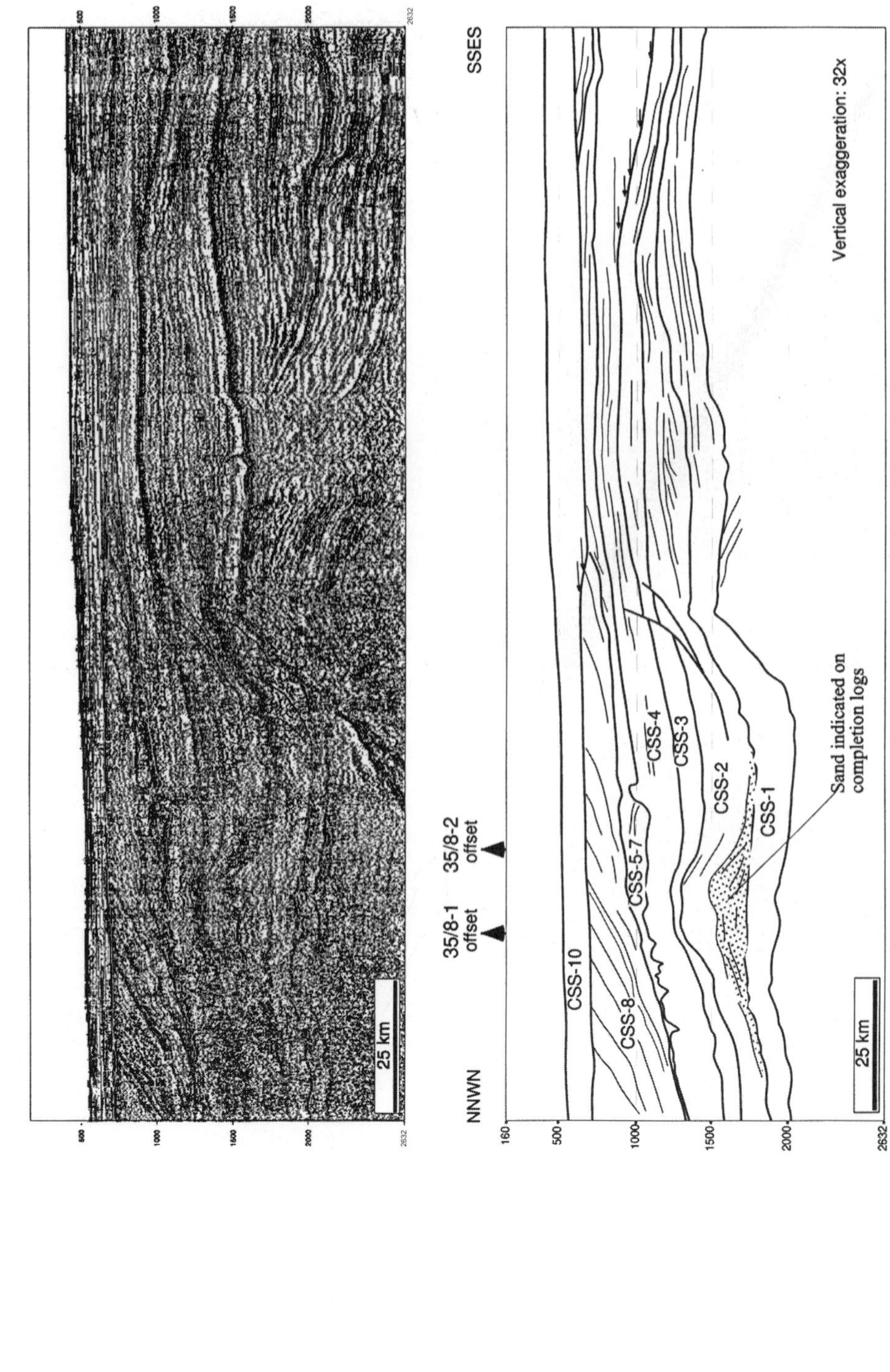

Fig. 12. N–S striking seismic section from the Horda Platform in the northern North Sea showing the mapped seismic sequences.

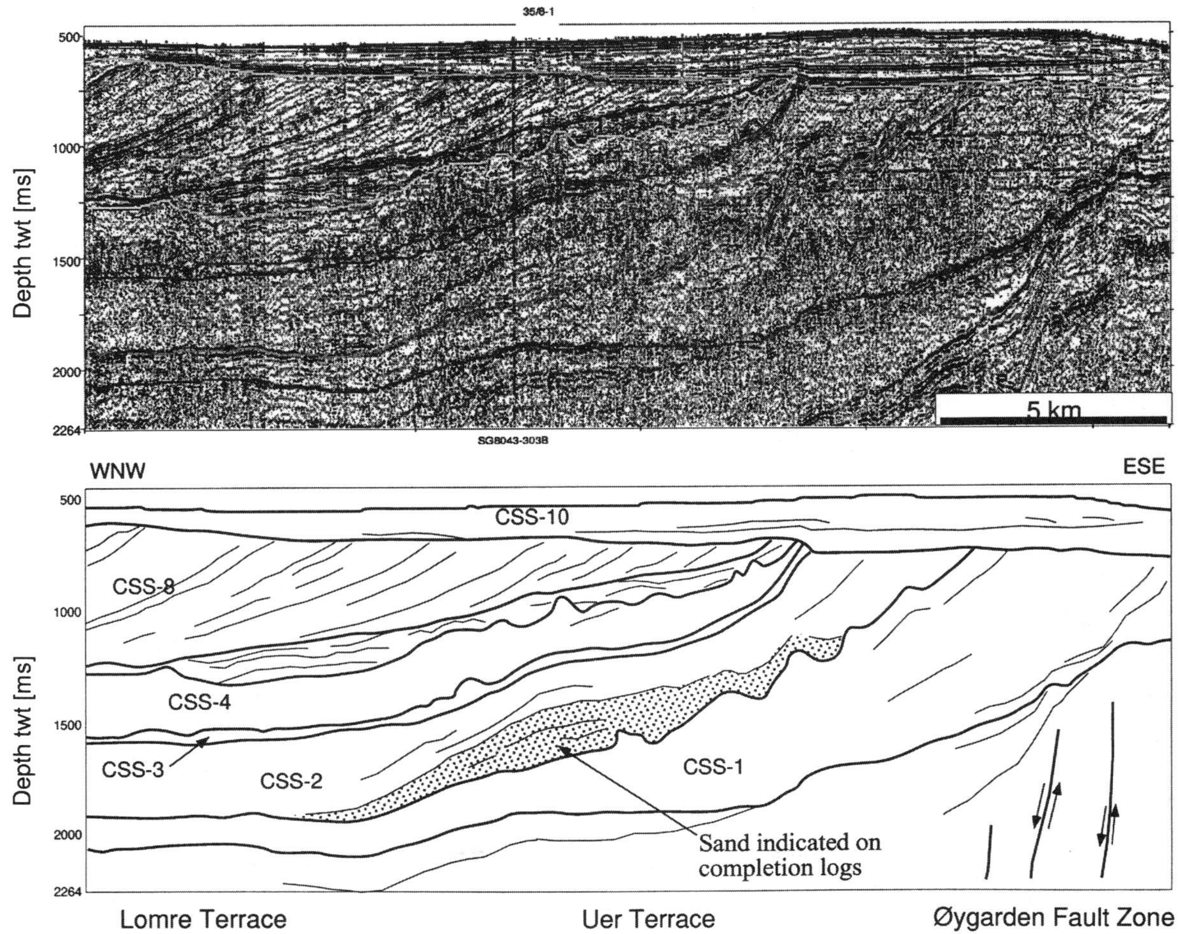

Fig. 13. NW–SE striking seismic section showing the mapped sequences in the northern North Sea. Line location is shown in Fig. 2.

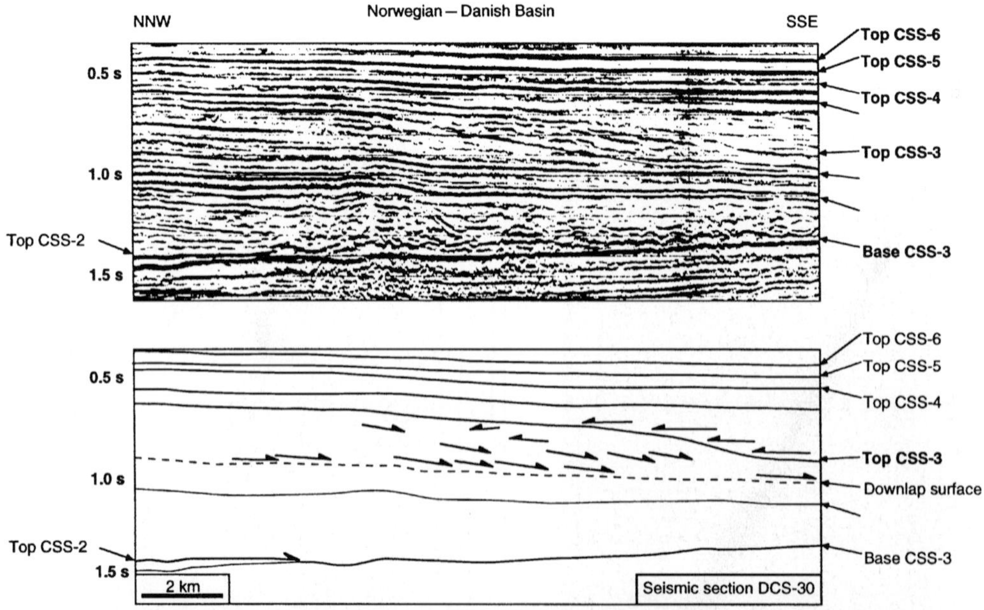

Fig. 14. DCS-30 N–S striking seismic section showing the Oligocene and Miocene sequences in the Norwegian–Danish Basin. Line location is shown in Fig. 2.

The sediment accumulation rates for the sequences CSS-6 and CSS-7 vary between 7 and 71 m Ma^{-1}, and 11 and 100 m Ma^{-1}, respectively (Table 1). The maximum net accumulation rates are all lower than for the CSS-1 and CSS-3 sequences, but higher than for sequences CSS-2 and CSS-4.

The Miocene sediments are characterized by an upward increase in feldspar content, reflecting a decrease in sedimentary maturity (Fig. 11). An upward increasing amount of clastic carbonate and chlorite reflects an increasing supply of clastic material.

Pliocene sequence

The CSS-8 sequence is dominated by a >1000 m thick clastic wedge north of the Horda Platform (Figs 16 and 17) which was built out from the south (Fig. 12) and east (Fig. 13). The base of sequence CSS-8 is marked by a prominent regional downlap surface (Fig. 17). In the Stord Basin a >400 m thick Pliocene wedge was built from north (Fig. 12) and east. The uppermost Pliocene sediment succession is eroded along the Øygarden Fault Zone, whereas further west the upper boundary of sequence CSS-8 is conformable, resulting from continuous progradation. Sequence CSS-8 onlaps basinward against Miocene sediments along the East Shetland Platform, generating a marked seismic unconformity (Jordt *et al.* 1995), and it aggrades and pinches out towards the Stavanger Platform (Fig. 16). The CSS-8 sequence builds northwards as well as southwards on the Horda Platform (Fig. 12). The overlying Quaternary sequence shows a thickening in the same area.

Pliocene sediment accumulation rates vary between 29 and 235 m Ma^{-1}, which is much higher than those of the Miocene (Table 1). The Pliocene maximum net accumulation rate has a similar magnitude to that for the CSS-1 and CSS-3 sequences.

The sediment composition of sequence CSS-8 is characterized by an immature mineralogy reflected by upward increasing amounts of feldspar, moderate amounts of clastic carbonate and detrital chlorite (Fig. 11).

Discussion

Outbuilding of the CSS-1 sequence along the basin margins indicates that the landmasses of southern Norway and the UK sector existed during Late Paleocene time. The sediment composition reflects erosion of areas with mixed lithologies. The high net accumulation rates (Table 1) and thickening of sequence CSS-1 into

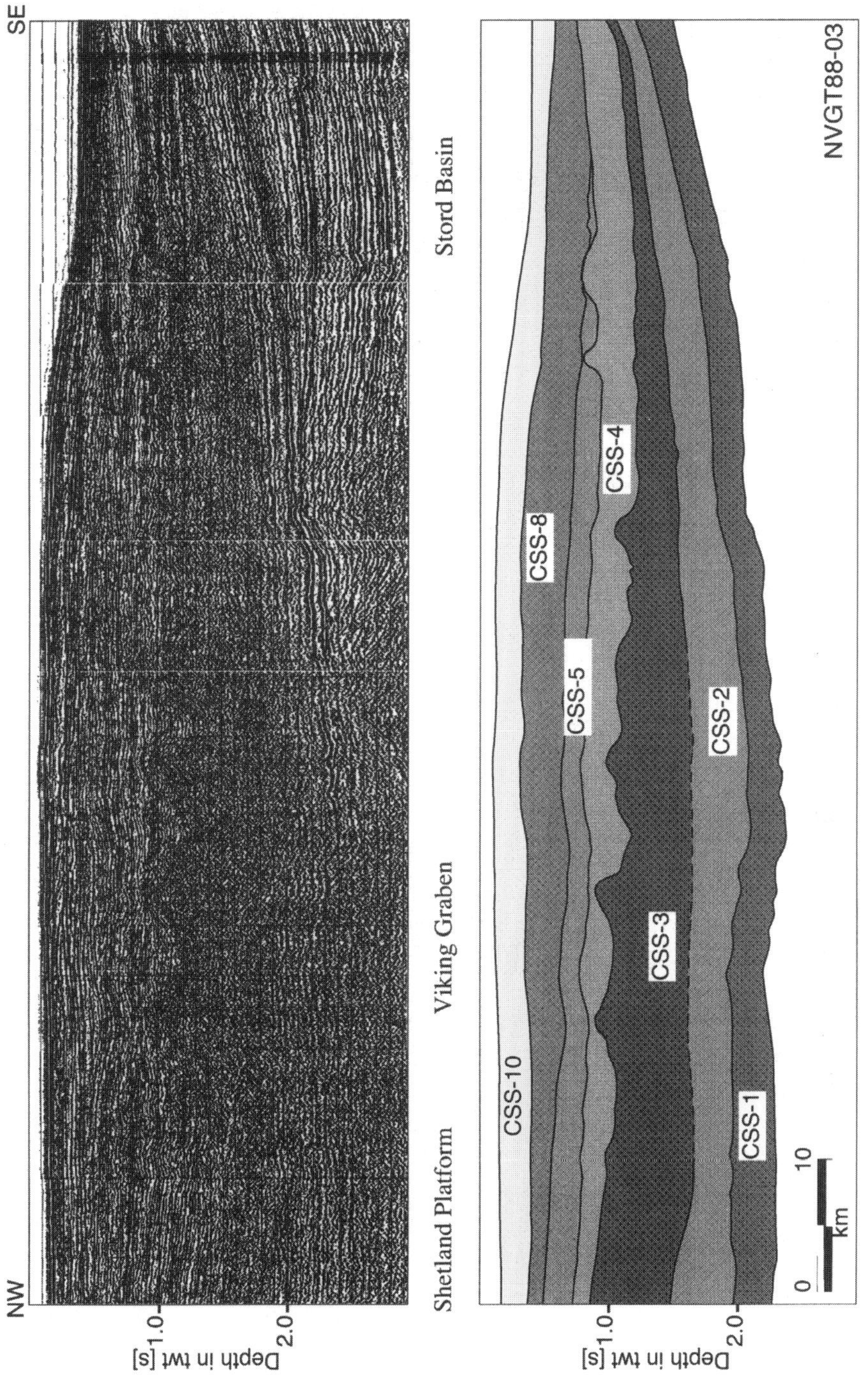

Fig. 15. NW–SE striking seismic section showing remobilized Oligocene clay.

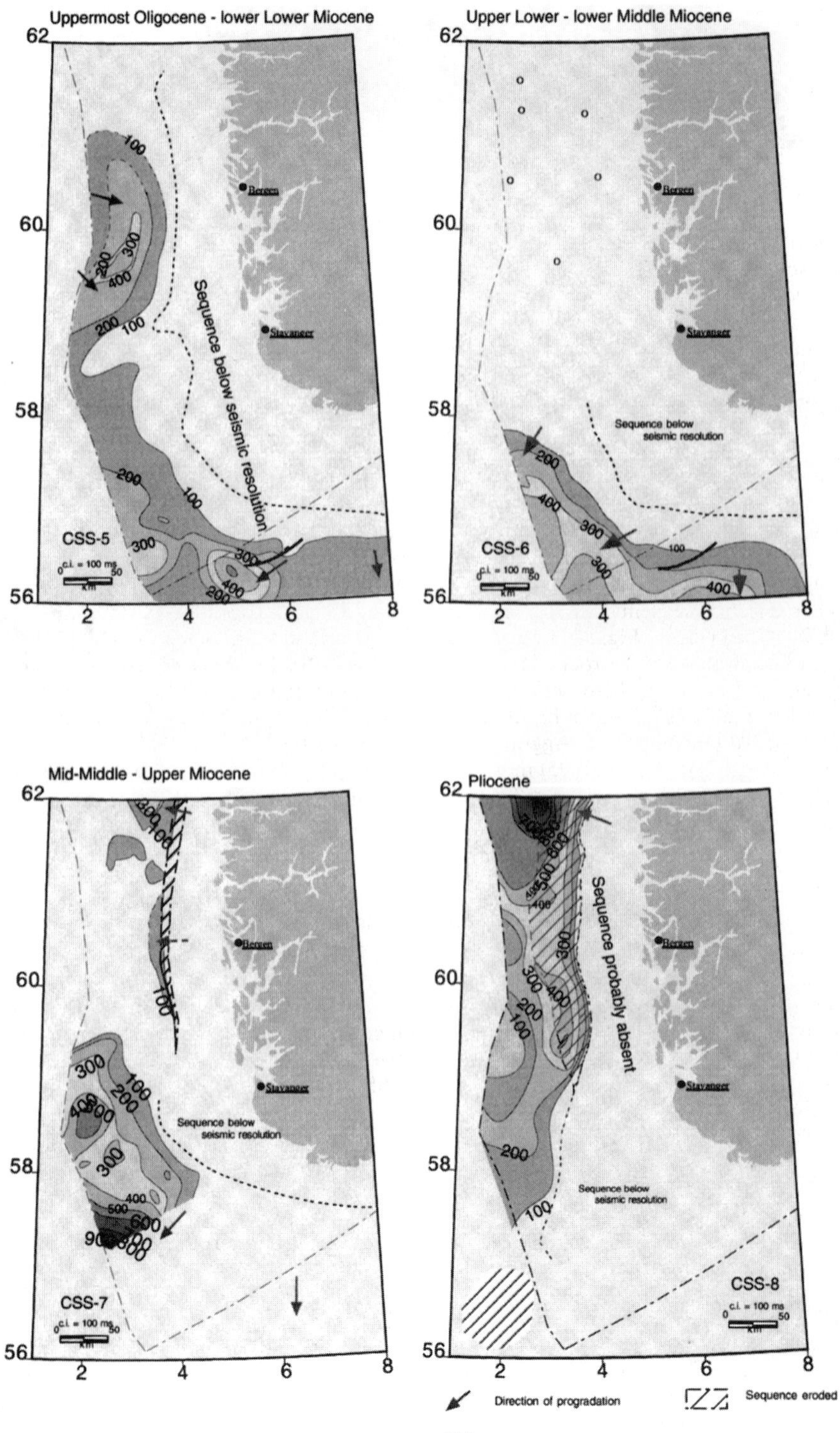

Fig. 16. Isopach maps of the Miocene (CSS-5 to CSS-7) and Pliocene (CSS-8) sequences. Outbuilding directions in the Norwegian–Danish Basin are based on Jordt (1995).

depocentres (Fig. 8) may suggest that sediment input was focused along older structure elements (i.e. Sognefjorden) into a basin possibly deeper than 1000 m. As the inherent post-Cretaceous bathymetry along the North Sea basin margin areas was not significant, the accommodation and development of the thick Upper Paleocene wedge probably resulted from subsidence of the basin floor combined with increased sediment supply related to tectonic uplift and subsequent erosion of southern Norway. Such uplift may have occurred in response to the opening of the North Atlantic rift zone, possibly related to spreading and emplacement of a mantle plume beneath the lithosphere (Skogseid *et al.* 1999), as the zone of uplift is concentrated along the flank of the rift. Rising and spreading of melt theoretically would have caused initial inversion of the central rift zone, followed by progressive uplift towards the surrounding flank areas. To what extent intra-plate stress induced by the Alpine collision played an important role is unknown.

The higher smectitic content in the upper part of the CSS-1 sequence (Fig. 11) indicates weathering of a basaltic source rock. The most likely provenance area for these sediments is the Late Paleocene–Early Eocene basalt province along the Atlantic rift zone. Aggradation of sequence CSS-1.2 (Jordt *et al.* 1995) indicates that supply from Norway was significantly reduced in earliest Eocene time, compared with Late Paleocene time. Reduced runoff caused by a climatic change, changes in land vegetation, or tectonic quiescence or erosional levelling of southern Norway, are possible mechanisms that could explain a sudden reduction in sediment supply. However, progradation of the lower part of sequence CSS-2 on the Horda Platform and deposition of sand along the eastern margin of the basin (Fig. 12) indicate continued sediment supply from the east following deposition of sequence CSS-1.2. This suggests that a temporary climatic deterioration, or a change in vegetation, may have led to the reduced supply from the east in the very earliest Eocene time. Subsequently, higher runoff and renewal of supply along the basin margins may have caused renewed outbuilding.

Eastward thinning of sequence CSS-2 and termination of seismic outbuilding from the east and northeast reflect a marked decrease in sediment supply from Norway into the Norwegian–Danish Basin during Early Eocene time (Fig. 8). The uniform lithology and composition of the Eocene sediments on the Horda Platform and northwards (Rundberg 1989; Thyberg *et al.* 2000), and the velocity distribution of sequence CSS-2 (Fig. 10), suggest a rather uniform source and sediment supply from an area dominated by basaltic rocks, i.e. the Atlantic basalt province to the northwest. Overall, it seems that sediment supply from basement erosion east of the study area was insignificant. It is suggested, therefore, that the reduced sediment supply into the Norwegian–Danish basin and North Sea from mainland Norway during Early Eocene time was probably a result of tectonic quiescence and erosional levelling of the provenance areas.

Middle Eocene sediments are overlain by Lower Oligocene sediments in the northern North Sea (Figs 5 and 6), reflecting a regional hiatus (Figs 3 and 4). A lower eustatic sea level during Late Eocene time (Abreu & Anderson 1998) may have enhanced erosion in the northern North Sea, whereas the Eocene succession in the Norwegian–Danish Basin appears unaffected by erosion. Lower Oligocene sediments were supplied from both the east and west and thick (>500 m) wedges were deposited in the northern Viking Graben and in the Norwegian–Danish Basin. A higher content of smectite in the CSS-3 sequence, compared with the sediments immediately below (Fig. 11), suggests input from a new source area in Early Oligocene time.

The observed changes in direction of outbuilding (Fig. 8) and increase in sediment accumulation rate (Table 1) from Eocene to Early Oligocene time, as well as the Upper Eocene hiatus in the northern North Sea, cannot be easily explained by eustatic sea-level changes or by changes in the sediment supply rate caused by climatic variations alone. It is our interpretation, therefore, that there was a significant uplift of southern Norway in Late Eocene–Early Oligocene times, which was accompanied by a lower eustatic sea level. A late or post-Eocene uplift of southern Norway is in accordance with the observations by Riis (1996) of an uplifted, deeply weathered Eocene erosional peneplain surface across Fennoscandia. It is suggested that lithospheric stress induced by ridge push forces from the North Atlantic rift zone and by the Alpine collision probably played an important role. To what extent continued depletion and movement of magma below the lithosphere along the Atlantic rift played a role is not known.

Subsequent rapid accumulation of Lower Oligocene sediments (Table 1) occurred during a relatively high eustatic sea level. However, the magnitude of the sea-level rise in Early Oligocene time was only *c.* 100 m (Abreu & Anderson 1998), compared with the >500 m of sediments

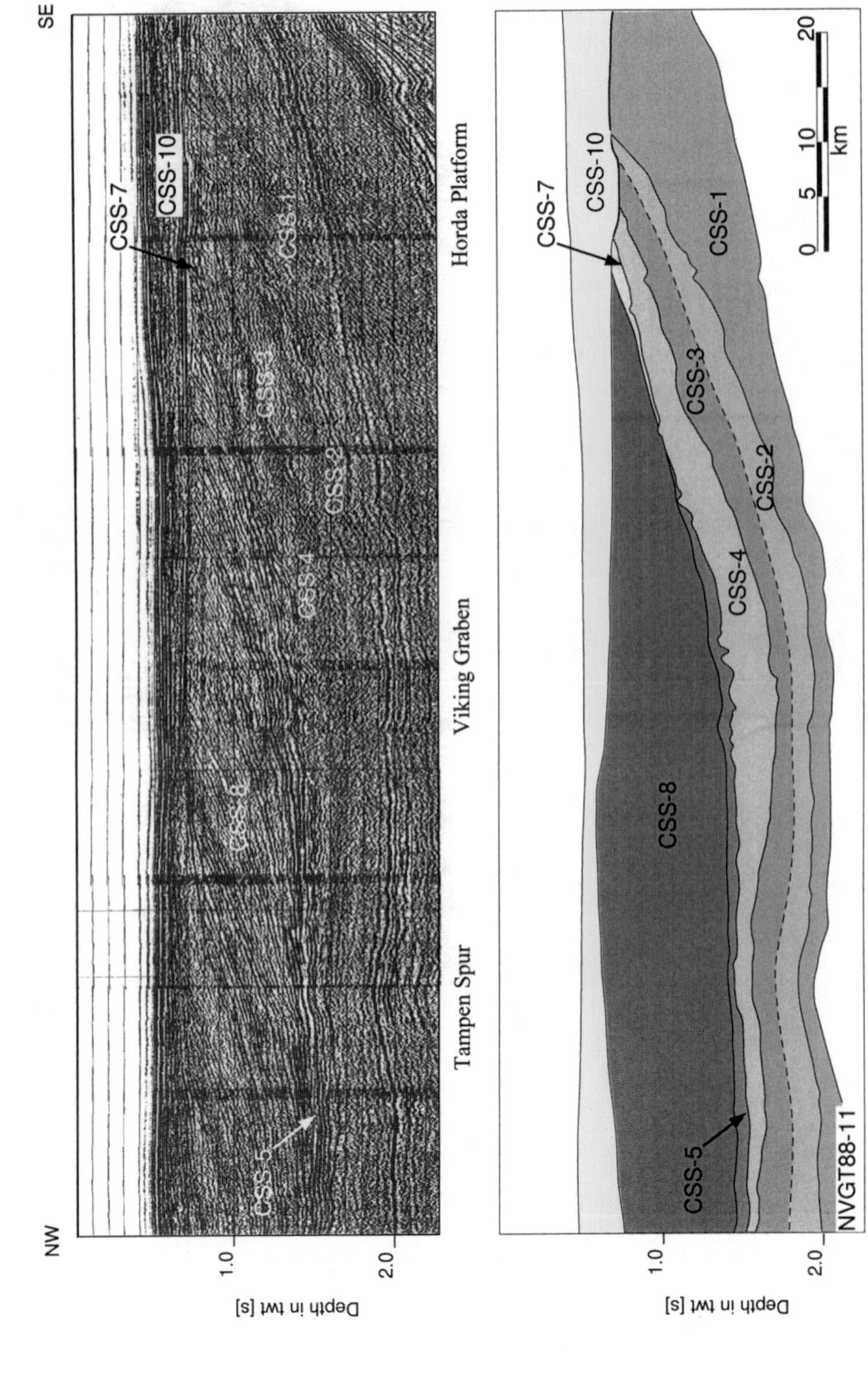

Fig. 17. NW–SE trending seismic section showing the mapped sequences in the northernmost North Sea. Line location is shown in Fig. 2.

that was deposited in the basin. The inherent Late Eocene bathymetry in the basinal areas of the northern North Sea was probably not significant, and as sediment loading can account for only part of the excess accommodation space needed, it is suggested that rapid tectonic subsidence occurred in the central and northern North Sea during Early Oligocene time.

No major changes in composition of the sediments occurred in the transition to Late Oligocene time, during deposition of the CSS-4 sequence (Fig. 11), and no major entry points have been identified. Sequence CSS-4 slowly filled in and smoothed the basin-floor topography generated by clay mobilization and deformation of sequence CSS-3 (Fig. 9).

Incision of Miocene channels into Oligocene strata along the eastern margin (Rundberg *et al.* 1995; Jordt *et al.* 1996; Gregersen 1998) indicates a shallowing of the northern North Sea during Miocene time. The shallowing may be related to uplift in the northern North Sea, as a continuous Oligocene to Miocene succession is preserved further to the south in the study area (Fig. 9). The presence of a >500 ms TWT thick CSS-5 sequence prograding from the west, a northward pinchout of sequence CSS-6 and subsequent deposition of sequence CSS-7 from the east (Fig. 16), indicate that an Early Miocene reduction in accommodation space was followed by an increase in accommodation space during late Miocene time. The eustatic sea level shifted from a relatively high level during Early and Mid-Miocene times to a significantly lower level in Late Miocene time (Abreu & Anderson 1998).

Dating of the Miocene succession in the northern North Sea is uncertain (Fig. 7), but, according to Gradstein & Bäckström (1996), the upper Miocene sequence appears to be absent in this area. Eustatic sea level was relatively low during deposition of the CSS-7 sequence (Abreu & Anderson 1998), supporting that the upper Miocene sequence may be absent and that eustatic variations may have controlled, or at least influenced, deposition and erosion of the Miocene sediments (i.e. low eustatic sea level v. deposition of sequence CSS-7). However, eustatic variations cannot adequately explain the change in outbuilding that occurred during Miocene time (Fig. 16), nor can they explain the change from a mature to an immature sediment (Fig. 11), which suggests a significant shift in provenance area. If, on the other hand, Upper Miocene sediments are present, as indicated by the mixed sediment composition (Fig. 11), then the eustatic sea-level variations were out of phase with deposition, and tectonic movements of the basin floor may have been a prime control on the Miocene development.

The Pliocene sequence is characterized by outbuilding from Norway and deposition of a major clastic wedge in the northern North Sea (Figs 16 and 17). It is, however, uncertain if Lower Pliocene sediments are present (Fig. 7). The thickness of the Pliocene wedge indicates a major increase in accommodation space before outbuilding. As deposition of sequence CSS-8 occurred during a general eustatic sea-level fall (Abreu & Anderson 1998), the increase in accommodation space cannot be explained purely by eustacy nor by sediment loading, suggesting that a significant tectonic subsidence of the basin floor occurred during (early) Pliocene times. The tectonic subsidence was accompanied by a dramatic increase in sediment supply, as a result of glacial erosion (Riis 1996).

The immature mineralogy of sequence CSS-8 (Fig. 11) reflects weathering of basement rocks that suggests contemporaneous uplift east of the study area, of the Norwegian mainland. It is debated in the literature to what extent the uplift was a result purely of glacial rebound, or if also had a tectonic component (Rohrman *et al.* 1995; Riis 1996).

Before deposition of the CSS-10 sequence, the northern Horda Platform was uplifted (Fig. 12), causing erosion into Pliocene and Miocene sediments. The timing of this relative uplift is uncertain; southward-dipping reflections in the Pliocene and Miocene sequences may indicate that it occurred before deposition of sequence CSS-8.

Summary and conclusions

In the present study, we have integrated seismic stratigraphic interpretation and velocity analyses with analyses of sediment composition. We propose that the bases of the CSS-1, CSS-3, CSS-7 and CSS-8 sequences correspond to periods of tectonic uplift in the northern North Sea. It appears that the northern North Sea was more affected by uplifts than the central North Sea, suggesting that the mechanisms causing uplift may have been related to the opening of the North Atlantic rift zone.

The unconformities at the base of the CSS-3 and CSS-8 sequences both were generated during periods with relatively lower eustatic sea levels. However, the subsequent sea-level rises were not of a sufficient magnitude (<100 m, Abreu & Anderson 1998) to allow the rapid deposition of thick (>500 m) overlying sequences during early Oligocene and Pliocene time, respectively. This suggests that, subsequent to the three or four uplift events, the basin subsided and the composition of the sediments changed as a result of uplift in the sediment source areas. The most significant subsidence event took place during Pliocene time, coincident with a general eustatic sea-level fall. However, deposition of the >500 m thick CSS-3 sequence (lower Oligocene) above an eroded Eocene succession in the northern North Sea may possibly reflect a subsidence event of similar magnitude.

This interplay, between tectonic uplift during eustatic lows and subsequent tectonic subsidence, cannot be explained in terms of passive thermal subsidence following the Jurassic rift event. Thus, a Mackenzie-type rifting model cannot adequately describe the Tertiary evolution of the central and northern North Sea. The mechanism causing the observed complex, tectonically controlled basin-floor movements is probably composite. However, lithospheric stress induced by ridge push forces from the Atlantic rift zone and by the Alpine collision probably played an important role, and depletion and movement of magma below the lithosphere along the Atlantic rift are possible other factors.

The research was funded by the Commission of the European Union and the Norwegian Research Council (NFR) in the framework of the DGXII-Joule Programme: Energy from fossil sources; Integrated Basin Studies (IBS) project. Norsk Hydro a.s., Saga Petroleum a.s. and Statoil a.s. provided data and stimulus for this study. We thank Nopec for giving permission to publish their seismic sections. We are also especially indebted to fellow colleagues within the above-mentioned oil companies who shared their knowledge and thoughts with us and gave valuable comments.

References

ABREU, V. S. & ANDERSON, J. B. 1998. Glacial eustasy during the Cenozoic: sequence stratigraphic implications. *AAPG Bulletin*, **82**, 1385–1400.

BERTRAM, G. T. & MILTON, N. J. 1988. Reconstructing basin evolution from sedimentary thickness; the importance of palaeobathymetric control with reference to the North Sea. *Basin Research*, **1**, 247–257.

BIJLSMA, S. 1981. Fluvial sedimentation from the Fennoscandian area into the north-west European basin during late Cenozoic. *In*: VAN LOON, A. J. (ed) *Quaternary Geology: a Farewell to, A. J. Wiggers. Geologie en Mijnbouw*, **60**, 337–345.

BONDE, N. 1973. Fiskefossiler, diatomiter og vulkanske askelag. *Dansk Geologisk Forening Årsskrift 1972*, 136–143.

CAMERON, T. D. J., BULAT, J. & MESDAG, C. S. 1993. High resolution seismic profile through a Late Cenozoic delta complex in the southern North Sea. *Marine and Petroleum Geology*, **9**, 591–599.

CARTWRIGHT, J. 1995. Seismic–stratigraphical analysis of large-scale ridge–trough sedimentary structures in the Late Miocene to Early Pliocene of the Central North Sea. *In*: PLINT, A. G. (ed.) *Sedimentary Facies Analysis*. Special Publications of the International Association of Sedimentologists, **22**, 285–303.

DEEGAN, C. E. & SCULL, B. J. 1977. *A Standard Lithostratigraphic Nomenclature for the Central and Northern North Sea*. Norwegian Petroleum Directorate Bulletin, **1**.

EIDVIN, T. & RIIS, F. 1992. En biostratigrafisk og seismostratigrafisk analyse av tertiære sedimenter i nordlige deler av Norskerenna, med hovedvekt på øvre pliocene vifteavsetninger. *Norwegian Petroleum Directorate Contribution*, **32**, 40.

GALLOWAY, W. E. 1989. Genetic stratigraphic sequences in basin analysis I: Architecture and genesis of flooding-surface bounded depositional units; II: Application to Northwest Gulf of Mexico Cenozoic Basin. *AAPG Bulletin*, **73**, 125–154.

GIBBARD, P. L. 1988. The history of the great northwest European rivers during the past three million years. *Philosophical Transactions of the Royal Society of London, Series B*, **317**, 559–602.

GLENNIE, K. W. 1990. Outline of the North Sea history and structural framework. *In*: GLENNIE, K. W. (ed.) *Introduction to the Petroleum Geology of the North Sea*. Blackwell, Oxford, 34–77.

GRADSTEIN, F. & BÄCKSTRÖM, S. 1996. Cainozoic biostratigraphy and palaeobathymetry, northern North Sea and Haltenbanken. *Norsk Geologisk Tidsskrift*, **76**, 3–32.

GREGERSEN, U. 1998. Upper Cenozoic channels and fans on 3D seismic data in the northern Norwegian North Sea. *Petroleum Geoscience*, **4**, 67–80.

HALL, B. D. & WHITE, N. 1994. Origin of anomalous Tertiary subsidence adjacent to North Atlantic continental margins. *Marine and Petroleum Geology*, **10**, 702–713.

HARLAND, W. B., ARMSTRONG, R. L., COX, A. V., CRAIG, L. E., SMITH, A. G. & SMITH, D. G. 1990. *A Geologic Time Scale 1989*. Cambridge University Press, Cambridge.

ISAKSEN, D. & TONSTAD, K. 1989. *A Revised Cretaceous and Tertiary Lithostratigraphic Nomenclature for the Norwegian North Sea*. Norwegian Petroleum Directorate Bulletin, **5**.

JORDT, H. 1995. Regional Cenozoic uplift and subsidence patterns in the southeastern North Sea. *In*: MICHELSEN, O. (ed.) *Proceedings of the 2nd Symposium on Marine Geology. Geology of the North Sea and Skagerrak, Aarhus University 1993*. Danmarks Geologiske Undersøgelse Serie C, **12**, 53–68.

——*The Cenozoic Evolution of the Central and Northern North Sea Based on Seismic Sequence Stratigraphy*. Doctoral thesis, University of Oslo.

——, ——, —— & IBRAHIM, M. T. 1995. Cenozoic sequence stratigraphy of the Central and Northern North Sea basin: tectonic development, sediment distribution and provenance areas. *Marine and Petroleum Geology*, **12**, 845–879.

KNOX, R. W. O. & HOLLOWAY, S. 1992. *Paleogene of the Central and Northern North Sea*. British Geological Survey, Keyworth.

MIALL, A. D. 1995. Sequence stratigraphy and chronostratigraphy: problems of definition and precision in correlation, and their implications for global eustacy. *Geoscience Canada*, **21**, 1–26.

MICHELSEN, O., DANIELSEN, M., HEILMANN-CLAUSEN, C., JORDT, H., LAURSEN, G. V. & THOMSEN, E. 1995. Occurrence of major sequence stratigraphic boundaries in relation to basin development in Cenozoic deposits of the southeastern North Sea. *In*: STEEL, R. J., FELT, V. L., JOHANNESEN, E. P. & MATHIEU, C. (eds) *Sequence Stratigraphy on the Northwest European Margin*. Norwegian Petroleum Society Special Publication, **5**, 415–427.

——, ——, ——, ——, —— & ——1998. Cenozoic sequence stratigraphy in the eastern North Sea. *In*: DE GRACIANSKY, P.-C., HARDENBOL, J., JACQUIN, T. & VAIL, P. R. (eds) *Mesozoic and Cenozoic Sequence Stratigraphy of Western European Basin*. SEPM Special Publication, **60**, 91–118.

MITCHUM, R. M. 1977. Seismic stratigraphy and global changes of sea level, Part 10: glossary of terms used in seismic stratigraphy. *In*: PAYTON, C. E. (ed.) *Seismic Stratigraphy – Applications to Hydrocarbon Exploration*. American Association of Petroleum Geologists, Memoir, **26**, 205–212.

ORTON, G. J. & READING, H. G. 1993. Variability of deltaic processes in terms of sediment supply, with particular emphasis on grain size. *Sedimentology*, **40**, 475–512.

READING, H. G. & RICHARDS, M. 1994. Turbidite systems in deep-water basin margins classified by grain size and feeder system. *AAPG Bulletin*, **78**, 792–822.

RIIS, F. 1996. Quantification of Cenozoic vertical movements of Scandinavia by correlation of morphological surfaces with offshore data. *Global and Planetary Change*, **12**, 331–357.

ROBERTS, A. M., YIELDING, G. & BADLEY, M. E. 1993. Tectonic and bathymetric controls on stratigraphic sequences within evolving half-graben. *In*: WILLIAMS, G. D. & DOBB, A. (eds) *Tectonics and Seismic Stratigraphy*. Geological Society, London, Special Publications, **71**, 87–121.

ROHRMAN, M., VAN DER BECK, P., ANDRIESSEN, P. & CLOETINGH, S. 1995. Meso-Cenozoic morphotectonic evolution of southern Norway: Neogene domal uplift inferred from apatite fission track thermochronology. *Tectonics*, **14**, 704–718.

RUNDBERG, Y. 1989. *Tertiary and sedimentary history and basin evolution of the Norwegian Sea between 60° and 62°N. An integrated approach*. Dr.ing. thesis, University of Trondheim.

——, OLAUSSEN, S. & GRADSTEIN, F. 1995. Incision of Oligocene strata; evidence for northern North Sea Miocene uplift and key to the formation of the Utsira sands. *Geonytt*, **22**, abstract.

SALOMONSEN, I. 1995. Origin of a deep buried valley system in Pleistocene deposits of the eastern North Sea. *In*: MICHELSEN, O. (ed.) *Proceedings of the 2nd Symposium on Marine Geology. Geology of the North Sea and Skaggerak, Aarhus University 1993*. Danmarks Geologiske Undersøgelse, Serie C, **12**, 7–19.

SHARMA, P. V. 1969. Early Tertiary field reversals recorded in volcanic ash layers in northern Denmark. *Meddelelser fra Dansk Geologisk Forening*, **19**, 218–223.

——, PLANKE, S., FALEIDE, J. I., PEDERSEN, T., ELDHOLM, O. & NEVERDAL, F. 1999. NE Atlantic continental rifting and volcanic margin formation. *This volume*.

SPJELDNÆS, N. 1975. Palaeogeography and facies distribution in the Tertiary of Denmark and surrounding areas. *In*: WHITEMAN, A., ROBERTS, D. & SELLEVOLL, M. A. (eds) *Petroleum Geology and Geology of the North Sea and Northwest Atlantic Continental Margin*. Norges Geologiske Undersøkelse Bulletin, **29**, 289–311.

STEURBAUT, E., SPIEGLER, D., WEINELT, M. & THIEDE, J. 1991. *Cenozoic Erosion and Sedimentation on the Northwest European Continental Margin*. Geomar, Kiel.

STUEVOLD, L. & ELDHOLM, O. 1996. Cenozoic uplift of Fennoscandia inferred from a study of the mid-Norwegian margin. *Global and Planetary Change*, **12**, 359–386.

——, SKOGSEID, J. & ELDHOLM, O. 1992. Post-Cretaceous uplift events on the Vøring continental margin. *Geology*, **20**, 919–922.

THYBERG, B. I., JORDT, H., BJØRLYKKE, K. & FALEIDE, J. I. 1999. Relationship between sequence stratigraphy, mineralogy and geochemistry in the Cenozoic sediments of the northern North Sea. *This volume*.

ZIEGLER, P. A. 1982. *Geological Atlas of Western and Central Europe*. Elsevier, Amsterdam.

——1990. Tectonic and palaeogeographic development of the North Sea rift system. *In*: BLUNDELL, D. J. & GIBBS, A. D. (eds) *Tectonic Evolution of the North Sea Rifts*. Clarendon Press, Oxford, 1–36.

Relationships between sequence stratigraphy, mineralogy and geochemistry in Cenozoic sediments of the northern North Sea

B. I. THYBERG,[1,2] H. JORDT,[1,2] K. BJØRLYKKE[1] & J. I. FALEIDE[1]

[1] *Department of Geology, University of Oslo, P.O. Box 1047 Blindern, N-0316 Oslo, Norway*
[2] *Present address: Statoil, Box 300, 4001 Stravanger, Norway*

Abstract: Cuttings of Cenozoic sediments from selected wells in the northern North Sea have been analysed both mineralogically and geochemically, and compared with seismic sequence stratigraphic units. Variations in the composition with respect to clay minerals, feldspars, major elements and selected trace elements have been used to establish changes in the provenance of the Cenozoic sequences. The Eocene sequence consists of a thickness of several hundreds of metres of almost exclusively smectitic clays with high Ni and Zn contents and little or no quartz. This represents a strong input from subaerial volcanism related to the rifting and initial spreading of the Norwegian–Greenland Sea. The Eocene–Oligocene smectitic mudstones have high porosity and low seismic velocity ($<2\,km\,s^{-1}$) compared with the underlying Paleocene sediments and overlying Miocene–Pliocene sequences. The Plio cene and Pleistocene mudstones are coarse-grained clays with an immature mineralogy, which compact much more rapidly than the underlying smectitic sediments. The seismic velocity may then reach $2.5–3.0\,km\,s^{-1}$ and thus cause a strong velocity inversion. This sequence is partly glacial and mineralogically immature, but may contain some kaolinite and smectite, which indicate reworking of lower Tertiary and Mesozoic sediments. Our obser vations indicate that sequence boundaries are closely related to tectonic movements and to changes in the sediment supply from adjacent areas. Both mineralogical and geochemical data can be used as additional tools for correlation in Cenozoic sediments in the northern North Sea. The mineralogy of these sediments also determines the rate of compaction and other factors relevant for basin modelling.

The Cenozoic succession in the North Sea Basin was one of the main examples used when seismic stratigraphy was introduced (Vail *et al.* 1977*a*, *b*) and several sequence stratigraphic studies from this area have subsequently been published (e.g. Nielsen *et al.* 1986; Stewart 1987; Rundberg 1989; Galloway *et al.* 1993; Jones & Milton 1994; Michelsen 1994; Jordt *et al.* 1995; Michelsen *et al.* 1995). The initial interpretation of these data related the seismic stratigraphic units to changes in eustatic sea level (Vail *et al.* 1977*a*, *b*). However, recent studies of the North Sea Basin suggest that the seismic sequences and sequence boundaries are mainly controlled by tectonic processes (e.g. Rundberg 1989; Galloway *et al.* 1993; Jordt *et al.* 1995, 1999). These tectonic processes cause subsidence (accommodation space), as well as uplift and erosion, and thus to a large extent control the sedimentation into a basin such as the North Sea. By contrast, sediment supply is not directly related to sea-level changes. It is dominantly influenced by environmental factors, including relief, climate, the drainage system and the biogenic productivity of the ocean. The clastic sediment supply is controlled by the rate of erosion and weathering in the source area, itself a function of the drainage and relief generated by tectonic uplift. The type of clay mineral found in a shale is thus a function of provenance (rock type), climate (weathering), transport (facies) and diagenetic history of the sequence (Weaver 1989).

The main objective of this paper is to analyse and establish the relationship between sediment composition and seismic sequence stratigraphic units. Special emphasis has been placed on the relationship between the sediment composition and the Cenozoic seismic sequence stratigraphic framework established by Jordt *et al.* (1995). The present paper also provides a contribution to the discussion of the origin of the stratigraphic sequences and their development.

The boundaries of seismic sequences are identified by physical variables, such as changes in the acoustic impedance, defined as the product of seismic velocity and bulk density, and by differences in the amplitude and frequency of the signal. As the seismic velocity is significantly influenced by primary mineral composition, texture and diagenetic processes, which are also

From: NØTTVEDT, A. *et al.* (eds) *Dynamics of the Norwegian Margin*. Geological Society, London, Special Publications, **167**, 245–272. 1-86239-056-8/00/$15.00

reflected in fine-grained sediments, the nature of seismic reflections is a direct function of the lithology (e.g. Rundberg 1989). In spite of this, sequence stratigraphic studies have almost exclusively been based on seismic profiles, biostratigraphic data and petrophysical well logs. Clay mineralogy has been used as a basis for the stratigraphy of barren units of Triassic, Permian and Devonian red beds in UK Quadrant 9 and as a stratigraphic tool to distinguish Mesozoic and Palaeozoic rock units (Glasmann & Wilkinson 1993). Fisher & Jeans (1982) divided the Permo-Triassic red beds into eight clay mineral zones, and dated and correlated a Permo-Triassic sequence exposed on the south Devon coast and a sequence of red beds penetrated in BNOC 72/10-1A, some 500 km away in the Western Approaches, by using facies-related clay mineral assemblages. In addition, Henson (1974) described various compositional changes in the clay mineral assemblages related to lithofacies. Rundberg (1989) integrated geophysical data and mineralogical analyses to reconstruct climatic and tectonic changes in the Cenozoic section in the northern North Sea. Triassic sediments from seven wells in the Central Graben also show clearly identifiable geochemical differences, which have been related to shift in sources and transport regimes (Racey *et al.* 1995). The present study compares seismic sequence stratigraphic units with the mineralogical composition of the fine-grained Cenozoic sediments.

Database and methods

Mineralogical and geochemical data obtained by X-ray diffraction (XRD) and X-ray fluorescence (XRF) of washed drill bit cuttings of wells 24/6-1, 26/4-1, 30/3-3, 34/7-1, 34/7-2 and 34/7-6 are presented in this paper (Fig. 1). They are selected from a very large mineralogical and geochemical database established during this study. The sampling interval was between 30 and 100 m. XRD analyses are based on the bulk samples as

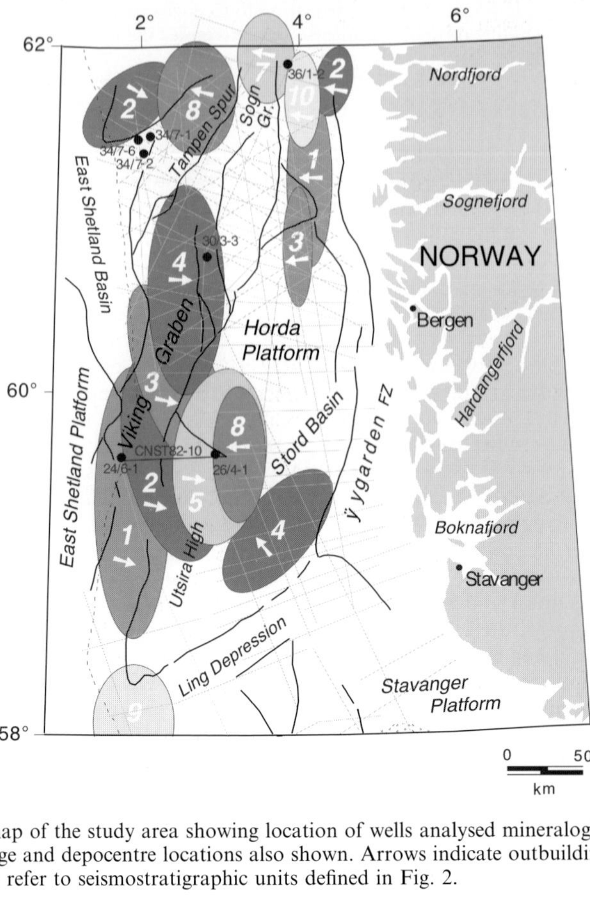

Fig. 1. Structural map of the study area showing location of wells analysed mineralogically and geochemically. Seismic data coverage and depocentre locations also shown. Arrows indicate outbuilding directions (Jordt *et al.* 1995). The numbers refer to seismostratigraphic units defined in Fig. 2.

well as the <2 μm fraction, which have been taken out to indicate the clay mineral composition of some selected wells. Unoriented powders of bulk samples were analysed on a Phillips PM1700/1710 diffractometer with CuKα radiation (1.54060). Diffraction diagrams were collected from 2 to 50°2θ. Additional slow scan runs (26.0–28.5°2θ) were carried out routinely, to obtain a better identification of feldspars. Peak intensities used for estimating weight percentages are from the diffracted intensities (DI) listings given by the Phillips APD software. Peak intensities for 'all clays', quartz, feldspars, carbonates and pyrites are weighted using intensity factors from Ramm (1991).

The clay fraction (<2 μm) was separated from the material by suspension settling following dispersal by ultrasonic disaggregation. Oriented samples of the <2 μm fraction were prepared by filtering the samples through a Millipore filter and then transferring the material to a silica slide by inverting. XRD diagrams from 2 to 50°2θ were obtained, after the following treatment; samples were Mg saturated and air-dried, then ethylene glycolated and heated to 550°C in 2 h. Slow-scan diagrams between 24 and 26°2θ on air-dried specimens were obtained routinely to separate the d(002) kaolinite and d(004) chlorite reflections. The clay minerals were categorized into 'expandable clays' (<14 Å minerals on ethylene glycolated runs; referred to below as smectite), illite, kaolinite and chlorite. The weight factors used were based on Pearson & Small (1988) and Pearson (1990).

The inorganic geochemical analyses obtained by XRF of the cuttings were also included in this study. The samples were analysed with respect to major elements such as SiO_2, Al_2O_3, MnO, MgO, CaO, Na_2O, K_2O, TiO_2 and P_2O_5, and loss of ignition (LOI) has been calculated, expressed in weight percent of the oxide. The sum of major elements is in most cases 100% but, mainly as a result of contamination of drilling mud, values of <100% have been obtained. Samples were analysed for trace elements (V, Cr, Co, Ni, Cu, Zn, Rb, Sr, Y Zr, Nb, Pb, Th and U), presented in parts per million (ppm). In this study we have presented the Co, Ni, Cu and Zn contents in selected wells. Major element data were obtained from crushed material dried at 100°C for at least 12 h, mixed with lithium tetraborate in the ratio of 9 : 1, and then fused to 1000°C (1 h). The trace element content was determined by using pressed pellets made by mixing 10 g finely ground material with 2 ml Elvacite solution. Analysis were performed in a Phillips PW2400 spectrometer. The USGS (US Geological Survey) standards were used for calibration.

Cenozoic seismic stratigraphy

The results of the geochemical analyses have been correlated with the Cenozoic seismic sequence stratigraphic framework established by Jordt *et al.* (1995), who divided the Cenozoic succession of the North Sea Basin into ten main sequences, CSS-1 to CSS-10 (Fig. 2). The CSS-1 sequence is of Late Paleocene–earliest Eocene age (60.5–56 Ma). The top CSS-2 is dated to the Eocene–Oligocene transition (35.4 Ma). Oligocene sediments correspond to the CSS-3 and CSS-4 sequences. The Miocene succession is divided into three seismic sequences: CSS-5, CSS-6 and CSS-7. Top CSS-5 is of late Early Miocene age and top CSS-6 is dated to early Mid-Miocene. A Mid- to Late Miocene age is assigned the CSS-7 seismic sequence. CSS-8 is of Pliocene age, and the Quaternary sediments correspond to CSS-9 and CSS-10. Approximate locations of the Cenozoic depocentres in the northern North Sea and corresponding outbuilding directions of the different seismic sequences, based on Jordt *et al.* (1995), are shown in Fig. 1.

Cenozoic sediment composition

The feldspar content of both shales and sandstones is a good indication of a rock's mineralogical maturity, which may be related to rock type, weathering and transport as detrital feldspar. K-feldspar in particular may, however, be dissolved by deep burial, which reduces its effectiveness as a provenance indicator.

The Al_2O_3/SiO_2 ratio is used as an indication of the clay/sand ratio (e.g. Bjørlykke 1974), as Al_2O_3 is mostly found in clays, feldspar and mica. The iron content of the samples can be related to iron-rich chlorite, pyrite or siderite. Iron-rich smectite has also been reported in Paleocene and Eocene sediments (Tyridal 1994). The amount of calcium present in silicate minerals is generally small in sediments from the North Sea, and sandstones without carbonate minerals contain less than 2–3% CaO.

The clay mineralogy at the time of deposition can be related to the sediment provenance and facies (e.g. Weaver 1989). This includes not only the nature of the erosion of the source rock, but also its rate and the processes of rock weathering acting in different climates. Kaolinite forms in soils developed under abundant rainfall, good drainage and in shallow or coastal marine sandstones during meteoric water flushing. In drier climates, illite and smectite become important, although volcanic ash and other volcanic rocks

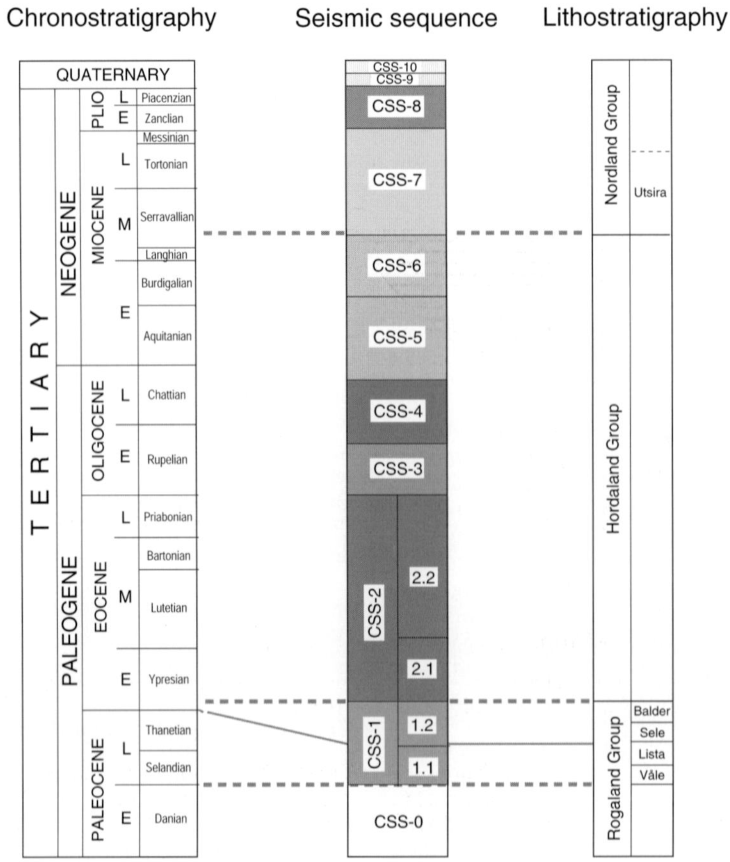

Fig. 2. Cenozoic seismic sequence stratigraphic units (Jordt *et al.* 1995, 2000) correlated with chronostratigraphy and lithostratigraphy from Deegan & Scull (1977).

are additional important sources of smectite. Illites in the basin, however, are probably derived from weathering of pre-existing muscovites or illites, but can also be formed diagenetically by the transformation of smectite to illite or kaolinite to illite. Chlorite forms from pre-existing chlorites but may be formed diagenetically with smectite or kaolinite as a precursor mineral.

The clay mineral size distribution relative to the proximity of the source has been used by several workers (e.g. Porrenga 1967; Millot 1970; Gibbs 1977). Some of the principles are summarized into a schematic and simplified representation of the theoretical mineralogical response to the sequence stratigraphic framework in mudrocks, reflecting periods of highstand and lowstand sea level in a sequence development (Fig. 3). This illustrates the distribution of the clay minerals during progradation, with periods of highstand ideally characterized by a higher concentration of fine-grained clay minerals, such as illite and smectite, whereas kaolinite is enriched in a more proximal facies, as the kaolinite crystals tend to be larger than those of illite and smectite. A starved interval in the basinward direction is evidenced in the most fine-grained part of the sequence. Condensed sections (e.g. Loutit *et al.* 1988; Galloway 1989) are laterally extensive thin layers. A condensed sequence may be represented by a dark, fine-grained shale or marl, often with a high content of biogenic carbonate and silica. Relative enrichments of phosphorus and manganese (Fig. 3) probably indicate very low sedimentation rates.

Upper Paleocene–lowermost Eocene sequence (CSS-1)

The clay mineralogy distribution of CSS-1 shows a high amount of smectite, but also chlorite, with smaller amounts of illite and kaolinite

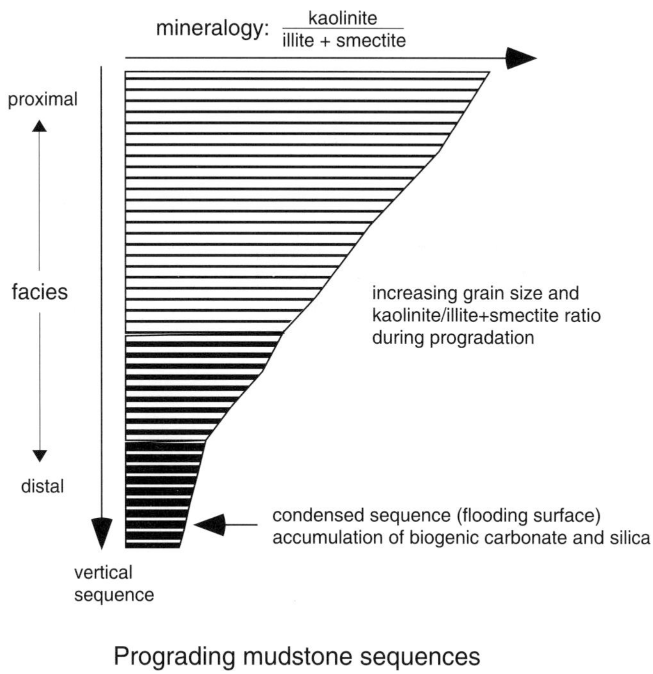

Fig. 3. Conceptual model for lateral variations in clay mineralogy in mudstones. Illite and smectite, being finer grained, are enriched in the distal facies compared with kaolinite.

detected by XRD (Figs 4 and 5). Rundberg (1989) also documented this in Paleocene sediments from the same region. The CSS-1 sediments are variable with respect to their feldspar content and the K-feldspar/plagioclase ratio, although plagioclase is dominant over K-feldspar in most of the analysed wells (Fig. 6), as is also reflected in the K_2O/Na_2O ratio (Fig. 7). Close to the top of CSS-1, an increase in plagioclase content is often observed. The erosion of a plagioclase-rich basement source could explain this. However, some plagioclase has volcanic origin, as Malm *et al.* (1984) and Morton & Knox (1990) have reported on plagioclase-rich xeno-crystals in the Balder tuff.

CSS-1 sediments have relatively high Fe_2O_3 and TiO_2 contents (Fig. 8); mean values of Fe_2O_3 and TiO_2 are 8.7 and 1.17 wt% respectively. The average MgO content is 2.6 wt%. Carbonate-cemented intervals are also rather characteristic (Figs 9 and 10), and some of the high iron content can be related to ankerite layers confirmed by XRD. Relatively high metal concentrations are seen in this sequence (Fig. 11), and the average values for Ni, Cu and Co are only lower than those of CSS-2 and CSS-3, and are much higher than in the uppermost sequences. An increase in the concentration of these trace elements close to the CSS-1–CSS-2 boundary, especially with respect to nickel and zinc, can be seen (Fig. 11). This is not found in well 26/4-1, however, as CSS-1 contains mainly sand and samples analysed are not from tuff-beds. The silica content in the shales mainly varies between 45 and 60 wt%, which is close to typical values for mafic rocks (e.g. Barth 1951). However, in the sand-rich units of CSS-1 in wells 26/4-1 and 24/6-1, a higher silica content is observed.

Paleocene sediments underlying the Balder Formation may have a relatively high smectite content, as well as high Fe, Mg, Cu, Ni and Zn content, indicating mafic source rocks. These sediments may be derived from basic basement rocks or erosion of Mesozoic sediments from surrounding areas. A supply of volcaniclastic material from the NW may also explain this distribution. The content of chlorite and a weak increase in illite in CSS-1 (Figs 4 and 5) can be attributed to diagenetic transformations with smectite as the precursor mineral, or the dissolution of rock fragments.

Top CSS-1 probably indicates a condensed section, with high concentrations of MnO, carbonate and a high concentration of fine-grained 'expandable' clays and sometimes P_2O_5 near the CSS-1–CSS-2 boundary. These data indicate periods of low clastic sediment supply

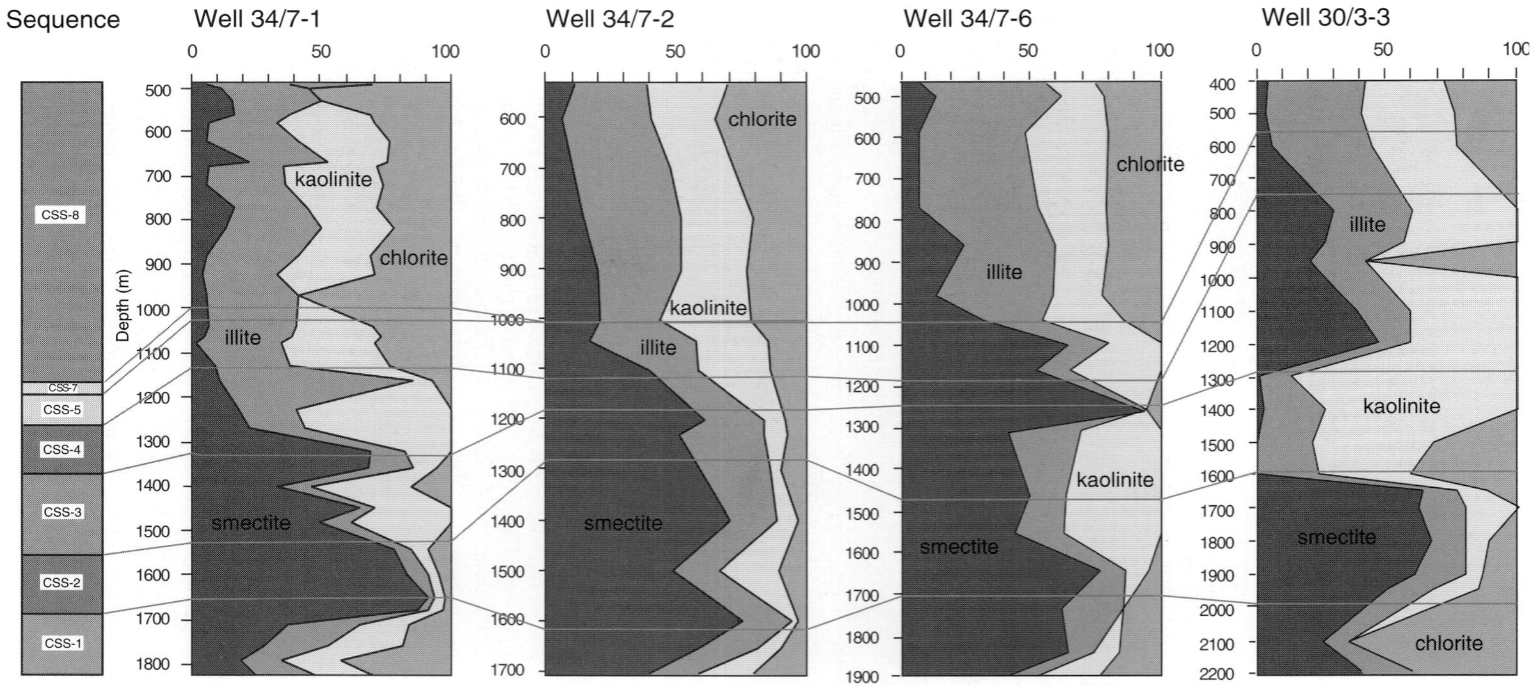

Fig. 4. Clay mineral distribution of some wells in the northern North Sea. The seismic sequence stratigraphic units are characterized by different clay mineral composition and particularly with respect to the smectite content. A low smectite and high kaolinite content in the Lower Oligocene sequence (CSS-3) in eastern well 30/3-3 may indicate a more proximal facies. Location of wells is shown in Fig. 1.

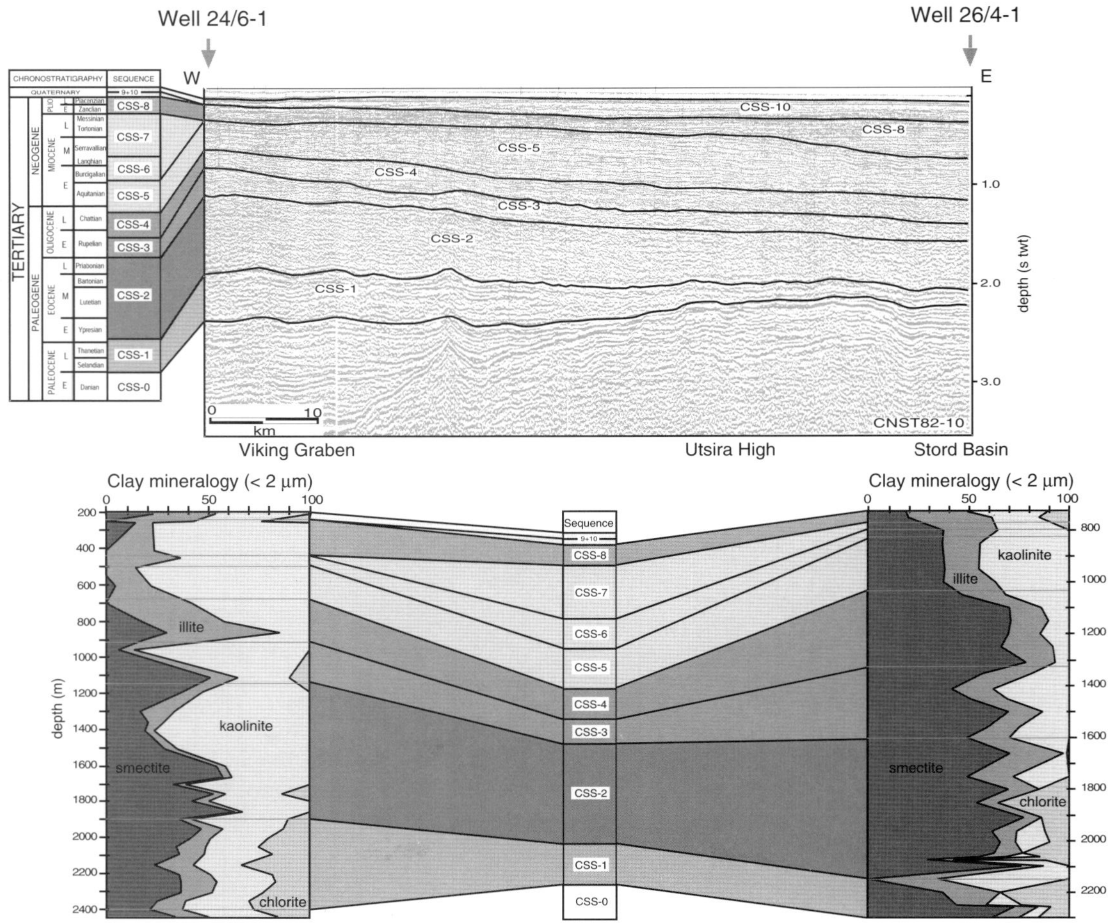

Fig. 5. Clay mineral distribution of wells 24/6-1 and 26/4-1, correlated with seismic section CNST82-10 and the Cenozoic seismic sequence stratigraphic framework. Depths are relative to mean sea level according to Jordt *et al.* (1995). Location of wells and seismic line is shown in Fig. 1.

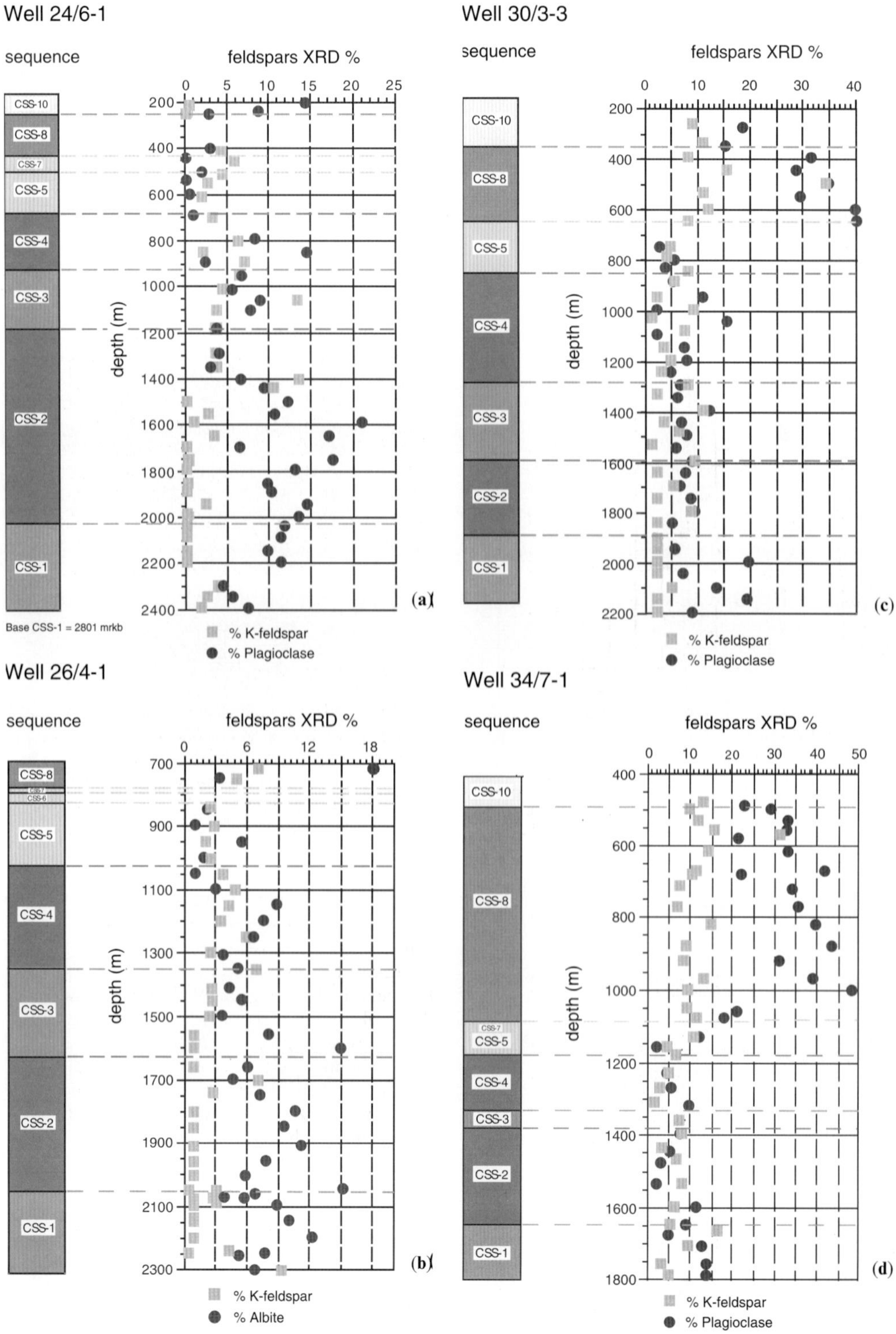

Fig. 6. Feldspar content analysed by XRD and related to the seismic sequence stratigraphic framework. Location of wells is shown in Fig. 1. Clastic feldspar may be used as an indicator of provenance and weathering. The K-feldspar content is low in all wells except in the Pliocene sediments, where it reflects a colder climate. The albite content is also low in Miocene and older sediments except in well 26/4-1. It is possible that albite to some extent may be authigenic in the Eocene sediments with a volcanic source. mrkb, metres below Kelly bushing.

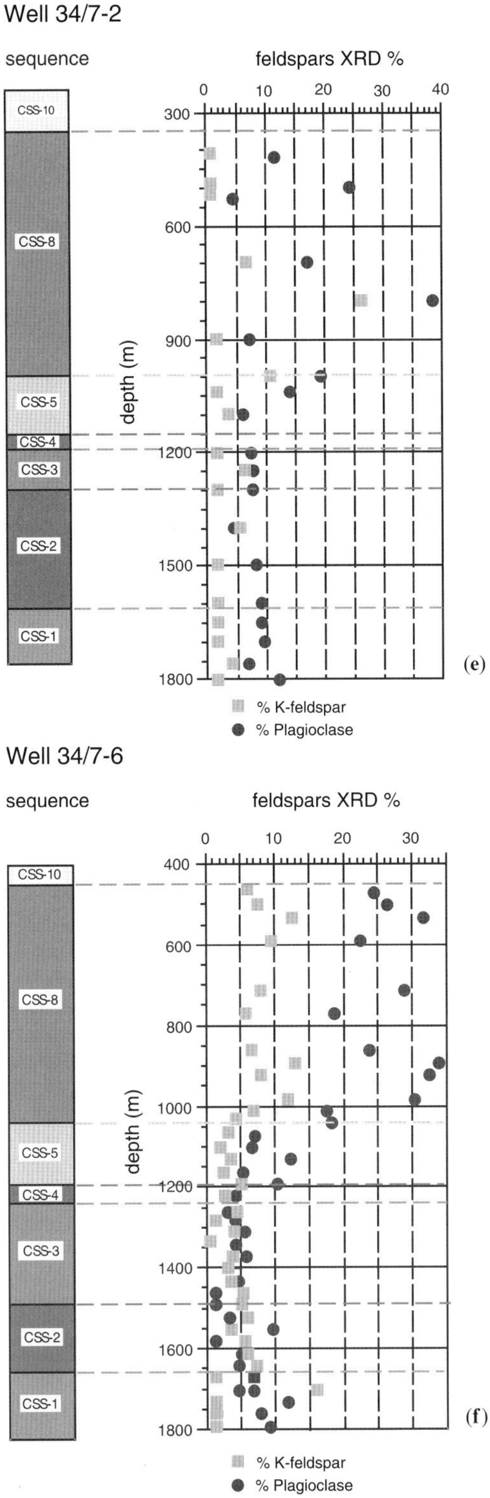

Fig. 6 (*continued*)

and a starved depositional environment (Fig. 3). The enrichment of manganese, phosphorus and zinc is therefore probably due to a high relative organic productivity, particularly of siliceous and calcareous organisms, though Sørensen & Nielsen (1981*a–c*) have interpreted the presence of rhodochrosite in Paleocene sediments as a result of the supply of Mn during volcanism.

The Balder Formation, corresponding to the upper part of sequence CSS-1, contains siliceous shales and numerous tuffaceous zones and layers (Jacqué & Thouvenin 1975; Malm *et al.* 1984). These interbedded shales and tuff layers contain a high amount of microcrystalline quartz and have a very low content of detrital minerals (Malm *et al.* 1984). Explosive volcanism in the British and the Faeroe–Greenland Tertiary volcanic provinces have been regarded as the source for the volcanic ashes found widespread in the North Sea (Malm *et al.* 1984; Knox & Morton 1988) and onshore Denmark (Spjeldnæs 1975; Nielsen & Heilmann-Clausen 1988).

Eocene sequence (CSS-2)

Eocene (CSS-2) sediments consist mostly of smectite in the study area (Figs 4 and 5). Petrographical observations show very low quartz content. Silica content has a mean value of 50 wt% (Fig. 7) and the XRD data show also little quartz (mainly less than 20 wt%) and feldspar (0–10 wt%). Magnesium has an average value of 2.5 wt% MgO. The Eocene mudrocks are rich in aluminium (15 wt% Al_2O_3), iron (9.5 wt% Fe_2O_3) and titanium (1.14 wt% TiO_2). This is typical of basic volcanic sediments, and the dry bulk composition is not far from that of Icelandic basalts (e.g. Barth 1951) and indicates that volcaniclastic material represents a dominant part of the sedimentation in CSS-2. The high content of the trace metals such as nickel, cobalt, zinc and copper (Fig. 11) is also consistent with a volcanic origin. Eocene sediments are sodium rich and this may possibly be because of the presence of authigenic albite formed in volcanic sediments. Enrichment of phosphorus and/or manganese is also seen close to the CSS-2–CSS-3 boundary or within the CSS-2 sequence in some wells, the best examples being wells 24/6-1 and 34/7-6. The fine-grained CSS-2 sediments and indications of low rates of non-volcanic clastic sediment supply indicate a relatively deep marine facies during CSS-2 times in the North Sea Basin.

The Paleocene and Eocene sediments in the North Sea are highly smectitic (Figs 4 and 5), and

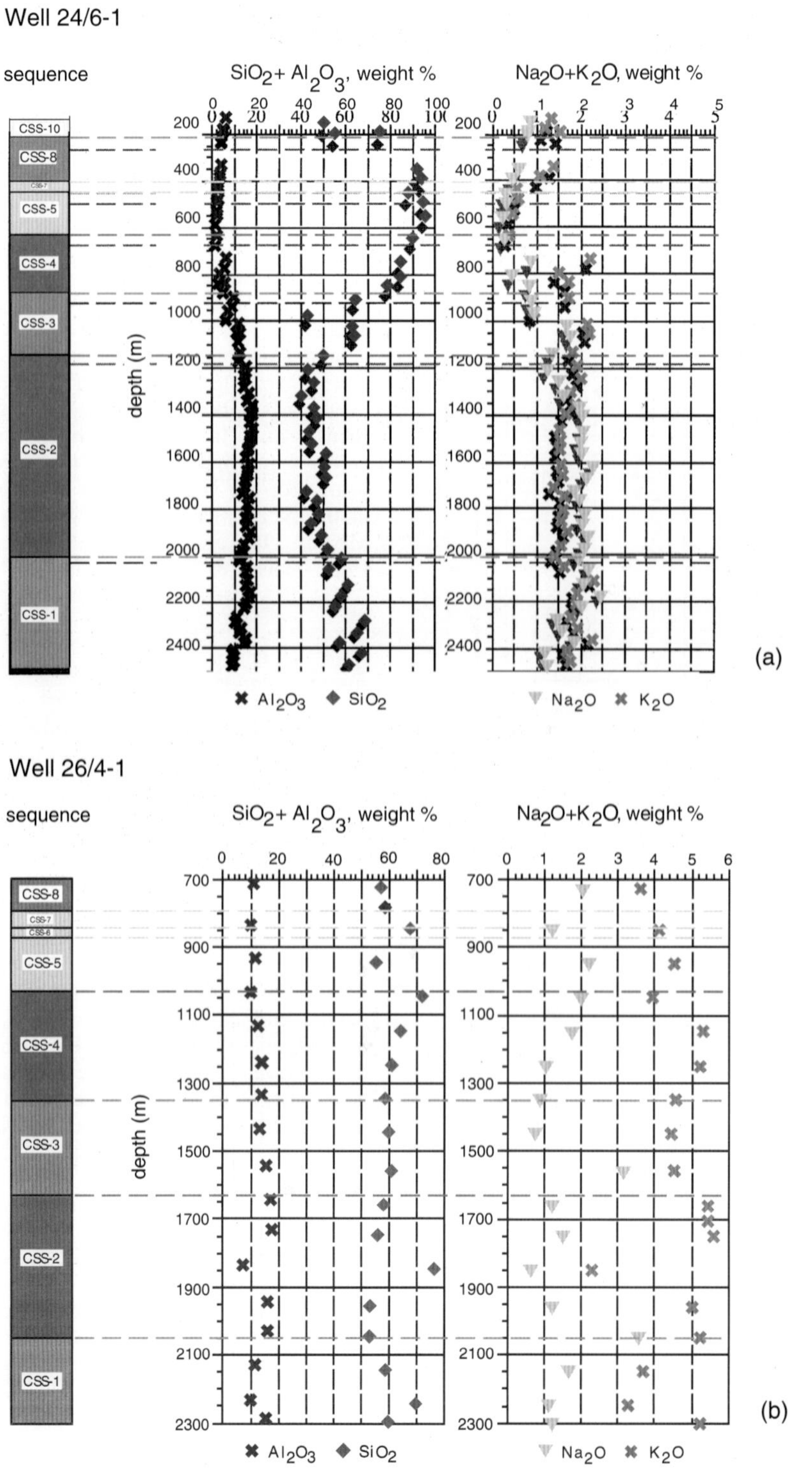

Fig. 7. Distribution of some major elements ($SiO_2 + Al_2O_3$ and $Na_2O + K_2O$) based on XRF analyses of cuttings and related to the seismic sequence stratigraphic framework. Location of wells is shown in Fig. 1. (Note the low silica content and high aluminium content in the smectite-rich Eocene sediments (CSS-2).)

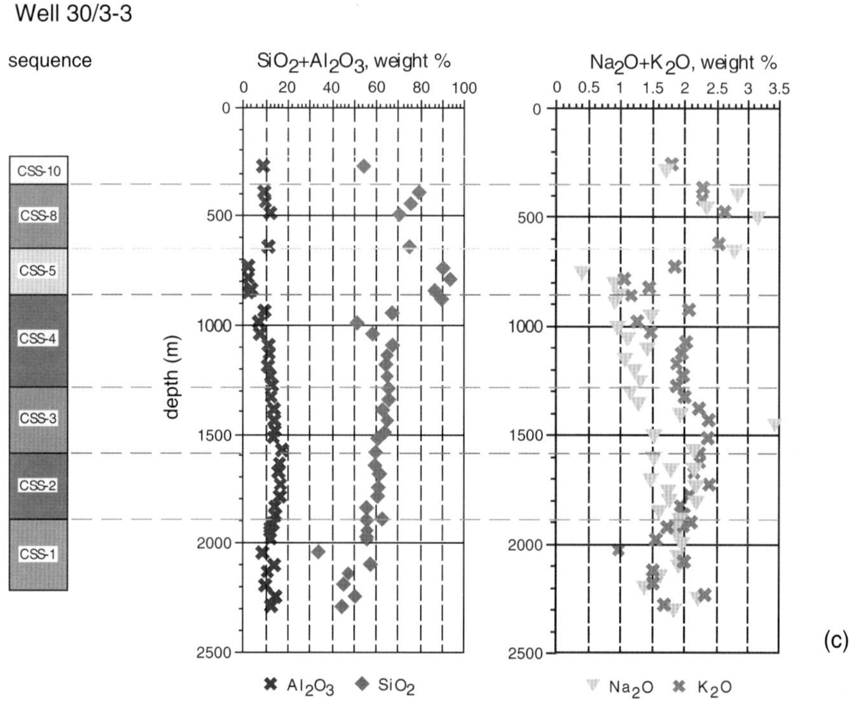

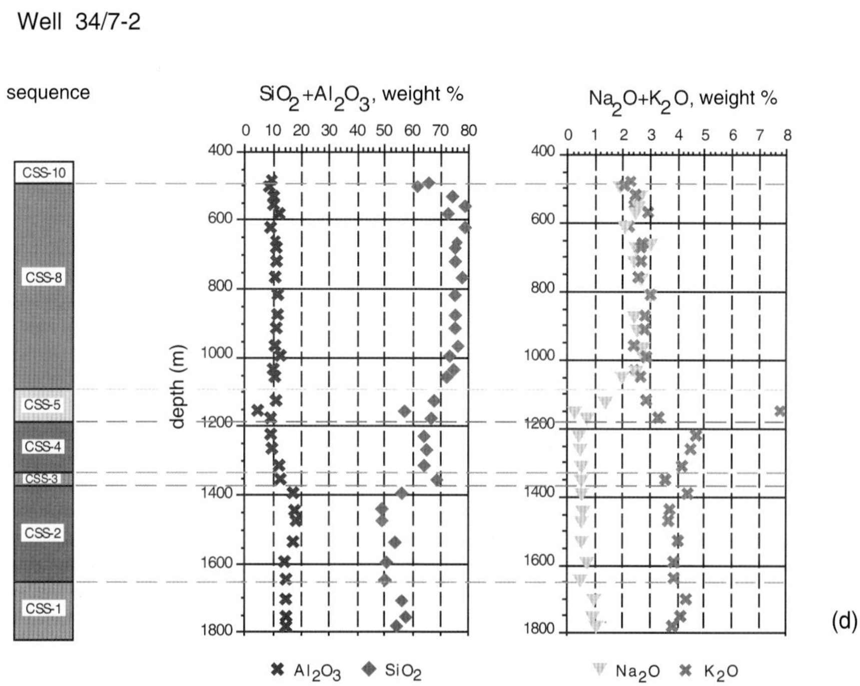

Fig. 7 (*continued*)

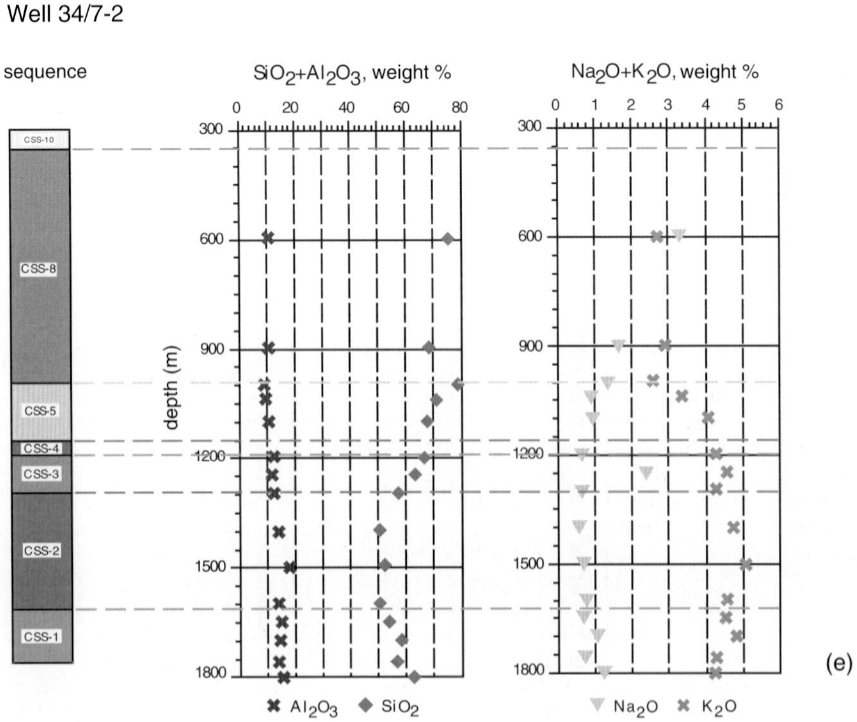

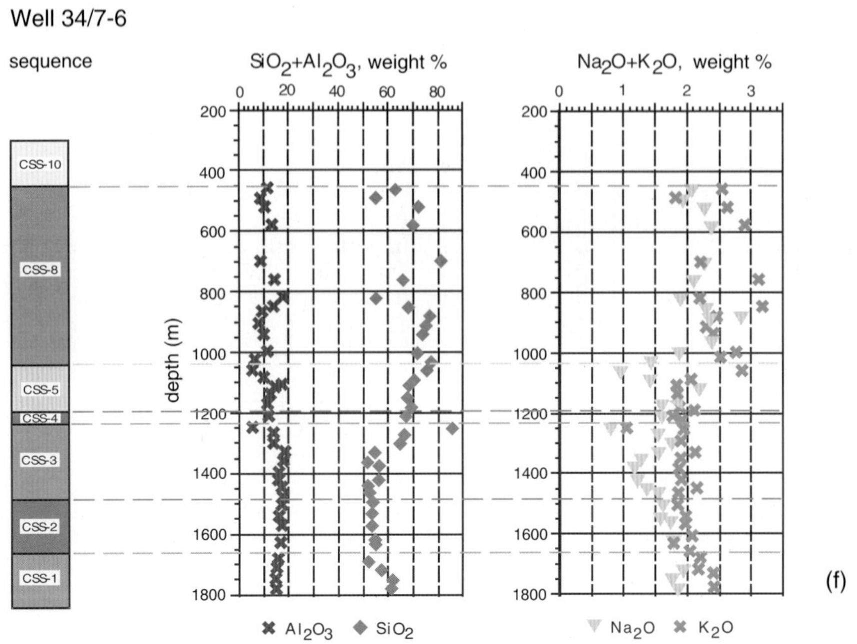

Fig. 7 (*continued*)

this has also by previous workers been related to a volcanic source (Nielsen 1974; Karlsson *et al.* 1979; Sørensen & Nielsen 1981*a–c*; Rundberg 1989; Pearson 1990; Hugget 1992). The quartz content would be expected to be higher if erosion of the adjacent landmasses was the main source of the basin sediments. The clay mineral analysis and the trace and main element data also show that this is not the case. A higher ratio of kaolinite–illite + smectite in sandstones in well 24/6-1 indicates this well to be located closer to the sediment source than the other wells in the study area (Ziegler 1982) (Figs 3 and 5).

Lower Oligocene sequence (CSS-3)

The clay mineral distribution (Figs 4 and 5) shows that smectite remains the main clay mineral component in the Lower Oligocene sediments (CSS-3). However, the content of kaolinite is higher, particularly in wells 30/3–3 and 24/6–1, which represent a more proximal facies close to an eastern and western source, respectively (Fig. 1). Well 26/4-1, located more in the distal part of the basin, has a correspondingly higher content of smectite compared with kaolinite. Thus, the regional clay mineral distribution in CSS-3 time is also controlled by the basin development and sediment transport direction as indicated in Fig. 3. In well 30/3-3 a pronounced shift in clay mineral assemblages near the CSS-2–CSS-3 boundary occurs, from a smectite-rich sediment to a clay mineral composition enriched in kaolinite above the sequence boundary (Fig. 4). However, Rundberg (1989) indicated a higher smectite content in well 30/3-3 in the depth interval corresponding to CSS-3 in Fig. 4.

The K-feldspar and plagioclase content is broadly similar and low throughout the sequence (Fig. 6). The silica content has values close to 59 wt% and Al_2O_3 content close to 13 wt%. A slight increase in silica, and corresponding decrease in alumina content, is usually seen close to the transition from Eocene to Oligocene sediments (e.g. well 30/3-3 (Fig. 7b) and 34/7-2 (Fig. 7d)), though this is not always the case, as illustrated by well 34/7-1 (Fig. 7c). The sediments in the lower part of CSS-3 in well 34/7-1 have silica content of about 48 wt%, which indicates a basic composition. In well 24/6-1 (Fig. 7a), the higher silica content is due to a more sandy lithology compared with the underlying section. The iron content in the investigated wells varies widely, between 4.4 and 8.27 wt% Fe_2O_3, and is enriched in smectitic mudstones (Fig. 8). This variation is also exhibited by titanium. Aluminium and titanium contents are, however, lower than in the Eocene sediments (Fig. 8).

The trace metal values (Ni, Co, Cu and Zn) are almost always lower than in the underlying CSS-2 (Eocene) sediments, but are still relatively high (Fig. 11). However, the presence of smectite, and the iron, titanium and trace element (Ni, Co, Cu and Zn) contents indicate a high volcanic component in the CSS-3 (Lower Oligocene) sediments. Continued input of volcanic ash falls or reworking of volcanic material is probably the explanation for these observed trends. The sedimentation rate increased during Early Oligocene time, probably in response to Oligocene uplifts and erosion of soft Eocene and Paleocene sediments. This also explains the high sedimentation rates found by Jordt *et al.* (1995, 1999) for this period. Ni and Zn sulphides are easily oxidized when the sediments are exposed, and will quickly be depleted. The reduced amount of the trace metals therefore may be an indication of reworked volcanic material. In well 34/7-1, trace elements, iron and titanium values are similar to those obtained from CSS-2 sequence (Fig. 11c). The higher content of kaolinite (Fig. 4) may indicate more sediment supply from land. Extensively reworked Early Paleocene and Early Eocene ash falls and lavas, yielding glass-rich sediments, thus occur throughout Late Eocene to Miocene time (Hugget 1992), and Pearson (1990) interpreted sediments from a volcanic source as far up as Lower Oligocene.

Lower–Upper Oligocene sequence (CSS-4)

Smectite is a major component in the clay fraction in wells 34/7-1, 34/7-2, 34/7-6, 34/7-2 (Fig. 4) and 26/4-1 (Fig. 5). Kaolinite is also an important constituent in these sediments, particularly in the proximal parts of the basin (well 24/6-1, Fig. 5) identified by progradation from the west in CSS-4 time (Figs 1 and 5). However, in well 34/7-1 these sediments have lower smectite and higher illite contents (Fig. 4). The feldspar content is approximately the same as in the underlying Oligocene and Eocene sediments (Fig. 6), with respect to both plagioclase and K-feldspar. The sandy lithology in well 24/6-1 has a relatively high silica content (Fig. 7a); otherwise the silica content is about 50 wt%. Differences in lithology and facies are also indicated in variables such as aluminium and iron contents, with the highest iron content found in well 34/7-6, which also has the highest accumulation of smectite (Fig. 4). The lowest iron content is in well 24/6-1. Otherwise average iron content for CSS-4 is about 4.4–4.9 wt%

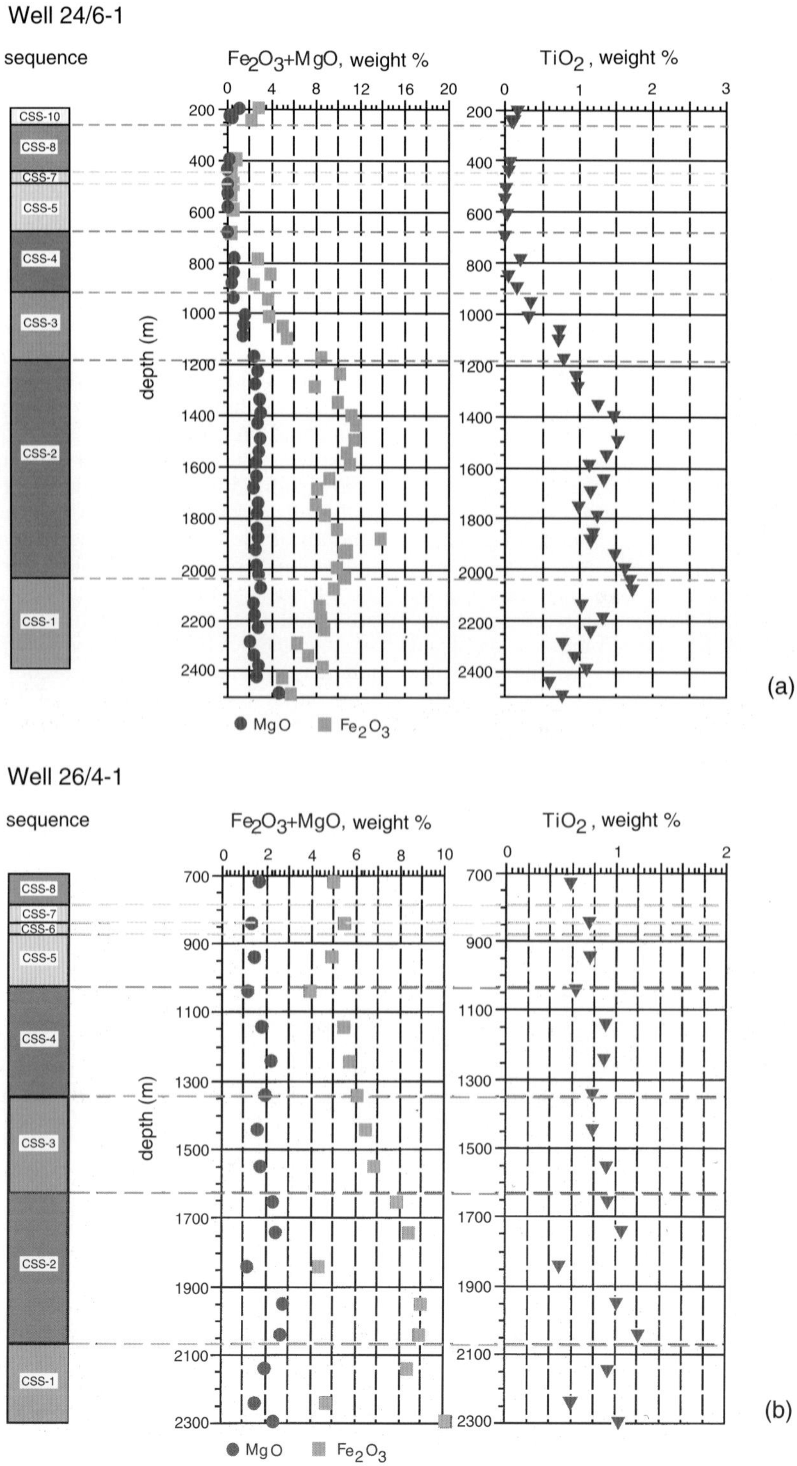

Fig. 8. Distribution of some major elements ($Fe_2O_3 + MgO$ and TiO_2) based on XRF analyses of cuttings and related to the seismic sequence stratigraphic framework. Location of wells is shown in Fig. 1. (Note the high iron content of the smectite-rich mud.)

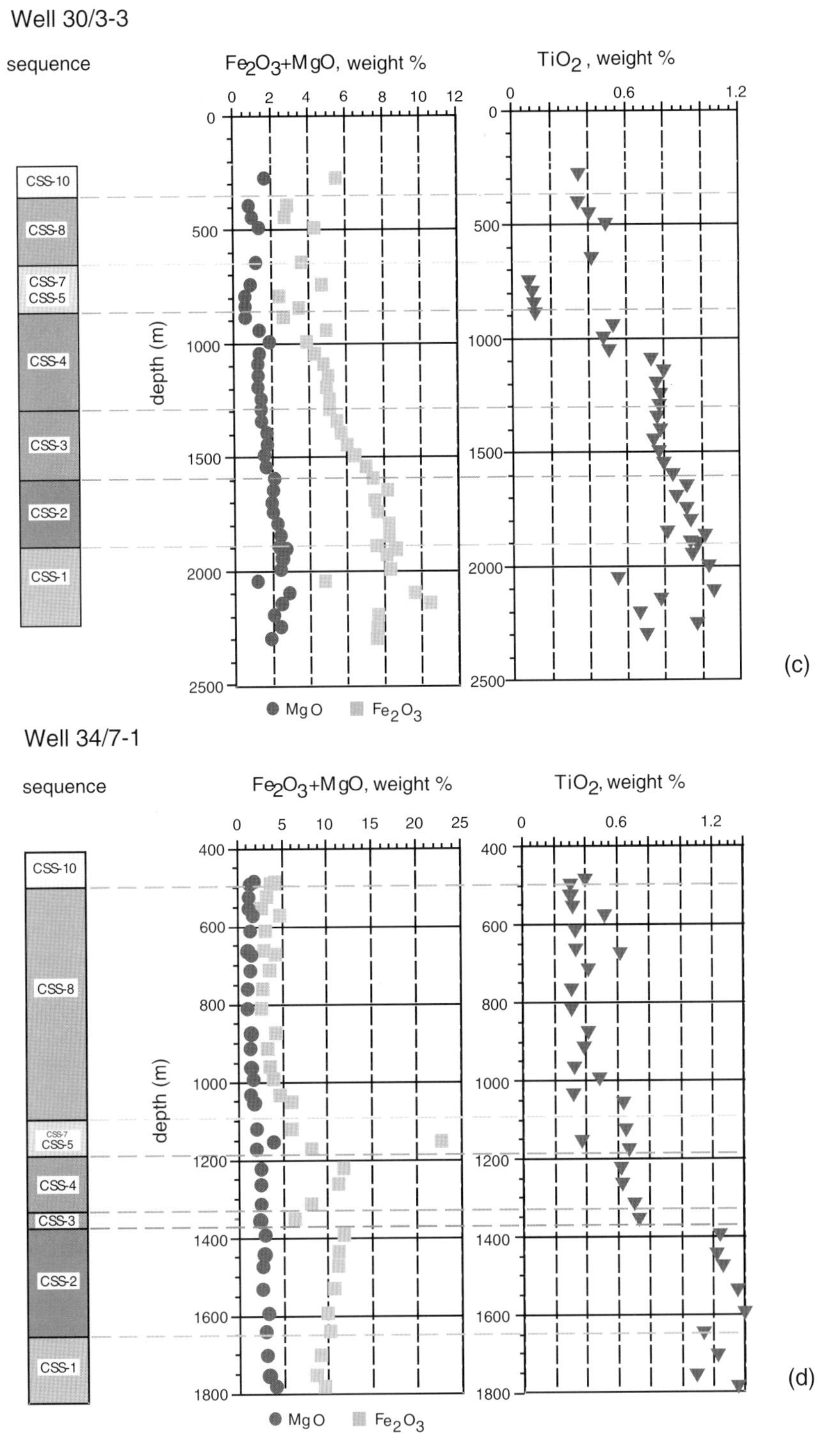

Fig. 8 (*continued*)

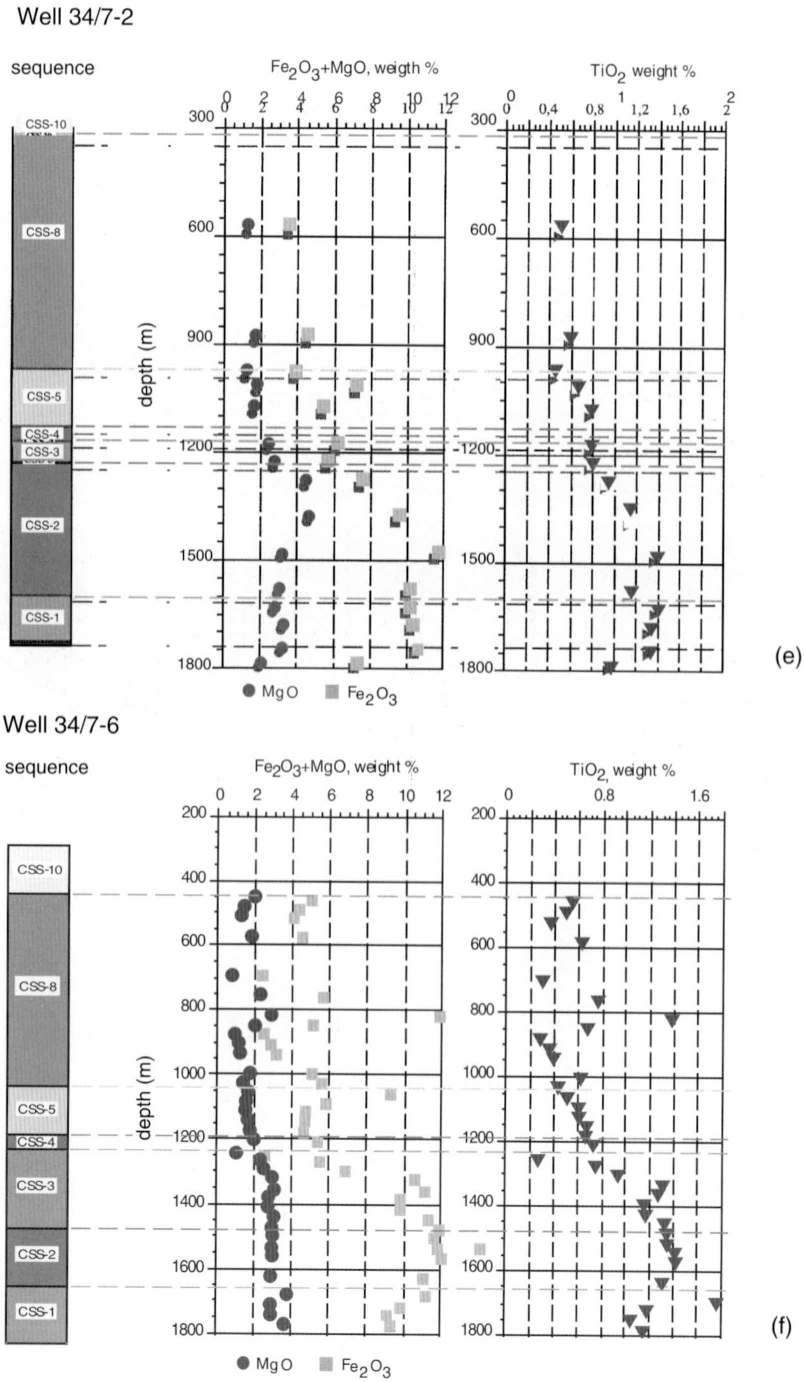

Fig. 8 (*continued*)

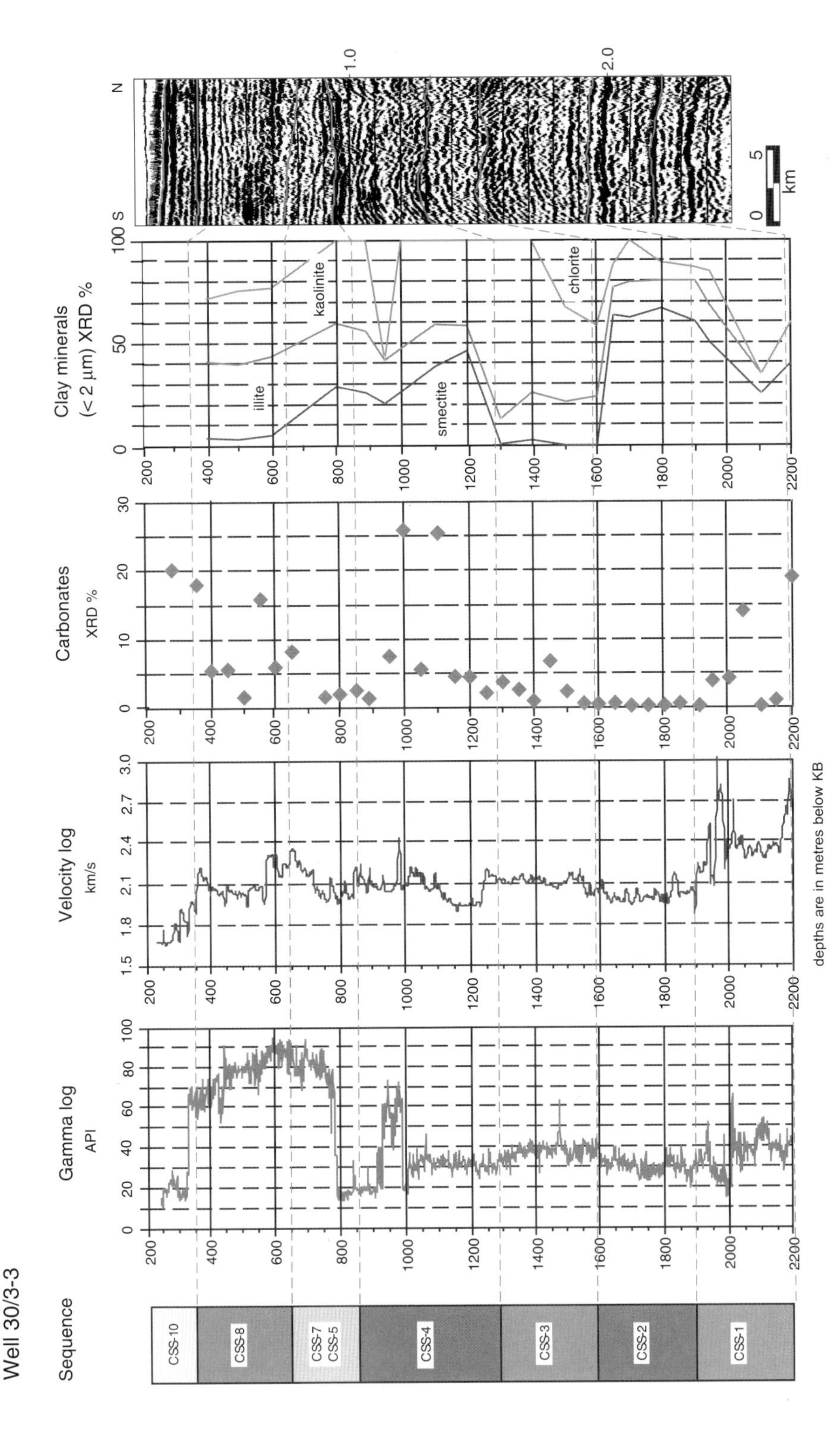

Fig. 9. Relationships between seismic stratigraphic units, velocity and clay mineralogy in well 30/3-3. Location of well and seismic line is shown in Fig. 1. (Note changes in velocity near the boundary between seismic units CSS-3 and CSS-4, probably caused by the differences in smectite content.)

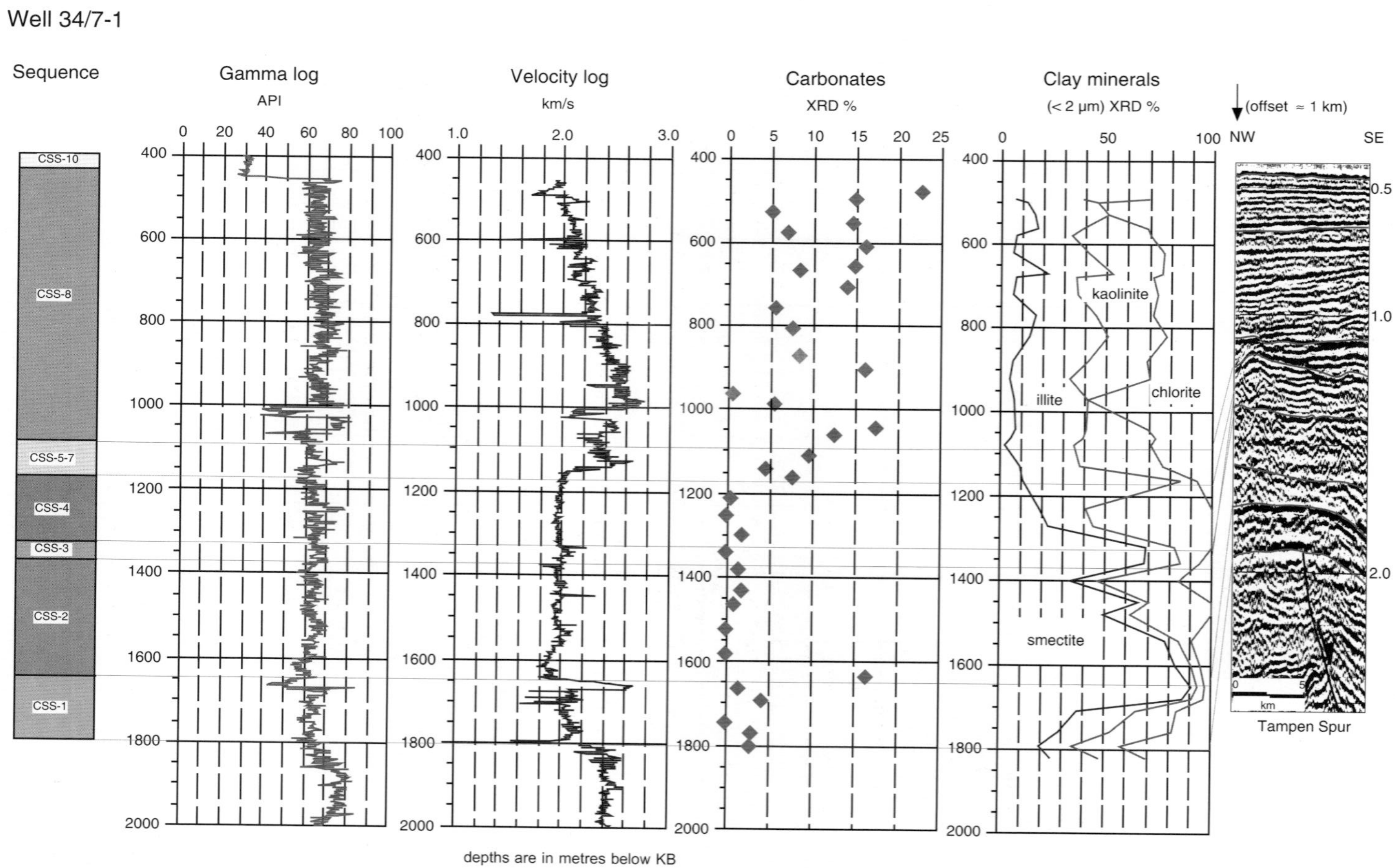

Fig. 10. Relationships between seismic stratigraphic units, velocity and clay mineralogy in well 34/7-1. Location of well and seismic line is shown in Fig. 1. It should be noted that the Pliocene sediments, which are mineralogically immature with low smectite content, show a strong increase in velocity as a result of progressive compaction. This is not the case with the Eocene and Oligocene smectite-rich sediments.

Fe_2O_3. The high CaO content in well 30/3-3 is due to carbonate-cemented intervals, as indicated by XRD (Fig. 9).

Both close to the CSS-3–CSS-4 sequence boundary and within the CSS-4 sequence there is an upward decrease in Fe, Ti and the trace metals Ni, Co, Cu and Zn, and increase in silica content (Figs 7, 8 and 11). The trace metals Co, Ni, Cu and Zn have generally slightly lower values in the CSS-4 than the underlying CSS-3 (Fig. 11); this finding also reflects that CSS-4 had a similar sediment composition to that of CSS-3 throughout the Oligocene epoch. However, the most pronounced reduction in trace metals such as Co, Ni, Cu and Zn can be seen in well 34/7-1 (Fig. 11c), which also shows a decreasing smectite content close to the CSS-3–CSS-4 boundary (Figs 4 and 10).

The highest concentration of biogenic silica is found in the basal part of sequence CSS-4 (wells 30/3-3, 34/7-1 and 34/7-6). The biogenic siliceous facies is mainly dominated by diatoms, but sponge spicules are also common and radiolaria are present as minor components (Thyberg *et al.* 1999). Thyberg *et al.* (1999), on the basis of detailed diatom flora in CSS-4, considered that the siliceous sediments may not represent the same stratigraphic horizons, indicated by the presence of diatoms both in Eocene in the northernmost studied well (36/1-2), and a low diatom content, mostly fragments, in CSS-5 (Lower Miocene) in well 26/4-1. An example of the diatom assemblages and commonly observed 'silica aggregates' is shown in Fig. 12.

The sequence CSS-4 has a rich diatom flora with the dominance of *Paralia* species, *Paralia thybergii* (Stabell 1996), possibly indicating a nearshore setting compared with the dominant *Paralia sulcata* found in coastal settings in general and in the North Sea during the present (Stabell 1985; Stabell & Lange 1990). Diatomaceous facies interbedded with glauconitic facies (Fig. 13) therefore suggest that shallower marine conditions prevailed in CSS-4 time than in earlier Cenozoic time. Poorly sorted, immature clastic and angular grains observed in well 34/7-1 also indicate a short-distance transport of clastic material, which interfingers with shallow marine biogenic and glauconitic sediments. This input of clastic fragments indicates structural local highs within, or close to, the basin. Rundberg (1989) suggested an eastern source for the sediments, on the basis of a sandy progradational system in the upper part of the Oligocene sequence in the Agat area.

An even sediment thickness in the northern North Sea Basin has been interpreted to indicate low clastic supply from the land and a reduced topographic relief onshore (Jordt *et al.* 1995, 1999). This is supported by the study of the sediment composition in the CSS-4 sequence, which indicates sediment-starved deposition in the basin. Thus, the mineralogy of CSS-4 and the aggrading seismic reflection pattern both indicate reduced clastic sediment supply. The basin configuration during Late Oligocene time was an important factor controlling the regional distribution of siliceous sediments, coupled with organic productivity, low dissolution rate, pH and a reduced clastic sediment supply.

Lower Miocene sequence (CSS-5)

A pronounced upward shift in the mineralogical and geochemical composition occurs close to the base of CSS-5 (Figs 4 and 5). Chlorite increases up-section, concomitant with a reduction in the concentration of smectite in the clay fraction (Fig. 4). Smectite-rich sediments are, however, characteristic in well 34/7-6, where CSS-5 sediments have similar clay mineralogy to the underlying CSS-4 (Fig. 4). The smectite content in this well, however, seems to decrease upwards, as well as in the CSS-5 sequence in well 30/3-3. In the southernmost wells, 24/6-1 and 26/4-1 (Fig. 5), upward enrichment of kaolinite and reduction of smectite close to the base of CSS-5 is also observed. In three wells, 34/7-1, 34/7-2 and 34/7-6, the plagioclase content in Lower Miocene sediments is relatively high (Fig. 6) compared with that in the underlying sequences. This is, however, not found in well 30/3-3 (Fig. 6b) or well 26/4-1 (Fig. 6a). The variations in K_2O/Na_2O ratio also reflect these regional variations with respect to the feldspars.

The silica content is high, with a mean value of 75 wt%, and the Al_2O_3 content is close to 6.85 wt% in CSS-5 (Fig. 7). The Ti, Mg and Fe contents are generally low (Fig. 8), with mean values of 0.4 wt% TiO_2, 1.0 wt% MgO and 4.3 wt% Fe_2O_3, respectively.

Trace metals (Co, Ni, Zn and Cu) often show a weak reduction in concentration close to the base of the Lower Miocene sequence (Fig. 11). However, this is not observed in well 26/4-1 (Fig. 11b). Low values suggest that volcanic influence is low in these sediments.

Eastward decreasing kaolinite content in wells 24/6-1 and 26/4-1 (Fig. 5) indicates that clastic sediments were sourced from the East Shetland Platform in CSS-5 time, in accordance with the seismic data marking outbuilding from the west in the area around 60°N, into the Viking Graben (Fig. 1).

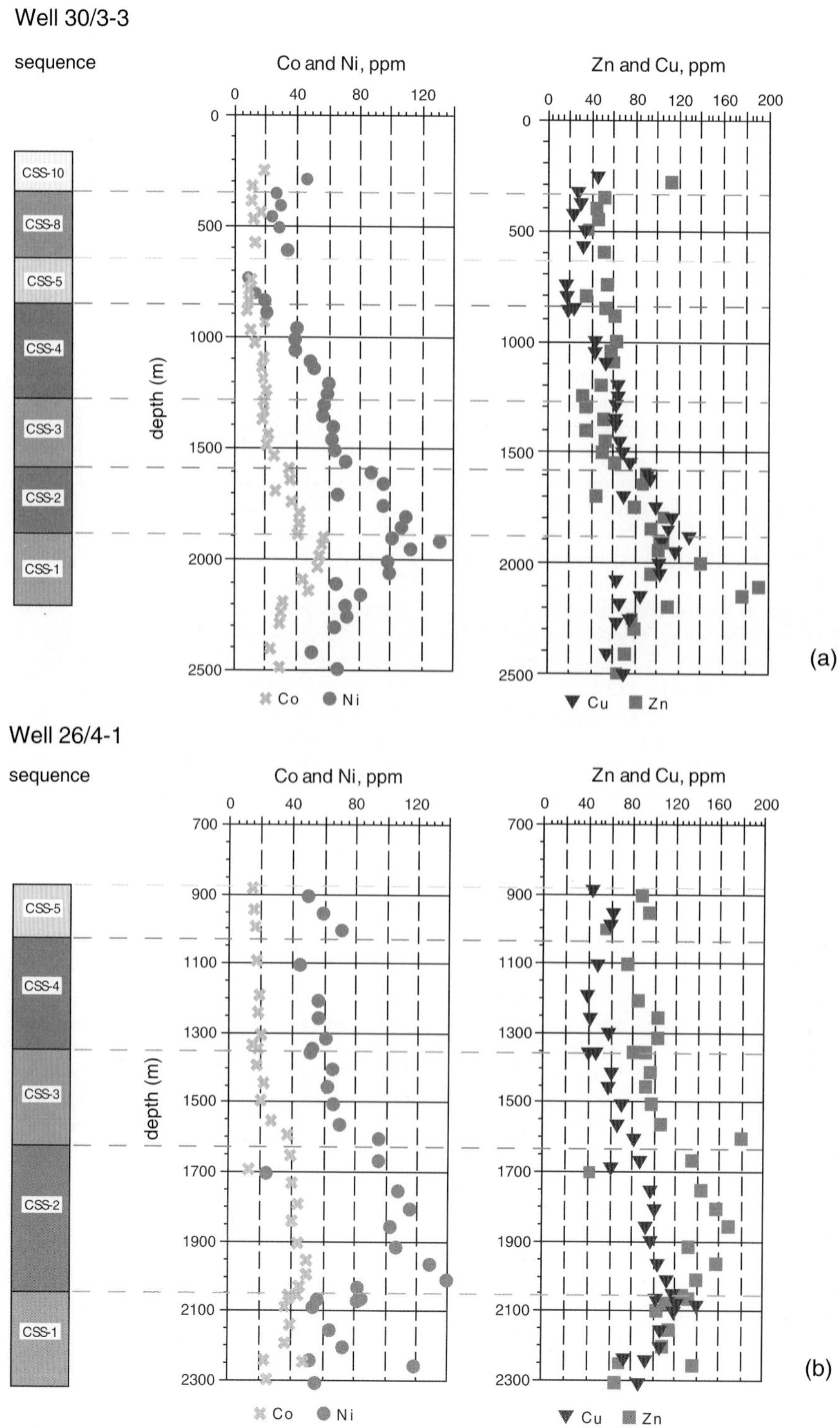

Fig. 11. Distribution of trace elements (Co, Ni, Cu and Zn) based on XRF analyses of cuttings. Location of wells is shown in Fig. 1. (Note the relatively high contents of most of these elements in the partly volcanic Eocene and Oligocene sediment.)

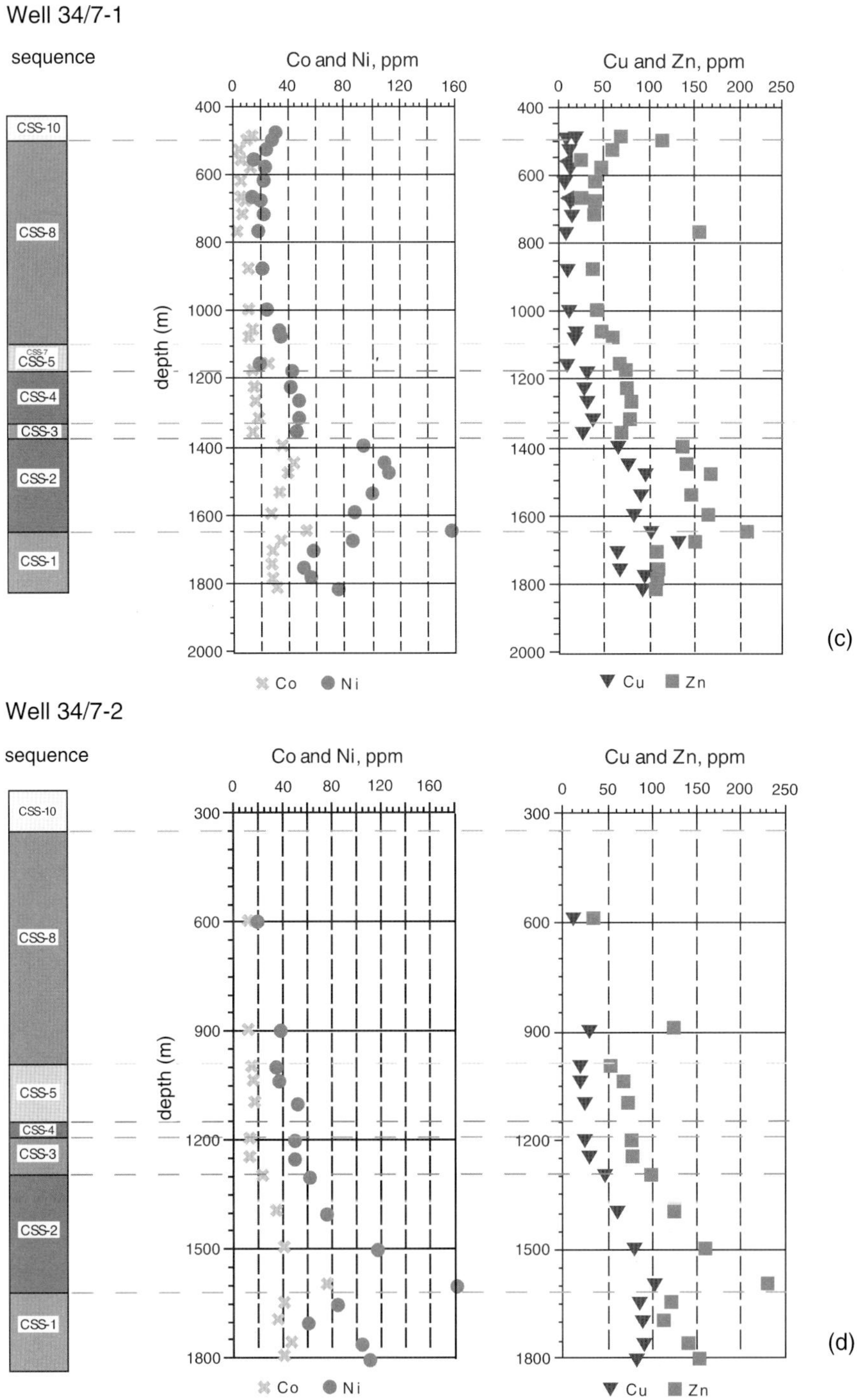

Fig. 11 *(continued)*

Fig. 12. Scanning electron micrographs of the most common diatoms in Upper Oligocene sediments in well 30/3-3.

Middle–Upper Miocene sequence (CSS-6 and CSS-7)

CSS-6 is mapped only in the southern part of the Norwegian North Sea, and thins below seismic resolution in a northward direction. Some biostratigraphic investigations have indicated that sediments of CSS-6 age are absent in the northern North Sea (e.g. Eidvin & Riis 1992); however, other studies (e.g. Gradstein & Bäckström 1996) have indicated that CSS-6 sediments actually are present there. The northward thinning of CSS-6 indicates that CSS-6 is absent in the northern North Sea; therefore, we have grouped possible CSS-6 sediments together with CSS-5.

The base of CSS-7 is characterized by a pronounced downlap surface generated by sediments building out from the east in the central North Sea. The thickness of CSS-7 is close to seismic resolution farther to the north, but biostratigraphic evidence from quadrant 34 (Steurbaut *et al.* 1991; Eidvin & Riis 1992) indicates that it is present locally. We have, however, not been able to distinguish between these sediments of CSS-7 age and the underlying CSS-5 on the seismic data in the northern North Sea. However, gamma-ray log response in the investigated wells often shows a marked shift in the upper part of the Miocene succession, suggesting a subdivision into two sequences (Fig. 10). We use this shift in log response to indicate the base of CSS-7. We have also used gamma-ray and sonic log data, in addition to biostratigraphic data, to identify the top of the Miocene sequence; this is often recognized by a marked upward increase in gamma-ray log response (Figs 9 and 10) and in some wells by an increase in the seismic velocity causing pronounced velocity inversion at the base Pliocene (Fig. 10). We have no mineralogical and geochemical data from CSS-7 in this study.

Pliocene (CSS-8) and Pleistocene sequences (CSS-9 and CSS-10)

The CSS-8 sequence has been investigated in wells 30/3-3, 34/7-1, 34/7-2 and 34/7-6. We therefore have only mineralogical and geochemical data from the northernmost wells (Fig. 1). The little Pleistocene sediment analysed appears to be similar to the mineralogical composition of the Pliocene sediments.

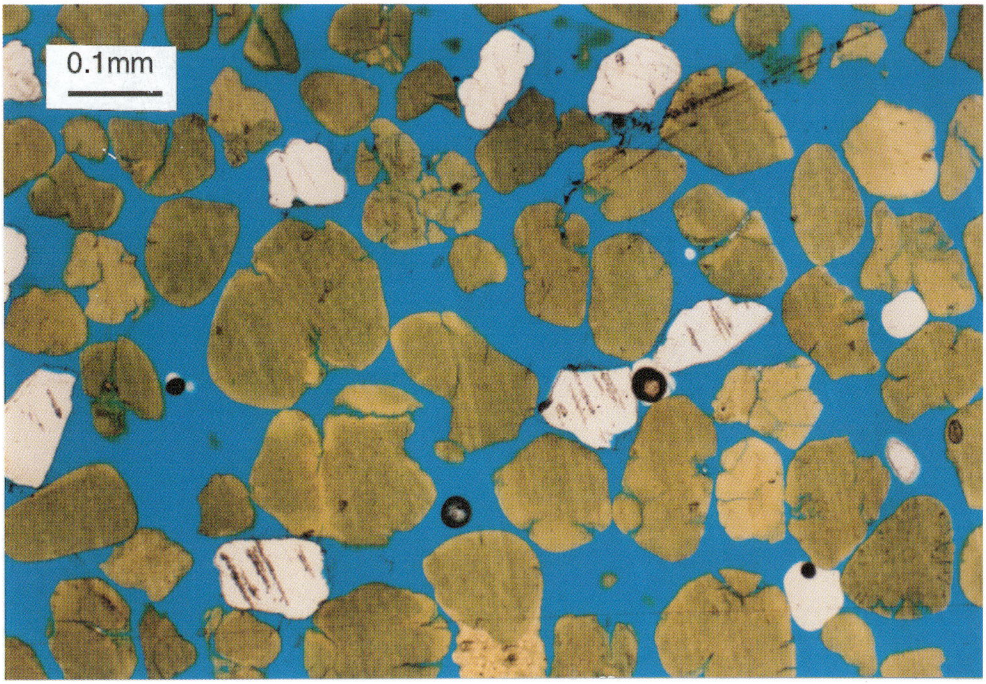

Fig. 13. Glauconite sand grains from well 34/7-1 indicating slow sedimentation, possibly on a submarine high.

The Pliocene and Pleistocene sequences are 'coarse-grained mudstones' with little smectite and enrichment of chlorite, illite and to a certain degree kaolinite (Fig. 4). Much of the kaolinite found may be taken as an indication of the degree of reworking of older sediments, as very little kaolinite would have been produced when the climate was cold and erosion rates were high. Much of what is recorded as illite by XRD in these layers is actually mica derived from relatively unweathered basement rocks. A relatively high chlorite content in the sediments is evidence of limited weathering in a cold climate in a region undergoing glacial erosion.

In the lower Tertiary sequence, much of the chlorite may be present as authigenic minerals formed by the alteration of volcanic glass and smectite or basic rock fragments, and can therefore be a possible source for CSS-8 sequences. However, the iron content in CSS-8 is normally relatively low compared with that in the lower part of the Cenozoic section (Fig. 8c–f), as was also observed by Rundberg (1989). Chlorite with a low iron and high Mg content may represent a metamorphic source (Curtis *et al.* 1985). The Pliocene and Pleistocene sediments have probably been derived from metamorphic basement rocks from western Norway. Karlsson *et al.* (1979) have also reported detrital chlorite in a study of well 2/11-1 in the central North Sea.

The highest silica content is found in the Pliocene and Pleistocene sediments and, coupled with low alumina (Fig. 7d–f), gives also strong indications that Pliocene and Pleistocene sediments have a different composition from those of the lower Tertiary sequence. Generally, the highest content (>20%) of plagioclase is almost always limited to CSS-8 (Fig. 6c–e). Lower amounts of K-feldspar (generally <10%) are also characteristic for CSS-8. Trace element (Ni, Co, Cu and Zn) values in the sequence are generally low (Fig. 11c and d), and also show an increasing trend towards the top of CSS-8. CSS-8 contains relatively fresh feldspar (e.g. albite), poorly sorted, crystalline rock fragments, volcanic rock fragments, mica, amphiboles, pyroxenes and hornblende gneisses within a muddy matrix and Danian chalk fragments (F. Gradstein, pers. comm.). Kaolinite and smectite indicate reworking of lower Tertiary and Mesozoic rocks or erosion of the basement. The Pliocene sediments are plagioclase rich, probably from albite gneisses and basic rocks from Western Norway.

Calcite is the dominant carbonate mineral, although siderite and ankerite–dolomite are also

Fig. 14. Clast of chalk in Pliocene sediment in well 36/1–2. The clast contains spherical calcareous fossils (calcisphera) typical of Danian chalk (from Singh 1996).

present in significant quantities. This is partly as contemporaneous biogenic carbonate but includes reworked Danian chalk fragments, microfossils (e.g. Foraminifera), shell debris and fragments. Occurrence of Danian chalk fragments (Fig. 14) as far north as well 36/1-2 shows that chalk facies extended into the northern North Sea and covered areas where late Neogene erosion reached into the basement. The high content of carbonate recorded by XRD (Figs 9 and 10) is therefore not mainly carbonate cement, but occurs mostly as reworked carbonate clasts.

The immature grains, and the angular and poorly sorted mineralogy of the CSS-8 to CSS-10 sequences indicate high erosion rates in a relatively cold climate prevailing in the northern North Sea. CSS-8 has a high content of glacial-induced eroded material and represents glacimarine deposits, with a proximal source area. Eidvin & Riis (1992) observed glacial material in Pliocene sediments in the northern North Sea, and suggested that these sediments are in close proximity to the provenance area. The high sediment input rate has been related to glacial erosion contemporaneous with tectonic uplift of Norway in Pliocene time (Eidvin & Riis 1992; Stuevold *et al.* 1992; Jordt *et al.* 1995, 2000).

Mineralogical changes associated with sequence boundaries

The mineralogical and geochemical data have been compared with seismic data for selected wells (Figs 5, 9 and 10); the clay mineralogy and carbonate distribution has been correlated with the gamma-ray and velocity logs of wells 30/3-3 and 34/7-1.

The Paleocene to Oligocene (CSS-1 to CSS-4) sediments in well 26/4-1 have a generally high content of smectite (Fig. 5). A marked shift in clay mineral composition occurs close to the CSS-4–CSS-5 sequence boundary, showing an upward reduction in smectite content. The CSS-1 to CSS-4 sequences are all smectite rich, although there are some minor variations in the clay mineral composition correlating with the seismic sequence boundaries (Figs 4 and 5). However, some minor shifts in clay mineral compositions also occur within the sequences, indicating a composite lithology of the source area.

Sediments corresponding to CSS-1, CSS-2 and partly CSS-3 in well 24/6-1 (Fig. 5) have a relatively high content of smectite. Kaolinite is, however, also an important constituent in the lower part of the Tertiary section (CSS-1 to CSS-4). A reduction in smectite content occurs close to the CSS-3–CSS-4 sequence boundary. The CSS-2–CSS-3 sequence boundary can to a certain degree also be correlated with a shift in clay mineral composition indicated by gradually increased kaolinite content.

A lateral decrease in the kaolinite/smectite ratio in CSS-1 to CSS-4 occurs from well 24/6-1 to well 26/4-1 (Fig. 5), indicating that the Upper Paleocene to Lower Miocene sediments of the most easterly well (26/4-1) were deposited in a more distal position. High kaolinite–smectite to illite ratios in the proximal facies are consistent with the proposed general relationship between clay mineralogy and facies and sea-level changes (Fig. 3), and with the outbuilding pattern indicated by the seismic data (Fig. 5). The clay mineral distribution of CSS-1 prograding from western Norway shows a similar, increasing kaolinite–smectite to illite ratio towards the depocentre (Rundberg 1989), indicating that these sediments were derived from the east.

An increased smectite and carbonate content is observed near the base of CSS-1 in well 30/3-3 (Fig. 9). Close to the sequence boundary between CSS-2 and CSS-3, there is an apparently marked decrease in smectite and an increase in the kaolinite content upwards. We find, however, less smectite in our samples from well 30/3-3 than reported by Rundberg (1989). A marked increase in smectite content is observed across the CSS-3–CSS-4 sequence boundary.

Marked changes in the clay mineral assemblages across seismic sequence boundaries are also observed in well 34/7-1 (Fig. 10). The top of CSS-1 correlates with a smectite peak and the top of CSS-2 is reflected by a decrease in smectite content in well 34/7-1. CSS-3 has a generally high content of smectite; a marked fall is, however, indicated near the base of the overlying CSS-4. The base of CSS-4 is reflected by a marked upward increase in the content of kaolinite.

Chlorite is found at several levels in the lower Tertiary sequence of the investigated wells (Figs 4, 5, 9 and 10), and occurs apparently independently of seismic sequence boundaries. However, the chlorite content of the Miocene to Pleistocene sequences cannot be explained by diagenetic changes, and the variations in the occurrence of chlorite correlate with the mapped sequence boundaries (Figs 4, 5, 9 & 10). An increase in clastic chlorite of the Miocene and Pliocene–Pleistocene sediments is probably related to provenance changes, high erosion rates and a reduced rate of weathering, reflecting a colder climate in the source area. For of the same reasons, CSS-8, and sometimes, the CSS-5 sequence boundaries can also be rather easily identified on the basis of the feldspar content, as increased plagioclase content close to base CSS-8–CSS-5 is seen in the northern North Sea.

A correlation between the major element distribution and the sequence boundaries is demonstrated by the Si, Al, Fe and Ti contents, and to a certain extent by the Na and K variations (Figs 7 and 8). In well 24/6-1 the increased concentration in Si is seen close to the CSS-2–CSS-3 and CSS-3–CSS-4 sequence boundaries (Fig. 7a). The upward increase in Si content corresponds to the CSS-3–CSS-4 and CSS-5–CSS-8 sequence boundaries in well 34/7-1 (Fig. 7d), and in well 30/3-3 (Fig. 7c) is close to the CSS-4–CSS-5 sequence boundary. The Al content shows an upward decrease in concentration, as exemplified by well 24/6-1 (Fig. 8a) and well 30/3-3 (Fig. 7c). A reduced concentration in Al content is seen close to the CSS-3–CSS-4 boundary in well 34/7-1 (Fig. 7d). The highest concentration of Fe and Ti is found in CSS-2, CSS-3 and sometimes also in CSS-4, and concentration variations in Fe and Ti are almost always following the same trend throughout the Cenozoic section and across sequence boundaries (Fig. 8). A reduction in Ti and also sometimes Fe content close to the CSS-3–CSS-4 boundary is demonstrated in well 34/7-1 (Fig. 8d) and close to the base CSS-5 boundary in well 30/3–3 (Fig. 8c). The differences in Na_2O and K_2O concentration also occur across the sequence boundaries, as demonstrated in well 34/7-1 (Fig. 7d) and well 30/3-3 (Fig. 7c).

Significant variations in trace elements, exemplified by Ni, Co, Cu and Zn, can be observed in the Tertiary sediments, as shown in Fig. 11, and they largely follow the same trend throughout the Cenozoic section, probably because these elements are enriched in volcanic material. Variations in the concentrations of Ni, Co, Cu and Zn occur across the sequence boundaries, and high Ni and Zn values are observed close to the CSS-1–CSS-2 boundary. Such very high Zn values may also be related to enrichment of silica-producing organisms. Enrichment of biogenic silica in the Balder Formation (top CSS-1) has been reported in the literature (e.g. Malm *et al.* 1984). Close to the top of the Balder Formation, accumulation of carbonate may be due to high organic productivity (Figs 9 and 10), particularly of siliceous and calcareous organisms. The upward decrease in trace element concentration (Co, Ni, Zn and Cu) is particularly demonstrated close to the CSS-2–CSS-3 boundary in well 30/3-3 (Fig. 11b) and well 26/4-1, and across the CSS-3–CSS-4 boundary in well 34/7-1 (Fig. 11c).

The lowermost siliceous facies in the Oligocene sequences close to the CSS-3–CSS-4 boundary represents a starvation with respect to clastic supply (Thyberg *et al.* 1999).

The Pliocene mudstones are coarse grained and poorly sorted with an immature mineralogy, and therefore probably rather permeable and have undergone rapid compaction compared with the Eocene and Oligocene smectitic mudstones. This is probably the main reason for the high sonic velocity (well 34/7-1, Fig. 10). Smectitic mudstones are extremely fine grained and have high tortuosity, high specific surface and low permeability, and compact very slowly (Weaver 1989). The low sonic velocities (Figs 9 and 10) may be partly due to the relatively high porosity and partly to the nature of the grain contacts in such fine-grained rocks. Overpressure primarily occurs where there is a thick sequence of smectitic mudstones without laterally extensive sandstone bodies to provide lateral drainage. This is also apparent from published pressure data from the North Sea (Buhrig 1989). The gamma-ray logs show a pronounced reduction, which correlates with dramatic increase in velocity close to the CSS-1 boundary. This shift has also been reported regionally (Nielsen *et al.* 1986; Isaksen & Tonstad 1989; Morton & Knox 1990). Carbonate cementation and the transformation of opal-A to opal-CT may cause the high seismic velocities.

Summary and conclusions

The main factors influencing the sediment composition in basins such as the North Sea are the clastic supply to the basin and the biogenic production. The volcaniclastic input is very dominant in the Eocene sediments and this is related to volcanism associated with the breakup and initial opening of the Norwegian–Greenland Sea, and a transgression over western Norway reducing the clastic sediment supply. Paleocene (CSS-1) sediments are probably, to a large extent, derived from the erosion of Mesozoic sediments. The volcanic component in the Eocene sequence (CSS-2) is rather distinctive and there is also some evidence for sediment starvation.

An increased rate of input of coarser-grained material in Early Oligocene (CSS-3) time occurred in the North Sea Basin, indicated by an increase in kaolinite and higher SiO_2 values. However, the volcanic input is still distinctive in CSS-3 and CSS-4 time, perhaps partly representing reworked volcanic Eocene material. During Oligocene time (CSS-4 and CSS-3), the basin configuration changed, and in late Oligocene time, a glauconitic facies in CSS-4 sediments indicates a transgression and sediment starvation.

Significant shifts in the sediment composition occurred at the Miocene–Pliocene transition, which coincides with a change towards a colder climate. Lower Miocene (CSS-5), Pliocene and Pleistocene sediments in the northern part of the study area possess a relatively immature mineralogy, including clastic chlorite and high content of plagioclase, reflecting a colder climate and a reduced rate of weathering in the source regions. A significant kaolinite and smectite content in these sediments is probably not due to contemporaneous weathering but may indicate the reworking of lower Tertiary or Mesozoic rocks during Late Cenozoic time. Prediction of regional variations in geophysical properties and rates of compaction can be carried out only if the primary mineralogical composition is known.

In this paper we have established a correlation between the seismic sequence stratigraphic framework of the Cenozoic succession in the northern North Sea and variations in the mineralogical and geochemical composition. Changes in the provenance areas and sediment supply, and selective erosion of different source rocks and sediments may explain the variations in sediment composition. This can be related to tectonic processes affecting the North Sea Basin and surrounding areas (Jordt *et al.* 1995, 2000). The regional clay mineral distribution is controlled by source and facies changes throughout the development of the North Sea Basin, which is probably controlled by tectonic movements and changes in sediment supply and distance from the source.

The mapped seismic sequences do not only reflect depositional processes. The primary mineralogical composition of the sediments and burial diagenetic processes strongly influence the seismic velocity and seismic attributes. The rate of compaction also influences the geometry of the seismic sequences.

An integration of mineralogy, geochemistry and sequence stratigraphy has therefore the potential for correlation in the North Sea Basin on local and regional scales.

This work was funded by the Commission of the European Union and the Norwegian Research Council (NFR) in the framework of the DGXII-Joule Programme, sub-programme: Energy From Fossil

Sources: Hydrocarbons, Integrated Basin Studies – The Dynamics of the Norwegian Margin. Norsk Hydro AS, Saga Petroleum AS, Statoil and Norwegian Petroleum Directorate are gratefully acknowledged for providing samples. R. G. Simmons provided interpretation of clay mineralogy of wells 26/4-1, 30/3-3 and 34/7-2. We would like to thank B. Stabell regarding discussion of diatoms. V. Cadman kindly corrected the manuscript.

References

BARTH, T. F. W. 1951. *Theoretical Petrology. A Textbook on the Origin and the Evolution of Rocks.*

BJØRLYKKE, K. 1974. *Depositional History and Geochemical Composition of Lower Paleozoic Epicontinental Sediments from the Oslo Region, Norway.* Norges Geologiske Undersøkelse Bulletin, **305**.

BUHRIG, C. 1989. Geopressured Jurassic reservoirs in the Viking Graben: modelling and geological significance. *Marine and Petroleum Geology*, **6**, 31–48.

CURTIS, C. D., HUGHES, C. R., WHITEMAN, J. A. & WHITTLE, C. K 1985. Compositional variations within some sedimentary chlorites and some comments on their origin. *Mineralogical Magazine*, **49**, 375–386.

DEEGAN, C. E. & SCULL, B. J. 1977. *A Standard Lithostratigraphic Nomenclature for the Central and Northern North Sea.* Norwegian Petroleum Directorate Bulletin, **1**.

EIDVIN, T. & Riis, F. 1992. *En biostratigrafisk analyse av sedimenter over og under basal pleistocen, regionale vinkeldiskordans i nordøstlige deler av Nordsjøen.* Norwegian Petroleum Directorate Contribution, **28**.

FISHER, M. J. & JEANS, C. V. 1982. Clay mineral stratigraphy in the Permo-Triassic Red Beds sequences of BNOC 72/10-1A, Western-Approaches, and the south Devon coast. *Clay Minerals*, **17**, 79–89.

GALLOWAY, W. E. 1989. Genetic stratigraphic sequences in basin analysis I: Architecture and genesis of flooding-surface bounded depositional units. *AAPG Bulletin*, **73**, 125–154.

——, GARBER, J. L., LIU, X. & SLOAN, J. 1993. Sequence stratigraphic and depositional framework of the Cenozoic fill, Central and Northern North Sea. *In*: PARKER, J. R. (ed.) *Petroleum Geology of Northwest Europe: Proceedings of the 4th Conference.* Geological Society, London, 33–43.

GIBBS, R. J. 1977. Clay mineral segregation in the marine environment. *Journal of Sedimentary Petrology*, **47**, 237–243.

GLASMANN, J. R. & WILKINSON, G. C. 1993. Clay mineral stratigraphy of Mesozoic and Paleozoic red beds, Northern North Sea. *In*: PARKER, J. R. (ed.) *Petroleum Geology of Northwest Europe: Proceedings of the 4th Conference.* Geological Society, London, 625–636.

GRADSTEIN, F. & BÄCKSTRÖM, S. 1996. Cainozoic biostratigraphy and paleobathymetry, northern North Sea and Haltenbanken. *Norsk Geologisk Tidsskrift*, **76**, 3–32.

HENSON, M. R. 1974. Clay minerals from the Lower New Red Sandstone of south Devon. *Proceedings of the Geologists' Association, London*, **84**, 429–445.

HUGGET, J. M. 1992. Petrography, mineralogy and diagenesis of over pressured Tertiary and Late Cretaceous mudrocks from the East Shetland Basin. *Clay Minerals*, **27**, 487–506.

ISAKSEN, D. & TONSTAD, K. 1989. *A Revised Cretaceous and Tertiary Lithostratigraphic Nomenclature for the Norwegian North Sea.* Norwegian Petroleum Directorate Bulletin, **5**.

JACQUÉ, M. & THOUVENIN, J. 1975. Lower Tertiary tuffs and volcanic activity in the North Sea. *In*: WOODLAND, A. W. (ed.) *Petroleum and the Continental Shelf of North-West Europe; Geology.* Wiley, New York, 455–465.

JONES, R. W. & MILTON, N. J. 1994. Sequence development during uplift: Paleogene stratigraphy and relative sea-level history of the Outer Moray Firth, UK North Sea. *Marine and Petroleum Geology*, **11**, 157–165.

JORDT, H., FALEIDE, J. I., BJØRLYKKE, K. & IBRAHIM, M. T. 1995. Cenozoic sequence stratigraphy in the Central and Northern North Sea: tectonic development, sediment distribution and provenance areas. *Marine and Petroleum Geology*, **12**, 845–879.

——, ——, THYBERG, B. I. & NØTTVEDT, A. 2000. Cenozoic evolution of the central and northern North Sea with focus on differential vertical movements of the basin floor and surrounding clastic source areas. *This volume.*

KARLSSON, W., VOLLSET, J., BJØRLYKKE, K. & JØRGENSEN, P. 1979. Changes in mineralogical compostion of Tertiary sediments from North Sea wells. *In*: MORTLAND, M. M & FARMER, V. C. (eds) *International Clay Conference 1978, Developments in Sedimentology*, **27**, 281–289.

KNOX, R. W. O'B. & MORTON, A. C. 1988. The record of early Tertiary N Atlantic volcanism in sediments of the North Sea Basin. *In*: MORTON, A. C. & PARSON, L. M. (eds) *Early Tertiary Volcanism and Opening of the NE Atlantic.* Geological Society, London, Special Publications, **39**, 407–419.

LOUTIT, T. S., HARDENBOL, J., VAIL, P. R. & BAUM, G. R. 1988. Condensed sections: the key to age determination and correlation of continental margin sequences. *In*: WILGUS, C. K., HASTINGS, B. S., KENDALL, C. G. ST. C., POSAMENTIER, H. W., ROSS, C. A. & VAN WAGONER, J. C. (eds) *Sea-level Changes – an Integrated Approach.* Society of Exploration Paleontologists and Mineralogists, Special Publication, **42**, 183–213.

MALM, O. A., CHRISTENSEN, O. B., FURNES, H., LØVLIE, R., RUESLÅTTEN, H. & ØSTBY, K. L. 1984. The Lower Tertiary Balder Formation: an organogenic and tuffaceous deposit in the North Sea region. *In*: SPENCER, A. M. *et al.* (eds) *Petroleum Geology of the North European Margin.* Graham and Trotman, London, 149–170.

MICHELSEN, O. 1994. Stratigraphic correlation of the Danish onshore and offshore Tertiary successions based on sequence stratigraphy. *Bulletin of Geological Society of Denmark*, **41**, 145–161.

——, DANIELSEN, M., HEILMANN-CLAUSEN, C., JORDT, H., LAURSEN, G. V. & THOMSEN, E. 1995. Occurrence of major sequence stratigraphic boundaries in relation to basin development in Cenozoic deposits of the southeastern North Sea. *In*: STEEL, R. J., VELT, V. L., JOHANNSESEN, E. P. & MATHIEU, C. (eds) *Sequence Stratigraphy on the Northwest European Margin*. Norwegian Petroleum Society, Special Publication, **5**, 415–427.

MILLOT, G. 1970. *Geology of Clays*. Springer, Berlin.

MORTON, A. C. & KNOX, R. W. O'B. 1990. Geochemistry of late Paleocene and early Eocene tephras from the North Sea Basin. *Journal of the Geological Society, London*, **147**, 425–437.

NIELSEN, O. B. 1974. Sedimentation and diagenesis of lower Eocene sediments at Ølst, Denmark. *Sedimentary Geology*, **12**, 25–44.

—— & HEILMANN-CLAUSEN, C. 1988. Early Tertiary volcanism and the opening of the NE Atlantic. *In*: MORTON, A. C. & PARSON, L. M. (eds) *Early Tertiary Volcanism and Opening of the NE Atlantic*. Geological Society, London, Special Publications, **39**, 395–405.

——, SØRENSEN, S., THIEDE, J. & SKARBØ, O. 1986. Cenozoic differential subsidence of North Sea. *AAPG Bulletin*, **70**, 276–298.

PEARSON, M. J. 1990. Clay mineral distribution and provenance in Mesozoic and Tertiary mudrocks of the Moray Firth and northern North Sea. *Clay Minerals*, **25**, 519–541.

—— & SMALL, J. S. 1988. Illite–smectite diagenesis and paleotemperatures in northern North Sea Quaternary to Mesozoic shale sequences. *Clay Minerals*, **23**, 109–132.

PORRENGA, D. H. 1967. *Clay Mineralogy and Geochemistry of Recent Marine Sediments in Tropical Area*. Geophysical Laboratory, University of Amsterdam.

RACEY, A., LOVE, M. A., BOBOLECKI, R. M. & WALSH, J. N. 1995. The use of chemical element analyses in the study of biostratigraphically barren sequences; an example from the Triassic of the central North Sea (UKCS). *In*: DUNNAY, R. E. & HAILWOOD, E. A. (eds) *Non-biostratigraphical Methods of Dating and Correlation*. Geological Society, London, Special Publications, **89**, 69–105.

RAMM, M. 1991. *Quantitative Mineral Analysis of Sandstone Samples by XRD*. Department of Geology, University of Oslo, Norway, Internal Publication **62**.

RUNDBERG, Y. 1989. *Tertiary and sedimentary history and basin evolution of the Norwegian North Sea between 60°N and 62°N. An integrated approach*. Dr.Ing. thesis, University of Trondheim.

SINGH, B. 1996. *Study of diagenesis and provenance of the sedimentary sequences in the northern North Sea*. Cand.scient. thesis, University of Oslo.

SPJELDNÆS, N. 1975. Paleography and facies distribution in the Tertiary of Denmark and surrounding areas. *In*: WHITEMAN, A., ROBERTS, D. & SELLEVOLL, M. A. (eds) *Petroleum Geology and Geology of the North Sea and Northwest Atlantic Continental Margin*. Norges Geologiske Undersøkelse Bulletin, **316**, 289–311.

STABELL, B. 1985. Diatoms in Upper Quaternary Skagerrak sediments. *Norsk Geologisk Tidsskrift*, **65**, 91–96.

——1996. *Paralia thybergii* sp. nov. – another fossil *Paralia*. *Diatom Research*, **11**, 155–163.

—— & LANGE, C. B. 1990. Diatom assemblages in upper Quaternary sediments from the southeastern Norwegian Sea, Core P76-17. *Beiheft zur Nova Hedwigia*, **100**, 289–300.

STEURBAUT, E., SPIEGLER, D., WEINELT, M. & THIEDE, J. 1991. *Cenozoic Erosion and Sedimentation on the Northwest European Continental Margin*. Geomar Research Centre for Marine Geosciences, Christian-Albrechts Universität, Kiel.

STEWART, I. J. 1987. A revised stratigraphic interpretation of the Early Paleogene of the Central North Sea. *In*: BROOKS, J. & GLENNIE, J. (eds) *Petroleum Geology of North West Europe*. Graham and Trotman, London, 557–576.

STUEVOLD, L. M., SKOGSEID, J. & ELDHOLM, O. 1992. Post-Cretaceous uplift events on the Vøring continental margin. *Geology*, **20**, 919–922.

SØRENSEN, S. & NIELSEN, O. B. 1981*a*. *Biostratigrafi, litostratigrafi og sedimentpetrografi av tertiærsedimenter fra den Norske Kontinentalsokkel, brønn 2/8-2*. Department of Geology, University of Oslo, Norway, Internal Publication **33**.

—— & ——1981*b*. *Biostratigrafi, litostratigrafi og sedimentpetrografi av tertiærsedimenter fra den Norske Kontinentalsokkel, brønn 15/6-2*. Department of Geology, University of Oslo, Norway, Internal Publication **34**.

—— & ——1981*c*. *Biostratigrafi, litostratigrafi og sedimentpetrografi av tertiærsedimenter fra den Norske Kontinentalsokkel, brønn 30/5-1*. Department of Geology, University of Oslo, Norway, Internal Publication **35**.

THYBERG, B. I., STABELL, B., FALEIDE, J. I. & BJØRLYKKE, K. 1999. Upper Oligocene diatomaceous silica deposits in the northern North Sea – silica diagenesis and paleogeographical implications. *Norsk Geologisk Tidsskrift*, in press.

TYRIDAL, D. S. 1994. *Litologisk og mineralogisk sammensetning av slamsteiner i relasjon til loggrespons*. Cand.scient. thesis, University of Oslo.

VAIL, P. R., MITCHUM, R. M. & THOMPSON, S. 1977*a*. Relative changes of sea level from coastal onlap. *In*: PAYTON, C. E. (ed.) *Seismic Stratigraphy –* Applications to Hydrocarbon Exploration. American Association of Petroleum Geologists Memoir, **26**, 63–82.

——, —— & ——1977*b*. Global cycles of relative changes of sea level. *In*: PAYTON, C. E. (ed.) *Seismic Stratigraphy – Applications to Hydrocarbon Exploration*. American Association of Petroleum Geologists Memoir, **26**, 83–98.

WEAVER, C. E. 1989. *Clays, Muds and Shales*. Elsevier, Amsterdam.

ZIEGLER, P. 1982. *Geological Atlas of Western and Central Europe*. Elsevier, Amsterdam.

Cenozoic tectonic subsidence from 2D depositional simulations of a regional transect in the northern North Sea basin

RUNE KYRKJEBØ,[1] MARTIN HAMBORG,[2] JAN INGE FALEIDE,[3] HENRIK JORDT[3] & PETER CHRISTIANSSON[3]

[1] *Geological Institute, University of Bergen, Allégaten 41, N-5007 Bergen, Norway*
[2] *SINTEF, Petroleum Research, N-7034 Trondheim, Norway*
[3] *Department of Geology, University of Oslo, P.O. Box 1047 Blindern, N-0316 Oslo, Norway*

Abstract: The Cenozoic depositional history along a regional E–W profile across the northern North Sea has been simulated using a forward process-based simulation program of dynamic-slope type. It involves a depth-dependent, dual-lithology diffusion equation that handles transport, erosion and deposition of sediments. The data used in the simulation were derived from a seismic line calibrated against wells, and from the regional literature concerning the northern North Sea. The most important of the factors used are: the initial basin form (Paleocene bathymetry), tectonic subsidence, isostatic variables, sediment supply (sand–shale), sediment compaction (porosity–depth relationships for sand-shale) and eustatic sea-level changes. The interaction between the data values extracted from the literature could not reproduce a cross-section similar to the observed cross-section from seismic data. Therefore, the subsidence pattern and the initial basin form were reconsidered. The resulting model gave an anomalous Cenozoic subsidence pattern, different from the expected post-rift thermal subsidence, with deviations corresponding to Paleocene and Late Miocene–Pliocene times. The model-derived Paleocene subsidence might have been overestimated by using an over-shallow palaeobathymetric value, although a deepening of the basin is also indicated by biostratigraphic data. The pronounced Neogene subsidence created accommodation space for a thick Pliocene sequence, derived from the uplifted eastern source area.

Computer simulations are increasingly important for the improved understanding of basin development and infilling, and several forward models that simulate sedimentary processes in basins are available. Kendall *et al.* (1991) summarized various approaches in forward modelling and concluded that the significance and utility of any particular model is a matter of need, computer hardware and programming resources. However, the quality of the simulation depends on the exactness of the input values (Aigner *et al.* 1990).

DEMOSTRAT (Rivenæs 1993), which is a forward computer-simulation model for siliciclastic basin-fill in two dimensions, has been used to simulate the Cenozoic seismic stratigraphic framework along a regional E–W transect in the northern North Sea region. The program DEMOSTRAT is a process-based dynamic-slope type model for clastic settings, with a dual-lithology, depth-dependent diffusion algorithm, which has successfully demonstrated realistic erosion and deposition (Syvitski *et al.* 1988; Flemings & Jordan 1989; Sinclair *et al.* 1991; Rivenæs 1992). The process-elements in the program are performed as logically linked events in a time-step loop (Helland-Hansen *et al.* 1988; Lawrence *et al.* 1990; Rivenæs 1992). The most important elements in the time-step loop that will be discussed here are: tectonic subsidence history (user-specified subsidence rates); eustatic sea-level variations (user-defined sea-level curve); erosion and deposition (user-defined transport coefficients that describe how easily erosion and transport occur, and a user-defined rate of sediment sand and shale supply); compaction of sediment (user-defined empirical porosity–depth relations); flexural isostatic subsidence caused by loading (or unloading) of sediment and water (user-defined size of the flexural rigidity describing the lithospheric strength). For detailed descriptions of the elements in the program, the reader is referred to Rivenæs (1993).

The present study is based on the seismic lines SG8043–101 and NSDP84–2 (transect 2) (Jordt *et al.* 1995, 2000; Christiansson *et al.* 2000; Odinsen *et al.* 2000), adjacent wells and a documented regional Cenozoic seismic sequence

From: NØTTVEDT, A. *et al.* (eds) *Dynamics of the Norwegian Margin*. Geological Society, London, Special Publications, **167**, 273–294. 1-86239-056-8/00/$15.00

analysis (Jordt *et al.* 1995, 2000). In addition, information has been extracted from the literature concerning the North Sea. The main objectives of the present simulation have been to reproduce the Cenozoic succession as seen in transect 2 by applying a set of realistic input values and to discuss the tectonic subsidence predicted by the simulation.

Geological setting

The North Sea Basin (Fig. 1) is an intracratonic basin situated at the northwestern margin of the European Platform. Since the crustal accretion completed during early Devonian time, the North Sea area has experienced several episodes of lithospheric extension (Devonian–Carboniferous, Permo-Triassic, late Jurassic–early Cretaceous), each followed by a stage of thermal subsidence (Ziegler 1982*a*; Glennie 1984; Giltner 1987; Badley *et al.* 1988; Gabrielsen *et al.* 1990). The Devonian–Carboniferous part of the structural history is not very well understood in the study area, mainly because of lack of data (Gabrielsen *et al.* 1990). The timing of the Permo-Triassic rift phase is still a matter of debate, and both Permian (Eynon

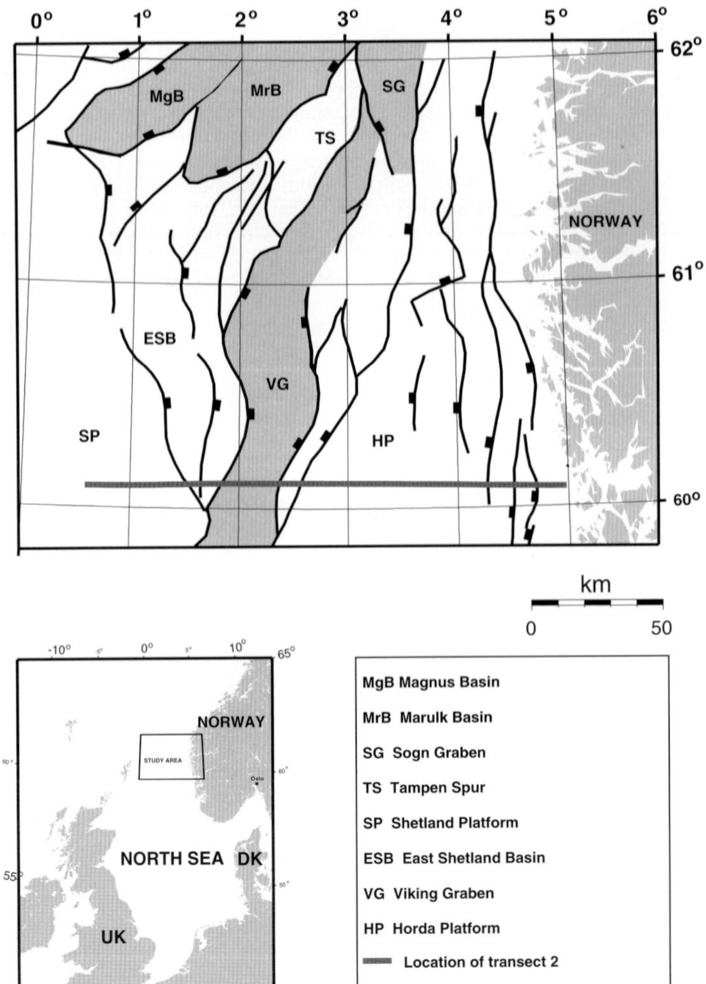

Fig. 1. Regional map of the northern North Sea with main structural elements and the location of transect 2. Transect 2 a combination of the conventional SG8043-101 line and the deep seismic line NSDP84-02, crosses the Horda Platform (HP), Viking Graben (VG), East Shetland Basin (ESB) and the Shetland Platform (SP), and has a total length of 240 km in this study (from Christiansson *et al.* 1999).

1981; Badley *et al.* 1988; Gabrielsen *et al.* 1990; Færseth *et al.* 1995) and Triassic (Beach *et al.* 1987; Giltner 1987; Roberts *et al.* 1995) ages have been suggested. Most of the Permo-Triassic stretching occurred between the Øygarden Fault Zone to the east and the Shetland Platform and the Hutton alignment to the west (Odinsen *et al.* 2000*b*). Modelling results (ter Voorde *et al.* 2000) show that no heat access from the broad extended Permo-Triassic rift event was left at the onset of the more focused late Jurassic–early Cretaceous rifting (Odinsen *et al.* 2000*b*), which led in turn to the development of the main structural elements of the Viking Graben area.

In early Cretaceous time, thermal subsidence commenced; this was marked by the cessation of fault block tilting. The following post-rift sedimentation (Cretaceous–Cenozoic) filled in and buried the rift topography. In the northern

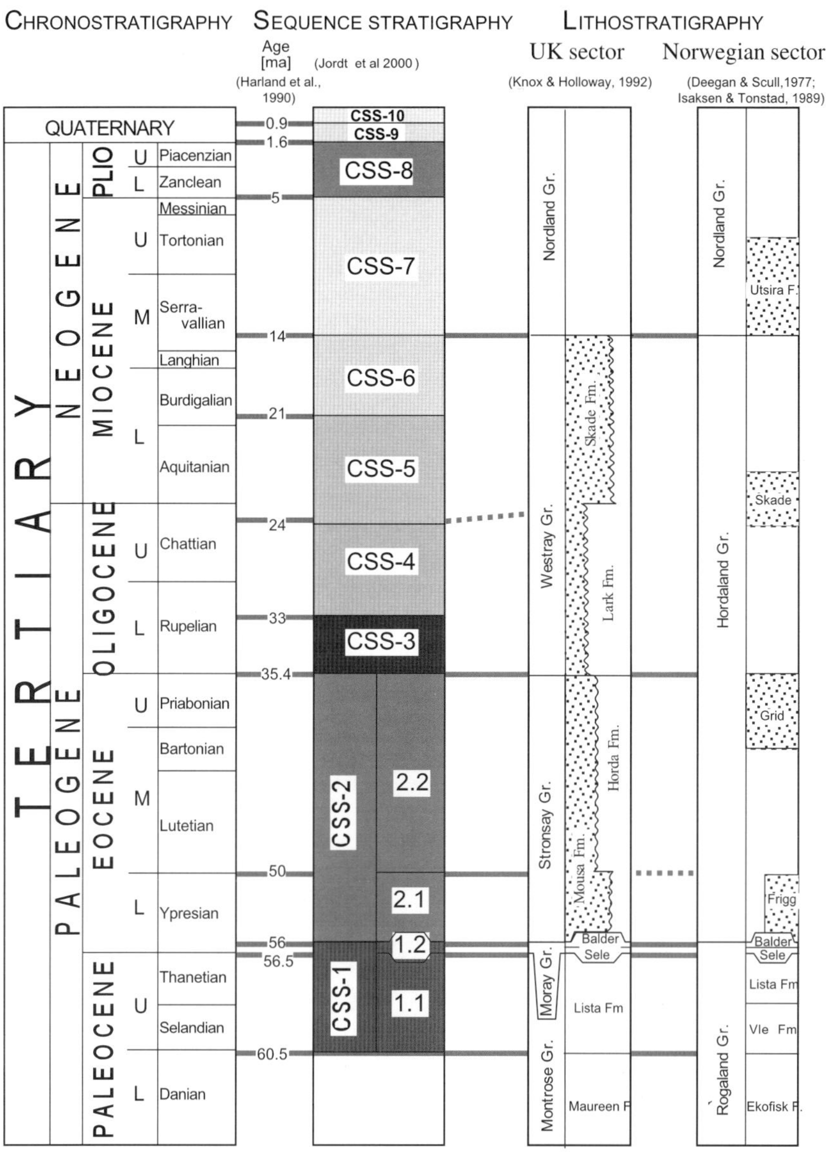

Fig. 2. Correlation chart showing the seismic sequence stratigraphic framework, connected to chronostratigraphy and lithostratigraphy (from Jordt *et al.* 1995, 2000).

North Sea, south of 61°N, early Cretaceous deposits are mainly restricted to the Viking Graben and mostly lap onto the basin margins, but lower Cretaceous sediments are also found in the Stord Basin and at the Horda Platform. Late Cretaceous sedimentation buried the rift topography and Cenozoic sedimentation overstepped the graben boundaries. Periods of increased subsidence associated with faulting are detected in the post-rift stage, particularly in late Cretaceous–early Tertiary time (Thorne & Watts 1989; White & Latin 1993; Hall & White 1994; Leperq & Gaulier 1996; Nadin & Kusznir 1996), but this cannot be considered a rift phase (Badley *et al.* 1988)

The Cenozoic basin-fill was derived from two main sources: the East Shetland Platform to the west and the Fennoscandian Shield to the east. Deposition in Tertiary time was controlled by several phases of uplift affecting the source areas, combined with relative sea-level changes and thermal subsidence of the basin (Jordt *et al.* 1995, 2000). The western source areas dominated the sediment influx from late Paleocene until early Miocene times (Stewart 1987; Galloway *et al.* 1993; Jordt *et al.* 1995, 2000). The basin shallowed from relative deep-marine conditions during the late Paleocene time, to shallow-marine and locally fluvial conditions in late Early Miocene time (Rundberg *et al.* 1995; Jordt *et al.* 1995, 2000). The sediment influx from the west ceased in Early Miocene time and most of Mid- and Late Miocene time, was characterized by non-depositional conditions. In Pliocene time, a major uplift of the Fennoscandian landmass caused a major sediment influx from the east. In late Pliocene and Pleistocene time, the glaciations of Fennoscandia caused the removal of proximal Tertiary and older deposits from the basin's eastern margin.

Cenozoic seismic stratigraphy

The seismic-stratigraphic framework used in this study was established by Jordt *et al.* (1995), on the basis of a regional analysis of the Cenozoic in the central and northern North Sea. The sequence descriptions are not repeated here: instead, the important constraints for the computer modelling are examined.

From east to west, transect 2 (Fig. 1) crosses the main structural features of the northern North Sea Basin, namely, the Øygarden Fault Zone, the Horda Platform, the Viking Graben and the eastern part of the East Shetland Platform. The Cenozoic succession in the North Sea has been subdivided into ten sequences, labelled CSS-1 to CSS-10 (Cenozoic Seismic Sequences). The correlation chart in Fig. 2 connects the sequence stratigraphic interpretation with the standard chrono- and lithostratigraphic subdivisions of Deegan & Scull (1977), revised by Isaksen & Tonstad (1989) for the Norwegian sector and by Knox & Holloway (1992) for the UK sector. A depocentre map with indicated outbuilding directions is shown in Fig. 3. Figure 4 shows the full cross-section, whereas the distribution and thickness (depth-converted) of the Cenozoic sequences are given in Fig. 5.

CSS-1 (of Late Paleocene–earliest Eocene age), which correlates with the Rogaland Group (Fig. 2), has its main outbuilding direction from the west, where it was sourced by the uplifted Shetland Platform. As can be seen from Fig. 5, westerly derived sediments reached far to the east. The sediments were deposited as submarine fan complexes in a relatively deep basin, with water depths suggested to be middle bathyal (500–1000 m) in well 3/25-1 (located close to transect 2 in the UK sector) (Gradstein *et al.* 1994). In the east, sedimentary wedges prograded westwards towards a depocentre probably situated north of the transect, west of Sognefjorden (Fig. 3). It is probable that CSS-1 covered a larger area, and that parts of western Norway were transgressed in late Paleocene time (Jordt *et al.* 1995). However, these deposits were removed by post Paleocene erosion during the Late Pliocene and Quarternary glaciations.

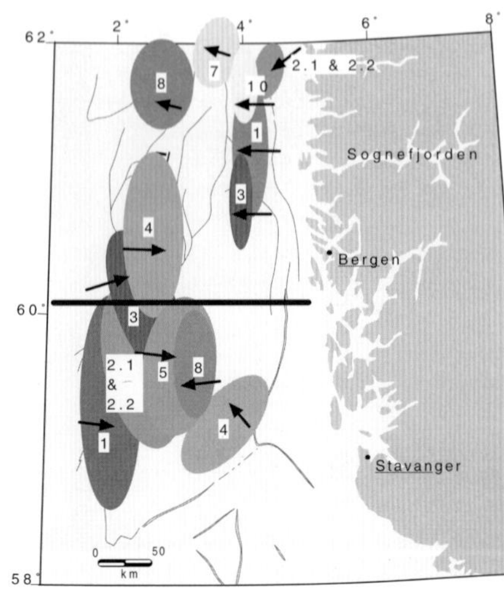

Fig. 3. Depocentre locations and outbuilding directions indicated by arrows for each of the Cenozoic seismic sequences given by numbers (from Jordt *et al.* 1999).

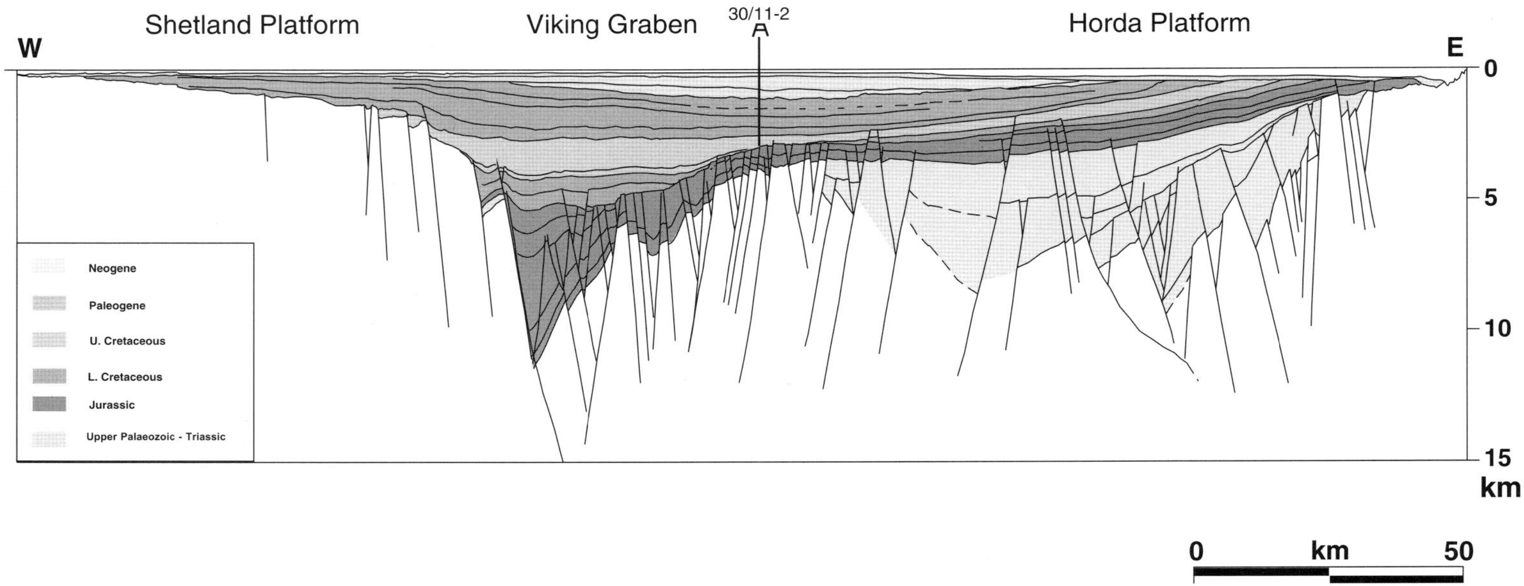

Fig. 4. Transect 2 in full basin scale and location of key well 30/11-2 (from Christiansson *et al.* 2000). The transect shows the Paleogene and Neogene successions relative to deeper structures and the main structural elements. Line location is indicated in Fig. 1.

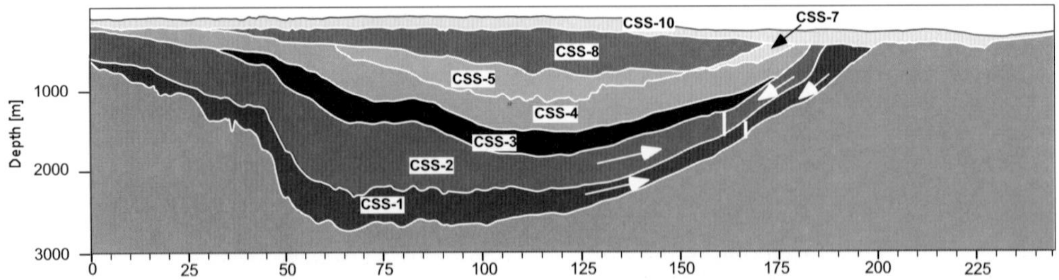

Fig. 5. Cenozoic basin fill (CSS-1 to CSS-10) along transect 2. Arrows show main outbuilding directions, and the interpreted border between easterly and westerly derived sediments is marked for sequences CSS-1 and CSS-2.

CSS-2 (of Eocene age) correlates with the lower part of the Hordaland Group (Fig. 2). This sequence has been subdivided into two units: a lower Eocene sequence, CSS-2.1, found mainly to the east, and a middle–upper Eocene sequence, CSS-2.2, with its depocentre in the Viking Graben (Fig. 3). CSS-2.1 is predominantly east derived, and its deposition was terminated with the corresponding submergence of Fennoscandia (Rundberg 1989; Sterbaut *et al.* 1991; Jordt *et al.* 1996). The depositional style of CSS-2 seems to be similar to that of CSS-1, both for westerly and easterly derived sediments.

CSS-3 (of Early Oligocene age) correlates with the middle part of the Hordaland Group (Fig. 2). At the Eocene–Oligocene transition, uplift of Fennoscandia caused a relative fall of sea level and an outbuilding of a sandy wedge from the east, south of transect 2. The depocentre of this sandy unit is located north of the transect (Fig. 3). Another depocentre is located in the Viking Graben, and was sourced from the western hinterland (Jordt *et al.* 1995).

CSS-4 (of late Early–Late Oligocene age) correlates with the middle part of the Hordaland Group (Fig. 2). It prograded eastwards, and has a depocentre along transect 2 (Fig. 3). Another depocentre is located south of the transect, where CSS-4 prograded to the northwest. An irregular surface on the top of CSS-4 (Fig. 5) was probably caused by the remobilization of clays (Jordt *et al.* 1995 and 1996).

CSS-5 (of Latest Oligocene–Early Miocene age) correlates with the upper part of the Hordaland Group (Fig. 2), and is the uppermost preserved strata of the group in the study area (see CSS-6). CSS-5 prograded from the west, but formed a north-trending sheet in the middle of the basin (Figs 3 and 5). No sediment influx from the east has been recognized in this sequence. The irregular top of CSS-5 is interpreted as an incised valley surface, probably formed by subaerial erosion. This latter feature is more distinct in the seismic sections taken further north in the North Sea. Rundberg *et al.* (1995) and Jordt *et al.* (1996) suggested that the subaerial exposure in this area was primarily controlled by tectonic uplift of the northernmost North Sea.

CSS-6 (of latest Early–early Mid-Miocene age) is the uppermost unit correlative with the Hordaland Group (Fig. 2). It is only found south of the Viking Graben and on the Horda Platform. In the present study area it is represented by a hiatus (or below seismic resolution).

CSS-7 (of Mid–Late Miocene age) correlates with the lower part of the Nordland Group (Fig. 2). This sequence prograded from the east as a result of the uplift of Fennoscandia relative to the subsiding basin in Mid-Miocene time. The deposition of this sequence marked a shift of the main source area from a westerly to an easterly domain. In the transect area, the sequence is represented by a relatively thin depositional wedge (Fig. 5), with the major depocentres located north and south of the transect (Fig. 3). This sequence also correlates with the Utsira Formation (Fig. 2). However, correlations of seismic and well data show that the sandy units of both CSS-5 and CSS-8 have also been interpreted as the Utsira Formation (Jordt *et al.* 1995).

CSS-8 (of Pliocene age) correlates with the Nordland Group (Fig. 2) and was deposited in response to erosion of a strongly uplifted Fennoscandian hinterland, together with North Sea basinal subsidence providing accommodation space (Jordt *et al.* 1995, 2000). The CSS-8 clastic wedge prograded to the west, forming two depocentres (Fig. 3): a smaller wedge was located on the transect in the Stord Basin, and a larger one in the Tampen Spur area.

CSS-9 (of early Quaternary age) correlates with the upper Nordland Group (Fig. 2), but is represented by a hiatus in transect 2.

CSS-10 (of late Quaternary age) correlates with the uppermost part of the Nordland Group

(Fig. 2). This sequence shows an angular unconformity at the lower boundary (Fig. 5), especially in the eastern part of the study area, where the older sequences are truncated by CSS-10.

Input factors

The modelling was performed by systematically changing input values to reproduce the geometry observed in the seismic reflection data. We started with a fixed set of values for initial basin form, amount of tectonic subsidence (uplift), rates of sediment supply, sea-level curve and values for isostatic calculations, whereas the values of the transport coefficients were varied. The fixed input values are summarized in Table 1 and the choice of values is discussed below.

Initial basin form

Water depth is a critical factor in subsidence and burial analysis, because it represents the amount of the basin's 'underfill'. Large palaeobathymetric values at the end of Cretaceous time provided much of the accommodation space for Tertiary sedimentation. On the basis of an analysis of faunal trends along palaeoslope transects in the northern North Sea, Gradstein *et al.* (1994) and Gradstein & Bäckström (1996) estimated conservatively that the early Palaeocene water depth was between 250 and 500 m, and that the Late Paleocene water depth was between 500 and 1000 m in the UK-well 3/25-1, which is close to the modelled transect. Other wells (Fig. 6) analysed in the area show that water depths in early Paleocene time were between 500 and 750 m in UK-well 9/23-1 and between 200 and 500 m in UK-well 9/13-1. Late Paleocene water depths were between 500 and 1000 m in UK-well 9/13-1, and between 750 and 1500 m in UK-well 9/23-1 (Gradstein *et al.* 1994, Gradstein & Bäckström, 1996). Palaeobathymetric maps (Barton & Wood 1984) based on information obtained from well samples, combined with the palaeogeographical maps of Ziegler (1981, 1982*a*), show a 500 m contour for the Paleocene water depths, indicating that water depths in the deepest part of the basin must have exeeded 500 m. Accordingly, a maximum initial basin depth in this study was estimated to be *c.* 550 m.

The initial basin configuration along the transect at the beginning of Late Paleocene time was constructed by combining the work of Barton & Wood (1984) (500 m contours), palaeogeographical maps of Rochow (1981) and Ziegler (1990) (Fig. 6), and the present base Tertiary relief. The initial form of the basin (Fig. 7) is believed to have been nearly symmetrical, with the western flank reaching sea level on the East Shetland Platform (Rochow 1981; Ziegler 1990), whereas parts of the eastern flank on the Horda Platform were submerged (Ziegler 1990).

Table 1. *Initial input factors summarized for the post-rift modelling*

Variable	Source
Initial basin form	Water depths along transect 2 based on work by Rochow (1981), Barton & Wood (1984), Ziegler (1990) and Gradstein *et al.* (1994)
Sea-level function	Haq *et al.* (1987) long-term curve; age based on Harland *et al.* (1990) time scale
Sediment supply	Derived from backstripping; supply from both sides
Transport coefficient function	According to the type curves shown in Fig. 7
Surface layer thickness	30 cm
Compaction	Sclater & Christie (1980) curves; the compaction of the base is measured and given together with tectonic subsidence
Isostasy	Flexural rigidity $D = 1.0 \times 10^{22}$ N m; no axial stress
Time span, number of time steps	60.5 Ma, 200 time steps, which gives each time step a duration of 302 500 a
Basin length, number of columns	240 km, 200 columns
Tectonic movements	Thorne & Watts (1989)

Sediment input

The depth-converted seismic profile was backstripped to determine the decompacted cross-sectional area of each sequence after removing the overburden. The backstripping process was carried out using the Balancing Section Program (XBSP; Midland Valley Exploration Ltd). The exponential porosity–depth relationships of Sclater & Christie (1980) for shales and sands were used in the backstripping procedure. The lithological composition of each sequence was established by investigating the completion log from well 30/11-2 (Fig. 4); the shale-to-sand-ratio was estimated from the gamma-ray log. The volumetric sand fraction (Table 2) appears

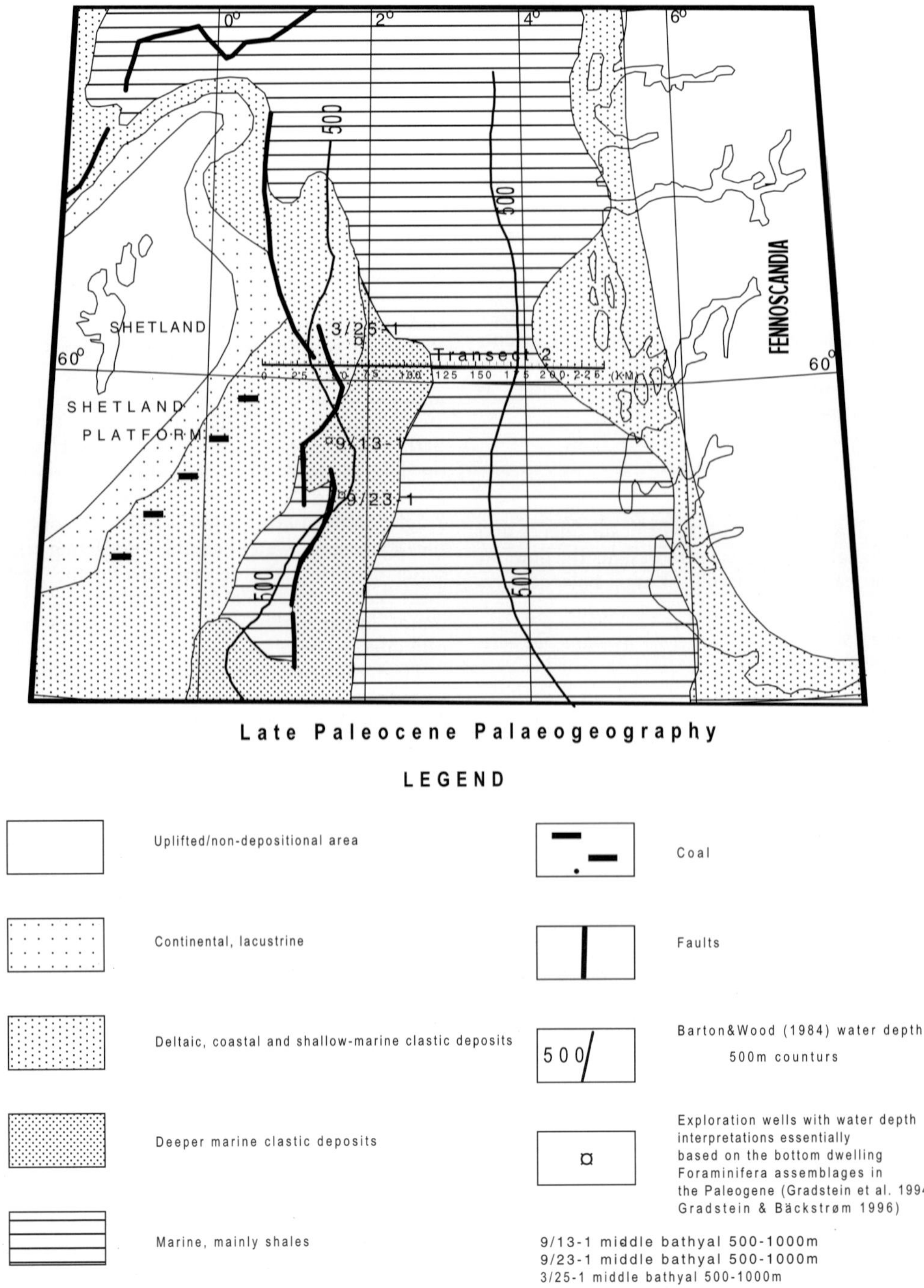

Fig. 6. Integration of Late Paleocene palaeogeographical maps by Rochow (1981), Barton & Wood (1984) and Ziegler (1990) with interpreted faunal trends from wells locations by Gradstein *et al.* (1994) and Gradstein & Bäckström (1996).

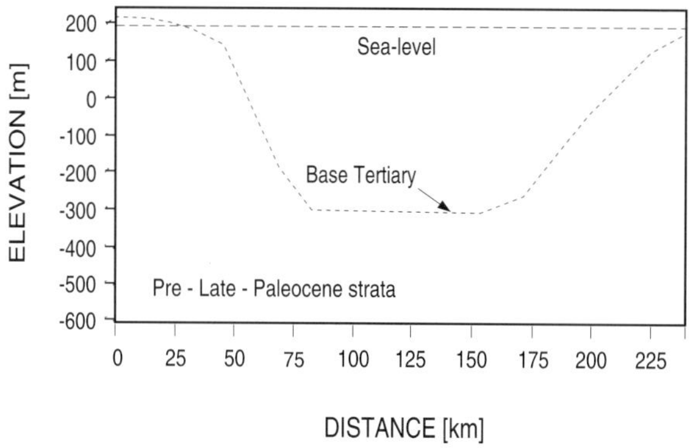

Fig. 7. The initial form of the basin in the beginning Late Paleocene time based on Fig. 6. The maximum water depth is suggested to be 550 m. The basin form is symmetrical with shallow water depths at the flanks.

to be relatively high for most of the sequences. The estimates are admittedly uncertain, because of limitations of the method and the uncertainty of the log-derived values.

Because the section is 2D, the decompacted sediment-flux rates ('2D volumes') for sand and shale are given in square metres per year ($m^2 a^{-1}$) (see Fig. 8). Changes in sediment-flux rates are closely related to tectonic movements and to changes in the sediment supply system; they are synchronous with the development of depositional sequence boundaries (Jordt *et al.* 1995).

To test the sensitivity of compaction to the lithological variations, a backstripping was done for 100% shale and 100% sand content. The difference between the shale and sand volumes in the basinal cross-section was found to range from 0 to 20%, with the greatest difference in the oldest (most compacted) units, and decreasing upwards. Hence, the sensitivity analysis indicates that the errors in bulk volume caused by departures in lithological composition are small.

The lack of information on the distribution of lithologies along the profile made it necessary to simplify this part of the problem. As the lithofacies architecture of the basinal cross-section was not the main objective of the simulation, the compaction-related errors caused by lateral lithological variations have been neglected. Instead, the modelling focused on the bulk geometrical relationships of each seismic sequence.

Table 2. *Interpreted sand fraction for each sequence applied in the simulation*

Sequence	Sand fraction
CSS-1	0.76
CSS-2	0.62
CSS-3	0.47
CSS-4	0.35
CSS-5	0.80
CSS-7	0.99
CSS-8	0.49
CSS-10	0.72

Some sequences, especially those derived from the eastern source, have been eroded from the flank. This erosional truncation is shown in Fig. 5. As the backstripping takes only the volume of sediment present in the profile (2D) into account and no out-of-plane transport of sediments is possible in the forward simulation, a redistribution pattern for the eroded sediment had to be suggested. The volume of the redistributed sediment was calibrated to match the overall thickness framework, and the total sediment bulk volume. The pattern of redistribution would obviously have to reflect the palaeogeographical situation. Sequences CSS-1, CSS-2 and CSS-4 therefore had to include additional sediment in the eastern part to fit the pre-Miocene palaeogeography. This was accomplished by extrapolating the observed sequences eastwards; the amount of volume added was calibrated to the related volumes of sequences CSS-7 and CSS-8. While CSS-7 and CSS-8 were being deposited, the additional sediment volume in the east is assumed to have been eroded and redeposited. These depositional events correspond to the uplift of the eastern source area.

The sediments are assumed to enter the basin at point sources on each side of the cross-section. On the seismic reflection data it is possible to distinguish between easterly and westerly derived sediments (see location of depocentres in Fig. 3) for some of the sequences (CSS-1, CSS-2, CSS-4 and CSS-10) (Jordt *et al.* 1995), and therefore calculate the decompacted cross-sectional area to enter from each side of the section. Because of restricted knowledge of the source

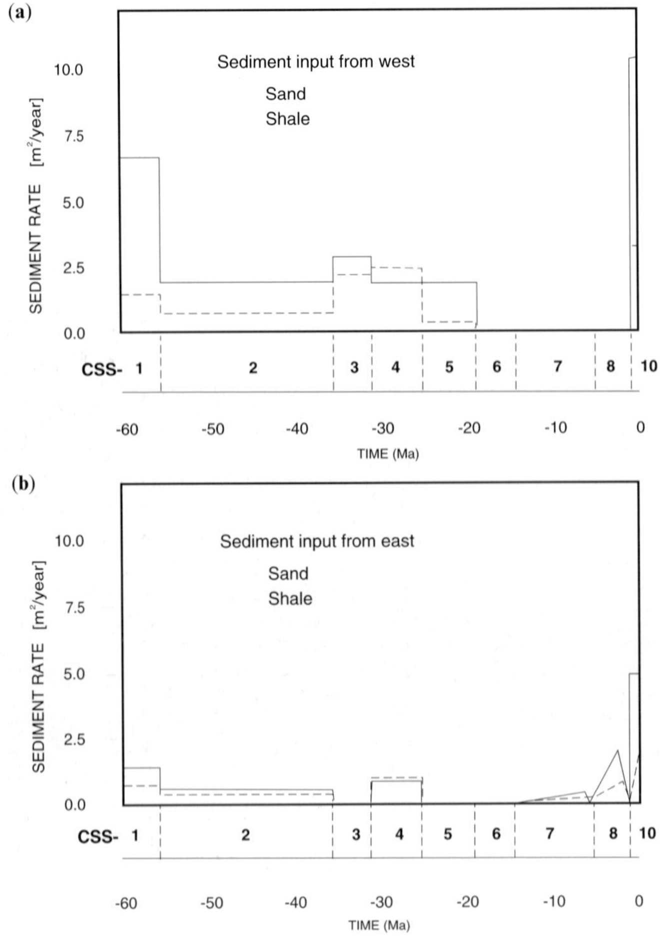

Fig. 8. Sediment supply ($m^2 a^{-1}$) of sand (continuous line) and shale (dashed line) from west (**a**) and east (**b**). Volumes are calculated from backstripping. These volumes have 0% porosity, which is the input form in DEMOSTRAT. The program will convert input volumes to maximum porosities, 0.49 for sand and 0.63 for shale, according to the empirical porosity–depth relations of Sclater & Christie (1980) for North Sea settings.

area, erosion of source area topography is not specified in detail in the simulation.

Compaction

The functions of Sclater & Christie (1980) have commonly been used to describe the porosity–depth relationship in the North Sea region. DEMOSTRAT uses one curve for each lithology, whether sand or shale, whereas XBSP (Midland Valley Exploration Ltd) uses a single lithology-dependent curve.

An important aspect is the volume control on the sediments belonging to each sequence. By using the same empirical porosity–depth relationship and sand fraction values both in the inverse and forward modelling, the volume is expected to be the same in the simulated and in the observed section. The porosity values in the simulated cross-section are not necessarily the same as in the observed section. These deviations are due to uncertainty related to the usage of empirical porosity–depth functions. However, as the error would be the same for the inverse and forward simulations, the uncertainty of the porosity–depth relationship is not relevant.

The underlying Cretaceous and older deposits have a non-uniform thickness distribution along the cross-section (Fig. 4), being thickest in the Viking Graben and thinning greatly towards the basin flanks. This probably caused differential compaction and base-Tertiary subsidence, which it is not possible to represent in the DEMOSTRAT simulation. In the decompaction process based on XBSP (Midland Valley Exploration Ltd), the Cretaceous strata were included and the resulting values were added to the tectonic subsidence in the DEMOSTRAT simulation. In all presentations of tectonic subsidence curves, these compaction values are subtracted. These are, how- ever, minimum values for the compaction of the underlying sediments, as no sediments older than Cretaceous have been taken into account in the decompaction process and in the depositional modelling.

Transport coefficients

In DEMOSTRAT (Rivenæs 1992, 1993), the depth-dependent diffusion dual-lithology algorithm handles erosion and deposition. Diffusion can be characterized as a time-dependent smoothing process. In geological terms this can be erosion of topographic highs and deposition of the erosional product in sedimentary basins. The model has several limitations. In the present case, the most important are lack of transverse sediment transport (out of the plane of the section) and inability to model turbidites.

Input data are values for transport coefficients, termed K for sand and shale. The transport coefficients give a measure of the transport efficiency of sediments, telling us how easily erosion and deposition occur (Rivenæs 1992). Sediments are supplied from point sources on each side of the cross-section. In general, depth variations of K values in the simulation are adapted as follows (Fig. 9): a higher transport efficiency above sea level, an exponential decrease near and below sea level when the load enters the basins, and below a certain level the transport efficiency is constant and 'low'.

Rivenæs (1993) discussed the quantification of the transport coefficients and summarized some measured and some synthetic K values proposed by various workers, showing that K values are a function of the scale of investigation and that K values are dependent on environment. Rivenæs (1992) suggested synthetic K values ranging from $300\,m^2\,a^{-1}$ in mountain environments to $1.2 \times 10^4\,m^2\,a^{-1}$ for nonmarine mud. Kenyon & Turcotte (1985) have measured K values for the Mississippi delta front of $5.3 \times 10^5\,m^2\,a^{-1}$.

Unfortunately, no values for the transport coefficients for the Cenozoic basin-fill in the North Sea are available, and in the present study it has been necessary to assume the transport coefficients representing the contributing processes invoked in sedimentation and erosion. The applied K values, therefore, are taken as those that reproduce the lateral distribution of sediments. The K curve pattern used in this simulation is complex, because of changes in depositional pattern both in time (differences in sediment supply for each sequence) and space (differences in sediment supply for the western and eastern part, and difference in depositional regime with depth).

The sequences CSS-1 to CSS-5 are mainly derived from a western source, with limited sediment supply from the east. This suggests a higher transport efficiency for the westerly sourced sediments. CSS-1, which was sourced from the

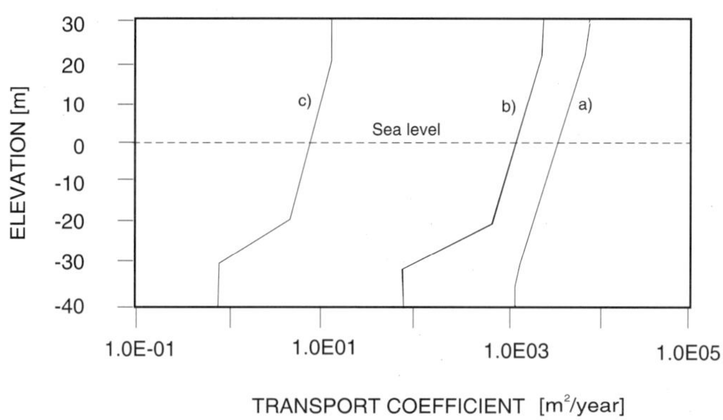

Fig. 9. The transport coefficient functions used in the simulation. The values decrease to a depth of 40 m and are constant at deeper levels. Differences in transport efficiency are needed to reproduce the observed geometries in Fig. 5. A high value above sea level, decreasing to a high value at 20 m below sea level, and constant high further down (line a) are adapted for the western derived CSS-1. A low value both above and below sea level, describing a period of non-deposition, and trying to avoid erosion, is used for CSS-6 to CSS-9 from the west. Intermediate values to those of line a and line c are applied for the rest of the sedimentary record.

west, has the largest sediment supply and a higher K value is assumed. Furthermore, CSS-1 consists partly of submarine fan deposits and therefore a high K value at great depths has been applied to match the observed geometries (Fig. 9, line a). In periods of erosion or non-deposition (CSS-6 to CSS-9 from the western source), K values are assumed to have been very low (*c.* 1.0) (Fig. 9, line c). This might not be correct considering the erosion observed in underlying sequences and the northward-directed transport of erosional products (Jordt *et al.* 1995), but because of the 2D limitations we are not able to transport sediments out of the plane. The CSS-8 sequence, sourced from the east, has high K values to simulate erosion of previously deposited CSS-1 to CSS-4 on the eastern flank (Fig. 9, line b).

We make no attempt to predict a detailed lithological facies distribution along the cross-section. Therefore, no particular sediment sorting (sand–shale) is applied to differentiate between the K values. This is due to restricted knowledge of sediment composition along the profile, time-step resolution (302.500 a), and to the necessarily limited scope of this study.

CSS-1 and CSS-2 are partly submarine fan deposits (turbidites) and each turbidite event cannot be regarded to have a diffusive nature, but the sum of turbidite events within each sequence can be assumed to be diffusive.

Tectonic movements

The general hypothesis of Cenozoic basin development in the northern North Sea is that the basin was dominated by thermal subsidence that followed the Late Jurassic–Early Cretaceous rifting (e.g. Sclater & Christie 1980; Badley *et al.* 1988; Gabrielsen *et al.* 1990). Several workers have quantified and explained the cause of this subsidence. Thorne & Watts (1989), White & Latin (1993), Hall & White (1994), Lepercq & Gaultier (1996) and Nadin & Kusznir (1996); have inferred a higher subsidence rate in Early Tertiary time and a lower rate in the rest of the Tertiary time. The subsidence data obtained by Thorne & Watts (1989) (Fig. 10) were applied in the initial simulation run. This is considered to be a minimum subsidence pattern, because it was a result of backstripping using Airy isostasy, which favours isostatic compensation. However, Thorne & Watts (1989) apparently did not consider the effects of uplift or subsidence of basin flanks. Therefore, tectonic movements on the flanks were connected (linked) to the qualitative interpretation of Rundberg (1989) and Jordt *et al.* (1995).

Only one fault in the basinal cross-section is thought to have substantially influenced the basin geometry. The fault is 46 km from the western end of the section and cuts sequence CSS-1, whereas sequence CSS-2 is unaffected. As this fault has affected the depositional pattern, it is included in the simulation. The relative tectonic drop of the eastern part of the section is taken to be about 350 m, and the faulting is assumed to have occurred in Late Paleocene time.

Eustasy

The model requires the input of sea-level variations through the geological time. The long-term curve of Haq *et al.* (1987), therefore, was chosen to reflect sea level. The curve was sampled and converted to the Harland *et al.* (1990) time scale.

However, eustatic sea-level changes are not within the main scope of the present study, as the rate of sea-level change in terms of the long-term curve (Haq *et al.* 1987) is too low for its effects to be represented in the depositional style. An exception is the large fall in sea level from 15 Ma to the present, which affects the sequences CSS-7 to CSS-10. To be comparable with the possible effects of the short-term eustatic curve of Haq *et al.* (1987), the simulation has been run with 200 time steps. The results confirm that there is no significant difference with respect to the main simulation based on the long-term eustatic curve, probably because of the combination of the depositional conditions in the deep basin and the large scale of the stratigraphic approach. An increase in the number of time steps and horizontal resolution (number of columns) might give greater detail of changes in a shallow-marine depositional system with respect to the short-term curve of Haq *et al.* (1987).

Isostasy

The lithosphere in the North Sea region was more stable in Cenozoic time than during Mid-Jurassic to Cretaceous times. The regional tectonic stress in Cenozoic times was apparently low, with the strain no greater than the development of minor features. Therefore, no axial stress has been applied in the simulation, and a flexural rigidity of the lithosphere of 1.0×10^{22} N m was used (Fjeldskaar 1994). The matrix densities for the isostatic compensation have been taken as follows: 2.65 g cm^{-3} for sand, 2.72 g cm^{-3} for shale, and 1.05 g cm^{-3} for water (the same

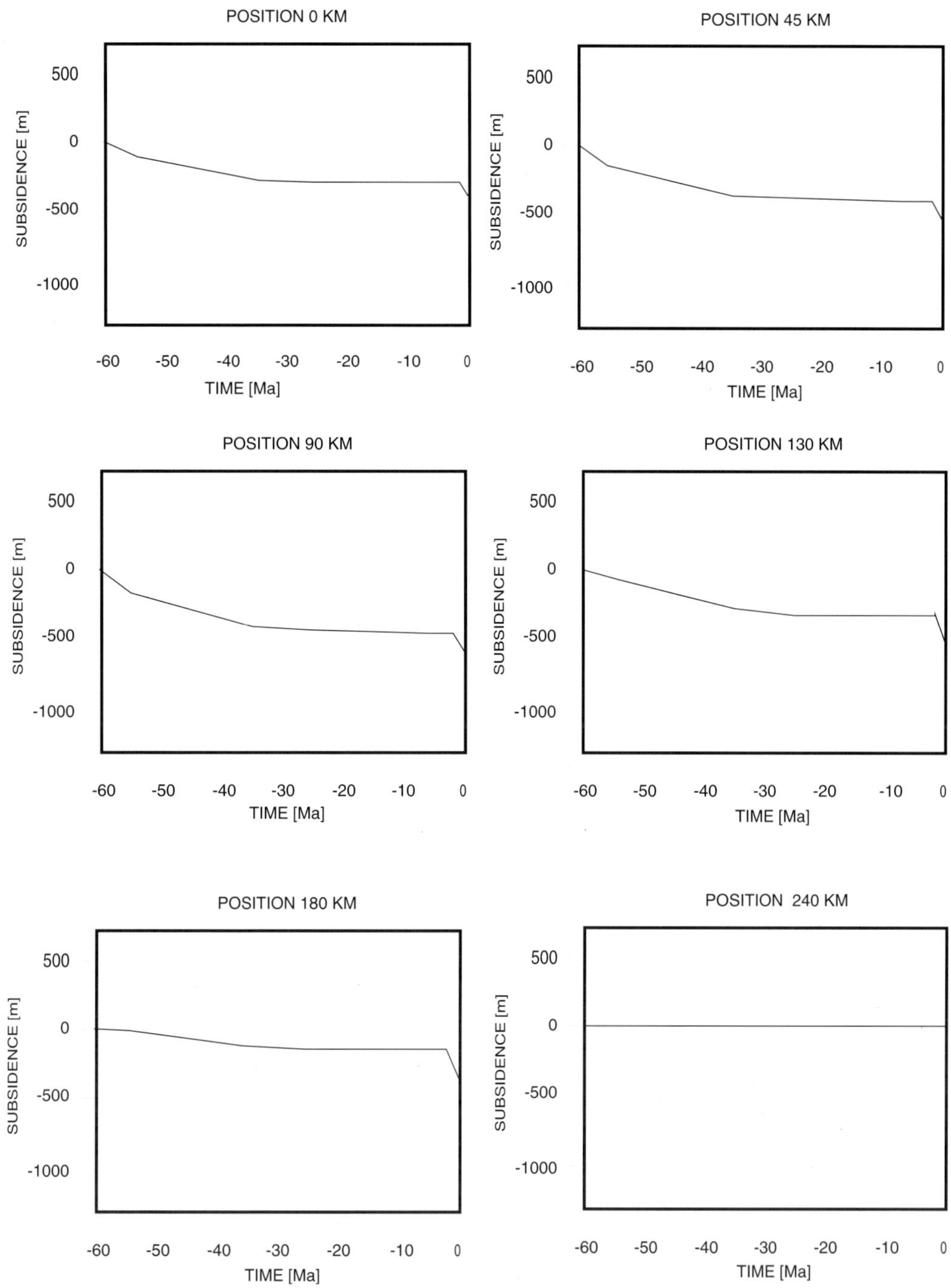

Fig. 10. Cenozoic tectonic subsidence pattern given by Thorne & Watts (1989). The subsidence analysis is based on backstripping with Airy isostasy (flexural rigidity zero) of wells in the northern Viking Graben area, and suggests an accelerated subsidence in Paleocene time followed by gentle subsidence rates in the rest of the Cenozoic period, reflecting post-rift thermal subsidence (McKenzie 1978). In the simulation, the subsidence data obtained by Thorne & Watts (1989) were applied for modelling case 1. A flexural rigidity of 1.0×10^{22} N m was used.

as used by Sclater & Christie (1980)). For the crust and the upper mantle, densities of 2.8 g cm^{-3} and 3.4 g cm^{-3}, respectively, have been assumed.

Modelling results, case 1

The use of the values specified in Table 1 resulted in an overfill of the basin by Early Oligocene time, and subaerial deposition of sequences younger than Early Oliogocene (see Fig. 11a). Therefore, it is not possible to compare the modelled sequence geometries with the observed geometries. Furthermore, the modelling results placed the base Tertiary too shallow at present time. The effect of introducing Airy isostasy (flexural rigidity zero) was tested, but was not sufficient to avoid early overfilling (see Fig 11b), which occurred in Mid-Oligocene times in this model, or to improve the position of the present depth level of the base Tertiary. It is evident from the simulated cross-section that the basin model underestimates either the total subsidence or the initial water depths along the cross-section. Accordingly, a new initial form of the basin (Fig. 12), with an asymmetrical form, enhanced initial water depths in its eastern part, and a slight increase in the maximum initial water depth was suggested.

The overall objective in this study was to carry out a simulation that could match the observed cross-section with respect to the geo-history related to the basin-filling processes and the observed geometrical relationships. The eustatic sea-level, isostasy and sediment input were considered to be documented values and it was assumed that the software handles erosion and deposition properly. The factors that are associated with the largest uncertainties were therefore tectonic subsidence and water depths. We could, by an increase in the maximum initial water depth of the basin to an order of 1000 m, still use the subsidence rates proposed by Thorne & Watts (1989). This is also in agreement with the water depths for early Paleocene time in the

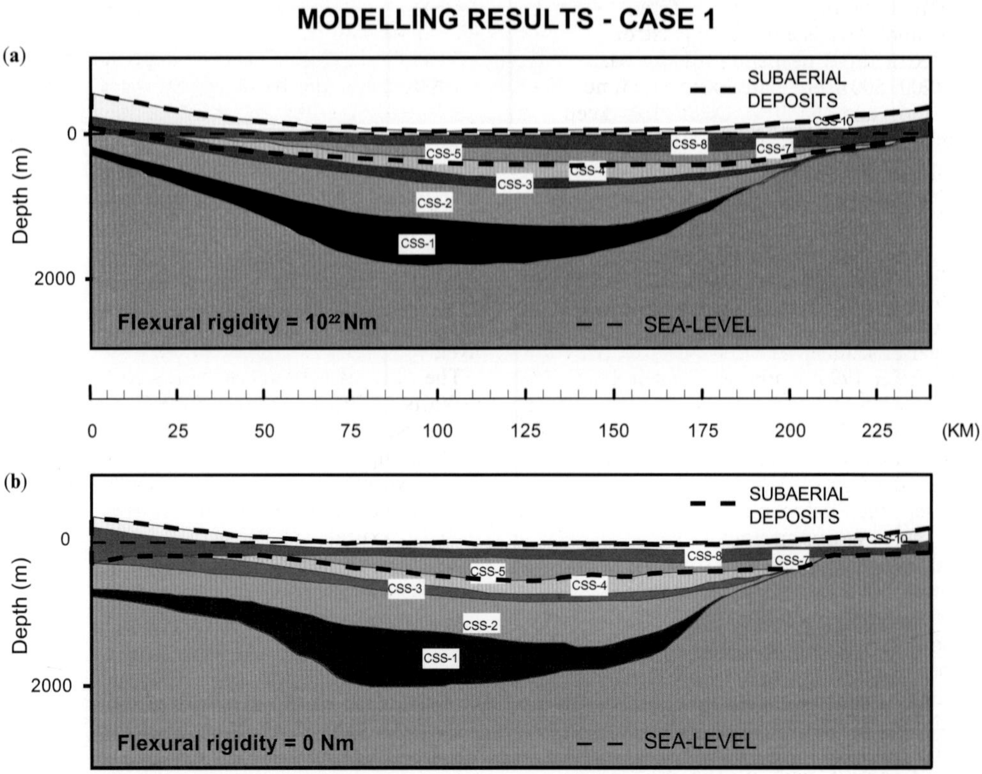

Fig. 11. (**a**) The interaction between the chosen values failed to reproduce the cross-section. The basin was overfilled in Early Oligocene time. (**b**) Use of $D = 0$, Airy isostasy, could not improve the results, as it also led to an overfilled basin in Late Oligocene time.

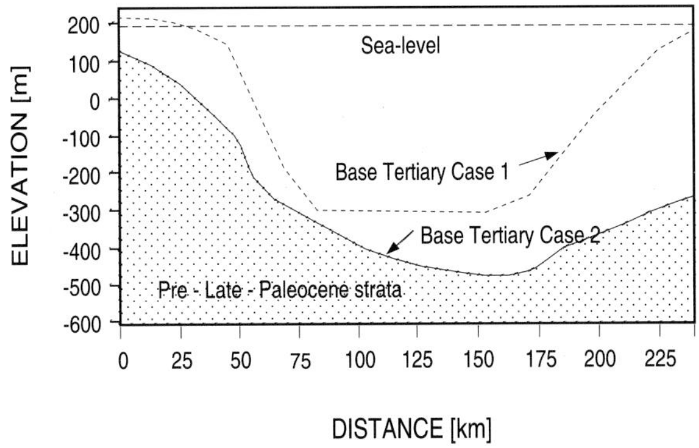

Fig. 12. On the basis of the assumption that the overfilled basin in Early Oligocene time in the model was mainly caused by an underestimated water depth of the initial basin, a revised initial form of the basin in Late Paleocene time with increased water depth along the cross-section was required. Also, the tectonic subsidence pattern was reconsidered to obtain agreement of the modelled and observed geometrical relationships.

northern North Sea proposed by Bertram & Milton (1989) and Nadin & Kusznir (1995), but differs from water depths proposed by Barton & Wood (1984) and Gradstein *et al.* (1994) for early Paleocene time. However, Gradstein *et al.* (1994) and Gradstein & Bäckström (1996) suggested a rapid deepening of the basin from Early (200–500 m) to Late Paleocene time (500–1000 m). It was therefore decided to keep the maximum initial water depth at a maximum of *c.* 700 m at the beginning of Late Paleocene time. Paleocene uplift in northwest Britain includes the shelf areas northwestward to Shetland, which were associated with the early Tertiary opening of the North Atlantic (Bott 1975; Rochow 1981). Delta-top lignite sequences in the Moray Firth Basin suggest marginal marine to subaerial sedimentation in this area (Rochow 1981; Ziegler 1990; Nadin & Kusznir 1995). No corresponding Paleocene uplift event has been detected for the Fennoscandian hinterland, which probably was partly submerged during this period (Ziegler 1990; Jordt *et al.* 1995). In principle, the sedimentation pattern should reflect the basin topography (Jordt *et al.* 1995). The depositional pattern of the first four sequences, CSS-1 to CSS-4, is characterized by sediment supply predominantly from the western source, with the sediments reaching far to the east. This gives reason to assume that the basin floor was dipping slightly to the east during the deposition of the first four sequences. Therefore, an asymmetrical basin form, where the westernmost part of the cross-section is characterized by shallow water depths as a result of the uplifted source area to the west, and water depths are about 700 m in the deepest part of the northern Viking Graben, is assumed for the beginning of Late Paleocene time. The eastern part of the basin is anticipated to have had relatively deep water depths at this time, indicating submerging of the westernmost part of the Norwegian mainland.

Several simulation runs using the revised initial form of the basin showed that an early overfill of the basin persisted. The rapid deepening of the basin from Early (200–500 m) to Late Paleocene time (500–1000 m) recognized on biostratigraphic data by Gradstein *et al.* (1994) and Gradstein & Bäckström (1996), is assumed to have had a tectonic origin. Therefore, it was decided to keep the maximum initial water depth at 700 m, and instead increase the tectonic subsidence rates compared with those given by Thorne & Watts (1989).

The tectonic subsidence suggested by Thorne & Watts (1989) and Hall & White (1994) is based on backstripping with the response to sedimentary loading assumed to be of local Airy-isostasy type (flexural rigidity $D = 0$), thus giving a minimum value. To improve the model, a lithospheric flexural rigidity of $D = 1.0 \times 10^{22}$ N m was applied. Therefore, subsidence rates should be increased compared with those given by Thorne & Watts (1989).

Geological studies (Jordt *et al.* 1995, 1996, 2000) suggest that the basin was filled at the end of sequence CSS-5 time and locally subjected to episodic subaerial erosion in Mid–Late Miocene time. This scenario has led to a modification of the tectonic subsidence (Fig. 13) pattern suggested by Thorne & Watts (1989), with the purpose of avoiding complete basin filling before the deposition of the CSS-5 sequence.

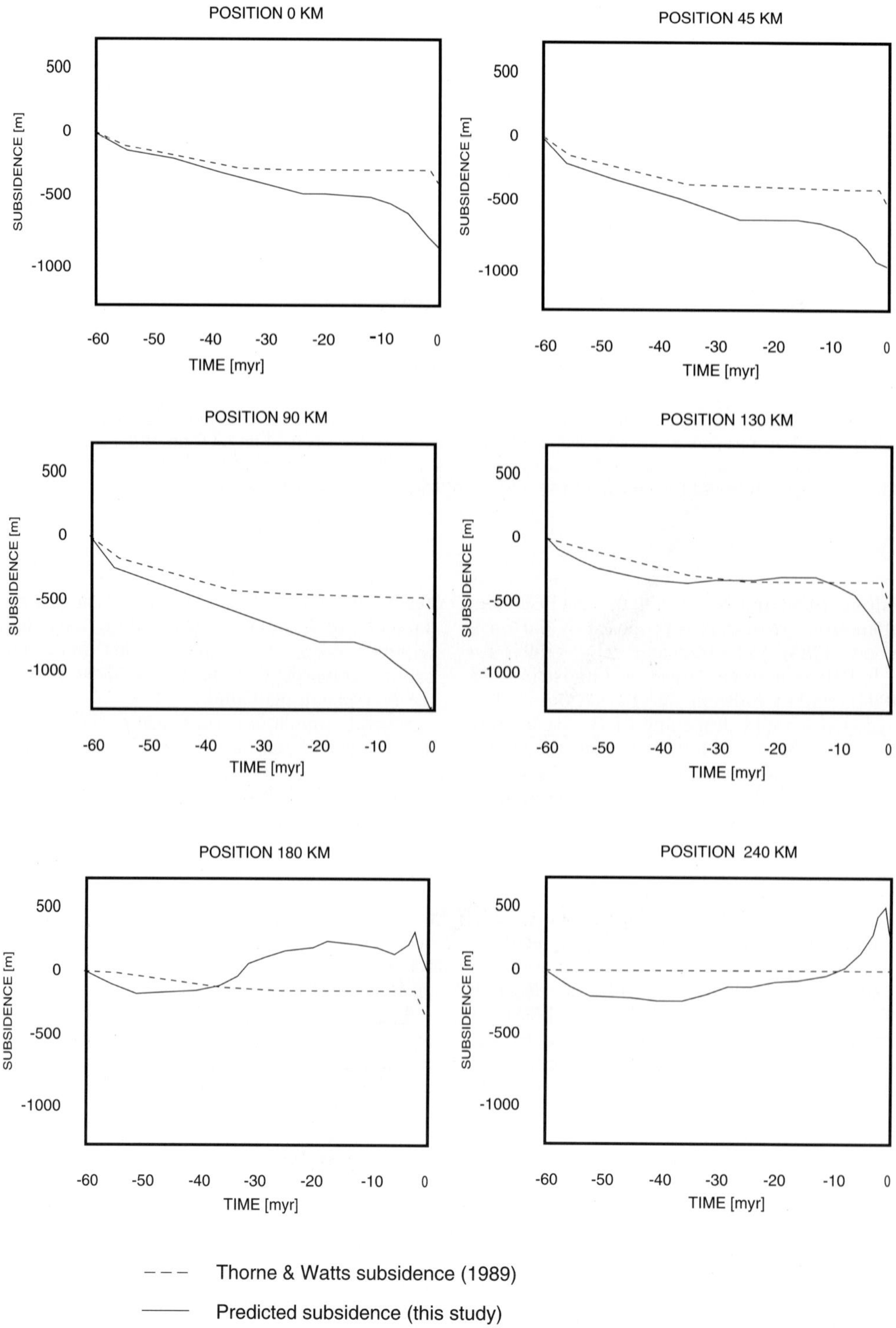

Fig. 13. Tectonic subsidence patterns with a flexural rigidity of 1.0×10^{22} N m at selected positions along the transect as predicted by the simulation. An accelerated tectonic subsidence pattern is detected for Paleocene time along the entire cross-section. In Late Miocene–Pliocene time, tectonic subsidence of the central and western part of the basin is detected, corresponding to the uplift of the eastern source area (Fennoscandia). Both of these subsidence events are anomalous relative to post-rift thermal subsidence (McKenzie 1978).

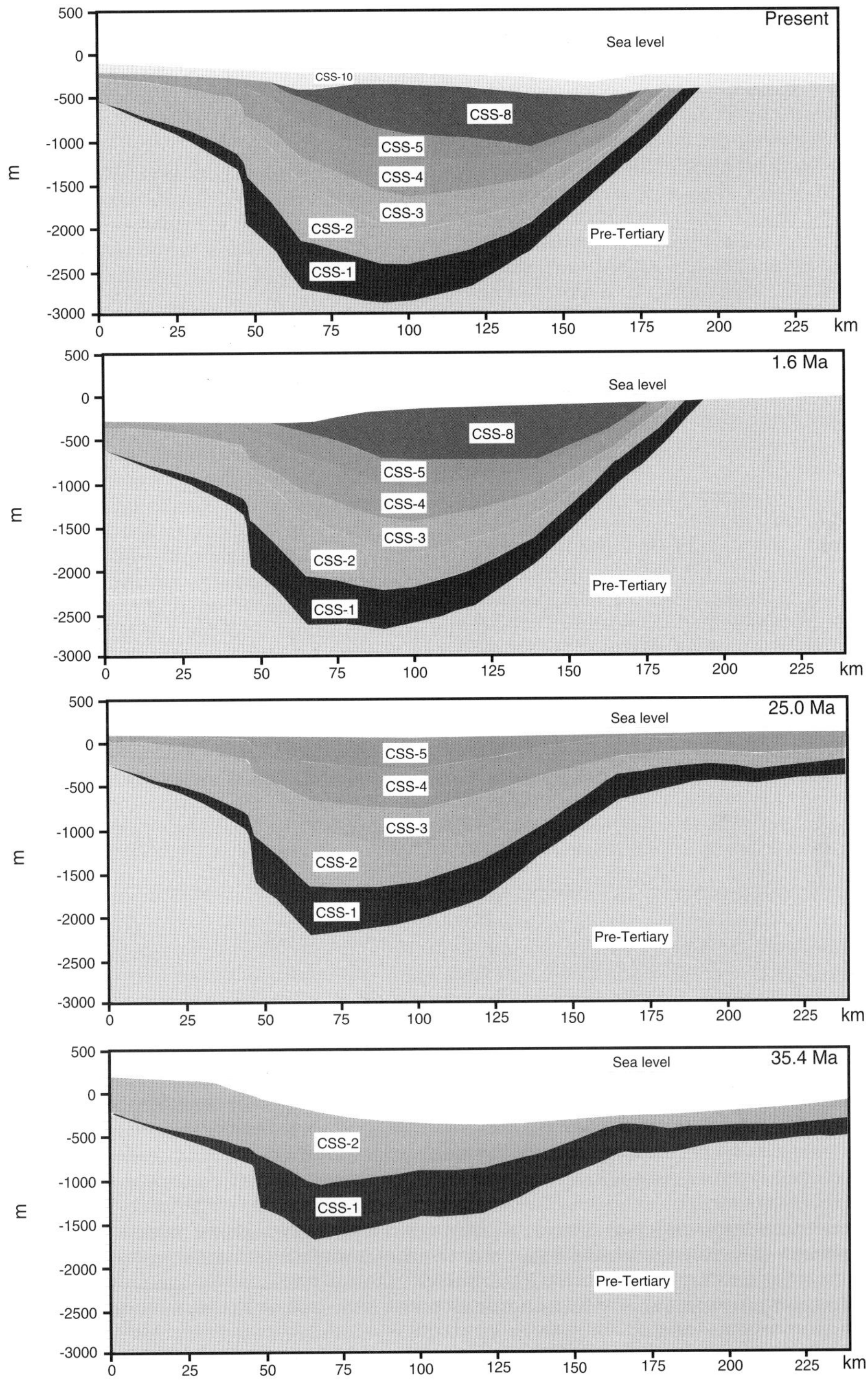

Fig. 14. Simulated basin geometry and stratigraphy at selected time steps in the Cenozoic basin evolution.

Modelling results, case 2

Figure 13 shows simulated temporal subsidence for selected localities along the cross-section resulting from the simulation. Accelerated subsidence is seen for the Paleocene CSS-1 sequence at all positions along the cross-section, and the subsidence rates are highest in the central part of the basin (at 90 km, Fig. 13). In the western (Fig. 13, 0 km and 45 km) and central parts of the basin (Fig 13, 90 km) subsidence rates in the Eocene–Oligocene CSS-2 to CSS-4 time show reduced, but still high rates. It is noted that sedimentation rates outpaced subsidence in the western part of the basin and the Eocene CSS-2 sequence built up to sea level (see Fig. 14). For the central part of the basin, Oligocene sedimentation rates also outpaced the subsidence rates and during Latest Oligocene–Mid-Miocene time, CSS-5 built up to sea level at the same time that subsidence ceased (see Fig. 14). During CSS-6 time, parts of the basin were subject to subaerial erosion. In the eastern central part of the basin (Fig. 13, 130 km), subsidence rates were low in Eocene time. In Late Eocene time, subsidence ceased. In the eastern part of the basin (Fig. 13, 180 km and 240 km), subsidence ceased in Eocene CSS-2 time. In the western (Fig. 13, 0 km and 45 km) and central parts of the basin, a new phase of subsidence started in Mid-Miocene time (CSS-7) and lasted throughout Pliocene time (CSS-8). Subsidence rates for the CSS-7 time are relatively gentle. During deposition of CSS-8, a considerable acceleration in subsidence occurred and provided accommodation for the Pliocene CSS-8 (Fig. 14) wedge, especially in the central part of the basin (Fig. 13, 90 km and 130 km). The eastern part of the cross-section (Fig. 13, 180 km and 240 km) shows uplift through Oligocene–Miocene time (CSS-3 to CSS-7), and a prominent uplift in Pliocene time (CSS-8). The uplift caused erosion of previously deposited sequences CSS-1 to CSS-4 from the flank and deposition in a rapidly subsiding basin in the west. The final cross-section with time-lines is shown in Fig. 14.

A comparison of the simulated and the observed basinal cross-sections (Fig. 15) shows some geometrical discrepancies, mainly caused by smoothing processes resulting from the diffusive algorithm used to handle erosion and deposition. Otherwise there is a striking similarity between the two. The similarity is sufficiently high to render the simulation realistic and justify the main conclusions derived from this study. This pertains particularly to the pattern of tectonic subsidence or uplift (Fig. 13) implied by the simulation and the pattern of source-area activity.

Discussion

The set of factors used in the simulation is not a unique solution. Other combinations of factors

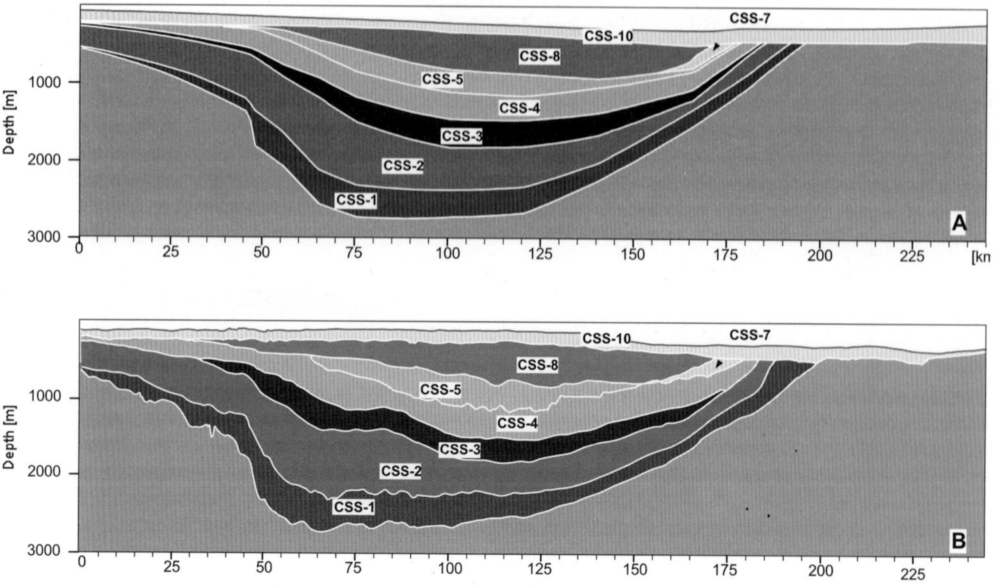

Fig. 15. Modelled (**a**) and observed (**b**) cross-section.

could undoubtedly give the same results. For instance, would inclusion of the source area reveal a totally different pattern of transport coefficients? A different sea-level fluctuation curve would affect the depositional pattern and the tectonic subsidence pattern, but among the published sea-level curves there are not curves of such a radical pattern to contradict our main results. An increased time-step resolution would make it possible to simulate a higher order of sea-level variations and to predict lithofacies distributions. We assume that our documentation and choice of factors is reasonable, and that the trends of the main results, therefore, are adequate.

The subsidence curves for different localities along the transect (Fig. 13) show an anomalous subsidence pattern, different from the expected curve of post-rift thermal subsidence (McKenzie 1978), both for Late Paleocene and Late Miocene–Pliocene times. Thorne & Watts (1989), White & Latin (1993) Hall & White (1994) and Lepercq & Gaultier (1996) have also pointed to the anomalous subsidence pattern in Paleocene time, with apparent subsidence acceleration. The Late Miocene–Pliocene subsidence increase, up to 1 km in magnitude, is anomalous, for it is not predicted by the existing model of lithospheric stretching, unless a late phase of Tertiary stretching is assumed to have occurred. However, there is little regional evidence of any significant normal faulting in late Tertiary time.

In this model, it was assumed that some of the Paleocene subsidence is incorporated in the initial basin form, because the simulation commences with the Late Paleocene development. In Fig. 16, the Cenozoic subsidence pattern for a pseudo-well, at 110.7 km from the section's left margin (west) (30/11-2), has been plotted in combination with the subsidence analysis by Hall & White (1994) from well 30/9-1, located relatively close to the pseudo-well site. The plot, showing the subsidence pattern from Jurassic time to the present, indicates anomalous subsidence rates in Paleocene (1) (CSS-1 sequence), and Late Miocene–Pliocene times (2) (CSS-7 to CSS-8 sequences).

The Paleocene subsidence rate may not be real, but rather an artefact of the backstripping procedure, which may have underestimated palaeobathymetries. According to Nadin & Kusznir (1995, 1996), no mechanism is required for the formation of Tertiary accommodation space other than post-Jurassic thermal subsidence 'buffered' by an event of transient Paleocene uplift. Bertram & Milton (1989) came to the same conclusion: if a water depth of 1 km is assumed for the Cretaceous and Tertiary development of the northern North Sea basin, then no post-Jurassic rifting needs to be invoked in the basin to explain its subsidence pattern. The latter workers also suggested an episode of uplift in the Paleocene which affected much of the basin's northwest margin. This uplift was subsequently eliminated by the margin collapse, except for the Inner Moray Firth area.

If the Paleocene subsidence rate from our simulation is considered, and we let it follow the curve of post-rift thermal subsidence (McKenzie 1978) (see upper dashed line in Fig. 16), the reduced subsidence creates the need for deeper water depths in the basin. In this case, a maximum basin depth of *c.* 1000 m would account for the anomalous subsidence pattern of Late Paleocene time. Water depths of that order have been suggested by Jones (1988) and Gradstein *et al.*

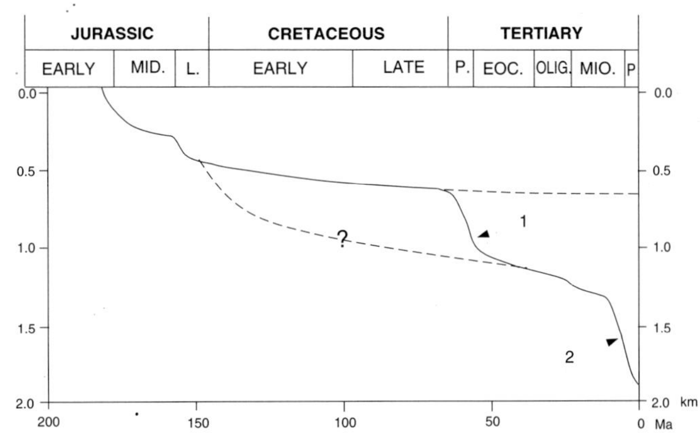

Fig. 16. Subsidence pattern for Early Jurassic time to the present from a pseudo-well on transect 2. Anomalous subsidence events, which do not follow the expected post-rift thermal subsidence pattern, are detected in Paleocene (1) and Late Miocene–Pliocene (2) times.

(1994), albeit those workers have also postulated a rapid deepening of the basin in Paleocene time.

The present simulation shows that the basin, after the Miocene episode of subaerial sedimentation, requires an anomalous amount of subsidence to provide sufficient accommodation space for Late Miocene–Pliocene sedimentation. To our knowledge, such an accelerated Pliocene subsidence has not been recognized in the northern North Sea. A possible coeval event has been suggested by Kooi *et al.* (1989) for the North Sea's Central Graben to the south, where the increased subsidence is attributed to a late-stage pulse of compression.

The present study does not reveal the causal mechanism to explain the Paleocene and Late Miocene–Pliocene anomalous subsidence, but indicates that such a subsidence pattern (Figs 13 and 16) is required according to the input values, which are considered to be realistic. The need for the anomalous Paleocene subsidence can be avoided only if the initial depth of the basin is taken to have been greater.

Conclusions

The main conclusions of this study are as follows.

The numerical simulation has produced a lithostratigraphic cross-section that matches fairly well the observed basinal cross-section (see Figs 14 and 15). The geometric departures are minor and do not significantly affect the similarity of the simulated cross-section to the observed one. The visual test renders the simulation result satisfactory.

The simulation indicates anomalous tectonic subsidence, different from the pattern of post-rift thermal subsidence predicted by general models, for Late Paleocene and Late Miocene–Pliocene time. This former episode of accelerated subsidence may partly be an artefact of the basin's underestimated initial water depth, but it is more difficult to attribute the latter episode to a mistaken input value. An application of a modified sea-level curve will not cause any major change of the simulation result. A decreased flexural rigidity will cause a lower degree of tectonic subsidence during the whole time interval considered.

In short, the result of the simulation seems to be valid and requires further lithostratigraphic analysis of the basin.

We thank O. Hansen of IKU for his valuable contribution to the computer simulation. We gratefully acknowledge the constructive comments made by A. Ryseth, J. C. Rivenæs, W. Nemec and R. Gabrielsen.

References

AIGNER, T., BRANDENBURG, A., VAN VLIET, A., DOYLE, M., LAWRENCE, D. & WESTRICH, J. 1990. Stratigraphic modelling of epicontinental basins: Two applications: *Sedimentary Geology*, **69**, 167–190.

BADLEY, M. E., PRICE, J. D., RAMBECH-DAHL, C. & AGDESTEIN, T. 1988. The structural evolution of the northern Viking Graben and its bearing upon extensional models of basin formation. *Journal of the Geological Society, London*, **145**, 455–472.

BARTON, P. & WOOD, R. 1984. Tectonic evolution of the North Sea basin: crustal stretching and subsidence. *Geophysical Journal of the Royal Astronomical Society*, **79**, 987–1022.

BEACH, A., BIRD, T. & GIBBS, A. 1987. Extensional tectonics and crustal structure: deep seismic reflection data from the northern North Sea Viking Graben. *In*: COWARD, M. P., DEWEY, J. F. & HANCOCK, P. L. (eds) *Continental Extensional Tectonics*. Geological Society, London, Special Publications, **28**, 467–476.

BERTRAM, G. T. & MILTON, N. J. 1989. Reconstructing basin evolution from sedimentary thickness; the importance of palaeobathymetric control, with reference to the North Sea. *Basin Research*, **1**, 247–257.

BOTT, M. H. P.1975. The structure and evolution of the North Sea Basin, the Faeroe Block and the intervening region. *In*: WOODLAND, A. W. (ed.) *Petroleum and the Continental Shelf of North West Europe, Vol. 1: Geology*. Applied Science, London, 105–116.

CHRISTIANSSON, P., FALEIDE, J. I. & BERGE, A. M. 1999. Crustal structure in the northern North Sea; an integrated geophysical study. *This volume*.

DEEGAN, C. E. & SCULL, B. J. 1977. *A Proposed Lithostratigraphic Nomenclature for the Central and Northern North Sea*. Norwegian Petroleum Directorate Bulletin **1**.

EYNON, G. 1981. Basin development and sedimentation in the Middle Jurassic of the northern North Sea. *In:* ILLING, L. V. & HOBSON, G. D. (eds) *Petroleum Geology of the Continental Shelf of NW Europe*. Heyden, London, 196–204.

FJELDSKAAR, W. 1994. Viscosity and thickness of the asthenosphere detected from the Fennoscandian uplift. *Earth and Planetary Science Letters*, **126**, 399–410.

FLEMINGS, P. B. & JORDAN, T. E. 1989. A synthetic stratigraphic model of foreland basin development. *Journal of Geophysical Research*, **94**, 430–434.

FÆRSETH, R. B., GABRIELSEN, R. H. & HURICH, C. A. 1995. Influence of basement in structuring of the North Sea Basin offshore southwest Norway. *Norsk Geologisk Tidsskrift*, **75**, 105–119.

GABRIELSEN, R. H., FÆRSETH, R. B., STEEL, R. J., IDIL, S. & KLØVJAN, O. S. 1990. Architectural styles of basin fill in the northern Viking Graben. *In*: BLUNDELL, D. J. & GIBBS, A. D. (eds) *Tectonic Evolution of the North Sea Rifts*. Oxford University Press, Oxford, 158–179.

GALLOWAY, W. E., GARBER, J. L., LIU, X. & SLOAN, B. J. 1993. Sequence stratigraphic and depositional framework of the Cenozoic fill, Central and northern North Sea. *In*: PARKER, J. R. (ed.) *Petroleum Geology of Northwest Europe: Proceedings of the 4th Conference*. Geological Society, London, 33–43.

GILTNER, J. P. 1987. Application of extensional models to the northern Viking Graben. *Norsk Geologisk Tidsskrift*, **67**, 339–352.

GLENNIE, K. W. (ed.) 1984. *Introduction to the Petroleum Geology of the North Sea*. Blackwell, Oxford, 1–236.

GRADSTEIN, F. M. & BÄCKSTRÖM, S. 1996. Cainozoic biostratigraphy and palaeobathymetry, northern North Sea and Haltenbanken. *Norsk Geologisk Tidsskrift*, **76**, 3–32.

——, KAMINSKI, M. A., BERGGREN, W. A., KRISTIANSEN, I. L. & D'IORIO, M. A. 1994. Cenozoic biostratigraphy of the North Sea and Labrador Shelf. *Micropalaeontology*, **40**(Suppl.), 1–152.

HALL, B. D., & WHITE, N. 1994. Origin of anomalous Tertiary subsidence adjacent to North Atlantic continental margins. *Marine and Petroleum Geology*, **11**, 702 – 714.

HAQ, B. U., HARDENBOL, J. & VAIL, P. R. 1987. Chronology of fluctuating sea levels since Triassic (250 Myr age to present). *Science*, **235**, 1156–1167.

HARLAND, W. B., ARMSTRONG, R. L., COX, A. V., CRAIG, L. E., SMITH, A. G. & SMITH, D. G. 1990. *A Geologic Time Scale 1989*. Cambridge University Press, Cambridge.

HELLAND-HANSEN, W., KENDALL, C. G. ST C., LERCHE, I. & NAKAYAMA, K. 1988. A simulation of continental basin margin sedimentation in response to crustal movements, eustatic sea level change, and sediment accumulation rates. *Mathematics and Geology*, **20**, 777–802.

ISAKSEN, D. & TONSTAD, K. (eds) 1989. *A Revised Cretaceous and Tertiary Lithostratigraphic Nomenclature for the Norwegian North Sea*. Norwegian Petroleum Directorate Bulletin **5**.

JONES, G. D 1988. A palaeoecological model of Late Palaeocene 'Flysch-type' agglutinated foraminifera using the palaeoslope transect approach, Viking Graben, North Sea. *Abhandlungen der Geologischen Bundesanstalt*, **41**, 143–153.

JORDT, H. 1996. *The Cenozoic geological evolution of the Central and Northern North Sea based on seismic sequence stratigraphy*. Dr. Scient. thesis, University of Oslo.

——, ——, —— & IBRAHIM, M. T. 1995. Cenozoic sequence stratigraphy of the Central and Northern North Sea basin: tectonic development, sediment distribution and provenance areas. *Marine Petroleum Geoleum*, **12**, 845–879.

——, THYBERG, B. I. & NØTTVEDT, A. 2000. Cenozoic evolution of the central and northern North Sea with focus on differential vertical movements of the basin floor and surrounding clastic source areas. *This volume*.

KENDALL, C. G. ST C., STROBEL, J., CANNON, R., BEZDEK, J. & BISWAS, G. 1991. The simulation of the sedimentary fill of basins. *Journal of Geophysical Research*, **96**, 6911–6929.

KENYON, P. M. & TURCOTTE, D. L. 1985. Morphology of a prograding delta by bulk sediment transport. *Geological Society of American Bulletin*, **96**, 1457–1465.

KNOX, R. W. & HOLLOWAY, S. 1992. Palaeogene of the Central and Northern North Sea. *In*: KNOX, R. W. & CORDEY, W. G. (eds) *Lithostratigraphic Nomenclature of the UK North Sea*. British Geological Survey, Keyworth.

KOOI, H., CLOETINGH, S. & REMMELTS, G. 1989. Intraplate stresses and the stratigraphic evolution of the North Sea Central Graben. *Geologie en Mijnbouw*, **68**, 49–72.

LAWRENCE, D. T., DOYLE, M. & AIGNER, T. 1990. Stratigraphic simulation of sedimentary basin: Concepts and calibration. *AAPG Bulletin*, **74**, 273–295.

LEPERCQ, J. Y & GAULIER, J. M. 1996. Two-stage rifting in the North Viking Graben area (North Sea): inferences from a new three-dimensional subsidence analysis. *Marine and Petroleum Geology*, **13**, 129–148.

MCKENZIE, D. P. 1978. Some remarks on the development of sedimentary basins. *Earth and Planetary Science Letters*, **40**, 25–32.

NADIN, P. A. & KUSZNIR, N. J. 1995. Palaeocene uplift and Eocene subsidence in the northern North Sea Basin from 2D forward and reverse stratigraphic modelling. *Journal of the Geological Society*, London, **152**, 833–848.

—— & ——1996. Forward and reverse stratigraphic modelling of Cretaceous–Tertiary post-rift subsidence and Palaeogene uplift in the Outer Moray Firth Basin, central North Sea. *In*: KNOX, R. W., CORFIELD, R. M. & DUNAY, R. E. (eds) *Correlation of the Early Palaeogene in Northwest Europe*. Geological Society, London, Special Publication, **101**, 43–62

ODINSEN, T., CHRISTIANSSON, P., GABRIELSEN, R. H., FALEIDE, J. I. & BERGE, A. M. 2000*a*. The geometries and deep structure of the northern North Sea rift system. *This volume*.

——, REEMST, P., VAN DER BEEK, P., FALEIDE, J. I., GABRIELSEN, R. H. 2000*b*. Permo-Triassic and Jurassic extension in the northern North Sea: results from tectonostratigraphic forward modelling. *This volume*.

RIVENÆS, J. C. 1992. Application of a dual-lithology, depth-dependent diffusion equation in stratigraphic simulation. *Basin Research*, **4**, 133–146.

——1993. *A computer simulation model for siliciclastic basin stratigraphy*. Dr.Ing. thesis, University of Trondheim.

ROBERTS, A. M., KUSZNIR, N. J., WALKER, I. & DORN-LOPEZ, D. 1995. Quantitative analysis of Triassic extension in the northern North Sea. *Journal of the Geological Society, London*, **152**, 15–26.

ROCHOW, K. A. 1981. Seismic stratigraphy of the North Sea 'Palaeocene' deposits. *In:* ILLING, L. V. & HOBSON, G. D. (eds) *Petroleum Geology of the Continental Shelf of NW Europe*. Heyden, London, 255–266.

RUNDBERG, Y. 1989. *Tertiary sedimentary history and basin evolution of the Norwegian North Sea between 60°–62°N – An integrated approach.* Dr.Ing. thesis, University of Trondheim.

——, OLAUSSEN, S. & GRADSTEIN, F. 1995. Incision of Oligocene strata; evidence for northern North Sea Miocene uplift and key to the formation of the Utsira Sands. *Geonytt*, 22, 62.

SCLATER, J. G. & CHRISTIE, P. A. F. 1980. Continental stretching: An explanation of the post-Mid Cretaceous subsidence of the Central North Sea Basin. *Journal of Geophysical Research*, **85**, 3711–3739.

SINCLAIR, H. D., COAKLEY, B. J., ALLEN, P. A. & WATTS, A. B. 1991. Simulation of foreland basin stratigraphy using a diffusion model for mountain belt uplift and erosion: an example form the Central Alps, Switzerland. *Tectonics*, **10**, 599–620.

STERBAUT, E., SPIEGLER, D., WEINELT, M. & THIEDE, J. 1991. *Cenozoic Erosion and Sedimentation on the Northwest European Continental Margin.* Geomar Research Centre for Marine Geosciences, Christian-Albrechts Universität Kiel.

STEWART, I. J. 1987. A revisited stratigraphic interpretation of the Early Palaeogene of the Central North Sea. *In*: BROOKS, J. & GLENNIE, K. (eds) *Petroleum Geology of North West Europe.* Graham and Trotman, London, 557–576.

SYVITSKI, J. P. M., SMITH, J. N., CALABRESE, E. A. & BOUDREAU, B. P. 1988. Basin sedimentation and the growth of prograding deltas. *Journal of Geophysical Research*, **93**, 6895–6908.

TER VOORDE, M., FÆRSETH, R. B., GABRIELSEN, R. H. & CLOETINGH, S. A., P. L. 1997. Repeated lithosphere extension in the northern Viking Graben: a coupled or a decoupled rheology? *This volume.*

THORNE, J. A. & WATTS, A. B. 1989. Quantitative analysis of the North Sea subsidence. *AAPG Bulletin*, **73**, 88–116.

WHITE, N. & LATIN, D. 1993. Subsidence analysis from the North Sea 'triple junction'. *Journal of the Geological Society, London*, **150**, 473–488.

ZIEGLER, P. A. 1981. Evolution of sedimentary basins in Northwest Europe. *In:* ILLING, L. V. & HOBSON, G. D. (eds) *Petroleum Geology of the Continental Shelf of NW Europe.* Heyden, London, 3–39.

——1982*a*. Faulting and graben formation in western and central Europe. *Philosophical Transactions of the Royal Society of London, Series A*, **305**, 113–143.

——1982*b*. *Geological Atlas of Western and Central Europe*, Elsevier, Amsterdam.

——1990. *Geological Atlas of Western and Central Europe.* Shell International, The Hague.

NE Atlantic continental rifting and volcanic margin formation

JAKOB SKOGSEID,[1,3] SVERRE PLANKE,[1] JAN INGE FALEIDE,[1] TOM PEDERSEN,[2] OLAV ELDHOLM[1] & FLEMMING NEVERDAL[1]

[1] *Department of Geology, University of Oslo, P.O. Box 1047 Blindern, N-0316 Oslo, Norway*
[2] *Institute for Energy Technology, P.O. Box 40, N-2007 Kjeller, Norway*
[3] *Present address: Saga Petroleum ASA, P.O. Box 490, N-1302 Sandvika, Norway (e-mail: Jakob.Skogseid@saga.com)*

Abstract: Deep seismic data from the Hatton–Rockall region, the mid-Norway margin and the SW Barents Sea provide images of the crustal structure that make it possible to estimate the relative amounts of crustal thinning for the Late Jurassic–Cretaceous and Maastrichtian–Paleocene NE Atlantic rift episodes. In addition, plate reconstructions illustrate the relative movements between Eurasia and Greenland back to Mid-Jurassic time. The NE Atlantic rift system developed as a result of a series of rift episodes from the Caledonian orogeny to early Tertiary time. The Late Palaeozoic rifting is poorly constrained, particularly with respect to timing. However, rifted basin geometries, inferred to be of this age, are observed at depth in seismic data on the flanks of the younger rift structures. Intra-continental rifting in Late Jurassic–Cretaceous times caused *c.* 50–70 km of crustal extension and subsequent Cretaceous basin subsidence from the Rockall Trough–North Sea areas in the south, to the SW Barents Sea in the north. In late Early to early Late Cretaceous times, new rifting occurred in the Rockall Trough and Labrador Sea associated with the northward propagation of North Atlantic sea-floor spreading. When sea-floor spreading was approached in the Labrador Sea the Rockall rift apparently became extinct. The final NE Atlantic rift episode was initiated near the Campanian–Maastrichtian boundary, lasted until continental separation near the Paleocene–Eocene transition, and caused *c.* 140 km extension. The late syn-rift and the earliest sea-floor spreading periods were affected by widespread igneous activity across a *c.* 300 km wide zone along the rifted plate boundary. The deep seismic data provide lower-crustal structural geometries that represent boundary conditions for a better mapping and understanding of the extensional thinning of the crust. The crustal geometries question extension estimates previously made from basin subsidence analysis, and aid in the definition of bodies of magmatic underplating beneath the outer volcanic margins.

The opening of the NE Atlantic Ocean at the Paleocene–Eocene transition marked the culmination of a *c.* 340 Ma history of extensional deformation and sediment basin formation since the end of the Caledonian orogeny (Fig. 1). In this paper we utilize previously published crustal transects between the Hatton–Rockall margin and the western Barents Sea to evaluate the degree of crustal thinning in the various regions with respect to the rifting history of the area.

The evaluation includes first a discussion of the observed structures and associated extension estimates made for separate rift episodes with respect to a relatively simple tectonic model. We then attempt to illustrate the spatial and temporal geological evolution by using extension estimates in plate reconstructions and transect restorations. The reconstruction to magnetic anomaly 24b time (53 Ma), the oldest magnetic spreading anomaly in the region, is based on interpretation of all available magnetic data for the NE Atlantic (Fig. 1). Finally, a presentation of different aspects of plume–lithosphere interaction and associated vertical motion during the margin formation is linked with a discussion of exploration challenges at volcanic margins in general.

Deep seismic reflection profiles provide images of both Moho and intra-crustal geometries, which in particular allow us to better constrain the dimensions of the Maastrichtian–Paleocene break-up-related rift, and the Late Jurassic–Cretaceous intra-continental rift. Integration of deep seismic reflection data with deep seismic refraction transects provides also the velocity distribution of the lower crust (Mutter *et al.* 1984, 1988; Gudlaugsson *et al.* 1987; Hinz *et al.* 1987; White *et al.* 1987; Jackson *et al.* 1990; Faleide *et al.* 1991, 1993*b*; Makris *et al.* 1991; Olafsson *et al.* 1991; Planke *et al.* 1991; Larsen &

From: Nøttvedt, A. *et al.* (eds) *Dynamics of the Norwegian Margin*. Geological Society, London, Special Publications, **167**, 295–326. 1-86239-056-8/00/$15.00

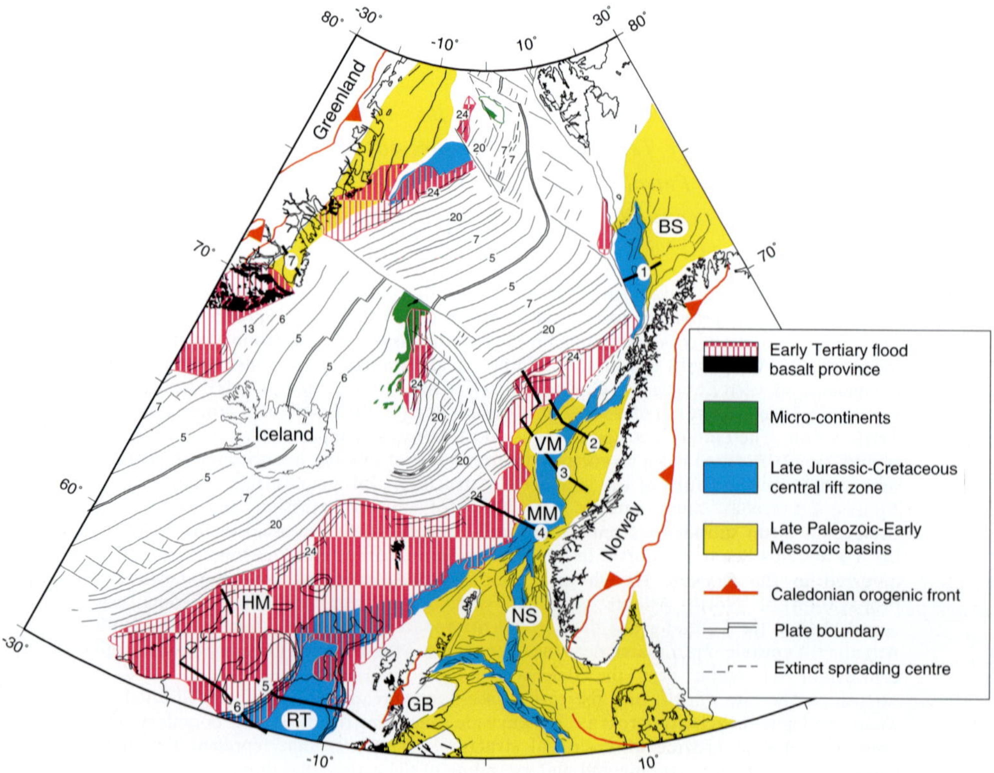

Fig. 1. Magnetic sea-floor spreading anomalies (numbered 5–7, 13, 20, 24) and structural setting in the NE Atlantic region. Transects 1–7 discussed in text are located by bold lines. The outline of the flood basalt province in the regions off Britain and in vicinity of the Faeroe Islands is based on Hitchen & Ritchie (1993) and Boldreel & Andersen (1994). BS, SW Barents Sea; GB, Great Britain; MM, Møre margin; NS, North Sea; RT, Rockall Trough; VM, Vøring margin; HM, Hatton-Rockall margin.

Marcussen 1992; Shannon *et al.* 1993; Eldholm & Grue 1994; Skogseid & Eldholm 1995; Mjelde *et al.* 1997; Christiansson *et al.* 1999). Such data reveal the basin geometries, and allow quantitative evaluation and modelling of the various rift episodes. None the less, the deep seismic database is limited. Moreover, multiple rifting, extensive flood basalts and sill intrusions along the outer margin make the interpretation of the crustal structure ambiguous in large parts of the NE Atlantic study area.

Regional geological framework

Late Palaeozoic–Early Mesozoic phase

The location and structural expression of the late Palaeozoic NE Atlantic rift system within the Caledonian orogenic domain was influenced by Caledonian, and possibly pre-Caledonian, structures. Between Norway and Greenland the rift system followed the NE-oriented Caledonides into the SW Barents Sea, where north-trending structures suggest a structural connection to the Arctic rift system. Between Britain and Greenland basin trends indicate a possible reactivation of the Appalachian deformation system.

A phase of late orogenic extensional collapse during latest Silurian–Early Devonian time apparently initiated the formation of a series of large half-graben basins, which subsequently were filled with thick successions of mainly intra-continental deposits (e.g. Ziegler 1988; Coward 1993; Hartz & Andresen 1995). An orogenic collapse of the Barents Sea Caledonides southeast of Bjørnøya has been suggested, on the basis of structures beneath the flanks of the SW Barents Sea Permo-Carboniferous rift system (Gudlaugsson *et al.* 1994). If we consider orogenic collapse mainly a syn-orogenic phenomenon, we assume that rifting in response to lithospheric extension between Eurasia and Greenland was initiated at the end of Devonian time. The main Late Palaeozoic rift episodes took place in Mid-Carboniferous, Carboniferous–Permian and Permian–Early Triassic times (e.g. Ziegler 1988).

Late Jurassic–Cretaceous phase

During Mid-Jurassic time the development of oceanic crust in the central Atlantic marked the onset of a new kinematic regime in the Atlantic domain (e.g. Larsen 1987; Ziegler 1988). Following the separation between Africa and North America, sea-floor spreading was confined to the region south of the Azores Fracture Zone for *c.* 50 Ma. Rifting persisted, however, into the North Atlantic domain, as documented by rift-associated sediments of Late Jurassic age commonly found in North Atlantic basins. Late Jurassic–Cretaceous rifting has in this context been considered a precursor to the progressive continental separation in the southern part of the North Atlantic, where the final break-up between Iberia and the Grand Banks has been given a Hauterivian age (*c.* 134 Ma) (Srivastava & Tapscott 1986; Keen & de Voogd 1988; Whitmarsh *et al.* 1993). Thus, at the end of the Jurassic period deep rift basins probably existed from the Rockall Trough to the SW Barents Sea, with a separate branch in the North Sea, with a possible linkage to spreading in the Tethys Ocean (Lundin & Doré 1997) (Fig. 1). Sea-floor spreading progressed northward, and the opening between Newfoundland and Goban Spur–Porcupine Bank began in late Aptian to early Albian (*c.* 112 Ma) times according to De Graciansky *et al.* (1985) and Srivastava *et al.* (1988).

Further north, the Labrador Sea and the Rockall Trough developed contemporaneously with rifting and drifting to the south. In the Labrador Sea gabbroic dykes of Berriasian–Valanginian (133–138 Ma) age in southwest Greenland are related to incipient rifting (Watt 1969; Larsen *et al.* 1999). On the basis of data from commercial wells off Canada, the oldest known rift deposits here are coarse-grained non-marine sandstones of Barremian age. These sediments lie, however, partly on Neocomian basalts (Berrasian–Barremian, Alexis formation) (Balkwill *et al.* 1990). The basalts are thus of the same age as the Greenland dykes and may suggest that even Upper Jurassic sedimentary rocks may occupy the deep, undrilled parts of some rift basins. The youngest rift deposits sampled in wells are Coniacian in age. Extension in the Labrador Sea continued to continental separation between North America and Greenland in Late Cretaceous (92 Ma, Roest & Srivastava 1989) to earliest Tertiary times (62 Ma, Chalmers *et al.* 1993).

The development of the Rockall Trough is more disputed (Roberts 1974, 1975*a*, *b*; Russel 1976; Kristoffersen 1978; Russel & Smythe 1978; Srivastava 1978; Roberts *et al.* 1981; Smythe 1989). Irrespective of the nature of the crust flooring the trough (oceanic or continental) it is clear that significant amounts of extension have been taken up along the structure (Joppen & White 1990; Makris *et al.* 1991; Doré 1992; Keser Neish 1993; Knott *et al.* 1993; O'Reilly *et al.* 1996). There seems to be consensus about a Cretaceous age for the last major extensional phase in the area. From plate reconstructions, Srivastava & Verhoef (1992) assumed that 65% of the trough's present width can be accounted for by Late Jurassic and Cretaceous extension. Further north the Late Jurassic boundary faults along the West Shetland Platform were mildly reactivated in Aptian time, and in the Faeroe Basin, which may be considered the northern extremity of the Rockall Trough, the Early Cretaceous period was characterized by increased subsidence rates (Duindam & Van Horn 1987; Hitchen & Ritchie 1987; Mudge & Rashid 1987; Nelson & Lamy 1987).

Off Norway the Møre and Vøring basins were formed primarily as a result of the Late Jurassic–Early Cretaceous rifting, whereas it may be questioned whether or not large-scale extension took place in this region contemporaneous with rifting to break-up in the Labrador Sea. According to Surlyk *et al.* (1981), coarse clastic submarine fans associated with deep-water shales suggest activity along the major boundary faults in central East Greenland persisting into Aptian–Albian time, and coeval activity has been reported along the flanks of main fault zones in the Møre and Vøring basins off Norway (Blystad *et al.* 1995; Doré *et al.* 1997*a*). Furthermore, according to Dalland (1981) Albian–Aptian tectonism is reported from Andøya, northern Norway, and three Early Cretaceous tectonic phases, Berriasian–Valanginian, Hauterivian–early Barremian and Aptian–Albian, have been reported from the SW Barents Sea (Faleide *et al.* 1993*a*,*b*).

Late Cretaceous–Tertiary phase

The final Maastrichtian–Paleocene rift episode lasted for *c.* 20 Ma, leading into continental separation and onset of sea-floor spreading at the Paleocene–Eocene transition. This rift episode formed a more than 300 km wide zone associated with lithospheric thinning and post-break-up subsidence (Skogseid 1994). In Paleocene time (63–62 Ma) the rift was affected by the impingement of the Iceland mantle plume beneath the thinned lithosphere, which induced considerable regional uplift. Subsequent igneous activity characterized the rift history. The magmatism was governed by the lithosphere relief, which provided pressure conditions (thin-spots) for voluminous melt generation (Thompson & Gibson 1991; Skogseid *et al.* 1992*a*). The continental separation and initial sea-floor spreading were associated with a 2–3 Ma period of massive volcanic activity along the more than 2600 km long new plate boundary (Eldholm & Grue 1994) (Fig. 1).

Stretching estimates

In this study the amount of crustal thinning is estimated from the present thickness of the continental crust along the transects in Fig. 1. The thinning is compared with results from independent estimates based on standard basin subsidence analyses, i.e. uniform extension models (McKenzie 1978; Steckler & Watts 1978), applied to the NE Atlantic margins (Skogseid *et al.* 1992*b*; Skogseid 1994; Roberts *et al.* 1997; Walker *et al.* 1997). All methods have limitations, and ideally require input data that are difficult to obtain along most transects. Nevertheless, we claim that the integration of subsidence modelling and crustal structure evaluation provides the best platform for analysing crustal extension and for furthering our understanding of the extensional development of the region.

With reference to the rifting history we subdivide the sedimentary succession in the crustal transects into three mega-sequences (Figs 2, 3 and 4). The pre-Cretaceous strata represent an unspecified tectonic and basin subsidence history for which little information exists in the offshore regions. The Cretaceous sequence outlines the basins formed after the Late Jurassic–Cretaceous rifting; whereas the Tertiary sequence and the present water

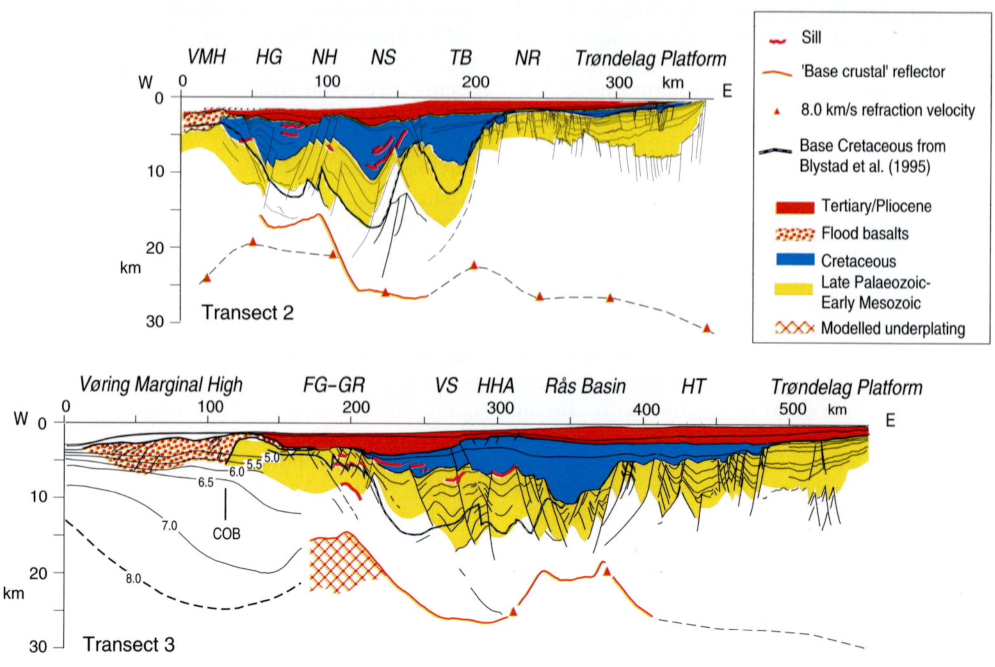

Fig. 2. Crustal transects 2 and 3 across the northern and southern Vøring margin (modified from Skogseid & Eldholm 1995) with an alternative interpretation of the base Cretaceous horizon from Blystad *et al.* (1995). Crustal velocities beneath the Vøring Marginal High (VMH) are annotated in km s^{-1}. COB, Continent–ocean boundary; FG–GR, Fenris Graben–Gjallar Ridge; HG, Hel Graben; HHA, Helland-Hansen Arch; HT, Halten Terrace; NH, Nyk High; NR, Nordland Ridge; NS, Någrind Syncline; TB, Trœna Basin; VS, Vigrid Syncline. Location shown in Fig. 1.

depth approximate the amount of subsidence since the Maastrichtian–Paleocene rift episode. It should be noted, however, that a large Plio-Pleistocene sedimentary wedge is part of the Tertiary mega-sequence in the eastern part of the transects. These deposits are related to continental uplift and glacial erosion of Fennoscandia (Riis & Fjeldskaar 1992; Stuevold *et al.* 1992).

The basin analysis depends on the definition of the base Cretaceous and the base Tertiary levels, separating the three mega-sequences. These horizons, which correspond to major hiatuses or condensed sequences on the basin flanks, may be bounding or found within thick sedimentary units reflecting rapid differential subsidence in the deepest basins. In fact, the base Cretaceous unconformity does not mark the transition from syn-rift to post-rift sediments in the North Atlantic rift, where rifting continued well into Early Cretaceous times. Ideally, tectono-structural considerations should have been made with reference to the top of the Middle Jurassic shallow marine sandstones, which marks the onset of Late Jurassic rifting and rapid subsidence. It is, however, difficult to ascertain this level because of the lack of stratigraphic control in the deep rift basins. Similarly, the Campanian–Maastrichtian horizon should be used to bound the last rift episode. None the less, here the near-base Tertiary rift unconformity is at present the best approximation in the seismic data. Erosion across the top of rotated fault blocks beneath this horizon is further used as an indicator of near sea-level conditions in Paleocene time.

The uncertainties in the seismic interpretation, which are subsequently implemented in the modelling, can be illustrated both in the SW Barents Sea and on the Vøring margin. In the SW Barents Sea the base Cretaceous horizon represents a main rift unconformity on the flanks of the deep Tromsø Basin, whereas it is located 1–2 s twt above the top Middle Juras-

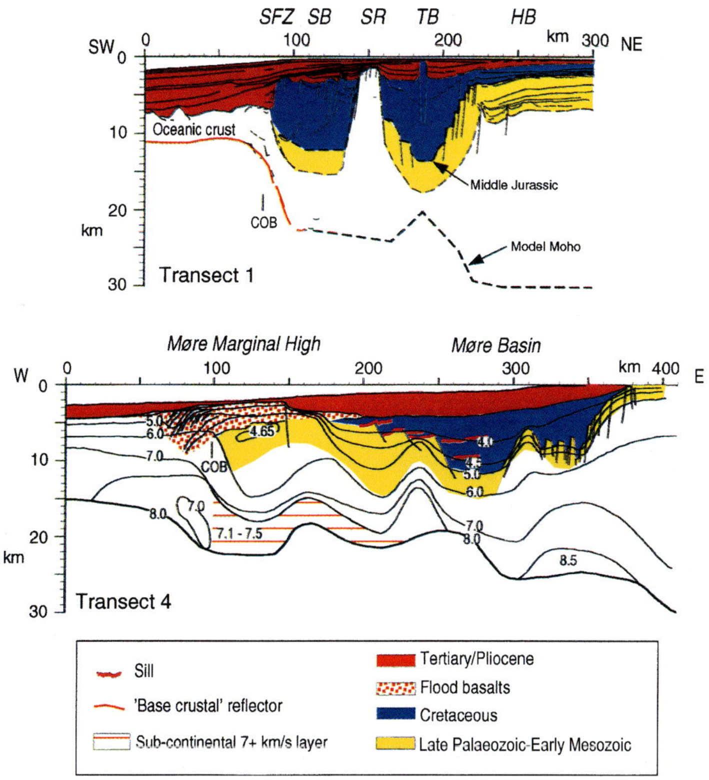

Fig. 3. Crustal transects 1 and 4 across the southwestern Barents Sea and Møre margins, respectively (modified from Olafsson *et al.* 1991; Faleide *et al.* 1993*b*; Breivik *et al.* 1998). Crustal velocities are annotated in km s^{-1}. COB, Continent–ocean boundary; HB, Hammerfest Basin; SB, Sørvestnaget Basin; SFZ, Senja Fracture Zone; SR, Senja Ridge; TB, Tromsø Basin. Location shown in Fig. 1.

sic level in the basin itself (Faleide *et al.* 1993*b*). On the Vøring margin the structural and stratigraphical continuity associated with the base Cretaceous horizon from the Rås and Træna basins westward into the Vigrid and Någrind synclines, respectively, is poorly constrained. The main reason is that the seismic data beneath the Tertiary Helland-Hansen Arch are generally poor, and that igneous intrusions make the structural setting difficult to interpret with confidence. To illustrate the variation in interpretation we have implemented the suggested base Cretaceous horizon from Blystad *et al.* (1995) (Fig. 2). However, our interpretation is aided by the observed lower-crustal geometries. The deep seismic reflection data show a *c.* 45–70 km wide zone of pronounced crustal thinning beneath the Rås Basin, whereas the Moho interface lies significantly deeper both eastward and westward of this zone. This configuration resembles the structural setting in the Rockall Trough, the Møre and Tromsø basins, and the northern North Sea, where a similar spatial correlation exists between the upper-crustal rift and areas of pronounced crustal thinning (Figs 2–4). In our opinion we find that a relatively narrow central rift basin therefore is more plausible than the very broad deep basin suggested by Blystad *et al.* (1995). Furthermore, our interpretation allows for thick units of pre-Cretaceous strata beneath the Vøring Basin, which is consistent with observed stratigraphical successions both beneath the adjacent Trøndelag Platform (Fig. 2) and in Greenland.

Vøring margin transects

We use the Vøring margin transects (Figs 2, 5 and 6) to illustrate the evaluation procedure for the crustal thinning derived extension. We interpret the *c.* 45 km long reflector segment at 8.5–9 s twt (*c.* 20 km) as Moho beneath the Rås Basin. On either side Moho deepens to 10–11 s twt, i.e. *c.* 25 km (Skogseid & Eldholm

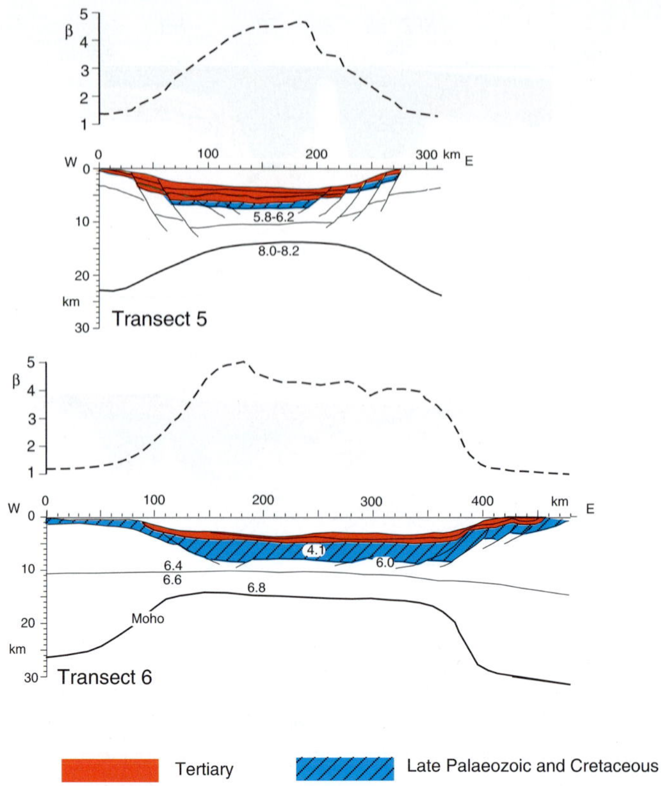

Fig. 4. Crustal transects 5 and 6 and associated degree of crustal thinning (β) across the Rockall Trough (modified from Makris *et al.* 1991; Keser Neish 1993; Shannon *et al.* 1993). Crustal velocities are annotated in km s^{-1}. (Note the relatively rapid decay in width of the Rockall Trough between the two transects, which are located *c.* 200 km apart.) Location shown in Fig. 1.

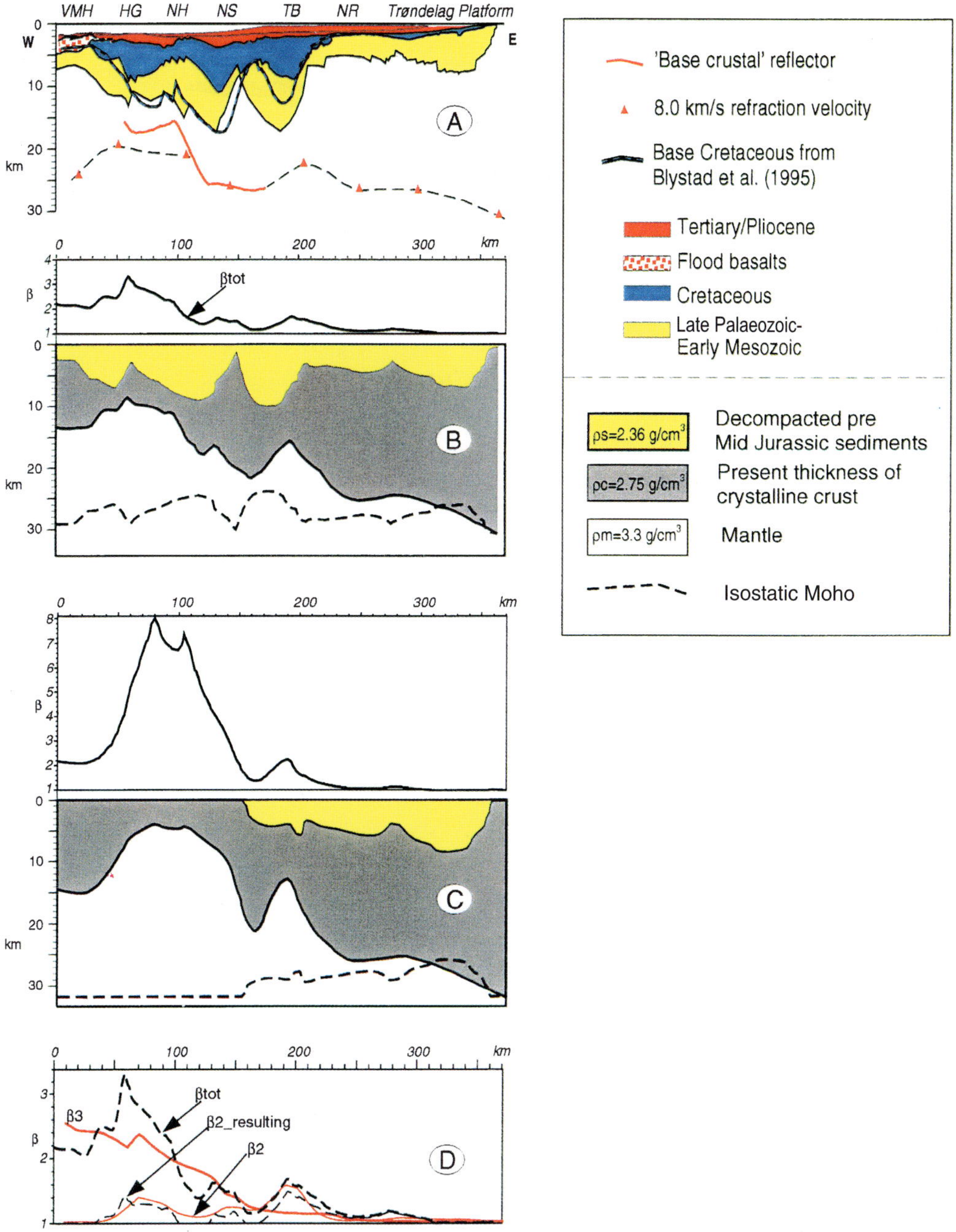

Fig. 5. (**a**) Transect 2 across the northern Vøring margin, where the three main tectono-stratigraphic units, the base Cretaceous interpretation from Blystad *et al.* (1995) and the lower crustal geometries are outlined. Abbreviations as in Fig. 2. (**b**) Transect backstripped to the base Cretaceous level and isostatically balanced. The stretching factor, βtot, illustrates the total post mid-Jurassic (base Cretaceous) thinning calculated as the ratio between depths to isostatic and backstripped Moho. (**c**) Same as in (**b**), but with the base Cretaceous interpretation from Blystad *et al.* (1995). (Note the extreme thinning beneath the western portions of the transect associated with this interpretation.) (**d**) A summary of extension estimates made along the transect. βtot, total post mid-Jurassic thinning; β2 and β3, previously made subsidence derived stretching estimates for the Late Jurassic–Cretaceous and the Maastrichtian–Paleocene rift episodes (Skogseid *et al.* 1992*b*); β2_resulting, new estimates of Late Jurassic–Cretaceous thinning derived as the ratio between βtot and β3. Where β3 is larger than βtot, β2_resulting is set to 1.0.

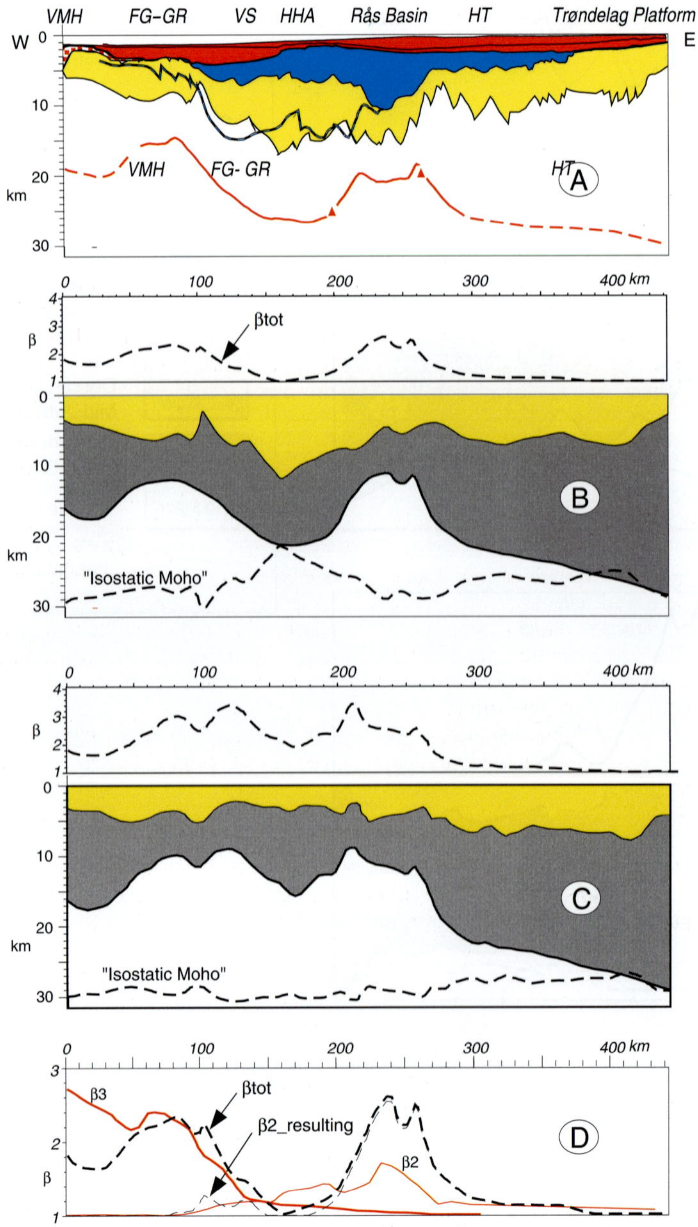

Fig. 6. (**a**) Transect 3 across the southern Vøring margin, where the three main tectono-stratigraphic units, the base Cretaceous interpretation from Blystad *et al.* (1995) and the lower-crustal geometries are outlined. Abbreviations as in Fig. 2 and label as in Fig. 5. (**b**) Transect backstripped to the base Cretaceous level and isostatically balanced. The stretching factor, β, illustrates the total post mid-Jurassic (base Cretaceous) thinning calculated as the ratio between depths to isostatic and backstripped Moho. (**c**) Same as in (**b**), but with the base Cretaceous interpretation from Blystad *et al.* (1995). (Note the broad (*c.* 250 km) area affected by high thinning factors in contrast to the two more distinct zones of thinning in (**b**).) (**d**) A summary of extension estimates made along the transect. βtot, total post mid-Jurassic thinning; $\beta 2$ and $\beta 3$, previously made subsidence derived stretching estimates for the Late Jurassic–Cretaceous and the Maastrichtian–Paleocene rift episodes (Skogseid *et al.* 1992*b*); $\beta 2$_resulting, new estimates of Late Jurassic–Cretaceous thinning derived as described in Fig. 5. (Note the large differences that exist between $\beta 2$_resulting and $\beta 2$ beneath the deep Rås Basin.)

1995). The overlying deep Cretaceous rift trough is between 50 and 70 km wide, whereas the Cretaceous Vøring Basin is *c.* 300 km wide. To the north, transect 2 reveals two Cretaceous depocentres, the Træna Basin and the Hel Graben. Between these depocentres, the Någrind Syncline contains a 3–7 km thickness of Cretaceous strata above a very thick pre-Late Jurassic section. Skogseid & Eldholm (1995) suggested a shallower base Cretaceous horizon in this region, resulting in an even thicker pre-Late Jurassic succession. Recent commercial drilling on the Nyk High, well 6707/10-1, demonstrated, however, that Late Cretaceous sediments still exist at 5039 m depth below sea level.

The base of the crust is determined by integrating deep seismic refraction and wide-angle reflection data (Planke *et al.* 1991; Mjelde *et al.* 1996, 1997). East of the Någrind and Vigrid synclines refraction Moho ($8\,km\,s^{-1}$) corresponds in depth to a deep crustal reflector (Fig. 2) (Skogseid & Eldholm 1995). Towards the west, however, refraction Moho is located deeper than this observed reflector. Here the two levels bound a $7+km\,s^{-1}$ velocity body interpreted to represent magmatic underplating emplaced during rifting and break-up. Thus, the reflector probably reflects the original base of the crust, a key crustal thinning factor. The underplating explains why the thinned crust has not subsided to form a deep early Tertiary rift basin, like the deep Cretaceous Rås Basin, which formed as a result of the Late Jurassic–Cretaceous extension episode. The crust again thickens beneath the Vøring Marginal High, presumably reflecting emplacement of an expanded igneous sequence approaching the continent–ocean boundary (COB).

The total amount of crustal thinning can be derived from the observed crustal thickness if we are able to determine a reasonable estimate of the initial, or pre-rift, crustal thickness. In standard basin analysis it is commonly assumed that a 32–35 km thick crystalline crust of density $2.8\,g\,cm^{-3}$, and a 125–130 km thick lithosphere is balanced at sea level (McKenzie 1978). Hence, sediments will be deposited below sea level associated with lithospheric thinning or flexural bending. In the NE Atlantic, however, eclogites in Greenland and Norway indicate syn-orogenic and maybe early post-orogenic crustal thicknesses of 80–90 km (Andersen & Jamtveit 1990; Brueckner *et al.* 1999). Furthermore, between the Caledonian fronts in Norway and Greenland the Late Palaeozoic–earliest Mesozoic sediments are generally of intra-continental and/or lacustrine facies. Marine deposits were regionally first introduced in Permian–early Triassic times. The eastern Barents Sea and the southern North Sea have a somewhat different palaeo-environmental history, both areas probably receiving large amounts of detritus from the high-standing Caledonian mountain belt (e.g. Coward 1993).

Our preferred margin stratigraphy includes very thick sequences of upper Palaeozoic and lowest Mesozoic sediments. In terms of the orogenic and early post-orogenic history the question of reference crustal thickness is important, and standard assumptions applied in basin modelling are considered not applicable to the pre-Late Jurassic basin history. More importantly, the so-called pre-rift sediments do not reflect amounts of rift-related tectonic subsidence, a key factor in calculation of extension factors from subsidence analysis also for the later rift episodes (Le Pichon & Sibuet 1981).

None the less, crustal thinning estimates can be made for the Late Jurassic–Cretaceous and Maastrichtian–Paleocene rift episodes, assuming: (1) a regional marine depositional environment in Mid-Jurassic time; (2) a correctly identified base Cretaceous horizon; (3) water depths were small and may be ignored; (4) the lithosphere had cooled over sufficiently long time since preceding rift episodes; (5) isostatic conditions. By backstripping the crust to the base Cretaceous (Middle Jurassic) horizon, an image of the pre-Late Jurassic basin and top basement configuration is achieved (Fig. 5b). Isostatic balancing of this section (assuming an average porosity of 20% in sediments with matrix density of $2.7\,g\,m^{-3}$, and crystalline crustal and mantle densities of 2.75 and $3.3\,g\,cm^{-3}$, respectively) results in an 'isostatic Moho', which is used as our reference crustal thickness. Thus, the total post-Mid-Jurassic crustal thinning is imaged by the difference between the isostatic Moho and the backstripped present Moho. This approach, like the traditional subsidence analysis, accounts for only compaction of the sediments, whereas it does not assume extensional thinning of the sedimentary column. The latter factor may not be important for small stretching factors ($\beta < 1.5$), but in the central rifts (e.g. Rås Basin), with factors between two and three, we may predict an extensional terrane with highly rotated fault blocks and possibly internally deformed pre-rift sediments. Beneath post-rift units of up to 8–10 km thickness we are not at present able to decipher this deformation. Our crustal thinning derived stretching estimates will therefore represent maximum estimates.

Figure 5c illustrates the same procedure, but with the base Cretaceous horizon of Blystad *et al.* (1995). The average stretching factors and the integrated amount of stretching (over the 325 km portion of the transect with stretching factors >1.0) for the two alternative interpretations are 1.7 and 127 km for our interpretation, v. 2.8 and 200 km for the interpretation of Blystad *et al.* (1995). Similarly, the southern Vøring margin transect (390 km portion with stretching factors >1.0) yields values of 1.75 and 170 km and 2.1 and 194 km, respectively (Fig. 6). These results do not alone exclude any of the interpretations. We should keep in mind, however, that also the conjugate east Greenland margin exhibits rifted basin geometries reflecting extension that will add to these estimates if they are related to the same rift episodes. In any case, the very high values of integrated stretching indicate that the width of the present margin was increased almost by a factor of two during the 170–55 Ma period. In addition to our previous comments on the upper- and lower-crust structural correlation, we believe the extreme maximum thinning (β between seven and eight) suggested by the Blystad *et al.* (1995) interpretation argues against the possibility that Cretaceous sediments are resting directly on basement.

To distinguish between the Maastrichtian–Paleocene and Late Jurassic–Cretaceous components of thinning we prefer to use the previously subsidence defined stretching estimates for the Maastrichtian–Paleocene rift episode. There is general consensus that the Maastrichtian–Paleocene stretching affected a 150–200 km wide zone landward of the continent–ocean boundary, with factors approaching 2.0–2.5 across the Fenris and Hel grabens on the outer margin (Skogseid *et al.* 1992*a*; Skogseid 1994; Roberts *et al.* 1997; Walker *et al.* 1997). The estimates from Skogseid *et al.* (1992*a*) and Skogseid (1994) were derived from subsidence analyses corrected for reduced subsidence because of magmatic underplating. The average stretching factors and amount of extension from the stretching distribution shown in Figs 5d and 6d ($\beta 3$) are 1.6 and 107 km and 1.6

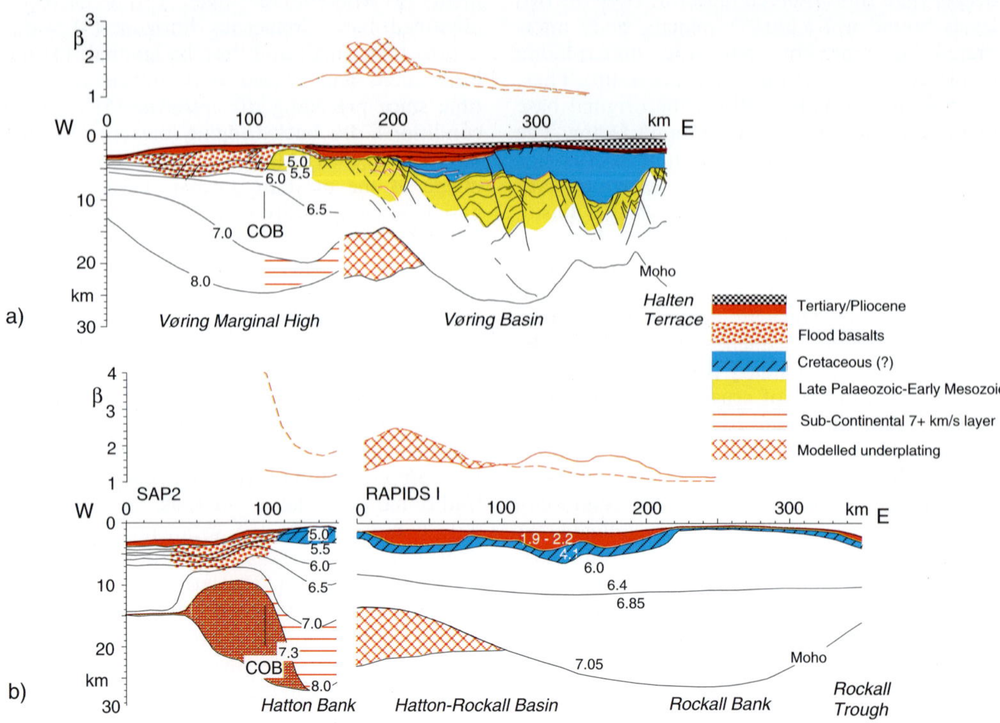

Fig. 7. Crustal transects across the Vøring (**a**) and Hatton–Rockall (**b**) margins illustrating crustal geometries, velocity structure (annotation in km s^{-1}) and modelled thickness of 'magmatic underplating'. COB, Continent–ocean boundary. Location as transects 3 and 6 in Fig. 1. Subsidence (continuous curve) and crustal thinning (stippled curve) derived stretching distribution for the Maastrichtian–Paleocene rifting episode are shown. The modelled thickness of underplating is derived from the difference between the thinning derived and the subsidence derived stretching factors as described by Skogseid (1994). The Hatton–Rockall transect is modified from White *et al.* (1987) and Makris *et al.* (1991).

and 104 km for the southern and northern transects, respectively. It should be noted that the thickness of underplating is the measured $7+\mathrm{km\,s^{-1}}$ velocity body on the northern Vøring margin transect, whereas the thickness is derived from modelling on the southern Vøring margin transect (Skogseid 1994) (Fig. 7). Although it may be questioned whether the $7+\mathrm{km\,s^{-1}}$ velocity body may contain some intruded lower continental crust, we consider these estimates as well constrained based on the present knowledge of the crustal structure, stratigraphic record and the subsidence history. In fact, they are comparable with stretching estimates on typical non-volcanic margins (Ginzburg *et al.* 1985), but tend to be high compared with estimates derived from amounts of extension from fault analysis (Roberts *et al.* 1997; Ren *et al.* 1999).

On both Vøring margin transects the Maastrichtian–Paleocene stretching estimates correlate with the total post Mid-Jurassic crustal thinning factors across the western provinces (Figs 5 and 6). Along the southern Vøring margin transect this comparison has no significance, as the thickness of modelled underplating (Fig. 7) was defined by the difference between observed and theoretical tectonic subsidence based on the crustal thickness (Skogseid 1994). On the northern transect, and the Møre margin transect (Fig. 8), however, the correlations indicate that most of the observed crustal thinning beneath the Nyk High–Hel Graben and western Møre Basin relates to the last rift episode only.

The Late Jurassic–Cretaceous component of the thinning, $\beta2$_resulting, can be compared with the subsidence-derived estimates of Skogseid *et al.* (1992*a*, *b*). We note that significantly greater thinning seems to have taken place beneath the Rås and Møre basins than previously suggested (Figs 6d and 8). In these deep basins, no notable difference is introduced by various interpretations of the base Cretaceous level, and it appears that less subsidence has taken place than would be expected from the crustal structure.

Our interpretation of only two main rift episodes, the Late Jurassic–Cretaceous and Maastrichtian–Paleocene phases, contrasts with those of Blystad *et al.* (1995) and Lundin & Doré (1997). These workers argued that additional extensional deformation affected the region also

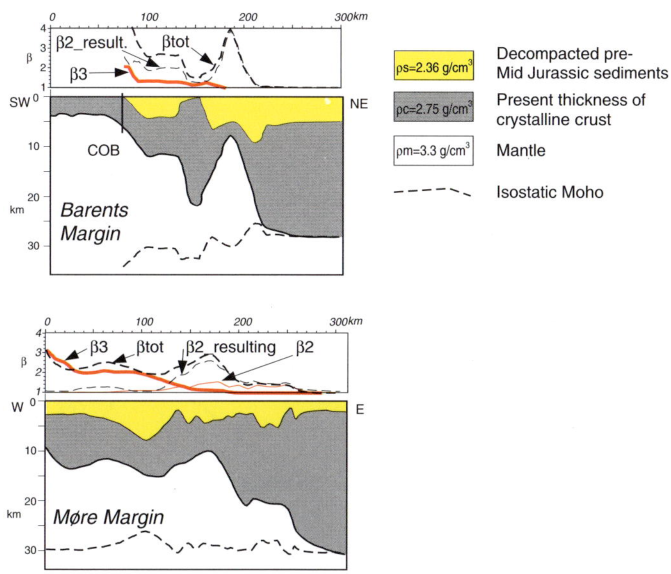

Fig. 8. Transects 1 and 4 across the Western Barents Sea and the Møre margins (modified from Faleide *et al.* (1993*b*) and Olafsson *et al.* (1991), respectively) backstripped to the mid-Jurassic–base Cretaceous level. Extension estimates are made with same procedure as described for Figs 5 and 6. βtot, total post mid-Jurassic thinning; $\beta2$ and $\beta3$, previously made subsidence derived stretching estimates for the Late Jurassic–Cretaceous and the Maastrichtian–Paleocene rift episodes (Skogseid *et al.* 1992*b*); $\beta2$_resulting, new estimates of Late Jurassic–Cretaceous thinning derived as described in Fig. 5. (Note large differences between $\beta2$_resulting and $\beta2$ beneath the Møre Basin, whereas βtot and $\beta3$ show very consistent values towards the west (continent–ocean boundary).)

in Late Cretaceous time, either as a separate rift episode or as continuous stretching. No distinct rift structures or basins of this age are identified, however. On the other hand, with reference to the Tromsø Basin we should use a broader definition of the Cretaceous rift basins, indicating that substantial amounts of Late Jurassic sediments may constitute the deepest parts of the rift structures, and that rifting may have continued also towards late Cretaceous time. Late lithospheric thinning can, in fact, partly explain both the relatively large discrepancy between subsidence and crustal thinning derived stretching factors beneath the deepest Cretaceous basins (i.e. the apparent lack of subsidence with reference to crustal thickness; $\beta 2$-resulting $-\beta 2$), as well as some of the discrepancy observed between the extension derived from normal faults and from subsidence for the last rift episode.

Hatton–Rockall, Møre and SW Barents Sea transects

The total crustal thinning has also been calculated for the other transects in Fig. 1, and it is shown that significantly more post Mid-Jurassic extension has affected the Hatton–Rockall margin, including the Rockall Trough, than is possible to envisage off Norway and in the western Barents Sea (Fig. 9).

The northern North Sea rift has been evaluated by Christiansson *et al.* (2000), using a

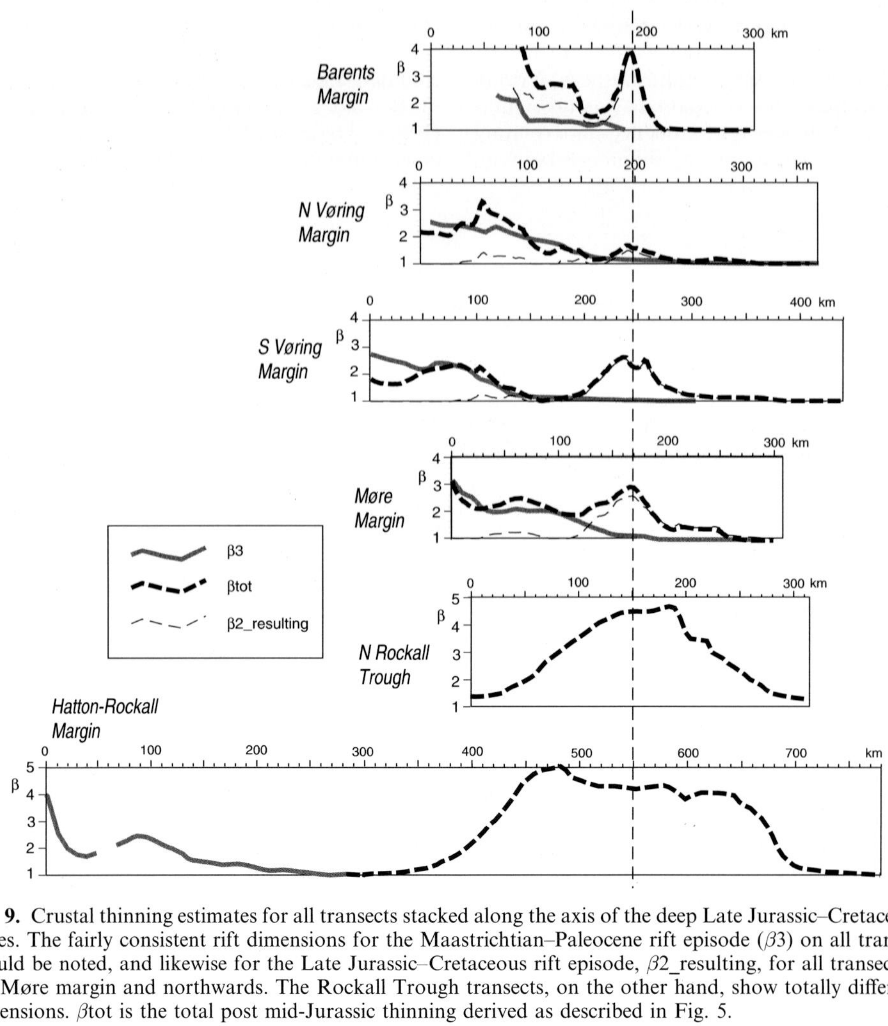

Fig. 9. Crustal thinning estimates for all transects stacked along the axis of the deep Late Jurassic–Cretaceous rift zones. The fairly consistent rift dimensions for the Maastrichtian–Paleocene rift episode ($\beta 3$) on all transects should be noted, and likewise for the Late Jurassic–Cretaceous rift episode, $\beta 2$_resulting, for all transects from the Møre margin and northwards. The Rockall Trough transects, on the other hand, show totally different rift dimensions. βtot is the total post mid-Jurassic thinning derived as described in Fig. 5.

comparable approach. They showed that the basin was mainly affected by Late Jurassic extension, and that it is characterized by a narrow central rift zone, 30–50 km, within a 250 km wide region affected by Cretaceous post-rift subsidence. The rift follows the Late Palaeozoic–Early Mesozoic basin trend, and is surrounded by rift flank terraces. The estimated stretching over the two profiles discussed by Christiansson *et al.* (2000) shows averages of 1.2 and 1.25, respectively, resulting in *c.* 40–50 km extension.

West of Britain, the wide Rockall Trough appears to follow the SW-oriented Appalachian trend. With reference to the above procedure, no backstripping is needed of the Rockall transects, as the existence of pre-Cretaceous sediments is speculative and accordingly not included in this analysis. In Fig. 4 a rapid decay in the magnitude of extension is observed from south to north in the trough. Over the distance of *c.* 200 km between the two transects the width of the Rockall Trough is reduced by *c.* 100 km. Extension estimates based on crustal thickness show that the average stretching factor and magnitude of extension are 2.9 and 315 km and 2.9 and 203 km for the southern and northern transects, respectively, assuming a 30 km pre-rift crustal thickness. Farther west the *c.* 300 km wide Hatton–Rockall Basin shows average thinning factors of *c.* 1.7, interpreted as 126 km of Maastrichtian–Paleocene extension according to Skogseid (1994) (Fig. 7).

The deep Møre and Vøring basins represent the northward continuation of the Rockall Trough and the Faeroe–Shetland Basin trend. On the Møre margin the average stretching factors and total amount of extension are 1.9 and 150 km, of which 87 km relates to the Maastrichtian–Paleocene extension (Fig. 8). Roberts *et al.* (1990) made similar extension estimates for the Viking Graben, Møre and Faeroe–Shetland basins. Although they only estimated Late Jurassic stretching (assuming it to be *c.* 25% of the total Late Jurassic–Early Cretaceous episode), they concluded that most of the extension in the Møre Basin had to have occurred west of Shetland.

In the SW Barents Sea the Late Jurassic–Cretaceous crustal thinning was severe, with maximum crustal thinning values approaching four in the Tromsø Basin (Fig. 8). Here the depth to the Middle Jurassic horizon is *c.* 13 km, and in the westernmost basin, the Sørvestnaget Basin (Fig. 3), the depth is comparable or greater. The extensional development of these two basins appears to be similar, giving rise to thick Upper Jurassic–Lower Cretaceous syn-rift sequences (Breivik *et al.* 1998). The final rift phase in the Late Jurassic–Cretaceous episode is well constrained to be of Aptian age in these western Barents Sea basins, whereas the subsequent period was characterized by rapid post-rift subsidence and sedimentation. Most of the structural relief was covered by Cenomanian time (Faleide *et al.* 1993*a*, *b*). The total crustal thinning derived stretching distribution shows an average value of 2.2, yielding 100 km extension over the ≤200 km long transect, of which the Maastrichtian–Paleocene episode may account for *c.* 30 km. In the Tromsø Basin removal of material below the Middle Jurassic horizon by salt diapirism has amplified the subsidence. Our thinning estimates are therefore considered maximum values.

Restored basin geometries and palaeogeography

In this study we have attempted to quantify the amount of relative movements between the continents associated with the respective extensional episodes, and spatially illustrate this on maps and along transects.

Break-up setting

A plate reconstruction to magnetic Chron 24n.3 (*c.* 53 Ma) represents the first time slice in this restoration, and shows a palaeogeographical view of the NE Atlantic configuration shortly after break-up (Fig. 10). The 18–20 Ma Maastrichtian–Paleocene rift episode pre-dating the final continental separation was characterized by normal faulting over broad regions, whereas during the late syn-rift period more focused extensional deformation was associated with uplift and erosion of the central rift zone (Skogseid *et al.* 1992*a*). Eventually, break-up was accompanied by massive volcanic activity, which covered very large provinces generally along the break-up axis. The distribution of flood basalts, as shown in Fig. 10, presumably outline the regions that both had a high potential for melt generation (e.g. thinnest lithosphere) and had dominantly a subaerial depositional environment during break-up.

The extension estimates for the Maastrichtian–Paleocene rift episode are very consistent from one margin segment to the other, and viewed in a conjugate margin setting the amount and distribution of stretching seem to resemble estimates made across typical non-volcanic margins (Skogseid 1994). Along the NE European

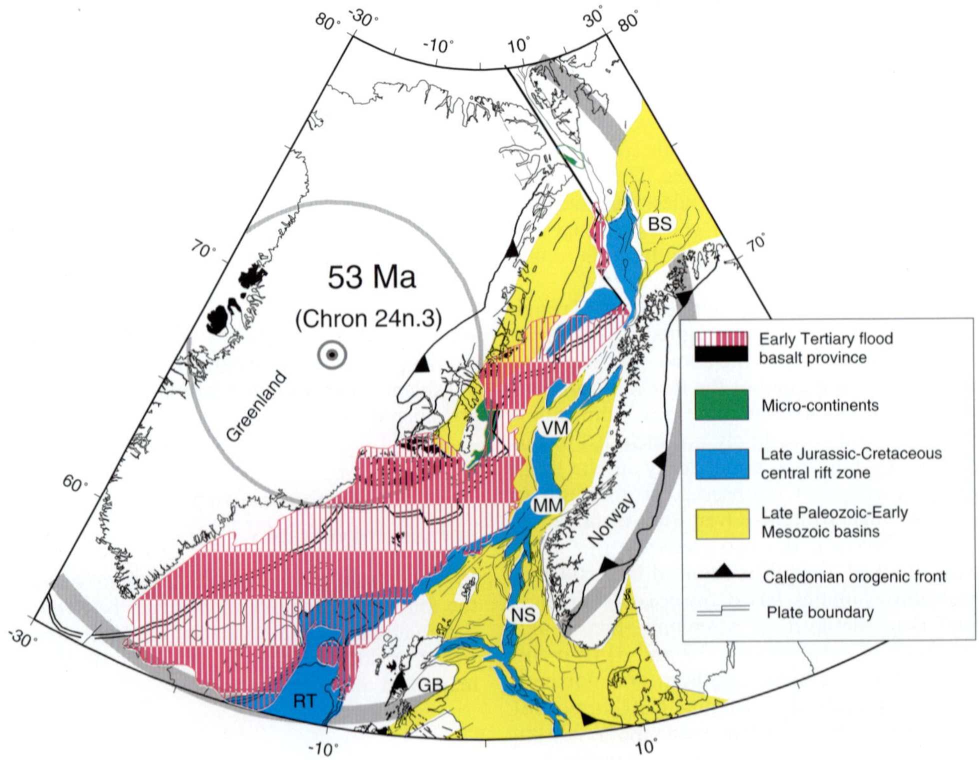

Fig. 10. Plate reconstruction at 53 Ma (magnetic Chron 24n.3) showing the NE Atlantic rift system shortly after continental break-up. Emphasis is placed on the Late Palaeozoic–Early Mesozoic, Late Jurassic–Cretaceous and the Maastrichtian–Paleocene rift zones. The last is generally illustrated by the Early Cenozoic volcanic overprint along the young plate boundary. The small and large circles, respectively, illustrate the suggested size of the Iceland mantle plume before and after spreading beneath the NE Atlantic lithosphere, assuming a symmetrical distribution. The circles are centred at the projected location of the Iceland mantle plume according to Lawver & Müller (1994). (For other abbreviations, see Fig. 1.)

margins the areas that have been affected by increased Cenozoic subsidence are *c.* 200–250 km wide, and coincide well with the region affected by extensional deformation by faulting and igneous activity (Skogseid *et al.* 1992*a*). The discrepancies that exist between extension estimates made from structural modelling on one hand, and from crustal structure and subsidence analysis on the other, may be ascribed to limitations in seismic resolution, which cause small faults to be left out (Roberts *et al.* 1997; Walker *et al.* 1997; Ren *et al.* 1999). More important, however, are the seismic imaging problems associated with the igneous rocks. If we assume break-up took place near the centre of the rift zone we can conclude that between 50 and 150 km of the central rift structure (on the European side, only) is at present covered by volcanic rocks. In this context, the Fenris and Hel grabens, on the Vøring margin, represent the exceptions within the NE Atlantic realm, being located in a region where the flood basalt province is particularly narrow (*c.* 50 km) (Fig. 10). Here, very well imaged extensional terranes are locally dominated by low-angle normal faults and associated lower-crustal extensional unroofing (Lundin & Doré 1997; Ren *et al.* 1999). With respect to the observed abrupt crustal thinning beneath these fault zones we interpret them as the main boundary faults of the central break-up-related rift zone.

Cretaceous setting; 75 Ma

The plate restoration to 75 Ma (Fig. 11) and the respective shortening implemented on the transects (Figs 12–14) compensates for the amount

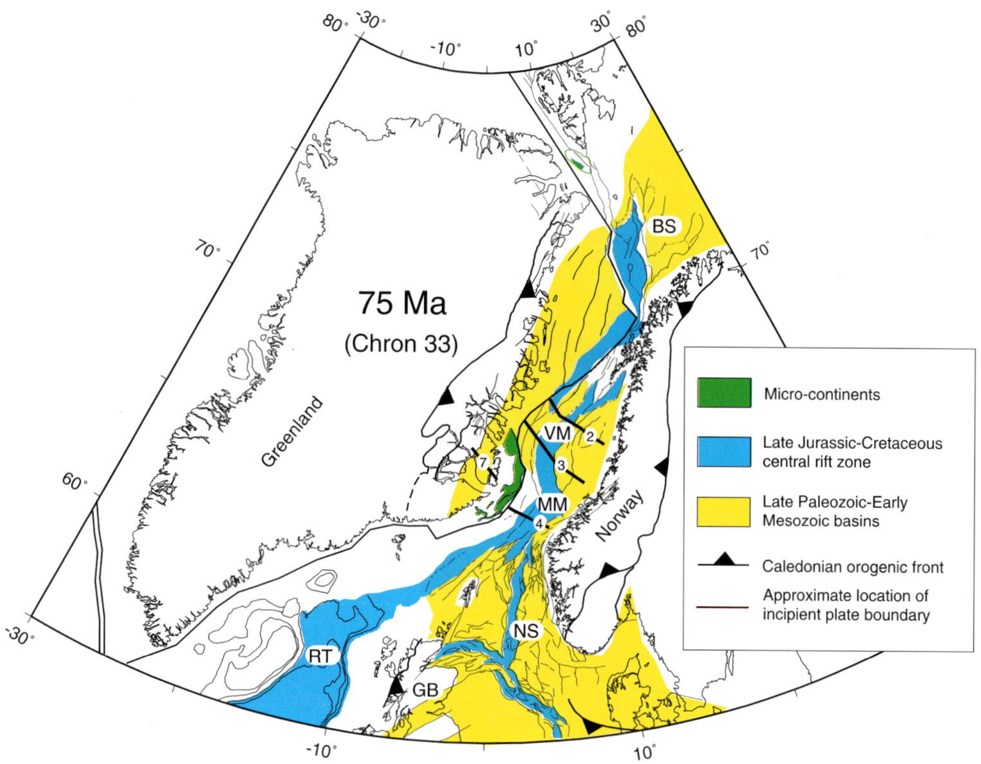

Fig. 11. Plate reconstruction at 75 Ma (*c.* magnetic Chron 33n) based on pre-break-up stretching estimates according to Fig. 9. Emphasis is placed on the Late Jurassic–Cretaceous central rift zones and the Late Palaeozoic–Early Mesozoic basins. (Note the close spatial correlation between the Late Jurassic-Cretaceous central rift zone on the present NE Greenland margin and those at the outer Vøring margin and in the SW Barents Sea. Also note the pronounced fan-shape of the Rockall Trough.) Locations of crustal transects (2, 3, 4 and 7) are shown. (For other abbreviations, see Fig. 1.)

of extension derived for the Maastrichtian–Paleocene rift episode (Fig. 9). The width of the Cretaceous basins has, thus, been reduced by *c.* 100 km. The reconstructed transects are generated assuming pure shear thinning of the crystalline crust, whereas the 'pre-rift' sediments are decompacted only, in general agreement with the way the stretching estimates were made. We realize that this restoration approach must be considered with caution, but believe that the restored sections show more realistic basin configurations and regional rift geometries than those observed on unrestored sections. In combination with the maps they may carry important information on provenance and basin fill relations in the area. The structural and stratigraphic details are unreliable, however.

The transects show that the Late Jurassic–Cretaceous central rift zones had widths between 30 and 60 km, whereas the post-rift regional subsidence formed Cretaceous basins 150–250 km wide. The Møre and southern Vøring basins are fairly symmetrical about their central rift zones, whereas the northern Vøring Basin reveals three main depocentres. Here, the central depocentre, the Någrind Syncline, is interpreted as a deep synform between two 'real' rift basins, the Træna Basin and the Hel Graben (Figs 2 and 14). In the Någrind Syncline post-Campanian sediments seem to represent a significant portion of the Cretaceous succession.

The reconstruction (Fig. 11) demonstrates our understanding of how the deep Late Jurassic–Cretaceous central rift structures were located, locally revealing overlapping rift zones. Off NE Greenland a 'NE Greenland margin basin' is interpreted from aeromagnetic data (Fig. 15). Here, a magnetic quiet zone obliquely truncates coastal parallel N–S trending anomalies. The latter are interpreted to reflect

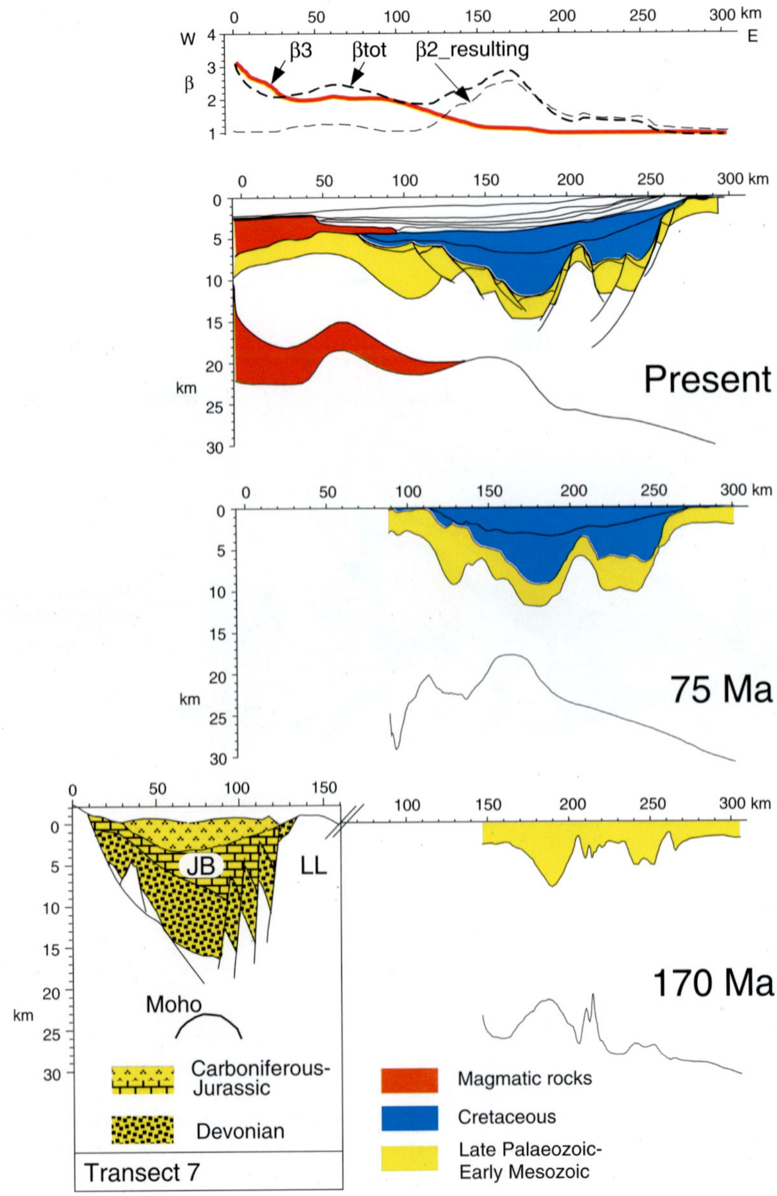

Fig. 12. Simple restoration of transect 4 across the Møre margin, where shortening and thickening of the crystalline crust are determined according to extension estimates for the Maastrichtian–Paleocene rift episode ($\beta 3$, 75 Ma), and the Late Jurassic–Cretaceous rift episode ($\beta 2$_resulting, 170 Ma), respectively. The sediment successions are decompacted, only. The 170 Ma restoration is combined with transect 7 across the Jameson Land Basin (JB) in Greenland (modified from Larsen & Marcussen 1992; Fig. 1 for location). Here, very thick Devonian to Jurassic deposits fill in a large half-graben basin bounded eastward by the Liverpool Land (LL) basement block.

a Palaeozoic–Early Mesozoic structural grain. The change in trend between the quiet zone and the coastal parallel structures is similar to what is observed between the Palaeozoic–Early Mesozoic grain and the Late Mesozoic basins off mid-Norway. The deep basins off Norway were, in fact, also first suggested based on their smooth magnetic signature (Åm 1971).

With respect to the dimension of the deep Late Jurassic–Cretaceous rift structures, it is

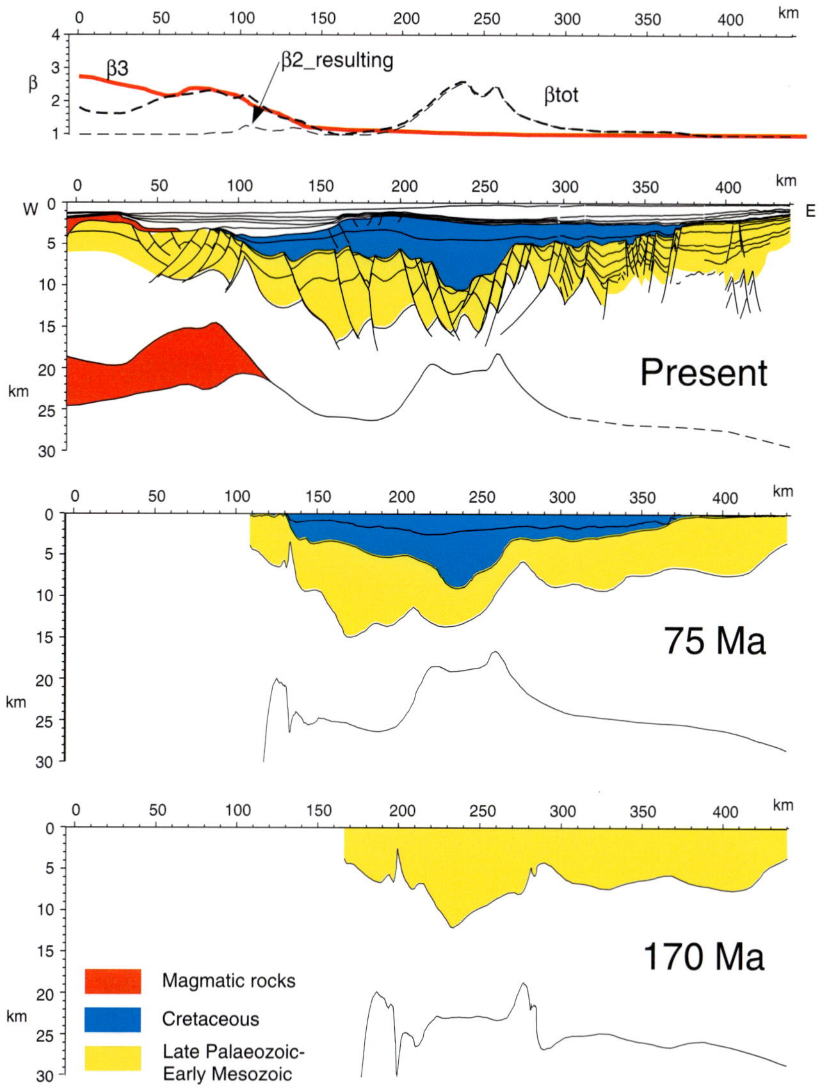

Fig. 13. Simple restoration of transect 3 across the southern Vøring margin, where shortening and thickening of the crystalline crust are determined according to extension estimates for the Maastrichtian–Paleocene rift episode ($\beta 3$, 75 Ma), and the Late Jurassic–Cretaceous rift episode ($\beta 2$_resulting, 170 Ma), respectively. The sediment successions are decompacted, only.

difficult to envisage that a northward continuation from the Vøring to the Barents Sea basins could take place without involving the present NE Greenland margin, and it therefore gives confidence that the relocation of the 'NE Greenland margin basin' merges with the Hel Graben in the northern Vøring Basin, the Harstad Basin off northern Norway, and the Tromsø Basin in the SW Barents Sea. The obliquely cutting line of continental break-up, thus, subsequently caused a potentially large Cretaceous basin to be found on the present NE Greenland margin.

With respect to the Labrador Sea and the Rockall Trough rift history, we propose a model where the southward fan-shape of the Rockall Trough reflects a failed rift formation. The trough was formed as a relatively short rift branch within the large-scale North Atlantic rift system, where the Labrador rift took over

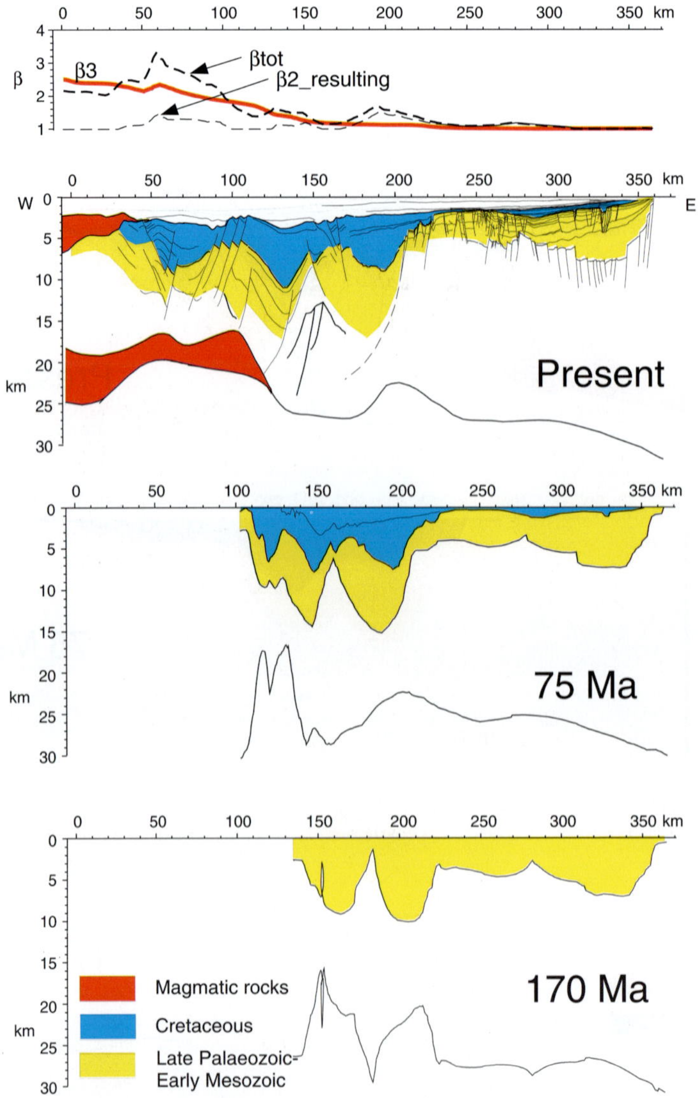

Fig. 14. Simple restoration of transect 2 across the northern Vøring margin, where shortening and thickening of the crystalline crust are determined according to extension estimates for the Maastrichtian–Paleocene rift episode ($\beta 3$, 75 Ma), and the Late Jurassic–Cretaceous rift episode ($\beta 2$_resulting, 170 Ma), respectively. The sediment successions are decompacted, only.

'all' tectonic activity between Coniacian and Campanian–Maastrichtian times. In this model the mid–Late Cretaceous period appears not to have represented an epoch of renewed large-scale basin-forming extension between Norway and Greenland north of the Rockall Trough. We recognize, as outlined above, that some tectonic activity occurred along some of the major fault zones (e.g. Faleide *et al.* 1993*b*; Blystad *et al.* 1995; Lundin & Doré 1997).

Pre-Late Jurassic setting; 170 Ma

In the reconstruction to 170 Ma we have compensated for *c.* 60 km of Late Jurassic–Cretaceous stretching (Fig. 16). We note that the remaining Late Palaeozoic–Early Mesozoic basin province is less than 400 km wide in east–west direction, located between the Caledonian fronts in Greenland and Norway within a range of 650–1100 km. These dimensional

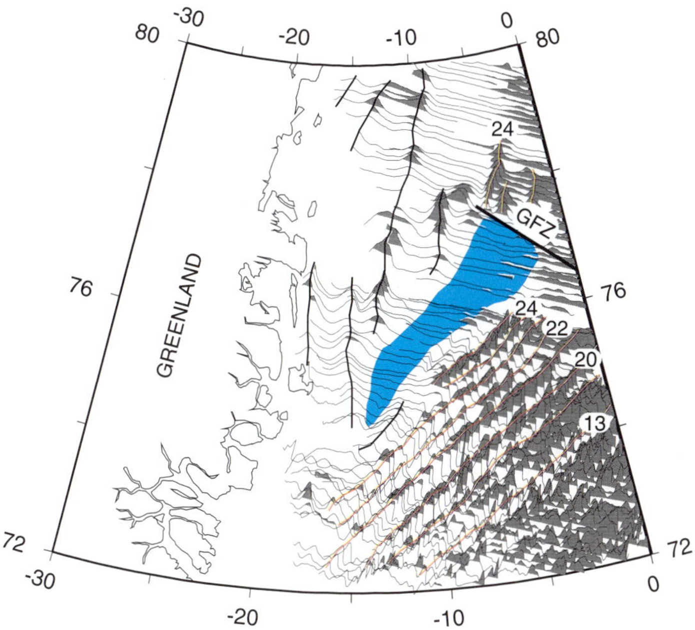

Fig. 15. Magnetic anomaly wiggle plot from the NE Greenland continental margin. Positive magnetic anomalies (grey), interpreted coastal parallel lineations (thick continuous lines) and a zone of 'smooth magnetic signature' (blue area) are identified beneath the shelf, whereas sea-floor spreading anomalies 13, 20, 22 and 24 are shown in the ocean basin (red lines). GFZ, Greenland Fracture Zone.

relations are important with respect to our understanding of the Caledonian orogeny itself, and to provenance relations both during the early stage non-marine environment and subsequently during the development of marine transgressions from both north and south.

Basin formation associated with the orogenic collapse of the Caledonides is generally reflected by deep half-graben structures of mainly Devonian age and regional deposition of continental red sandstones (e.g. Hossach 1986; Duindam & van Hoorn 1987; Marcussen *et al.* 1987; Ziegler 1988; Hartz & Andresen 1995). Subsequent episodes of lithospheric extension have been reported, of Late Carboniferous–Early Permian, Permo-Triassic, Mid–Late Triassic and Triassic–Early Jurassic ages (e.g. Surlyk *et al.* 1984; Ziegler 1988; Surlyk 1991). However, the early non-marine depositional setting, the general lack of good stratigraphic correlation, and the totally unknown structure of the orogenic crust and lithosphere make basin modelling and extension estimates for these basins a speculative task.

In the SW Barents Sea the Late Palaeozoic tectonic development gave rise to a 300 km wide rift zone, the Nordkapp Basin, suggested by Gudlaugsson *et al.* (1994) to extend at least 600 km northeastward into the Barents Sea. The rift formed mainly during Mid-Carboniferous times, probably linking up with Carboniferous–Early Permian extension as far southwestward as the Jameson Land Basin in Greenland (Surlyk 1991), and causing the establishment of a marine depositional environment along this axis during Late Permian time.

There is only little direct information on the pre-Mid-Jurassic basin history beneath the deep SW Barents Sea basins, although the deep seismic reflection and refraction data indicate a considerable section of old sediments (Fig. 3). In the Tromsø Basin the deep seismic data indicate a crystalline basement at 18–20 km (Gudlaugsson *et al.* 1987; Jackson *et al.* 1990; Faleide *et al.* 1993*b*). Regional considerations combined with the presence of salt in the Tromsø and Sørvestnaget basins suggest that the pre-Middle Jurassic sequence probably includes thick Triassic and Jurassic clastic sediments and Permo-Carboniferous mixed carbonates, evaporites and clastic deposits (Faleide *et al.* 1993*a*, *b*; Gudlaugsson *et al.* 1994). Basin characteristics, sediment thicknesses and the salt diapirism can be compared with observations on the NE Greenland shelf (Hinz *et al.* 1991; Escher & Pulvertaft 1995), whereas similar salt diapirs are not reported off mid-Norway.

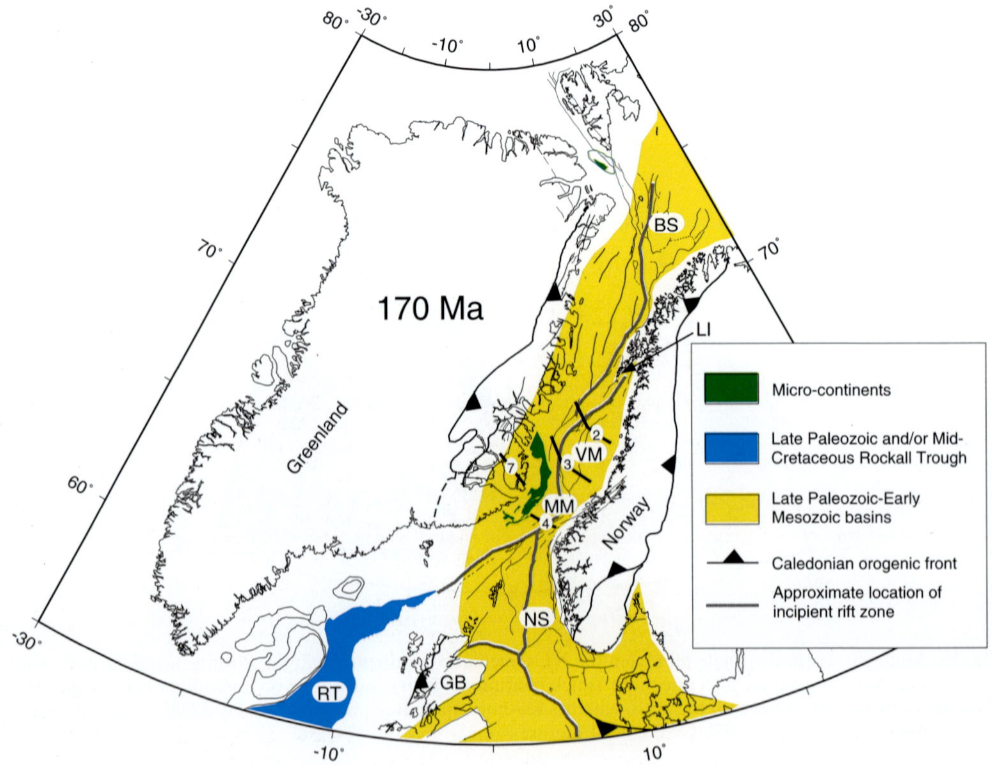

Fig. 16. Reconstruction at 170 Ma based on total post mid-Jurassic stretching estimates according to Fig. 9 revealing a *c.* 400 km wide basin province between Greenland and Eurasia. Location of crustal transects (2, 3, 4 and 7) discussed in this respect are shown. The remaining fan-shaped Rockall Trough is left shaded partly to illustrate the size of the region we believe may have a separate history in terms of rifting and/or drifting as described in text. LI, Lofoten Islands. (For other abbreviations, see Fig. 1.)

In contrast to the Late Carboniferous–Early Permian marine transgression in the Barents Sea, and its presumable continuation off- and onshore East Greenland, a generally non-marine depositional environment persisted off mid-Norway and in the northern North Sea until Late Triassic time (e.g. Eide 1989; Steel 1993). With respect to Fig. 16, we may thus infer that a structural and/or depositional boundary existed west or southwest of the Lofoten Islands, separating a Carboniferous–Permian marine transgression and associated evaporite basin north and west from the areas to the south and east characterized by continuous non-marine conditions.

In the transect restorations (Figs 12–14) removal of the Late Jurassic–Cretaceous extension provides images of the pre-Late Jurassic basin configuration, which demonstrates very thick local depocentres as parts of a regional basin province. The Møre margin transect (Fig. 12) is relocated to be almost conjugate to the Jameson Land Basin in Greenland (where the Jan Mayen Ridge appears to have represented an intermediate basement high). Along the transect only moderate thicknesses of pre-Late Jurassic sediments can be imaged, whereas in the Jameson Land Basin as much as 16 km of Devonian to Jurassic sediments have been suggested, on the basis of seismic reflection data (Larsen & Marcussen 1992) (a significantly smaller thickness of 8–10 km is, however, inferred from seismic refraction data according to Fechner & Jokat (1996)). According to Marcussen *et al.* (1987), the succession generally consists of Devonian braided river and aeolian sediments found deposited in a rapidly subsiding intra-mountain basin setting, underlying Carboniferous to Lower Permian coarse-grained sandstones, red siltstone, organic-rich black shale and coal beds of fluviatile and lacustrine origin. The Upper Permian sequence shows conglomerate and black shale, and thin interbedded layers of evaporite are noted. The Lower Triassic formations consist of finely laminated marine silt and sandstone, whereas Middle and Upper Triassic sequences are continental to shallow marine in origin.

Along the southern Vøring margin transect (Fig. 13) very thick successions of pre-Late Jurassic sediments exist beneath the Trøndelag Platform area close to the Norwegian mainland, and beneath the Vigrid Syncline, in the central parts of the transect (Figs 2 and 13). The stratigraphic interpretations assume a substantial accumulation of Permo-Triassic sediments, but Devonian sediments may also be present in a number of deep half-graben basins. The transect connects to the Greenland margin between the Triall Ø and Hold with Hope, an area that is characterized by thick sequences of Carboniferous to Jurassic units overlain by thinner squences of Cretaceous shales generally found on the footwall side of major Late Jurassic fault blocks (e.g. Surlyk 1991; Price & Whitham 1997).

Further north, the northern Vøring margin transect still reveals thick units of pre-Late Jurassic sediments in what may appear to be deep half-graben structures (Figs 2 and 14). In view of the plate reconstructions (Figs 10, 11 and 16) it is tempting to use the deep Jameson Land Basin (Fig. 12) as an analogue for this basin configuration. The eastern flank of the Någrind Syncline, the Utgard High, exhibits a crystalline basement high adjacent to the deepest pre-Late Jurassic basin (Planke *et al.* 1991; Skogseid *et al.* 1992*b*). We interpret this to indicate a similar deep half-graben basin as on East Greenland, where the Utgard High in this context resembles the Liverpool Land basement block. Westward the northern Vøring margin transect connects to Greenland where the shelf starts to become broader, but where Tertiary lavas inhibit good seismic imaging of the substratum (Larsen 1990; Escher & Pulvertaft 1995). Onshore, in the Wollastone Foreland province, however, the stratigraphy is characterized by Upper Mesozoic sediments deposited in Late Jurassic–Cretaceous extensional half-grabens directly onto Caledonian basement (Surlyk 1991).

The magnetic lineations on the NE Greenland margin (Fig. 15) may be correlated with basement structures, which, furthermore, may resemble metamorphic rocks such as those of the Lofoten peninsular islands, the Utrøst and Utgard highs off mid-Norway, and the pronounced basement-related structures along the eastern Møre margin and beneath the Møre Basin (i.e. the last generally reflected by gravity and magnetic lineations, e.g. Doré *et al.* 1997*b*).

The interpretation of the Någrind Syncline (Fig. 2) as primarily a pre-Jurassic basin contrasts severely with the interpretation presented by Blystad *et al.* (1995). In their interpretation no pre-Cretaceous strata have existed in this region, indicating as much as 15 km of Cretaceous sediments in the syncline. We argue against this interpretation primarily on the basis of the crustal configuration combining reflection and refraction data as discussed previously. We further think the observation of pre-Jurassic strata both beneath the Trøndelag Platform and offshore Greenland indicates that the Vøring Basin has also been part of the pre-Jurassic basin province (Fig. 16). The possibility of a somewhat deeper level for the base Cretaceous reflector than presented in our interpretation can, however, not be ruled out. With reference to the Tromsø Basin in the western Barents Sea, it may be that this inter-basinal syncline trapped significant thicknesses of late Jurassic and possibly early Cretaceous sediments.

In the northern North Sea Late Palaeozoic rifting is best documented in marginal areas with respect to the Viking Graben. The seismic data display a tilted fault-block geometry where sediments probably of Permian age fill in the tectonic relief and are subsequently buried by Triassic deposits. The initiation of this extensional tectonism is not dated, but from observations made from the seismic data combined with more regional information, it is generally assumed to have been during Permian time (Færseth *et al.* 1995; Faleide *et al.* 2000). Furthermore, seismic and gravity data indicate thick Upper Palaeozoic strata of probable Devonian age to exist beneath both the East Shetland and the Horda platforms (Platt 1995; Christiansson *et al.* 2000; Faleide *et al.* 2000).

The reconstruction (Fig. 16) shows a remaining fan-shaped region in the Rockall Trough, which we have left shaded partly to illustrate the size of the area we believe may have a separate history in terms of rifting and/or drifting (i.e. Smythe 1989; Srivastava & Verhoef 1992; Knott *et al.* 1993). The rapid decay in width of the trough from south to north is interpreted to suggest that 'extra' crustal thinning affected this rather short rift segment, i.e. *c.* 800–1000 km. It should be noticed, however, that further north neither the Late Jurassic–Cretaceous 'major' rift episode nor the Maastrichtian–Paleocene rifting to continental break-up were able to create a similarly wide area of thin crust.

As an alternative to previously presented interpretations for the formation of the trough (i.e. Permian to Cretaceous), or rather as a possible addition to these, there are indications that the Maastrichtian–Paleocene extension episode and subsequent sea-floor spreading may have influenced the area. This assumption comes from the plate tectonic reconstruction between Greenland and Eurasia (Fig. 10). To generate good closure between conjugate pairs of magnetic Chrons 21 to 24n.3 south of Iceland, the Hatton–Rockall 'plate' has to be rotated anti-clockwise relative to Eurasia, yielding post-break-up (i.e. post 53–46 Ma) movements possibly centred along the trough. With respect to the significant Early Tertiary igneous activity in both the Rockall Trough and the Porcupine Basin (Tate *et al.* 1993) this could be an attractive explanation.

Volcanic margin formation and exploration challenges

Plume–lithosphere interaction and Paleocene vertical movements

The Maastrichtian–Paleocene rift episode in the NE Atlantic region was strongly influenced by the arrival of the Iceland mantle plume at lithospheric levels. Regionally, Early Tertiary igneous activity initiated in West Greenland and in the British Tertiary Igneous Province (BTIP) about 62–63 Ma (Mussett *et al.* 1988; Larsen *et al.* 1992; Hitchen & Ritchie 1993). The ages from the BTIP have, however, recently been questioned by Jolley (1997), who proposed an age of *c.* 58 Ma, on the basis of new radiometric dating. The massive volcanism along the Maastrichtian–Paleocene NE Atlantic rift zone started some million years later, at *c.* 55–54 Ma, using the Cande & Kent (1992) time scale (Roberts *et al.* 1984; Larsen & Watt 1985; Eldholm *et al.* 1989; Skogseid *et al.* 1992*a*). The late introduction of the igneous activity with respect to the initiation of rifting has been related to the timing of the emplacement of the Iceland mantle plume beneath the proto NE Atlantic lithosphere. On the basis of our present understanding, early rifting (75–63 Ma) occurred without the plume-related thermal anomaly at lithospheric levels, and thus without much melt formation. The key to this interpretation is found in the Vøring Basin, where a thick Paleocene sediment package away from the central rift zone allows us to resolve the time relations between initiation of rifting, uplift and magmatism as described by Skogseid *et al.* (1992*a*).

Thermal plumes, which transport energy towards the Earth's surface because of buoyancy, contain a large head of hot material that ascends through the mantle. Upon impinging on the lithosphere the plume head spreads laterally. The arrival of plumes therefore results in an episode of surface uplift over large areas. According to Griffiths & Campbell (1990) the average temperature anomaly carried by the plume head with respect to the adjacent mantle is *c.* 100°C. This magnitude of the thermal anomaly is in general agreement with the asthenospheric temperatures derived from depth and gravity anomalies (Cochran & Talwani 1978) and analysis of crustal thickness in the realm of the Iceland plume (i.e. 10.3 ± 1.7 km, from White *et al.* 1992). It is presumed that the hot plume material preferentially fills traps at the base of the lithosphere created either by previous plate deformations or by continuing lithospheric thinning (e.g. Artyshkov *et al.* 1980; Sleep 1997). It is, however, not obvious from the various plume models at what velocity the plume material flows laterally beneath the lithosphere. In terms of dimensions, Hill (1991) has estimated that an original plume head, with a diameter of 1000 km, will form an ellipsoidal disc 2000 km wide, still with a maximum thickness of 160 km centrally.

The initiation of magmatism at 62–63 Ma in the NE Atlantic region, suggested to be correlated with the impingement of the plume head on the lithosphere, was centred beneath Greenland at *c.* 63 Ma according to Lawver & Müller (1994). A plume 3000 km in diameter (after spreading) is required to encircle the North Atlantic Volcanic Province, including the British Tertiary Volcanic Province and the Labrador Sea volcanic provinces (Fig. 10). The shape and thickness of the plume body underneath the lithosphere is still speculative, but should in principle be possible to estimate from observations of relative vertical movements in the area.

Here, we want to outline the first-order patterns of uplift expected in relation to the plume emplacement. To achieve this, thickness models of the lithosphere and the plume body at the time of plume emplacement (*c.* 63 Ma) are constructed. The lithosphere thickness model is based on our break-up reconstruction and knowledge about the regional geology in the NE Atlantic region, but cannot be considered well constrained (Fig. 17). The model generally reflects the main rift configurations within the area, where in particular the Late Jurassic–Cretaceous rift zones and the Maastrichtian–Paleocene rift along the subsequent break-up

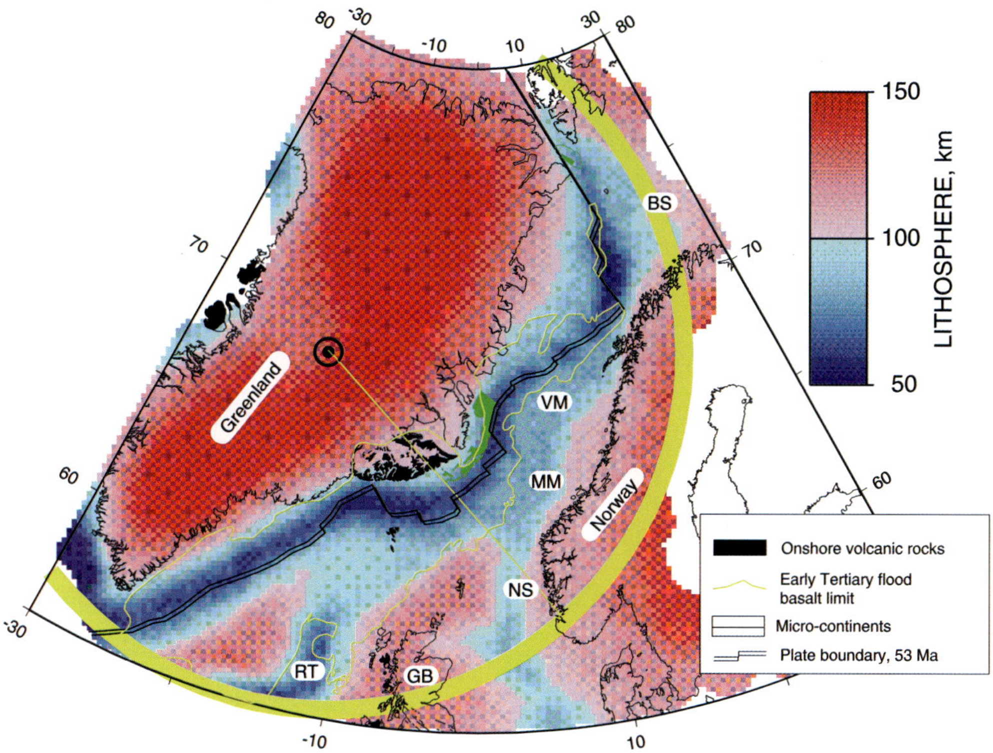

Fig. 17. NE Atlantic lithosphere thickness model suggested for *c.* 63 Ma, but implemented on the map showing the plate reconstruction to time of break-up (Fig. 10). The circle with a 1500 km radius illustrates the size of a flattened plume head large enough to encircle the entire Early Tertiary NE Atlantic volcanic province, and centred approximately at the projected Iceland plume location according to Lawver & Müller (1994). The yellow straight line show the location of the plume profile in inset frame in Fig. 18. (For other abbreviations, see Fig. 1.)

axis have thinnest lithosphere (i.e. 55–100 km). In contrast, the more stable cratonic areas may have thicknesses as high as 160 km (Suhadolc *et al.* 1990).

The shape of the plume body beneath the lithosphere is based on the assumption that the volume of a spherical plume head, 1200 km in diameter, spreads out radially and that the thickness of plume materials is a function of both the distance from the plume centre and the structure of the overlying lithosphere (Fig. 18). It is, thus, evident that the thin lithosphere areas govern a significantly thicker plume body than the cratonic regions.

The emplacement of a thermal plume beneath the lithosphere results in an isostatically defined uplift, where the magnitude of movements is a function of the thickness of plume body and density difference between the plume material and the original asthenosphere. We have assumed an average thermal anomaly of $\Delta T = 100^\circ C$ in the plume material, and that a thermal expansion coefficient of $\alpha = 0.00003\,K^{-1}$ is applicable for plume-related materials. $\Delta T \cdot \alpha$ thus characterize the density difference between the plume body and the adjacent asthenosphere. The plume–lithosphere interaction demonstrates the probability of the establishment of an elongated uplift area along the active rift zones, as well as inversion of regions with thin lithosphere. The same regions are those that by nature are prone to decompression melt generation, and from Fig. 18 we note that there is a close spatial relationship between the uplift region and the distribution of subaerially emplaced flood basalts. Formation, migration and eventually ponding of igneous melts at the base of the crust is a secondary process of central rift inversion (Pedersen *et al.* 1996), and the ascending melts, causing significant additional weakening of the lithosphere preferentially along the rift zone, may have driven the final continental separation (*c.* 55 Ma) (Skogseid *et al.* 1992*a*).

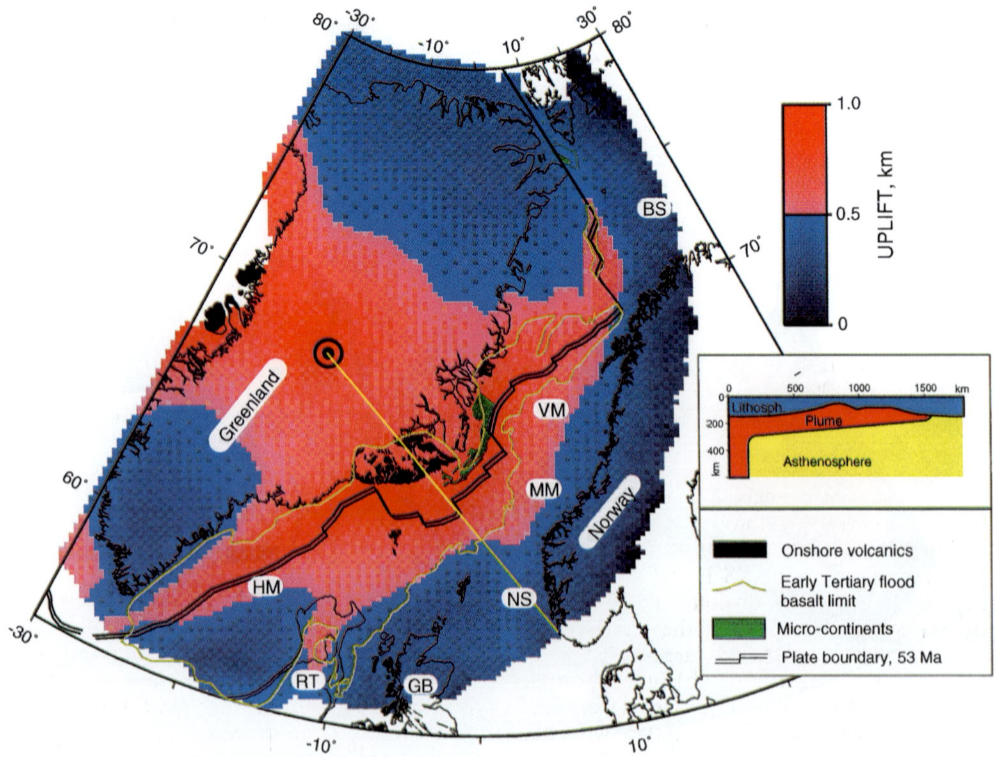

Fig. 18. Theoretical isostatic lithosphere uplift related to an average temperature anomaly of 100°C in a plume head where the thickness distribution is a function of both the distance from the plume centre and the relief at the base of the overlying lithosphere (see plume profile in inset frame located by yellow straight line on map). (For other abbreviations, see Fig. 1.)

The central rift inversion described above may cause the development of a relatively short-lived land area along the rift axis, which because it is prone to erosion may represent a provenance region for sediments being shed into adjacent shallow-water basins. Erosion is therefore a third mechanism of amplified uplift. Thick Paleocene sedimentary sequences are mapped along the outer margins off mid-Norway and west of the Shetland Islands (e.g. Hitchen & Ritchie 1987; Skogseid & Eldholm 1989).

According to Griffiths *et al.* (1989), the emplacement of mantle plumes beneath continents appears to show less uplift than in areas where plumes are located beneath oceanic lithosphere. The experiments of Griffiths *et al.* demonstrate that thicker surface layers result in reduced swell heights. The effect is described as a buoyancy response, and applies to a dynamic system only. The spreading of the plume beneath the lithosphere should also be considered dynamically, which during the period of plume head collapse yields a situation where the thickness, and thus weight, of the lithosphere plays an important role with respect to the magnitude of vertical movements. Thus, in regions where the lithosphere is thick with respect to the thickness of the plume, i.e. Greenland and regions around the margin of the plume, such as the Canadian margin of the Labrador Sea (Balkwill *et al.* 1990), the North Sea (e.g. White & Latin 1993; Jordt *et al.* 1999), mainland Norway and Fennoscandia (e.g. Tynni 1982; Riis 1996), the Western Barents Sea (Faleide *et al.* 1993*a*, *b*), Svalbard (e.g. Nøttvedt *et al.* 1992), etc., significant and rapid subsidence ($\leq$500 m) may be expected during the plume emplacement, i.e. during late Paleocene–early Eocene times. This subsidence should theoretically be followed by a rapid rebound resulting in net uplift as the plume movement ceases.

The *c.* 20 Ma rift development associated with break-up in the NE Atlantic, apparently starting before a plume establishment at lithospheric levels, contrasts to some degree with the active rifting model of Campbell & Griffiths (1990) and

Hill (1991). Base lithosphere erosion during plume emplacement may, on the other hand, be one mechanism that can explain observed discrepancies in magnitude of upper-crustal v. lithospheric thinning. Melt generation and subsequent migration through the lithosphere will further cause dramatic mantle–lithosphere weakening and probably facilitate break-up. The apparently low temperature anomaly (50–100°C) required to explain volumes of 'observed' igneous rocks across the Vøring margin (Pedersen & Skogseid 1989), the geochemistry of the extrusive complexes at some distance from the central plume trail (e.g. Zehnder *et al.* 1990), and the thickness of the adjacent oceanic crust (White *et al.* 1992) within the NE Atlantic large igneous province (LIP), should also open a broader discussion on tectono-magmatic models v. plume and non-plume hypotheses. Holbrook *et al.* (1994*a*, *b*) suggested a non-plume mechanism for the igneous activity on the US East Coast margin primarily because no obvious plume trail is present. Wilson (1997), on the other hand, indicated that a *c.* 200 Ma large-scale plume head caused a widespread tholeiitic magmatic event on the West African margin, which both temporally and spatially correlates with the Central Atlantic rifting and opening. Throughout the past 200 Ma the centres of the Cape Verde, Canary and Madeira plumes have been located on the African plate that have moved relatively little in terms of absolute movements (Watts & Marr 1995; Demets *et al.* 1990). From the assumption that mantle plumes are fixed points with respect to the Earth and that the lithospheric plates are moving, a plume trail should not exist on the North American plate. In this discussion it may also be important to note that, according to Brooks *et al.* (1996), volcanic rocks at distal regions of the Iceland plume show a small degree of partial melting and provide samples from the mantle uncontaminated by reaction with continental crust, partly in contrast to some of the tholeiitic volcanism. These rocks show a mantle rich in trace elements and volatiles; this finding may be of significance also to the much more voluminous volcanism of the region, where such characteristics are masked by large degrees of partial melting.

Exploration challenges

Volcanic margins are considered to be more common then previously thought (Coffin & Eldholm 1994). To broaden our knowledge of passive margins, in general, and volcanic passive margins, in particular, comparative studies between the NE Atlantic, the Namibian, the US Atlantic and the NW Australian margins are considered important as references or analogues (Eldholm *et al.* 2000). The main results from this comparative work are that all margins seem to have had a protracted rift development before continental separation, that significant crustal thinning is observed adjacent to the continent–ocean boundary, and that the main pulse of igneous activity is centred around the time of break-up.

One important observation is the recognition of that magmatic underplating (characterized by $7+\mathrm{km\,s^{-1}}$ velocity; Furlong & Fountain 1986) is found up to 100–150 landward from the COB (Fig. 7). These rocks represent actual growth in crustal thickness, which leads to rift inversion during magma emplacement and significantly reduced post-rift subsidence compared with a-magmatic rifts. Thus, the width of volcanic margins, in terms of the area affected both by lithospheric thinning and by much magmatism, may be increased significantly compared with the 'narrow rift' hypothesis initially applied on volcanic margins by Mutter *et al.* (1988). This hypothesis is often repeated when subsidence analysis across the volcanic constructions is made without application of a more regional perspective (e.g. Clift *et al.* 1995), or where the lack of a plume trail in the vicinity of the volcanic margin makes it difficult to point to a plume source for the excess magmatism (Holbrook *et al.* 1994*a*, *b*; Keen & Boutilier 1995).

Thus, in addition to the large variability and complexity associated with continental rifting in general, a number of tectono-magmatic events characterize volcanic margins and make them particularly interesting both for scientific research and as exploration targets. The expression of these tectono-magmatic events is mostly associated with vertical motions during extension; first, by alteration of the temperature structure (plume–lithosphere interaction) during plume emplacement; second, by central rift inversion, erosion and associated sediment transport and redeposition primarily linked with the magma generation, migration and emplacement (White & McKenzie 1989); third, by relatively little post-rift subsidence compared with non-volcanic margins (e.g. Bott 1992). Rift-related footwall uplift and possible vertical motion related to differential lithospheric stretching are additional factors that may play important roles locally. We stress that from a palaeo-environmental, an ocean circulation and a basin history point of view it is important to recognize that during rifting and break-up the present outer margins formed parts of a wide land-bridge stretching along the rift axis. In the NE Atlantic region this land area probably reached from the Charlie Gibbs Fracture Zone, at *c.* 55°N, to the SW Barents margin, at *c.* 75°N (Fig. 18).

With respect to exploration on volcanic margins we have to understand the complexity in the geological development and to be able to evaluate the heat-flow history and temperature effects associated with the igneous rocks at crustal levels. Here, the igneous characteristics of volcanic margins represent a special challenge by means of seismic definition. On seismic reflection data the flood basalt province is normally characterized by a series of seaward-dipping reflectors forming huge wedges of volcanic rocks and interbedded sediments ($\leq$10 km thick) (Hinz 1981; Mutter *et al.* 1984; Skogseid & Eldholm 1987; White *et al.* 1987; Hopper *et al.* 1992; Holbrook *et al.* 1994*a*, *b*; Planke 1994; Planke & Eldholm 1994). From a tectono-magmatic point of view we assume them, in general, to be associated with the actual continental break-up and first phase of sea-floor spreading when full asthenosphere decompression is reached. As a result of the syn-rift uplift referred to above, the rift zone may be at or above sea level at time of break-up, resulting in a significant increase in basalt flowability. The basalts may, thus, overflow large portions of the central rift zone and associated extensional structures. The internal seismic characteristics of the flood basalt wedges are generally resolvable, whereas the interpretation of the sub-basaltic parts of the seismic section is more challenging (Planke *et al.* in press).

Landward of the main wedges of volcanic rocks a wide (50–100 km) zone of the margin may be affected by sills and dykes in the pre-break-up sedimentary units (Skogseid *et al.* 1992*a*). The sills are often easy to interpret by their seismic signature and abrupt lateral termination. They hamper, however, the interpretation of the sedimentary sequences they are intruded in, and make structural interpretation more difficult as they partly are crossing faults without reflecting the fault offsets.

The third type of igneous features associated with volcanic margins are the bodies of magmatic underplating, which from an exploration point of view may be considered the most important factor in heat-flow history and hydrocarbon maturation (Pedersen *et al.* 1996). Factors influencing the volumes of melt to be formed are primarily the magnitude of the temperature anomaly (McKenzie & Bickle 1988) and the magnitude and duration of lithospheric stretching (Pedersen & Ro 1992). The molten rocks are emplaced beneath the crust, within the crust as dykes or sills, or at the surface as flood basalts. A typical thickness of 3–5 km underplating will result in uplift or reduced subsidence in the range between 0.8 and 1.5 km. This uplift, which is ascribed to a permanent re-creation of crustal thickness, is subsequently identical with the relative lack of thermal subsidence associated with volcanic margins with respect to non-volcanic margins.

The uplift associated with rifting, plume emplacement, melt formation and melt emplacement will take place over the same period of time during margin formation, will be focused along the central rift zones, and may sum to magnitudes of several kilometres.

Exploration on volcanic margins needs to consider the problem related to the thermal effects from the igneous units on maturation of organic matter. The tectono-magmatic history, including the various surface uplift–erosion mechanisms and the temporal development of the igneous complexes, needs to be treated as input to the calculation of heat flow (Pedersen *et al.* 1996). Traditionally the heat-flow history is related to extensional thinning of the lithosphere. At a passive margin, with high extension factors, this background heat flow will be substantially increased over relatively long periods. However, the heat flow from a magmatic body at the base of the crust can yield a three-fold increase in heat flow over a relatively short ($\leq$15 Ma) time interval, thus spiking the temperatures in the sediments by an amount mainly depending on the thickness of the underplate, the duration of melt emplacement, the distance from the underplate to the sediments, and the sediment conductivity (Pedersen *et al.* 1996). On the basis of models, with the mid-Norwegian margin as the study area, the present temperatures are generally lower within the pre-rift section than they were shortly after the igneous emplacement (Pedersen *et al.* 1996).

Additionally, sills intruded in the sedimentary units produce positive temperature anomalies. Thin sills ($\leq$10 m) will quickly die out thermally and increase the level of maturation of organic matter within distances no more than their own thickness (Jaeger 1964). Thick sills ($\geq$100 m), on the other hand, may yield substantial increase in maturation at distances up to 3–4 times their thickness. Flood basalts, on the other hand, are probably deposited as several batches of magma (<50 m thick). Their most important effect is that they cause burial of the underlying sediments.

Conclusions

The integrated approach, using deep seismic data and plate restoration, has allowed a definition of the rift zones, the degree of crustal thinning associated with rifting, and an evaluation of the spatial relations between basins at the time of formation. We believe that the lower-crustal geometries together with the upper-crustal structural setting demonstrates the value of our approach. We reiterate in particular the overall spatial correlation between Moho relief and upper-crustal structure beneath the main (deep) rift basins, as well as the extreme crustal thinning under the deep central rift troughs.

In the conjugate margin perspective, *c.* 140 km of stretching occurred between Greenland and Eurasia during the 75–55 Ma Maastrichtian–Paleocene rifting and continental break-up, whereas the *c.* 170–130 Ma Late Jurassic–Cretaceous rift episode caused *c.* 60 km of stretching. In addition, the Rockall Trough seems to have been affected by 100–200 km additional stretching with a southward fan-shaped signature. After plate restoration accounting for the above stretching, the pre-Late Jurassic basins province is less than 400 km wide. The general lack of detailed stratigraphic information and knowledge about the post-Caledonian crustal and lithospheric structure does not allow further quantification of the degree of stretching associated with these basins.

The crustal structure beneath the Late Jurassic–Cretaceous rift zones, i.e. the Møre and southern Vøring basins, shows degrees of crustal thinning which by far overstep the stretching estimates derived from subsidence analysis. This discrepancy may occur because we have underestimated the duration of the Late Jurassic–Cretaceous extension episode, and thus not included the full subsidence history in the region, and/or because the relatively narrow rift basins resist subsidence owing to the elastic strength of the lithosphere. A similar discrepancy does not exist beneath the Maastrichtian–Paleocene rift zones to the west. Here, subsidence and crustal thinning estimates are fairly consistent, whereas they are significantly larger than the results from structural analysis. These difference may be related to resolution limits in the seismic data, to the fact that most of the central rift zone is covered by volcanic rocks and therefore was not included in the structural analysis, and finally, because of differential lithosphere thinning by base lithosphere erosion during plume emplacement.

Theoretical estimates of uplift associated with the Iceland mantle plume show a clear correlation with the regions that subsequently were affected by subaerial volcanic activity. The formation of this short-lived central rift land area may have caused conditions for drainage systems and subsequent deposition of sand sequences of Late Paleocene age (63–55 Ma), and is, thus, considered particularly important for hydrocarbon exploration along the outer margins.

The geometric definition and an understanding of the emplacement history of the units of igneous materials at crustal levels have relevance for calculation of heat-flow history. In particular, the bodies of underplated materials can spike heat flow by a factor of three over time spans of 5–15 Ma, and according to models from the mid-Norwegian margin, may have resulted in palaeo-temperatures in the pre-rift sequences that were higher than present temperatures. The magnitude and duration of the temperature increase is able to cause maturation of organic matter, which cannot be predicted from present heat flow measurements or from traditional basin modelling.

We acknowledge data made available by the Norwegian Petroleum Directorate, Bundesanstalt für Geowissenschaften und Rohstoffe, and Statoil AS. We thank the PLATES Project, Institute for Geophysics, University of Texas at Austin, for access to their plate rotation software. We also thank R. England, P. Haremo and L. N. Jensen for comments and critical review of the manuscript and L. M. Larsen for adding details and for fruitful discussions. The work has been supported by the Norwegian Research Council, the IBS (Integrated Basin Studies) project, part of Joule II research programme funded by the Commission of European Communities (Contract JOU2-CT-92-0110) and by 'University of Oslo – Passive Margin Research Group' funded by Statoil.

References

Andersen, T. B. & Jamtveit, B. 1990. Uplift of deep crust during orogenic extensional collapse; a model based on field studies in the Sogn–Sunnfjord region of western Norway. *Tectonics*, **9**, 1097–1111.

Artyushkov, E. V., Schlesinger, A. E. & Yansin, A. L. 1980. The origin of vertical movements within lithospheric plates. *In*: Bally, A. W., Bender, P. L., McGetchin, T. K. & Walcott, R. I. (eds) *Dynamics of Plate Interiors*. American Geophysical Union, Geodynamics Series, **1**, 37–51.

Balkwill, H. R., McMillan, H. J., MacLean, B., Williams, G. L. & Srivastava, S. P. 1990. Geology of the Labrador Shelf, Baffin Bay and Davis Strait. *In*: Keen, M. J. & Williams, G. L. (eds) *Geology of the Continental Margin of Eastern Canada*. Geological Society of America, The Geology of North America, **I-1**, 293–348.

BLYSTAD, P., FÆRSETH, R. B., LARSEN, B. T., SKOGSEID, J. & TØRUDBAKKEN, B. 1995. *Structural Elements of the Norwegian Continental Margin between 62° and 69°N*. Norwegian Petroleum Directorate Bulletin, **8**.

BOLDREEL, L. O. & ANDERSEN, M. S. 1994. Tertiary development of the Faeroe–Rockall Plateau based on reflection seismic data. *Geological Society of Denmark, Bulletin*, **41**, 162–180.

BOTT, H. M. P. 1992. Passive margins and their subsidence. *Journal of the Geological Society, London*, **149**, 805–812.

BREIVIK, A., FALEIDE, J. I. & GUDLAUGSSON, S. T. 1998. Pre-Tertiary sediment distribution and crustal extension: SW Barents Sea margin. *In*: BREIVIK, A. 1998. *Crustal structure and development of the SW Barents Sea and the adjacent continental margin*. Dr.Scient. thesis, University of Oslo.

BROOKS, K., BERNSTEIN, S. & TEGER, C. 1996. Alkaline volcanism along North Atlantic passive margins (abstract). *Tectonic, Magmatic and Depositional Processes at Passive Continental Margins*. Geological Society, London, Meeting, 20–22 February.

BRUECKNER, H. K., GILOTTI, J. A. & NUTMAN, A. 1999. Caledonian eclogite-facies metamorphism of Early Proterozoic protoliths from the northeast Greenland eclogite province. *Contributions to Mineralogy and Petrology*, **1301**, 103–120.

CAMPBELL, I. H. & GRIFFITHS, R. W. 1990. Implication of mantle plume structure for the evolution of flood basalts. *Earth and Planetary Science Letters*, **99**, 79–93.

CANDE, S. C. & KENT, D. V. 1992. A new geomagnetic polarity time scale for the Late Cretaceous and Cenozoic. *Journal of Geophysical Research*, **97**, 13 917–13 951.

CHALMERS, J. A., PULVERTAFT, T. C. R., CHRISTIANSEN, F. G., LARSEN, H. C., LAURSEN, K. H. & OTTESEN, T. G. 1993, The southern West Greenland continental margin: rifting history, basin development and petroleum potential. *In*: PARKER, J. R. (ed.) *Petroleum Geology of Northwest Europe: Proceedings of the 4th Conference*. Geological Society, London, 915–931.

CHRISTIANSSON, P., FALEIDE, J. I. & BERGE, A. M. 2000. Crustal structure in the northern North Sea; an integrated geophysical study. *This volume*.

CLIFT, P. D., TURNER, J. & OCEAN DRILLING PROGRAM LEG 152 SCIENTIFIC PARTY 1995. Dynamic support by the Iceland plume and vertical tectonics of the northeast Atlantic continental margins. *Journal of Geophysical Research*, **100**, 24 473–24 486.

COCHRAN, J. R. & TALWANI, M. 1978. Gravity anomalies, regional elevation, and the deep structure of the North Atlantic. *Journal of Geophysical Research*, **83**, 4907–4924.

COFFIN, M. F. & ELDHOLM, O. 1994. Large Igneous Provinces: crustal structure, dimensions, and external consequences. *Reviews of Geophysics*, **32**, 1–36.

COWARD, M. P. 1993. the effect of Late Caledonian and Variscan continental escape tectonics on basement structure: Paleozoic basin kinematics and subsequent Mesozoic basin development in NW Europe. *In*: PARKER, J. R. (ed.) *Petroleum Geology of Northwest Europe: Proceedings of the 4th Conference*. Geological Society, London, 1095–1108.

DALLAND, A. 1981. Mesozoic sedimentary succession at Andøya, northern Norway, and relation to structural development of the North Atlantic. *In*: KERR, J. W., MACHAN, L. C. & FERGUSSON, A. J. (eds) *Geology of the North Atlantic Borderlands*. Canadian Society of Petroleum Geologists, Memoir, **7**, 563–584.

DE GRACIANSKY, P. C., POAG, C. W. *et al.* 1985. The Goban Spur Transect: geological evolution of a sediment-starved passive continental margin. *Geological Society of America Bulletin*, **96**, 58–76.

DEMETS, C., GORDON, R. G., ARGUS, D. F. & STEIN, S. 1990. Current plate motions. *Geophysical Journal International*, **101**, 425–478.

DORÉ, A. G. 1992. Synoptic palaeogeography of the Northeast Atlantic Seaway: late Permian to Cretaceous. *In*: PARNELL, J. (ed.) *Basins of the Atlantic Seaboard: Petroleum Geology, Sedimentology and Basin Evolution*. Geological Society, London, Special Publications, **62**, 421–446.

——, LUNDIN, E. R., BIRKELUND, Ø., ELIASSEN, P. E. & JENSEN, L. N. 1997*a*. The NE Atlantic Margin: implications of late Mesozoic and Cenozoic events for hydrocarbon prospectivity. *Petroleum Geoscience*, **3**, 117–131.

——, ——, FICHLER, C. & OLESEN, O. 1997*b*. Patterns of basement structures and reactivation along the NE Atlantic margin. *Journal of the Geological Society, London*, **154**, 85–92.

DUINDAM, P. & VAN HORN, B. 1987. Structural evolution of the West Shetland continental margin. *In*: BROOKS, J. & GLENNIE, K. W. (eds) *Petroleum Geology of North-West Europe*. Graham and Trotman, London, 765–773.

EIDE, F. 1989. Biostratigraphic correlation within the Triassic Lunde Formation in the Snorre area. *In*: COLLINSON, J. D. (ed.) *Correlation in Hydrocarbon Exploration*. Graham and Trotman, London, 291–297.

ELDHOLM, O. & GRUE, K. 1994. North Atlantic volcanic margins: dimensions and production rates. *Journal of Geophysical Research*, **99**, 2955–2968.

——, GLADCZENKO, T. P., SKOGSEID, J. & PLANKE, S. 2000. Atlantic volcanic margins: a comparative study. *This volume*.

——, THIEDE, J. & TAYLOR, E. 1989. Evolution of the Vøring volcanic margin. *In*: ELDHOLM, O., THIEDE, J. & TAYLOR, E. *et al.* (eds) *Proceedings of the Ocean Drilling Program, Scientific Results*, **104**. Ocean Drilling Program, College Station, TX, 1033–1065.

ESCHER, J. C. & PULVERTAFT, T. C. R. 1995. *Geological map of Greenland*, 1:2 500 000. Geological Survey of Greenland, Copenhagen,

FALEIDE, J. I., GUDLAUGSSON, S. T., ELDHOLM, O., MYHRE, A. M. & JACKSON, H. R. 1991. Deep seismic transects across the sheared western Barents Sea–Svalbard margin. *Tectonophysics*, **189**, 73–89.

——, VÅGNES, E. & GUDLAUGSSON, S. T. 1993*a*. Late Mesozoic–Cenozoic evolution of the southwestern Barents Sea in a rift–shear tectonic setting. *Marine and Petroleum Geology*, **10**, 186–214.

——, —— & ——1993*b*. Late Mesozoic–Cenozoic evolution of the southwestern Barents Sea. *In*: PARKER, J. R. (ed.) *Petroleum Geology of Northwest Europe: Proceedings of the 4th Conference*. Geological Society, London, 933–950.

——, MATHISEN, G., CHRISTIANSON, P., MYKLEBUST, R. & SANDVIK, J. 2000. Deep basin geometry and crustal structure in the northern North Sea; constrains form gravity and magnetic data. *This volume*.

FECHNER, N. & Jokat, W. 1996. Seismic refraction investigations on the crustal structure of the western Jameson Land Basin, East Greenland. *Journal of Geophysical Research*, **101**, 15 867–15 881.

FURLONG, K. P. & FOUNTAIN, D. M. 1986. Continental crustal underplating: thermal considerations and seismic petrological consequences. *Journal of Geophysical Research*, **91**, 8285–8294.

FÆRSETH, R. B., GABRIELSEN, R. H. & HURICH, C. A. 1995. Influence of basement in structuring of the North Sea Basin, offshore Southwest Norway. *Norsk Geologisk Tidsskrift*, **75**, 105–119.

GINZBURG, A., WHITMARSH, R. B., ROBERTS, D. G., MONTADERT, L. CAMUS, A. & AVEDIK, F. 1985. The deep seismic structure of the northern continental margin of the Bay of Biscay. *Annales Geophysicae*, **3**, 499–510.

GRIFFITHS, R. W. & CAMPBELL, I. H. 1990. Stirring and structure in mantle starting plumes. *Earth and Planetary Science Letters*, **99**, 66–78.

——, GURNIS, M. & EITELBERG, G. 1989. Holographic measurements of surface topography in laboratory models of mantle hotspots. *Geophysical Journal*, **96**, 477–495.

GUDLAUGSSON, S. T., FALEIDE, J. I., FANAVOLL, S. & JOHANSEN, B. 1987. Deep seismic reflection profiles across the western Barents Sea. *Geophysical Journal of the Royal Astronomical Society*, **89**, 273–278.

——, ——, JOHANSEN, S. E. & BREIVIK, A. 1994. Late Palaeozoic structural development of the southwestern Barents Sea. *In*: GUDLAUGSSON, S. T. 1994. Structural and tectonic history of the western Barents Sea. PhD thesis, University of Oslo.

HARTZ, E. & ANDRESEN, A. 1995. Caledonian sole thrust of central East Greenland: a crustal scale Devonian extensional detachment. *Geology*, **23**, 637–640.

HILL, R. I. 1991. Starting plumes and continental break-up. *Earth and Planetary Science Letters*, **104**, 398–416.

HINZ, K. 1981. A hypothesis on terrestrial catastrophes: wedges of very thick oceanward dipping layers beneath passive continental margins – their origin and palaeoenvironment significance. *Geologische Jahrbuch, Serie E*, **22**, 3–28.

——, MEYER, H. & MILLER, H. 1991. North-east Greenland Shelf north of 79°N: results of a reflection seismic experiment in sea ice. *Marine and Petroleum Geology*, **8**, 461–467.

——, MUTTER, J. C., ZEHNDER, C. M. & THE NGT STUDY GROUP 1987. Symmetric conjugation of continent–ocean boundary along the Norwegian and East Greenland margins. *Marine and Petroleum Geology*, **4**, 166–187.

HITCHEN, K. & RITCHIE, J. D. 1987. Geological review of the West Shetland area. *In*: BROOKS, J. & GLENNIE, K. W. (eds) *Petroleum Geology of North-West Europe*. Graham and Trotman, London, 737–749.

—— & ——1993. New K–Ar ages, and a provisional chronology, for the offshore part of the British Tertiary Igneous Province. *Scottish Journal of Geology*, **29**, 73–85.

HOLBROOK, W. S., PURDY, G. M., SHERIDAN, R. E., GLOVER, L., TALWANI, M., EWING, J. & HUTCHINSON, D. 1994*a*. Seismic structure of the US Mid-Atlantic continental margin. *Journal of Geophysical Research*, **99**, 17 871–17 891.

——, REITER, E. C., PURDY, G. M. *et al.* 1994*b*. Deep structure of the, US Atlantic continental margin, offshore South Carolina, from coincident ocean bottom and multichannel data. *Journal of Geophysical Research*, **99**, 9155–9178.

HOPPER, R. J., MUTTER, J. C., LARSON, R. L., MUTTER, C. Z. & NW AUSTRALIA STUDY GROUP 1992. Magmatism and rift margin evolution: evidence from northwest Australia. *Geology*, **20**, 67–73.

HOSSACH, J. R. 1986. The geometry of listric faults in the Devonian Basins of SUNNFJORD, W. Norway. *Journal of the Geological Society, London*, **141**, 629–637.

JACKSON, H. R., FALEIDE, J. I. & ELDHOLM, O. 1990. Crustal structure of the sheared southwestern Barents Sea continental margin. *Marine Geology*, **93**, 119–146.

JAEGER, J. C. 1964. Thermal effects of intrusions. *Reviews of Geophysics*, **2**, 443–466.

JOLLEY, D. W. 1997. Palaeosurface palynofloras of the Skye lava field and the age of the British Tertiary volcanic province. *In*: WIDDOWSON, M. (ed.) *Palaeosurfaces: Recognition, Reconstruction and Palaeoenvironmental Interpretation*. Geological Society, London, Special Publications, **120**, 67–94.

JOPPEN, M. & WHITE, R. S. 1990. The structure and subsidence of Rockall Trough from two-ship seismic experiments. *Journal of Geophysical Research*, **95**, 19 821–19 837.

JORDT, H., THYBERG, B. I. & NØTTVEDT, A. 2000. Cenozoic evolution of the central and northern North Sea with focus on differential vertical movements of the basin floor and surrounding clastic source areas. *This volume*.

KEEN, C. E. & BOUTILIER, R. R. 1995. Lithosphere–asthenosphere interactions below rifts. *In*: BANDA, E., TORNÉ, M. & TALWANI, M. (eds) *Rifted Ocean–Continent Boundaries*. NATO ASI Series **C463**, Kluwer, Dordrecht, 17–30.

—— & DE VOOGD, B. 1988. The continent–ocean boundary at the rifted margin off eastern Canada: new results from deep seismic reflection studies. *Tectonics*, **7**, 107–124.

KESER NEISH, J. 1993. Seismic structure of the Hatton–Rockall area: an integrated seismic/modeling study from composite data sets. *In*: PARKER,

J. R. (ed.) *Petroleum Geology of Northwest Europe: Proceedings of the 4th Conference*. Geological Society, London, 1047–1056.

KNOTT, S. D., BURCHELL, M. T., JOLLEY, E. J. & FRASER, A. J. 1993. Mesozoic to Cenozoic plate reconstructions of the North Atlantic and hydrocarbon plays of the Atlantic margins. *In*: PARKER, J. R. (ed.) *Petroleum Geology of Northwest Europe: Proceedings of the 4th Conference*. Geological Society, London, 953–974.

KRISTOFFERSEN, Y. 1978. Sea-floor spreading and the early opening of the North Atlantic. *Earth and Planetary Science Letters*, **38**, 273–290.

LARSEN, H. C. 1990. The East Greenland Shelf. *In*: GRANTZ, A., JOHNSON, L. & SWEENEY, J. F. (eds) *The Arctic Ocean Region*. Geological Society of America, The Geology of North America, **L**, 185–210.

—— & MARCUSSEN, C. 1992. Sill-intrusions, flood basalt emplacement and deep crustal structure of the Scoresby Sund region, East Greenland. *In*: STOREY, B. C., ALABASTER, T. & PLANKHURST, R. J. (eds) *Magmatism and the Causes of Continental Break-up*. Geological Society, London, Special Publications, **68**, 365–386.

LARSEN, L. M. & WATT, W. S. 1985. Episodic volcanism during break-up of the North Atlantic: evidence from the East Greenland plateau basalts. *Earth and Planetary Science Letters*, **73**, 105–116.

——, PEDERSEN, A. K., PEDERSEN, G. K. & PIASECKI, S. 1992. Timing and duration of Early Tertiary volcanism in the North Atlantic: new evidence from West Greenland. *In*: STOREY, B. C., ALABASTER, T. & PANKHURST, R. J. (eds) *Magmatism and the Causes of Continental Break-up*. Geological Society, London, Special Publications, **68**, 321–333.

——, REX, D. C., WATT, W. S. & GUISE, P. G. 1999. ^{39}Ar–^{40}Ar dating of alkali basaltic dykes along the south-west coast of Greenland: Cretaceous and Tertiary igneous activity along the eastern margin of the Labrador Sea. *Geology of Greenland Survey Bulletin*, **184**, 15–25.

LARSEN, V. B. 1987. A synthesis of tectonically-related stratigraphy in the North Atlantic–Arctic region from Aalenian to Cenomanian time. *Norsk Geologisk Tidsskrift*, **67**, 281–293.

LAWVER, L. A. & MÜLLER, D. R. 1994. Iceland hotspot track. *Geology*, **22**, 311–314.

LE PICHON, X. & SIBUET, J.-C. 1981. Passive margins: a model of formation. *Journal of Geophysical Research*, **86**, 3708–3720.

LUNDIN, E. R. & DORÉ, A. G. 1997. A tectonic model for the Norwegian passive margin with implications for the NE Atlantic: Early Cretaceous to break-up. *Journal of the Geological Society, London*, **154**, 545–550.

MAKRIS, J., GINZBURG, A., SHANNON, P. M., JACOB, A. W. B., BEAN, C. J. & VOGT, U. 1991. A new look at the Rockall region, offshore Ireland. *Marine and Petroleum Geology*, **8**, 410–416.

MARCUSSEN, C., CHRISTIANSEN, F. G., Larsen, P.-H. *et al.* 1987. *Studies of the onshore hydrocarbon potential in East Greenland 1986–1987: field work from 72° to 74°N*. Greenland Geological Survey, Report, **135**, 72–81.

MCKENZIE, D. P. 1978. Some remarks on the development of sedimentary basins. *Earth and Planetary Science Letters*, **40**, 25–32.

—— & BICKLE, M. J. 1988. The volume and composition of melt generated by extension of the lithosphere. *Journal of Petrology*, **29**, 625–679.

MJELDE, R., BERG, E. W., STRØM, A. *et al.* 1996. An extensive Ocean Bottom Seismograph survey in the Vøring BASIN, N. Norway. *First Break*, **14**, 247–256.

——, KODAIRA, S., SHIMAMURA, H., KANAZAWA, T., SHIOBARA, H., BERG, E. W. & RIISE, O. 1997. Crustal structure of the central part of the Vøring Basin, mid-Norway margin, from ocean bottom seismographs. *Tectonophysics*, **277**, 235–257.

MUDGE, D. C. & RASHID, B. 1987. The geology of the Faeroe Basin area. *In*: BROOKS, J. & GLENNIE, K. W. (eds) *Petroleum Geology of North-West Europe*. Graham and Trotman, London, 751–763.

MUSSETT, A. E., DAGLEY, P. & SKELHORN, R. R. 1988. Time and duration of activity in the British Tertiary Igneous Province. *In*: MORTON, A. C. & PARSON, L. M. (eds) *Early Tertiary Volcanism and the Opening of the NE Atlantic*. Geological Society, London, Special Publications, **39**, 337–348.

MUTTER, J. C., BUCK, W. R. & ZEHNDER, C. M. 1988. Convective partial melting. 1. A model for the formation of thick basaltic sequences during the initiation of spreading. *Journal of Geophysical Research*, **93**, 1031–1048.

——, TALWANI, M. & STOFFA, P. L. 1984. Evidence for a thick oceanic crust adjacent to the Norwegian margin. *Journal of Geophysical Research*, **89**, 483–502.

NELSON, P. H. H. & LAMY, J.-M. 1987. The Møre/West Shetland Area: a review. *In*: BROOKS, J. & GLENNIE, K. W. (eds) *Petroleum Geology of North-West Europe*. Graham and Trotman, London, 775–784.

NØTTVEDT, A., CECCHI, M., GJELBERG, J. G. *et al.* 1992. Svalbard–Barents Sea correlation: a short review. *In*: VORREN, T. O., BERGSAKER, E., DAHL-STAMNES, Ø. A., HOLTER, E., JOHANSEN, B., LIE, E. & LUND, T. B. (eds) *Arctic Geology and Petroleum Potential*. Norwegian Petroleum Society, Special Publications, **2**, 363–375.

OLAFSSON, I., SUNDVOR, E., ELDHOLM, O. & GRUE, K. 1991. Møre Margin: crustal structure from analysis of Expanded Spread Profiles. *Marine Geophysical Research*, **14**, 137–162.

O'REILLY, B. M., HAUSER, F., JAKOB, B. A. W. & SHANNON, P. M. 1996. The lithosphere below the Rockall Trough: wide-angle seismic evidence for extensive serpentinisation. *Tectonophysics*, **255**, 1–23.

PEDERSEN, T. & RO, H. E. 1992. Finite duration extension and decompression melting. *Earth and Planetary Science Letters*, **113**, 15–22.

—— & SKOGSEID, J. 1989. Vøring Plateau volcanic margin: extension, melting and uplift. *In*: ELDHOLM, O., THIEDE, J. & TAYLOR, E. *et al.* (eds) *Proceedings of the Ocean Drilling Program, Scientific Results*, **104**. Ocean Drilling Program, College Station, TX, 985–991.

——, ——, JOHANSEN, H. & THRONDSEN, T. 1996. Integrated tectonomagmatic and basin modeling at the Vøring Basin volcanic margin: magmatism, source and reservoir rocks (abstract). *Tectonic, Magmatic and Depositional Processes at Passive Continental Margins*. Geological Society, London, Meeting, 20–22 February.

PLANKE, S. 1994. Geophysical response of flood basalts from analysis of wireline logs: ODP Site 642, Vøring volcanic margin. *Journal of Geophysical Research*, **99**, 9279–9296.

—— & ELDHOLM, O. 1994. Seismic response and construction of seaward dipping wedges of flood basalts: Vøring volcanic margin. *Journal of Geophysical Research*, **99**, 9263–9278.

——, SKOGSEID, J. & ELDHOLM, O. 1991. Crustal structure off Norway, 62°–70°N. *Tectonophysics*, **189**, 91–107.

——, SYMONDS, P. A., ALVESTAD, E. & SKOGSEID, J. 1999. Seismic volcanostratigraphy of large-volume basaltic extrusives on rifted margins. *Journal of Geophysical Research*, in press

PLATT, N. H. 1995. Structure and tectonics of the northern North Sea: new insight from deep penetration regional seismic data. *In*: LAMBIASE, J. J. (ed.) *Hydrocarbon Habitat in Rift Basins*, Geological Society, London, Special Publications, **80**, 103–113.

PRICE, S. P. & WHITHAM, A. G. 1997. Exhumed hydrocarbon traps in East Greenland: analogs for Lower–Middle Jurassic Play of Northwest Europe. *AAPG Bulletin*, **81**, 196–221.

REN, S., SKOGSEID, J. & ELDHOLM, O. 1999. Late Cretaceous–Paleocene extension on the Vøring volcanic margin. *Marine Geophysical Research*, **20**, 343–369.

RIIS, F. 1996. Quantification of Cenozoic vertical movements of Scandinavia by correlation of morphological surfaces with offshore data. *Global and Planetary Change*, **12**, 331–357.

—— & FJELDSKAAR, W. 1992. On the magnitude of the late Tertiary and Quaternary erosion and its significance for the uplift of Scandinavia and Barents Sea. *In*: LARSEN, R. M., BREKKE, H., LARSEN, B. T. & TALLERAAS, E. (eds) *Tectonic Modelling and its Application to Petroleum Geology*. Norwegian Petroleum Society, Special Publications, **1**, 163–185.

ROBERTS, A. M., LUNDIN, E. & KUSZNIR, N. 1997. Subsidence of the Vøring Basin and the influence of the Atlantic continental margin. *Journal of the Geological Society, London*, **154**, 551–557.

——, YIELDING, G. & BADLEY, M. E. 1990. A kinematic model for the orthogonal opening of the late Jurassic North Sea rift system, Denmark–Mid Norway. *In*: BLUNDELL, D. J. & GIBBS, A. D. (eds) *Tectonic Evolution of the North Sea Rifts*. Oxford University Press, Oxford, 180–199.

ROBERTS, D. G. 1974. Structural development of the British Isles, the continental margin, and the Rockall Plateau. *In*: BURKE, C. A. & DRAKE, C. L. (eds) *The Geology of Continental Margins*. Springer, New York, 343–359.

——1975*a*. Marine geology of the Rockall Plateau and trough. *Philosophical Transactions of the Royal Society of London, Series A: Mathematical and Physical Sciences*, **278**, 447–509.

——1975*b*. Tectonic and stratigraphic evolution of the Rockall Plateau and trough. *In*: WOODLAND, A. W. (ed.) *Petroleum and the Continental Shelf of Northwest Europe, Vol. 1, Geology*. Applied Science, London, 77–91.

——, BACKMAN, J., MORTON, A. C., MURRAY, J. W. & KEENE, J. B. 1984. Evolution of volcanic rifted margins: synthesis of Leg 81 results on the west margin of Rockall Plateau. *Initial Reports of the Deep Sea Drilling Project*, **81**, US Government Printing Office, Washington, DC, 883–911.

——, MASSON, D. G. & MILES, P. R. 1981. Age and structure of the southern Rockall Trough, new evidence. *Earth and Planetary Science Letters*, **52**, 115–128.

ROEST, W. R. & SRIVASTAVA, S. P. 1989. Sea-floor spreading in the Labrador Sea: a new reconstruction. *Geology*, **17**, 1000–1003.

RUSSEL, M. J. 1976. A possible lower Permian age for the onset of ocean floor in the northern North Atlantic. *Scottish Journal of Geology*, **12**, 315–323.

—— & SMYTHE, D. K. 1978. Evidence for an early Permian oceanic rift in the northern North Atlantic. *In*: NEUMANN, E.-R. & RAMBERG, I. B. (eds) *Pertology and Geochemistry of Continental Rifts*. Reidel, Dordrecht, 173–179.

SHANNON, P. M., MOORE, J. G., JACOB, A. W. B. & MAKRIS, J. 1993. Cretaceous and Tertiary basin development west of Ireland. *In*: PARKER, J. R. (ed.) *Petroleum Geology of Northwest Europe: Proceedings of the 4th Conference*. Geological Society, London, 1057–1066.

SKOGSEID, J. 1994. Dimensions of the Late Cretaceous–Palaeocene Northeast Atlantic rift derived from Cenozoic subsidence. *Tectonophysics*, **240**, 225–247.

—— & ELDHOLM, O. 1987. Early Cenozoic crust at the Norwegian continental margin and the conjugate Jan Mayen Ridge. *Journal of Geophysical Research*, **92**, 11 471–11 491.

—— & ——1989. Vøring Plateau continental margin: seismic interpretation, stratigraphy and vertical movements. *In*: ELDHOLM, O., THIEDE, J. & TAYLOR, E. *et al.* (eds) *Proceedings of the Ocean Drilling Program, Scientific Results*, **104**. Ocean Drilling Program, College Station, TX, 993–1030.

—— & ——1995. Rifted continental margin off mid-Norway. *In*: BANDA, E., TORNÉ, M. & TALWANI, M. (eds) *Rifted Ocean–Continent Boundaries*. NATO ASI Series **C463**, Kluwer, Dordrecht, 147–153.

——, PEDERSEN, T., ELDHOLM, O. & LARSEN, B. T. 1992*a*. Tectonism and magmatism during NE Atlantic continental break-up: the Vøring Margin. *In*: STOREY, B. C., ALABASTER, T. & PLANKHURST, R. J. (eds) *Magmatism and the Causes of Continental Break-up*. Geological Society, London, Special Publications, **68**, 305–320.

——, —— & LARSEN, V. B. 1992*b*. Vøring Basin: subsidence and tectonic evolution. *In*: LARSEN,

R. M., BREKKE, H., LARSEN, B. T. & TALLERAAS, E. (eds) *Structural and Tectonic Modeling and its Application to Petroleum Geology*. Norwegian Petroleum Society, Special Publications, **1**, 55–82.

SLEEP, N. H. 1997. Lateral flow and ponding of starting plume material. *Journal of Geophysical Research*, **102**, 10 001–10 012.

SMYTHE, D. K. 1989. Rockall Trough – Cretaceous or late Palaeozoic? *Scottish Journal of Geology*, **25**, 5–43.

SRIVASTAVA, S. P. 1978. Evolution of the Labrador Sea and its bearing on the early evolution of the North Atlantic. *Geophyical Journal of the Astronomical Society*, **52**, 313–357.

—— & TAPSCOTT, C. R. 1986. Plate kinematics of the North Atlantic. *In*: VOGT, P. R. & TUCHOLKE, B. E. (eds) *The Western North Atlantic Region*. Geological Society of America, The Geology of North America, **M**, 379–404.

—— & VERHOEF, J. 1992. Correlation of sedimentary basins across the North Atlantic as obtained from gravity and magnetic data, and its relation to the early evolution of the North Atlantic. *In*: TANKARD, A. J. & BALKWILL, H. (eds) *Extensional Tectonics and Stratigraphy of the North Atlantic Margins*. American Association of Petroleum Geologists, Memoir, **46**, 131–147.

——, —— & MACNAB, R. 1988. Results from a detailed aeromagnetic survey across the northeast Newfoundland margin, Part II: early opening of the North Atlantic between the British Isles and Newfoundland. *Marine and Petroleum Geology*, **5**, 324–337.

STECKLER, M. S. & WATTS, A. B. 1978. Subsidence of the Atlantic-type continental margin off New York. *Earth and Planetary Science Letters*, **41**, 1–13.

STEEL, R. J. 1993. Triassic–Jurassic megasequence stratigraphy in the Northern North Sea: rift to post-rift evolution. *In*: PARKER, J. R. (ed.) *Petroleum Geology of Northwest Europe: Proceedings of the 4th Conference*. The Geological Society, London, 299–315.

STUEVOLD, L., SKOGSEID, J. & ELDHOLM, O. 1992. Post-Cretaceous uplift events on the Vøring continental margin. *Geology*, **20**, 919–922.

SUHADOLC, P., PANZA, G. F. & MUELLER, S. 1990. Physical properties of the lithosphere–asthenosphere system in Europe. *Tectonophysics*, **176**, 123–135.

SURLYK, F. 1991. Sequence stratigraphy of the Jurassic–lowermost Cretaceous of Greenland. *AAPG Bulletin*, **75**, 1468–1488.

——, CLEMMENSEN, L. B. & LARSEN, H. C. 1981. Post-Palaeozoic evolution of the East Greenland continental margin. *In*: KERR, J. W. & FERGUSSON, A. J. (eds) *Geology of the North Atlantic Borderlands*. Canadian Society of Petroleum Geologists, Memoir, **7**, 611–645.

——, PIASECKI, S., ROLLE, F., STEMMERIK, L., THOMPSON, E. & WRANG, P. 1984. The Permian basin of East Greenland. *In*: SPENCER, A. M. *et al.* (eds) *Petroleum Geology of the North European Margin*. Graham and Trotman, London, 303–315.

TATE, M., WHITE, N. & CONROY, J.-J. 1993. Lithospheric extension and magmatism in the Porcupine Basin west of Ireland. *Journal of Geophysical Research*, **98**, 13 905–13 923.

THOMPSON, R. N. & GIBSON, S. A. 1991. Subcontinental mantle plumes, hotspots and pre-existing thinspots. *Journal of the Geological Society, London*, **148**, 972–977.

TYNNI, R. 1982. *The Reflection of Geological Evolution in Tertiary and Interglacial Diatomes and Silico-Flagellates in Finnish Lapland*. Bulletin of the Geological Survey of Finland, **320**.

WALKER, I. M., BERRY, K. A., BRUCE, J. R., BYSTØL, L. & SNOW, J. H. 1997. Structural modeling of regional depth profiles in the Vøring Basin: implications for the structural and stratigraphical development of the Norwegian passive margin. *Journal of the Geological Society, London*, **154**, 537–544.

WATT, W. S. 1969. The coast-parallel dyke swarm of Southwest Greenland in relation to the opening of the Labrador Sea. *Canadian Journal of Earth Science*, **6**, 1320–1321.

WATTS, A. B. & MARR, C. 1995. Gravity anomalies and the thermal and mechanical structure of rifted continental margins. *In*: BANDA, E., TORNÉ, M. & TALWANI, M. (eds) *Rifted Ocean–Continent Boundaries*. NATO ASI Series **C463**, Kluwer, Dordrecht, 65–94

WHITE, N. & LATIN, D. 1993. Lithospheric stretching from subsidence analyses in the North Sea 'triple junction'. *Journal of the Geological Society, London*, **150**, 473–488.

WHITE, R. S. & MCKENZIE, D. P. 1989. Magmatism at rift zones: the generation of volcanic continental margins and flood basalts. *Journal of Geophysical Research*, **94**, 7685–7729.

——, MCKENZIE, D. & O'NIONS, K. R. 1992. Oceanic crustal thickness from seismic measurements and rare earth element inversions. *Journal of Geophysical Research*, **97**, 19 683–19 715.

——, SPENCE, G. D., FOWLER, S. R., MCKENZIE, D. P., WESTBROOK, G. K. & BOWEN, A. N. 1987. Magmatism at rifted continental margins. *Nature*, **330**, 439–444.

WHITMARSH, R. B., PINHEIRO, L. M., MILES, P. R., RECQ, M. & SIBUET, J.-C. 1993. Thin crust at the western Iberia ocean–continent transition and ophiolites. *Tectonics*, **12**, 1230–1239.

WILSON, M. 1997. Thermal evolution of the Central Atlantic passive margins: continental break-up above a Mesozoic super-plume. *Journal of the Geological Society, London*, **154**, 491–495.

ZEHNDER, C. M., MUTTER, J. C. & BUHL, P. 1990. Deep seismic and geochemical constraints on the nature of rift-induced magmatism during breakup of the North Atlantic. *Tectonophysics*, **173**, 545–565.

ZIEGLER, P. (ed.) 1988. *Evolution of the Arctic–North Atlantic and the Western Tethys*. American Association of Petroleum Geologists, Memoir, **43**.

ÅM, K. 1971. *Aeromagnetic Investigation on the Continental Shelf of Norway, Stad-Lofoten (62–69°N)*. Norges Geologiske Undersøkelser, **266**, 49–61.

The tectonic evolution of the Norwegian Sea Continental Margin with emphasis on the Vøring and Møre Basins

HARALD BREKKE

Norwegian Petroleum Directorate, P.O. Box 600, 4001 Stavanger, Norway (e-mail: harald.brekke@npd.no)

Abstract: The Norwegian Sea continental margin is dominated by two major basins with a very thick Cretaceous basin fill: the Vøring and Møre Basins. The basins are flanked by the uplifted mainland and the Cretaceous Trøndelag Platform to the east and by the Møre and Vøring Marginal Highs capped by Eocene lavas to the west. The tectonic development of the area is controlled by two structural trends: NE–SW and NW–SE. The area has been tectonically active from Carboniferous to Late Pliocene time with the main tectonic phases in Late Palaeozoic, late Mid-Jurassic–Early Cretaceous and Late Cretaceous–Early Tertiary time. The general tectonic development comprised a long period of extension and rifting that ended in Early Eocene time by continental separation, major volcanism and subsequent sea-floor spreading in the Norwegian–Greenland Sea. In Carboniferous to Early Cretaceous time the extensional tectonics were related to within-plate continental rifting. The tectonics of the Late Cretaceous and the Tertiary periods were controlled by the relative movements along plate boundaries. The overall NE–SW structural grain is constituted by faults and basin axes that probably originated in Late Palaeozoic time and were active during all subsequent tectonic phases. The transverse NW–SE trend is expressed as major lineaments that probably reflect the old, Precambrian grain of the basement. These lineaments, two of which are the continuation into the continental crust of major oceanic fracture zones, controlled the tectonic activity throughout Cretaceous and Tertiary time and constitute the boundaries between the major structural provinces of the area. The differentiation into the Cretaceous basins and the bounding platforms and marginal highs started by the late Mid-Jurassic–Early Cretaceous extensional phase. The subsequent Cretaceous subsidence history, where the basin flanks formed by flexuring rather than faulting, resulted in an exceptionally thick basin fill. In the Vøring Basin the Cretaceous development comprised an early thermal subsidence phase and a post-Cenomanian phase of tectonically driven subsidence involving intermittent phases of normal faulting and compression and folding. The Vøring Basin was tectonically active also during Tertiary time with the main phases of strike-slip–compression coinciding with the Alpine orogenies in Late Eocene and Mid-Miocene time. Within the Vøring Basin there is evidence of the formation of a fossil opal A–opal-CT transition and extensive regional marine erosion in Mid-Miocene and Late Pliocene times. In contrast, the Møre Basin was generally tectonically quiet throughout the Cretaceous and Tertiary periods, experiencing mainly continuous subsidence.

The continental margin of the Norwegian Sea has been the subject of accelerating geological and geophysical studies ever since the Norwegian authorities acquired the first seismic data for exploration purposes in 1969. During the next decade the Norwegian Petroleum Directorate (NPD) steadily increased its database of seismic, gravimetric and magnetic data. In the same period the Deep Sea Drilling Project (DSDP) and academic institutions started to unravel the details of the plate tectonics and the sea-floor spreading history of the Norwegian–Greenland Sea (e.g. Johnson & Heezen 1967; Avery *et al.* 1968; Meyer *et al.* 1972; Talwani & Eldholm 1972, 1977; Bott 1973; Huseby *et al.* 1975; Talwani *et al.* 1976; Eldholm & Thiede 1980).

The first exploration licences in the area were allocated in 1980 and the next 15 years saw an increased activity of exploration and scientific studies from both the oil industry and academic institutions. The industry focused on the Late Palaeozoic and Mesozoic geological development and based many of its regional tectonic models on the shallow shelf areas off Norway and the mainland geology of East Greenland (e.g. Rønnevik & Navrestad 1977; Bukovics *et al.* 1984; Bøen *et al.* 1984; Gabrielsen *et al.* 1984; Price & Rattey 1984; Surlyk *et al.* 1984; Brekke

From: Nøttvedt, A. *et al.* (eds) *Dynamics of the Norwegian Margin*. Geological Society, London, Special Publications, **167**, 327–378. 1-86239-056-8/00/$15.00 © The Geological Society of London 2000.

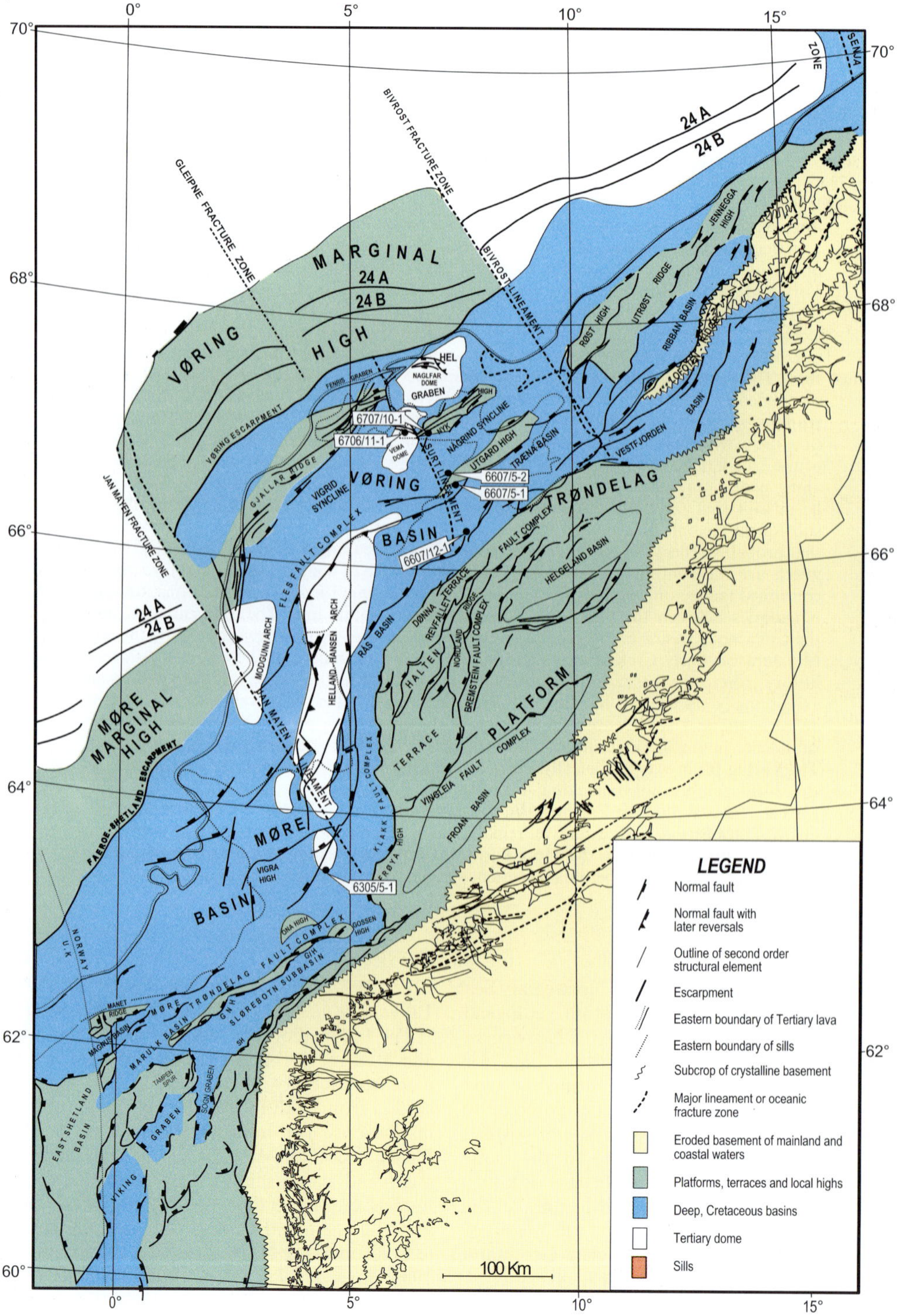

Fig. 1. Simplified structural map of the Norwegian Sea continental margin. GIH, Giske High; GNH, Gnausen High; SH, Selje High.

& Riis 1987; Surlyk 1990; Doré 1991; Torske & Prestvik 1991). The tectonic framework, crustal structure and the tectono-magmatic development of the outer parts of the continental margin were mainly treated by academic institutions (Hinz *et al.* 1982, 1984; Mutter *et al.* 1984; Skogseid & Eldholm 1987, 1989, Eldholm *et al.* 1989; Planke *et al.* 1991; Skogseid *et al.* 1992; Skogseid 1994).

From 1994 the industry also changed its focus to the deep-water outer part of the continental margin as a modern seismic grid was made available and the first concession round in those areas was imminent. The intention of the present author is to give a description based on an updated interpretation of the present regional seismic grid with an emphasis on the outer parts of the continental margin, but with due reference to the present state of knowledge of the geology from the shallow part of the shelf. Less emphasis is placed on trying to make an overall regional tectonic model to explain all observations.

Tectonic setting

The overall tectonic framework of the continental margin between 62° and 69°N consists of a central area of NE–SW trending deep Cretaceous basins, the Vøring and Møre Basins, flanked by palaeo-highs and -platforms and the elevated mainland (Figs 1 & 2). The platforms to the west are termed the Møre and Vøring Marginal Highs and are characterized by thick, Early Eocene basalt flows overlying an unknown substrate. The boundary between the marginal highs and the basin area is formed by the Faeroe–Shetland Escarpment to the south and the Vøring Escarpment to the north.

The central part of the eastern flanks is dominated by the Late Jurassic–Early Cretaceous Trøndelag Platform. North and south of the Trøndelag Platform the basin area is flanked to the east by the eroded mainland.

To the north, the main basin area is bounded by the NW–SE trending Bivrost Lineament, which separates the wide and deep Vøring Basin from the narrow and tectonically uplifted continental margin around the Lofoten Ridge.

To the south, the southeastern boundary of the Møre Basin is located where the NE–SW trending Møre–Trøndelag Fault Complex truncates the N–S and NNE–SSW trends of the northern North Sea. To the southwest, the Møre Basin borders upon the Faeroe–Shetland Basin along the NW–SE trending Erlend Platform (Pharaoh *et al.* 1996).

Apart from the main NE–SW structural grain, the continental margin is subdivided by two NW–SE trending lineaments into three structural provinces: a northern, a middle and a southern province. These lineaments, the Bivrost and Jan Mayen Lineaments, continue into the oceanic crust as the Bivrost and Jan Mayen Fracture Zones, respectively.

The tectonic activity that has given the present structural configuration of the Norwegian Sea continental margin may be traced back into Permo-Carboniferous time (Fig. 3) (Bukovis *et al.* 1984; Gabrielsen *et al.* 1984; Price & Rattey 1984; Surlyk *et al.* 1984; Brekke & Riis 1987; Blystad *et al.* 1995; Doré & Lundin 1996). Three main rifting episodes are identified: in Carboniferous to Permian, in late Mid-Jurassic to Early Cretaceous and in Late Cretaceous to Early Eocene times. In Carboniferous to Early Cretaceous time the extensional tectonics were related to continental rifting. The tectonics of the Late Cretaceous phase and, in particular, the Tertiary extensional regime were more directly influenced by the relative movements along plate boundaries just before and during the continental break-up and onset of sea-floor spreading in the North Atlantic. Locally, this is expressed as prominent compressional structures.

Database, stratigraphy and seismic ties

The present mapping and description of the Vøring and Møre Basins and adjacent areas are mainly based on the interpretation of the seismic data acquired by the NPD from 1985 to 1992, supplemented by well data and literature on the geology of East Greenland (Larsen 1984; Surlyk *et al.* 1984; Surlyk 1990, 1991; Larsen & Marcussen 1992; Larsen *et al.* 1999).

Most of the wells in this part of the Norwegian shelf have been drilled on the Trøndelag Platform and the Halten Terrace, and the post-Lower Triassic stratigraphy of the platform and terrace sequences is well known (Dalland *et al.* 1988). By analogy to East Greenland it is assumed that the platform and terraces also comprise sedimentary rocks of Carboniferous and Permian age on top of crystalline basement.

So far, only seven conventional exploration wells have tested parts of the Cretaceous basin fill sequences. These wells, 6505/10-1, 6607/5-1, 6607/5-2, 6607/12-1, 6704/12-1, 6706/11-1 and 6707/10-1, were drilled in the northeastern part of the Vøring Basin and terminated in the Upper Cretaceous sequence, proving a very thick Upper Cretaceous sequence of marine shales and intebedded sandstone. Parts of the sandstone

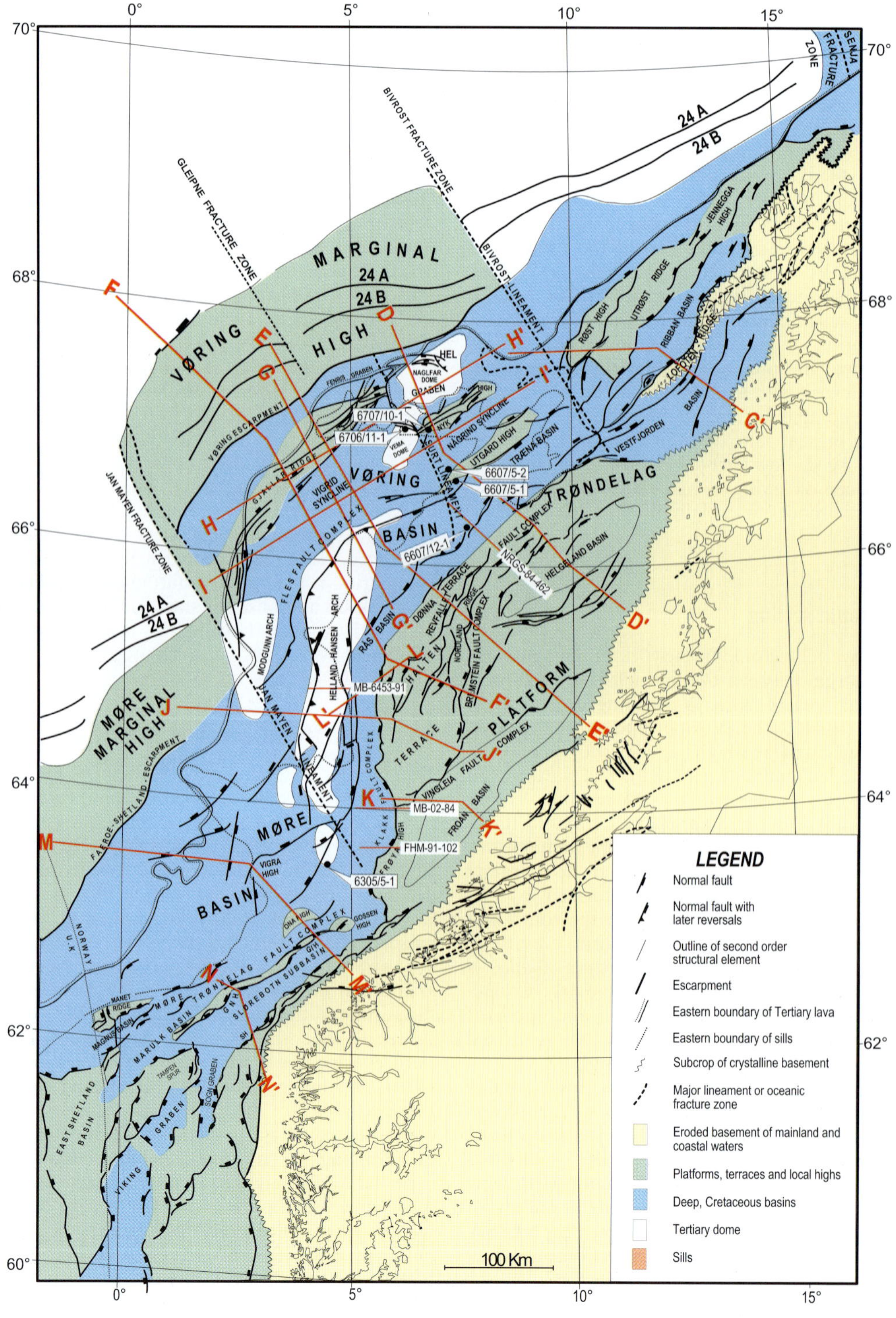

Fig. 2. Simplified structural map of the Norwegian Sea continental margin with line locations for seismic and geoseismic sections in Figs 5, 6, 8–12, 14, 17, 18, 22, 23, 25, 26 and 30.

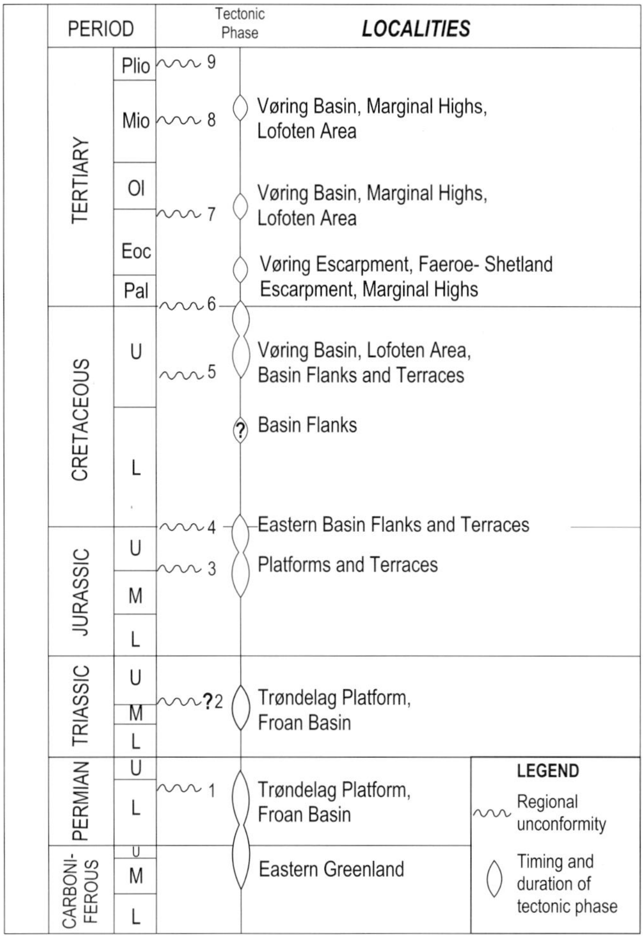

Fig. 3. Tectonic history of the Norwegian Sea continental margin.

sequences are thick and of good reservoir quality (Kittilsen *et al.* 1999; Brekke *et al.* 1999). The pre-Cenomanian sequences within the Vøring and Møre Basins are still unknown.

The Paleocene and Eocene units of the western part of the Vøring Basin show an expansion in thickness compared with the eastern part (Fig. 5). DSDP and Ocean Drilling Program (ODP) wells in the area have only penetrated the Miocene sequence (Talwani *et al.* 1976; Eldholm *et al.* 1987), and the lithologies of the Lower Tertiary sequence are left to speculation.

From wells and seismic surveys a total of nine regional unconformities are identified in the stratigraphy of the area (Fig. 3). The three oldest unconformities may be identified only in the platform areas: these are of late Early Permian, Mid-Triassic and late Mid-Jurassic age. The dating of these unconformities is based on exploration well data and analogies with East Greenland (Surlyk *et al.* 1984).

The six younger regional unconformities in the stratigraphy are identified in both the basin and platform–terrace areas: these are of base Cretaceous, top Cenomanian, base Tertiary, Upper Eocene–Lower Oligocene, Middle Miocene and intra Upper Pliocene positions in the sequence.

The base Cretaceous unconformity is well defined as an onlap surface in the eastern basin flanks, platform and terrace areas. Within the basin areas itself the base Cretaceous position lies very deep and is difficult to interpret and tie up because of lateral variation in the quality of the seismic data. However, in the Rås Basin and partly in the Træna Basin seismic data clearly show the base Cretaceous unconformity as a strong reflector between 6 and 8 s twt (see Figs 6 and 7, and Figs 24, 25, 28 and 30, below). Other

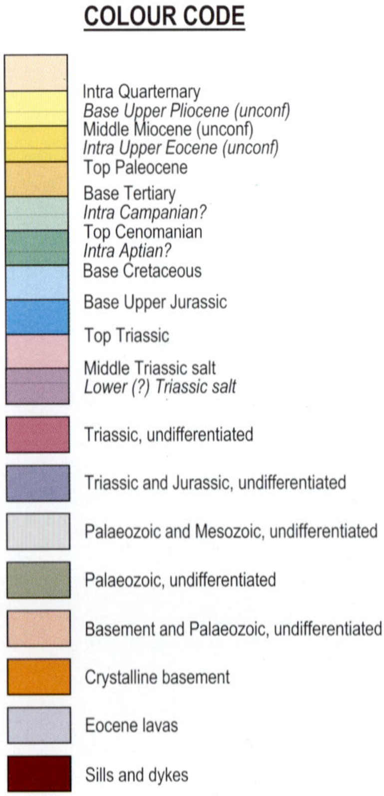

Fig. 4. Colour legend of geoseismic sections in Figs 5, 6, 8–12, 14, 17, 22 and 23.

workers in the area (Skogseid & Eldholm 1989) have placed the base Cretaceous position at shallower depths west of the Fles Fault Complex, corresponding to the top Cenomanian unconformity in the present interpretation. In our interpretation the base Cretaceous position remains deep also to the west of the Fles Fault Complex. This is based both on a clear onlap surface (taken to be the base Cretaceous level) dipping eastwards along the flanks of the Gjallar Ridge in the southwestern Vøring Basin, and on a very good, continuous double reflector of probable Aptian age which puts limits to the position of the base Cretaceous level. This Aptian reflector is found throughout the Vøring and Møre Basins. A deep position of the top Cenomanian unconformity (and hence the deep position of the base Cretaceous unconformity) west of the Fles Fault Complex is now confirmed by the recent well 6707/10-1 on the Nyk High (Kittilsen *et al.* 1999).

The top Cenomanian unconformity is identified as a tilted and faulted surface onlapped by younger strata along both flanks of the Vøring Basin (e.g. seismic lines NPD-VB-2-87 and NPD-VB-4-87), and in the southern parts of the Trøndelag Platform as noted by Brekke & Riis (1987).

Seismic data show that the base of the Tertiary sequence is an erosional unconformity in the Vøring Basin (e.g. seismic line NPD-VB-15-90, Fig. 8) and along the flanks of the Møre Basin. This is confirmed by well data in the basin (e.g. well 6607/5-1) and basin flank position (Gradstein & Backstrøm 1996).

The Upper Eocene–Lower Oligocene unconformity is identified as an onlap surface along the Vøring Escarpment (e.g. seismic line NPD-VB-15-90 as indicated in Fig. 8), on the flanks of the main compressional structures in the Vøring Basin, and as a hiatus or erosion surface all over the eastern part of the Vøring Basin, Halten Terrace, Trøndelag Platform and the North Sea (Eidvin & Riis 1991; Gradstein & Backstrøm 1996).

The Middle Miocene unconformity is seen as an important erosional break all over the Norwegian shelf. Much effort has been put into dating this unconformity both in the Norwegian Sea and in the North Sea (Eidvin & Riis 1991, 1992; Eidvin *et al.* 1993; Rundberg *et al.* 1995; Gradstein & Backstrøm 1996). Gradstein & Backstrøm (1996) argued that this unconformity is of Late Miocene age, but pointed out that this is based on a difference in opinion with Eidvin & Riis (1991, 1992) on the time-transgressive nature of one of the critical biozones.

The intra Upper Pliocene unconformity is the base of the thick, regional sediment wedge that built out westward from the elevated mainland across the entire shelf during late Pliocene time (Eidvin & Riis 1991, 1992; Riis & Fjeldskaar 1992).

The mapping of the Vøring and Møre Basins is based on the interpretation of ten seismic reflectors. All the interpreted reflectors are tied to wells in the area. The most important wells for the Vøring and Møre Basins are 6607/5-1, 6607/5-2, 6707/10-1 (Fig. 1) and 6506/11-1 (Fig. 9), and wells on northern Tampen Spur and along the Møre–Trøndelag coast. In several places references are made to specific seismic lines that are not included as figures as they are not yet released.

The bounding lineaments

The Jan Mayen Lineament separates the Møre Basin to the south from the Vøring Basin to the north. This lineament is defined by a distinct

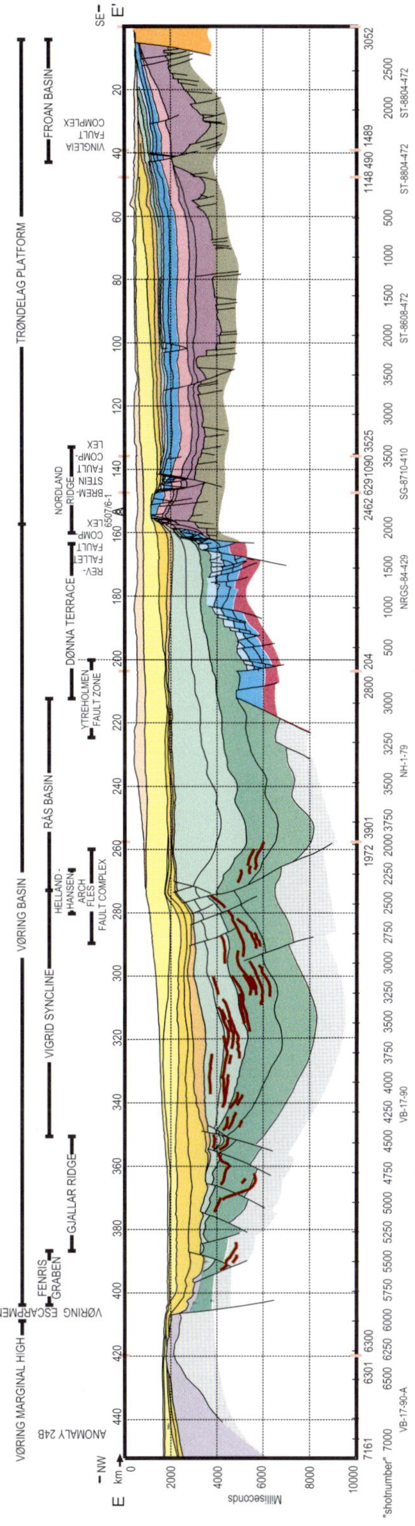

Fig. 5. Geoseismic section EE′. (See Fig. 2 for line location and Fig. 4 for colour legend.) Modified from Blystad *et al.* (1995).

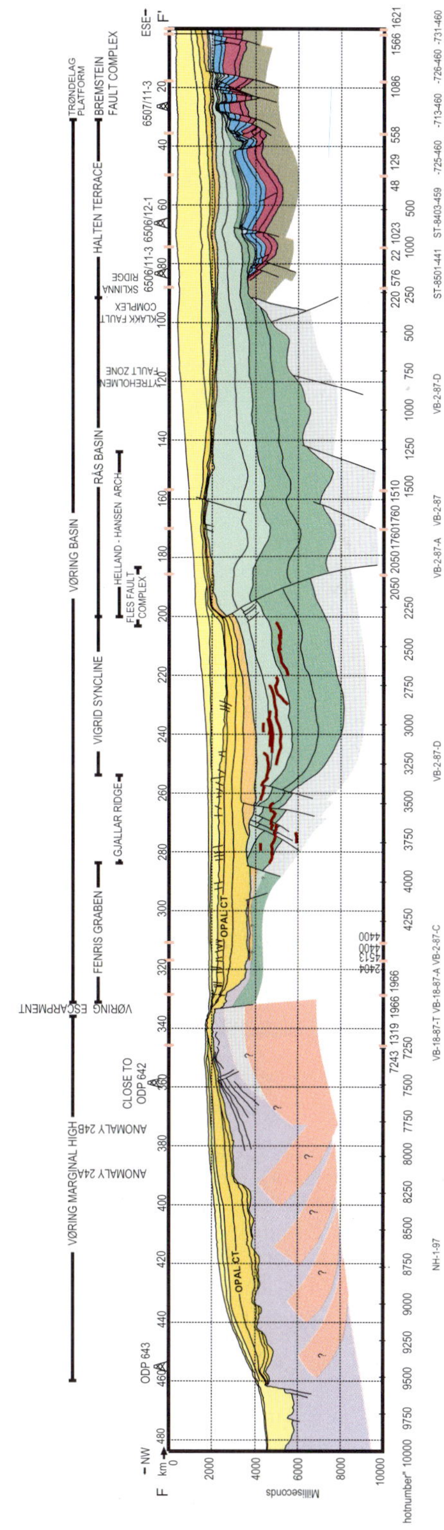

Fig. 6. Geoseismic section FF′. (See Fig. 2 for line location and Fig. 4 for colour legend.) Modified from Blystad *et al.* (1995).

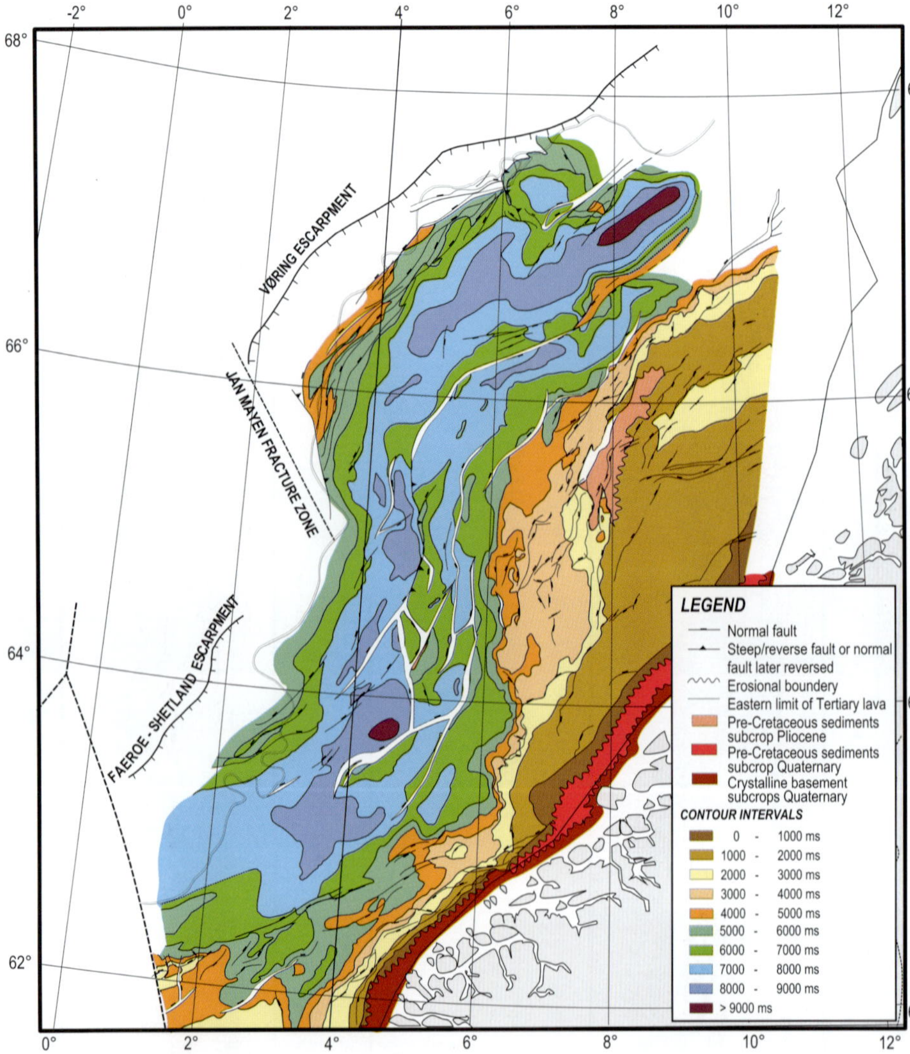

Fig. 7. Two-way travel time map of the base Cretaceous unconformity of the Vøring and Møre Basins.

sinistral shift of basin axes and basin flanks. It also coincides with the southern boundary of the Trøndelag Platform. A similar sinistral shift is seen in the magnetic spreading anomalies in the ocean crust along the Jan Mayen Fracture Zone (Fig. 1). The Bivrost Lineament separates the wide and deep Vøring Basin from the narrow and tectonically uplifted continental margin around Lofoten to the north. The lineament is further defined as an apparent dextral shift in basin axes and flanks and coincides with the northern termination of the Trøndelag Platform. On the other hand, the magnetic spreading anomalies in the ocean crust, as interpreted by Skogseid & Eldholm (1987), along the Bivrost Fracture Zone display a sinistral shift (Fig. 1). In recent compilations of magnetic data (Olesen *et al.* 1997) this shift is not obvious and the Bivrost Fracture Zone rather appears to be the tectonic boundary between the thick crust of the Vøring Marginal High and the oceanic crust to the northeast.

The two lineaments probably reflect an old structural grain in the crystalline basement. This is substantiated by the NW–SE strike of the fjords and major fractures where the lineaments meet the mainland. In the regional gravity anomaly pattern the Jan Mayen Lineament continues as the most prominent NW–SE trending gravity low observed in the Scandinavian

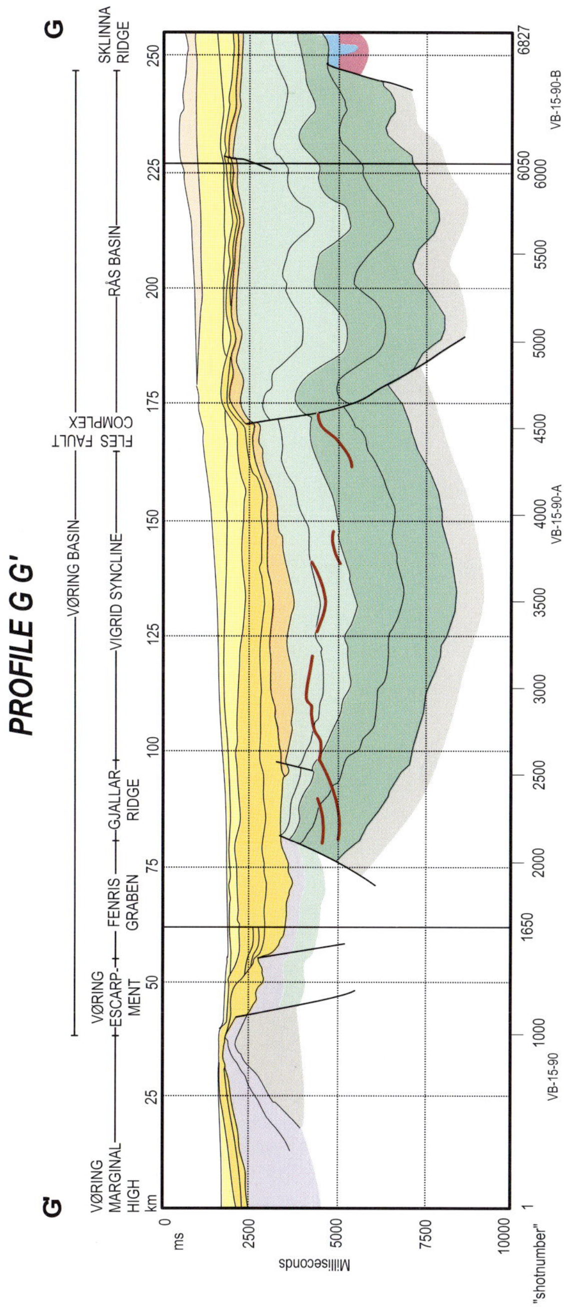

Fig. 8. Geoseismic section GG′. (See Fig. 2 for line location and Fig. 4 for colour legend.)

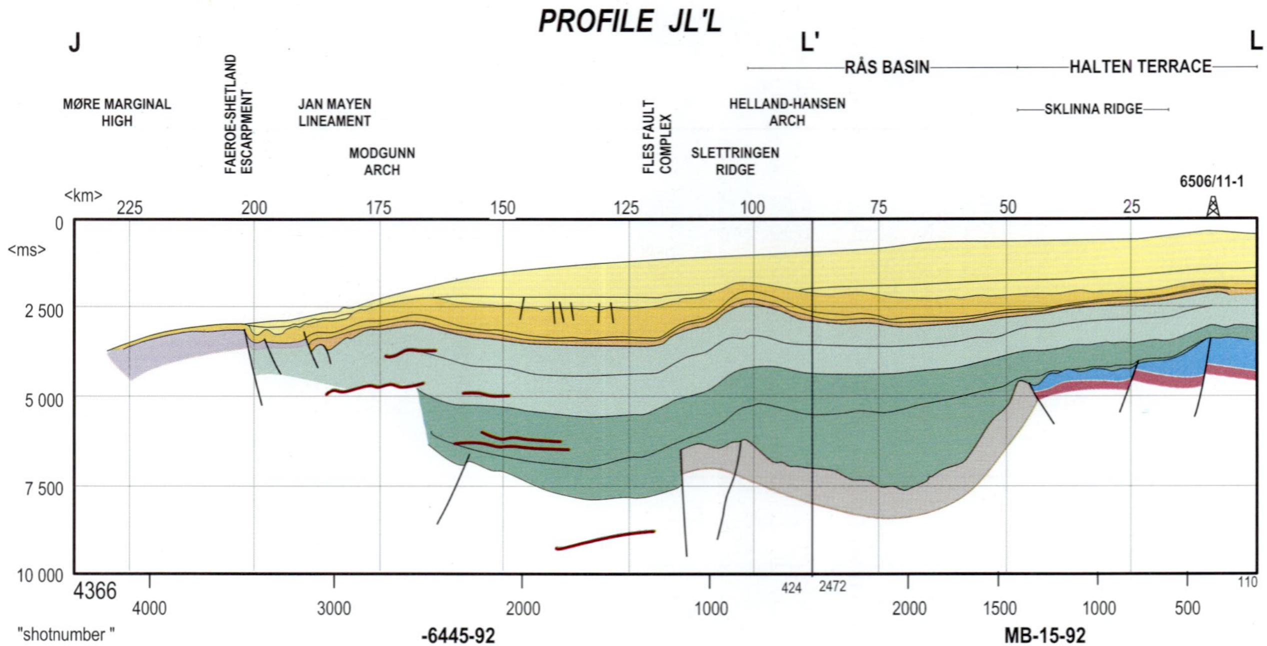

Fig. 9. Geoseismic section JL′L. (See Fig. 2 for line location and Fig. 4 for colour legend.)

Peninsula (Fairhead *et al.* 1994). The observed apparent lateral displacements of main structures across the lineaments cannot be matched with similar displacements in the structural grain of the mainland Caledonian geology. The lateral shifts are hence not due to any large-scale post-Caledonian transcurrent tectonic movements. However, the way in which these inherited NW–SE lineaments divide the continental margin into structural provinces strongly indicates that they have had a major control on the post-Caledonian tectonic development. Torske & Prestvik (1991) described a model where these lineaments were the location of the controlling transfer zones in a simple shear extension at crustal scale between Norway and Greenland.

The position, and possibly the formation, of the Trøndelag Platform were probably controlled by the Bivrost and Jan Mayen Lineaments. Whether similar platform areas existed across the now deeply eroded mainland to the north and south of the two lineaments is unknown. The old NW–SE structural grain probably also controlled the Eocene spreading geometry of the Norwegian–Greenland Sea, where the fracture zones perpendicular to the early spreading ridges initiated along these trends. The Jan Mayen Lineament–Fracture Zone pair is the most important of these structures, as it has been the location of the prominent ridge offset that has bisected the Norwegian–Greenland Sea throughout its spreading history (see Figs 33 and 34, below). Through the whole of Late Cretaceous and Tertiary time the Jan Mayen Lineament was a tectonic barrier between the tectonically very active Vøring Basin and the quiet and non-active Møre Basin, confirming the significance of this lineament.

The middle province between the Jan Mayen and Bivrost Lineaments

The major structural elements in the middle province are the Trøndelag Platform, the Vøring Basin and the Vøring Marginal High.

The Trøndelag Platform and adjacent terraces

In the internal parts of the Trøndelag Platform the main tectonic episode took place in the Carboniferous to Late Permian time (Fig. 5). The horsts and half-grabens at depth are of that age. The major faults were also active through much of Triassic time, giving rise to several en echelon, NE–SW trending basins filled with Triassic and Upper Palaeozoic sediments. The southernmost and best defined of these basins is termed the Froan Basin. The Vingleia Fault Complex forms the northwestern boundary of this basin and was reactivated in both Jurassic and Cretaceous times. The Frøya High must have been a basement high at least from Late Permian times (Fig. 10). In the high, a condensed succession of probable Triassic and Late Permian age unconformably overlies deeply eroded basement and remnants of an older basin sequence. This sequence in turn lies unconformably on the basement and shows internal angular unconformities, testifying to the complex Late Palaeozoic tectonic history. The boundary fault between the Frøya High and the Froan Basin was active through much of Triassic time, as can be seen by the marked expansion of the Triassic sequence in the hanging wall (Fig. 10). Towards the south the Froan Basin becomes progressively shallower, as a result of a combination of an original thinning of basin sequences and a later uplift and erosion in late Mid- to Late Jurassic time.

During the tectonic episode in late Mid-Jurassic to Early Cretaceous time the interior parts of the Trøndelag Platform were subject to minor faulting partly by reactivations of older faults in the Vingleia Fault Complex, and by faulting along the flanks of the Helgeland Basin in Early Cretaceous time. The uplift of the Nordland Ridge on the platform edge was initiated in this period, as may be seen by the onlap of the Early Cretaceous basin fill of the Helgeland Basin (Fig. 11). Along the platform edges and in the future terrace areas of the Halten and Dønna Terraces there was intense fault activity in this tectonic episode (Fig. 5). The subdivision into the deep basins to the west and the platform area to the east was probably initiated in the early stage of the tectonic episode. The terraces as individual elements were initiated by faulting in Late Jurassic time, which resulted in an expanded Upper Jurassic succession along the Vingleia, Bremstein and Revfallet Fault Complexes along the eastern margin of the terraces. However, it is important to note that throughout the late Mid-Jurassic to Early Cretaceous tectonic episode, the Halten Terrace stayed close to the same elevation as the Trøndelag Platform relative to the basins to the west, and the main subsidence of the terraces relative to the Trøndelag Platform did not take place until later in Cretaceous time. Hence, in Late Jurassic and earliest Cretaceous time, the high along the western edge of the Halten Terrace, the Sklinna Ridge, was at the same

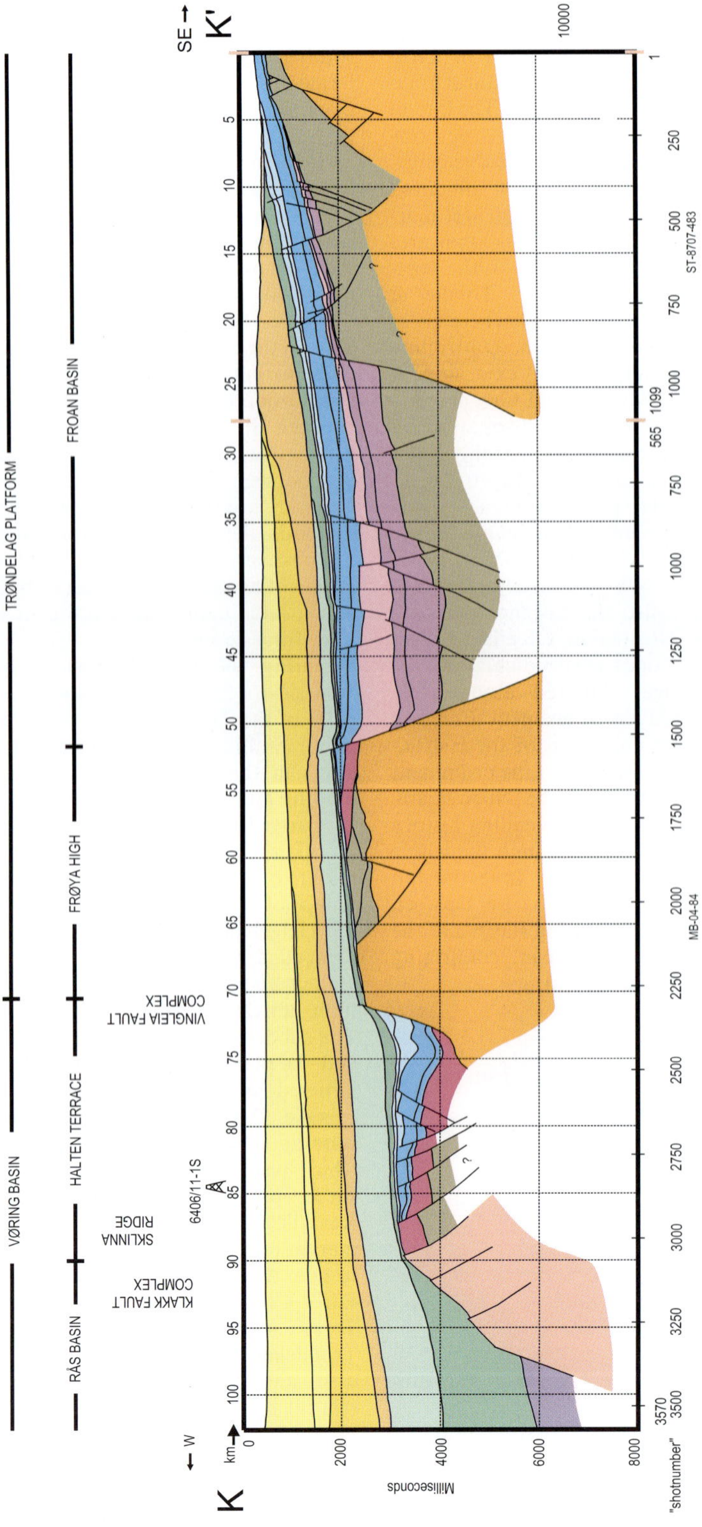

Fig. 10. Geoseismic section KK′. (See Fig. 2 for line location and Fig. 4 for colour legend.) Modified from Blystad *et al.* (1995).

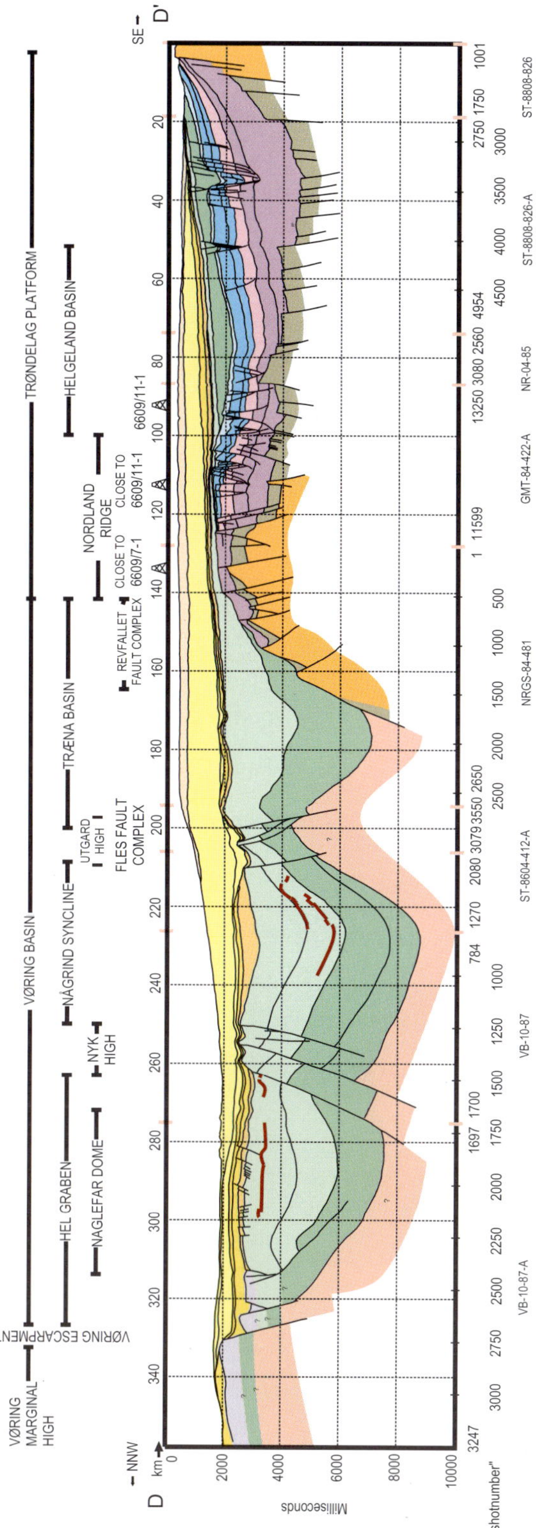

Fig. 11. Geoseismic section DD′. (See Fig. 2 for line location and Fig. 4 for colour legend.) Modified from Blystad *et al.* (1995).

elevation as the highs along the edge of the Trøndelag Platform, i.e. the Nordland Ridge and the Frøya High. All these elevated areas along the platform edge were deeply eroded in Late Jurassic time, leaving the crystalline basement at shallow depths and in places even subcropping the uppermost Jurassic and/or the Cretaceous units (Figs 6, 10 and 11). The flat tops of these eroded highs were then all part of the common Late Jurassic peneplain of the Trøndelag Platform (see also Riis 1996).

The expansion of the Cretaceous sequence across the Vingleia, Bremstein and Revfallet Fault Complexes along the eastern boundaries of the Halten and Dønna Terraces shows that these faults controlled the final separation of the terraces from the platform during two phases of subsidence; one in Early Cretaceous time and the last, and major one, in post-Cenomanian time (Figs 5 and 6). The latter is part of the main Late Cretaceous tectonic episode seen in the Vøring Basin (see below). During this phase, the southern part of the Nordland Ridge was subjected to uplift and faulting, which destroyed the Late Jurassic peneplain in that area. This part of the Nordland Ridge was further uplifted during several phases in Tertiary time so that Triassic to Jurassic rocks now subcrop Paleocene to Pliocene strata.

The Vøring Basin area

The typical feature of both the Vøring Basin area and the Møre Basin is the enormous thickness of the Cretaceous sequence (Figs 5 and 7). In places, the base of the Cretaceous units is as deep as >9s twt, and the axial parts of the basins are generally deeper than 7 s twt. This means that the base of the Cretaceous sequence in the central parts of the basins lies at depths between 9000 and 13 000 m.

The Vøring Basin area is bounded to the west by the Vøring Escarpment along the Vøring Marginal High, and to the east by the fault complexes along the edge of the Trøndelag Platform. The basin area is bisected by the Fles Fault Complex, which runs along the basin axis from the Jan Mayen Lineament in the south to the Bivrost Lineament in the north. The other important tectonic boundary within the basin is the Surt Lineament, including the Rym Fault Zone, in the northern half of the basin area. This feature runs parallel to the Jan Mayen and Bivrost Lineaments and is obviously controlled by the same ancient NW–SE structural grain of the basement. The western part of the basin area is dominated by the basin parallel Gjallar Ridge. The development of the Vøring Basin area should be described in terms of three main phases; from late Mid-Jurassic to Late Cenomanian time, from Late Cenomanian to Early Paleocene time, and from Early Paleocene time to the present.

Late Mid-Jurassic to Late Cenomanian phase. The Fles Fault Complex probably originated as a Mid- to Late Jurassic zone of normal faults with an eastward polarity along its northern NE–SW trending segment, and a westward polarity along its southern N–S trending segment. By several phases of reactivations, the Fles Fault Complex has played a central role in the tectonic development of the area since Early Cretaceous time. The Rås and Træn Basins were established as deep depocentres between the Fles Fault Complex to the west and the terraces to the east (Figs 5, 6 and 11). The area to the west of the Fles Fault Complex, comprising the present Vigrid and Någrind Synclines, the Hel Graben and the Nyk High, probably constituted a single broad Early Cretaceous basin somewhat shallower than the Rås and Træna Basins (e.g. Fig. 6). Researchers who prefer a shallower correlation for the base of the Cretaceous sequence to the west of the Fles Fault Complex have a regional high, informally termed the 'Fulla Ridge', in place of the present author's western Early Cretaceous basin (Skogseid & Eldholm 1989; Skogseid *et al.* 1992).

The location of the western flank of the western Early Cretaceous basin is uncertain, but the present author's interpretation of the base of the Cretaceous sequence implies that the Lower Cretaceous units thin and onlap older rocks in the southern parts of the Gjallar Ridge (Figs 6 and 12). To the north of the Gjallar Ridge, along the western margin of the Hel Graben, the Lower Cretaceous sequence is interpreted to thin in a similar way (Fig. 11). This is taken to indicate that the western flank of the Early Cretaceous basin was situated more or less parallel to the present Gjallar Ridge, but probably offset to the west of the ridge along its central segments. This Early Cretaceous basin formation is believed to be the result of thermal subsidence after the major rifting episode in late Mid-Jurassic to Early Cretaceous time.

Late Cenomanian to Early Paleocene phase. Through Cenomanian time the different depocentres of Early Cretaceous time and the Halten and Dønna Terraces merged into one large basin, the Vøring Basin *sensu stricto*. This basin was mainly formed by the renewed tectonics that

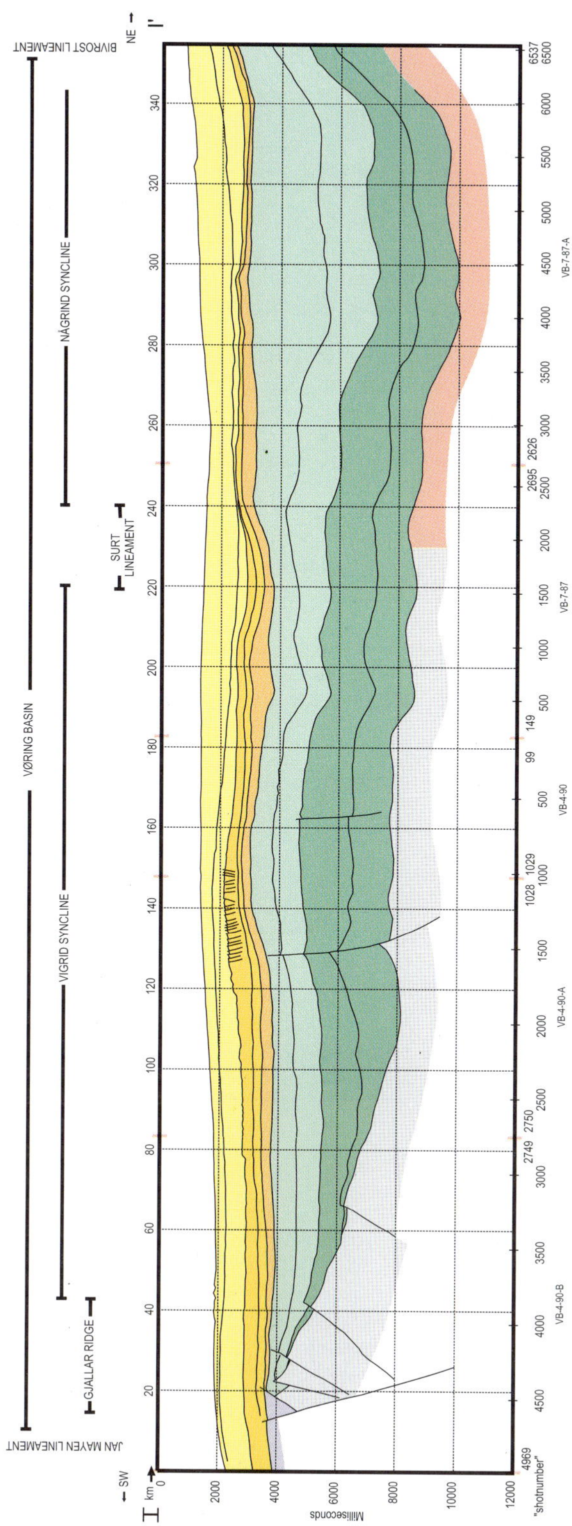

Fig. 12. Geoseismic section II′. (See Fig. 2 for line location and Fig. 4 for colour legend.) Modified from Blystad *et al.* (1995).

started in latest Cenomanian to earliest Turonian time and the associated subsidence that reached its maximum in Campanian time. Hence, the Vøring Basin *sensu stricto* is defined at the post-Cenomanian Cretaceous levels (Blystad *et al.* 1995).

At the transition into Turonian time, the Gjallar Ridge started to rise and formed the western flank of the Vøring Basin. This is demonstrated by the thinning and onlap of the post-Cenomanian strata towards the tilted Top Cenomanian units on the eastern flank of the Gjallar Ridge (Figs 5 and 6). This Late Cenomanian–Early Turonian eastward tilting of the Gjallar Ridge is mirrored by a simultaneous westward tilt of the southern part of the Trøndelag Platform; across the Frøya High and Froan Basin the post-Cenomanian sequence thins and onlaps the tilted Top Cenomanian units (Fig. 10). This symmetrical basinward tilting of the basin flanks indicates that the development of the Vøring Basin started with a phase of renewed or accelerated subsidence. At the same time, or just afterwards, faulting activity started in the Gjallar Ridge and continued through Campanian time, creating the complex pattern of strongly rotated fault blocks that characterizes the ridge (Figs 5, 6, 7 and 12). In the same period, renewed faulting activity took place in the Revfallet Fault Complex along the Nordland Ridge (Fig. 5) and on the eastern boundary faults of the Halten Terrace (the Bremstein and Vingleia Fault Complexes) (Fig. 6), across which the post-Cenomanian sequence demonstrates a marked growth. These fault complexes then acted as the eastern flank of the Vøring Basin, which implies that the Halten and Dønna Terraces must be regarded as parts of the Vøring Basin *sensu stricto*.

In the northern part of the Vøring Basin, the Surt Lineament acted as a tectonic hinge zone in Late Cretaceous time. The post-Cenomanian sequence expands greatly to the north of the lineament. The lineament, including the Rym Fault Zone, constituted the southwestern flank of the Hel Graben in post-Cenomanian times (Fig. 12) and also marks the transition zone between the very thick post-Cenomanian deposits in the Någrind Syncline and the thinner post-Cenomanian succession of the Vigrid Syncline. The post-Cenomanian to Campanian thickness is more or less constant across the Någrind Syncline, Nyk High and the Hel Graben. This indicates that during most of the period from Cenomanian to latest Campanian–Maastrichtian time the area north of the Surt Lineament was a major depocentre before the formation of the synclines and highs (Fig. 11).

The fact that the Late Cretaceous phase of subsidence in the Vøring Basin started slightly before the observed faulting activity, and that the maximum subsidence rate (in Campanian time) seems to have coincided with the maximum faulting activity, indicates that the formation of the Late Cretaceous Vøring Basin was not the result of a simple sequence of rifting and consequent thermal subsidence. The Late Cretaceous subsidence was probably driven by active tectonic stresses rather than passive thermal relaxation (see discussion below).

South of the Surt Lineament, the Vøring Basin seems to be regionally folded into a central anticline along the Fles Fault Complex, flanked by two synclines (Fig. 5). The western syncline is termed the Vigrid Syncline, whereas the eastern syncline coincides with the Rås Basin. This regional folding suggests a probably regional compressional tectonic phase involving reactivation along the Fles Fault Complex. The role of the Fles Fault Complex is well demonstrated in section in Fig. 8 by the more intense internal folding of the Rås Basin to the east of the fault complex compared with the broad, open fold of the Vigrid Syncline to the west. The smaller-scale internal folds of the Rås Basin have a NE–SW strike and may be seen on the map of the Top Cenomanian units (Fig. 13). The folds are best developed in the regional bend where the strike of the fault complex changes from N–S to NE–SW, indicating space problems during the reactivation of the Fles Fault Complex.

North of the Surt Lineament the Vøring Basin is folded into three NE–SW trending synclines (the Træna Basin, the Någrind Syncline and the Hel Graben) separated by two anticlines (the Utgard and Nyk Highs) (Fig. 11). The folds affect the whole of the Cretaceous sequence, but in the uppermost part of the Upper Cretaceous (Maastrichtian?) sequence the Någrind Syncline becomes progressively less tight, indicating a latest Campanian–Maastrichtian age for the folding. The Nyk and Utgard Highs show no internal onlaps and, therefore, must also have been formed in latest Cretaceous time. A possible minor growth of the uppermost Cretaceous sequence in the Hel Graben indicates that the faulting activity in the Nyk High and along the Rym Fault Zone that established the Hel Graben initiated just before the folding. This indicates that the overall anticlinal shape of the Nyk High and its position between the two synclines is the result of changing phases of both extension and compression. Thinning of the Lower Cretaceous sequence indicates that the Utgard High was probably initiated as a marginal high along the Træna Basin already in

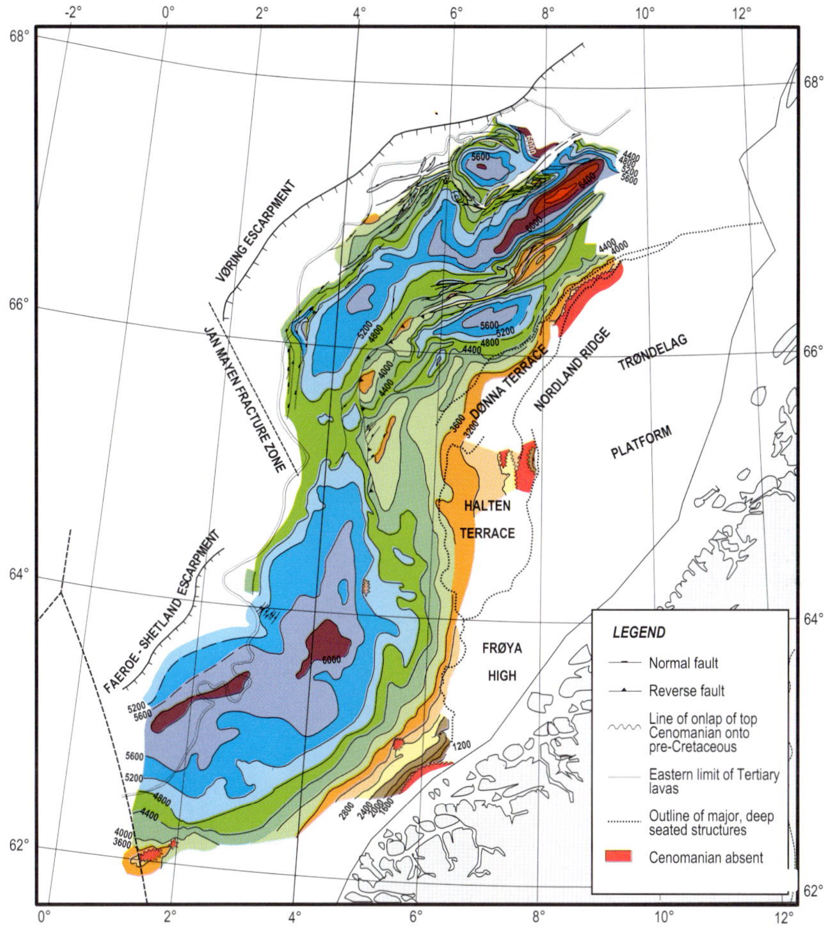

Fig. 13. Two-way travel time map of the top Cenomanian unconformity in the Vøring and Møre Basins. Contours labelled in milliseconds.

Early Cretaceous time (Fig. 11). However, its present shape as a faulted anticline between two synclines is the result of a combination of latest Cretaceous extension and subsequent compression similar to the development of the Nyk High.

In addition to being a tectonic transfer and hinge zone, the Surt Lineament also probably marks a change in the substrate to the Cretaceous and latest Jurassic basin fill. South of the lineament the basin fill probably overlies Upper Palaeozoic to Jurassic strata, whereas to the north the basin fill rests directly on the crystalline basement or deeply eroded Upper Palaeozoic rocks (Figs 12 and 14). In the north, this statement is based on a westward extrapolation of the geological relationships revealed by wells and seismic mapping of the northern part of the Trøndelag Platform (e.g. wells 6609/7-1 and 6609/10-1 in Fig. 11). A westward thinning, caused by pre-Cretaceous erosion, of the basement cover across the platform edge and basin flanks indicates a very shallow basement also beneath the northern Vøring Basin (Fig. 11). To the south, one may also argue for a shallow basement under the basin area from the erosional thinning of the basement cover in the Frøya High and Sklinna Ridge along the platform and terrace edges (Figs 6 and 10). Influx of sand from the west into the western Halten Terrace during Mid-Jurassic time also indicates that the Vøring Basin area was subject to strong uplift and erosion then (Gjelberg *et al.* 1987). However, sections such as Figs 5 and 14 show that Jurassic and Triassic rocks are present in many parts of the basin margin along the terrace areas, indicating that Triassic, and perhaps

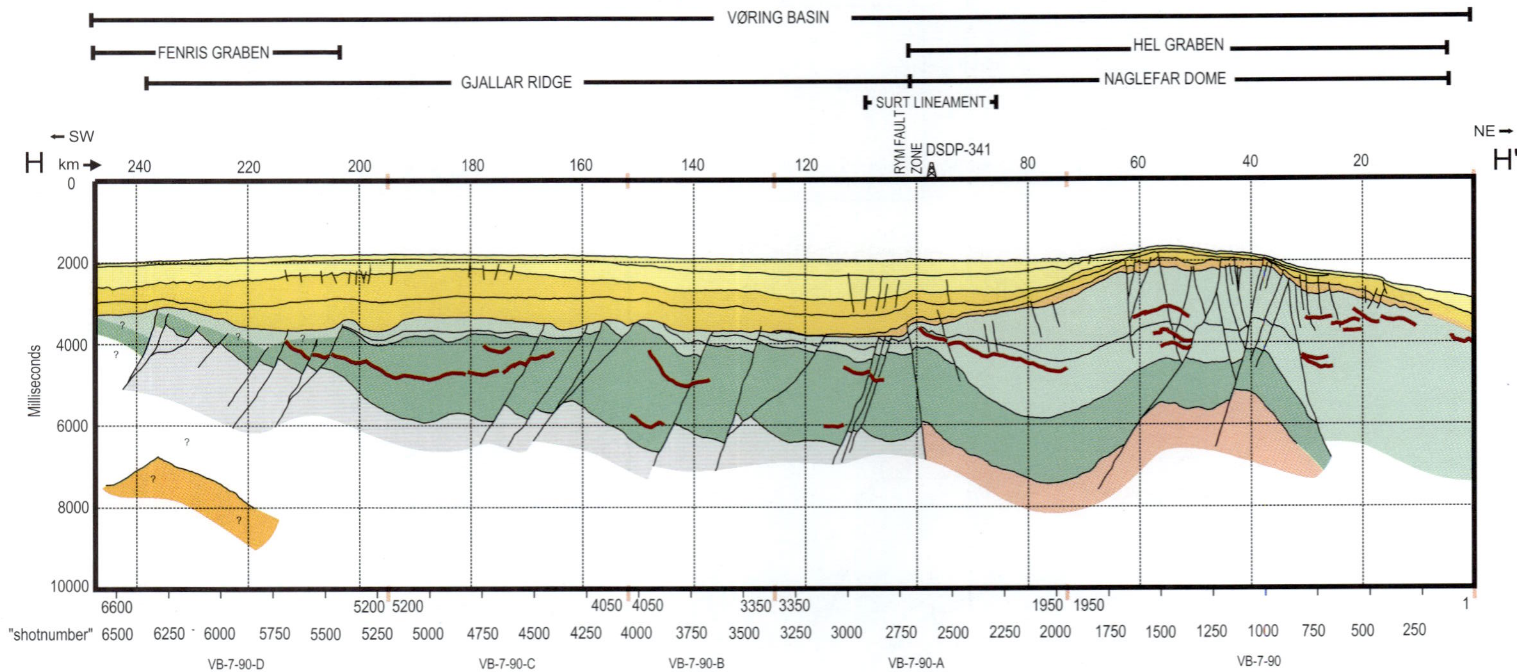

Fig. 14. Geoseismic section HH′. (See Fig. 2 for line location and Fig. 4 for colour legend.) Modified from Blystad *et al.* (1995).

Jurassic, deposits may also be present within the adjacent parts of the basin (see also Brekke at al. 1999). Sub-basin fill internal parallel reflectors in the Slettringen and Gjallar Ridges probably represent sediment layers.

Across the whole of the Vøring Basin, the base of the Tertiary sequence is a regional angular unconformity constituting a generally even surface not involved in the folding of the underlying Cretaceous strata (Figs 5 and 8), substantiating the latest Cretaceous age for the folding. The major highs, such as the Gjallar and Nordland Ridges, the Nyk and Utgard Highs and the central anticline along the Fles Fault Complex, are seen to have been eroded (Figs 5 and 11). This means that the whole basin was uplifted at the end of the Cretaceous and beginning of the Tertiary period.

Early Paleocene time to the present. In the two large synclines, the Vigrid and Någrind Synclines, weak folding probably persisted into Paleocene time and the synclines were infilled by a Paleocene wedge that onlaps the syncline limbs (Figs 5 and 11). This indicates that the whole area stayed uplifted, giving erosion of the most elevated highs through most of Paleocene time. The uplift increased westwards so that the Gjallar Ridge was deeply eroded and was subsequently onlapped by the Paleocene sequence and progressively overstepped by early Eocene deposits (Figs 4 and 15). Even to the east some of the highest areas were also onlapped by the Paleocene sequence and overstepped by younger sediments; these areas include the Nyk High, the Nordland Ridge and an area in the northern Rås Basin close to the Surt Lineament

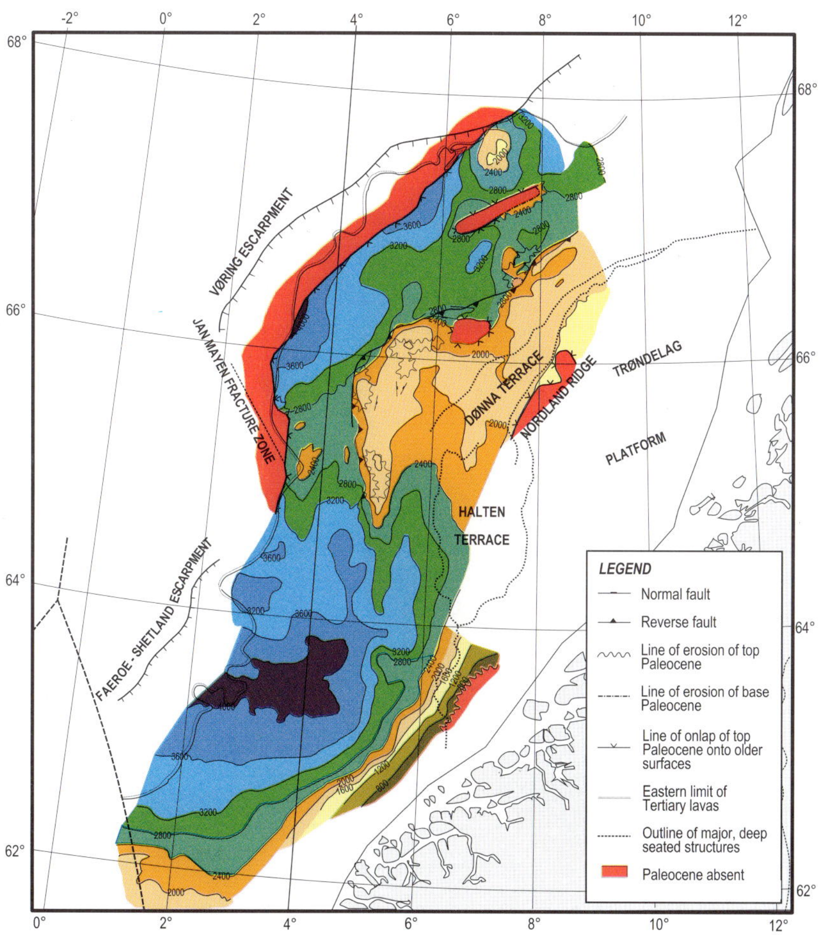

Fig. 15. Two-way travel time map of the near top Paleocene unconformity in the Vøring and Møre Basins. Contours labelled in milliseconds.

(Fig. 15). This area must have been a dome during Paleocene times, and subsequently became inactive and subsided.

During Tertiary time, the Cretaceous eastern basin margin, the entire Trøndelag Platform and probably large parts of the present mainland were transgressed. The main Paleocene to Pliocene depocentre seems to have been to the west, bounded by the Surt and Jan Mayen Lineaments, the Vøring Escarpment and the reversed Fles Fault Complex (Figs 6 and 12). In Neogene time, the mainland and adjacent regions were strongly uplifted, and huge amounts of Plio-Pleistocene sediments were transported westwards, building the shelf edge to its present position (Figs 6 and 11).

The tectonic activity in latest Cretaceous time continued into Paleocene time and was probably part of the rifting that was centred west of the Vøring Escarpment and led to the break-up of the continent and separation of Norway and Greenland (Skogseid & Eldholm 1989; Skogseid 1994). Later tectonic phases were probably linked to major relative plate boundary movements. As in Late Cretaceous time, the Tertiary tectonics were controlled by reactivations of the major lineaments and fault complexes. The Vøring Escarpment, however, was a new element, which was initiated in Eocene time and established itself as an important additional feature of tectonic control throughout Tertiary time. The Vøring Escarpment is a combination of flexuring and faulting (Fig. 8). The main faulting took place in Early Eocene time and established the separation between the Vøring Marginal High and the Vøring Basin area. The marginal high was additionally uplifted (relative to the basin) during two phases of

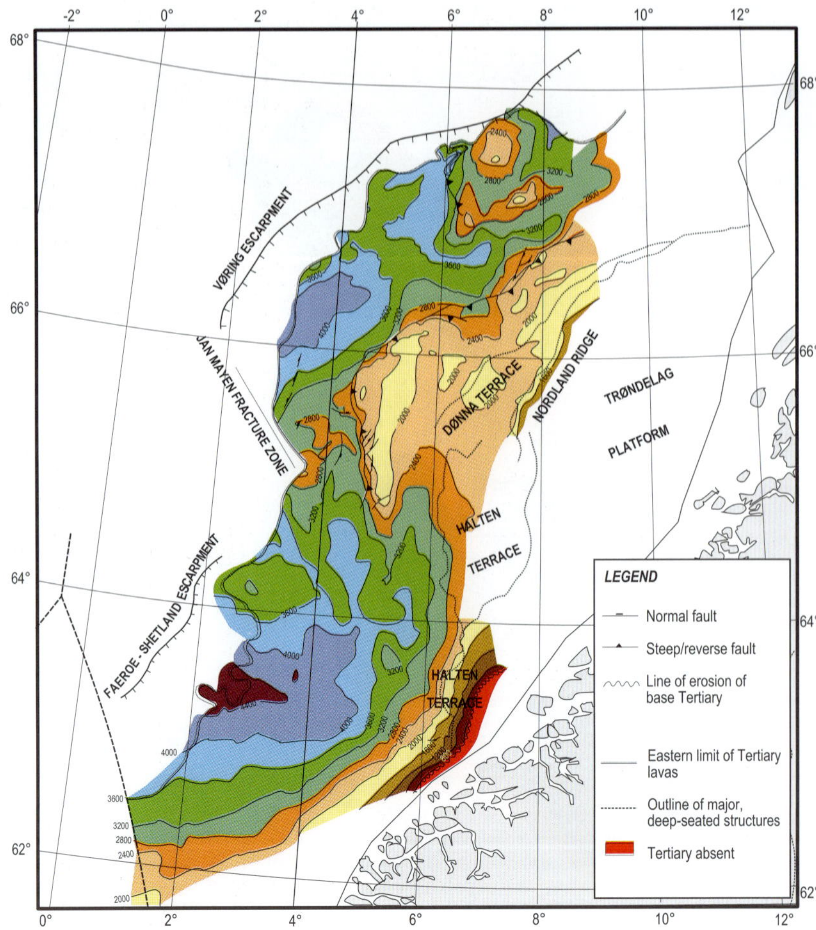

Fig. 16. Two-way travel time map of the base Tertiary unconformity in the Vøring and Møre Basins. Contours labelled in milliseconds.

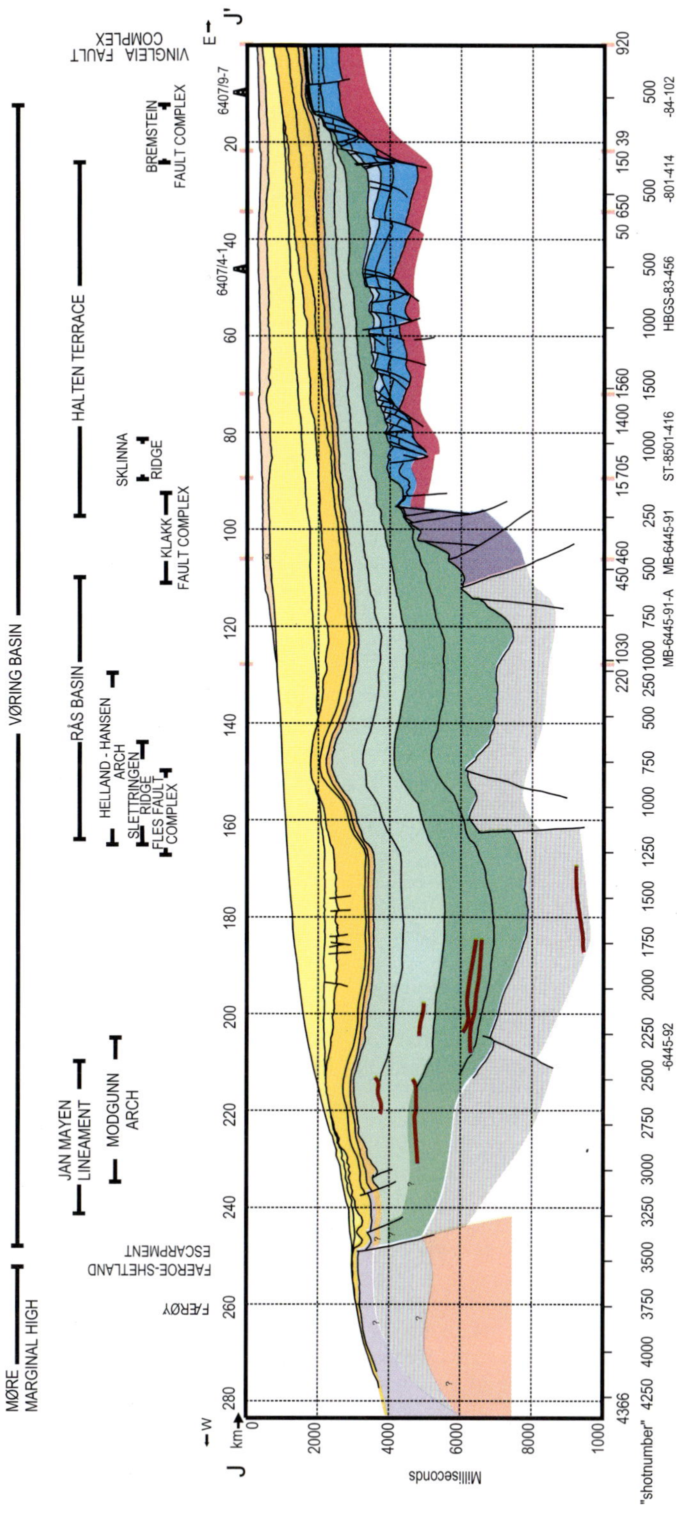

Fig. 17. Geoseismic section JJ′. (See Fig. 2 for line location and Fig. 4 for colour legend.) Modified from Blystad *et al.* (1995).

flexuring along the Vøring Escarpment in the latest Eocene–Early Oligocene and Late Miocene times, respectively. This is indicated by the onlap of post-Eocene strata onto the flexured top of the Eocene sequence, and a similar relationship between the middle Miocene unconformity and Late Miocene strata (Fig. 8).

The same phâses of movement are also recorded as the most important Tertiary reactivation phases on the major tectonic lineaments: the Jan Mayen and Bivrost Lineaments, the Fles Fault Complex and the Surt Lineament including the Rym Fault Zone. The activity resulted in large domes and arches that are well defined on the maps of the base Tertiary unconformity and of the Middle Miocene unconformity (Figs 16 and 19).

Reverse reactivations of the Fles Fault Complex governed the growth of the very prominent Helland-Hansen Arch (Figs 6 and 17). A similar compressional structure, the Modgunn Arch, was controlled by reactivations on the Jan Mayen Lineament (Fig. 17). South of 65°N the strata across the arch are continuous; this makes it possible to correlate the Cretaceous reflectors from well control in the east across the Fles Fault Complex into the western parts of the basin area (Fig. 9). A study of the internal stratigraphic relationships of the Helland-Hansen Arch in this area supports the dating of the tectonic events. The Middle Miocene unconformity defines the present arch surface of the Helland-Hansen Arch (Fig. 18). The Upper Miocene and Pliocene strata onlap and overstep this surface. The unconformity is probably erosional and truncates the underlying strata. The oldest strata overlying this unconformity have been dated to latest Mid-Miocene time in nearby wells, and the youngest strata beneath the unconformity have been dated to Early Miocene time (Eidvin & Riis 1991). This implies that the latest phase of folding of the arch took place in Mid-Miocene time, according to the dating of Eidvin & Riis (1991, 1992). (The dating of Gradstein & Backstrøm (1996) would place the event in Late Miocene time.)

Flattening the unconformity still leaves an arch structure at the Upper Eocene–Lower Oligocene surface. There are some indications of thinning and onlap onto this surface on the flanks of the arch, but these relationships are not conclusive (Fig. 18). However, combined with the clear onlaps onto the flexure of the same surface along the Vøring Escarpment (see above), these relationships indicate onset of folding of the arch already in latest Eocene to earliest Oligocene time.

In the northern part of the Vøring Basin, the Eocene and Miocene tectonics are controlled by the reverse reactivations along the Fles Fault Complex and the Surt Lineament, which initiated folding and doming of the Middle Miocene Unconformity. The prominent Naglfar Dome is an inversion of the Hel Graben controlled by the Rym Fault Zone and the faults along the Nyk High (Figs 11 and 14). The Vema Dome is situated on the intersection between the Surt Lineament and the faults along the Nyk High (Fig. 1). The Træna Basin was inverted along the Fles Fault Complex, giving rise to minor domes and arches (Figs 11 and 16). In addition, the tectonic activity gave rise to further elevation of the Nordland Ridge, the Nyk and Utgard Highs (Figs 5 and 11).

The Lower to Middle Miocene hiatus is recorded all over the Norwegian Continental Shelf (Eidvin & Riis 1989, 1991, 1992; Eidvin *et al.* 1993; Jordt *et al.* 1995; Gradstein & Backstrøm 1996). This implies that just before the folding in Miocene time, there must have been a period of erosion or non-deposition to give the observed hiatus. As only deep-water faunas are recorded immediately below and above the unconformity, this must have been a submarine erosion, as an uplift of the whole basin area to above wave base seems excluded. However, Gradstein & Backstrøm (1996) stated that the basin flanks seem to have reached wave base and been considerably eroded. The mechanism behind this regional event of submarine erosion is not known. However, there is evidence of a regional uplift, lasting for about 7 Ma, of the whole continental margin and the surrounding mainlands in Early to Mid-Miocene time (Anderton *et al.* 1979; Jordt *et al.* 1995).

The Mid-Miocene folding created a significant submarine topography that lasted through Late Miocene time. The map of the Middle Miocene unconformity shows that the summit of the large domes and arches lies 1000–1800 ms twt above the adjacent lows (Fig. 19), corresponding to a sea-bottom topography of the order of 1000–2000 m at that time. Substantial deposits of flat-lying Upper Miocene sediments filled in the lows, onlapping and burying the tectonically created topography. A regional hiatus is also demonstrated at the base of the Upper Pliocene sequence along the continental margin (Eidvin & Riis 1989; Riis & Fjeldskaar 1992; Eidvin *et al.* 1993). On the mainland and in the shallow parts of the continental margin this hiatus is an unconformity related to uplift and deep erosion. In the deeper-water parts of the margin it is mainly a surface of non-deposition, downlapped by the deposits of a huge sedimentary wedge

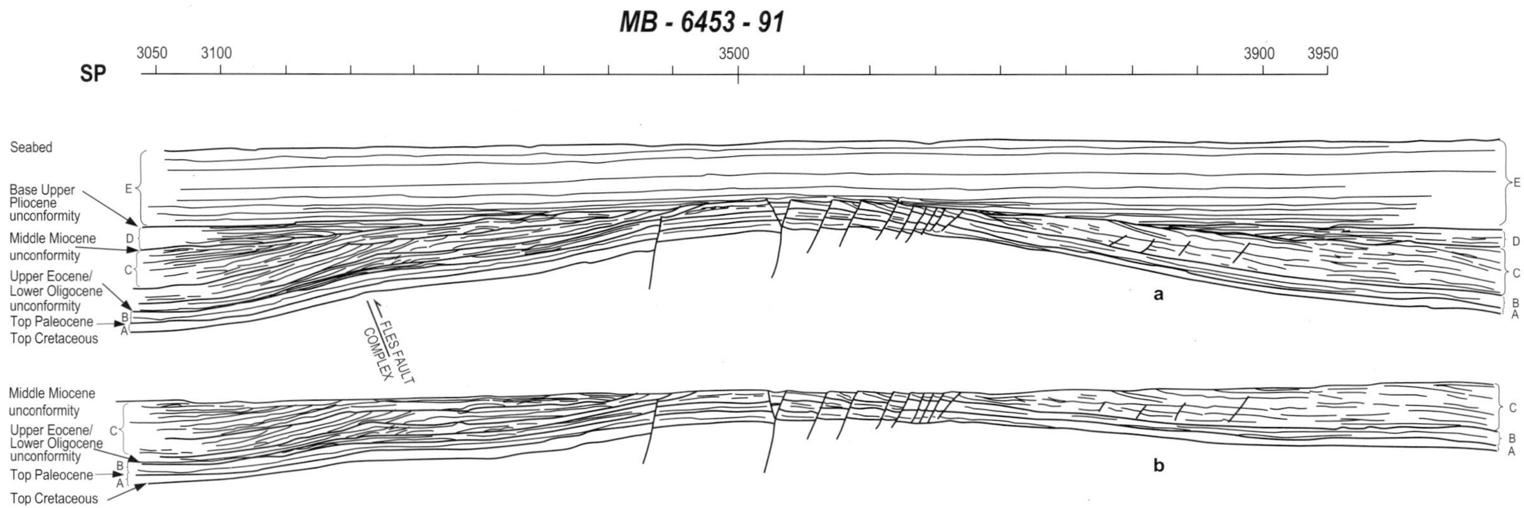

Fig. 18. Line interpretation of seismic line MB-6453-91. (**a**) The present structural geometry. (**b**) The Middle Miocene unconformity surface restored to horizontal (flattening), not compensated for compaction. (See Fig. 2 for line location.) Seismic sequences: A, Paleocene; B, Eocene; C, Oligocene, and Lower Miocene; D, Upper Miocene and Lower Pliocene; E, Upper Pliocene and Pleistocene.

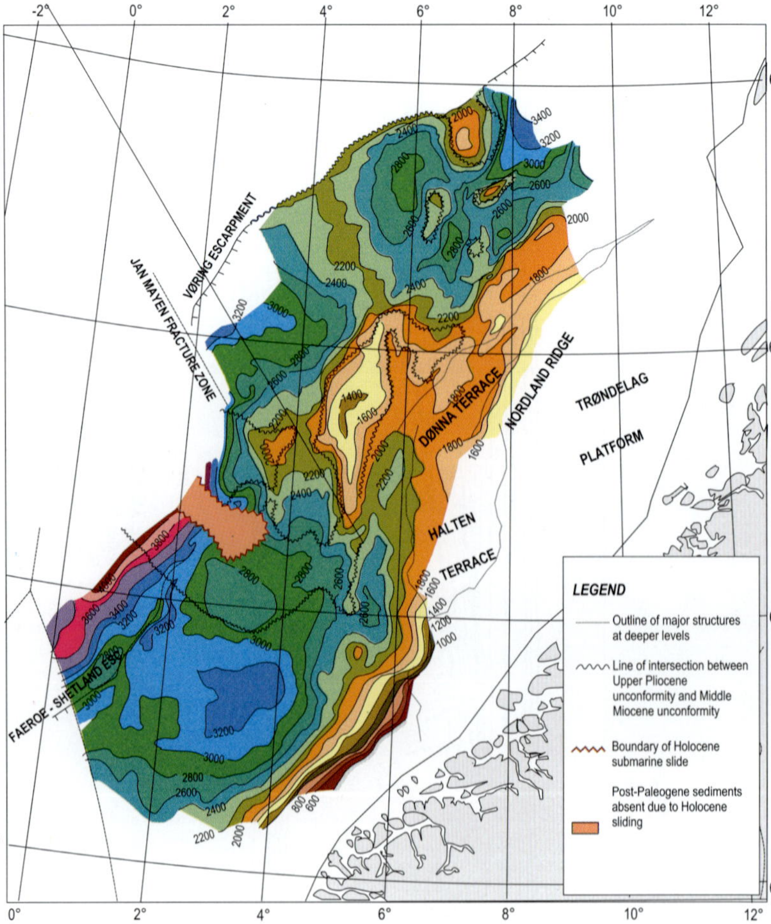

Fig. 19. Two-way travel time map of the Middle Miocene unconformity in the Vøring and Møre Basins. Line of intersection between Upper Pliocene and Middle Miocene unconformities delineates areas where sediments of Middle and Upper Miocene and Lower Eocene are missing. Within these areas contours are on the base Upper Pliocene unconformity. Contours labelled in milliseconds.

prograding from the eroded mainland and shelf. However, in places this hiatus is an erosional unconformity also within the basin area. In the Vøring Basin the Pliocene unconformity is relatively flat, and Upper Pliocene sediments overstep the Middle Miocene unconformity on the summit of the large domes and arches (Figs 6 8 and 9). In these elevated areas the Middle and Upper Miocene sequence is missing because of tectonic uplift, erosion and non-deposition: these areas include the Helland-Hansen Arch, the Naglfar and Vema Domes, the Nyk and Utgard Highs, the eastern edge of the Vøring Marginal High, and areas involved in the Jan Mayen Lineament (Fig. 19). The mechanism behind the regional, submarine erosion in Pliocene time is also unknown, but is clearly related to the onset of glaciations and rapid uplift of the surrounding mainland (Riis & Fjeldskaar 1992; Riis & Jensen 1992; Riis 1996).

The Vøring Marginal High

The marginal high is situated to the west of the Vøring Escarpment between the Jan Mayen and Bivrost Lineaments. The high comprises Tertiary sediments on top of thick Lower Eocene flood basalts, which are probably underlain by continental crust that progressively thins and becomes transitional to oceanic crust towards the west. Early workers placed the boundary between continental and thickened oceanic crust beneath the observed set of seaward-dipping

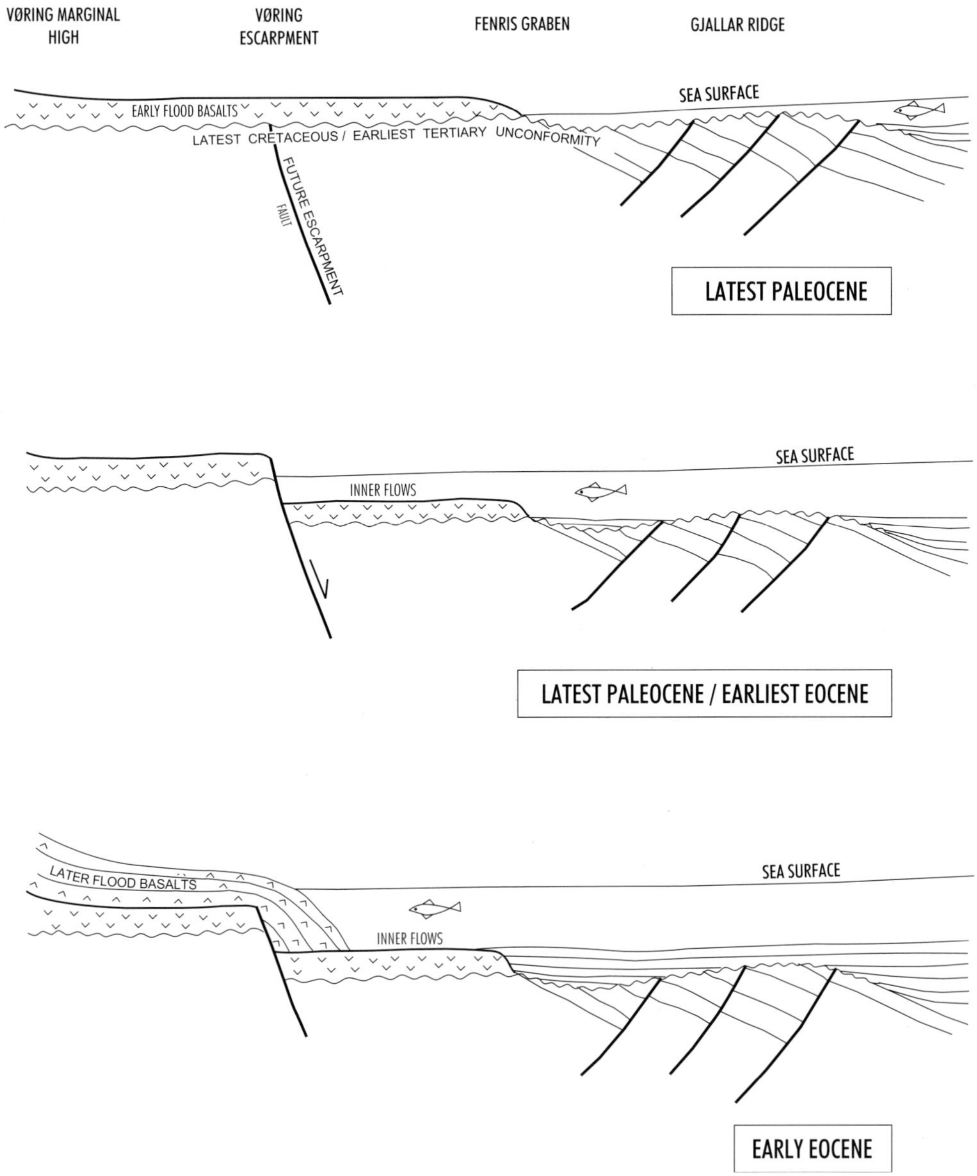

Fig. 20. Model for the development of the Vøring Escarpment and the formation of the 'inner flows'. The same model is inferred for the Faeroe–Shetland Escarpment. (See text for further discussion.)

reflectors that was identified as volcanic flows by well ODP-642 (Talwani *et al.* 1983; Hinz *et al.* 1984; Eldholm *et al.* 1989) (Fig. 6). However, the position of this ocean–continent boundary has not been substantiated by relevant detailed geophysical data. This transition is probably a wide complex zone involving intrusions and tectonic thinning and fragmentation of the continental crust. The magnetic anomalies 24A and 24B in the Vøring Marginal High, as defined by Skogseid & Eldholm (1987), do not appear as typical sea-floor spreading anomalies in current compilations (Olesen *et al.* 1997). The magnetic anomaly pattern in the high rather resembles the signature of the shallow basement ridges that are common in the continental margins around the Norwegian–Greenland Sea (e.g. the Lofoten and Utrøst Ridges).

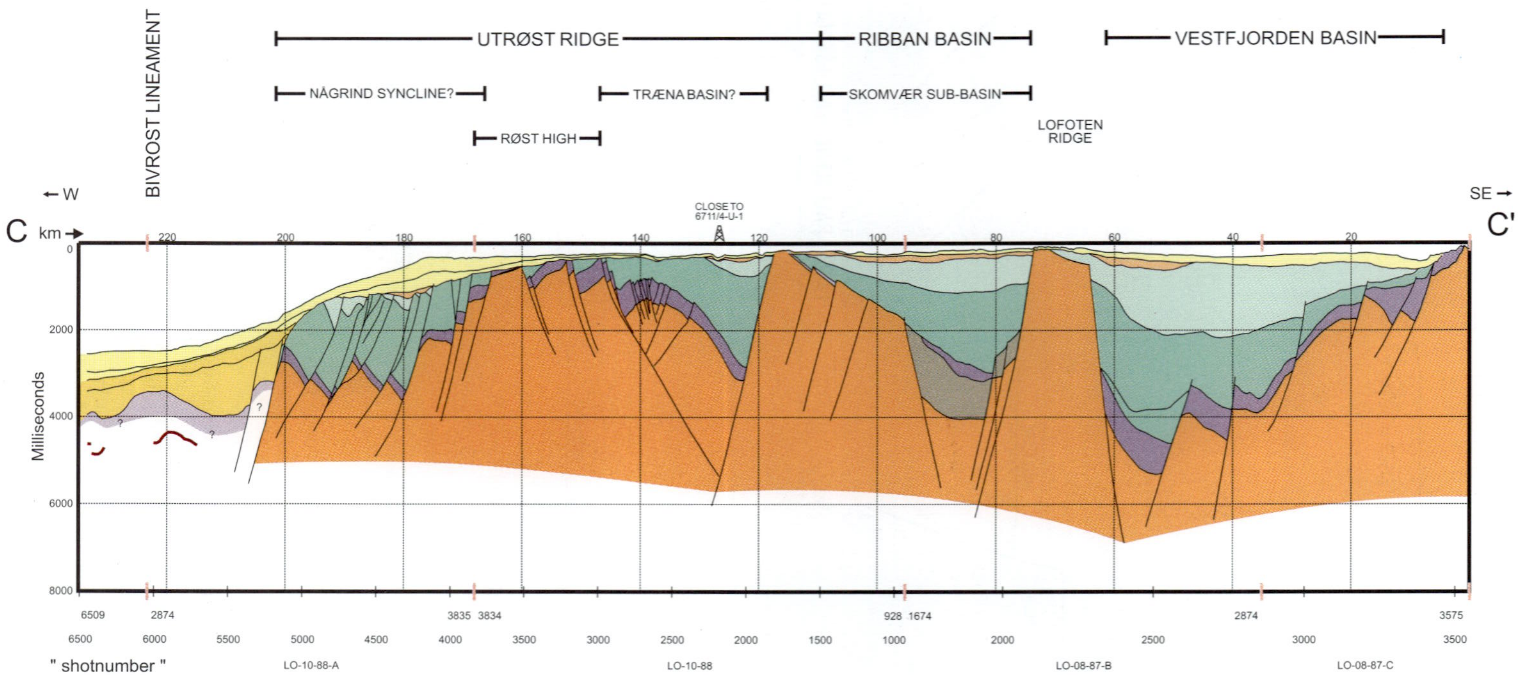

Fig. 21. Geoseismic section CC′. (See Fig. 2 for line location and Fig. 4 for colour legend.) Modified from Blystad *et al.* (1995).

As described above, the present eastern flank of the Vøring Marginal High was probably close to the western margin of the Vøring Basin throughout Cretaceous time. Therefore, the marginal high probably was the western bounding platform and sediment source area to the Cretaceous Vøring Basin area (see also Brekke *et al.* 1999). As the present boundary element, the Vøring Escarpment, is a post-Paleocene feature, the regional base Tertiary erosion surface deeply truncating the Gjallar Ridge and Fenris Graben in the western Vøring Basin must have continued as an unbroken surface into the area of the present marginal high (Fig. 20). This implies that, in latest Cretaceous time, the western platform included the Gjallar Ridge and the Fenris Graben. In latest Paleocene to Early Eocene time, large volumes of tholeiitic flood basalts flowed across the eroded platform and in places covered both the Fenris Graben and the Gjallar Ridge. When the Vøring Escarpment was initiated, the early flows east of the escarpment were separated from the emerging marginal high and were probably submerged and are now termed 'the inner flows' (Hinz *et al.* 1982; Talwani *et al.* 1983) (Fig. 20). Because of the water-front then established along the Vøring Escarpment, the younger flows of flood basalt from the fissure eruptions to the west stopped on reaching the escarpment and a prominent coastline lava front was built, enhancing the topography of the escarpment. The substrate to the flood basalts in the marginal high is unknown, but probably consists of deeply eroded Cretaceous or older rocks.

The northern province between the Bivrost Lineament and the Jennegga High

The continental margin in the northern province is very narrow. The main elements are the Vestfjorden and Ribban Basins and the elevated area of the Lofoten and Utrøst Ridges (Fig. 1). The Eocene flood basalt lavas onlap the flank of the Utrøst Ridge and cover the adjacent deep-water area, making it very difficult to map the underlying geology (Mjelde *et al.* 1993).

The area around the Jennegga High on the northernmost part of Utrøst Ridge constitutes a deeply eroded transverse ridge separating the Norwegian Sea continental margin to the south from the Barents Sea margin to the north. The Utrøst Ridge, Ribban Basin, Lofoten Ridge and Vestfjorden Basin constitute a characteristic regional horst-and-graben system (Fig. 21). The Ribban Basin and the Vestfjorden Basin are NE–SW striking half-grabens with 4–5 s twt of Cretaceous basin fill and with mutually opposite polarities. The basin boundary faults run along both flanks of the narrow basement horst of the southern Lofoten Ridge, which separates the two basins (Fig. 21). The Utrøst Ridge and the mainland are the rotated, non-faulted flanks of the Ribban and Vestfjorden Basins, respectively. In the very north of the Ribban Basin the bounding faults change to an easterly polarity. The Ribban Basin is divided into two sub-basins, the Skomvær Sub-basin in the south and the Havbåen Sub-basin in the north. The general horst and graben geometry was initiated because of late Mid-Jurassic–Early Cretaceous crustal stretching and later thermal subsidence and minor reactivations. Seismic mapping shows that there was a phase of rotation and subsidence in the Ribban Basin in the Mid-Cretaceous time. Truncation of the Cretaceous succession (Fig. 21) indicates that the whole of this area was uplifted and deeply eroded in latest Cretaceous time, a phase of uplift also recorded in the Vøring Basin to the south (see above). However, the Lofoten–Utrøst area experienced a stronger uplift than the Vøring Basin, and the Bivrost Lineament acted as the hinge line between the two areas. Phases of renewed uplift and erosion probably took place in Tertiary time, but the details are not known because of the erosion of the Tertiary sequences. However, by reference to the tectonic development in the Vøring Basin to the south (see above) and in the adjacent Barents Sea (Brekke & Riis 1987; Gabrielsen *et al.* 1990), it seems likely that the main phases took place in Eocene, Miocene and Pliocene time. As a result of the erosion, crystalline basement rocks are at present exposed in the Lofoten Ridge and at the sea bed on local highs on the Utrøst Ridge (the Røst and Jennegga Highs).

The southern province between 62°N and the Jan Mayen Lineament

The main elements of the southern province are the Møre Basin, the Møre Marginal High and the Møre–Trøndelag Fault Complex (Fig. 1). The southern boundary of the province is the area of intersection between the NE–SW trend of the Møre–Trøndelag Fault Complex and the N–S to NNE–SSW trends of the northern North Sea at *c.* 62°N (Fig. 1).

The Møre Basin

The Møre Basin is bounded to the north by the Jan Mayen Lineament, to the southeast by

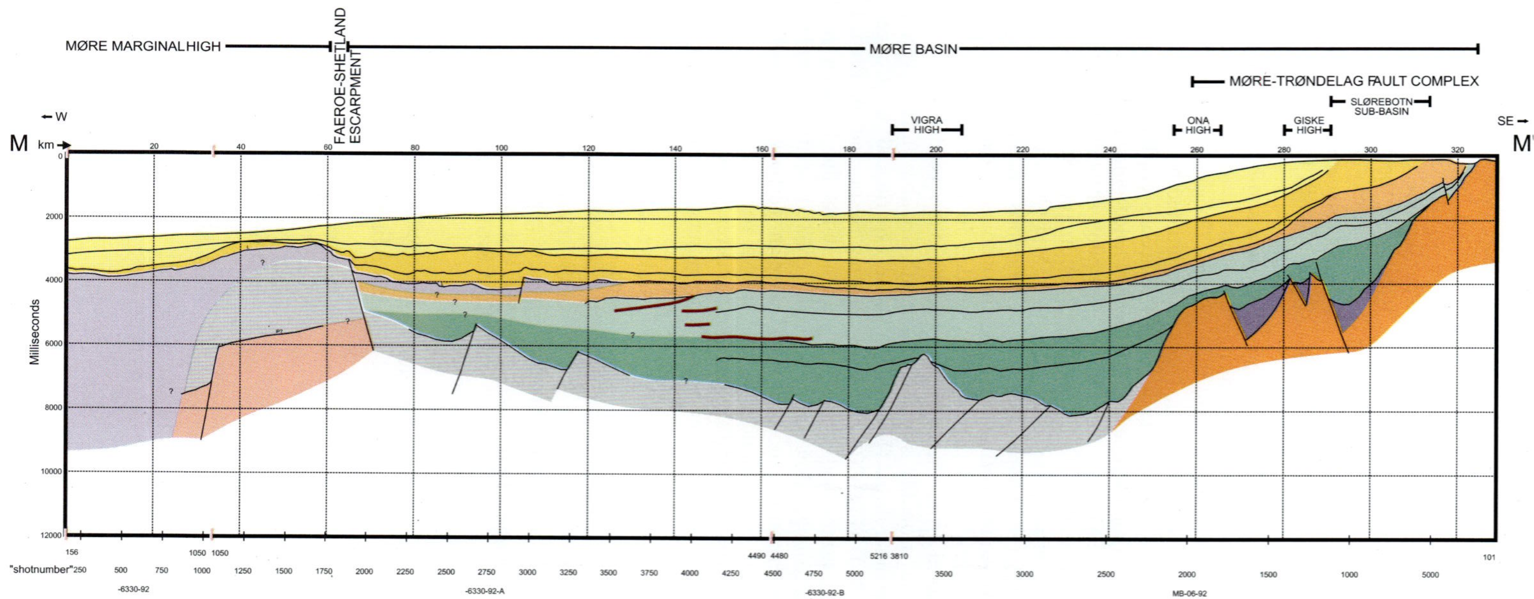

Fig. 22. Geoseismic section MM′. (See Fig. 2 for line location and Fig. 4 for colour legend.) Modified from Blystad *et al.* (1995).

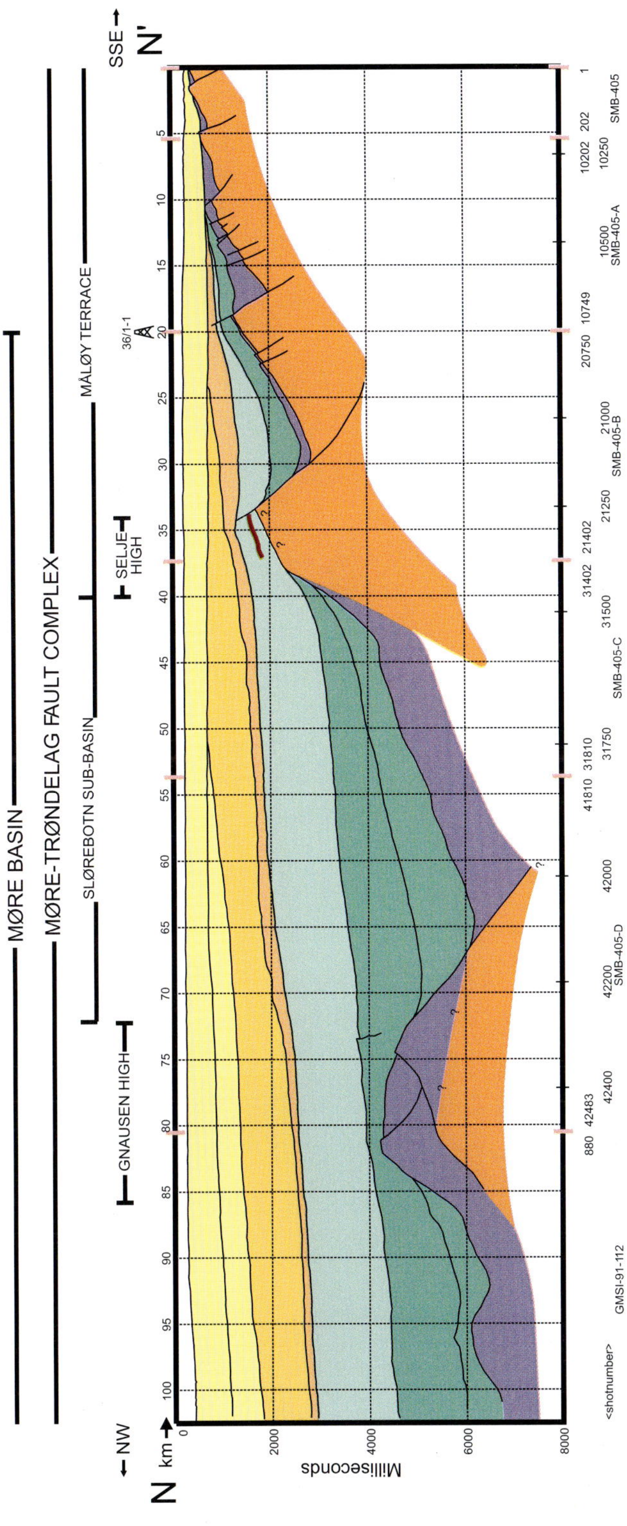

Fig. 23. Geoseismic section NN′. (See Fig. 2 for line location and Fig. 4 for colour legend.) Modified from Blystad *et al.* (1995).

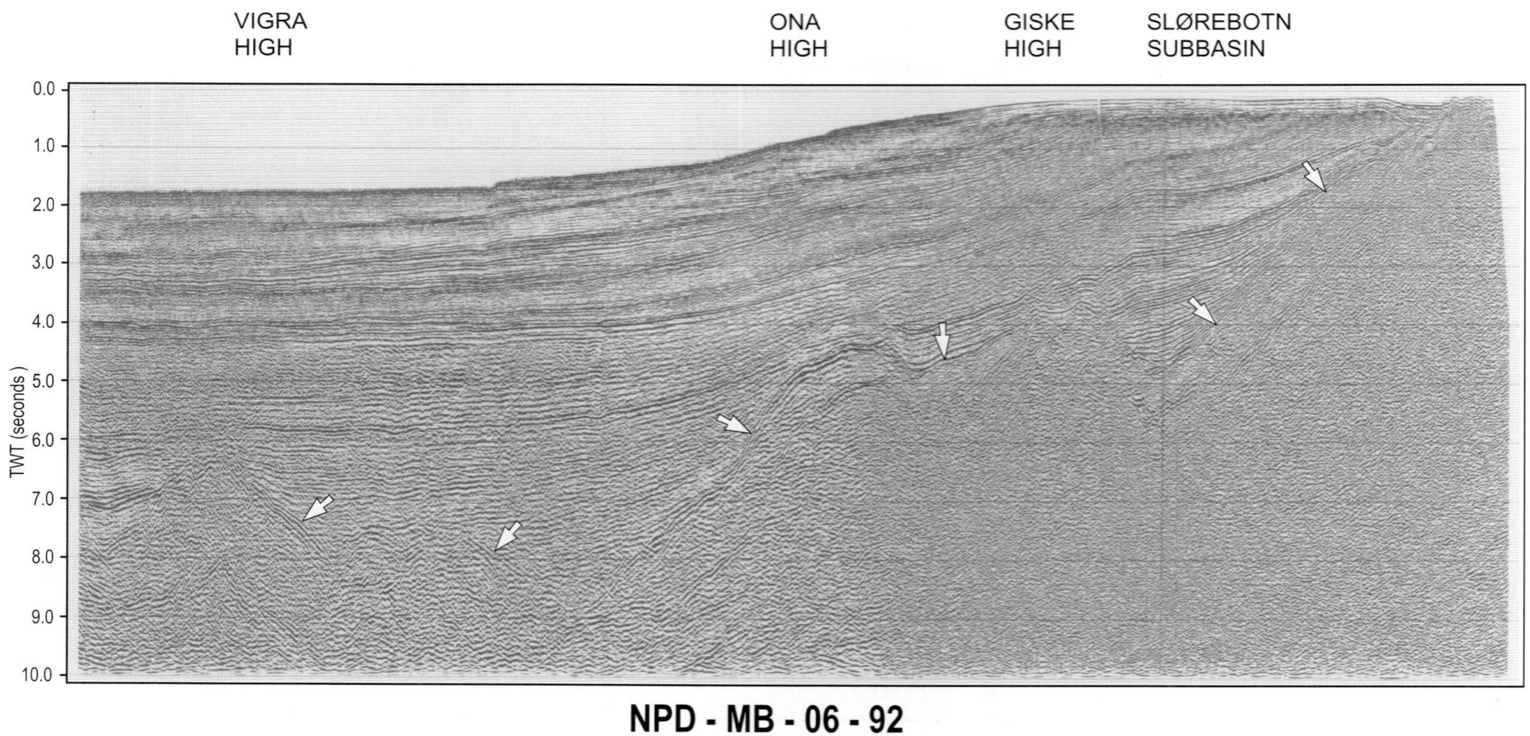

Fig. 24. Seismic line MB-06-92. The overall downflexed nature of the eastern flank of the Møre Basin. (Note the general eastward polarity of the major faults in the flank.) The line is the basis for eastern part of geoseismic section MM′ in Fig. 22 (see Fig. 2 for line location). White arrows point at the base Cretaceous unconformity.

the Møre–Trøndelag Fault Complex and to the west by the Faeroe–Shetland Escarpment. The basin is defined at the base Cretaceous unconformity and by a greatly expanded Cretaceous sequence. The basin has an overall NE–SW trend and in the axial parts of the basin the Cretaceous interval may be up to 6 km thick and the base of the Cretaceous sequence lies between 7 and 9 s twt (Fig. 7).

The late Mid-Jurassic–Early Cretaceous rifting was the main tectonic episode in the Møre Basin (Fig. 22). The Cretaceous sediments onlap the base Cretaceous unconformity all along both the eastern and western margins, indicating that the basin was mainly formed by downwarping of its flanks subsequent to the rifting (Figs 22 and 23). The rift faults show polarity away from the basin axis. The dip of the eastern flank is a composite of the downflexing in Early Cretaceous times and the Late Tertiary uplift of the adjacent mainland. Removing the effect of the Tertiary uplift leaves the Cretaceous Møre Basin as a perfectly symmetrical sag-basin. Influx of coarse clastic deposits from the west into a coastal or shallow marine environment in the Slørebotn Sub-basin (Fig. 1) in Mid-Jurassic times (Jongpier *et al.* 1996) indicates that the deepest parts of the Cretaceous Møre Basin were highly elevated and subject to erosion at that time. This is in agreement with the observation that the pre-Cretaceous sequences are thin wherever recognized within the Møre Basin.

The prominent Vigra High, close to the central axis of the Møre Basin, is a large, NE–SW trending rotated fault block onlapped by the Lower Cretaceous sequence (Fig. 22). The bounding fault along its northeast margin has a maximum throw of about 2500 m. It dies out towards the south, and ends at the Jan Mayen Lineament to the north. The high was formed by large-scale faulting in the late Mid-Jurassic–Early Cretaceous rifting episode. Internal reflectors parallel to its eastward sloping summit indicate the existence of a thin interval of pre-Cretaceous sediments within the high (Fig. 24).

In sharp contrast to the Vøring Basin north of the Jan Mayen Lineament, the Møre Basin seems to have been tectonically quiet, and experienced passive Cretaceous subsidence subsequent to the late Mid-Jurassic–Early Cretaceous rifting, and no observable tectonic activity in Late Cretaceous time and only minor activity in Tertiary time. The Tertiary tectonics were mainly associated with reactivations of the Jan Mayen Lineament and minor faulting along the Faeroe–Shetland Escarpment.

The thinning and onlap of the entire Cretaceous sequence onto the sloping basin flanks strongly indicate the existence of Cretaceous bounding platforms and sediment source areas both across the now eroded mainland to the east and across the Møre Marginal High to the west. Much of the western flanks of the Møre Basin are hidden under the Eocene lavas so that it is not possible to produce continuous maps of details of the Cretaceous features. However, in some areas where the lavas are reasonably seismically transparent, data show that the Cretaceous sequence thins by onlap onto the western basin flank (Fig. 22). Close to the present Faeroe–Shetland Escarpment the Cretaceous interval is as thin as on the upper parts of the eastern basin flank (Fig. 22) and on the edge of the Trøndelag Platform (e.g. Figs 6 and 11). Thus, the Faeroe–Shetland Escarpment is probably close to the Cretaceous boundary between the Møre Basin and the western platform (Fig. 22).

The Møre–Trøndelag Fault Complex

The southeastern margin of the Møre Basin consists of the Møre–Trøndelag Fault Complex, comprising a broad NE–SW to ENE–WSW trending system of fault-controlled ridges, highs and minor basins. In two of these highs wells 6305/12-2 and 6306/10-1 have proven crystalline basement overlain by a thin Middle Jurassic sequence underneath post-Albian strata (Jongpier *et al.* 1996). The Manet Ridge (south), and Gnausen, Giske, Ona and Gossa Highs (north) make up a row of such basement-cored highs that separate the mainly Cretaceous Magnus Basin, Marulk Basin and Slørebotn Sub-basin from the main, deeper part of the Møre Basin (Brekke & Riis 1987; Blystad *et al.* 1995). This set of highs and minor basins possibly originated as a system of horsts and half-grabens already in Triassic times, but was greatly accentuated during the late Mid-Jurassic–Early Cretaceous rifting episode. The basement-cored Selje High to the southeast is also part of this system. The relatively thin Jurassic and Triassic cover to the crystalline basement along the Møre–Trøndelag Fault Complex, is in agreement with the notion of a shallow basement beneath the Cretaceous and latest Jurassic basin fill in the Møre Basin (Fig. 22).

On the mainland, the fault complex can be seen to affect the Precambrian basement and rocks of Lower Palaeozoic, Devonian and Jurassic ages. The ENE–WSW structural grain in northwestern Norway can be related to Caledonian deformations, but the fault complex has

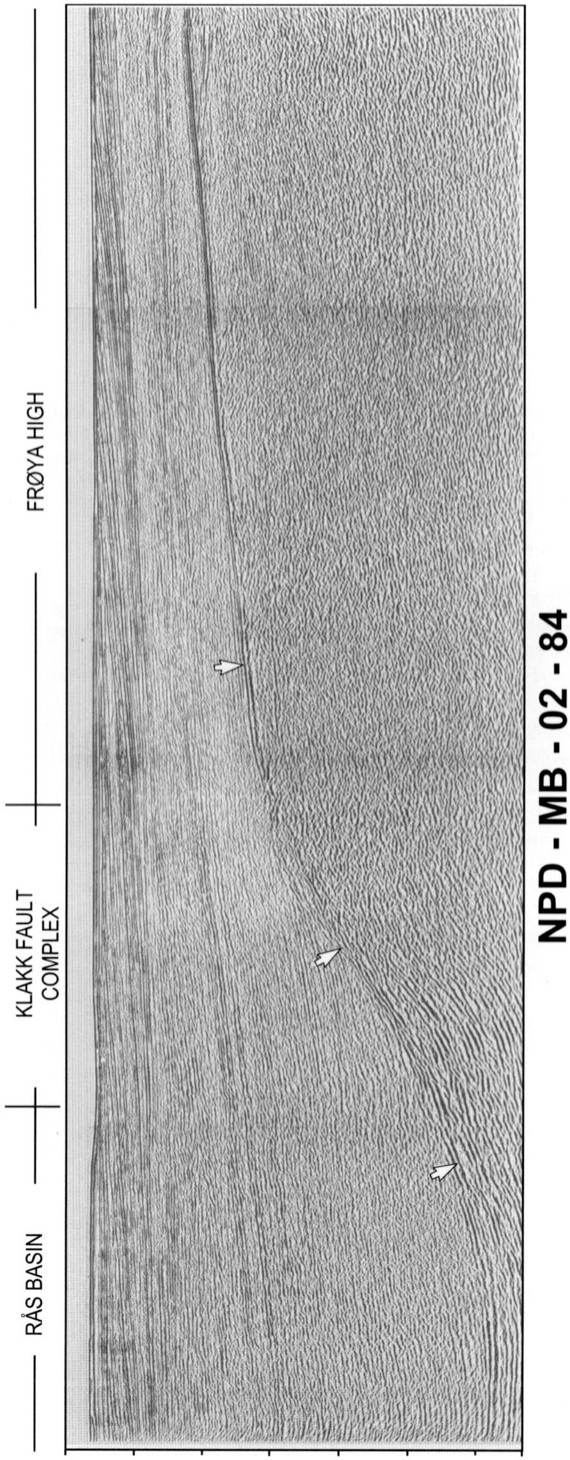

Fig. 25. Seismic line MB-02-84. Downflexed western slope of the Frøya High. White arrows point at the base Cretaceous unconformity. (See Fig. 2 for line location.)

been reactivated in several episodes (Bering 1992; Grønlie *et al.* 1994).

The Møre Marginal High

The marginal high is situated to the west of the Faeroe–Shetland Escarpment and bounded to the north by the Jan Mayen Fracture Zone. To the south, the Møre Marginal High merges with the Faeroe Plateau, which includes the Faeroe Islands. The high comprises Tertiary sediments on top of thick early Eocene flood basalts, and, like the Vøring Marginal High, probably contains continental crust that westwards becomes increasingly more intruded, faulted, thinned and stretched (Fig. 22).

The substrate to the flood lavas in the continental part of the marginal high is unknown. The westward thinning and onlap of the Cretaceous units of the Møre Basin strongly indicate that the Cretaceous sequence is very thin or missing beneath the lavas on the high. The Paleocene sequence in the Møre Basin seems to thicken westwards towards the Faeroe–Shetland Escarpment, suggesting early fault activity along the escarpment and differential uplift and erosion of the marginal high at that time. The fact that the early lavas flowed far eastward of the present escarpment indicates that the area was above sea level in Late Paleocene–Early Eocene time. Subsequent faulting along the escarpment probably submerged and separated the 'inner flows' from the marginal high (Fig. 20). The present morphology of the Faeroe–Shetland Escarpment seems to be mainly the result of the build-up of a volcanic front along that fault-related shoreline (see also Smythe *et al.* 1983). However, in the northern segments it is enhanced by minor faulting in Late Eocene time (Fig. 22).

Probably because of a relatively thin cover of lava in some places, it is possible to identify a deep-seated, faulted angular unconformity within the marginal high. As it is juxtaposed by the Cretaceous sequence of the Møre Basin in the hanging wall, it must be older than the Cretaceous sequence. In Fig. 22 it is tentatively suggested that it may be correlated with the Upper Permian unconformity identified in the Trøndelag Platform.

The eastern basin flanks

The eastern flank of the Møre Basin obviously formed by the downflexing of the rift–unconformity of late Mid-Jurassic–Early Cretaceous time to a western dip along the basin margin. A majority of the rift faults have an eastern polarity and their fault planes were rotated to a shallow dip during the flexuring (Figs 22 and 23). In the present interpretation the same flexure geometry also dominates the eastern flanks of the Rås and Træna Basins in the Vøring Basin (Figs 10, 11 and 17). This means that the traditional maps showing these basins as the result of down-to-the-basin faulting along the Klakk, Revfallet and Ytreholmen Fault Complexes are simplifications and do not reveal the true nature of the crustal mechanics behind the basin formation. In an attempt to substantiate the present interpretation, a series of seismic dip-lines are presented from the Møre Basin in the south to the Træna Basin in the north (Figs 24–30).

The data show the general westward slope of the base Cretaceous reflector and that even the apparently steep western flank of the Ona High is a flexure and not a fault (seismic data are plotted with a 2–4 times vertical exaggeration) (Fig. 24). Along the Frøya High the basin flank is considerably steeper, but Fig. 25 indicates it to be the limb of a monoclinal flexure rather than a fault scarp. However, the flexural steepening of the slope in some places has resulted in a subsequent gravitational collapse of the slope, giving detached and highly rotated fault blocks (involving even the basement) hanging in the slope (Fig. 26). The flank of the Frøya High is part of what is termed the Klakk Fault Complex, which also runs along the steep flank of the Halten Terrace further north. The data quality along the terrace flank is generally poor, but the present author believes the flexuring and rotation of early faults as shown in Figs 10 and 17 to be correct (compare original data in Figs 27 and 28, respectively).

The southern segment of the Nordland Ridge was strongly reactivated and uplifted during Late Cretaceous and Tertiary time, overprinting much of the late Mid-Jurassic–Early Cretaceous tectonic features. Along this segment of the ridge, the Late Cretaceous basin-facing faults are clearly an important element of the basin flank (Fig. 5). However, a close look at the original data reveals that the underlying, early structure of the Revfallet Fault Complex is a monoclinal flexure (Fig. 29). In the northern segment of the Nordland Ridge this early (Late Jurassic–earliest Cretaceous) flexuring is better preserved, being less affected by later faulting (Figs 11 and 30). The characteristic back-rotated Jurassic faults facing away from the basin are also clearly observable.

Accepting that the underlying structure of the Late Jurassic–Early Cretaceous eastern basin

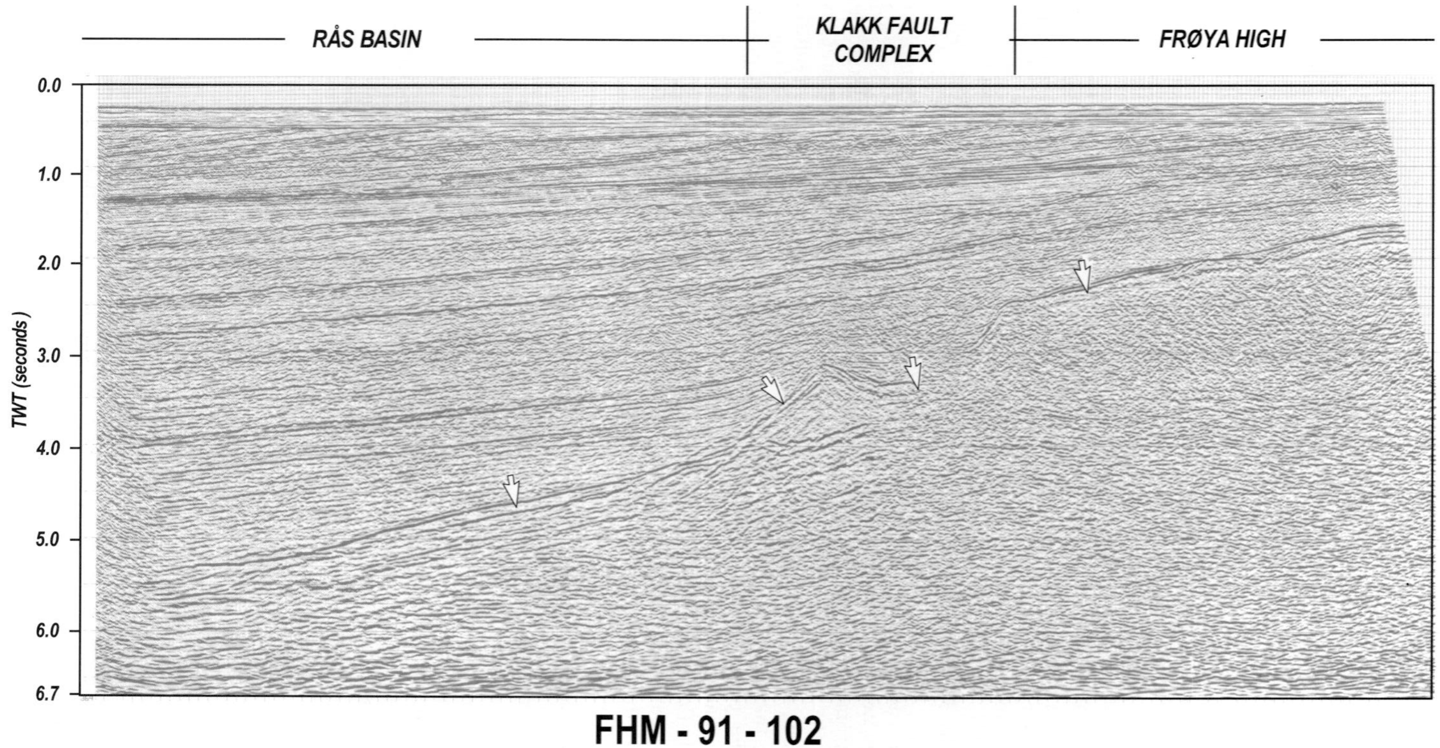

Fig. 26. Seismic line FHM-91-102. Gravitational collapse of western slope of the Frøya High. White arrows point at the base Cretaceous unconformity. (See Fig. 2 for line location.) By courtesy of Norsk Hydro a.s.

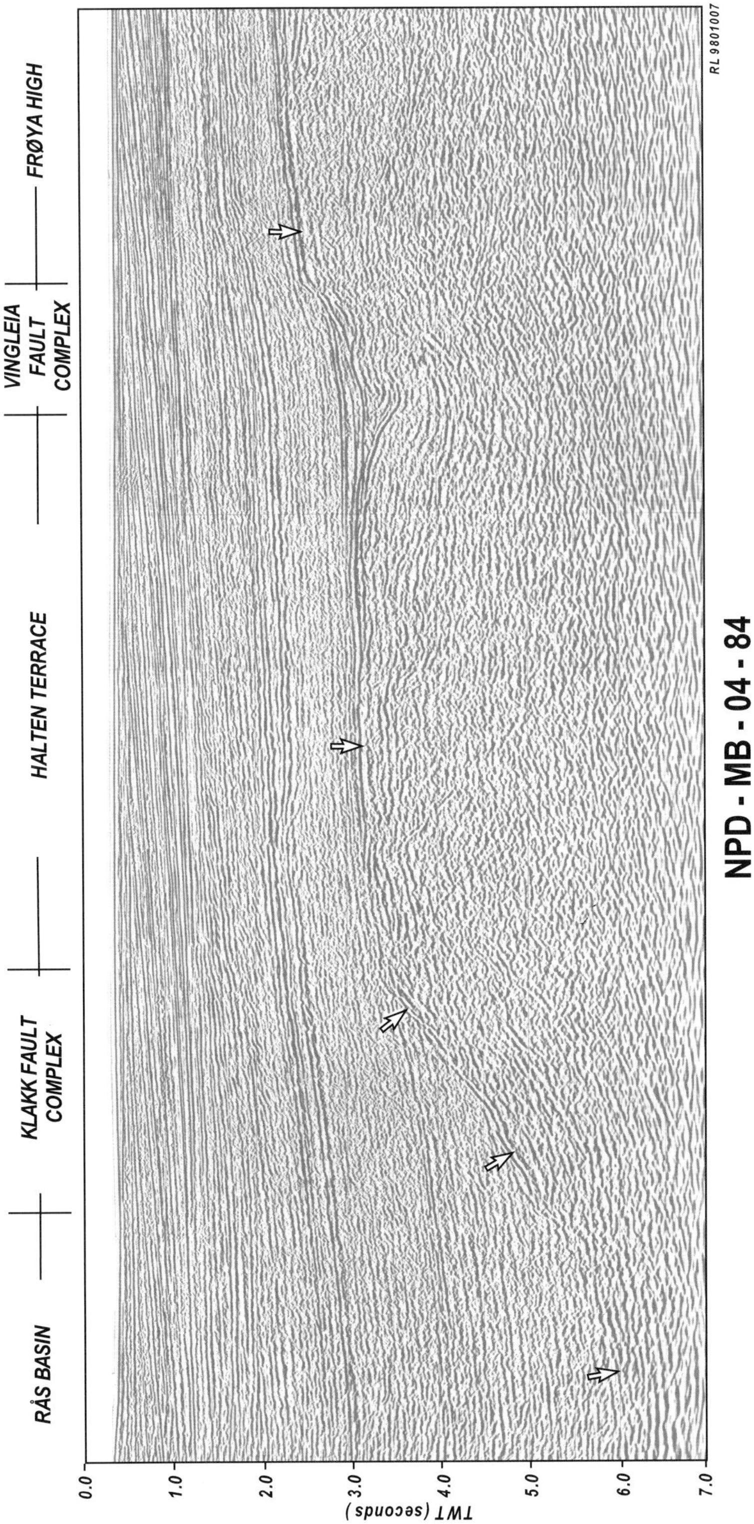

Fig. 27. Seismic line MB-04-84. The overall downflexed nature of the western flank of the Føya High. (Note the indications of Jurassic faults with a general eastward polarity in the upper part of the flank.) The line is the basis for the western part of geoseismic section KK′ in Fig. 10 (see Fig. 2 for line location). White arrows point at the base Cretaceous unconformity.

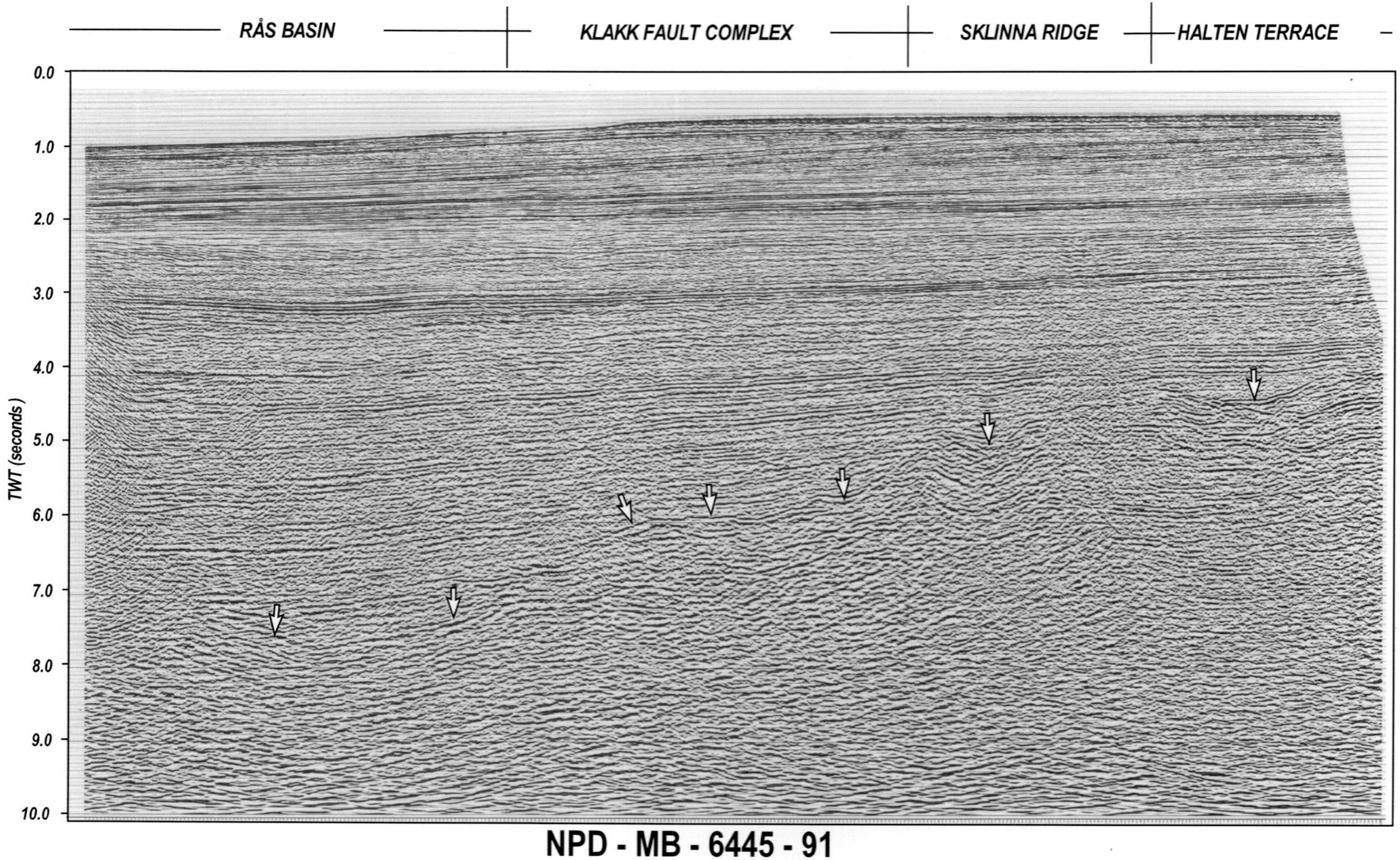

Fig. 28. Seismic line MB-6445-91, which is the basis for the interpretation of the overall downflexed nature of the western flank of the Halten Terrace in geoseismic section JJ′ in Fig. 17 (see Fig. 2 for line location). White arrows point at the base Cretaceous unconformity.

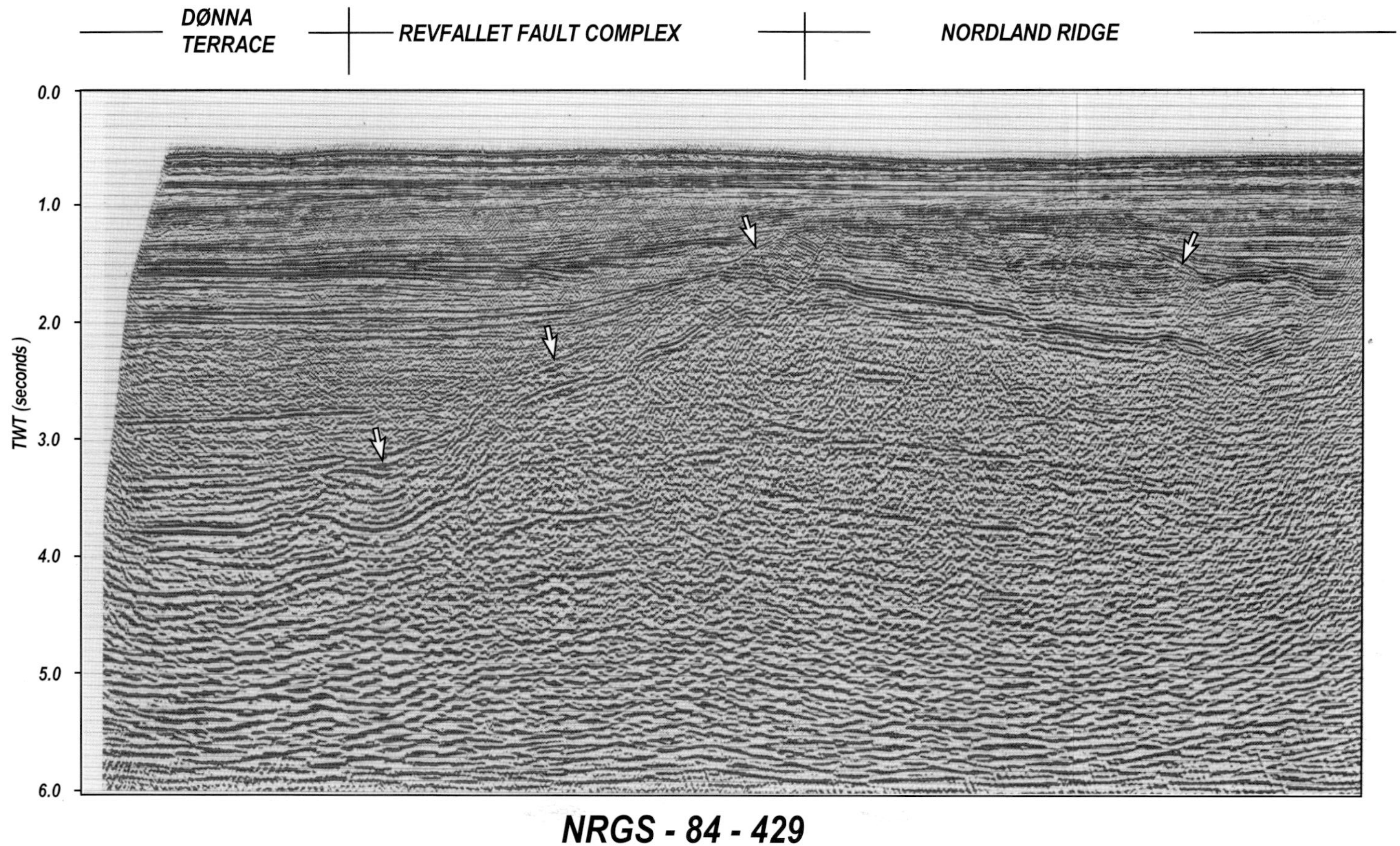

Fig. 29. Seismic line NRGS-84-429, reprocessed by Saga Petroleum a.s, which illustrates the underlying, Late Jurassic monoclinal nature of the western flank of the Nordland Ridge cut by younger, Cretaceous faults. The line is the basis for the interoperation of that part of geoseismic section EE′ in Fig. 5 (see Fig. 2 for line location). White arrows point at the base Cretaceous unconformity. By courtesy of Saga Petroleum a.s.

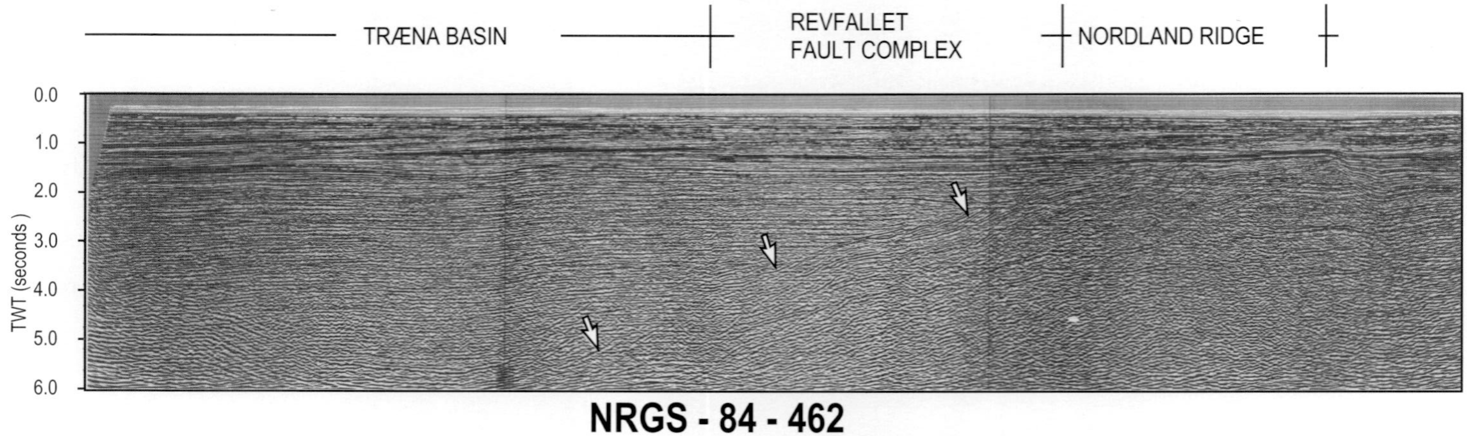

Fig. 30. Seismic line NRGS-84-462. The overall downflexed nature of the western flank of the northern part of the Nordland Ridge. (Note the general eastward polarity of the major Jurassic faults and the general lack of later, Cretaceous faults in this part of the Nordland Ridge.) White arrows point to the base Cretaceous unconformity. (See Fig. 2 for line location.) By courtesy of Statoil.

flanks is a flexure implies that both the Møre Basin and the Vøring Basin area were symmetrical sag-like basin areas in Early Cretaceous time (e.g. Figs 6 and 22). Such a symmetry, combined with the great subsidence of the axial parts of the basin area, strongly indicates that during the late Mid-Jurassic–Early Cretaceous rifting phase the crust of central parts of the rift system was highly stretched and attenuated. The accompanying heat flow must have been high, causing a significant uplift and erosion of the future basin areas. An early onset of increased heat flow in Mid-Jurassic time would explain the enigmatic influx of sand from the west in the western parts of the Halten Terrace and Slørebot Sub-basin during Jurassic time (Gjelberg *et al.* 1987; Jongpier *et al.* 1996). This could also explain the indications of a generally thin pre-Cretaceous basement cover all over both the Vøring and Møre Basins. As already argued, north of the Surt Lineament the Cretaceous (or uppermost Jurassic) units may even rest directly on the basement.

A possible fossil opal-CT transition

An anomalous reflector of possible diagenetic origin is observed in the Lower Miocene sequence throughout the Vøring Basin west of the Fles Fault Complex. The same reflector is also seen within the Neogene sequence on the Vøring Marginal High, where it was penetrated by well ODP-463. According to the data from well ODP-643, opal A is converted to opal-CT at depths of 275–310 m (Roaldset & He 1995). This depth correlates well with the diagenetic seismic reflector (C. Magnus, pers. comm.) (see also Fig. 6).

This reflector crosscuts the bedding reflectors where these are involved in small-scale compaction faulting and Late Miocene doming. The reflected surface is not parallel to the sea floor, nor to the Late Pliocene bedding, and is deeper and covered by considerably greater thicknesses of Late Pliocene sediments to the east than to the west (Fig. 6). The transition surface is parallel to bedding reflectors just below the Upper Pliocene unconformity and is situated at a more or less constant depth of 550 m below the Upper Pliocene unconformity, both in the Vøring Basin as well as on the Vøring Marginal High. On the flanks of the large Late Miocene arches and domes the reflector clearly crosscuts the flexured bedding, but is also seen to be itself involved in the latest phase of the flexuring. The same relationship is observed at the flexured limbs of the Vøring Escarpment. These relationships indicate that the phase transition surface is fossil and was formed probably in latest Miocene or Early Pliocene time. The fossil nature of the phase transition surface was also suggested by Roaldset & He (1995), on the basis of present temperature relationships. The estimated transition temperature of 55°C in the ODP-643 well (see Fig. 6) corresponds to a thermal gradient of 180°C km^{-1}, as opposed to the present gradient of 50–65°C. The seismic data (line NH-1-79: the western end of geoseismic section in Fig. 6) indicates a missing section of about 250 m as a result of slumping at the well location (ODP 643), but allowing for this will only reduce the calculated thermal gradient from 180 to 100°C km^{-1}, still strongly indicating a fossil process. Such fossil opal-A to opal-CT phase transition surfaces are also reported from the North Sea (Rundberg 1989) and the Barents Sea (Roaldset & He 1995).

As the opal-A to opal-CT transition is temperature dependent (Roaldset & He 1995), the fossil transition surface is a record of a peak heat flow at that time or an event of rapid reduction of overburden by erosion all over the western parts of the continental margin, or a combination of the two.

At what sediment depths the phase transition took place is not known, as the depth of erosion in Late Pliocene time is not known. Seismic data show that, in a restricted area above the Vigrid Syncline, an extra 150 m thick section of Upper Miocene–Lower Pliocene units is preserved as an erosional remnant in the shape of a large, gentle mound-like feature (line NPD-VB-17-90: western end of geoseismic section in Fig. 5). The seismic correlations indicate that this 'mound' contains the youngest Upper Miocene–Lower Pliocene sediments preserved in the Vøring Basin. Everywhere else in the Vøring Basin this 150 m thick section seems to have been removed by erosion, giving a minimum estimate of missing section beneath the Upper Pliocene unconformity. However, as the process was submarine and of regional extent this also seems to be close to a likely maximum of eroded section. This implies a maximum burial depth of about 700–800 m in pre-erosion times. According to Roaldset & He (1995), the expected transition temperature in siliceous sediments of ages less than 20 Ma is about 40°C. This would imply a Late Miocene–Early Pliocene thermal gradient of at least 50–60°C km^{-1} if the phase transition was due to higher thermal gradients in the past. The existence of a past thermal event is also indicated by the high thermal gradient as calculated from the data in ODP-643 well (Roaldset & He 1995). If such a thermal event originated from mantle processes,

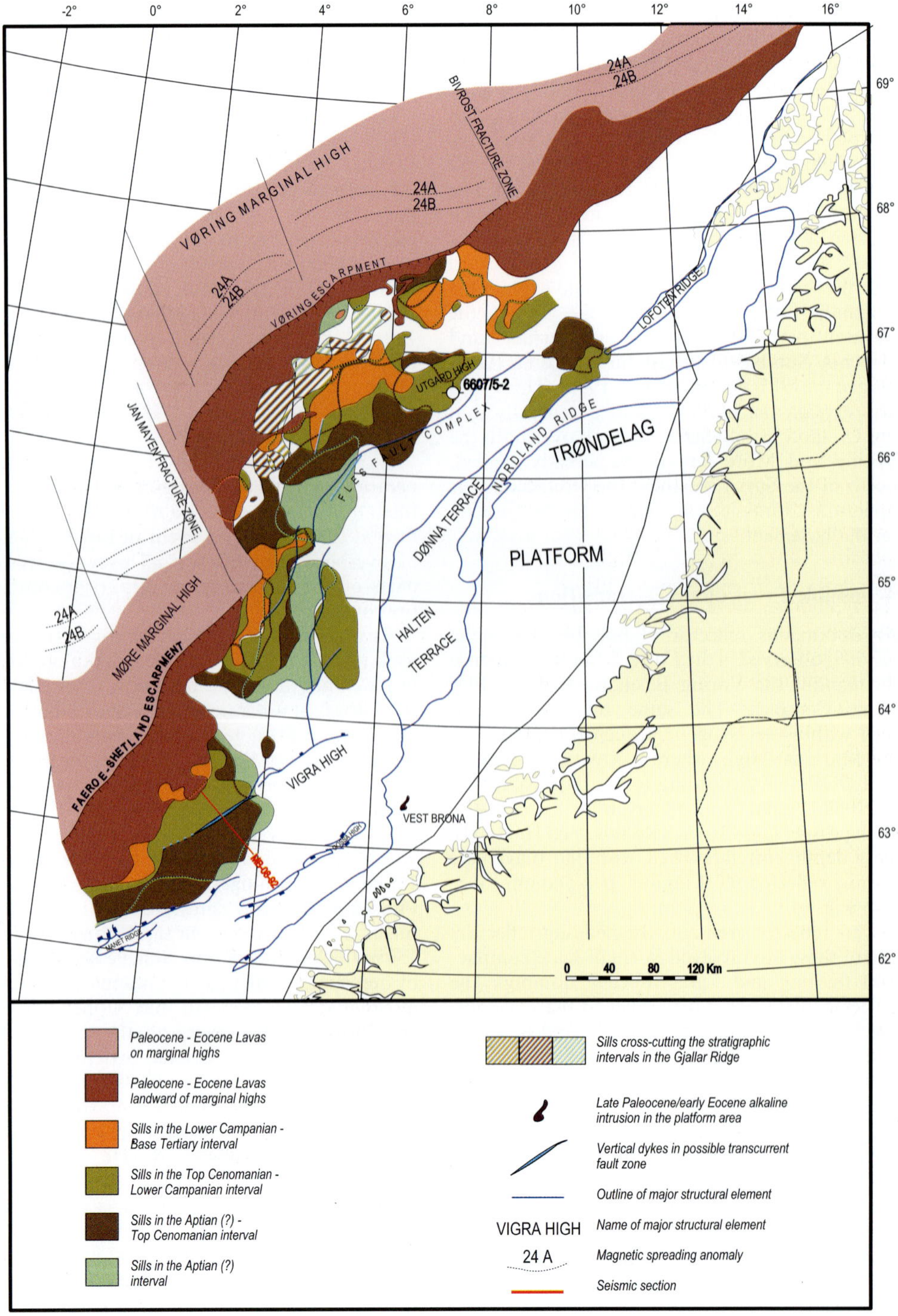

Fig. 31. Map of the distribution of magmatic rocks in the Norwegian Sea continental shelf based on seismic interpretation. The distribution of the sill intrusions in the Møre and Vøring Basins is given in terms of both area and stratigraphy.

it must have been associated with a phase of thermal uplift of the whole continental margin. This could fit with the 7 Ma period of regional Miocene uplift described earlier, but the time since then (10–15 Ma) is far too short to cool the mantle and crust by normal conduction to its present temperature gradient and depth of subsidence. This calls for either an unknown mechanism of extremely rapid cooling of the crust and underlying mantle or a mechanism for the heating of the upper parts of the crust and cover only, in which case the heating and the uplift are caused by separate processes.

Alternatively, the 'freezing' of the transition surface may be due to Pliocene erosion. The cited age–temperature relationship for the opal-A–opal-CT transition gives an expected burial depth of 1100–1200 m at a normal thermal gradient of $35^{\circ}C\,km^{-1}$ in the Vøring Basin, and *c.* 1000 m at the present $55^{\circ}C\,km^{-1}$ on the Vøring Marginal High. This would imply a regional submarine erosion of the order of 700–500 m in the Vøring Basin and 400–500 m on the Vøring Marginal High, if the present position of the transition surface is due to Early Pliocene erosion. Riis & Fjeldskaar (1992) suggested that the mechanism to initiate the coincident major erosional event onshore could be the change to a colder climate and the onset of glaciation. The consequent increase in erosion rates would cause uplift of the mainland and adjacent offshore areas by unloading. However, the model does not indicate that this uplift mechanism would affect the outer parts of the continental margin as far as the Vøring Marginal High, and the required erosion in that area must have been purely an effect of a change in the submarine currents. To the present author, it seems highly unlikely that submarine currents have been responsible for removing a 500–700 m sediment column all over such a wide area as the Vøring Basin and the Vøring Marginal High. One would expect such submarine erosion to be much more channelled. This leaves us with the option that the fossil opal-A–opal-CT surface records a higher temperature gradient in Late Miocene–Early Pliocene time than at present. The author finds it most likely that this was due to an event of increased heat flow in Mid-Miocene times, though caused by an unknown mechanism.

Sill intrusions

East of the 'inner flows' the Cretaceous basin fill is intruded by numerous sills of probably latest Paleocene–earliest Eocene age (Figs 5, 6, 17, 22 and 31). In the Vøring Basin the eastern limit of the sills generally coincides with the Fles Fault Complex. However, along the Jan Mayen and Bivrost Lineaments the intrusions are seen to extend further east, even into the flanks of the Trøndelag Platform (Fig. 31). In the Møre Basin the general eastern limit of the sills coincides with the western faults in the broad Møre–Trøndelag Fault Complex along the eastern flanks of the basin.

In both basins the sills are generally found at their deepest position and lowest stratigraphic levels (Aptian?) to the east. From this position they typically step upwards towards the west to their highest levels (Maastrichtian) close to the front of the 'inner flows' (Fig. 31); this feature indicates that the sills originated from feeder dykes in the central parts of the basins. In the Gjallar Ridge in the western parts of the Vøring Basin, the semi-horizontal intrusions cut across the eastward tilted bedding (see Figs 8 and 31). These intrusions in the Gjallar Ridge may have been fed from dykes in the Fenris graben to the west or from the base of the Gjallar Ridge itself.

Narrow, vertical zones of deteriorated seismic data, which are interpreted as wrench fractures, are thought to be the loci of the feeder dykes to the sill system. These are distributed over large parts of the basins and in many places coincide with locations of shifts in stratigraphic level of the sills, indicating that they are linked to zones of dilation accommodating the vertical steps in the dyke–sill system. The wrench movements have resulted in 'bulges', which appear as 'eye structures' in cross-section (Fig. 32). The largest eye structure is identified in the eastern part of the Møre Basin, and is believed to be the location of a major feeder dyke to the sills (Fig. 32). A row of strong magnetic anomalies coincides with the location and strike of this feature.

Such wrench-related eye structures (mainly in the Paleocene sequence) occur on a variety of scales and the transcurrent tectonic movements on these fractures are probably minor, but clearly affect all strata up to the uppermost Paleocene–lowermost Eocene units. This would indicate a Paleocene–Early Eocene age for the magmatic activity. There are no seismic indications that the magma ever reached the surface along these 'eye structures' in Paleocene–Eocene time. Many of the wrench zones have been reactivated later through Tertiary time without accompanying magmatism.

Discussion and plate tectonic relationships

The very thick Cretaceous sediment pile and the present great depths to the base of the

Fig. 32. Seismic line MB-08-92 showing a large 'eye structure' typical of the Møre and Vøring Basins (central part of figure). Here 'eye structure' affects mainly Paleocene and Eocene strata, with some later reactivation. Within the 'eye' the base of the Tertiary sequence is downflexed, whereas the Upper Paleocene and Eocene strata are upward bulging. The disturbed zone beneath 'eye structure' is interpreted to be major feeder dyke to the sill intrusions seen to protrude from it. Thin, slanted arrows point at near top Eocene. Small, vertical arrows point at near top Paleocene. Thick, vertical arrows point at base Tertiary. Open, wide arrows point at sills in the Cretaceous section. White triangles point at top of Eocene lavas of the 'inner flows'. (See Fig. 31 for line location.)

Cretaceous sequence in the Møre and Vøring Basins show that the crust beneath the basins was stretched and thinned significantly over an exceptionally wide rift area in the late Mid-Jurassic–Early Cretaceous rifting episode. The stretching was accompanied by normal faulting and fault-block rotation across the whole basin area, but without any major, basinward boundary faults. Instead, the basin margins apparently formed by symmetrical tilting and flexuring during Early Cretaceous time, indicating a less brittle behaviour of the crust than seems usual for narrower rifts. Together with the indications of a high elevation of the central basin areas during Mid- and Late Jurassic time, these features point towards a very high heat flow during the stretching phase. No attempt has been made to model how this mode of basin formation would fit with the models of crustal behaviour based on the recent concept of 'depth of necking' (Cloetingh *et al.* 1985; Braun & Beaumont 1989; Kooi & Cloetingh 1992). However, such models predict cases where post-rifting compression would add to the subsidence of a basin depending on the 'depth of necking' (Kooi & Cloetingh 1992). The post-rift Cretaceous history of the Vøring and Møre Basins involves subsidence to great depths of the base of the Cretaceous sequence, the development of relatively steep, polyphasally tilted (flexured) basin flanks, and internal folding, all of which indicate that intermittent phases of intraplate compression have played a role in the subsidence history of the two basins. This is discussed in the following.

The tectonic history of the Cretaceous and Tertiary periods seems to be closely linked to the plate tectonic evolution of Europe and the North Atlantic (Table 1). In Cretaceous times the Møre and Vøring Basins were part of a system of continental rift basins between Norway and Greenland (Fig. 33). Before break-up the Jan Mayen Lineament lined up with Kong Oscar Fjord in Greenland, and the southern part of the Cretaceous platform of the Vøring Marginal High was close to Liverpool Land and Jameson Land (Fig. 33). In the area between Liverpool Land and the Møre Marginal High, which was later rifted off to become the Jan Mayen Microcontinent, there was room for a Cretaceous basin. However, recent data from the Jan Mayen Microcontinent indicate only a moderately thick Mesozoic section there (Kuvaas & Kodaira 1997). These reconstructions are taken to imply the existence of a central platform area between Greenland and Norway throughout the Cretaceous period (Brekke *et al.* 1999). Svalbard was situated north of Greenland and the proto-Senja Fracture Zone was probably already established as a crustal tectonic lineament (Brekke & Riis 1987).

The tectonic pulses in Late Cretaceous times (post-Cenomanian–Paleocene) were probably coupled to the opening of the Labrador Sea and the associated anti-clockwise rotation of Greenland that was initiated in Cenomanian time (Srivastava & Tapscott, 1986) (Table 1). In the Vøring Basin this was reflected as tilting of basin margins, increased subsidence, faulting and regional folding. The different structural expressions indicate that the intra-plate stress regime, set up by the rotation of Greenland, changed through time relative to the Vøring Basin. It is possible that the end-Cenomanian tilting of the basin margins and the increase in basin subsidence were due to an initial phase of weak intraplate compression, inducing further crustal flexuring (Cloetingh *et al.* 1985; Kooi & Cloetingh 1992).

Faulting on the basin margin (Revfallet and Bremstein Fault Complexes) and associated growth of the Campanian sequence indicate a change to an extensional regime that lasted into latest Cretaceous (Maastrichtian?) time (Figs 1, 5, 6 and 17).

The folding of the Vøring Basin in latest Cretaceous time involved folds of wavelengths much less than the width of basin, and apparently did not contribute significantly to the overall deepening of the basin. However, the large Vigrid and Någrind Synclines constituted local depocentres and sites of further deepening by downwarp during this phase. The change in the scale of folding may have been due to the reactivation of the Fles Fault Complex creating a regional anisotropy along the basin axis, which controlled the distribution of folds. Except for the tilting and steepening of the basin flanks there are no signs of this Late Cretaceous tectonics in the Møre Basin, which implies that the Jan Mayen Lineament must have acted as a regional transfer zone during the whole of Late Cretaceous time. The Maastrichtian fold axes in the Vøring Basin are generally NE–SW, indicating an orientation of the principal compressional stress axis normal to the basin axis. This implies that the Jan Mayen Lineament at that time must have acted as a dextral transfer zone, whereas the southern, N–S trending segment of the Fles Fault Complex acted as a conjugate sinistral zone. The northern segment of the Fles Fault Complex is parallel to the fold axes, implying compression and reverse movements on reactivation.

Tectonic activity and uplift are also evident north of the Bivrost Lineament and in the

Table 1. *A regional comparison of plate tectonics with relevance to the evolution of the Møre and Vøring Basins*

MAGNETIC ANOMALY	TIME IN Ma	PERIOD	SERIES	MØRE BASIN	VØRING BASIN	NORWEGIAN-GREENLAND SEA AND RELATED AREAS	ALPS
	5	TERTIARY		LANDWARD UPLIFT AND WESTWARD TILTING	LANDWARD UPLIFT AND WESTWARD TILTING MINOR FAULTING ON NYK HIGH	GLACIATIONS, UPLIFT AND EROSION OF CIRCUM-BASINAL LAND AREAS	MAIN COMPRESSION IN JURA MOUNTAINS
5 6	24	TERTIARY	MIO	RAPID SUBSIDENCE TRANSPRESSIONAL DOMING ALONG JAN MAYEN LINEAMENT ? UPLIFT, EROSION ALONG FLANKS ONLY ?	RAPID SUBSIDENCE AND WESTWARD TILTING MINOR EROSION OF HIGHS TRANSPRESSIONAL DOMING ALONG JAN MAYEN AND SURT LINEAMENT, FLES FAULT ZONE, VØRING ESCARPMENT AND NORDLAND RIDGE ? UPLIFT AND EROSION ?	SUBSIDENCE OF NORWEGIAN SEA AND ICELAND-FÆROE RIDGE PRONOUNCED CHANGE IN OCEANIC CIRCULATION AND CLAIMATE ACTIVE SPREADING ON PRESENT KOLBEINSEY RIDGE ESTABLISHED JAN MAYEN RIDGE SEPARATES FROM GREENLAND PERIOD OF NO DISTINC SPREADING RIDGE WEST OF JAN MAYEN FRACTURE ZONE	"NEOGENE OROGENY" MAIN COMPRESSIONAL PHASE RELATED TO TERMINATION OF NORTHERN SUBDUCTION (HELVETIC NAPPES)
7 10 13	36	TERTIARY	OL	TRANSPRESSION FORMING DOMES AND FLEXURES (ALONG JAN MAYEN LINEAM, ONLY)	TRANSPRESSION REACTIVATING VØRING ESCARPMENT, FLES FAULT ZONE, JAN MAYEN LINEAMENT AND NORDLAND RIDGE. GROWTH OF DOMES AND FLEXURES	SPREADING IN NORWEGIAN SEA (ÆGIR RIDGE) TERMINATES SVALBARD SEPARATES FROM GREENLAND. EXTINCTION OF SEA-FLOOR SPREADING IN LABRADOR SEA AND BAFFIN BAY. GREENLAND PART OF NORTH AMERICA. SEA-FLOOR SPREADING CONTINUES IN NORWEGIAN-GREENLAND SEA	"PALEOGENE OROGENY" MAIN COMPRESSIONAL PHASE RELATED TO TERMINATION OF SOUTHERN SUBDUCTION (WESTERN, CENTRAL & EASTERN ALPS)
17 21 24	55	TERTIARY	EO	FAULTING ALONG FÆROE-SHETLAND ESCARPMENT VOLCANISM ON MØRE MARGINAL HIGH	FAULTING ALONG VØRING ESCARPMENT AND NYK HIGH VOLCANISM ON VØRING MARGINAL HIGH	FORMATION OF ICELAND-FÆROE RIDGE, STRIKE-SLIP BETWEEN SVALBARD AND GREENLAND. CHANGE IN DIRCTION OF MOTION OF GREENLAND. (?POSSIBLE INITIATION OF SPREADING TO THE WEST OF THE ÆGIR RIDGE?) ACTIVE SEA-FLOOR SPREADING IN NORWEGIAN-GREENLAND SEA AND DAVIS STRAIT. CHANGE IN DIRECTIONOF MOTION OF GREENLAND. STRIKE-SLIP BETWEEN SVALBARD AND GREENLAND	COMPRESSION (ULTRAHELVETIA, VALAIS, BRIANC.)
	66	TERTIARY	PAL	MINOR FAULTING	POSSIBLE GRABEN FORMATION ALONG VØRING ESCARPMENT. ELSEWHERE ONLY MINOR FAULTING UPLIFT AND EROSION OF WHOLE BASIN AREA	VOLCANISM IN DAVIS STRAIT, EASTERN GREENLAND, VØRING PLATEAU AND FÆROE ISLANDS. CHANGE IN DIRECTION OF MOTION BETWEEN GREENLAND AND NORTH AMERICA. STRIKE-SLIP BETWEEN SVALBARD AND GREENLAND	MINOR COMPRESSION? (VALAIS, BRIANC.)
31	71	CRETACEOUS	MAAST	STRIKE-SLIP REACTIVATION (ALONG JAN MAYEN LINEAM. ONLY)	SYNCLINES AND ANTICLINES FORMED DUE TO TRANSPRESSION ALONG MAIN STRUCTURES	SEAFLOOR SPREADING IN NORTHERN LABRADOR SEA RIFTING IN NORWEGIAN-GREENLAND SEA	
	83	CRETACEOUS	CAMP		SUBSIDENCE MAIN CRETACEOUS RIFTING BLOCK FAULTING ALONG NYK HIGH, GJALLAR RIDGE, REACTIVATION OF FLES FAULT ZONE	ACTIVE SEA-FLOOR SPREADING IN SOUTHERN LABRADOR SEA	? COMPRESSION (EASTERN & CENTRAL & ALPS) ?
34	91	CRETACEOUS	SANTON CONIAC TURON		TILTING OF TRØNDELAG PLATFORM AND GJALLAR RIDGE	RIFTING IN NORTHERN LABRADOR SEA AND NORWEGIAN-GREENLAND SEA. COUNTER-CLOCKWISE ROTATION OF GREENLAND	? MAIN CRETACEOUS COMPRESSIONAL PHASE (EASTERN & CENTRAL ALPS) ?
	97	CRETACEOUS	CEN			RIFTING BETWEEN GREENLAND AND LABRADOR AND RELATED VOLCANISM	
	109	CRETACEOUS	ALB		SUBSIDENCE		? COMPRESSION (EASTERN ALPS, VALAIS) ?
140		CRETACEOUS	APT	MINOR FAULTING ASSOCIATED WITH FLES FAULT ZONE?	MINOR FAULTING ON FLES FAULT ZONE?	INITIAL, RIFT RELATED VOLCANISM ON LABRADOR SHELF RIFTING IN ROCKALL TROUGH	
	122	CRETACEOUS	BAR		SUBSIDENCE		
	140	LATE JURASSIC		GRAVITATIONAL SLIDING ON MARGINS MAIN RIFTING PHASE, BLOCK FAULTING BASIN MARGIN FLEXURE	GRAVITATIONAL SLIDING ON MARGINS MAIN RIFTING PHASE. BLOCK FAULTING BASIN MARGIN FLEXURE		

Based on of Trumpy (1973), Talwani & Eldholm (1977), Eldholm & Thiede (1980), Srivastava & Tapscott (1986), and Dewey & Windley (1988).

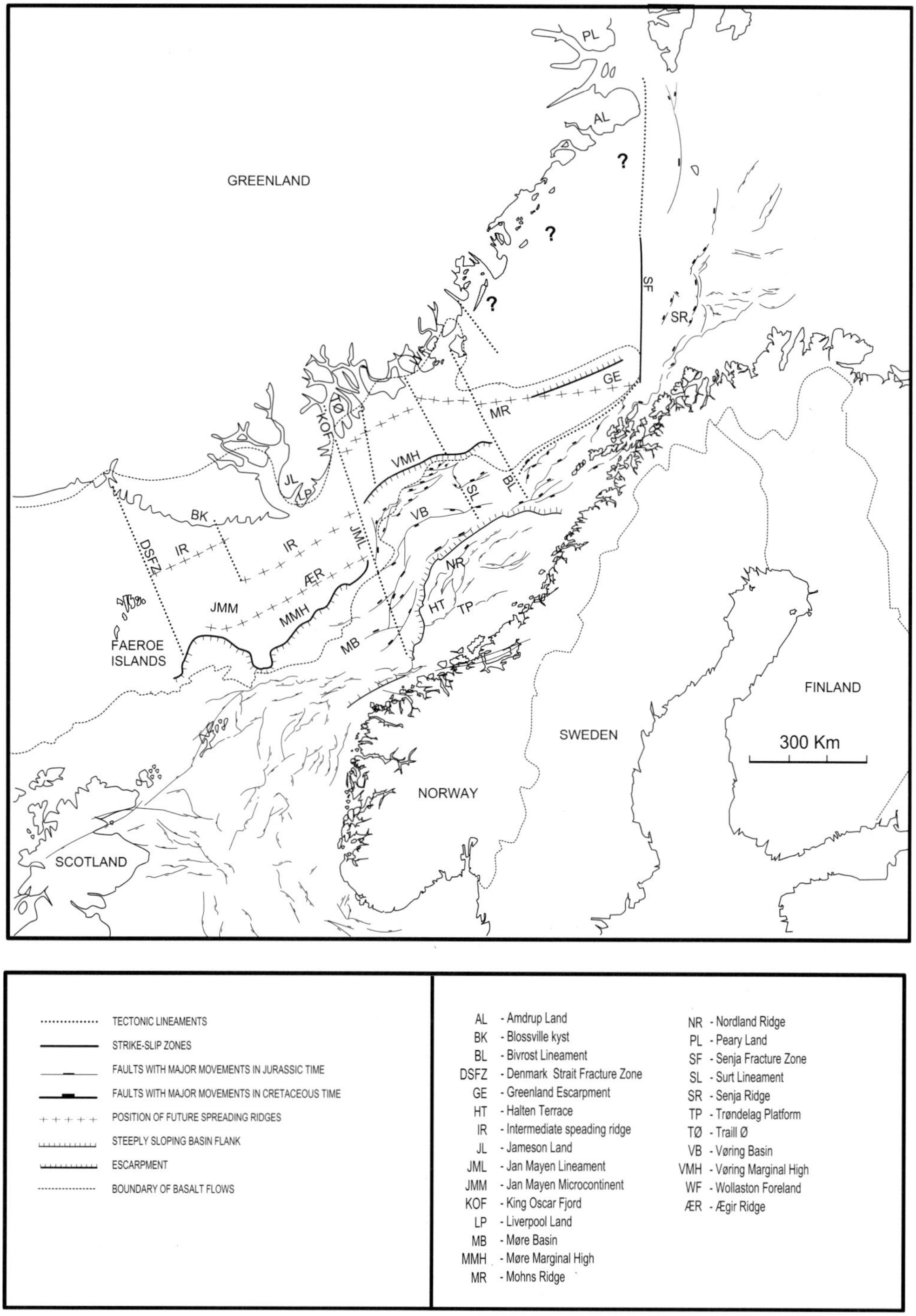

Fig. 33. Palinspastic reconstruction of the Norwegian Sea continental shelf at anomaly 24 time (Ypresian). Plate reconstruction taken from Srivastava & Tapscott (1986). Positions of the future intermediate spreading ridges compiled from Talwani & Eldholm (1977) and Nunns (1983). Other details of map based on author's own mapping and Talwani & Eldholm (1977), Larsen (1984), Stewart *et al.* (1992) and White (1992).

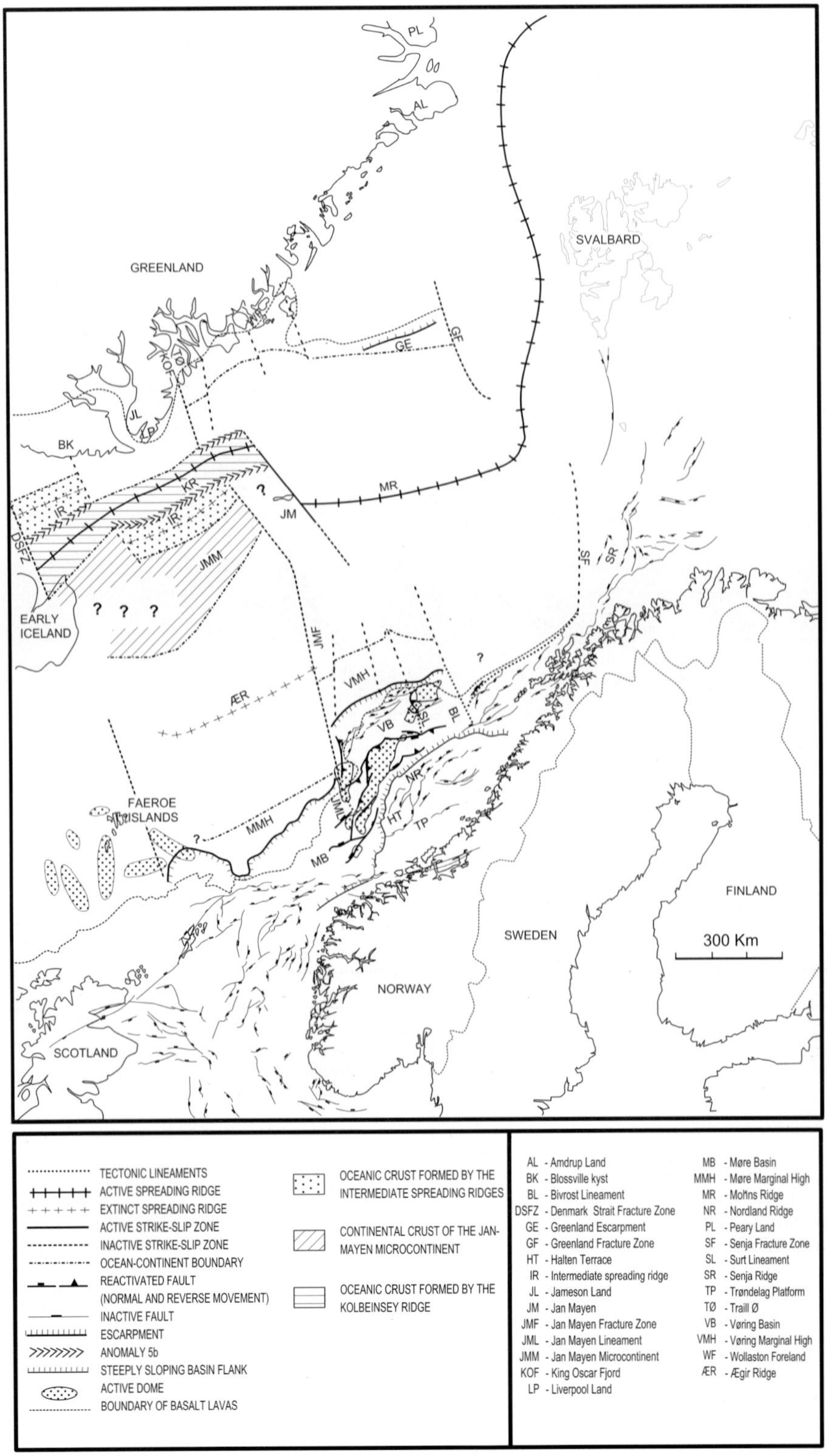
PL
AL
GREENLAND
SVALBARD
GE
GF
KOF
TØ
WF
JL
LP
BK
MR
KR
IR
IR
DSFZ
JM
JMM
? ? ?
EARLY ICELAND
SF
SR
JMF
VMH
ÆR
VB
SL
BL
NR
HT
TP
FAEROE ISLANDS
MMH
JML
MB
FINLAND
SWEDEN
300 Km
NORWAY
SCOTLAND
TECTONIC LINEAMENTS
ACTIVE SPREADING RIDGE
EXTINCT SPREADING RIDGE
ACTIVE STRIKE-SLIP ZONE
INACTIVE STRIKE-SLIP ZONE
OCEAN-CONTINENT BOUNDARY
REACTIVATED FAULT (NORMAL AND REVERSE MOVEMENT)
INACTIVE FAULT
ESCARPMENT
ANOMALY 5b
STEEPLY SLOPING BASIN FLANK
ACTIVE DOME
BOUNDARY OF BASALT LAVAS
OCEANIC CRUST FORMED BY THE INTERMEDIATE SPREADING RIDGES
CONTINENTAL CRUST OF THE JAN-MAYEN MICROCONTINENT
OCEANIC CRUST FORMED BY THE KOLBEINSEY RIDGE
AL - Amdrup Land
BK - Blossville kyst
BL - Bivrost Lineament
DSFZ - Denmark Strait Fracture Zone
GE - Greenland Escarpment
GF - Greenland Fracture Zone
HT - Halten Terrace
IR - Intermediate spreading ridge
JL - Jameson Land
JM - Jan Mayen
JMF - Jan Mayen Fracture Zone
JML - Jan Mayen Lineament
JMM - Jan Mayen Microcontinent
KOF - King Oscar Fjord
LP - Liverpool Land
MB - Møre Basin
MMH - Møre Marginal High
MR - Mohns Ridge
NR - Nordland Ridge
PL - Peary Land
SF - Senja Fracture Zone
SL - Surt Lineament
SR - Senja Ridge
TP - Trøndelag Platform
TØ - Traill Ø
VB - Vøring Basin
VMH - Vøring Marginal High
WF - Wollaston Foreland
ÆR - Ægir Ridge

Barents Sea (Brekke & Riis 1987). This indicates that the Late Cretaceous stress regime reactivated a system controlled by the Jan Mayen Lineament, the Fles Fault Complex and the proto-Senja Fracture Zone. Associated structures such as the Bivrost and Surt Lineaments were integral parts of the reactivation system. To link this system to the opening of the Labrador Sea, and to accommodate the difference between the quiet Møre Basin and the tectonically very active Vøring Basin, there must have been a tectonically active belt west of the Møre High, perhaps along the future break-up line and future Ægir Ridge (Figs 33 and 34).

The regional uplift and erosion of the continental margin at the beginning of Tertiary time was due to the increased heat flow of the area associated with the intense rifting centred west of the marginal highs, which led to the continental separation between Norway and Greenland in latest Paleocene–earliest Eocene time (Skogseid & Eldholm 1989). The severe extension and rifting also induced widespread intrusion activity along the axis and in western parts of the Vøring and Møre Basins. Although associated with the old zones of weakness, the intrusion activity was accompanied by minor, probably transtensional, tectonic movements, which were widespread and evenly distributed throughout the basins west of the Møre–Trøndelag and Fles Fault Complexes. This style of tectonic deformation and the volume of intrusive rocks indicate that at this stage the crust beneath the basins was generally weakened through heating. However, it was not sufficiently weakened to be the location of the continental break-up. This may indicate that the Jurassic–Cretaceous basin-forming process itself had changed the physical properties of the underlying crust to make it more resistant to rupture than the crust beneath the western basin bounding platforms (i.e. the present Vøring and Møre Marginal Highs).

The continental break-up was accompanied by enormous volumes of flood basalts, and the Ægir and Mohns Spreading Ridges were established (Talwani & Eldholm 1977; White 1992) (Table 1 Fig. 33). The regional uplift in Paleocene to earliest Eocene times, just before the break-up, affected areas even far away (900–1000 km) from the centre of the upwelling hot mantle material underneath the future spreading ridges. These distant areas (e.g. the Trøndelag Platform and the North Sea and surrounding mainland) experienced an extremely rapid subsidence in early Eocene times. Nadin & Kusznir (1995) explained this by a dynamic uplift of the crust by lateral fluid flow in the cold asthenosphere, which was induced by the dynamic flow-field set up by the upwelling hot mantle material in the distant upwelling centre before break-up. Subsequent to break-up the mantle flow was focused on the ridge axis, diverting cold asthenosphere back from the distal parts of the uplifted region, and thus explaining the rapid subsidence in those areas.

Up to Late Eocene time, there seems to be no obvious link between the tectonics of the Norwegian Sea and the compressional phases in the Alps (Table 1). But from Late Eocene time onwards the main orogenic phases in the Alps coincide with the timing of events in the Norwegian Sea. This may be due to the fact that the Norwegian Sea was now established as the plate boundary between the American and Eurasian plates, and that the relative motions between the two were changed because of the continuing collision between the Eurasian and Gondwanian plates. In particular, the subduction-related collisions of the Alps in Late Eocene and Miocene time caused changes in the relative plate motions, which were expressed as tectonic adjustments at the new plate boundary. A link between these events and the Alpine Orogeny and associated plate reorganization has been suggested also by Cloetingh *et al.* (1990) and Gradstein & Backstrøm (1996). As documented by Gradstein & Backstrøm (1996), the Late Eocene–Early Oligocene phase of flexuring and doming north of the Jan Mayen Lineament coincides with a regional hiatus on the eastern basin margins across major parts of the North Sea and on the Halten Terrace. This Late Eocene Alpine orogenic phase may be correlated with the end of the sea-floor spreading in the Labrador Sea and the final passage of Svalbard past northern Greenland, events that left Greenland as an integral part of the American plate (Eldholm & Thiede 1980). This also corresponds in time to the change in azimuth of the Jan Mayen Fracture Zone from NNW (eastern branch) to NW (western branch) (Fig. 34),

Fig. 34. Palinspastic reconstruction of the Norwegian Sea continental shelf at anomaly 5 time (Late Miocene). Plate reconstruction taken from Srivastava & Tapscott (1986). Positions of the intermediate spreading ridges and the Jan Mayen Microcontinent compiled from Talwani & Eldholm (1977) and Nunns (1983). Tertiary domes around Faeroe Islands based on anticlinal axes of Boldreel & Andersen (1993). Other details of map based on author's own mapping and Talwani & Eldholm (1977), Larsen (1984), Stewart *et al.* (1992) and White (1992).

implying a shift in the relative plate motions (Talwani & Eldholm 1977).

At the end of Oligocene time the sea-floor spreading along the Ægir Ridge slowed down and stopped (between anomalies 20 and 7). According to Eldholm & Thiede (1980), the present Kolbeinsey Ridge was established only just before anomaly 5 (10 Ma). In the period between anomaly 7 (26 Ma) and anomaly 5 (i.e. Early to late Mid-Miocene time) there were no distinct spreading axis south of the Jan Mayen Fracture Zone to replace the Ægir Ridge (Talwani & Eldholm 1977; Eldholm & Thiede 1980). However, Talwani & Eldholm (1977) suggested the existence of an intermediate spreading axis forming a local oceanic basin SW of Jan Mayen in the interval between anomalies 6A (21 Ma) and 5D (18 Ma), bracketed by intermittent periods of no spreading. Vogt *et al.* (1980) postulated spreading on the Kolbeinsey Ridge since anomaly 6C (23.5 Ma), and Nunns (1983) proposed a model with conjugate wedges of sea-floor spreading on both sides of the Jan Mayen Microcontinent in the interval between anomalies 20 and 7. However, the magnetic data shown by Nunns (1983) do not document any well-defined symmetry along the Kolbeinsey Ridge before anomaly 5B (15 Ma), nor any anomaly older than anomaly 7 (26 Ma) west of the Jan Mayen Microcontinent. Between what is interpreted by Nunns as anomalies 5B and 7 the pattern seems to be more in line with an intermediate axis SSW of Jan Mayen as proposed by Talwani & Eldholm (1977). In their model the intermediate ridge became extinct along their anomaly 5D, which corresponds to anomaly 6 of Nunns (1983). The implication of this model is a local basin with a set of anomalies with a symmetrical age distribution around that central anomaly. The identification of anomalies 6 and 6B by Nunns (1983) further SW, close to the present mainland of Greenland, may then be explained by the postulation (by the author) of another, coincident intermediate spreading ridge segment in that area. This model implies the existence of two en echelon local oceanic spreading basins west of the Jan Mayen Ridge in the period between anomalies 6A and 5D (as correlated by Talwani & Eldholm (1977)) (Fig. 34). The offset between the two segments of the intermediate ridge was taken up by a transverse zone parallel to the Jan Mayen Fracture Zone (Fig. 34). This offset was later cut by the Kolbeinsey Ridge, which was established by a new ridge shift sometime between anomalies 5B and 5 (Fig. 34).

The fan-shaped anomaly pattern around the Ægir Ridge was explained by Nunns (1983) by an anti-clockwise rotation of the Jan Mayen Microcontinent. This seems to be impossible, as the microcontinent was separated from Greenland only subsequent to the extinction of the Ægir Ridge. Such a rotation would have to take place between anomalies 22 and 7 to explain the fan. A more likely explanation is that that was a period of rifting and severe stretching and extension of the continental crust in the southern part of the future Jan Mayen Microcontinent, which compensated for the fan geometry in the ocean crust. The model with two en echelon intermediate spreading ridges implies a substantial size of the southern part of the Jan Mayen Microcontinent (Fig. 34), which would fit with such an early stretching episode. This is also in agreement with the mapping of the microcontinent based on seismic reflection data (Gudlaugsson *et al.* 1988): the present water depth and the intensity of past block-faulting increase considerably southwards on the microcontinent, indicating increased thinning and stretching of the crust in that direction. Another implication of such a model is that Iceland may be underlain by continental crust or fragments of such crust (Fig. 34).

From this discussion of the literature it seems safe to conclude that the interval between anomalies 7 and 5B was one of slow or no spreading at the Ægir Ridge and a complex pattern of temporary spreading centres west of Jan Mayen. The picture is complicated by the uncertain timing of the rifting east of Liverpool Land and the subsequent separation of the Jan Mayen Microcontinent (Gudlaugsson *et al.* 1988). It may be possible that the rifting in the continental crust compensated for periods of no spreading in the interval between the extinction of the Ægir Ridge and the final establishment of the Kolbeinsey Ridge.

This reorganization of the spreading system coincided with the onset of the Neogene Orogeny of the Alps (Table 1; Trumpy 1973; Dewey & Windley 1988) and the Mid-Miocene phase of compression in the Vøring Basin. This indicate that the Alpine Neogene Orogeny probably affected the relative plate motions between the Eurasian and American plates. This event was also accompanied with the earlier mentioned 7 Ma period of regional uplift, erosion, non-deposition and subsequent rapid subsidence seen all over the Norwegian Continental Shelf and adjacent mainland. In the absence of evidence of a regional heat pulse, the regional rapid uplift and subsidence may be explained by the dynamic fluid mantle flow model of Nadin & Kusznir (1995). After the extinction of the Ægir Ridge, the centre of mantle upwelling shifted to the area beneath the future Kolbeinsey Ridge. The spreading activity between the Denmark Strait Fracture Zone and Jan Mayen Fracture Zone in the intermittent period (Early to Late Miocene time) was slow and discontinuous (see above), and may have induced central uplift just as in the situation immediately before the continental break-up in Eocene time. This may in turn have induced a dynamic lateral flow in the cold asthenosphere outwards from the upwelling centre, resulting in uplift in the distant flanks: the Vøring and Møre Basins and the marginal highs. The subsequent rapid subsidence followed from the final establishment of the Kolbeinsey Ridge at the end of the overall plate motion adjustment in late Mid-Miocene time.

A phase of strong and widespread Miocene compression was also documented around the Faeroe Islands, by Boldreel & Andersen (1993, 1994) and Andersen & Boldreel (1995). Those workers have also identified phases of compression in Early Eocene and Oligocene time, which fit with the timing of tectonic activity in the Vøring Basin. As there is no evidence of Tertiary compressional tectonics in the Møre Basin, the Jan Mayen Fracture Zone and the Jan Mayen Lineament must have acted as a transfer zone between the Faeroe Islands area and the Vøring Basin during Tertiary time (as also inferred for Late Cretaceous time).

After late Mid-Miocene time the whole region again started to subside and the tectonically created ocean-bottom topography was filled in until the period of non-deposition in late Early–early Late Pliocene time. The regional Late Pliocene hiatus or unconformity has been well documented by Eidvin & Riis (1989, 1991, 1992) and Eidvin *et al.* (1993). A phase of considerable uplift and deep erosion of the mainland started in Late Pliocene time, resulting in the accumulation and progradation of a huge sedimentary wedge all along the subsiding continental shelf (Riis & Fjeldskaar 1992). However, in the mean time the thermal gradient in the shelf has continued to decrease, and the overburden accumulated since Late Pliocene time has still not brought the fossil opal transition level back down to the temperature of its formation.

The mechanism behind the strong Pliocene uplift of the mainland and the symmetrical subsidence of the continental shelf is still poorly understood (see discussion by Van der Beek (1995) and Riis (1996)).

The author is grateful to Norsk Hydro, Saga Petroleum, Statoil and the Norwegian Petroleum Directorate for granting permission to publish critical seismic data. L. O. Boldreel and T. Torske are acknowledged for their thorough review of the manuscript.

References

ANDERTON, R., BRIDGES, P. H., LEEDER, M. R. & SELLWOOD, B. W. 1979. *A Dynamic Stratigraphy of the British Isles. A Study in Crustal Evolution.* George Allen & Unwin, London.

ANDERSEN, M. S. & BOLDREEL, L. O. 1995. Tertiary compression structures in the Faeroe–Rockall area. *In:* SCRUTTON, R. A., SCHIMMIELD, G. B. & TUDHOPE, A. W. (eds) *The Tectonics, Sedimentation, and Palaeoceanography of the North Atlantic Region.* Geological Society, London, Special Publications, **90**, 215–216.

AVERY, O. E., BURTON, G. D. & HEIRTZLER, J. R. 1968. An aeromagnetic survey of the Norwegian Sea. *Journal of Geophysical Research*, **73**, 4583–4600.

BERING, D. 1992. The orientation of minor fault plane striae and the associated deviatoric stress tensor as a key to the fault geometry in part of the Møre–Trøndelag Fault Zone, onshore central Norway. *In:* LARSEN, R. M., BREKKE, H., LARSEN, B. T. & TALLERAAS, E. (eds) *Structural and Tectonic Modelling and its Application to Petroleum Geology.* Norwegian Petroleum Society Special Publication, **1**, 83–90.

BLYSTAD, P., BREKKE, H., FÆRSETH, R. B., LARSEN, B. T., SKOGSEID, J. & TØRUDBAKKEN, B. 1995. *Structural Elements of the Norwegian Continental Shelf. Part II: The Norwegian Sea Region.* Norwegian Petroleum Directorate Bulletin, **8**.

BOLDREEL, L. O. & ANDERSEN, M. S. 1993. Late Palaeocene to Miocene compression in the Faeroe–Rockall area. *In*: PARKER, J. R. (ed.) *Petroleum Geology of Northwest Europe: Proceedings of the 4th Conference.* Geological Society, London, 1025–1034.

—— & ——1994. Tertiary development of the Faeroe–Rockall Plateau based on reflection seismic data. *Bulletin of the Geological Society of Denmark*, **41**, 162–180.

BOTT, M. H. P. 1973. The evolution of the Atlantic north of the Faeroe Islands. *In*: TARLING, D. H. & RUNCORN, S. K. (eds) *Implications of Continental Drift to the Earth Sciences*, **1**, Academic Press, New York, 175–189.

BRAUN, J. & BEAUMONT, C. 1989. A physical explanation of the relation between the flank uplifts and the breakup unconformity at rifted continental margins. *Geology*, **17**, 760–764.

BREKKE, H. & RIIS, F. 1987. Tectonics and basin evolution of the Norwegian shelf between 62°N and 72°N. *Norsk Geologisk Tidsskrift*, **67**, 295–322.

——, DAHLGREN, S., NYLAND, B. & MAGNUS, C. 1999. The prospectivity of the Vøring and Møre Basins on the Norwegian Sea continental margin. *In*: FLEET, A. & BOLDY, S. A. R. (eds) *Petroleum Geology of Northwest Europe: Proceedings of the 5th Conference.* Geological Society, London, 261–274.

BUKOVICS, C., SHAW, N. D., CARTIER, E. G. & ZIEGLER, P. A. 1984. Structure and development of the Mid-Norway continental margin. *In*: SPENCER, A. M., JOHNSEN, MØRK, A., NYSÆTHER, E., SONGSTAD, P. & SPINNANGR, Å. (eds) *Petroleum Geology of the North European Margin*. Graham and Trotman, London, 407–423.

BØEN, F., EGGEN, S. & VOLLSET, J. 1984. Structures and basins of the margin from 62°N to 69°N and their development. *In*: SPENCER, A. M., JOHNSEN, MØRK, A., NYSÆTHER, E., SONGSTAD, P. & SPINNAGR, Å. (eds) *Petroleum Geology of the North European Margin*. Graham and Trotman, London, 271–284.

CLOETINGH, S., GRADSTEIN, F. M., KOOI, H., GRANT, A. C. & KAMINSKI, M. A. 1990. Plate reorganization. A cause of rapid late Neogene subsidence and sedimentation around the North Atlantic? *Journal of the Geological Society, London*, **147**, 495–506.

——, MCQUEEN, H. & LAMBRECK, K. 1985. On a tectonic mechanism for regional sealevel variations. *Earth and Planetary Science Letters*, **75**, 157–164.

DALLAND, A., WORSLEY, W. & OFSTAD, K. (eds) 1988. *A Lithostratigraphic Scheme for the Mesozoic and Cenozoic Succession Offshore Mid- and Northern Norway*. Norwegian Petroleum Directorate Bulletin, **4**.

DEWEY, J. F. & WINDLEY, B. F. 1988. Palaeocene–Oligocene tectonics of NW Europe. *In*: MORTON, A. C. & PARSON, L. M. (eds) *Early Tertiary Volcanism and the Opening of the NE Atlantic*. Geological Society, London, Special Publications, **39**, 25–31.

DORÉ, A. G. 1991. The structural foundation and evolution of Mesozoic seaways between Europe and the Arctic. *Palaeogeography, Palaeoclimatology and Palaeoceanography*, **87**, 441–446.

—— & LUNDIN, E. R. 1996. Cenozoic compressional structures of the NE Atlantic margin: nature, origin and potential significance for hydrocarbon exploration. *Petroleum Geoscience*, **2**, 299–311.

EIDVIN, T. & RIIS, F. 1989. *Nye dateringer av de tre vestligste borehullene i Barentshavet. Resultater og konsekvenser for den tertiære hevingen*. Norwegian Petroleum Directorate Contribution, **27**.

—— & ——1991. *En biostratigrafisk analyse av tertiære sedimenter på kontinentalmarginen av Midt-Norge, med hovedvekt på øvre pliocene vifteavsetninger*. Norwegian Petroleum Directorate Contribution, **29**.

—— & ——1992. *En biostratigrafisk og seimostratigrafisk analyse av tertiære sedimenter i nordlige deler av Norskerenna, med hovedvekt på øvre pliocene vifteavsetninger*. Norwegian Petroleum Directorate Contribution, **32**.

——, ——, BEYER, C. & RUNDBERG, Y. 1993. *Et multidisiplinært, stratigrafisk studie av neogene sedimenter i sentrale deler av Nordsjøen (Ekofiskfeltet)*. Norwegian Petroleum Directorate Contribution, **34**.

ELDHOLM, O. & THIEDE, J. 1980. Cenozoic continental separation between Europe and Greenland. *Palaeogeography, Palaeoclimatology, Palaeoecology*, **30**, 243–259.

——, —— & TAYLOR, E. 1989. Evolution of the Vøring volcanic margin. *In*: ELDHOLM, O., THIEDE, J., TAYLOR, E. *et al.* (eds) *Proceedings of the Ocean Drilling Program, Scientific Results*, **104**. Ocean Drilling Program, College Station, TX, 1033–1065.

——, ——, —— *et al.* 1987. *Proceedings of the Ocean Drilling Program, Initial Reports*, **104**. Ocean Drilling Program, College Station, TX.

FAIRHEAD, J. D., ASPLAND, S. W., CAMPBELL, S. J. *et al.* 1994. *West–East European Gravity Project*. Geophysical Exploration Technology Ltd, University of Leeds.

GABRIELSEN, R. H., FÆRSETH, R., HAMAR, G. & RØNNEVIK, H. C. 1984. Nomenclature of the main structural features on the Norwegian Continental Shelf north of the 62nd parallel. *In*: SPENCER, A. M., JOHNSEN, MØRK, A., NYSÆTHER, E., SONGSTAD, P. & SPINNANGR, Å. (eds) *Petroleum Geology of the North European Margin*. Graham and Trotman, London, 41–60.

——, ——, JENSEN, L. N., KALHEIM, J. E. & RIIS, F. (eds) 1990. *Structural Elements of the Norwegian Continental Shelf. Part I: The Barents Sea Region*. Norwegian Petroleum Directorate Bulletin, **6**.

GJELBERG, J., DREYER, T., HØIE, A., TJELLAND, T. & LILLENG, T. 1987. Late Triassic to Mid-Jurassic sandbody development on the Barents and Mid-Norwegian shelf. *In*: BROOKS, J. & GLENNIE, K. (eds) *Petroleum Geology of North West Europe*. Graham and Trotman, London, 1105–1129.

GRADSTEIN, F. M. & BACKSTRØM, S. A. 1996. Cainozoic biostratigraphy and palaeobathymetry, northern North Sea and Haltenbanken. *Norsk Geologisk Tidsskrift*, **76**, 3–32.

GRØNLIE, A., NAESER, C. W., NAESER, N.D, MITCHELL, J. G., STURT, B. A. & INESON, P. R. 1994. Fission-track and K–Ar dating of tectonic activity in a transect across the Møre–Trøndelag Fault Zone, central Norway. *Norsk Geologisk Tidsskrift*, **74**, 24–34.

GUDLAUGSSON, S. T., GUNNARSSON, K., SAND, M. & SKOGSEID, J. 1988. Tectonic and volcanic events at the Jan Mayen Ridge microcontinent. *In*: MORTON, A. C. & PARSON, L. M. (eds) *Early Tertiary Volcanism and Opening of the NE Atlantic*. Geological Society, London, Special Publications, **39**, 85–93.

HINZ, K., DOSTMAN, H. J. & HANISCH, J. 1982. Structural framework of the Norwegian Sea. *In*: *Offshore Northern Seas*. Norwegian Petroleum Society, Oslo.

——, —— & ——1984. Structural elements of the Norwegian Sea continental margin. *Geologisches Jahrbuch*, **A75**, 193–211.

HUSEBYE, E. S., GJØYSTDAL, H., BUNGUM, H. & ELDHOLM, O. 1975. The seismicity of the Norwegian–Greenland Sea. *Tectonophysics*, **26**, 55–70.

JOHNSON, G. L. & HEEZEN, B. G. 1967. The morphology and evolution of the Norwegian–Greenland Sea. *Deep-Sea Research*, **14**, 755–771.

JONGPIER, K., RUI, J. C. & GRUE, K. 1996. Triassic to Early Cretaceous stratigraphic and structural development of the northeastern Møre Basin margin, off Mid-Norway. *Norsk Geologisk Tidsskrift*, **76**, 199–214.

JORDT, H., FALEIDE, J. I., BJØRLYKKE, K. & IBRAHIM, M. T. 1995. Cenozoic sequence stratigraphy of the central and northern North Sea Basin; tectonic development, sediment distribution and provenance areas. *Marine and Petroleum Geology*, **12**, 845–879.

KITTELSEN, J. E., OLSEN, R. R., MARTEN, R. F., HANSEN, E. K. & HOLLINGSWORTH, R. R. 1999. The first deepwater well in Norway and its implications for the Upper Cretaceous Play, Vøring Basin. *In*: FLEET, A. & BOLDY, S. A. R. (eds) *Petroleum Geology of Northwest Europe: Proceedings of the 5th Conference*. Geological Society, London.

KOOI, H. & CLOETINGH, S. 1992. Lithospheric necking and regional isostasy at extensional basins 2. Stress-induced vertical motions and relative sea level changes. *Journal of Geophysical Research*, **97**, 17573–17591.

KUVAAS, B. & KODAIRA, S. 1997. The formation of the Jan Mayen microcontinent: the missing piece in the continental puzzle between the Møre–Vøring Basins and East Greenland. *First Break*, **15**(7), 239–247.

LARSEN, H. C. 1984. Geology of the East Greenland Shelf. *In*: SPENCER, A. M., JOHNSEN, MØRK, A., NYSÆTHER, E., SONGSTAD, P. & SPINNANGR, Å. (eds) *Petroleum Geology of the North European Margin*. Graham and Trotman, London, 329–339.

—— & MARCUSSEN, C. 1992. Sill-intrusion, flood basalt emplacement and deep crustal structure of the Scoresby Sund region, East Greenland. *In:* STOREY, B. C., ALABASTER, T. & PANKHURST, R. J. (eds) *Magmatism and the Causes of Continental Break-up*. Geological Society, London, Special Publications, **68**, 365–386.

——, HAMBERG, L., OLAUSSEN, S., PREUSS, T. & STEMMERIK, L. 1999. Sandstone wedges of Cretaceous–Lower Teriary Kangerlassuaq Basin, East Greenland – outcrop analogues to the offshore North Atlantic. *In*: FLEET, A. & BOLDY, S. A. R. (eds) *Petroleum Geology of Northwest Europe: Proceedings of the 5th Conference*. Geological Society, London.

MEYER, O., VOPPEL, D., FLEISCHER, U., CLOSS, H. & GERKE, K. 1972. Results of bathymetric, magnetic and gravimetric measurements between Iceland and 70°N. *Deutsche Hydrographische Zeitschrift*, **25**, 193–201.

MJELDE, R., SELLEVOLL, M. A., SHIMAMURA, H., IWASAKI, T. & KANAZAWA, T. 1993. Ocean bottom seismographs used in a crustal study of an area covered with flood-basalt off Lofoten, N. Norway. *Terra Nova*, **5**, 76–84.

MUTTER, J. C., TALWANI, M. & STOFFA, P. L. 1984. Evidence for a thick oceanic crust adjacent to the Norwegian margin. *Journal of Geophysical Research*, **89**, 483–502.

NADIN, P. A. & KUSZNIR, N. J. 1995. Palaeocene uplift and Eocene subsidence in the northern North Sea Basin from 2D forward and reverse stratigraphic modelling. *Journal of the Geological Society of London*, **152**, 833–848.

NUNNS, A. 1983. The structure and evolution of the Jan Mayen Ridge and surrounding regions. *In*: WATKINS, J. S. & DRAKE, C. L. (eds) *Continental Margin Geology*. American Association of Petroleum Geologists, Memoirs, **34**, 193–208.

OLESEN, O., BRÆKSTAD, G., GELLEIN, J., HÅBREKKE, H., KIHLE, O., SKILBREI, J. R. & SMETHURST, M. A. 1997. *Magnetic anomaly map, Norway and adjacent ocean areas. Scale 1:3 million*. Geological Survey of Norway.

PHARAOH, T. C., MORRIS, J. H., LONG, C. B. & RYAN, P. D. (compilers) 1996. *Tectonic map of Britain, Ireland and adjacant areas. 1:1 500 000*. British Geological Survey, Keyworth.

PLANKE, S., SKOGSEID, J. & ELDHOLM, O. 1991. Crustal structure off Norway, 62°–70°N. *Tectonophysics*, **189**, 91–107.

PRICE, I. & RATTEY, R. P. 1984. Cretaceous tectonics off Mid-Norway: implications for the Rockall and Faeroe–Shetland troughs. *Journal of the Geological Society, of London*, **141**, 985–992.

RIIS, F. 1996. Quantification of Cenozoic vertical movements of Scandinavia by correlation of morphological surfaces with offshore data. *Global and Planetary Change*, **12**, 331–357.

—— & FJELDKSAAR, W. 1992. On the magnitude of the Late Tertiary and Quaternary erosion and its significance for the uplift of Scandinavia and the Barents Sea. *In*: LARSEN, R. M., BREKKE, H., LARSEN, B. T. & TALLERAAS, E. (eds) *Structural and Tectonic Modelling and its Application to Petroleum Geology*. Norwegian Petroleum Society Special Publication, **1**, 163–185.

—— & JENSEN, L. N. 1992. Introduction: measuring uplift and erosion – proposal for a terminology. *Norsk Geologisk Tidsskrift*, **72**, 223–228.

ROALDSET, E. & HE, W. 1995. *Silica-phase transformation of opal-A to opal-CT to quartz – an experimental approach*. Report, Department of Geology and Mineral Resources Engineering, Norwegian Institute of Technology, Trondheim.

RUNDBERG, Y. 1989. *Tertiary and sedimentary history and basin evolution of the Norwegian North Sea between 60°N and 62°N. An integrated approach*. PhD thesis, Norwegian Institute of Technology, Trondheim.

——, OLAUSSEN, S. & GRADSTEIN, F. M. 1995. Incision of Oligocene strata, clue for northern North Sea Miocene uplift and formation of the Utsira sands (abstract). *Geonytt*, **22**, 62.

RØNNEVIK, H. C. & NAVRESTAD, H. 1977. Geology of the Norwegian shelf between 62°N and 69°N. *GeoJournal*, **1**(1), 33–46.

SKOGSEID, J. 1994: Dimensions of the Late Cretaceous–Paleocene Northeast Atlantic rift derived from Cenozoic subsidence. *Tectonophysics*, **240**, 225–247.

—— & ELDHOLM, O. 1987, Early Cenozoic crust at the Norwegian continental margin and the conjugate Jan Mayen Ridge. *Journal of Geophysical Research*, **92**, 11 471–11 491.

—— & ——1989. Vøring Continental Margin: seismic interpretation, stratigraphy and vertical movements. *In*: ELDHOLM, O., THIEDE, J., TAYLOR, E. *et al.* (eds) *Proceedings of the Ocean Drilling Program, Scientific Results*, **104**. Ocean Drilling Program, College Station, TX, 993–1030.

——, PEDERSEN, T. & LARSEN, V. B. 1992. Vøring Basin: subsidence and tectonic evolution. *In*: LARSEN, R. M., BREKKE, H., LARSEN, B. T. & TALLERAAS, E. (eds) *Structural and Tectonic Modelling and its Application to Petroleum Geology*. Norwegian Petroleum Society Special Publication, **1**, 55–82.

SMYTHE, D. K., CHALMERS, J. A., SKUCE, A. G., DOBINSON, A. & MOULD, A. S. 1983. Early opening of the North Atlantic – I. Structure and origin of the Faeroe–Shetland Escarpment. *Geophysical Journal of the Royal Astronomical Society*, **72**, 373–398.

SRIVASTAVA, S. P. & TAPSCOTT, C. R. 1986. Plate kinematics of the North Atlantic. *In*: VOGT, P. R. & TUCHOLKE, B (eds). *The Western North Atlantic Region*. The Geological Society of America, The Geology of North America, **M**, 379 – 404.

STEWART, I. J., RATTEY, R. P. & VANN, I. R. 1992. Structural style and the habitat of hydrocarbons in the North Sea. *In*: LARSEN, R. M., BREKKE, H., LARSEN, B. T. & TALLERAAS, E. (eds) *Structural and Tectonic Modelling and its Application to Petroleum Geology*. Norwegian Petroleum Society Special Publication, **1**, 197–220.

SURLYK, F. 1990. Timing, style and sedimentary evolution of Late Palaeozoic–Mesozoic extensional basins of East Greenland. *In*: HARDMAN, R. F. P. & BROOKS, J. (eds) *Tectonic Events Responsible for Britain's Oil and Gas Reserves*. Geological Society, London, Special Publications, **55**, 107–125.

——1991. Sequence stratigraphy of the Jurassic–Lowermost Cretaceous of East Greenland. *The AAPG Bulletin*, **75**, 1468–1488.

——, PIASECKI, S., ROLLE, F., STEMMERIK, L., THOMSEN, E. & WRANG, P. 1984. The Permian basin of East Greenland. *In*: SPENCER, A. M., JOHNSEN, MØRK, A., NYSÆTHER, E., ONGSTAD, P. & SPINNAGR, Å. (eds) *Petroleum Geology of the North European Margin*. Graham and Trotman, London, 303–315.

TALWANI, M. & ELDHOLM, O. 1972. The continental margin off Norway: a geophysical study. *Geological Society of America Bulletin*, **83**, 3375–3608.

—— & ——1977. Evolution of the Norwegian–Greenland Sea. *Geological Society of America Bulletin*, **88**, 969–999.

——, MUTTER, J. C. & HINZ, K. 1983. Ocean–continent boundary under the Norwegian continental margin. *In*: BOTT, M. H. P., SAXOV, S., TALWANI, M. & THIEDE, J. (eds) *Structure and Development of the Greenland–Scotland Ridge* – New Methods and Concepts. Plenum, New York, 121–131.

——, UDINTSEV, G. *et al.* 1976. *Initial Reports of the Deep Sea Drilling Project*, **38**, US Government Printing Office, Washington, DC.

TORSKE, T. & PRESTVIK, T. 1991. Mesozoic detachment faulting between Greenland and Norway: inferences from Jan Mayen Fracture Zone system and associated alkalic volcanic rocks. *Geology*, **19**, 481–484.

TRUMPY, R. 1973. The timing of orogenic events in the Central Alps. *In*: DE JONG, K. A. & SCHOLTEN, R. S. (eds) *Gravity and Tectonics*. Wiley, New York, 229–252.

VAN DER BEEK, P. A. 1995. Mechanisms of Neogene tectonic uplift in southern Norway. *In*: *Tectonic evolution of continental rifts – inferences from numerical modelling and fission track thermochronology*. PhD thesis, Research School of Sedimentary Geology (NSG), Amsterdam.

VOGT, P. R., JOHNSON, G. L. & KRISTJANSSON, L. 1980. Morphology and magnetic anomalies north of Iceland. *Journal of Geophysics*, **47**, 67–80.

WHITE, R. S. 1992. Crustal structure and magmatism of North Atlantic continental margins. *Journal of the Geological Society, London*, **149**, 841–854.

Late Cretaceous and Tertiary structural evolution of the northeastern part of the Vøring Basin, Norwegian Sea

TOMMY EGEBJERG MOGENSEN, RUNE NYBY, RIDVAN KARPUZ & PÅL HAREMO

Norsk Hydro a.s. Exploration, Harstad, P.O. Box 31, N-9401 Harstad, Norway (e-mail: Tommy.Mogensen.Egebjerg@hydro.com)

Abstract: The Late Cretaceous–Tertiary structural evolution of the northeastern part of the Vøring Basin, mid-Norway, is highly complicated. Although tectonic activity occurred throughout Cretaceous time in much of the Vøring Basin, including the Gjallar Ridge and along the Fles Fault Zone, in the Vema Dome–Nyk High area evidence of such activity is not observed until the latest Maastrichtian time. In the Vema Dome–Nyk High area, several faults with both a NW–SE orientation and a NE–SW orientation have experienced lateral movements. The NE–SW orientation is the old Caledonian trend. The complex structural evolution of the Vema Dome–Nyk High area is probably related to the existence of continental weakness zones, which are the onshore extension of known oceanic fracture zones. The two lineaments that delineate these weakness zones, and that delineate the Vema Dome–Nyk High area, namely the Bivrost and Surt Lineaments, diverge slightly from one another toward the SE. This complex structural framework of NW–SE oriented lineaments and NE–SW oriented deep-seated basement structures of Caledonian compressional and/or Mesozoic rift origin, facilitated minor clockwise rotation of the area between the two lineaments during the extensional regime before the break-up of the North Atlantic, as well as during the compressional regime that has been proposed after the break-up.

Several regional tectonic studies have been carried out around the Vøring Basin on the mid-Norway continental margin (Fig. 1) (Brekke & Riis 1987; Doré & Gage, 1987; Planke *et al.* 1991; Skogseid *et al.* 1992; Hinz *et al.* 1993; Skogseid 1994). More detailed local structural studies of the area have been hampered by the lack of closely spaced reflection seismic surveys. Before the Norwegian 15th licensing round, a large number of reflection seismic data were acquired in most of the Vøring Basin, which made it possible to carry out a more detailed structural analysis in several key tectonic areas, such as the Vema Dome–Nyk High area (Fig. 2) (Blystad *et al.* 1995). Several papers have dealt with the structural evolution of this area (Bjørnseth *et al.* 1997; Roberts *et al.* 1997; Walker *et al.* 1997) and some studies, such as that by Lundin & Doré (1997) have focused on the structural role of the lineaments seen in the Norwegian Sea. All studies agree on a complicated tectonic evolution of the Vema Dome–Nyk High area and that the structural evolution varies in several ways from that of the rest of the Vøring Basin. This paper addresses the structural setting before and after break-up in and around the Vema Dome–Nyk High, with its basis in depocentre configuration and fault orientations, and the structural influence by the Surt and Bivrost Lineaments.

Late Cretaceous–Paleocene structural evolution (pre-break-up)

Most of the structural elements in the Vøring Basin, e.g. the Gjallar Ridge and Fles Fault Zone, seem to be controlled by fault systems that have been tectonically reactivated throughout most of Cretaceous time (Fig. 3). During most of Cretaceous time, however, the Vema Dome–Nyk High area were in a basin-floor position, as shown by the thick parallel Cretaceous sequences across the area (Fig. 4), similar to the depositional setting of the nearby Vigrid Syncline (Fig. 3). Minor faulting and block rotation could have taken place in the Vema Dome–Nyk High area in Early Cretaceous time, but it is difficult to differentiate the deepest sequences because of limited seismic resolution at depth. The first clear evidence of fault activity in the Vema Dome–Nyk High area is recorded in the latest part of Maastrichtian or Early Paleocene time, when a fault-related roll-over structure was eroded at the crest of the roll-over (Fig. 5 at A). From that time, this slowly emerged as a result of tectonic activity, especially in the area between the present Vema Dome and Nyk High (Fig. 6). Most of the Nyk High area continued to be a site of deposition throughout most of the Paleocene period (Fig. 6), and only in the latest

From: NØTTVEDT, A. *et al.* (eds) *Dynamics of the Norwegian Margin.* Geological Society, London, Special Publications, **167**, 379–396. 1-86239-056-8/00/$15.00

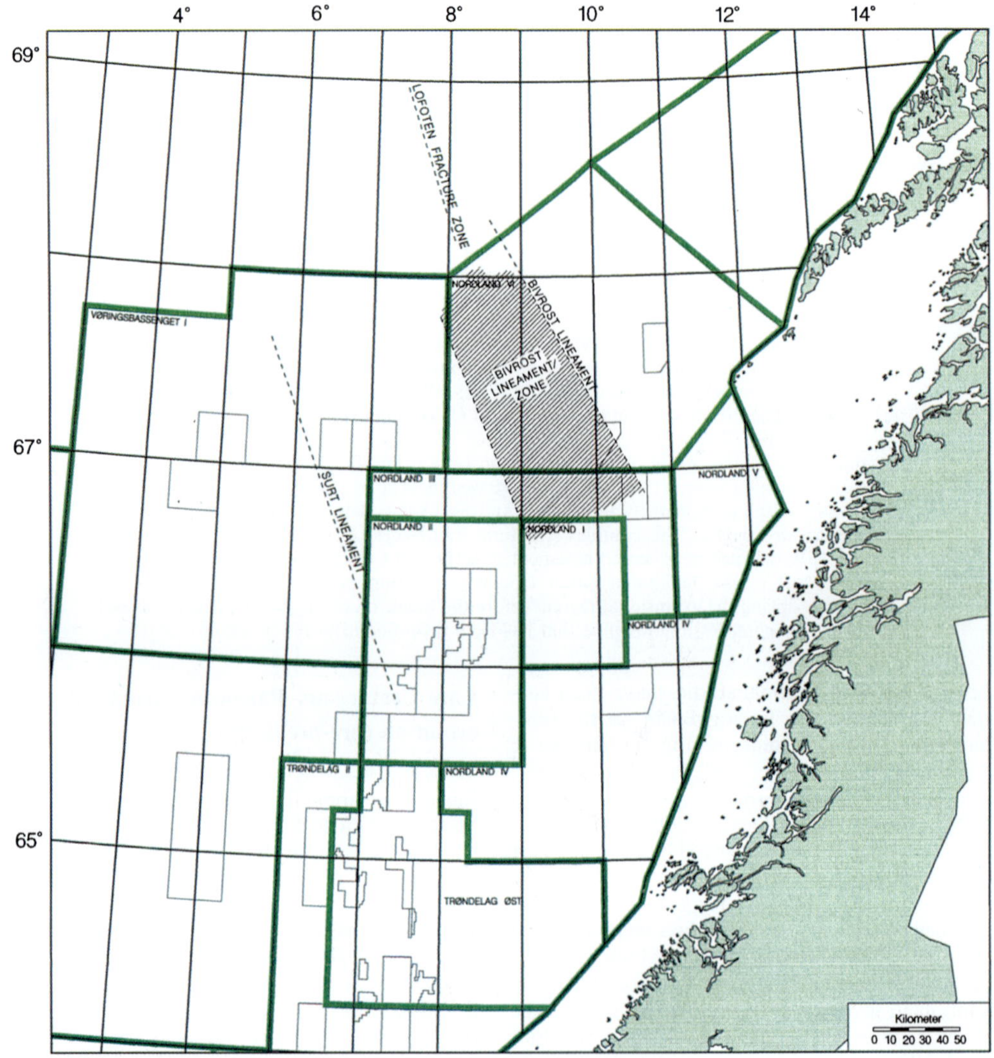

Fig. 1. Location of the Surt and Bivorst Lineaments in the Northern part of the Vøring Basin, offshore Mid Norway.

Paleocene or early Eocene time did this area become structurally high as a result of tectonic activity (Fig. 7 at A).

The Base Cretaceous is deeply buried in most of the Vøring Basin, and fault activity in pre-Cretaceous time is difficult to identify. Closer to the Norwegian mainland, on the Trøndelag Platform, where the Jurassic sequence is well imaged, the main Late Jurassic fault orientation as well as the old Caledonian structural grain is oriented in a NE–SW direction. We therefore assume that the observed NE–SW oriented extensional faulting in Late Maastrichtian–Paleocene time took place by reactivation of older NE–SW oriented Mesozoic and/or Palaeozoic faults, because of increased extensional forces before the North Atlantic break-up in this area (see also Doré *et al.* 1997).

From Intra Paleocene to the Near Top Paleocene time faulting was intensified and several transtensional features evolved (Fig. 8). Evidence of transtension is the apparent lack of extension at Near Top Cenomanian level, whereas there has been clear extension at shallower Cretaceous

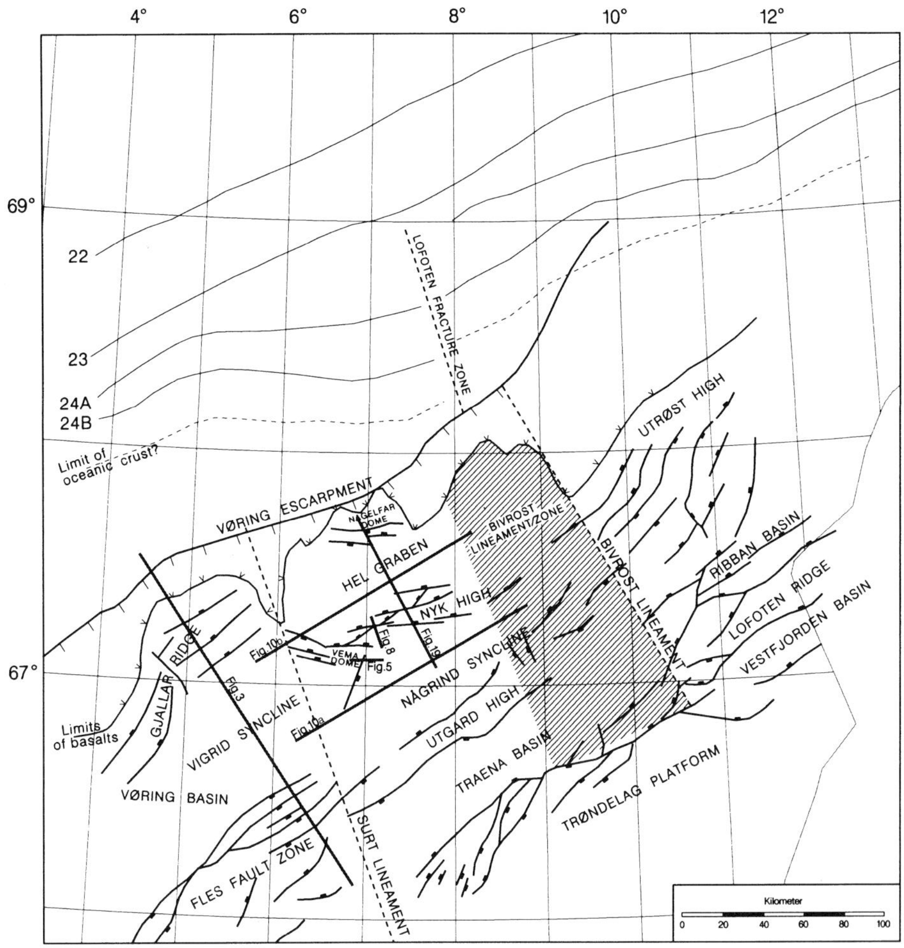

Fig. 2. Structural element map of the northern part of the Vøring Basin, showing location of seismic sections and geosections, in Figs 3, 5, 8, 10 and 19.

levels (Fig. 8 at A). Depocentre configuration along faults that have experienced lateral movements can be used as an indicator of the orientation of the lateral movement, with increasing subsidence in releasing bends and less subsidence or erosion in restraining bends (Mogensen 1994). Using this method the depocentre configuration along the NE–SW trending transtensional features in the Vema Dome–Nyk High area indicates a sinistral fault movement, from Intra Paleocene to the Near Top Paleocene time (Fig. 7 at B). Another indication of sinistral movement is the configuration of the isochrons across a steep fault, which we assume might be lateral (Fig. 6 at A). The non-eroded part of the uppermost Maastrichtian to Lower Paleocene sequence shows clearly an 'apparent' sinistral movement after the deposition of the sequence of the order of *c.* 2 km. There is no structural or stratigraphic indication that this fault was active in Latest Maastrichtian or Early Paleocene times, so the 2 km of sinistral movement post Intra Paleocene might be a good estimate of sense and amount of displacement.

Contemporaneous with the lateral movements along the NE–SW trending fault system, a relatively thick depocentre evolved in the western part of the Vema Dome along the NW–SE oriented Rym Fault Zone–Surt Lineament, where this structural element meets the NE–SW oriented fault system (Fig. 7 at C, and Fig. 9 at A). The position and configuration of this

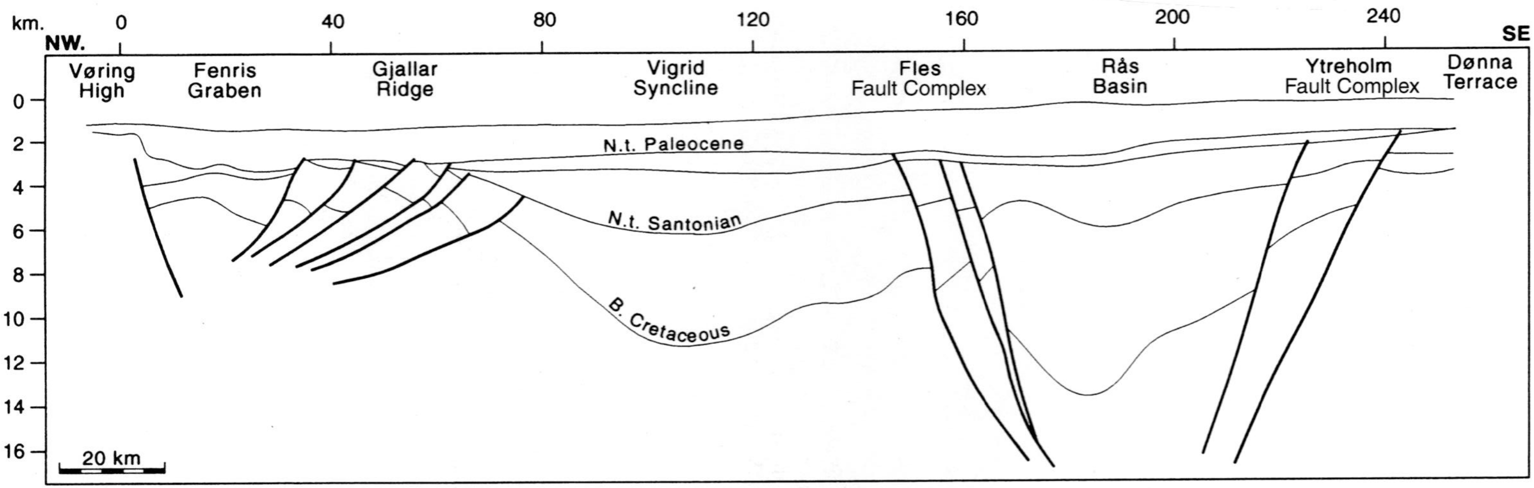

Fig. 3. Geosection across the northern part of the Vøring Basin from the Vøring Escarpment to the Dønn Terrace, showing repeated tectonic activity on the Gjallar Ridge and the Fles Fault Zone (thinning) throughout most of Cretaceous time. The total gravity as well as the residual gravity field for the section is also presented. (For location, see Fig. 2.)

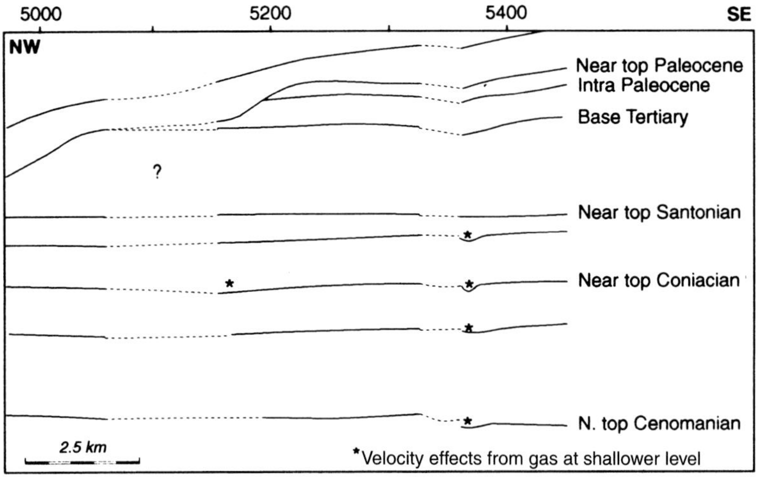

Fig. 4. Thick parallel Upper Cretaceous sequences on top of the Nyk High before Late Maastrichtian time, indicating that this area was in a basin-floor position in early Late Cretaceous time. This section is the same as in Fig. 8 flattened on the Near Top Santonian surface. Location, as for Fig. 8 is shown in Fig. 2.

depocentre compared with the configuration of the NW–SE trending fault indicates a releasing bend depocentre and sinistral fault movement. The faults transecting the former Vema Dome with an E–W orientation (Fig. 9) are interpreted as transfer faults between the NE–SW and the NW–SE fault system.

Influence of bounding crustal weakness zones

The different structural evolution of the Vema Dome–Nyk High compared with most of the rest of the Vøring Basin seems to be governed by two lineaments or zones that delineate the Vema Dome–Nyk High area, the Surt and Bivrost Lineaments (Blystad *et al.* 1995; Doré *et al.* 1997) (Figs 2 and 10a and b). The existence of these lineaments or zones has hitherto not been well constrained, but it is thought their orientation is NW–SE to NNW–SSE. Our interpretation suggests that the orientations of the two continental lineaments or zones diverge slightly, with the Surt Lineament having a NNW–SSE orientation whereas the Bivrost Lineament is oriented NW–SE (Fig. 2). The Bivrost Lineament could more correctly be named the Bivrost Zone, as there is no obvious single lineament, as for the Surt Lineament, but probably a series of lineaments acting as a hinge zone towards the elevated areas of the Utrøst High (Fig. 2). This is clearly seen on the Base Cretaceous time map (Fig. 11), which here is shown only with the 5.5 s twt contour to highlight the structure. Also, the total magnetic field (Fig. 12), the total gravity anomaly field (Fig. 13) and the best structural image (filtered free air) (Fig. 14) identify the location of the lineament or zone. The Base Cretaceous 5.5 s twt map is superimposed on all the potential field maps. The continental Bivrost Zone and Surt Lineament link up with the oceanic Lofoten Fracture Zone and a major bend in the magnetic anomalies 24A and 24B (Fig. 15). The earliest opening of the Norwegian Sea, north of the Jan Mayen Fracture Zone, was probably affected by these weakness zones in the continental crust. The magnetic anomalies 24B and 24A between the Lofoten Fracture Zone and the kink in the magnetic anomalies, oceanward of the Surt Lineament, have an almost E–W orientation (Figs 12 and 15) (Hagevang *et al.* 1983; Karpuz *et al.* in press *a*, *b*), indicating that the initial opening within this compartment might have been N–S. The orientation of Late Cretaceous faults on the continental side also seems to reflect a local N–S extensional field within this segment, with E–W trending individual faults (Fig. 11). We suggest that such a local N–S extensional field in late Late Cretaceous time could reactivate old crustal weakness zones, such as both the NNW–SSE oriented Surt Lineament and NW–SE oriented Bivrost Zone, with a sinistral component, setting up a clockwise rotation of the crustal block delineated by these lineaments or zones and the Vema Dome–Nyk High fault system (Fig. 9).

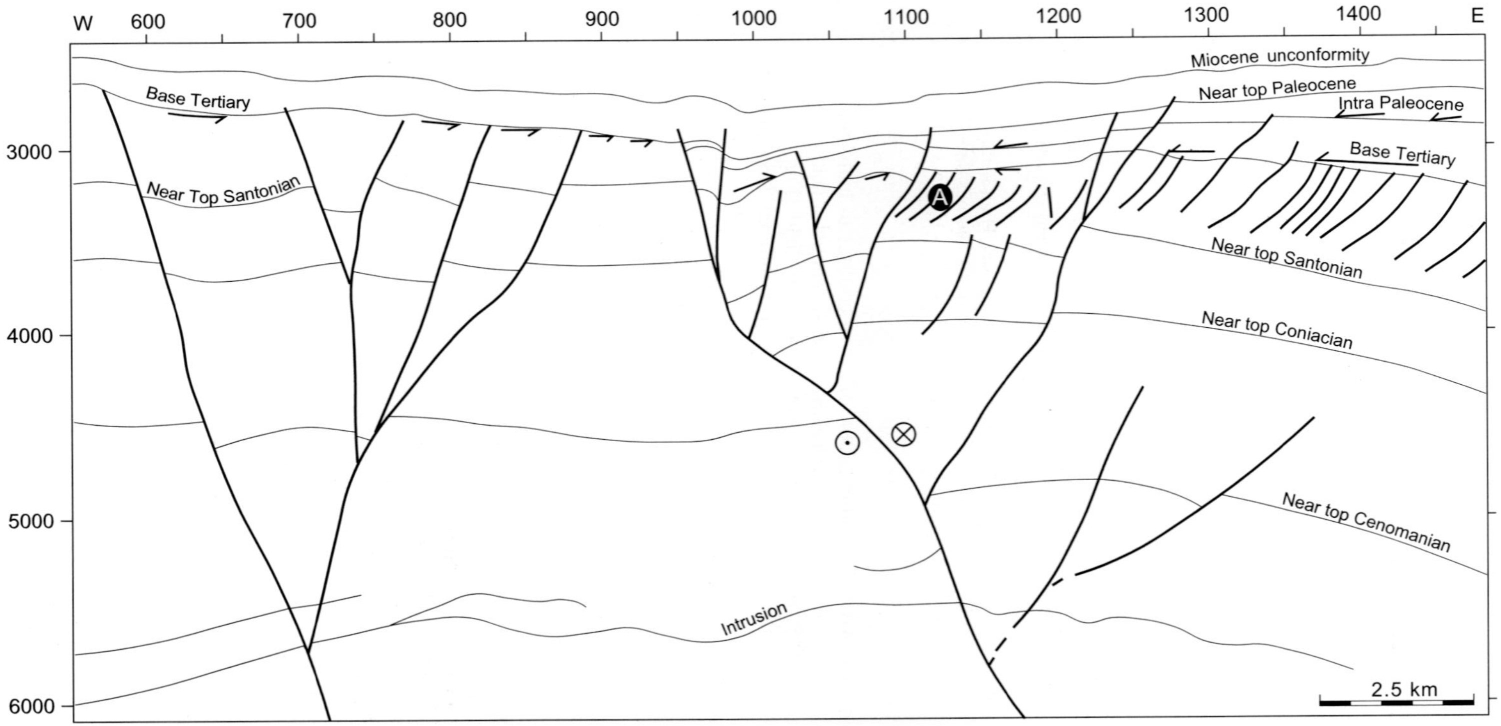

Fig. 5. Line drawing from seismic section. (Note erosion at A indicating that the roll-over structure was eroded at this level, which is in the latest part of Maastrichtian time.) (For location, see Fig. 2.)

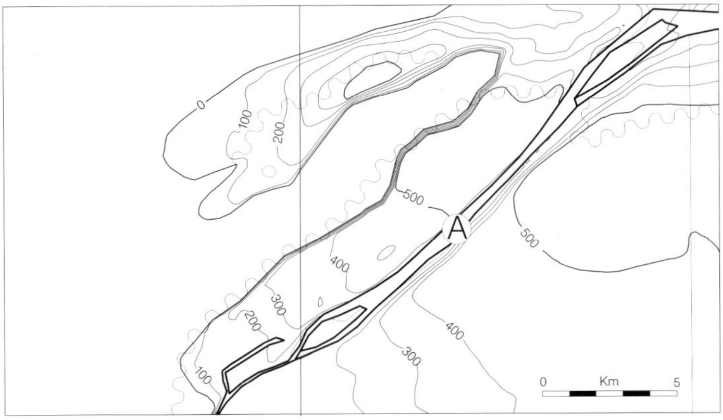

Fig. 6. Time difference map for the Base Tertiary to the Intra Paleocene sequence. Nyk High was at this time still in a depositional setting. Indication of 2 km left lateral movement in a pull-apart graben at A. Wavy line shows the post-Base Tertiary erosional limit. (See also Fig. 7.)

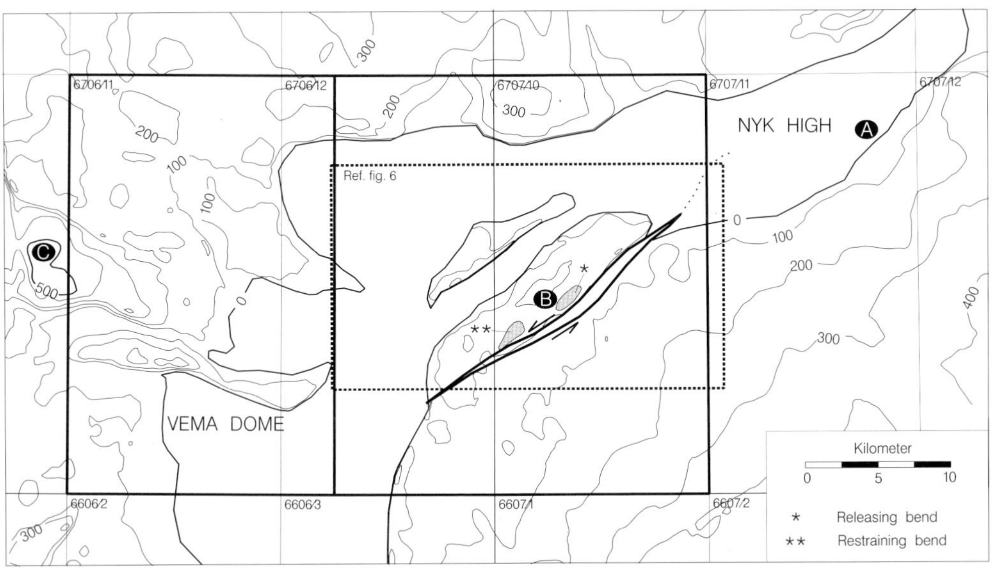

Fig. 7. Time difference map for the Intra Paleocene to Near Top Paleocene sequence. Nyk High evolved as a high in this period, at A. Depocentre configuration with transtensional (releasing) and transpressional (restraining) bends indicate left lateral movement on the fault at B. (Note the local depocentre at C.)

Such a clockwise block rotation would explain the observed sinistral movements along both the NE–SW trending Vema Dome–Nyk High and the NNW–SSE trending Surt Lineament. Although the amount of rotation is relatively small (sinistral movements of the order of 2 km), it would cause varying types of deformation that could explain the observed complex fault patterns (Figs 8 and 9), especially in the Vema Dome area.

Latest Late Cretaceous structural deformation seems to diminish away from the oceanic–continental transition within the crustal compartment between the Surt Lineament and the Bivrost Zone, and is less at the Utgard High than on the Vema Dome–Nyk High. This indicates that the Vema Dome–Nyk High fault system has taken up most of the latest Late Cretaceous structural deformation, both dip slip and strike slip, and it also implies that the crustal block

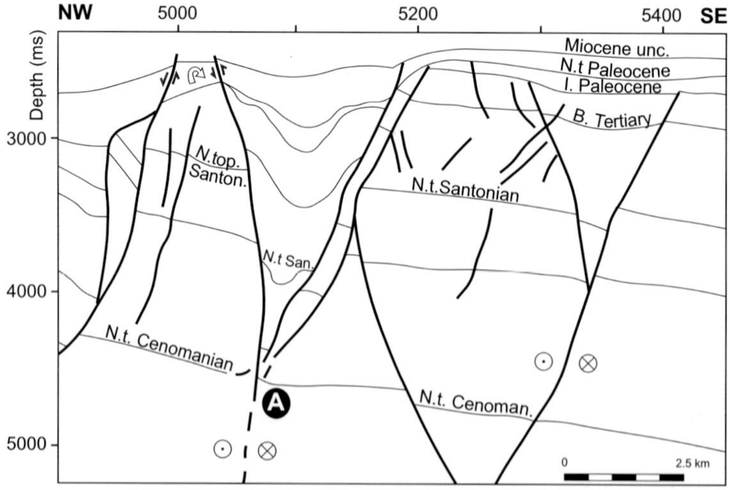

Fig. 8. Line drawing of seismic section. Extension on the Near Top Santonian level, but no apparent extension on the Near Top Cenomanian level is a good indication of lateral movements on the Nyk High. (For location, see Fig. 2.)

rotation between these two crustal weakness zones primarily is restricted to the outer parts of the continental crust, closest to the initial break-up (shown schematically in Fig. 16), which is in good agreement with the stretching configuration of Roberts *et al.* (1997, fig. 6).

Eocene to present structural evolution (post-break-up)

The clockwise rotation that seems to have been effective during the pre-break-up extensional regime might have continued during the post-break up compressional regime.

With the break-up of the North Atlantic in Eocene time a different tectonic setting was introduced to the Vøring Basin and the Vema Dome–Nyk High area. The whole area became exposed to compressional forces, and several phases of inversion have been identified from Eocene to the Late Miocene time. Earthquake focal mechanisms indicate that the Vema Dome–Nyk High area is today under compressional forces (Bungum *et al.* 1991), but the last major inversion tectonic activity took place in Mid- to Late Miocene time.

The orientation of the compressional forces today is NW–SE based on earthquake focal mechanisms (Bungum *et al.* 1991), whereas reactivated faults in the southern part of the Vøring Basin and on the Trøndelag Platform indicate that the maximum compressional force during the first compressional phases in Paleogene time might have had a more WNW–ESE orientation. The origin of the compressional forces is still under debate, but might be a combination of several opposing plate tectonic events (Lundin & Doré 1997). Whatever origin these compressional forces had, they seem to have reactivated old lineaments, and also the Surt and Bivrost Lineaments or Zones, once again with a sinistral component (Fig. 17). It is difficult to identify any lateral movement in the reactivation of these zones, but left lateral movements can explain the configuration and pattern of the inversion structures in the area (Figs 17 and 18). Sinistral movements would again induce a clockwise rotation of the crustal block between the Surt and Bivrost Lineaments or Zones, but this time in a compressional setting. The Nyk High and Vema Dome would therefore become structures that evolved in the corners of this crustal block (Fig. 17). Depending on the configuration of the underlying fault system these two structures would evolve differently. The Vema Dome would evolve as a normal inversion structure, whereas the Nyk High, being a former rotated footwall, was over-rotated and finally locked when the main faults became too steep (Fig. 19). This could also explain the inversion structure formed in the former Någrind Syncline area (Figs 10a, 17 and 19), indicating that the whole compartment between the Surt Lineament and Bivrost Zone

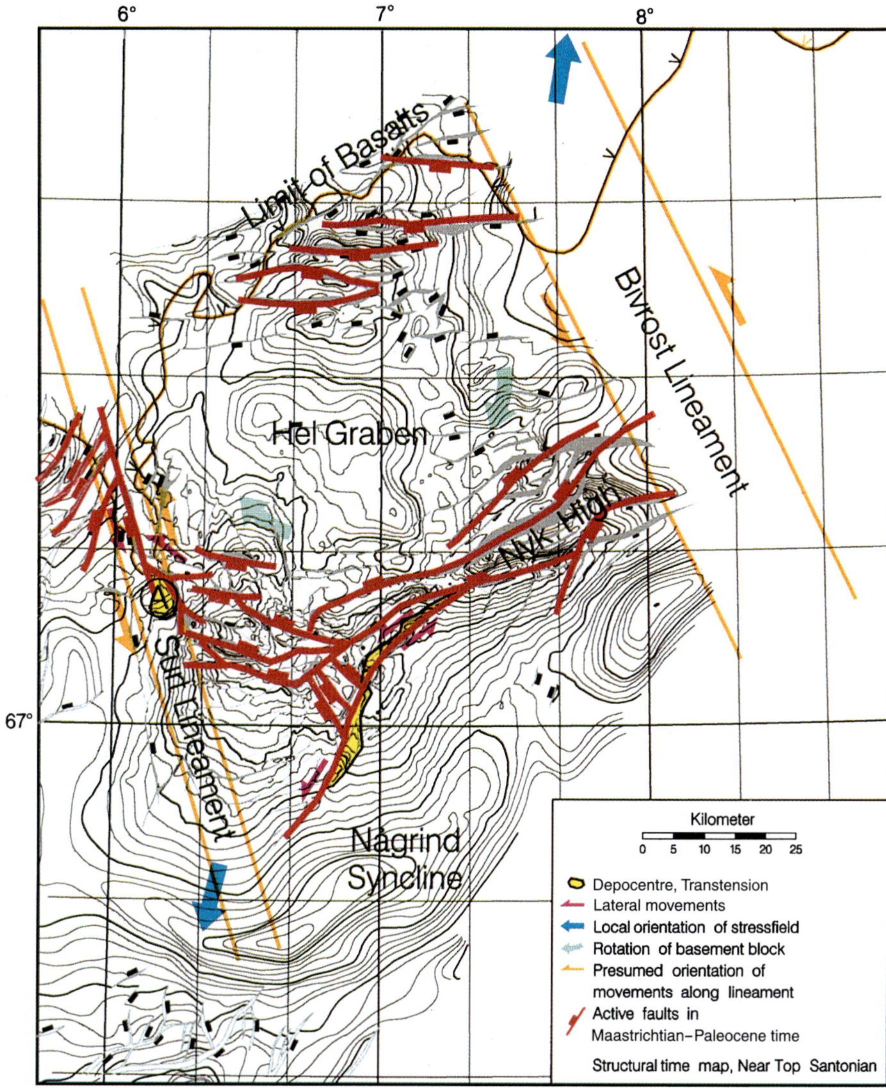

Fig. 9. Structural development in Late Maastrichtian–Paleocene time. As a result of local extensional forces a clockwise rotation and sinistral fault movements of the area between the Surt Lineament and the Bivrost Lineament or Zone can explain the depocentre configuration and the complex structural evolution of the area. Depocentre A under influence of sinistral shear.

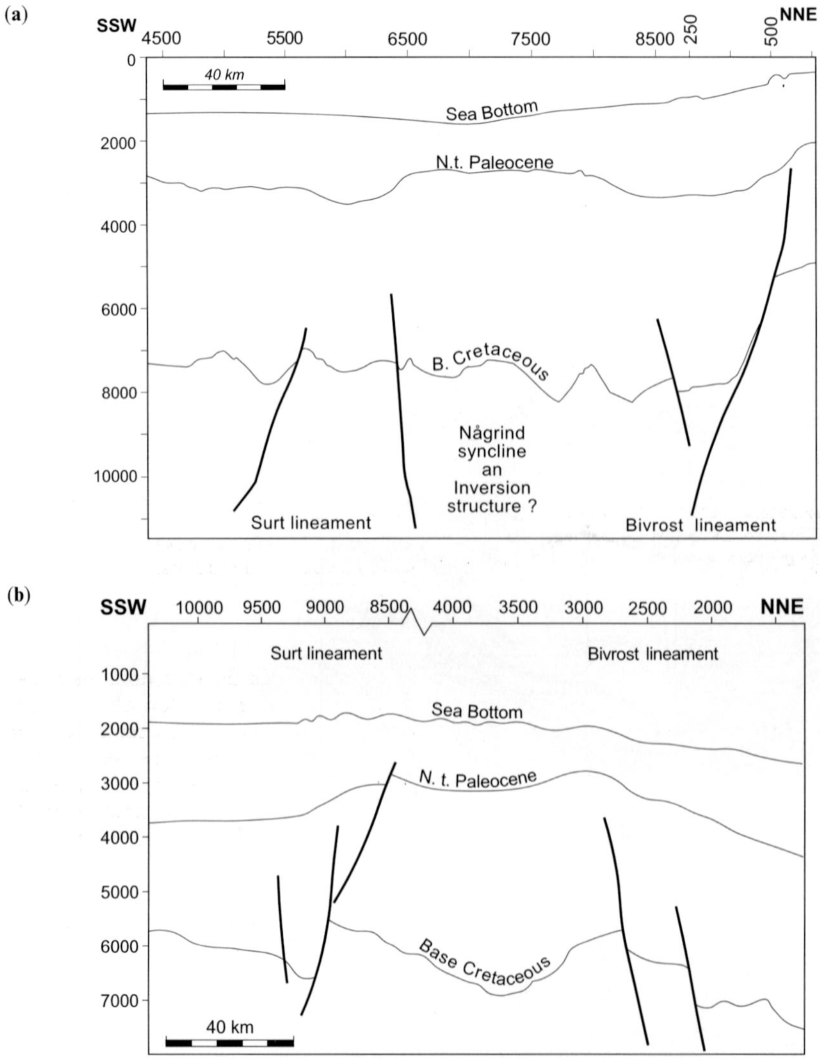

Fig. 10. (**a**) Section across the Surt Lineament and Bivrost Zone south of the Vema Dome–Nyk High. (**b**) Section across the Surt Lineament and Bivrost Zone north of the Vema Dome–Nyk High. (For location, see Fig. 2.)

SE of the Vema Dome–Nyk High area became locked.

The primary structural evolution during the compressional stages in Tertiary time (Doré & Lundin 1996) took part in the outermost part of the continental crust, as during the pre-break-up structural evolution, probably because of the weakened nature of the continental crust close to the oceanic–continental transition.

Since Late Miocene time, large Plio-Pleistocene clinoforms are evidence of progradation from the Norwegian mainland, gradually burying the inversion structures, and only the most oceanward inversion structures such as the Nyk High are imaged today on the sea bottom. The hiatus on the apex of the Nyk High includes the Campanian to Quaternary period, so that we have no exact information about when the Nyk High submerged. The Nyk High has clearly been exposed to erosional forces from at least Maastrichtian time until possibly Eocene time (Figs 7 and 19), but whether the Nyk High submerged as early as Late Eocene or as late as Late Miocene time is still an unsolved question.

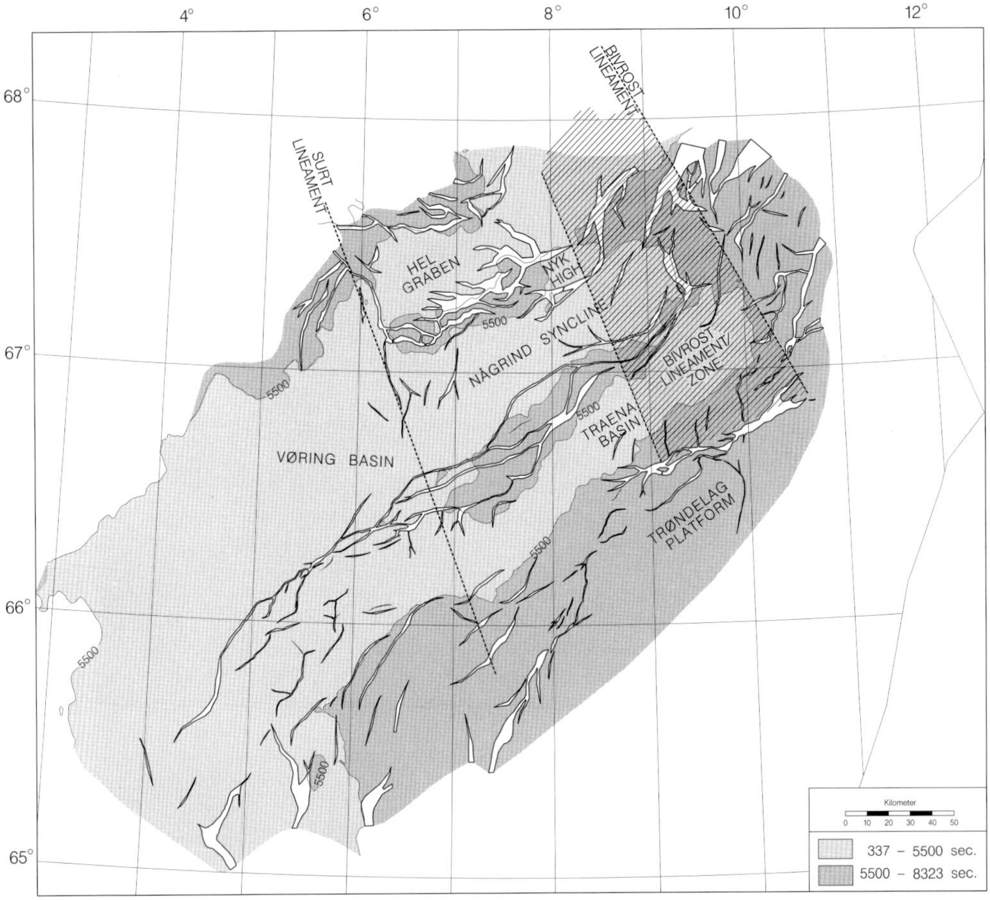

Fig. 11. A Base Cretaceous twt map with only the 5.5 s twt contour, showing the Surt Lineament and the Bivrost Zone.

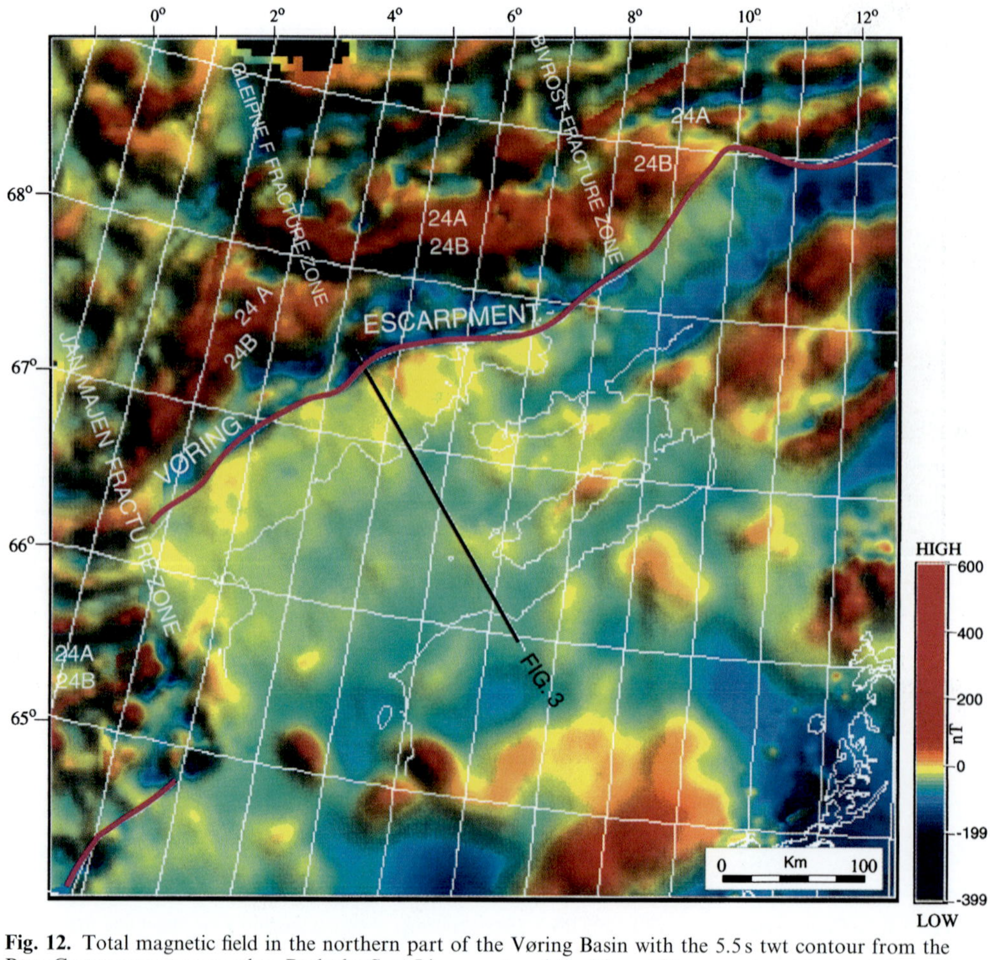

Fig. 12. Total magnetic field in the northern part of the Vøring Basin with the 5.5 s twt contour from the Base Cretaceous as an overlay. Both the Surt Lineament and the Bivrost Zone can be picked out. Location of Fig. 3 is indicated.

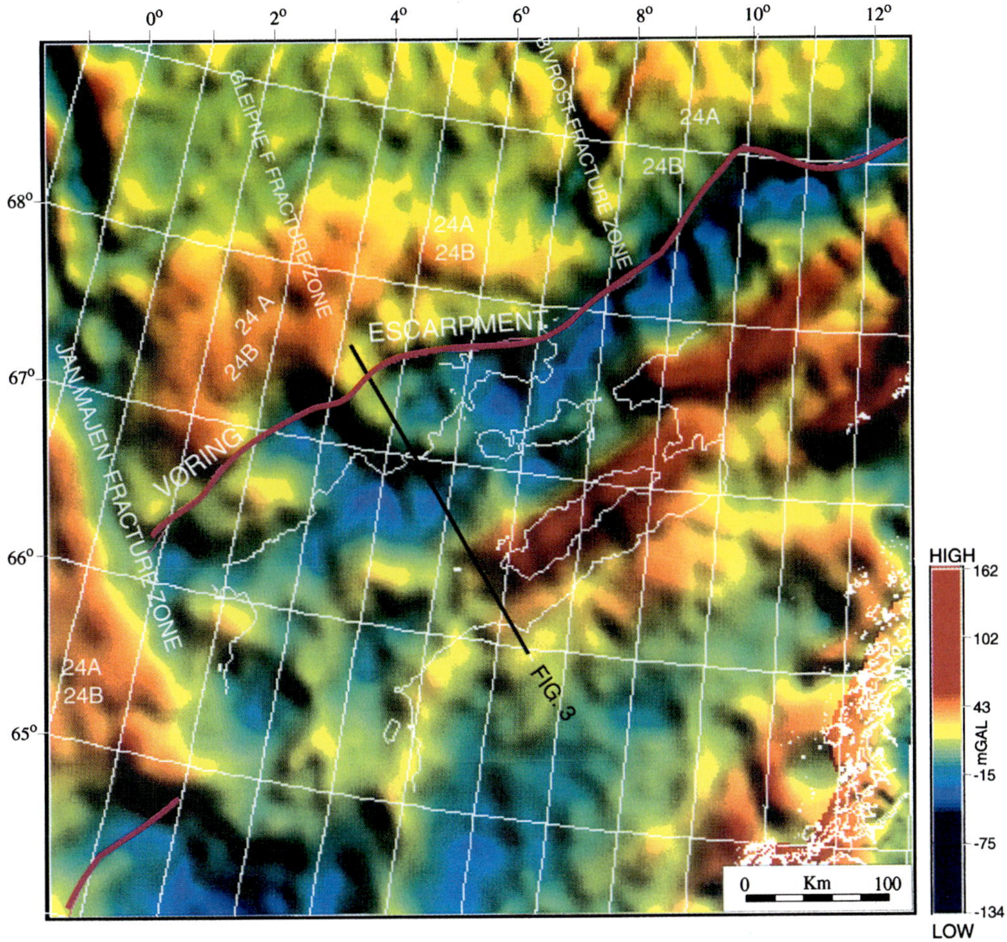

Fig. 13. Total free-air gravity field in the northern part of the Vøring Basin with the 5.5 s twt contour from the Base Cretaceous as an overlay. Both the Surt Lineament and the Bivrost Zone can be picked out. Location of Fig. 3 is indicated.

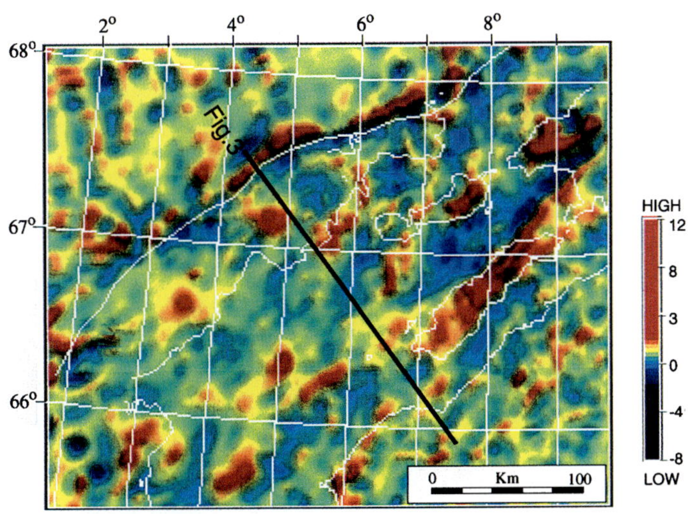

Fig. 14. Best structual Image (filtered free air gravity) in the northern part of the Vøring Basin with the 5.5 s twt contour from the Base Cretaceous as an overlay.

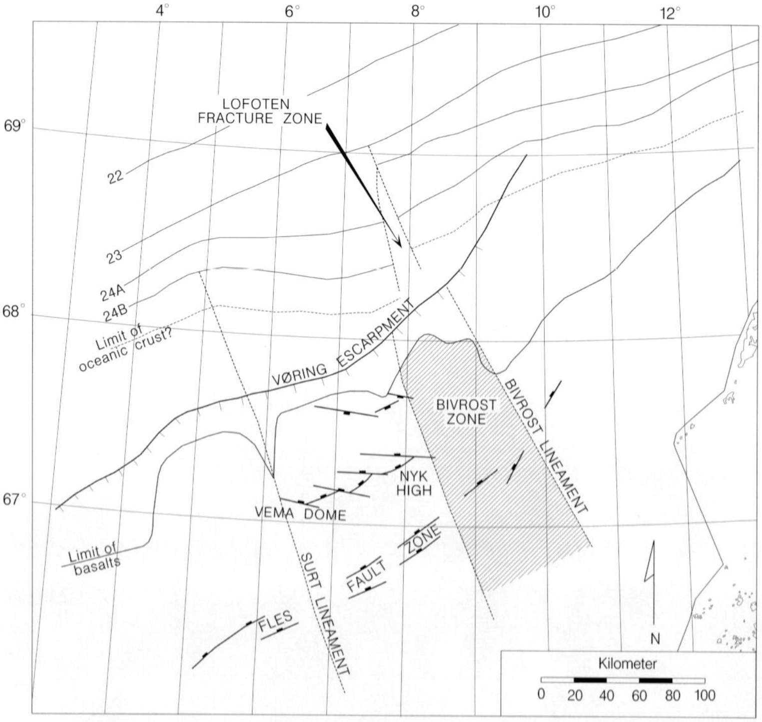

Fig. 15. Location of the Surt Lineament and Bivrost Zone, and the relationship to the magnetic oceanic spreading anomalies (note the kink of anomaly 24B and 24A in prolongation of the Surt Lineament.)

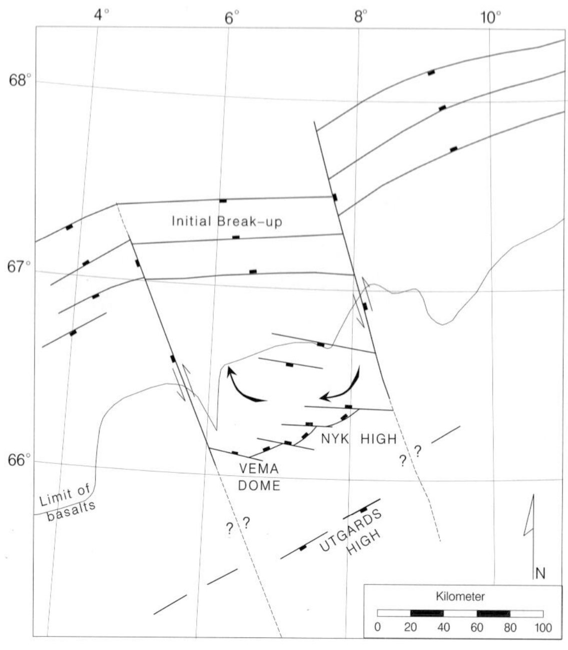

Fig. 16. Schematic illustration of the structural evolution in the outermost part of the area between the Surt Lineament and the Bivrost Zone. Local extensional forces give rise to basement block rotation.

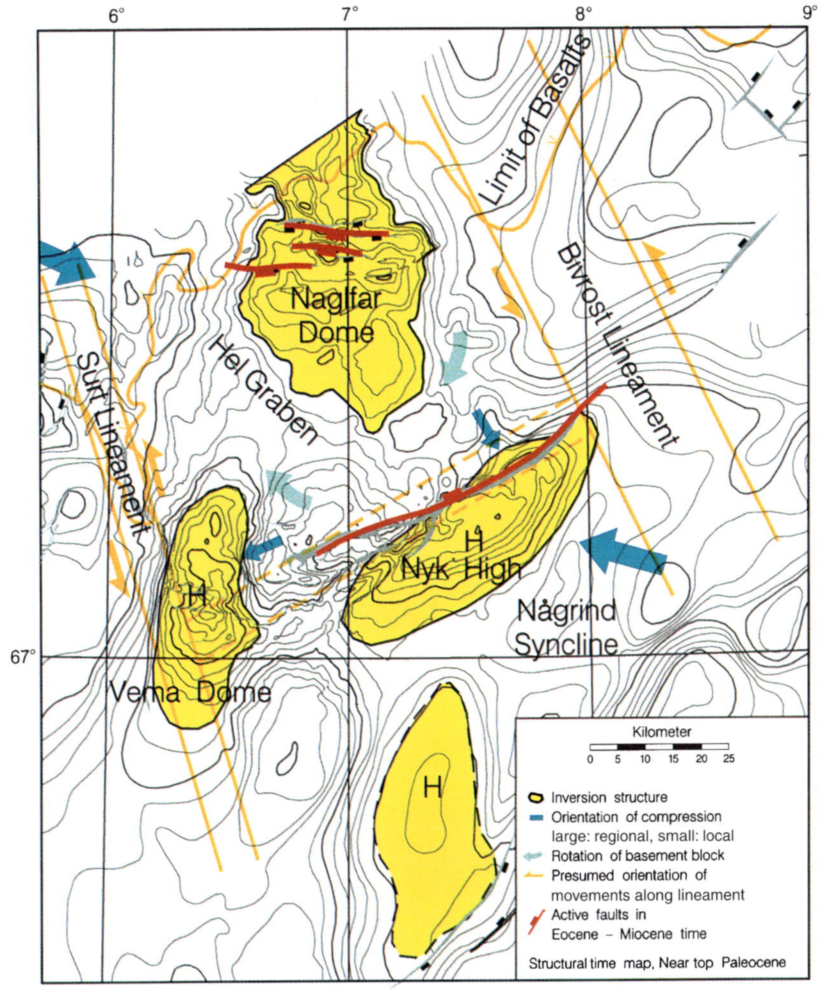

Fig. 17. Structural development in Eocene–Miocene time. Because of the regional compressional forces, clockwise rotation of the area between the Surt Lineament and the Bivrost Lineament or Zone once again can explain the structural evolution of the area. The inversion structures, the Vema Dome and the Nyk High, evolve in the corners of the rotating basement block.

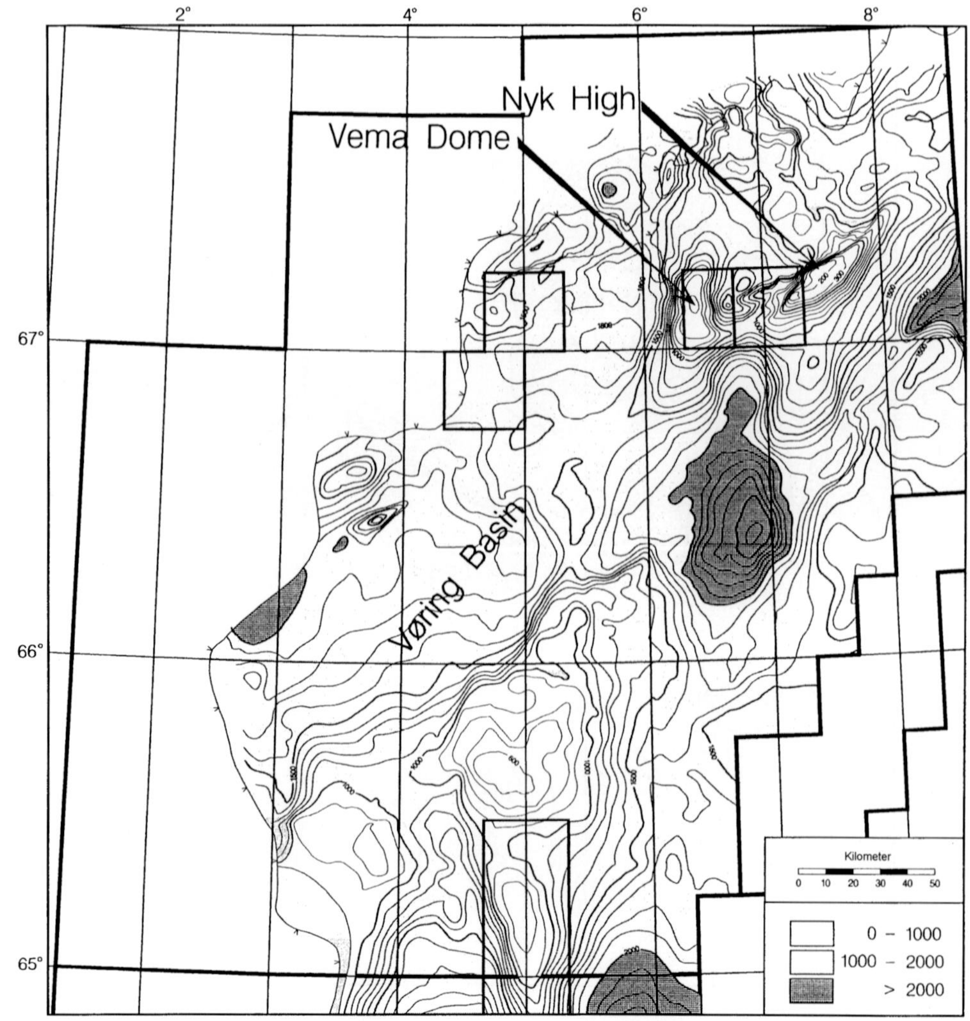

Fig. 18. The time difference map from the sea bottom to the Near Top Paleocene position in the Vøring Basin outlines the western and northernmost inversion structures related to the limited post-Miocene deposition in these areas.

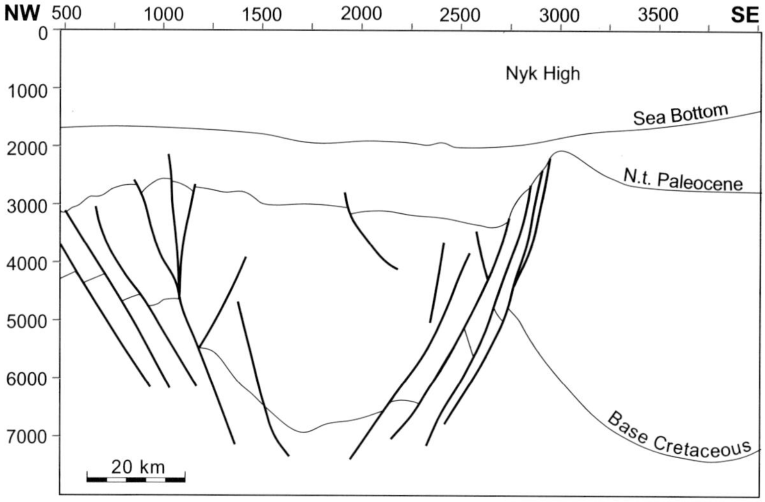

Fig. 19. Over-steepening of the normal faults on the Nyk High during the compression in Tertiary time could lock the faults. This would give rise to a transfer of the compressional forces to the whole basement block southeast of the Nyk High, which could explain the inversion of the former Någrind Syncline. (For location, see Fig. 2.)

The authors wish to thank B. Tørudbakken, R. Scrutton and Ø. Birkeland for their constructive comments during the review process, and E. M. Olsen for her excellent drafting work. We also want to thank Norsk Hydro a.s.a. for permission to publish these studies.

References

BJØRNSETH, H. M., GRANT, S. M., HANSEN, E. K., HOSSACK, J. R., ROBERTS, D. G. & THOMSON, M. 1997. Structural evolution of the Vøring basin, Norway, during the Late Cretaceous and Paleogene. *Journal of the Geological Society, London*, **154**, 545–550.

BLYSTAD, P., BREKKE , H., FÆRSETH, R. B., LARSEN, B. T., SKOGSEID, J. & TØRUDBAKKEN, B. 1995. *Structural Elements of the Norwegian Continental Shelf*. Norwegian Petroleum Directorate Bulletin, **8**, 1535–1541.

BREKKE, H. & RIIS, F. 1987. Tectonic and basin evolution of the Norwegian Shelf between 62°N and 72°N. *Norsk Geologisk Tidsskrift*, **67**, 295–321.

BUNGUM, H., ALSAKER, A., KVAMME, L. B. & HANSEN, R. A. 1991. Seismicity and seismotectonics of Norway and nearby continental shelf areas. *Journal of Geophysical Research*, **24**, 8249–8265.

DORÉ, A. G. & GAGE, M. S. 1987. Crustal alignments and sedimentary domains in the evolution of the North Sea, North-east Atlantic margin and Barents Shelf. *In*: BROOKS, J. & GLENNIE, K. (eds) *Petroleum Geology of North West Europe*. Graham and Trotman, London, 1131–1148.

—— & LUNDIN, E. R. 1996. Cenozoic compressional structures on the NE Atlantic margin: nature, origin and potential significance for hydrocarbon exploration. *Petroleum Geoscience*, **2**, 299–311.

——, ——, FICHLER, C. & OLESEN, O. 1997. Patterns of basement structure and reactivation along the NE Atlantic margin. *Journal of the Geological Society, London*, **154**, 85–92.

HAGEVANG, T., ELDHOLM, O. & AALSTAD, I. 1983. Pre-23 magnetic anomalies between Jan Mayen and Greenland–Senja Fracture Zones in the Norwegian Sea. *Marine Geophysical Research*, **5**, 345–363.

HINZ, K., ELDHOLM, O., BLOCK, M. & SKOGSEID, J. 1993. Evolution of North Atlantic volcanic continental margins. *In*: PARKER, J. R. (ed.) *Petroleum Geology of North West Europe*. Geological Society, London, 901–913.

KARPUZ, M. R., FLECHE, H. & FARELLY, B. 2000*a*. Magnetic total field anomalies. *Norsk Geologisk Tidsskrift*, in press.

——, —— & ——2000*b*. Free air (offshore)–Bouguer (land) anomalies. *Norsk Geologisk Tidsskrift*, in press.

LUNDIN, E. R. & DORÉ, A. G. 1997. A tectonic model for the Norwegian passive margin with implications for the NE Atlantic: Early Cretaceous to break-up. *Journal of the Geological Society, London*, **154**, 545–550.

MOGENSEN, T. E. 1994. Paleozoic structural development along the Tornquist Zone, Kattegat area, Denmark. *Tectonophysics*, **240**, 191–214.

PLANKE, S., SKOGSEID, J. & ELDHOLM, O. 1991. Crustal structure off Norway, 62°–70°N. *Tectonophysics*, **189**, 91–107

ROBERTS, A. M., LUNDIN, E. R. & KUSZNIR, N. J. 1997. Subsidence of the Vøring Basin and the influence of the Atlantic continental margin. *Journal of the Geological Society, London*, **154**, 551–557.

SKOGSEID, J. 1994. Dimensions of the Late Cretaceous–Paleocene Northeast Atlantic rift derived from Cenozoic subsidence. *Tectonophysics*, **240**, 225–247.

——, PEDERSEN, T., ELDHOLM, O. & LARSEN, B. T. 1992. Tectonism and magmatism during NE Atlantic continental break-up: the Vøring margin. *In*: STOREY, B. C., ALABASTER, T. & PANKHURST, R. J. (eds) *Magmatism and the Causes of Continental Break-up*. Geological Society, London, Special Publications, **68**, 305–320.

WALKER, I. M., BERRY, K. A., BRUCE, J. R., BYSTØL, L. & SNOW, J. H. 1997. Structural modelling of regional depth profiles in the Vøring Basin: implications for the structure and stratigraphic development of the Norwegian passive margin. *Journal of the Geological Society, London*, **154**, 537–544.

Norwegian–Greenland Sea thermal field

EIRIK SUNDVOR,[1] OLAV ELDHOLM,[2] TADEUSZ P GLADCZENKO[2] & SVERRE PLANKE[2]

[1] *Institute of Solid Earth Physics, University of Bergen, Allegt. 41, N-5007, Bergen, Norway (e-mail: eirik.sundvor@ifif.uib.no)*

[2] *Department of Geology, University of Oslo, Pb. 1047 Blindern, N-0316 Oslo, Norway*

Abstract: Thermal gradient and thermal conductivity measurements carried out in the Norwegian–Greenland Sea and in the western Eurasia Basin from 1967 to 1996 have yielded 436 heat-flow values which have been stored on a UNIX-platform database, HEAT, to facilitate retrieval, documentation and display. HEAT contains all public domain data for the region, and the database and application programs are freely available. The heat flow on oceanic crust reveals a clear, first-order heat-flow–crustal-age relationship. The thermal conductivity is relatively uniform and consistent, with values in the 0.85–1.15 $W m^{-1} K^{-1}$ range. The lower values are from the ocean basins and the higher values from the continental margins. Local, very high heat flow along the present spreading axis may imply active volcanism or venting. We also document local continental slope maxima on the Møre and western Barents Sea margins, contrasting greatly with the typical low heat flow, generally <75 $mW m^{-2}$, on old oceanic and thinned continental crust. The abnormally high heat flow, exceeding 1000 $mW m^{-2}$, on the 'Håkon Mosby' gas-seeping mud volcano on the Barents Sea margin suggests that also other local slope maxima may be related to active seeps.

Most of the early heat-flow measurements in the Norwegian–Greenland Sea were collected by Lamont–Doherty Geological Observatory (LDGO) during coring operations in the period 1967–1973. Despite an uneven geographical distribution, these measurements outlined the first-order thermal properties of the oceanic crust (Langseth & Zielinski 1974).

Few additional measurements were made until the acquisition of a Norwegian heat-flow probe dedicated to high-latitude work in 1983. At the same time, petroleum industry interest in the sedimentary basins along the Norwegian continental margin led to a French–Norwegian programme, from 1983 to 1986, to assess the thermal and burial history of sediment basins on the continental margin by mapping the crustal structure and the thermal field along selected margin transects. The joint programme, between Elf Petroleum Norge and the universities of Bergen and Oslo, included seismic wide-angle measurements by the expanded spread profiling technique (NORMAR Project), and heat-flow measurements (FLUNORGE Project) in co-operation with Institut Français de Recherche pour l'Exploitation de la Mer (IFREMER). A total of 161 heat-flow values were acquired during FLUNORGE (Sundvor *et al.* 1989; Sundvor & Eldholm 1992). These data, as well as measurements during many Norwegian cruises, have greatly expanded our understanding of the thermal properties of the Norway–Svalbard continental margin and the Jan Mayen Ridge. In addition, a number of values have been measured on oceanic crust in the Norwegian–Greenland Sea and the Eurasia Basin.

In this study, we present the heat-flow and thermal conductivity data from the Norwegian–Greenland Sea, western Eurasia Basin and northern Svalbard margin (Figs 1 and 2; Table 1), and discuss the regional character of the thermal field over oceanic crust and along transects across the Eurasian margin.

Data and database

The heat-flow surveys within the study area are listed in Table 1. The majority of temperature gradients have been measured by three instrument packages. The LDGO cruises applied the Ewing thermograd consisting of up to five thermistors mounted on the Ewing piston corer (Gerard *et al.* 1962; Haenel 1979). The Norwegian LDGO-manufactured instrument, designed for fast operations, employs thermistor probes mounted on outriggers several centimetres away from a solid lance that is attached to a core-head weight. It also monitors the bottom-water temperature and the tilt of the instrument

From: NØTTVEDT, A. *et al.* (eds) *Dynamics of the Norwegian Margin*. Geological Society, London, Special Publications, **167**, 397–410. 1-86239-056-8/00/$15.00

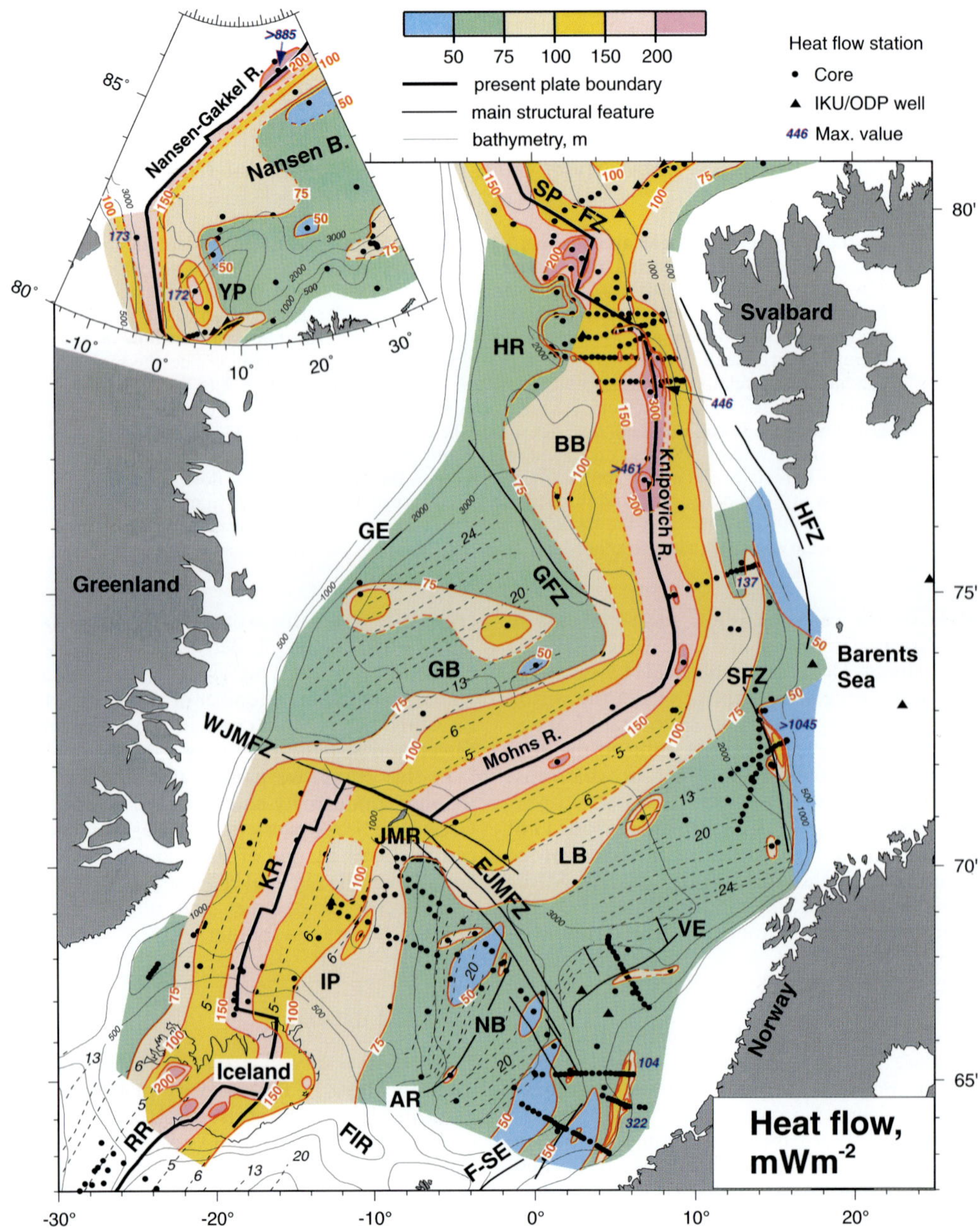

Fig. 1. Heat-flow measurements and selected heat-flow contours. Simplified bathymetry in m. Main structural features and sea-floor spreading anomalies from Talwani & Eldholm (1977), Skogseid & Eldholm (1987) and Eldholm *et al.* (1990). Iceland heat-flow contours from Flovenz & Saemundsson (1993). RR, Reykjanes Ridge; KR, Kolbeinsey Ridge; AR, Aegir Ridge; FIR, Faeroe–Iceland Ridge; JMR, Jan Mayen Ridge; HR, Hovgaard Ridge; JMFZ, Jan Mayen Fracture Zone (WJMFZ and EJMFZ); SFZ, Senja Fracture Zone; GFZ, Greenland Fracture Zone; SPFZ, Spitsbergen Fracture Zone; HFZ, Hornsund Fault Zone; IP, Iceland Plateau; NB, Norway Basin; LB, Lofoten Basin; GB, Greenland Basin; BB, Boreas Basin; F-SE, Faeroe–Shetland Escarpment; VE, Vøring Escarpment; GE, Greenland Escarpment; YP, Yermak Plateau.

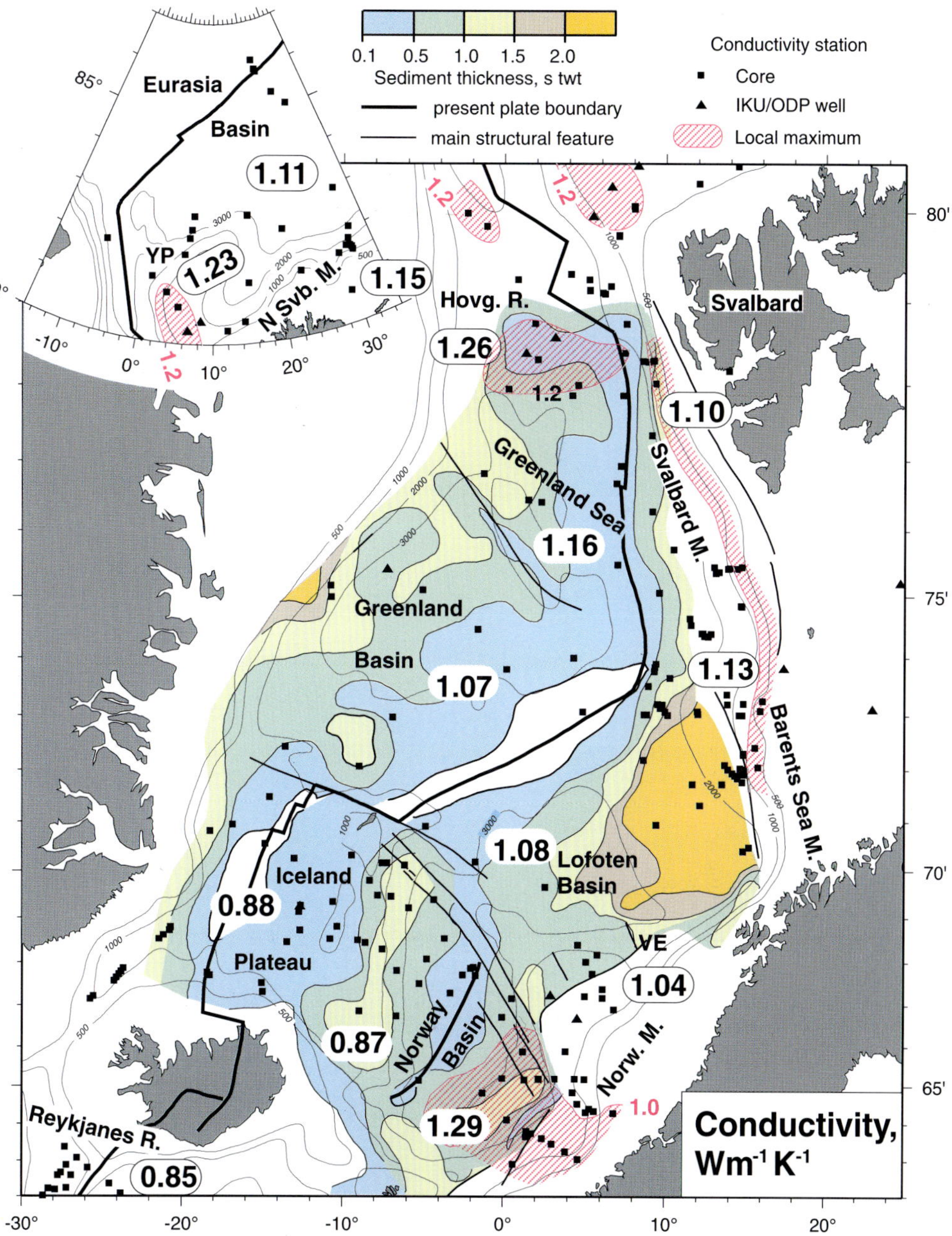

Fig. 2. Distribution of thermal conductivity measured in near sea-floor sediments, and average values for separate physiographic provinces (Table 2). Gridded, 1.5° × 0.5°, sediment thickness contours in s twt from Johansen (1989). Other references and acronyms in Fig. 1.

package. The IFREMER heat-flow instrument is based on a modified Aanderaa thermistor string, refurbished for enhanced resolution, attached to the outer part of the piston core barrel.

Thermal conductivities (Table 2) have been measured on sediment cores using the needle-probe method (von Herzen & Maxwell 1959; Haenel 1979). During the FLUNORGE and Norwegian cruises a Fenwal probe consisting of thermistor and heating wire was used, and the thermal conductivity was measured at 10 cm intervals along the cores and corrected to *in situ*

Table 1. *Marine heat-flow surveys in Norwegian–Greenland Sea and Eurasia Basin*

Survey		Area	Number of measurements			References	Comments
Institution	Year		*T* grad.	Cond.	Heat flow		
USGS; ONR	1965	Denmark Strait	10	11	10	1	Ice island Arlis II
LDGO	1966–1973	Norwegian–Greenland Sea	56	48	56	2	RV *Vema* 23, 27, 28, 29, 30
Shirshov Inst., Moscow	1971	Norwegian–Greenland Sea	11	11	11	3	RV *Akademik Kurchatov*
BGR	1969, 1972	Vøring Plateau; Lofoten Basin	7	7	7	4	RV *Planet* 1969, RV *Meteor* 1972
LDGO, UiB, UiO	1980	Norwegian–Greenland Sea, Yermak Plateau	16	15	16	5	Ymer-80
CGC, ONR, UiB, UiO	1981	Yermak Plateau	4	4	4	6	Ice island Fram III
Elf, IFREMER, UiB, UiO	1983	Norway–Svalbard Margin	71	18	71	7–8	FLUNORGE
UiB	1984	Svalbard Margin, Greenland Sea, Yermak Plateau	96	2	96	9	RV *Håkon Mosby*
Elf, IFREMER, UiB, UiO	1985	Møre Margin	65	16	65	8	FLUNORGE
Elf, IFREMER, UiB, UiO	1986	Jan Mayen Ridge	26	11	26	10	FLUNORGE
IKU	1985	Barents Sea	4	4	4	11–13	Shallow drilling
UiB, UiO	1986	Norwegian–Greenland Sea	23	7	23	10	RV *Håkon Mosby*
ODP	1985	Norwegian–Greenland Sea	1	1	1	14	Leg 104
AWI, Geomar, UiB	1987	Eurasia Basin	17	17	17	15	RV *Polarstern*
ODP	1994	Norwegian–Greenland Sea, Yermak Plateau	6	7	6	16	Leg 151
NRL, UiB, UiO	1995	Svalbard margin	25	77	25	17	RV *Håkon Mosby*
ODP	1995	Svalbard, Greenland margins	2	2	2	18	Leg 162
NRL, UiB, UiO, VNIIO	1996	Norway–Svalbard margin	16	13	16	17	RV *Profesor Logachev*
Total			456	271	456		

AWI, Alfred Wegener Institute for Polar Research; BGR, Bundesanstalt fur Geowissenschaften und Rohstoffe; Elf, Elf Petroleum Norge; GSC, Geological Survey of Canada; IKU, IKU Petroleum Research; LDGO, Lamont–Doherty Geological Observatory; NRL, Naval Research Laboratory; ONR, Office of Naval Research; UiB, University of Bergen; UiO, University of Oslo; USGS, US Geological Survey; VNIIO, VNII-okeangeologia Institute St Petersburg.

1, Lachenbruch & Marshall (1968); 2, Langseth & Zielinski (1974); 3, Udintsev & Lubimova (1973); 4, Haenel (1974); 5, Crane *et al.* (1982); 6, Jackson *et al.* (1984); 7, Crane *et al.* (1988); 8, Sundvor *et al.* (1989); 9, Sundvor (1986); 10, Sundvor & Myhre (1987); 11, Sættem (1988); 12, Zielinski *et al.* (1986); 13, Løseth *et al.* (1992); 14, Eldholm *et al.* (1989); 15, Sundvor & Torp (1987); 16, Myrhe *et al.* (1995); 17, Eldholm *et al.* (1999); 18, Jansen *et al.* (1996).

Table 2. *Thermal conductivities measured in near-surface sediment*

Physiographic province	Number of values	Average conductivity ($W m^{-1} K^{-1}$)	Standard deviation
North Svalbard margin	5	1.15	0.33
Yermak Plateau	11	1.23	0.23
Svalbard margin (74–80°N)	25	1.10	0.26
Barents Sea margin (70–74°N)	36	1.13	0.15
Norwegian margin (62–70°N)	22	1.04	0.28
Margin total	105	1.11	0.24
Eurasia Basin	17	1.11	0.23
Greenland Sea	25	1.16	0.16
Greenland Basin	11	1.04	0.13
Lofoten Basin	27	1.08	0.09
Iceland Plateau	36	0.88	0.16
Norway Basin	20	0.87	0.12
Reykjanes Ridge (north of 61°N)	27	0.85	0.20
Basins total	163	0.99	0.20
Hovgaard Ridge region	5	1.26	0.04
NE Norway Basin	7	1.29	0.27
Overall total	280	1.05	0.23

Stations, Y11 and Y18 (Crane *et al.* 1982) and 15F83 (Crane *et al.*1988), interpreted as linearly increasing conductivity with depth, yield average values >2.0 $W m^{-1} K^{-1}$. These values contrast greatly with those of adjacent stations and have been excluded in the averages.

conditions (Ratcliffe 1960). The LDGO and IFREMER core barrel method allows conductivity measurements on the recovered core, whereas the Norwegian lance probe data were complemented by gravity cores at selected probe locations. At locations without sediment cores, conductivities from neighbouring stations have been used.

Normally, the average core conductivity is used to calculate the heat flow. However, increased conductivity with depth measured in some cores on the Svalbard margin made Crane *et al.* (1982, 1988) introduce a linear conductivity–depth relationship at these stations. To obtain comparable heat-flow values, we have calulated the average conductivity over the length of core, and recalculated the heat flow.

Heat flow has also been determined at scientific drill sites (Figs 1 and 2; Table 1). Near-continuous thermal conductivity profiles exist at most Ocean Drilling Program (ODP) sites, and a limited number of temperature gradients have been measured at Sites 642 (Eldholm *et al.* 1987*b*), 907–912 (Myhre *et al.* 1995), and 981, 984, 986 and 987 (Jansen *et al.* 1996). In general, the temperature gradients are not considered well established, although there is a first-order consistency between heat flow in ODP holes and in core or probe locations (A. M. Myhre, pers. comm.). In addition, a few heat-flow values have been published from shallow exploration holes in the Barents Sea (Zielinski *et al.* 1986; Sættem 1988; Løseth *et al.* 1992).

Marine heat-flow data are affected by instrumental errors and bottom-water instability. Moreover, geology (sedimentation, heat refraction, sea-floor topography, uplift and erosion, etc.) may introduce local anomalies affecting data reliability and interpretation potential (Langseth & Zielinski 1974; Haenel 1979). Ideally, studies of lithospheric properties based on heat flow require a number of corrections. None the less, the common absence of such corrections in regional studies may explain part of the relatively large standard deviations in Table 3 (below).

All heat-flow data available to us have been included in a database, HEAT, to facilitate retrieval, documentation and display. HEAT also includes some surveys and measurements outside the region shown in Fig. 1. The database and application programs to make listings, maps and sections have been documented by Planke (1989). The database accepts, and converts between, c.g.s. and SI heat flow units. Here, we use SI units, i.e. thermal gradient in $m K m^{-1}$, thermal conductivity of sediments in $W m^{-1} K^{-1}$, and sea-floor heat flow in $mW m^{-2}$. The software is designed for a UNIX-platform and may be linked with the marine geophysical GMT system (Wessel & Smith 1991), as well as with other display and interpretation packages.

Table 3. *Table 1. Average heat flow for various crustal provinces compared with the averages of Haenel (1979)*

Region	Number of values	Mean value	Standard deviation	Reference
Earth	5417	74.3	90.9	Haenel (1979)
All oceans	3718	79.8	105.9	Haenel (1979)
All continents	1699	62.3	40.1	Haenel (1979)
Northeast Atlantic Ocean	124	82.8	73.7	Haenel (1979)
Mid-Atlantic Ridge	64	100.0	90.0	Haenel (1979)
Non-ridge areas	54	52.2	22.4	Haenel (1979)
Norwegian–Greenland Sea	64	96.4	48.3	Haenel (1979)
Norwegian Sea*	94	93.4	54.6	This study
Oceanic basins >15 Ma*	72	73.1	23.9	This study
Oceanic crust <10 Ma*	18	165.8	79.0	This study
Near-spreading axis <2 Ma†	19	234.7	93.3	This study
Norwegian margin				
Møre–Vøring	43	56.6	15.6	This study
Møre	29	51.5	13.9	This study
Vøring	14	67.3	12.4	This study
Jan Mayen Ridge	11	75.2	10.5	This study

Distribution of measurements in Fig. 1.
* Includes only values on identified oceanic crust, i.e. south of Greenland–Senja Fracture Zone.
† Includes Knipovich and Molloy ridges.

The database and the HEAT software are freely available from Department of Geology, University of Oslo.

Regional setting

The early Tertiary break-up between Eurasia and Greenland during Chron 24r was accompanied by transient, large-scale, subaerial break-up volcanism (Eldholm *et al.* 1989). The subsequent plate tectonic evolution north of the Faeroe–Iceland Ridge may spatially be divided into four main regions separated by the Jan Mayen, Greenland–Senja and Spitsbergen fracture zone systems (Fig. 1); and temporally by the change in relative plate motion during Chron 13 time (Talwani & Eldholm 1977). In particular, the Norway Basin–Iceland Plateau and Greenland Sea reveal a complex sea-floor spreading history with unstable plate boundaries and formation of the Jan Mayen and Hovgaard microcontinents, respectively. The plate geometry delayed crustal extension in the northern Greenland Sea until early Oligocene times, and generation of oceanic crust was probably initiated as late as mid- to late Miocene time (Myhre & Eldholm 1988; Eldholm *et al.* 1994). The plate boundary configuration through time also developed a number of rifted and sheared continental margin segments.

The sediment distribution on oceanic crust (Fig. 2) reflects the plate tectonic framework (Eldholm & Windisch 1974; Johansen 1989). However, late Neogene continental uplift and the subsequent Northern Hemisphere glaciation led to erosion of the inner shelf and adjacent landmass, and to deposition of thick, terrigenous sediment wedges along the eastern margin (Eldholm *et al.* 1994). In some areas, huge fan complexes have been constructed in front of prominent continental drainage systems. The distal fans commonly extend onto oceanic crust, exerting a significant, excess sedimentary load (Myhre *et al.* 1992; Faleide *et al.* 1996).

Thermal conductivity

The 271 thermal conductivity measurements show uniform and consistent values mostly in the 0.85–1.15 $W m^{-1} K^{-1}$ range. Except for local areas, it is not possible to make meaningful contours. Therefore, we have illustrated the regional features by calculating average conductivity values for different physiographic provinces (Fig. 2), showing that the conductivity is slightly higher on the margins than in the ocean basins. Moreover, the conductivity map outlines local areas of relatively high conductivity: the East Norway Basin and Møre margin, a region that has been affected by huge Quaternary submarine slides from the upper slope (Bugge *et al.* 1987); the North Greenland Sea in the vicinity of the Hovgaard microcontinent; the Barents Sea–Svalbard shelf and upper slope; the western Yermak Plateau.

With the exception of two cores sampling Eocene mud diapirs on the Vøring Plateau, the cores sampled the Quaternary sequence. The thermal conductivity of marine sediments correlates with bulk density and inversely with water content (Hamilton 1974), and we note that the ODP cores on the Vøring margin revealed high densities and low water contents in the dark, glacial samples compared with other glacial or interglacial samples (Eldholm *et al.* 1987*b*). Therefore, the variability in thermal conductivity is probably related to a dominance of pelagic components in the ocean basins, whereas low-conductivity sediments deposited during the present interglacial may locally have been removed on the continental margin. Thus, the local areas of high conductivity on the margin may be indicators of present erosion or non-deposition.

Heat flow

The regional distribution of the 456 heat-flow stations north of the Faeroe–Iceland Ridge varies considerably. Most stations are located along transects across the eastern margin, in the North Greenland Sea and in the North Norway Basin–Jan Mayen Ridge region. None the less, we have been able to contour the regional heat-flow field (Fig. 1). We also illustrate the spatial and temporal heat-flow distribution by a series of regional transects (Figs 3–5). The transects are located in regions of relatively dense heat-flow coverage, and adjacent stations are projected onto the transects. The minimum water depth and maximum projection distance are annotated on each transect. Most transects approximate plate tectonic flowlines or run perpendicular to main structural trends.

Oceanic crust

Both the contour map and the transects reveal a clear first-order correspondence of heat flow and age of the oceanic crust (Figs 1 and 4). The mid-oceanic ridges province and adjacent crust is characterized by heat flow $>100\,\mathrm{mW\,m^{-2}}$, whereas most stations on crust older than *c.* 30 Ma yield values in the $50–75\,\mathrm{mW\,m^{-2}}$ range. The marine data correspond fairly well to onshore measurements in Iceland (Flovenz & Sæmundsson 1993). On the other hand, the data coverage resolves neither the thermal signature of the extinct Aegir spreading axis in the Norway Basin nor the question of an extinct axis west of the the Jan Mayen Ridge (Talwani & Eldholm 1977). Previously, heat-flow–age relationships for various regions have been discussed based on sub-sets of the data in Fig. 1 related to lithospheric cooling models (Langseth & Zielinski 1974; Zielinski 1979; Langseth *et al.* 1990). We restrict our analysis to stations on reasonably well-identified oceanic crust (Talwani & Eldholm 1977; Hagevang *et al.* 1983; Nunns 1983; Skogseid & Eldholm 1987), i.e. the crust south of the Greenland–Senja Fracture Zone. The empirical function, heat flow $= 420\,\mathrm{age}^{-1/2}$, appears to fit the observed data (Fig. 3).

For reference to the global and northeast Atlantic mean heat flow of Haenel (1979), we have calculated average values. We attribute the higher averages and lower standard deviations in Table 3 to the relatively young age of the ocean, a hotter than normal upper mantle, and data of generally good quality.

The heat flow along the present spreading axis south of the Spitsbergen Fracture Zone is shown in Table 3 for crusts less then *c.* 2 and 10 Ma. We also observe some local, very high near-axis heat-flow values. In fact, the $461\,\mathrm{mW\,m^{-2}}$ value on the Knipovich Ridge (Langseth & Zielinski 1974) and the 855 and $1164\,\mathrm{mW\,m^{-2}}$ values on the Nansen Ridge (Sundvor & Torp 1987) are minima because the measured temperatures exceeded the calibrated thermistor ranges. We suggest that these extremal values reflect active volcanism and/or venting along the plate boundary.

Microcontinents

The Jan Mayen Ridge microcontinent was separated from East Greenland by rifting and sea-floor spreading since from *c.* 44 Ma, and complete separation was achieved at *c.* 24 Ma (Myhre *et al.* 1984). Four transects cross the main ridge block, which is flat topped in the north and narrows and plunges southward. The ridge has a relatively steep slope towards the Iceland Plateau and a more gentle slope towards the Norway Basin (Fig. 1). The transects show that the ridge has no heat-flow signature and that there is a regional, gently decreasing heat-flow trend from the younger Iceland Plateau towards the older Norway Basin reflecting the crustal ages (Fig. 3). None the less, the average heat flow on the ridge block proper, $75\,\mathrm{mW\,m^{-2}}$, is greater than that measured on extended continental crust off Norway (Table 3). We attribute the relatively high heat flow to the small size of the continental fragment and the fact that its crust has

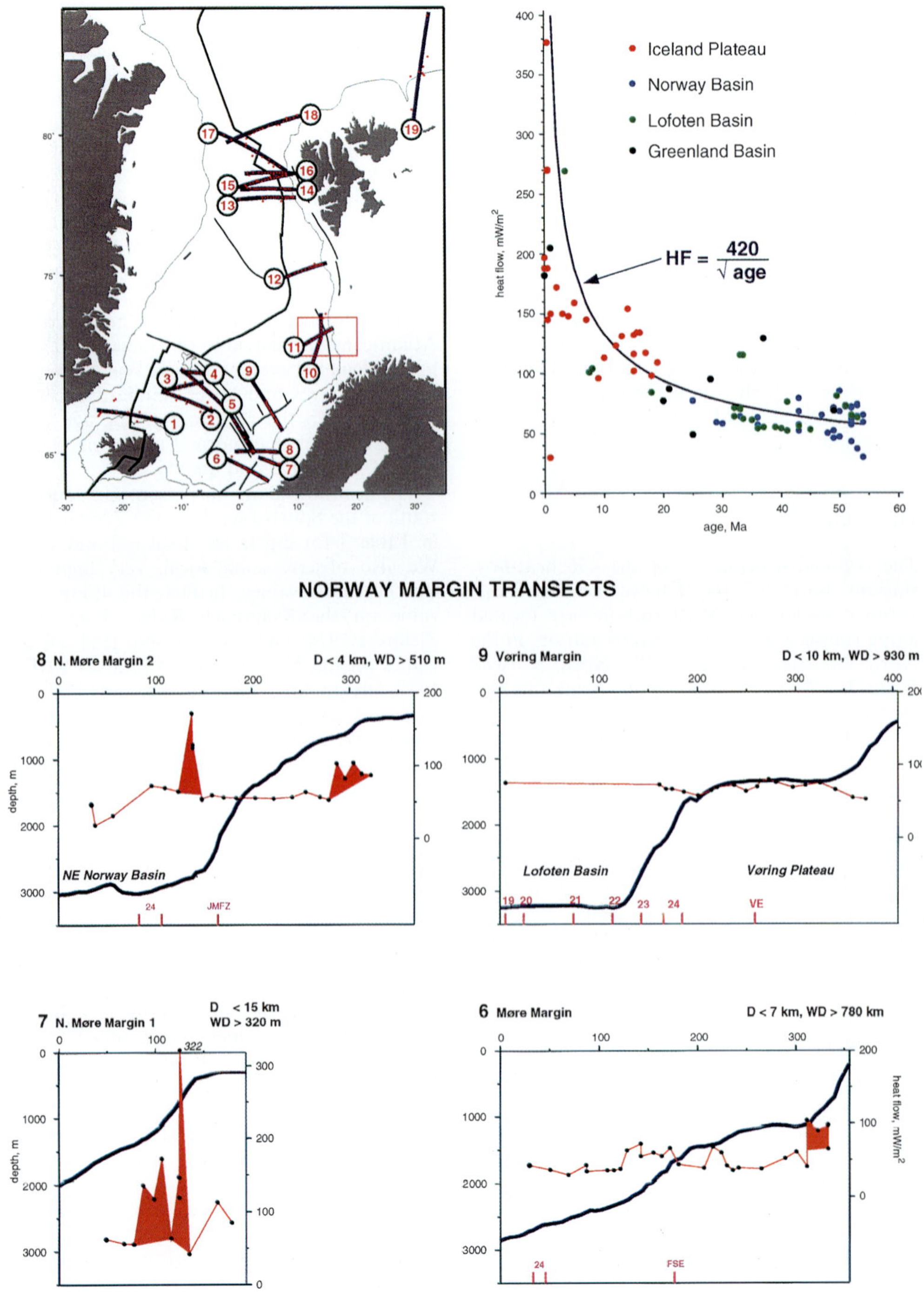

Fig. 3. Top: location of heat-flow transects (Figs 3–5), and heat flow v. age of oceanic crust for areas of well-identified crust south of the Greenland–Senja Fracture Zone. Bottom: Norway margin heat-flow transects. Red shading refers to local continental slope maxima. WD, water depth; D, distance normal to transect, 20 magnetic anomaly.

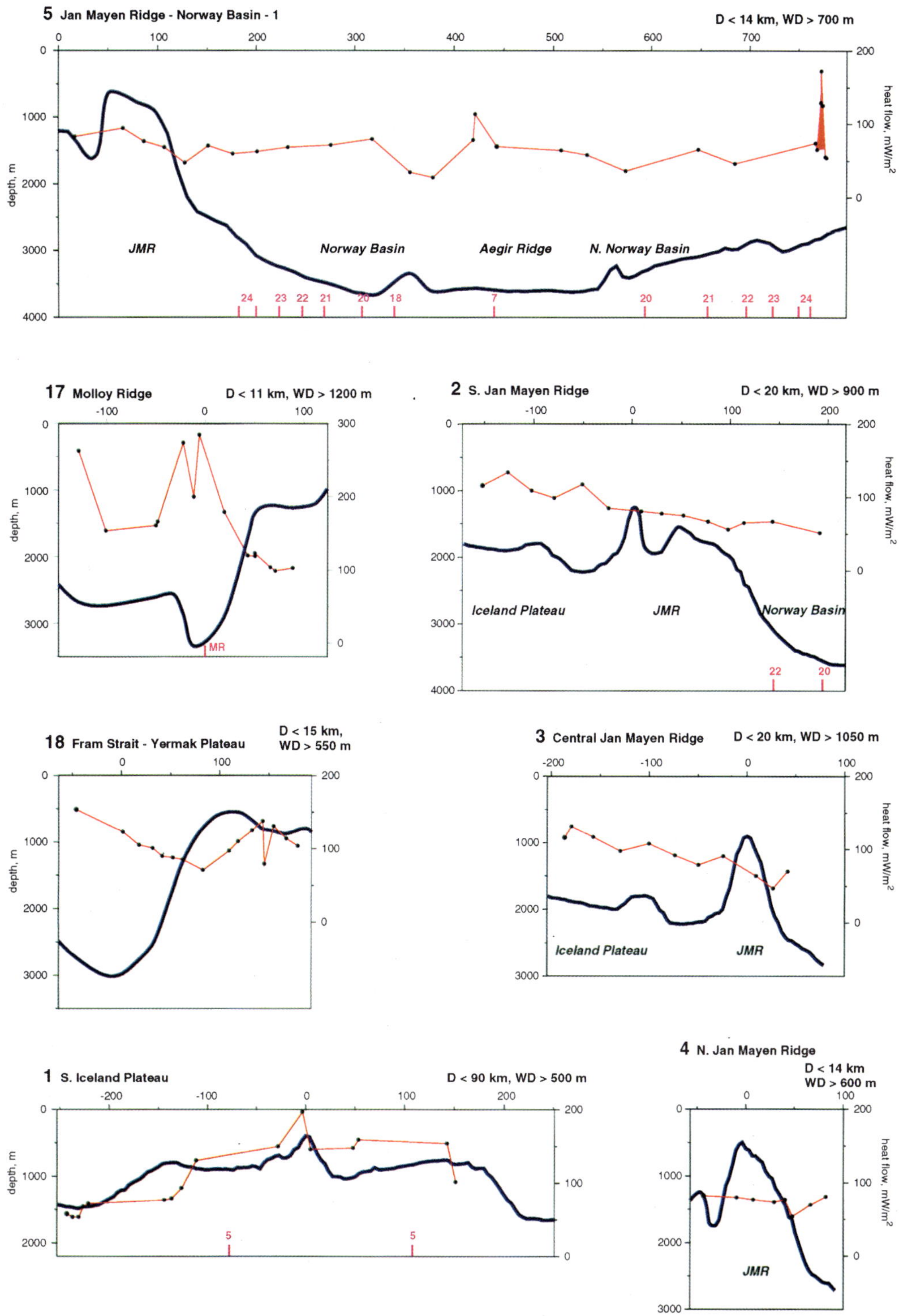

Fig. 4. Heat-flow transects over predominantly oceanic crust and microcontinents (Fig. 3). Red shading refers to local continental slope maxima.

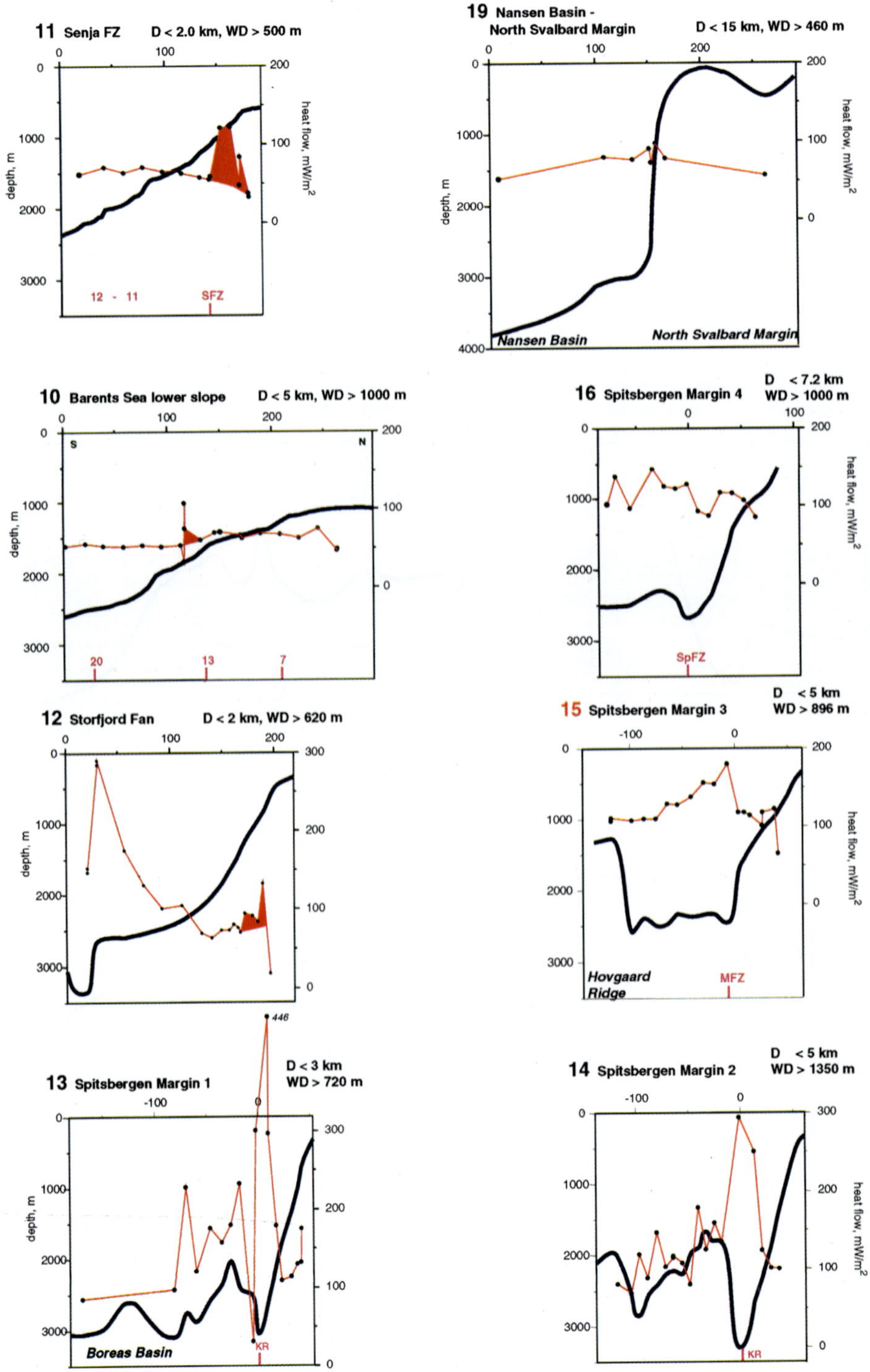

Fig. 5. Barents Sea and Svalbard margin heat-flow transects (Fig. 3). Red shading refers to local continental slope maxima.

been thinned, and intruded by two separate rift episodes (Gudlaugsson *et al.* 1988; Johansen *et al.* 1988).

The Hovgaard Ridge microcontinent (Fig. 1) separated from the Svalbard–Barents Sea margin by the readjustment of the plate boundary in early Oligocene time (Eldholm *et al.* 1987*a*). Recently, a continuation of the continental crust into the Hovgaard Basin northwest of the ridge proper has been suggested by A. M. Myhre (pers. comm.). The heat flow over the interpreted continental fragment is again similar to that in the adjacent basins (transects 14-15, Fig. 5), and there is excellent correspondence of the shallow measurements and those in ODP holes on top of the ridge and in the Hovgaard Basin. The average ridge heat flow is $91\,mW\,m^{-2}$, increasing to $117\,mW\,m^{-2}$ on the thinner crust in the Hovgaard Basin.

Norway–Svalbard continental margin

The heat-flow map reveals local highs on the continental slope off Norway and the Barents Sea (Fig. 1). We discuss the slope maxima separately, and these heat-flow values are not included in the regional statistics (Table 3).

The measurements on the Møre and Vøring margin cover both the marginal highs and the continental crust, which underwent extension before the early Tertiary break-up. Transects 6–9 exhibit a very small regional decrease in heat flow toward the continent; however, neither the marginal highs nor the Faeroe–Shetland and Vøring marginal escarpments have any appreciable heat-flow signature (Fig. 3). Modelling of the Vøring margin by Zielinski (1979) shows that the data fit a continent–ocean boundary in the vicinity of the Vøring Escarpment, and that effects of lateral heat flow may occur in this setting. The average heat flow landward of the marginal escarpments is $57\,mW\,m^{-2}$, but the heat flow is distinctly higher on the Vøring margin than farther south, i.e. 67 v. $52\,mW\,m^{-2}$ (Table 3). As first noted by Langseth & Zielinski (1974), the low Møre margin heat flow is surprising in view of the conductivity maximum in the northern part (Fig. 3). Our expanded database confirms the abnormal, and still enigmatic, heat flow and conductivity in this region. We also note that transect 9 which strongly affects the Vøring average, crosses the Vema Dome, which underwent both Paleocene and Late Oligocene–early Pliocene deformation, as well as the active Vema mud diapir field initiated in late Pliocene time (Hjelstuen *et al.* 1997).

The measurements on oceanic crust just west of the Senja Fracture Zone are internally consistent, exhibiting a very gentle increase with crustal age from 53 to $75\,mW\,m^{-2}$ along the north-trending transect 10, whereas the thermal field is stable, averaging $63.5 \pm 5.3\,mW\,m^{-2}$, along transect 11, which runs approximately along the magnetic anomaly 13 lineation (Fig. 5). The few measurements east of the fracture zone are not statistically significant, but several values are less than $50\,mW\,m^{-2}$.

The Senja Fracture Zone, a well-defined structural boundary demarcating a very narrow continent–ocean transition zone (Faleide *et al.* 1993), does not exhibit an appreciable heat-flow signature in transect 10; whereas reliable values of $121–122\,mW\,m^{-2}$, decreasing to $<40\,mW\,m^{-2}$ *c.* 20 km farther east, appear in transect 11 (Fig. 5). In view of several other local highs in this region, the anomaly may not relate to the fracture zone and we classify it as a slope maximum.

Both transect 12, on the post-late Pliocene Storfjord Fan (Hjelstuen *et al.* 1996), and transects 13–16 between 78 and 79°N are located west of the Hornsund Fault Zone (Fig. 5). This structure on the outer shelf south of 79°N demarcates the westernmost extent of the continental crust (Myhre & Eldholm 1988). The region between the fault zone and the Knipovich Ridge province is characterized by thick sediments of which the main part comprises prograded late Pliocene and Quaternary glacial fans and wedges (Faleide *et al.* 1996). In general, there is a larger scatter in heat flow than farther south, as well as a high regional level outside the ridge province. We attribute the former to the larger number of stations within the partly sediment-covered ridge province, and the latter to the young crust, particularly in the northern Greenland Sea.

The thermal evolution of the Svalbard margin and the northern Greenland Sea–Fram Strait region has been discussed and modelled by Crane *et al.* (1982, 1988, 1991). The results are consistent with the regional plate tectonic framework of Eldholm *et al.* (1987*a*), although a propagating rift and a migrating plate boundary have been advocated for the northern Greenland Sea.

North of Svalbard, the crustal nature of the margin is still largely unresolved (Sundvor & Austegard 1990). The regional heat-flow level west of 35°E is generally less than $75\,mW\,m^{-2}$; however, the western part of the Yermak Plateau constitutes an area of anomalously high heat flow and thermal conductivity (Figs 1 and 2). The high heat flow, and the late Miocene volcanism and present thermal activity in northern Svalbard, have been related to crustal dyke injection from 16 Ma to the present (Crane *et al.* 1982; Okay & Crane 1993). On the other hand, there is no direct spatial correspondence between the onshore thermal activity and the offshore heat-flow anomaly.

Slope maxima

Sundvor *et al.* (1989) pointed out several heat-flow maxima on the continental slope on the Møre and Barents Sea margins. Such highs, shown in red in Figs 3–5 and as elongate maxima in Fig. 1 are characterized by local, greatly enhanced heat flow with respect to the surrounding regional field. For example, in transect 7 (Fig. 3) we observe maxima of 322 mW m^{-2} and 118–172 mW m^{-2} at depths of 695 and 1100–1350 m respectively. Subsequently, Naval Research Laboratory SeaMARC II sidescan sonar and swath bathymetry cruises (Crane *et al.* 1995) revealed two circular back-scatter objects at of 1 km diameter *c.* 1250 m depth on the Barents Sea margin just south of transect 11. A core from the flank of the main anomaly contained gas and gas hydrate, and the heat flow was calculated to be 377 mW m^{-2} (Vogt & Sundvor 1996), whereas Eldholm *et al.* (1999), using a lower thermal conductivity, arrived at 298 mW m^{-2}. This discovery led to a detailed survey in 1996 confirming that the main structure, named the Håkon Mosby Mud Volcano, oozed mud and seeped gas, and that it is probably capped by gas hydrate at the sea floor (Vogt *et al.* 1997). Furthermore, the heat-flow maximum measured on the structure exceeds 1000 mW m^{-2} (Eldholm *et al.* 1999).

The mud volcano is located within a large slide scar on the post-late Pliocene Bear Island Fan, and is underlain by *c.* 33 Ma old oceanic crust (Hjelstuen *et al.* 1999). Farther south, the heat flow maximum in transect 7 and the outer maximum in transect 8 (Fig. 3) are measured within the Storegga slide (Bugge *et al.* 1987). Although the mechanism causing the heat-flow anomalies is still discussed (e.g. Vogt & Sundvor 1996; Eldholm *et al.* 1999), a relationship with large-scale, rapid mass movement inducing changes in sediment temperature and pressure as well as providing new pathways for fluids and gas needs to be explored. None the less, the discovery of the Håkon Mosby Mud Volcano and its associated gas seep suggests that the other local slope anomalies on the Norway–Svalbard margin may also be caused seeping mud, fluid and gas. However, detailed local surveys are required to confirm this inference.

Summary

All published thermal gradient, thermal conductivity and heat-flow values from the Norwegian–Greenland Sea and in the western Eurasia Basin have been stored on a freely available UNIX-platform database. The compilation of heat-flow and thermal conductivity data shows the following. There is a clear, first-order heat-flow–crustal age relationship on dated oceanic crust. Thermal conductivities are relatively uniform and consistent in the 0.85–1.15 W m^{-1} K^{-1} range; low values are measured in the ocean basin and higher values on the continental margin. Local areas of high conductivity may reflect present erosion or non-deposition. Local very high heat flow along the present spreading axis implies probable active volcanism or venting. Continental slope maxima on the Møre and western Barents Sea margins contrast greatly with the typical low heat flow, generally less than 75 mW m^{-2}, on old oceanic and thinned continental crust. Abnormally high heat flow, exceeding 1000 mW m^{-2}, has been measured on the Håkon Mosby gas-seeping mud volcano on the Barents Sea margin. Continental slope maxima elsewhere may also be related to active seeps.

We are particularly grateful to G. L. Johnson, for supporting the acquisition in 1983 of a Norwegian heat-flow probe dedicated to high-latitude studies, and to Elf Petroleum Norge for initiating and supporting the French–Norwegian FLUNORGE Project. We also thank P. R. Vogt and K. Crane for their efforts and advice during a long period of active collaboration.

References

BUGGE, T., BEFRING, S., BELDERSON, R. H. *et al.* 1987. A giant three-stage submarine slide off Norway. *Geo-Marine Letters*, **7**, 191–198.

CRANE, K., ELDHOLM, O., MYHRE, A. M. & SUNDVOR, E. 1982. Thermal implications for the evolution of the Spitsbergen transform fault. *Tectonophysics*, **89**, 1–32.

——, SUNDVOR, E., BUCK, R. & MARTINEZ, F. 1991. Rifting in the northern Norwegian–Greenland Sea: thermal tests of asymmetric spreading. *Journal of Geophysical Research*, **96**, 14 529–14 550.

——, ——, FOUCHER, J. P., HOBART, M., MYHRE, A. M. & LEDOUARAN, S. 1988. Thermal evolution of the western Svalbard margin. *Marine Geophysical Researches*, **9**, 165–194.

——, VOGT, P. R., SUNDVOR, E., SHOR, A. & REED, T., IV 1995. SeaMARC II investigations in the northern Norwegian–Greenland Sea. *Meddelelser Norsk Polarinstitutt*, **137**, 132–140.

ELDHOLM, O. & WINDISCH, C. C. 1974. Sediment distrbution in the Norwegian–Greenland Sea. *Geological Society of America Bulletin*, **85**, 1661–1676.

——, FALEIDE, J. I. & MYHRE, A. M. 1987*a*. Continent–ocean transition at the western Barents Sea/Svalbard continental margin. *Geology*, **15**, 1118–1122.

——, KARASIK, A. M. & REKSNES, P. A. 1990. The North American plate boundary. *In*: GRANTZ, A., JOHNSON, G. L. & SWEENEY, J. (eds) *The Arctic Ocean Region*. Geological Society of America, North America, **L**, 171–184.

——, MYHRE, A. M. & THIEDE, J. 1994. Cenozoic tectono-magmatic events in the North Atlantic: potential paleoenvironmental implications. *In*: BOULTER, M. C. & FISCHER, H. C. (eds) *Cenozoic Plants and Climates of the Arctic*. NATO ASI Series I27. Springer, Heidelberg, 35–55.

——, SUNDVOR, E., VOGT, P. R., HJELSTUEN, B. O., CRANE, K., NILSEN, A. K. & GLADCZENKO, T. P. 1999. SW Barents Sea continental margin heat flow and Håkon Mosby Mud Volcano. *Geo-Marine Letters*, **18**, in press.

——, THIEDE, J. & TAYLOR, E. 1989. Evolution of the Vøring Volcanic Margin. *Proceedings of the Ocean Drilling Program, Scientific Results*, **104**. Ocean Drilling Program, College Staion, TX, 1033–1065.

——, —— & —— *et al.* (eds) 1987*b*. *Proceedings of the Ocean Drilling Program, Initial Reports*, **104**. Ocean Drilling Program, College Station, TX.

FALEIDE, J. I., SOLHEIM, A., FIEDLER, A., HJELSTUEN, B. O., ANDERSEN, E. S. & VANNESTE, K. 1996. Late Cenozoic evolution of the western Barents Sea–Svalbard continental margin. *Global and Planetary Change*, **318**, 53–74.

——, VÅGNES, E. & GUDLAUGSSON, S. T. 1993. Late Mesozoic–Cenozoic evolution of the south-western Barents Sea in a regional rift–shear tectonic set-ting. *Marine and Petroleum Geology*, **10**, 186–214.

FLOVENZ, O. G. & SÆMUNDSSON, K. 1993. Heat flow and geothermal processes in Iceland. *Tectonophysics*, **225**, 123–138.

GERARD, R. S., LANGSETH, M. G. & EWING, M. 1962. Thermal gradient measurements in the water and bottom sediment of the western Atlantic. *Journal of Geophysical Research*, **67**, 785–803.

GUDLAUGSSON, S. T., GUNNARSON, K., SAND, M. & SKOGSEID, J. 1988. Tectonic and volcanic events at the Jan Mayen Ridge microcontinent. *In*: MORTON, A. C. & PARSON, L. M. (eds) *Early Tertiary Volcanism and the Opening of the NE Atlantic*. Geological Society, London, Special Publications, **39**, 85–93.

HAENEL, R. 1974. Heat flow measurements in the Norwegian Sea. *Meteor Forschungsergebnisse, Reihe C*, **17**, 74–78.

——1979. A critical review of heat flow measurements in sea and lake bottom sediments. *In*: CERMAK, V. & RYBACH, L. (eds) *Terrestrial Heat Flow in Europe*. Springer, Berlin, 49–70.

HAGEVANG, T., ELDHOLM, O. & AALSTAD, I. 1983. Pre-23 magnetic anomalies between Jan Mayen and Greenland–Senja fracture zones in the Norwegian Sea. *Marine Geophysical Researches*, **5**, 345–363.

HAMILTON, E. L. 1974. Prediction of deep sea sediment properties: state of the art. *In*: INDERBITZEN, A. L. (ed.) *Deep Sea Sediments: Physical and Mechanical Properties*. Plenum, New York, 1–42.

HJELSTUEN, B. O., ELDHOLM, O., FALEIDE, J. I. & VOGT, P. R. 1999. Regional setting of the Håkon Mosby Mud Volcano, SW Barents Margin. *Geo-Marine Letters*, **18**, in press.

——, —— & SKOGSEID, J. 1997. Vøring Plateau diapirs and their structural and depositional settings. *Marine Geology*, **144**, 33–57.

——, ELVERHØI, A. & FALEIDE, J. I. 1996. Cenozoic erosion and sediment yield in the drainage area of the Storfjorden Fan. *Global and Planetary Change*, **12**, 95–117.

JACKSON, H. R., JOHNSON, G. L., SUNDVOR, E. & MYHRE, A. M. 1984. The Yermak Plateau: formed at a triple junction. *Journal of Geophysical Research*, **89**, 3223–3232.

JANSEN, E., RAYMO, M. E., BLUM, P. *et al.* (eds) 1996. *Proceedings of the Ocean Drilling Program, Initial Reports*, **162**. Ocean Drilling Program, College Station, TX.

JOHANSEN, B. 1989. *Marine geophysical studies in the North Atlantic with emphasis on sedimentation, crustal structure and subsidence*. Dr.scient. thesis, University of Oslo.

——, ELDHOLM, O., TALWANI, M., STOFFA, P. L. & BUHL, P. 1988. Expanding spread profile at the northern Jan Mayen Ridge. *Polar Research*, **6**, 95–104.

LACHENBRUCH, A. H. & MARSHALL, B. V. 1968. Heat flow and water temperature fluctuations in the Denmark Strait. *Journal of Geophysical Research*, **73**, 5829–5842.

LANGSETH, M. G. & ZIELINSKI, G. W. 1974. Marine heat flow measurements in the Norwegian–Greenland Sea and in the vicinity of Iceland. *In*: KRISTJANSON, L. (ed.) *Geodynamics of Iceland and the North Atlantic Area*. Reidel, Dordrecht, 277–295.

LANGSETH, M. G., LACHENBRUCH, A. H. & MARSHALL, B. V. 1990. Geothermal observations in the Arctic region. *In*: GRANTZ, A., JOHNSON, G. L. & SWEENEY, J. (eds) *The Arctic Ocean Region*. Geological Society of America, The Geology of North America, **L**, 113–151.

LØSETH, H., LIPPARD, S. J., SÆTTEM, J. *et al.* 1992. Cenozoic uplift and erosion of the Barents Sea – evidence from the Svalis Dome area. *In*: VORREN, T. O., BERGSAKER, E., DAHL-STAMNES, Ø. A., HOLTER, E., JOHANSEN, B., LIE, E. & LUND, T. B. (eds) *Artic Geology and Petroleum Potential*. Norwegian Petroleum Society Special Publication, **2**, 643–664.

MYHRE, A. M. & ELDHOLM, O. 1988. The western Svalbard margin (74–80°N). *Marine and Petroleum Geology*, **5**, 134–156.

——, ——, FALEIDE, J. I. *et al.* 1992. Norway–Svalbard Margin: structural and stratigraphical styles. *In*: POAG, C. W. & DE GRACIANSKY, P. C. (eds) *Geologic Evolution of Atlantic Continental Rises*. Van Nostrand Reinhold, New York, 157–185.

——, —— & SUNDVOR, E. 1984. The Jan Mayen Ridge: present status. *Polar Research*, **2**, 47–59.

——, THIEDE, J., FIRTH, J. *et al.* (eds) 1995. *Proceedings of the Ocean Drilling Program, Initial Reports*, **151**. Ocean Drilling Program, College Station, TX.

NUNNS, A. G. 1983. Plate tectonic evolution of the Greenland–Scotland Ridge and surrounding regions. *In*: BOTT, M. H. P. *et al.* (eds) *Structure and Development of the Greenland–Scotland Ridge*. Plenum, New York, 11–30.

OKAY, N. & CRANE, K. 1993. Thermal rejuvenation of the Yermak Plateau. *Marine Geophysical Researches*, **15**, 243–263.

PLANKE, S. 1989. *HEAT – A heat flow data base program*. Computer Program/Database Documentation Series, Marine and Applied Geophysics Research Group, Department of Geology, University of Oslo, **2**.

RATCLIFFE, A. E. 1960. The thermal conductivities of ocean sediments. *Journal of Geophysical Research*, **65**, 1535–1541.

SKOGSEID, J. & ELDHOLM, O. 1987. Early Cenozoic crust at the Norwegian continental margin and the conjugate Jan Mayen Ridge. *Journal of Geophysical Research*, **92**, 11 471–11 491.

SUNDVOR, E. 1986. *Heat Flow Measurements on the Western Svalbard Margin*. University of Bergen, Seismological Observatory, Internal Report.

—— & AUSTEGARD, A. 1990. The evolution of the Svalbard margins: synthesis and new results. *In*: BLEIL, U. & THIEDE, J. (eds) *Geological History of the Polar Oceans: Arctic versus Antarctic*. Kluwer Dordrecht, 77–96.

—— & ELDHOLM, O. 1992. Norway: offshore and north-east Atlantic. *In*: HURTIG, E., CERMAK, V., HAENEL, R. & ZUI, V. (eds) *Geothermal Atlas of Europe*. GeoForschungZentrum Potsdam, Publication, **1**, 63–66.

—— & MYHRE, A. M. 1987. *Heatflow Measurements: Jan Mayen Ridge and Norway Basin*. University of Bergen, Seismological Observatory, Seismo-series, **9**.

—— & TORP, J. E. 1987. *Measurements of Heat Flow and Thermal Conductivity in the Nansen Basin, Arctic Ocean*. University of Bergen, Seismological Observatory: Cruise Report.

——, MYHRE, A. M. & ELDHOLM, O. 1989. *Heat Flow Measurements on Norwegian Continental Margin during the Flunorge Project*. University of Bergen, Seismological Observatory, Seismo-series, **27**.

SÆTTEM, J. 1988. Varmestrømsmålinger i Barentshavet. *18. Nordiske Geologiske Vintermøde, Geological Survey of Denmark* (extended abstract), 406–408.

TALWANI, M. & ELDHOLM, O. 1977. Evolution of the Norwegian–Greenland Sea. *Geological Society of America Bulletin*, **88**, 969–999.

UDINTSEV, G. B. & LUBIMOVA, E. A. 1973. Teplovye potoki vblizi Islandii. *Izvestiya Akademii Nauk SSSR, Fizika Zemli*, **11**.

VOGT, P. R. & SUNDVOR, E. 1996. Heat flow highs in the Norwegian–Barents–Svalbard continental slope; deep crustal fractures, dewatering, or 'memory in the mud'. *Geophysical Research Letters*, **23**, 3571–3574.

—— *et al.* 1997. Haakon Mosby mud volcano provides unusual example of venting. *EOS Transactions, American Geophysical Union*, **78**, 549–557.

VON HERZEN, R. P. & MAXWELL, A. E. 1959. The measurements of thermal conductivity of deep-sea sediments by a needle probe method. *Journal of Geophysical Research*, **64**, 1557–1563.

WESSEL, P. & SMITH, W. H. F 1991. Free software helps map and display data. *EOS Transactions, American Geophysical Union*, **72**, 441–446.

ZIELINSKI, G. W. 1979. On the thermal evolution of passive continental margins, thermal depth anomalies and the Norwegian–Greenland Sea. *Journal of Geophysical Research*, **84**, 7577–7588.

——, GUNLEIKSRUD, T., SÆTTEM, J., ZUIDBERG, H. M. & GEISE, J. M. 1986. Deep heat flow measurements in Quaternary sediments on the Norwegian continental shelf. *Offshore Techology Conference Houston, TX 1986*, 277–282.

Atlantic volcanic margins: a comparative study

O. ELDHOLM, T. P. GLADCZENKO, J. SKOGSEID & S. PLANKE

Department of Geology, University of Oslo, P.O. Box 1047 Blindern, N-0316 Oslo, Norway (e-mail: olav.eldholm@geologi.uio.no)

Abstract: Volcanic margins in the Atlantic Ocean reveal a series of common crustal units and structural features developed during continental extension and break-up. We suggest that four main crustal zones can be recognized on volcanic margins. This tectono-magmatic zonation implies a history of development where tectonic and magmatic styles and dimensions depend on the interaction of lithospheric and asthenospheric properties and dynamics. The amount of excess igneous activity depends on the temperature and fluid content of the asthenosphere along the incipient plate boundary and the dynamic history of the lithosphere during the rift phase. An adequate understanding of the margin history requires studies of the entire rift, i.e. the conjugate margins. We also note that the spectacular wedges of seaward-dipping reflectors observed along many rifted margins are only one of many igneous features originating during the process of break-up and initial sea-floor spreading. Probably, most passive rifted margins represent intermediate cases relative to the volcanic and non-volcanic end-members. A mantle plume impinging on lithosphere already under extension emplacing Large Igneous Province-type initial oceanic crust, including an extensive extrusive cover, is considered the most likely explanation for volcanic margins. Hydrocarbon resource evaluations of volcanic margins have to include their characteristic tectono-magmatic features and their consequences for vertical motion, erosion, sedimentation, thermal and burial histories, and maturation.

The main Norwegian contribution to the European Union-Joule II research project 'Integrated Basin Studies' comprises the module 'Dynamics of the Norwegian Margin' (IBS-DNM), focusing on the development and evolution of sediment basins in intraplate and passive margin settings (Nøttvedt *et al.* 2000). The North Atlantic and North Sea Mesozoic rift systems merge on the mid-Norway continental margin where the subsequent Late Cretaceous–Paleocene rift episode led to sea-floor spreading, accompanied by massive igneous activity, near the Paleocene–Eocene transition.

Evidence of massive, transient igneous activity during the final break-up of continents and the initial period of sea-floor spreading exists from many other of the world's passive continental margins (Fig. 1). To evalute the processes that govern the inception and evolution of such margins it is necessary to compare crustal structure, tectono-magmatic relations and the history of vertical motion. Consequently, the theme 'comparative volcanic margin studies' became part of IBS-DNM. Here, we present results mainly from Atlantic margins with an emphasis on crustal structure and tectono-magmatic style and dimensions.

We have compiled a global volcanic margin database from the scientific literature and studies of selected margins such as the North Atlantic conjugate margins north of Charlie Gibbs Fracture Zone, the North Namibia margin and other South Atlantic margin segments south of the Abutment and São Paulo plateaux, the US East Coast margin, the western margin off India, and the western Australia margin (see Fig. 1). Globally, the North Atlantic volcanic margins (Fig. 2) are the best explored both by geophysical surveys and drilling (see Eldholm *et al.* 1995; Skogseid *et al.* 2000), and we use this region to define typical volcanic margin tectono-magmatic features.

Volcanic margins

The massive extrusive complexes along the rifted margin segments off Norway (Fig. 2) were first recognized by an exceptionally smooth acoustic basement surface near the continent–ocean boundary (COB) (Talwani & Eldholm 1972, 1977). Later, multichannel seismic (MCS) lines imaged wedges of seaward-dipping, intrabasement reflectors (Hinz & Weber 1976; Hinz & Schlüter 1978; Talwani 1978; Eldholm *et al.* 1979). Similar wedges were also recognized elsewhere (e.g. Hinz 1981), and the number of observations has steadily increased with the

From: NØTTVEDT, A. *et al.* (eds) *Dynamics of the Norwegian Margin*. Geological Society, London, Special Publications, **167**, 411–428. 1-86239-056-8/00/$15.00

	Physiography	Age	Structural framework	Overburden	Pre-rift geology	Magmatism
End-member	Narrow (steep slope)	Young	Rift	Starved	Craton	Volcanic
	Marginal plateau	Intermediate	Shear-rift			
End-member	Wide (gentle slope)	Mature	Shear	Major basin	Basin	Non-volcanic

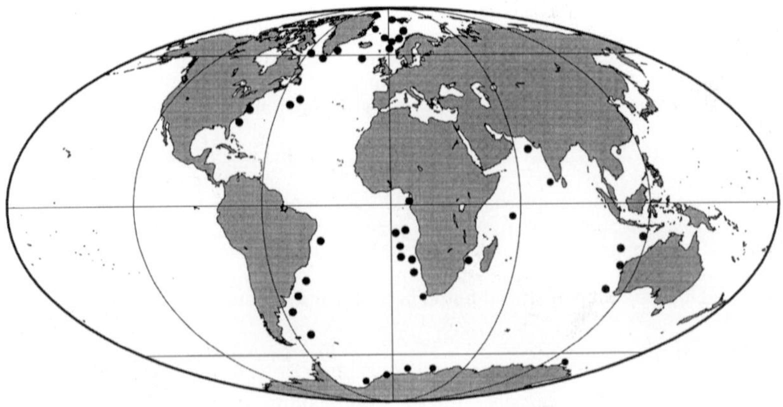

NE NORTH ATLANTIC
- Hatton Bank Margin
- Møre Margin
- Vøring Margin
- Lofoten Margin
- Vestbakken Volc. Prov.
- Jan Mayen Ridge
- NE and SE Greenland Margins

SW GREENLAND/LABRADOR SEA

US/CANADA EAST COAST
- Baltimore Trough
- Inner Blake Plateau
- Carolina Trough
- Georgia Embayment
- Georges Bank
- Newfoundland Basin
- SE Newfoundland Ridge

AUSTRALIA
- Scott Plateau
- Cuvier Margin
- Wallaby Plateau
- Exmouth Plateau
- Naturaliste Plateau

SOUTH ATLANTIC
- S Cape Basin
- Namibia Margin
- Abutment Plateau
- Angola Basin
- Ceara Rise
- Sierra Leone Rise
- Pelotas Basin
- São Paulo Plateau
- Argentina Margin
- Falklands Margin

ANTARCTICA
- Weddell Sea Margin
- Wilkes Land Margin
- Georgia Rise

INDIAN OCEAN
- Mozambique Margin
- Kathiawar Margin
- Cochin Margin
- Bay of Bengal
- NE Seychelles Margin

Fig. 1. Top: passive continental margin classifications (Eldholm *et al.* 1995). Middle and bottom: distribution of volcanic margins based on reported intrabasement and seaward-dipping reflectors.

acquisition of high-quality seismic lines on the outer margins (Fig. 1). Outside the North Atlantic, extensive and voluminous extrusive units exist along the US East Coast (e.g. Talwani *et al.* 1995) and in the South Atlantic south of the Abutment and São Paulo plateaus (Hinz *et al.* 1995; Gladczenko *et al.* 1997*b*). Furthermore, wide-angle seismic experiments commonly reveal a high-velocity lower-crustal body (LCB) beneath the extrusive cover. During the past decade these margins have been classified as volcanic margins (e.g. Eldholm *et al.* 1995).

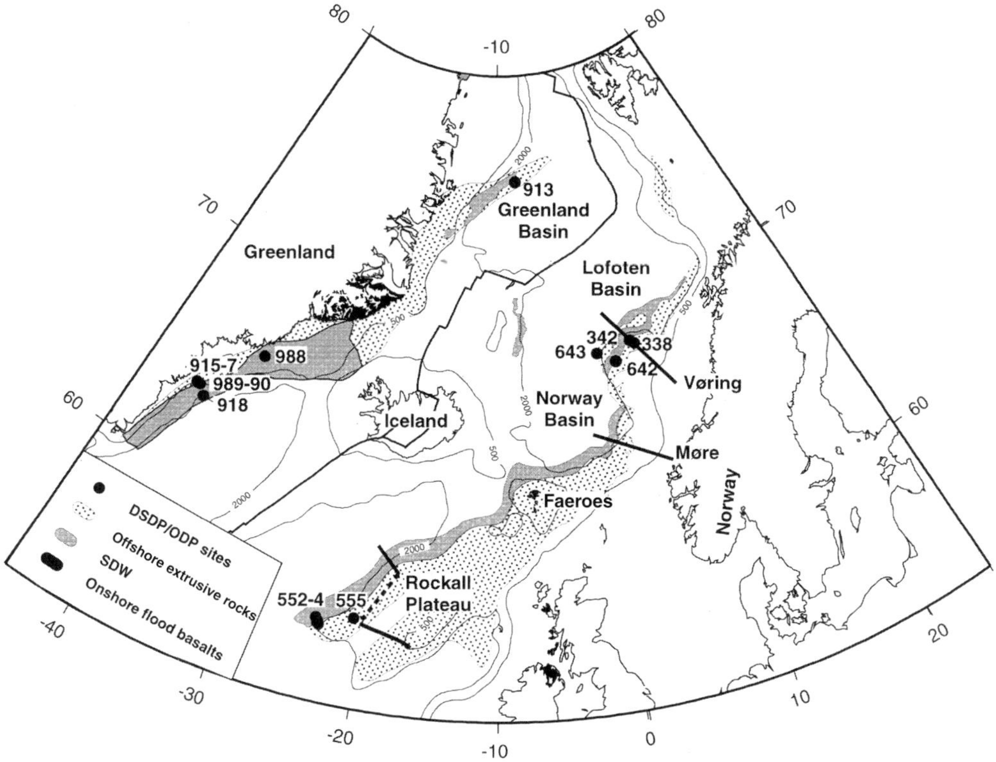

Fig. 2. North Atlantic volcanic margins with distribution of flood basalts (updated from Eldholm & Grue 1994), and locations of DSDP and ODP drill sites sampling igneous basement rocks (see Table 1). SDW, main wedges of seaward-dipping reflectors.

Volcanic margins, continental flood basalts (CFB), oceanic plateaux and ocean basin flood basalts constitute the main categories of transient large igneous provinces (LIP) composed of voluminous constructions of predominantly mafic igneous rocks that have not been emplaced by normal sea-floor spreading. The transient, large-scale volcanism is commonly attributed to mantle plumes (e.g. White & McKenzie 1989; Duncan & Richards 1991; Larson 1991). The dimensions and emplacement rates for some volcanic margins show that they contribute significantly to the global crustal production budget and that they may induce environmental change (Coffin & Eldholm 1994).

North Atlantic conjugate rifted margins

The Early Tertiary continental break-up and onset of sea-floor spreading between Eurasia and Greenland, *c.* 55 Ma, was accompanied by massive transient volcanism emplacing onshore flood basalts (Dickin 1988) and massive coeval extrusive and intrusive rock complexes on the rifted margins (Fig. 2). The break-up volcanism took place, in part subaerially, along more than 2600 km of the early Tertiary plate boundary, with most of the lavas extruded during Chron 24r. The transient event contrasts with persistent subaerial volcanism for *c.* 60 Ma along the Iceland plume trail leaving the Greenland–Iceland–Faeroe ridge between the conjugate Faeroe and East Greenland CFBs (e.g. Eldholm *et al.* 1989) (Fig. 2).

Several scientific drill holes have recovered rocks from the seaward-dipping wedges on the Hatton Bank, Vøring and SE Greenland margins (Fig. 2, Table 1). The wedges consist of mainly tholeiitic basalt and thin interbedded sediments reflecting a subaerial and/or shallow-water constructional environment. The maximum penetration was achieved at ODP Site 642, which drilled through *c.* 800 m of basalts and *c.* 130 m into underlying dacitic–andesitic lavas and interbedded sediments.

Although the seaward-dipping wedges yield a characteristic seismic image, which is commonly

Table 1. *Summary of scientific volcanic margin drill sites sampling basement rocks in the North Atlantic (Fig. 2)*

Site	Water depth (m)	Sediment thickness (m)	Basement penetration (m)	Reference
DSDP 338	1297.0	400.8	0.95	Talwani *et al.* (1976)
DSDP 342	1303.0	153.2	17.3	Talwani *et al.* (1976)
DSDP 553	2301	282.7	31.3	Roberts *et al.* (1984)
DSDP 553	2329	499.35	183	Roberts *et al.* (1984)
DSDP 554	2574	126.6	82.4	Roberts *et al.* (1984)
DSDP 555	1659	927.32	37	Roberts *et al.* (1984)
ODP 642	1286	315.2	914.2	Eldholm *et al.* (1987)
ODP 643	2753	565.2+	–	Eldholm *et al.* (1987)
ODP 913	3318.6	770+	–	Myhre *et al.* (1995)
ODP 915	533.1	196.8	12.6	Larsen *et al.* (1994)
ODP 917	508.1	41.9	833.0	Larsen *et al.* (1994)
ODP 918	1868.2	1189.4	15.0	Larsen *et al.* (1994)
ODP 988	262.6	10	22.0	Duncan *et al.* (1996)
ODP 989	554.6	4	80.2	Duncan *et al.* (1996)
ODP 990	541.5	211.9	130.8	Duncan *et al.* (1996)

ODP sites 643 and 913, which terminated just above basement, are included because they provide data on volcanic margin subsidence.

taken as a criterion for volcanic margins, several other igneous features relate to the transient break-up event (Table 2). In particular, the basaltic lavas may extend for large distances landward of the wedges, and there is an apparent spatial correlation between the LCB and the most voluminous extrusive rocks. The 10–20 km thick initial oceanic crust west of the COB thins to normal oceanic thickness over a distance of 50–150 km (Fig. 3). Moreover, sills and dykes intrude the pre-Eocene continental crust. These observations document that the volcanic margin encompasses intrusive and extrusive features far beyond the dipping wedges (Eldholm *et al.* 1989; Skogseid & Eldholm 1989). Thus, the North Atlantic LIP includes flood basalts both onshore and on the rifted margin; in fact, the volcanic margin constitutes its major component. The LCB and crustal intrusive companions to flood basalt volcanism make also significant contributions to the crustal volume. Minimum extrusive and total igneous crustal volumes are estimated to be 1.8×10^6 km^3 and 6.6×10^6 km^3, respectively (Eldholm & Grue 1994) (Table 3).

Table 2. *Geological features associated with transient, igneous activity during break-up in the North Atlantic LIP (Skogseid & Eldholm 1995)*

- Continental flood basalts
  - associated intrusive rocks
- Extrusive complexes along continent–ocean transition
  - seaward-dipping wedges and sub-horizontal units
  - associated introsive rocks
- Sills and low-angle dykes
- Volcanic vents
- Regional tephra horizons
- High-velocity lower-crustal bodies (LCB)
- Thick initial oceanic crust

The tectonic dimensions, based on structural features, subsidence modelling and crustal thickness variations, have been discussed by Skogseid *et al.* (1992*a*) and Skogseid (1994). They suggest that North Atlantic volcanic margin formation was preceded by a rift phase lasting for about 18–20 Ma before break-up (Table 3). The lithospheric extension, which affected a 300 km wide region, separated Eurasia and Greenland by about 140 km (Skogseid 1994).

Namibia and other South Atlantic margins

The South Atlantic break-up occurred at *c.* 130 Ma off South Africa and progressed northward. The oldest identified anomaly off Namibia is M4, *c.* 127 Ma (Rabinowitz & LaBrecque 1979). The Paraná and Etendeka CFBs (Fig. 4) have been dated to 137–127 Ma (Turner *et al.* 1994) and 130–125 Ma (Erlank *et al.* 1984; Milner *et al.* 1992), respectively. They are linked by the Walvis Ridge–Rio Grande Rise along the Tristan plume trail (O'Connor & Duncan 1990). Large extrusive constructions exist on both conjugate margins south of the plume trail (Table 3) (e.g. Hinz *et al.* 1995; Abreu *et al.* 1996; Condi *et al.* 1996; Gladczenko *et al.* 1997*b*).

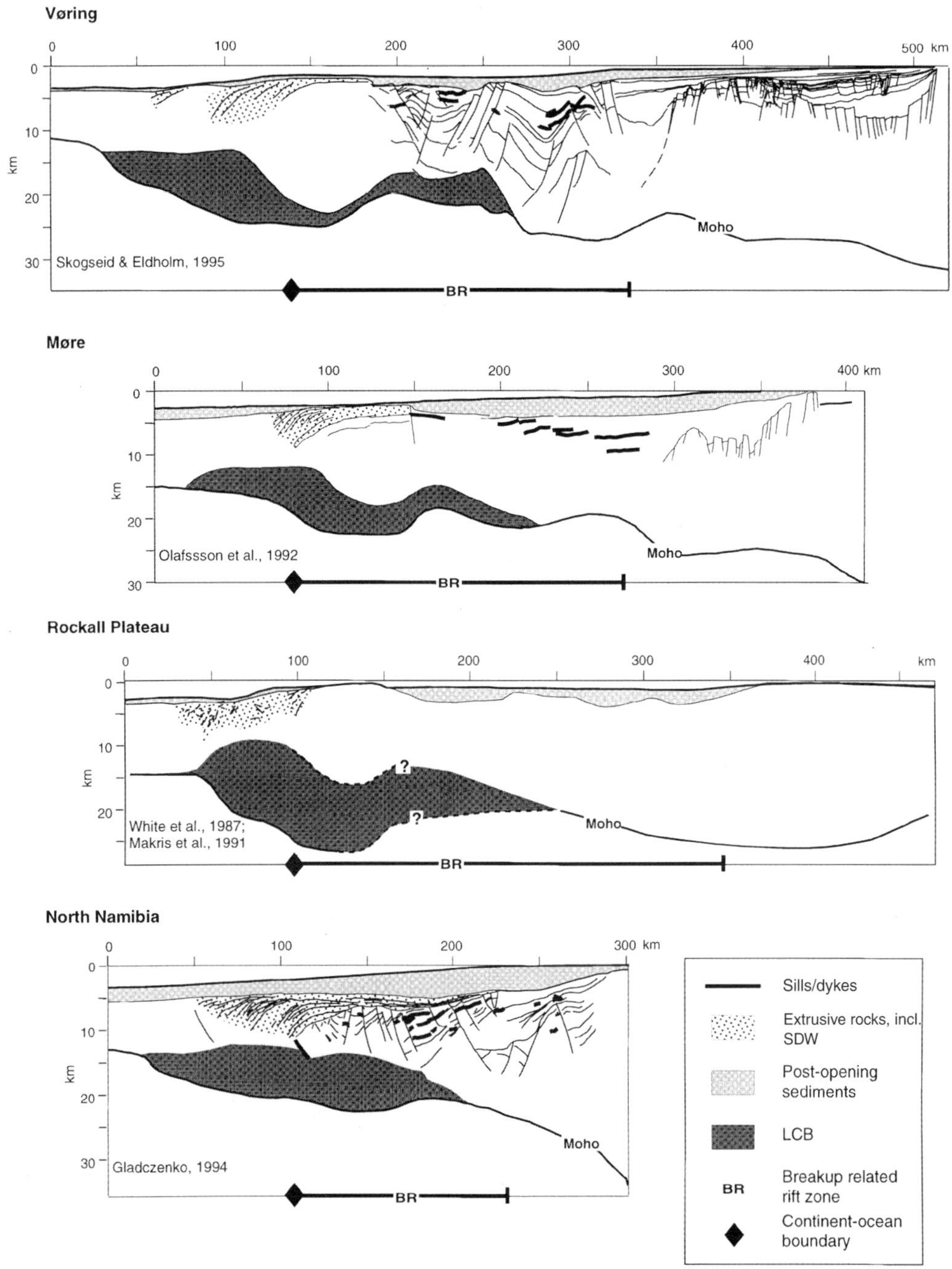

Fig. 3. Simplified crustal margin transects. SDW, main wedges of seaward-dipping reflectors; LCB, lower-crustal high-velocity, high-density body. The continent–ocean boundary is placed at the seaward termination of the base reflector below the inner dipping wedge. Transect locations in Figs 2 and 4–6.

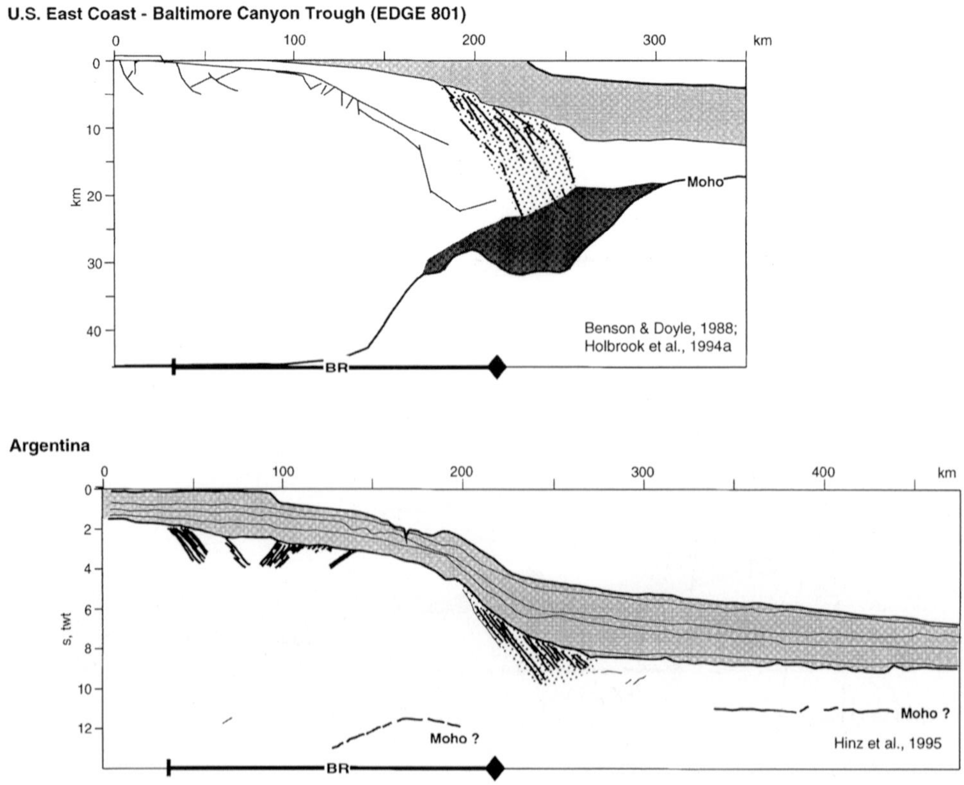

Figure 3. (*continued*)

Table 3. *Tectono-magmatic dimensions for volcanic margin LIPs*

Volcanic margin or LIP	Tectonic dimensions		Magmatic dimensions			
	Rift width (km)	Rift duration (Ma)	Length (km)	Area ($\times 10^6$ km^2)	Extrusive volume ($\times 10^6$ km^3)	Total crustal volume ($\times 10^6$ km^3)
North Atlantic*	300	18–20	2600	1.3	2.0	6.6
South Atlantic†	280	25	2400		2.4	
US East Coast Margin‡	200	70	1000		0.19	0.72

* Eldholm & Grue (1994).
† Gladczenko *et al.* (1997*b*).
‡ Gladczenko *et al.* (1994).

We have interpreted a grid of commercial MCS lines on the outer North Namibia margin, which reveal prominent seaward-dipping wedges and other coeval igneous features. The margin is divided into four tectono-magmatic zones (Figs 3 and 5): (1) oceanic crust; (2) thickened oceanic crust covered by seaward-dipping wedges; (3) a *c.* 150 km wide break-up related Late Jurassic–Early Cretaceous rift (BR, Fig. 3), partly covered by the dipping wedges in the west and lava flows and intrusions to the east; (4) thicker continental crust. The crust in zones (3) and (4) has undergone earlier, Late Palaeozoic extension.

Faulting in zone (3) sediments records the Late Jurassic–Early Cretaceous rifting which culminated with break-up. Central rift uplift led to an erosional rift unconformity. The subsequent break-up volcanism caused initial, subaerial seafloor spreading, seaward-dipping wedges, as well

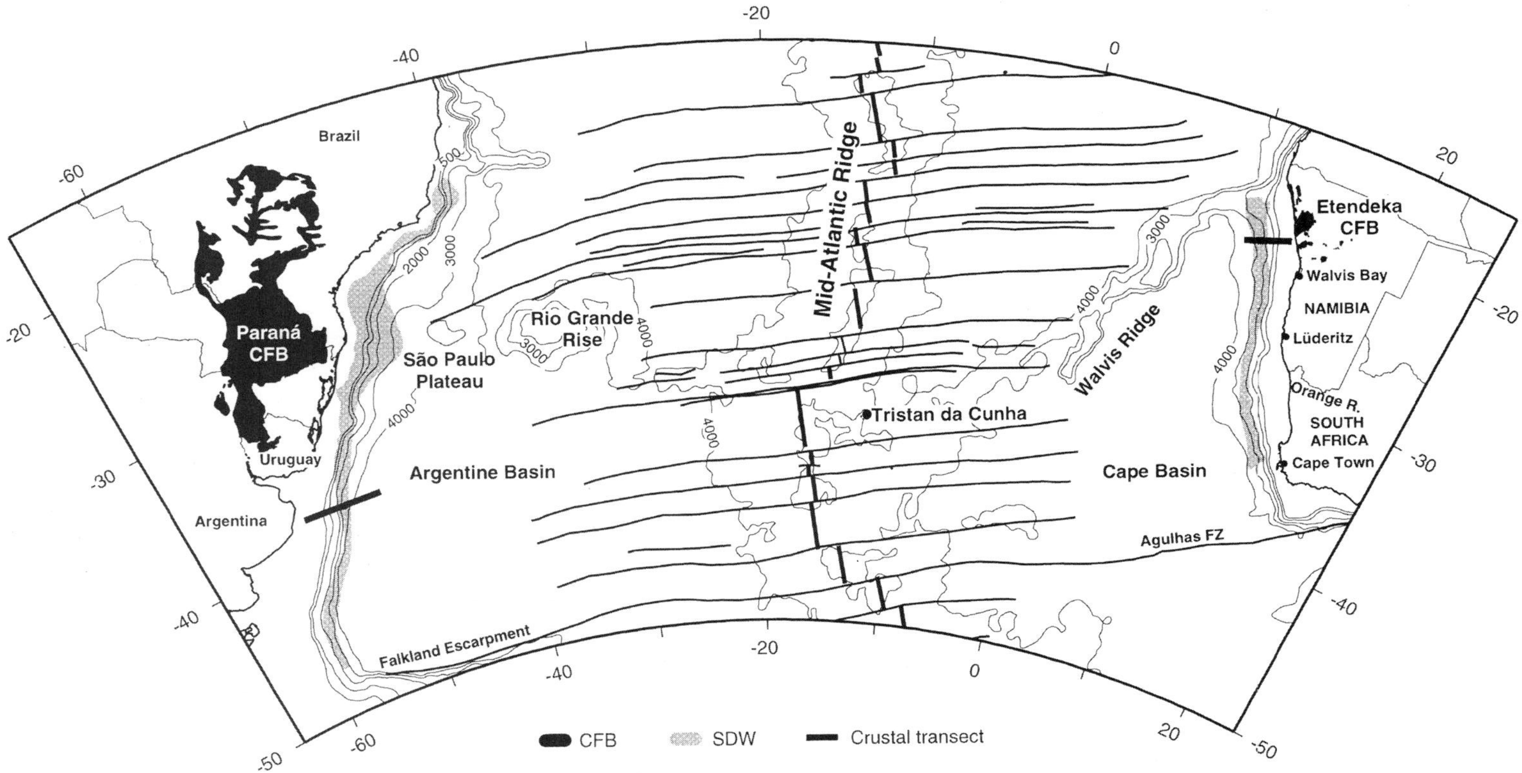

Fig. 4. Main South Atlantic structural features (Cande *et al.* 1989), Etendeka and Paraná CFBs (Milner *et al.* 1992; Turner *et al.* 1994), and distribution of seaward-dipping reflectors. Bathymetry in m (ETOPO-5 1988).

as lavas, abundant sills, and low-angle dykes east of the COB. Moreover, the Early Cretaceous sequence may also contain lavas from the Etendeka CFB. The most voluminous volcanism took place on the Abutment Plateau, reflecting the proximity of the plume and the persistent volcanism along the plume trail.

The COB is placed at the western termination of the rift; we note that the rift unconformity defines a base of the innermost dipping wedges and is absent farther west. Thus, the boundary is landward of magnetic anomaly G of Rabinowitz & LaBrecque (1979), which lies over the main wedge (Fig. 5). A deep continuous intracrustal reflector that may image the top of an LCB (Fig. 3) is observed below the Late Jurassic–Early Cretaceous rift zone. Newly acquired refraction seismic data show a *c.* 5 km thick 7.1–7.5 km s^{-1} LCB below the outer margin (Bauer & Schulze 1996).

The volcanic margin off North Namibia continues to the south (Figs 4 and 5). Seaward-dipping reflectors have been reported between Walvis Bay and Lüderitz (Austin & Uchupi 1982), and off the Orange River (Gerrard & Smith 1983) and Capetown (Hinz 1981). Thus, the entire >2400 km long eastern margin has a volcanic signature. Hinz *et al.* (1995) showed that the conjugate margin off South America, from the São Paulo Plateau to the Falkland Escarpment, has a similar character. In particular, the Uruguay and Argentine margins (Fig. 3) have a tectono-magmatic zonation similar to that of Namibia, and the cross-sectional dimensions of the dipping wedges are also similar.

Assuming that the North Namibia margin transect in Fig. 3 is representative, the extrusive volume is *c.* 0.2×10^6 km^3 for the margin segment in Fig. 5. Volume estimates farther south are uncertain, but appear smaller per length

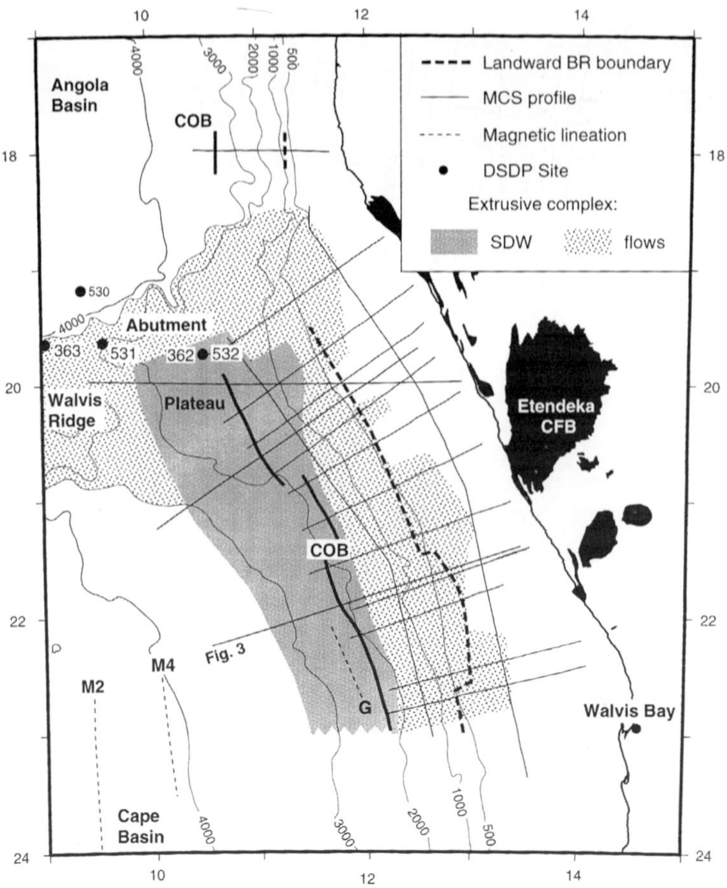

Fig. 5. Namibia margin tectono-magmatic zonation. MCS profiles from Intera-ECL89/91 and PGS Nopec surveys. Magnetic anomalies from Rabinowitz & LaBrecque (1979). Bathymetry in metres (GEBCO 1994). BR, break-up related rift zone; COB, continent–ocean boundary.

unit. We estimate a volume of $c. 0.58 \times 10^6\,km^3$ for the entire margin, and $c. 0.5 \times 10^6\,km^3$ for its conjugate. The entire South Atlantic LIP, including the Paraná–Etendeka CFBs (Milner *et al.* 1992; Peate *et al.* 1992), has an extrusive volume of at least $2.35 \times 10^6\,km^3$ (Table 3).

The onset of rifting leading to break-up is proposed at *c.* 160 Ma (Uliana *et al.* 1989; Nürnberg & Müller 1991), and dynamic modelling of lithospheric extension in the Paraná–Etendeka region suggests a rift duration of *c.* 25Ma (Harry & Sawyer 1992). The latter period is consistent with seismic data on the Namibian shelf (Light *et al.* 1993). We estimate rift widths of 120 and 150 km off Namibia and Argentina, respectively; and that the up to 300 km and 2400 km long rift underwent extension for *c.* 25 Ma before break-up.

US East Coast margin

The US East Coast margin (Fig. 6) was initiated by break-up of North America and Africa following a Late Triassic–Early Jurassic rift episode (Klitgord *et al.* 1988). It is covered by very thick sediments limiting seismic resolution in the deep basins and the underlying crust. The

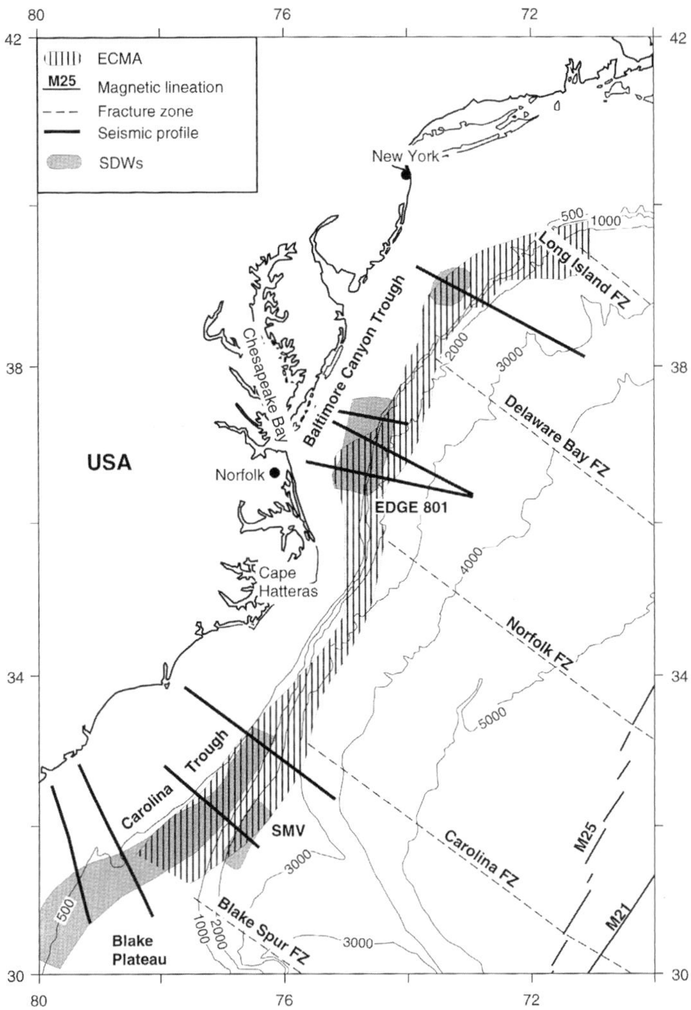

Fig. 6. Distibution of seaward-dipping wedges on the the US East Coast margin (Oh *et al.* 1995; Talwani *et al.* 1995) with selected seismic profiles used for volume estimates in Table 3. GEBCO (1994) bathymetry in metres, East Coast Magnetic Anomaly (ECMA) from Talwani *et al.* (1995), and fracture zones and sea-floor spreading anomalies from Klitgord *et al.* (1988). SMV, submarine volcanic rocks interpreted by Austin *et al.* (1990).

on- and offshore rift basins extend over a 200 km wide zone across Chesapeake Bay into the Baltimore Trough (Benson & Doyle 1988). Distinct magnetic and gravity anomaly belts delineate crustal features on the margin (e.g. Rabinowitz 1974; Alsop & Talwani 1984).

Seaward-dipping reflectors were imaged by Klitgord *et al.* (1988) and Austin *et al.* (1990), and a 7.2–7.5 km s^{-1} LCB was mapped by wide-angle profiles in the Baltimore Canyon (LASE Study Group 1986) and Carolina troughs (Tréhu *et al.* 1989). Recent surveys have led to improved mapping of geometries and distribution of these rock complexes (Holbrook & Kelemen 1993; Sheridan *et al.* 1993; Holbrook *et al.* 1994*a*, *b*; Oh *et al.* 1995; Talwani *et al.* 1995).

The *c.* 35 km thick continental crust on the inner margin thins seaward into transitional crust with 10–15 km thick and 25–70 km wide seaward-dipping wedges with velocities of 6.5–6.9 km s^{-1} (e.g. Austin *et al.* 1990). Some dipping reflectors extend into the up to 15 km thick, 7.1–7.5 km s^{-1} LCB (Holbrook & Kelemen 1993) (Fig. 3). A layer of flood basalt, extending from the seaward-dipping wedge across the inner margin, correlates with a basalt–diabase layer in onshore wells. It has been compared to CFBs because of the areal extent (Behrendt *et al.* 1983; Austin *et al.* 1990) and the 184 Ma age (Lanphere 1983). Talwani *et al.* (1995) argued that the 60–100 km wide transitional crust is the oldest oceanic crust, and was partly accreted subaerially.

Austin *et al.* (1990) estimated the extrusive volume of the 450 km long Carolina Trough segment to be 0.17×10^6 km^3 (Table 3), including the dipping wedges and crust with less clear intrabasement reflectors farther east interpreted as submarine volcanic rocks (Figs 3 and 6). The dipping wedge correlates spatially with the East Coast Magnetic Anomaly, and Talwani *et al.* (1995) suggested that the wedge gives rise to the anomaly, which they used to estimate the lateral extent of the extrusive rocks. By assuming a 100 km wide, 10 km thick and 1000 km long body from the Blake Spur Fracture Zone to the northern Baltimore Canyon Trough, they estimated a volume of 1×10^6 km^3. In view of the poorly developed intrabasement reflectors east of the main seaward-dipping wedge, we consider this a maximum value. We estimate a volume of 0.19×10^6 km^3, including extrusive rocks landward of the main wedge (Oh 1993), using average cross-sectional areas from the Carolina and Baltimore Canyon troughs and a 15% reduction for anisotropy (Planke & Eldholm 1994). The volume increases if we include dipping wedges on the Blake Plateau, where the nature of the underlying crust is uncertain (Oh *et al.* 1995). Holbrook & Kelemen (1993) calculated the total igneous crust emplaced during break-up to be 1.6×10^6 km^3. On the other hand, if we apply the same criteria as in the North Atlantic (Eldholm & Grue 1994), we arrive at a more conservative estimate of 0.72×10^6 km^3.

Although no seaward-dipping wedges have been reported on the conjugate margin, extrusive rocks may explain the linear magnetic anomaly off Morocco (Steiner & Roeser 1996). Furthermore, a 7.1–7.4 km s^{-1} LCB appears to replace typical Layer 3 velocities in the oldest oceanic crust (Holik *et al.* 1991).

Discussion

The volcanic margin history depends on lithospheric and asthenospheric properties before, during and after continental break-up. Therefore, one has to study the entire tectono-magmatic break-up history, i.e. consider the lithospheric setting before the onset of continental extension, the history of magmatism and tectonism during rifting and break-up, and the subsequent margin subsidence. This implies consideration of the entire crust at conjugate margins; however, the database to achieve this goal is as yet meagre, even at the best explored margins.

We observe changes in tectono-magmatic style and dimensions along single margin segments and among different margins. These are ascribed primarily to the lithospheric configuration before rifting, mode of rifting, magnitude of the mantle thermal anomaly, and distance from the plume. None the less, we note gross similarities in tectono-magmatic style and dimensions, and in main crustal units (Fig. 3, Table 3). In particular, the continental crust undergoes extension before break-up, forming a wide rift zone (BR, Fig. 3). Hence, we apply a crustal zonation comprising: (1) normal oceanic crust; (2) expanded oceanic crust; (3) pre- and syn-rift sediments and continental basement rocks that are extended, intruded and locally covered by flood basalt; (4) normal continental crust. Zones (2) and (3) are underlain by a high-velocity LCB, and zones (3) and (4) may have undergone previous tectonic events.

The COB is placed at the zone (2)–(3) boundary, seaward of which there is no base to the dipping wedge. Whether a distinct COB exists on rifted margins is debatable. We infer a narrow COB on volcanic margins corresponding to rapid, lateral compositional changes in

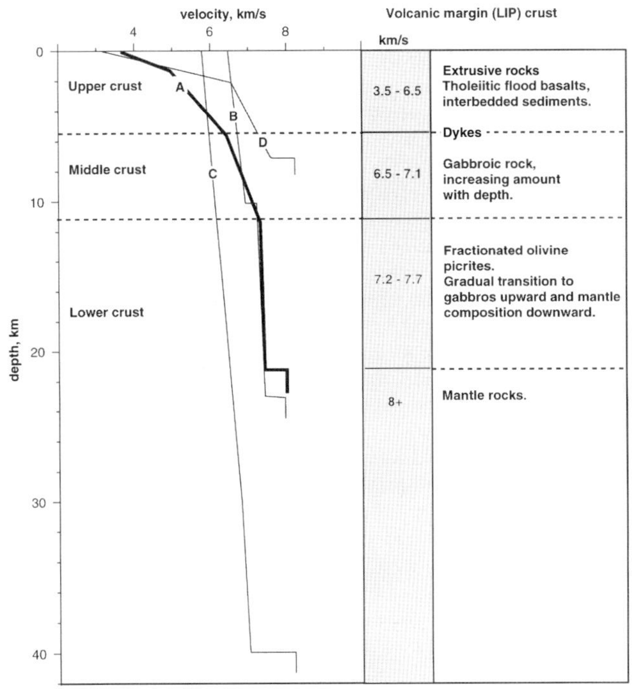

Fig. 7. Velocity–depth function for expanded volcanic margin oceanic crust in the North Atlantic (line A) (Eldholm & Grue 1994) and off the US East Coast (line B) (Holbrook *et al.* 1994*b*) compared with normal continental (line C) (Christensen & Mooney 1995) and oceanic (line D) (White *et al.* 1992) crusts. Suggested compositional column on right.

the uppermost crystalline crust below the extrusive rocks. Thus, the COB (Fig. 3) separates intruded, thinned continental crust from rocks emplaced entirely after break-up. Hence, magnetic profiles may delineate the COB if break-up occurred during periods of frequent reversals. On the other hand, if the entire crust is included, zone (2) and the most intruded part of zone (3) may be considered a transition zone.

The velocity distribution, the thick extrusive cover largely emplaced subaerially, and the LCB distinguish zone (2) from normal oceanic crust. The three-layer North Atlantic zone (2) crust of Eldholm & Grue (1994) consists of an upper extrusive layer, a mid-crustal layer and the LCB (Fig. 7). This type of crust is similar to the giant Ontong Java Plateau LIP (Gladczenko *et al.* 1997*a*) and to other oceanic LIPs (Coffin & Eldholm 1994). The consistent LIP velocity structure may suggest common emplacement and compositional elements.

Extrusive cover

The dipping wedges in the North Atlantic consist of up to 6 km thick flood basalt and very thin interbedded sediments. The velocity increases rapidly from *c.* 3.5 to >5.0 km s^{-1} in the uppermost part with a gentler velocity gradient at depth (Fig. 7). Velocities of 6.0–6.5 km s^{-1} near the base of the thickest dipping wedges may suggest an increasing proportion of dykes with depth. Integration of log, core and seismic data from ODP Sites 642 (Planke 1994; Planke & Eldholm 1994) and 917 (Planke & Cambray 1998) (Fig. 2) yields rock properties of lavas and interbedded sediments. Planke (1994) determined 0.6–18.5 m flow thicknesses at Site 642, whereas most sediment layers are <1 m. Physical properties at flow and composite-flow scale, and seismic modelling show that most lavas are too thin and without the physical property distribution required to produce reflectors resolved by standard MCS surveys. The dipping reflectors appear to originate from extensive, thick individual flows and from seismic interference. Therefore, MCS data are not suited for interpreting the detailed internal stratigraphy of the seaward-dipping wedges. We also note that Sites 642 and 917 as well as vertical seismic profiling (VSP) experiments in Iceland suggest that lava velocities recorded on the surface may be 10–20% too high as a result of transverse isotropic properties of the lavas (Planke & Eldholm 1994; Planke & Flovenz 1994), and that available seismic profiles record 2D images whereas volcano and fissure eruptions construct 3D features (Eldholm *et al.* 1995).

Sites 642 and 917 were drilled near the feather edge of a dipping wedge, i.e. landward of the most typical dipping reflectors. Though there is seismic continuity from the sites to the main wedges, there are few distinct, extensive reflectors at the sites proper. Therefore, thick lava series may exist also without distinct intrabasement reflectors, i.e. flood basalts may extend far beyond the prominent wedges. In fact, the dipping wedge is only one of several igneous features caused by the break-up event (Table 2) (e.g. Andersen 1988; Wood *et al.* 1988; Eldholm *et al.* 1989), and rifted margins may comprise extrusive constructions not imaged by the seismic record. The variety in seismic style is related to volume and rate of magma production, constructional environment, and syn- and post-constructional deformation and subsidence. For example, Planke *et al.* (1999) have proposed that the varying seismic characteristics reflect a change from subaerial flood basalt through shallow-water hyaloclastic mounds to deep-water flows.

Middle and lower crust

Zone 2 middle crust (Fig. 7) has a 6.5–6.7 km s^{-1} velocity at the top and a gentle velocity gradient, resembling a thickened oceanic layer 3A (Ewing & Houtz 1979; White *et al.* 1992). It probably consists of dykes at the transition with the extrusive cover and gabbro below (Zehnder *et al.* 1990). The 7+ km s^{-1} LCB velocity is not typical for normal oceanic or continental crusts (Meissner 1986; Christensen & Mooney 1995), but is characteristic of LIPs (Coffin & Eldholm 1994). There is still uncertainty in LCB geometry and velocity, partly because of data quality, acquisition and interpretation techniques. Thus, one has to be careful in using seismic velocity alone to distinguish crustal type and composition. Moreover, linearly scaled models of 'normal' oceanic (Zehnder *et al.* 1990; Mutter & Mutter 1993) and continental crusts have obvious genetic implications which may not be valid in view of melt volume and emplacement setting for the initial oceanic crust. There is similarity of the upper and middle crust in zone (2) with Icelandic crust (Mutter *et al.* 1984), thus the term Icelandic oceanic crust has been applied (e.g. Eldholm *et al.* 1989; Hinz *et al.* 1993).

We relate zone (2) to LIP-type crustal emplacement (Coffin & Eldholm 1994), characterized by increased decompressional partial melting during break-up emplacing the LCB, the middle crust in zone (2) and the extrusive cover. High-quality expanded spread profile (ESP) and ocean bottom seismograph (OBS) experiments (e.g. Eldholm & Mutter 1986; Hinz *et al.* 1987; Fowler *et al.* 1989; Olafsson *et al.* 1992; Mjelde *et al.* 1993; Holbrook *et al.* 1994*a*) yield a range of velocities, 7.1–7.7 km s^{-1}, for the LCB. The fact that increased MgO content in ponded decompressional basaltic melts at the base of the crust yields only 7.1–7.2 km s^{-1} velocities (White & McKenzie 1989) suggests that the LCB velocity range relates to a varying degree of melt fractionation. The upper LCB may represent a transition from gabbro to olivine cumulates derived from picritic melts (Fig. 7). On the other hand, the 7+ km s^{-1} velocity may also represent a secondary, metamorphic facies boundary (Eldholm & Grue 1994; Eldholm *et al.* 1995), probably the gabbro–garnet-granulite transition. However, this process requires the presence of substantial amounts of fluids shortly after emplacement, for which a viable source is not obvious (Gladczenko *et al.* 1997*a*). LCBs are commonly described as magmatic underplating (e.g. LASE Study Group 1986; White *et al.* 1987), a process that refers to accumulation of mantle-derived material below continental crust requiring a melt–crust density contrast (Herzberg *et al.* 1983; Furlong & Fountain 1986). Because a density filter is not applicable during oceanic crust formation only the LCB below the extended continental crust in zone (3) is truly underplated (Fig. 3).

Tectono-magmatic dimensions

For LIPs globally, our database allows only first-order volume estimates of the offshore extrusive component and of the total igneous crust. Most volumes are considered minimum values (Table 3). It is notable that the main contribution to the igneous volumes at volcanic margin LIPs with coeval CFBs, such as the North and South Atlantic (Table 3) and the Deccan–Seychelles, is found offshore, showing these margins contribute significantly to the global LIP inventory.

At margins associated with mantle plumes there is some evidence of narrowing, and less voluminous wedges away from the plume. The wedge is thickest, *c*. 15 km, off the US East Coast (Holbrook *et al.* 1994*a*), where it extends down to the LCB. Individual reflectors have also been interpreted to this level on the Hatton Bank margin (Spence *et al.* 1989). These reflectors may originate within gabbroic rocks and not from extrusive rock.

The margins in Fig. 3 show breakup related rift zones with 150–200 km half-widths, and rifting appears to have lasted for 50–20 Ma before break-up (Table 3). The Namibia and Vøring margins experienced one or more rift episodes pre-dating the break-up rift. Hence, the continental crystalline crust may be thinned over a wide region, whereas the break-up related rift is less that 350 km wide. Flow of lavas onto continental crust and pervasive intrusions inhibit seismic resolution and commonly hide rift structures. The apparent lack of extensional features has led to models of very rapid break-up of the continental lithosphere without significant rifting (Mutter *et al.* 1984; Larsen 1990; Hopper *et al.* 1992). In contrast, we show that a protracted rift phase is compatible with data from many margins. Thus, a separate tectonic framework for volcanic margins is not required. The similarity in structural style and dimensions of volcanic margins, non-volcanic margins and continental rifts makes us suggest that the principal difference between volcanic and non-volcanic margins is derived from the melt potential of the asthenosphere during rifting and break-up.

Margin asymmetry and rifting style

There is magmatic and/or tectonic asymmetry on many conjugate volcanic margins. The magmatic asymmetry, expressed by the on- and off-shore extrusive and LCB distribution and volume, may exist along and across the initial plate boundary. In the North Atlantic, the area and volume of basalts on continental crust are greatest south of Iceland, becoming smaller to the north (Eldholm & Grue 1994). This configuration may reflect stepwise propagation of the plate boundary during break-up resulting in diminished melt potential northward.

Extrusive across-plate-boundary asymmetry, commonly shown by distribution of CFBs and dipping wedges, may reflect the position of a mantle plume with respect to the line of break-up. The prominent wedges off the US East Coast without obvious equivalents on the conjugate Morocco margin may be another example. The present distribution of basaltic lavas has been related to the combined effects of variable melt production, vulnerability of the continental crust to melt penetration, multiple transient feeders, lateral melt migration, constructional environment and erosion (e.g. Eldholm *et al.* 1995).

The tectonic style of the margin is determined by the pre-rift lithospheric setting and the style of the rift deformation outlined by fault and detachment distributions and geometries, and by conjugate transfer systems (Lister *et al.* 1991). Crustal break-up away from the rift axis (Keen 1987) will create asymmetric margin structures, as does simple-shear extension proposed for the US East Coast margin (Benson & Doyle 1988; Klitgord *et al.* 1988). Simple-shear extension and associated syn-constructional listric faults may also explain the abrupt seaward termination sometimes observed at large dipping wedges (Eldholm *et al.* 1989).

Volcanic margins and mantle plumes

A relationship between most LIP emplacements and mantle plumes, recognized by hotspots, is well documented (e.g. White & McKenzie 1989; Duncan & Richards 1991). The igneous activity related to the transient break-up event is caused by decompressional melting. If a plume reaches the base of the lithosphere in a region under extension, or in a region with pre-existing thinned lithosphere, melting will be amplified and the excess melts may result in a volcanic margin. The variability in both extrusive cover and total igneous crustal volume emplaced during break-up lead to the inference that volcanic margins are expressions of asthenospheric melt anomalies of different magnitudes. Similar relations apply to transient LIPs in general (Coffin & Eldholm 1994). Noting the range in size, Eldholm *et al.* (1995) pointed out that some observations may challenge the plume model as the only mechanism for volcanic margin initiation. For example, the lengths of the volcanic rifted margins in the North Atlantic (Fig. 2) and in the South Atlantic (Fig. 4) require very large diameters for collapsed plume heads.

There are no obvious plumes to explain the US East Coast (Holbrook & Kelemen 1993; Talwani *et al.* 1995) and West Australia margins (Mutter *et al.* 1984; Hopper *et al.* 1992; Colwell *et al.* 1994), although a plume relationship was inferred by Wilson (1997) for the US East Coast and the conjugate West Africa margins. The inferred volcanic margin–plume relationship is commonly based on less excessive, persistent volcanism caused by the tail of the plume, and recognized by a submarine ridge or seamount chain such as the Iceland and Tristan plume trails expressed by the Greenland–Iceland–Faeroe ridge (Fig. 3) and the Walvis Ridge–Rio Grande Rise, respectively (Fig. 4). On the other hand, the plume concept may be retained if the plume source generates a mantle 'blob' (Griffiths & Campbell 1991) rather than a persistent plume, or if the plume is located beneath a plate remaining relatively stationary with respect to the asthenosphere. Thus, a volcanic margin may still have a deep mantle thermal source without being located in the vicinity of a hotspot.

In view of the variety in size and distribution of igneous volumes on volcanic margins, we prefer to treat the mantle plume as a sufficient, but not necessary, condition for excess igneous activity during complete plate separation. The asthenospheric melt potential is determined by the thermal state and fluid content in the asthenosphere and the dynamic state of the lithosphere, i.e. magnitude and duration of rifting. Consequently, the combined effect of small regional asthenospheric temperature and fluid content anomalies in the asthenosphere and lithospheric extension may induce excess melting during break-up. This concept does not depend on a plume or a specific mantle circulation model, although the existence of a plume will greatly facilitate large-scale melting, particularly if it impinges on lithosphere that is already under extension or has been thinned by previous rift episodes (Thompson & Gibson 1991). The similarity in tectonic style of volcanic or non-volcanic margins suggests that the classification in Fig. 1 refers to end-member types and that most margins are intermediate types. The margins off W Australia that have seaward-dipping wedges and LCBs of relatively small volumes are examples of intermediate margins (Mutter *et al.* 1989, Hopper *et al.* 1992; Planke *et al.* 1996).

Table 4. *Tectono-magmatic and depositional effects of volcanic margin formation having potential resource implications*

Pre- amd syn-rift (pre-opening) basins

- Syn-rift uplift
  - erosion
  - redeposition
  - restricted basins
- Thermal imprint
- Faulting
- Intrusive activity

Post-opening (early opening) basins

- Along- and across-margin barriers
  - restricted basin
  - high biogenic productivity
- Central sediment source
- Thermal imprint
- Flood basalts
- Margin subsidence
  - LCB influence

Primarily based on studies of the margin off Norway (Fig. 2).

Implications for resource evaluation

Large-scale transient geological events influence the palaeoenvironment by changing oceanographic and atmospheric circulation patterns and compositions. In particular, the syn-rift uplift and subsequent massive, transient volcanism during break-up affects environments on local, regional and possibly global scales by modification of basin geometries, new depositional and erosional environments, and changes in the composition of the biosphere. The effects have been discussed by Coffin & Eldholm (1994) for LIPs in general, and by Eldholm & Thomas (1993) and Eldholm *et al.* (1995) for volcanic margins in particular.

The potentially global environmental impact of volcanic margin formation has been suggested for the North Atlantic, where sediments show that the flood basalt emplacement near the Paleocene–Eocene boundary was accompanied by regional ashfalls. There is also an apparent temporal correspondence between this boundary event and the global Paleocene–Eocene extinction event. Subsequently, the Earth entered the early Eocene greenhouse, the warmest period over the past 70 Ma (Eldholm & Thomas 1993).

In terms of hydrocarbon exploration, the crustal movements and thermal regime associated with volcanic margin formation influence the resource potential of the pre-opening sedimentary basins, i.e pre- and syn-rift basins, as well as the post-opening margin basins (Table 4). In particular, the combination of transfer faults, fracture zones and central rift uplift during the syn-rift and early post-rift periods may form across- and along-margin barriers and thus develop a series of restricted basins, which in some cases existed tens of millions of years after break-up. The correspondence of restricted basins and periods of global warming may, in fact, induce favourable conditions for source rock formation.

Few studies have yet addressed these questions, except on the margin off Norway, where it has been shown that the LCB causes a significant, quantifiable reduction in the subsidence of the outer margin (Skogseid 1994). Thus, the LCB must be included during modelling of relative vertical motion and subsidence-derived lithospheric extension (Skogseid *et al.* 2000). The uplifted central region was eroded and acted as a main source of Paleocene and Eocene sediments into the regional Vøring and Møre basins, whereas sediments from the east first reached the highs in mid-Eocene time. The extrusive rocks became completely sediment covered as late as mid-Oligocene to early Miocene time, i.e. *c.* 30 Ma after break-up (Skogseid & Eldholm 1989; Skogseid *et al.* 1992*a*, *b*). The thermal imprint is almost entirely restricted to the part of basins overlying the LCB. Here, there is up to 200% increase in heat flow, and potential source rocks reach their maximum temperature a few million years after break-up, and return to normal thermal conditions 15–20 Ma later (Pedersen *et al.* 1996). Moreover, modelling of 100 m or thicker sills shows considerable maturation increase at distances 3–4 times the sill thickness (Pedersen *et al.* 1996). This effect, documented by wells on the Exmouth Plateau on the western Australia margin (Reeckmann & Mebberson 1984), will become even more important if convective heat transport is achieved.

This study has benefited from results and comments from a number of colleagues and students involved in continental margin studies at the University of Oslo. In addition to the IBS institutional and industry partners, we are grateful for advice and data support from the Australian Geological Survey Organization, Bundesanstalt für Geowissenschaften und Rohstoffe, Germany, and PGS Nopec, Norway. The work has been supported by the Research Council of Norway as part of the IBS (Integrated Basin Studies) project under the JOULE II research programme funded by the Commission of European Communities (Contract JOU2-CT 92-0110), and in part by the ProPetro research program. This paper is an IBS Contribution.

References

ABREU, V. S., CONDI, F. J., BALLY, A. W., SAWYER, D. S. & DROXLER, A. W. 1996. Tectono-stratigraphic evolution of a volcanic rifted margin – offshore Namibia (abstract). *EOS Transactions, American Geophysical Union*, **77**, F825.

ALSOP, L. E. & TALWANI, M. 1984. The East Coast Magnetic Anomaly. *Science*, **226**, 1189–1191.

ANDERSEN, M. S. 1988. Late Cretaceous and early Tertiary extension and volcanism around the Faeroe Islands. *In*: MORTON, A. C. & PARSON, L. M. (eds) *Early Tertiary Volcanism and the Opening of the NE Atlantic*. Geological Society, London, Special Publications, **39**, 115–122.

AUSTIN, J. A. & UCHUPI, E. 1982. Continental–oceanic transition off southwest Africa. *AAPG Bulletin*, **66**, 1328–1347.

——, STOFFA, P. L., PHILLIPS, J. D. *et al.* 1990. Crustal structure of the Southeast Georgia embayment–Carolina Trough: preliminary results of a composite seismic image of a continental suture (?) and a volcanic passive margin. *Geology*, **18**, 1023–1027.

BAUER, K. & SCHULZE, A. 1996. Seismic investigations of the passive continental margin of Namibia from wide-angle off-shore data (abstract) *EOS Transactions, American Geophysical Union*, **77**, F668.

BEHRENDT, J. C., HAMILTON, R. M., ACKERMANN, H. D., HENRY, V. J. & BAYER, K. C. 1983. Marine multichannel seismic-reflection evidence for Cenozoic faulting and deep crustal structure near Charleston, SC. *In:* GOHN, G. S. (ed.) *Studies Related to the Charleston, South Carolina Earthquake of 1886 – Tectonics and Seismicity*. US Geological Survey Professional Paper, **1313**, J1–J29.

BENSON, R. N. & DOYLE, R. G. 1988. Early Mesozoic rift basins and the development of the United States middle Atlantic continental margin. *In*: MANSPEIZER, W. (ed.) *Triassic–Jurassic Rifting Continental Breakup and the Origin of the Atlantic Passive Margins*. Elsevier, Amsterdam, 99–127.

CANDE, S. C., LA BRECQUE, J. L., LARSON, R. L. PITMAN, W. C. III, GOLOVCHENKO, X. & HAXBY, W. F. 1989. *Magnetic Lineations of the World's Ocean Basins*. AAPG Map Series, Tulsa.

CHRISTENSEN, N. I. & MOONEY, W. D. 1995. Seismic velocity structure and composition of the continental crust: a global view. *Journal of Geophysical Research*, **100**, 9761–9788.

COFFIN, M. F. & ELDHOLM, O. 1994. Large Igneous Provinces: crustal structure, dimensions, and external consequences. *Reviews of Geophysics*, **32**, 1–36.

COLWELL, J. B., SYMONDS, P. A. & CRAWFORD, A. J. 1994. The nature of the Wallaby (Cuvier) Plateau and other igneous provinces of the west Australia margin. *AGSO Journal of Australian Geology and Geophysics*, **15**, 137–156.

CONDI, F. J., ABREU, V. S., BALLY, A. W. & SAWYER, D. S. 1996. Evolution of a volcanic margin: case study from Pelotas Basin, Brasil (abstract). *EOS Transactions, American Geophysical Union*, **77**, F825.

DICKIN, A. P. 1988. The North Atlantic Tertiary Province. *In*: MACDOUGALL, J. D. (ed.) *Continental Flood Basalts*. Kluwer, Hingham, MA, 111–149.

DUNCAN, R. A. & RICHARDS, M. A. 1991. Hotspots, mantle plumes, flood basalts, and true polar wander. *Reviews of Geophysics*, **29**, 31–50.

ELDHOLM, O. & GRUE, K. 1994. North Atlantic volcanic margins: dimensions and production rates. *Journal of Geophysical Research*, **99**, 2955–2968.

—— & MUTTER, J. C. 1986. Basin structure structure on the Norwegian margin from analysis of digitally recorded sonobuoys. *Journal of Geophysical Research*, **91**, 3763–3783.

—— & THOMAS, E. 1993. Environmental impact of volcanic margin formation. *Earth and Planetary Science Letters*, **117**, 319–329.

——, SKOGSEID, J., PLANKE, S. & GLADCZENKO, T. P. 1995. Volcanic margin concepts. *In*: BANDA, E., TALWANI, M. & TORNÉ, M. (eds) *Rifted Ocean–Continent Boundaries*. NATO ASI Series. Kluwer, Dordrecht, 1–16.

——, SUNDVOR, E. & MYHRE, A. M. 1979. Continental margin off Lofoten–Vesterålen, Northern Norway. *Marine Geophysical Research*, **4**, 3–35.

——, THIEDE, J. & TAYLOR, E. 1989. Evolution of the Vøring Volcanic Margin. *In*: ELDHOLM, O., THIEDE, J., Taylor, E. *et al.* (eds) *Proceedings of the Ocean Drilling Program, Scientific Results*, **104**. Ocean Drilling Program, College Station, TX, 1033–1065.

ERLANK, A. J., MARSH, J. S., DUNCAN, A. R., MILLER, R. M., HAWKESWORTH, C. J., BETTON, P. J. & REX, D. C. 1984. Geochemistry and petrogenesis of the Etendeka volcanic rocks from SWA/Namibia. Geological Society of South Africa Special Publication, **13**, 195–245.

ETOPO-5 1988. *National Geophysical Data Center, ETOPO-5 bathymetry/topography data.* National Oceanic and Atmospheric Administration, US Department of Commerce, Boulder, CO, Data Announcements, **88-MGG-02**.

EWING, J. & HOUTZ, R. 1979. Acoustic stratigraphy and structure of the oceanic crust. *In*: TALWANI, M. (ed.) *Deep Drilling Results in the Atlantic Ocean: Ocean Crust.* American Geophysical Union, Maurice Ewing Series, **2**, 1–14.

FOWLER, S. R., WHITE, R. S., WESTBROOK, G. K. & SPENCE, G. D. 1989. The Hatton Bank continental margin, II. Deep structure from two-ship expanded spread seismic profiles. *Geophysical Journal*, **96**, 295–309.

FURLONG, K. P. & FOUNTAIN, D. M. 1986. Continental crustal underplating: thermal considerations and seismic petrological consequences. *Journal of Geophysical Research*, **91**, 8285–8294.

GEBCO 1994. Sheet 5.12. *In*: HUNTER, P., CHERKIS, N., BRENNER, C., UDINTSEV, G. B., FALCONER, R. & HANDLEY, J. (compilers) *GEBCO Digital Atlas.* British Oceanographic Data Centre on behalf of the IOC and IHO.

GERRARD, I. & SMITH, G. C. 1983. Post Palaeozoic succession and structure of the south-western African continental margin. *In:* WATKINS, J. S. & DRAKE, C. L. (eds) *Studies in Continental Margin Geology.* American Association of Petroleum Geologists Memoir, **34**, 49–74.

GLADCZENKO, T. P. 1994. *Crustal structure and composition of selected transient large igneous provinces.* Cand.scient. thesis, University of Oslo.

——, COFFIN, M. F. & ELDHOLM, O. 1997*a*. Crustal structure of the Ontong Java Plateau: modeling of new gravity and existing seismic data. *Journal of Geophysical Research*, **102**, 27711–27730.

——, HINZ, K., ELDHOLM, O., MEYER, H., NEBEN, S. & SKOGSEID, J. 1997*b*. South Atlantic volcanic margins. *Journal of the Geological Society, London*, **154**, 465–470.

GRIFFITHS, R. W. & CAMPBELL, I. H. 1991. Interaction of mantle plume heads with the Earth's surface and onset of small-scale convection. *Journal of Geophysical Research*, **96**, 18295–18310.

HARRY, D. L. & SAWYER, D. S. 1992. Basaltic volcanism, mantle plumes, and the mechanics of rifting: the Paraná flood basalt province of South America. *Geology*, **20**, 207–210.

HERZBERG, C. T., FYFE, W. S. & CARR, M. J. 1983. Density constraints on the formation of the continental Moho and crust. *Contributions to Mineralogy and Petrology*, **84**, 1–5.

HINZ, K. 1981. A hypothesis on terrestrial catastrophes: wedges of very thick oceanward dipping layers beneath passive margins; their origin and paleoenvironment significance. *Geologisches Jahrbuch*, **E22**, 345–363.

—— & SCHLÜTER, H.-U. 1978. Der Nordatlantik – Ergebnisse geophysikalischer untersuchungen der Bundesanstalt für Geowissenschaften und Rohstoffe an Nordatlantischen Kontinentalrändern. *Erdöl, Erdgas Zeitschrift*, **94**, 271–280.

—— & WEBER, J. 1976. Zum geologischen Aufbau des Norwegischen Kontinentalrandes und der Barents-See nach reflexionsseismischen Messungen. *Erdöl und Kohle, Erdgas, Petrochemie*, 3–29.

——, ELDHOLM, O., BLOCK, M. & SKOGSEID, J. 1993. Evolution of N Atlantic volcanic continental margins. *In*: PARKER, J. R. (ed.) *Petroleum Geology of Northwest Europe: Proceedings of the 4th Conference.* Geological Society, London, 901–913.

——, GEONCALVES DE SOUZA, K., HEMMERICH, M., NEBEN, S., RONDA, C. L., ROESER, H. A. & SCHRECKENBERGER, B. 1995. *The Argentine continental margin north of 48°S: sedimentary successions and structural elements.* Internal report, Bundesanstalt für Geowissenschaften und Rohstoffe, Hannover, **47**.

——, MUTTER, J. C., ZEHNDER, C. M. & NGT Study Group 1987. Symmetric conjugation of continent–ocean boundary structures along the Norwegian and East Greenland margins. *Marine and Petroleum Geology*, **3**, 166–187.

HOLBROOK, W. S. & KELEMEN, P. B. 1993. Large igneous province on the US Atlantic margin and implications for magmatism during continental breakup. *Nature*, **364**, 433–436.

——, PURDY, G. M., SHERIDAN, R. E. *et al.* 1994*a*. Seismic structure of the US Mid-Atlantic continental margin. *Journal of Geophysical Research*, **99**, 17871–17891.

——, REITER, E. C., PURDY, G. M. *et al.* 1994*b*. Deep structure of the US Atlantic continental margin, offshore South Carolina, from coincident ocean bottom and multichannel data. *Journal of Geophysical Research*, **99**, 9155–9178.

HOLIK, J. S., RABINOWITZ, P. D. & AUSTIN, J. A. 1991. Effects of Canary hotspot volcanism on structure of oceanic crust off Morocco. *Journal of Geophysical Research*, **96**, 12039–12067.

HOPPER, J. R., MUTTER, J. C., LARSON, R. L., MUTTER, C. Z. & NW AUSTRALIA STUDY GROUP 1992. Magmatism and rift margin evolution: evidence from northwest Australia. *Geology*, **20**, 853–857.

KEEN, C. E. 1987. Some important consequences of lithospheric extension. *In*: COWARD, M. P., DEWEY, J. F. & HANCOCK, P. L. (eds) *Continental Extensional Tectonics.* Geological Society, London, Special Publications, **28**, 67–73.

KLITGORD, K. D., HUTCHINSON, D. R. & SCHOUTEN, H. 1988. US Atlantic continental margin: structural and tectonic framework. *In*: *The Atlantic Continental Margin*. Geological Society of America, The Geology of North America, **I-2**, 19–55.

LANPHERE, M. A. 1983. $^{40}Ar/^{39}Ar$ ages of basalt from Clubhouse Crossroads test hole 2 near Charleston, South Carolina. *In*: GOHN, G. S. (ed.) *Studies Related to the Charleston, South Carolina Earthquake of 1886 – Tectonics and Seismicity*. US Geological Survey Professional Paper, **1313**, B1–B8.

LARSEN, H. C. 1990. The East Greenland Shelf. *In*: *The Arctic Ocean Region*. Geological Society of America, The Geology of North America, **L**, 185–210.

LARSON, R. L. 1991. Geological consequences of superplumes. *Geology*, **19**, 963–966.

LASE STUDY GROUP 1986. Deep structure of the US East Coast passive margin from large aperture seismic experiments (LASE). *Marine and Petroleum Geology*, **3**, 234–242.

LIGHT, M. P. R., MASLANYJ, M. P., GREENWOOD, R. J. & BANKS, N. L. 1993. Seismic sequence stratigraphy and tectonics offshore Namibia. *In*: WILLIAMS, G. D. & DOBB, A. (eds) *Tectonics and Seismic Sequence Stratigraphy*. Geological Society, London, Special Publications, **71**, 163–191.

LISTER, G. S., ETHERIDGE, M. A. & SYMONDS, P. A. 1991. Detachment models for the formation of passive continental margins. *Tectonics*, **10**, 1038–1064.

MAKRIS, J., GINZBURG, A., SHANNON, P. M., JACOB, A. W. B., BEAN, C. J. & VOGT, U. 1991. A new look at the Rockall region, offshore Ireland. *Marine and Petroleum Geology*, **8**, 410–416.

MEISSNER, R. 1986. *The Continental Crust. A Geophysical Approach*. Academic, New York.

MILNER, S. C., DUNCAN, A. R. & EWART, A. 1992. Quartz latite rheoignimbrite flows of the Etendeka Formation, north-western Namibia. *Bulletin of Volcanology*, **54**, 200–219.

MJELDE, R., SELLEVOLL, M. A., SHIMAMURA, H., IWASAKI, T. & KANAZAWA, T. 1993. Ocean bottom seismographs used in a crustal study of an area covered with flood-basalt off Lofoten, N. Norway. *Terra Nova*, **5**, 76–84.

MUTTER, C. Z. & MUTTER, J. C. 1993. Variations in thickness of layer 3 dominate oceanic crustal structure. *Earth and Planetary Science Letters*, **117**, 295–317.

——, LARSON, R. L., BUHL, P. *et al.* 1989. Extension of the Exmouth Plateau, offshore northwestern Australia; deep seismic reflection/refraction evidence for simple and pure shear mechanisms. *Geology*, **17**, 15–18.

——, TALWANI, M. & STOFFA, P. 1984. Evidence for a thick oceanic crust adjacent to the Norwegian margin. *Journal of Geophysical Research*, **89**, 483–502.

NØTTVEDT, A. *et al.* 2000. *This volume*.

NÜRNBERG, D. & MÜLLER, R. D. 1991. The tectonic evolution of the South Atlantic from Late Jurassic to present. *Tectonophysics*, **191**, 27–53.

O'CONNOR, J. M. & DUNCAN, R. A. 1990. Evolution of the Walvis Ridge–Rio Grande Rise hot spot system: implications for African and South American plate motions over plumes. *Journal of Geophysical Research*, **95**, 17475–17502.

OH, J. 1993. *Basement structures associated with Mesozoic continental breakup along the Southeastern United States Continental Margin from multichannel seismic profiles*. PhD dissertation, University of Texas at Austin.

——, AUSTIN, J. A., PHILLIPS, J. D., COFFIN, M. F. & STOFFA, P. L. 1995. Seaward-dipping reflectors offshore the southeastern United States: seismic evidence for extensive volcanism accompanying sequential formation of the Carolina trough and Blake Plateau basin. *Geology*, **23**, 9–12.

OLAFSSON, I., SUNDVOR, E., ELDHOLM, O. & GRUE, K. 1992. Møre Margin: crustal structure from analysis of Expanded Spread Profiles. *Marine Geophysical Research*, **14**, 137–162.

PEATE, D. W., HAWKESWORTH, C. J. & MANTOVANI, M. S. M. 1992. Chemical stratigraphy of the Paraná lavas (South America): classification of magma types and their spatial distribution. *Bulletin of Volcanology*, **55**, 119–139.

PEDERSEN, T., SKOGSEID, J., JOHANSEN, H. & THRONDSEN, T. 1996. Integrated tectonomagmatic and basin modelling at the Vøring Basin volcanic margin: magmatism, source and reservoir rocks (abstract). *In*: *Tectonic, Magmatic and Depositional Processes at Passive Continental Margins*. Geological Society, London, 20–22 February.

PLANKE, S. 1994. Geophysical response of flood basalts from analysis of wireline logs: ODP Site 642, Vøring volcanic margin. *Journal of Geophysical Research*, **99**, 9279–9296.

—— & CAMBRAY, H. 1998. Seismic properties of flood basalts from Hole 917A downhole data, southeast Greenland volcanic margin. *In*: LARSEN, H. C., SAUNDERS, A. D., CLIFT, P. D. *et al.* (eds) *Proceedings of the Ocean Drilling Program, Scientific Results*, **152**. Ocean Drilling Program, College Station, TX, 453–462.

—— & ELDHOLM, O. 1994. Seismic response and construction of seaward dipping wedges of flood basalts: Vøring volcanic margin. *Journal of Geophysical Research*, **99**, 9263–9278.

—— & FLOVENZ, O. G. 1994. Integration of downhole and surface seismic data in flood basalt terrains: implications for seismic imaging and crustal structure (abstract). *EOS Transactions, American Geophysical Union*, **75**, 316.

——, ALVESTAD, E. & ELDHOLM, O. 1999. Seismic characteristics of basaltic extrusives and intrusives. *Leading Edge*, **18**, 342–348.

——, SYMONDS, P. A. & SKOGSEID, J. 1996. Seismic volcanostratigraphy on rifted volcanic margins: implications for vertical movements (abstract). *In*: *Tectonic, Magmatic and Depositional Processes at Passive Continental Margins*. Geological Society, London, 20–22 February.

RABINOWITZ, P. D. 1974. The boundary between oceanic and continental crust in the western

North Atlantic. *In*: BURK, C. A. & DRAKE, C. L. (eds) *The Geology of Continental Margins*. Springer, New York, 67–84.

—— & LABRECQUE, J. 1979. The Mesozoic South Atlantic Ocean and evolution of its continental margins. *Journal of Geophysical Research*, **84**, 5973–6002.

REECKMANN, S. A. & MEBBERSON, A. J. 1984. Igneous intrusion in the North-West Canning Basin and their impact on oil exploration. *In:* PURCELL, P. G. (ed.) *The Canning Basin W.A.* Geological Society Australia, 389–399.

SHERIDAN, R. E., MUSSER, D. L., GLOVER, L., III *et al.* 1993. Deep seismic reflection data of EDGE, US mid-Atlantic continental-margin experiment: implications for Appalachian sutures and Mesozoic rifting and magmatic underplating. *Geology*, **21**, 563–567.

SKOGSEID, J. 1994. Dimensions of the Late Cretaceous–Paleocene Northeast Atlantic rift derived from Cenozoic subsidence. *Tectonophysics*, **240**, 225–247.

—— & ELDHOLM, O. 1989. Vøring Plateau continental margin: seismic interpretation, stratigraphy and vertical movements. *In*: ELDHOLM, O., THIEDE, J., TAYLOR, E., *et al.* (eds) *Proceedings of the Ocean Drilling Program, Scientific Results*, **104**. Ocean Drilling Program, College Station, TX, 993–1030.

—— & ——1995. Rifted continental margins off mid-Norway. In: BANDA, E., TALWANI, M. & TORNÉ, M. (eds) *Rifted Ocean–Continent Boundaries*. NATO ASI Series. Kluwer, Dordrecht, 147–153.

——, PEDERSEN, T., ELDHOLM, O. & LARSEN, B. T. 1992*a*. Tectonism and magmatism during NE Atlantic continental break-up: the Vøring Margin. *In*: STOREY, B. C., ALABASTER, T. & PARKHURST, R. J. (eds) *Magnetism and the Causes of Continental Break-up*. Geological Society London, Special Publications, **68**, 305–320.

——, —— & LARSEN, V. B. 1992*b*. Vøring Basin: subsidence and tectonic evolution. *In*: LARSEN, R. M., BREKKE, H., LARSEN, B. T. & TALLERAAS, E. (eds) *Tectonic Modelling and its Application to Petroleum Geology*. Norwegian Petroleum Society Special Publication, **1**, 55–82.

——, PLANKE, S., FALEIDE, J. I., PEDERSEN, T., ELDHOLM, O. & NEVERDAL, F. 2000. NE Atlantic continental rifting and volcanic margin formation. *This volume*.

SPENCE, G. D., WHITE, R. S., WESTBROOK, G. K. & FOWLER, S. R. 1989. The Hatton Bank continental margin, I Shallow structure from two-ship expanded spread seismic profiles. *Geophysical Journal*, **96**, 273–294.

STEINER, C. & ROESER, H. A. 1996. On the early history of the North Atlantic off Morocco (abstract). EuroColloquium *Ocean Drilling Program/Deep Sea Drilling Project*, 28 February–1 March, Oldenburg.

TALWANI, M. 1978. Distribution of basement under the eastern North Atlantic Ocean and Norwegian Sea. *Geological Journal Special Issue*, **5**, 25–57.

—— & ELDHOLM, O. 1972. The continental margin off Norway: a geophysical study. *Geological Society of America Bulletin*, **83**, 3575–3608.

—— & ——1977. Evolution of the Norwegian–Greenland Sea. *Geological Society of America Bulletin*, **88**, 969–999.

——, EWING, J., SHERIDAN, R. E., HOLBROOK, W. S. & GLOVER, L., III 1995. The EDGE experiment and the US East Coast Magnetic Anomaly. *In*: BANDA, E., TALWANI, M. & TORNÉ, M. (eds) *Rifted Ocean–Continent Boundaries*. NATO ASI Series. Kluwer, Dordrecht, 155–181.

THOMPSON, R. N. & GIBSON, S. A. 1991. Subcontinental mantle plumes, hotspots and pre-existing thinspots. *Journal of the Geological Society, London*, **148**, 973–977.

TRÉHU, A. M., KLITGORD, K. D., SAWYER, D. S. & BUFFLER, R. T. 1989. Regional investigations of the crust and upper mantle, Atlantic and Gulf of Mexico continental margins. *In:* PAKISER, L. & MOONEY, W. (eds) *Geophysical Framework of Continental United States*. Geological Society of America Memoir, **72**, 349–382.

TURNER, S., REGELOUS, M., KELLEY, S., HAWKESWORTH, C. & MANTOVANI, M. 1994. Magmatism and continental break-up in the South Atlantic: high precision $^{40}Ar–^{39}Ar$ geochronology. *Earth and Planetary Science Letters*, **121**, 333–348.

ULIANA, M. A., BIDDLE, K. T. & CERDAN, J. 1989. Mesozoic extension and the formation of Argentine sedimentary basins. *In*: TANKARD, A. J. & BALKWILL, H. (eds) *Extensional Tectonics and stratigraphy of the North Atlantic Margins*. American Association of Petroleum Geologists Memoir, **46**, 599–614.

WHITE, R. S. & MCKENZIE, D. 1989. Magmatism at rift zones: the generation of volcanic continental margins and flood basalts. *Journal of Geophysical Research*, **94**, 7685–7729.

——, —— & O'NIONS, R. K. 1992. Oceanic crustal thickness from seismic measurements and rare earth element inversions. *Journal of Geophysical Research*, **97**, 19683–19715.

——, WESTBROOK, G. K., FOWLER, S. R. *et al.* 1987. Hatton Bank (northwest, UK) continental margin structure. *Geophysical Journal of the Royal Astronomical Society*, **89**, 265–272.

WILSON, M. 1997. Thermal evolution of the Central Atlantic passive margins: continental break-up above a Mesozoic super-plume. *Journal of the Geological Society, London*, **154,** 491–495.

WOOD, M. V., HALL, J. & DOODY, J. J. 1988. Distribution of early Tertiary lavas in the NE Rockall Trough, *In*: MORTON, A. C. & PARSON, L. M. (eds) *Early Tertiary Volcanism and the Opening of the NE Atlantic*. Geological Society, London, Special Publication, **39**, 283–292.

ZEHNDER, C. M., MUTTER, J. C. & BUHL, P. 1990. Deep seismic and geochemical constraints on the nature of rift-induced magmatism during breakup of the North Atlantic. *Tectonophysics*, **173**, 545–565.

Crustal stress and tectonics in Norwegian regions determined from earthquake focal mechanisms

CONRAD D. LINDHOLM,[1] HILMAR BUNGUM,[1] ERIK HICKS[1] & MARIO VILLAGRAN[2]

[1] *NORSAR, Box 51, 2007 Kjeller, Norway*
[2] *Present address: NOC-SOREQ, Nuclear Research Centre of Israel, 81800 Yavne, Israel*

Abstract: A database of 109 earthquake focal mechanisms for Scandinavia and Svalbard has been compiled, comprising 29 new solutions. Stress directions extracted from these earthquake focal mechanisms have been calculated for southeastern and southwestern Norway, the Norwegian Sea and the Barents–Finnmark region. The principal horizontal stress is *c.* N45°W, and indicates a clockwise rotation as one moves from south to north, in accordance with stress trajectories that can be expected from ridge push. The data indicate that in southern Norway normal faulting regimes are more dominant in onshore regions, whereas thrust faulting dominates the offshore seismic activity. Offshore mid-Norway a preference for normal faulting is also found, whereas the shallow earthquakes in Finnmark, northern Norway, are all reverse. Focal depth distributions were studied in the northern North Sea region, revealing substantial differences with depths in the range 5–30 km.

In the last part of the 1980s the World Stress Map (WSM) project has made large amounts of crustal stress data available and accessible (Zoback 1992). Of all the data in the WSM database more than 50% stem from earthquakes, indicating how important that source of information is on a global scale. The WSM data have now convincingly demonstrated that stresses, not only at the plate margins, but also in intraplate areas, are closely connected to global tectonics in general and to plate motions in particular. The European part of the WSM has been discussed in detail by Müller *et al.* (1992) and for Fennoscandia by Gregersen (1992). For the northern North Sea area Lindholm *et al.* (1995) compared shallow borehole inferred stress directions with directions extracted from earthquakes.

Stresses in the crust are described by a stress tensor where the three principal stresses σ_1 (maximum), σ_2 (intermediate) and σ_3 (minimum) form an orthogonal set. The stress tensor describes completely the stresses in a body by magnitude and direction of principal axes (e.g. Aki & Richards 1980; Turcotte & Schubert 1982). The magnitude of the principal stresses can be inferred by *in situ* methods such as hydrofracturing and overcoring measurements (Hast 1958, 1969; Stephanson *et al.* 1987). In this study we have limited the description of the stress tensor to the direction of the principal axes.

The understanding of the present-day stress patterns and the sources and mechanisms behind these have a direct bearing on the understanding of the present-day tectonics (Bungum *et al.* 1991) as well as practical applications in petroleum exploitation (Lindholm *et al.* 1995).

The present-day crustal deformation (the brittle part) is reflected in the seismic activity, which is shown in Fig. 1. It can be seen that Fennoscandia may be divided in three seismotectonic active regions: the western Norway and northern North Sea region; the Norwegian Sea with coastal areas; the weakly indicated zone from the Oslofjord and Skagerrak areas towards the northeast along the Swedish coast of the Gulf of Bothnia and further northwards from the Bothnian Bay. The largest known earthquake in Fennoscandia was the 31 August 1819 earthquake in the Rana area, with an estimated magnitude of 5.8 (Muir Wood 1989).

The purpose of this study has been to utilize earthquakes to compile a database of old and new focal mechanisms, and to synthesize regional stress directions from that database. The criteria used for the definition of principal horizontal compression (σ_H) are the same as used in the WSM database (Zoback 1992). The study also includes an analysis of focal depths for northern North Sea earthquakes.

Wave-form data have been used in some of the investigations, together with readings and

From: Nøttvedt, A. *et al.* (eds) *Dynamics of the Norwegian Margin*. Geological Society, London, Special Publications, **167**, 429–439. 1-86239-056-8/00/$15.00

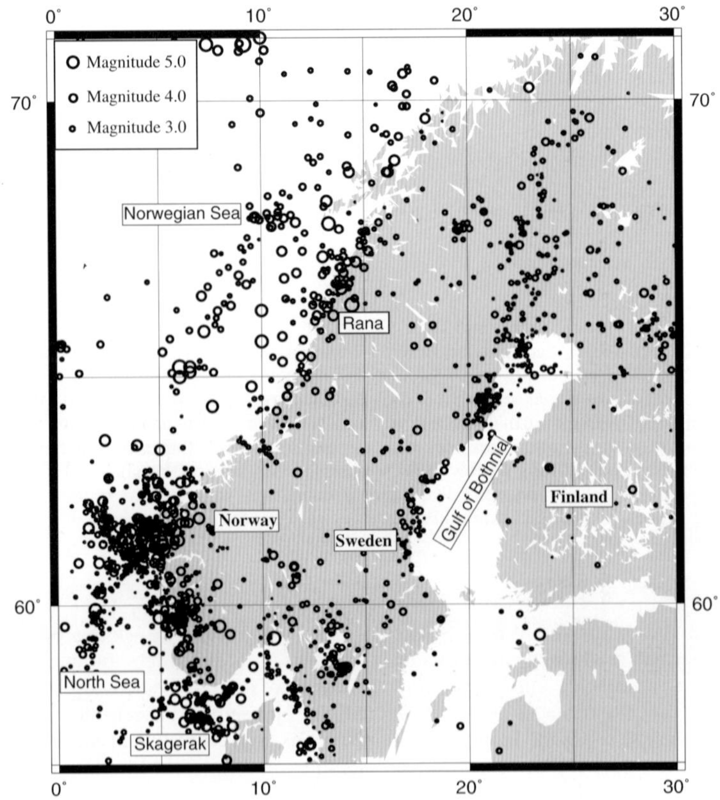

Fig. 1. Earthquakes in Scandinavia from 1800 to 1991. Data from FENCAT (Ahjos & Uski 1992).

locations of events in the northern North Sea. The data used stem from the Norwegian Seismic Network (NSN) and from NORSAR arrays.

Sources of data

A focal mechanisms database has been established that currently comprises 109 focal mechanism solutions for the Fennoscandian region. For Swedish territory only the largest earthquakes were included, as taken from Arvidsson & Kulhanek (1994). The bulk of the data have otherwise been published earlier by Bungum *et al.* (1991), and are also mostly included in the WSM database. Other sources that have been used are Lindholm *et al.* (1995), Bungum & Lindholm (1997) and Hicks (1996). The last study included 29 new solutions, which have been detailed and discussed together with all older focal mechanisms by Hicks *et al.* (in prep.), and only the 14 that were compiled within the frame of the IBS project are detailed below. Figure 2 gives in this respect an overview of all the focal mechanisms used in this study (see also Lindholm *et al.* (in prep.)), and the 14 solutions are listed in Table 1 (of which most are shown in Fig. 3).

Most of the focal mechanism solutions were obtained only by means of first motion polarity information, some of the older ones were constrained by amplitude ratio information, and some of the new solutions (Table 1) have been constrained by wave-form modelling. All of the solutions have been evaluated with respect to quality on a scale from A (good) to D (poor). The quality scheme suggested by the WSM project was applied, but there is also a significant amount of expert judgement in this evaluation.

Results

From the large database with 109 Fennoscandian and Svalbard focal mechanism solutions, 77 were located within the four regions shown in

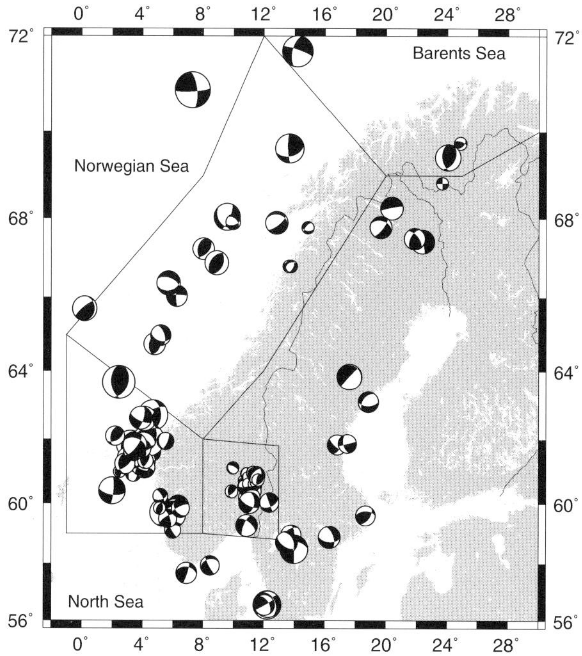

Fig. 2. Earthquake focal mechanisms in Scandinavia used in this study. The four regions discussed in the text are indicated. Structural elements from Blystad *et al.* (1989), Brekke *et al.* (1989) and Gabrielsen *et al.* (1990).

Table 1. *Recent earthquake focal mechanism solutions*

Date	Lat.	Lon.	D	M	T_1	T_2	P_1	P_2	Q
1990 2 26	57.67	6.91	6.4	3.4	252	39	139	26	C
1990 5 16	66.04	6.26	30	3.4	225	1	135	10	C
1991 9 1	79.02	3.59	10	5.0	84	0	174	0	C
1991 12 31	61.98	4.23	15	3.3	210	5	30	85	C
1992 2 19	59.27	10.88	10	3.6	258	9	160	39	C
1992 4 14	59.50	5.66	12	3.0	202	30	292	0	B
1992 6 30	60.88	11.53	12	2.8	299	5	182	78	B
1993 1 20	64.75	4.81	15	3.5	175	90	285	0	C
1993 6 26	62.61	4.14	17	3.9	280	37	180	12	B
1993 9 13	66.37	5.72	20	3.9	7	9	246	72	B
1993 12 27	61.25	2.84	20	3.3	0	58	121	18	C
1994 7 27	62.63	3.90	10	3.7	66	2	334	30	C
1994 11 19	60.17	11.06	13	3.5	185	18	68	54	B
1995 2 6	59.84	6.51	10	3.0	6	19	115	44	C

Date, year, month, day; Lat., latitude; Lon., longitude; D, depth (km); M, magnitude; T_1, azimuth of T-axis; T_2, dip of T-axis; P_1, azimuth of P-axis; P_2, dip of P-axis; Q, quality evaluation (A–D, where A is best).

Fig. 2. Of these, 38 and 21 solutions were in southwestern and southeastern Norway, respectively, and 13 and 5 solutions were in the two northern areas as indicated in Fig. 2.

For all the regions rose diagrams of the directions of the principal horizontal compressions (σ_H) have been prepared. The stress direction has been calculated by following the rules used in the WSM project (Zoback 1992). For each region a rose diagram of σ_H has been prepared (using the Generic Mapping Tools; Wessel & Smith 1991). The data have been

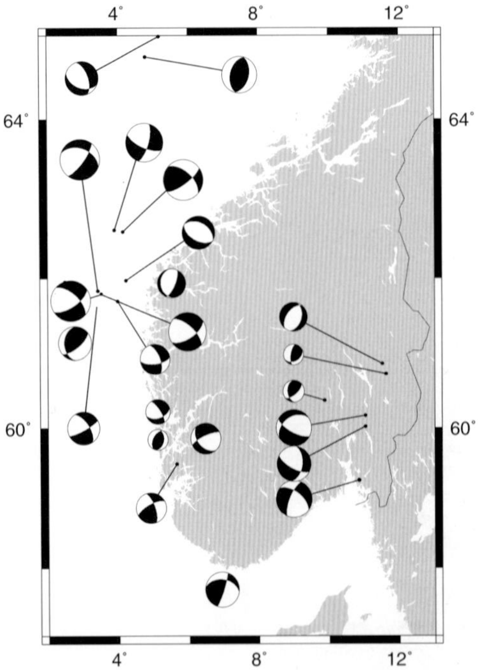

Fig. 3. The 23 southernmost of the 29 most recent earthquake focal mechanisms for southern Norway.

grouped in bins of 10°, with the number of data points within each group easily countable (see also Fejerskov *et al.* in prep.).

To illustrate the type of faulting involved we have used triangle plots as introduced by Frohlich & Apperson (1992).

Southeastern Norway

Southeastern Norway experienced in 1904 the most damaging earthquake in this century in Fennoscandia. Located in the Oslofjord region, it is thought to have occurred on the flank of the Oslo Graben. Three earthquakes have been felt in this region, on 19 February 1992 (M_L 3.3 in Fredrikstad), on 30 June 1992 (M_L 2.1 near Odalen), and on 19 November 1994 (M_L 3.2 near Kløfta). The activity rate in the Oslo Graben area is, however, low compared with west coast seismicity, even though these new events substantiate the earlier observation of Bungum *et al.* (1991) that this area is more active than its surroundings. Twenty-one focal mechanisms (Fig. 4) have until now been calculated for this area, largely because the area is particularly well covered with seismic instruments.

From the triangle diagram inset in Fig. 4 it is evident that normal and strike-slip faulting processes are relatively more abundant in southeastern Norway than reverse faulting, and that the maximum principal horizontal stress is between N100°E and N150°E.

Southwestern Norway and Adjacent seas

As pointed out in earlier papers (e.g. Husebye *et al.* 1978; Bungum *et al.* 1991), this region is among the seismically most active in northwestern Europe, with frequent magnitude 4 and stronger earthquakes. The largest recent one was a magnitude 5.3 earthquake that occurred on 23 January 1989 (Hansen *et al.* 1989). Nevertheless, the calculation of focal mechanisms in this region are often difficult because the station coverage is in general limited to the coastal stations in western Norway.

Earthquake focal depths. Earthquake focal depth is one of the most difficult factors to determine with sufficient precision. This is a well-recognized seismological problem, and is valid as much for the northern North Sea area as it is for other parts of the world. The reason for this is that most of the stations are onshore, and often at distances between 200 and 400 km from the epicentre, whereas accurate depths generally require stations within distances of twice the focal depth. Some attempts have been made to study earthquake depth in western Norway (e.g. Engell-Sørensen & Havskov 1986; Hansen *et al.* 1989; Bungum *et al.* 1991), but the depths found so far have largely been only indicative. In southeastern Norway, earthquake focal depths could in many cases be determined with accuracy to the lower crust. This higher location accuracy for the smaller earthquakes in the NORSAR siting area is a result of the dense instrumental coverage.

Since a reliable focal depth estimate is a prerequisite for a reliable focal mechanism solution, we have in this study analysed 500 earthquakes with respect to focal depth in the northern North Sea, covering the period 1984–1994. The events were binned in roughly 100 km × 100 km bins as shown in Fig. 5 and then located with fixed depth in 1 km depth intervals (using different location methods and crustal velocity models). The root mean square (r.m.s.) values of the travel time residuals were then calculated and summed for each bin. Only first P and S arrival times (i.e. no guided phases) were used, and because of the station–hypocentre distances these do in almost all cases correspond to Moho refracted phases.

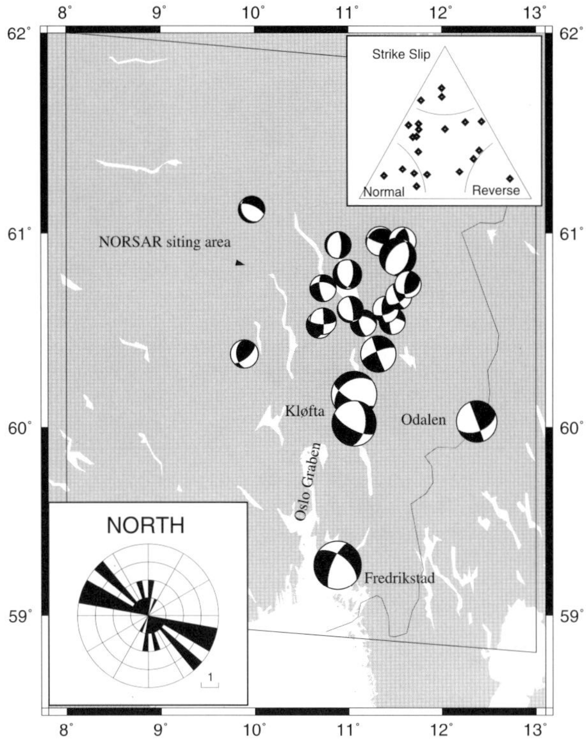

Fig. 4. Epicentres with focal mechanisms in southeastern Norway. The rose diagram shows the directions of the principal horizontal compressions.

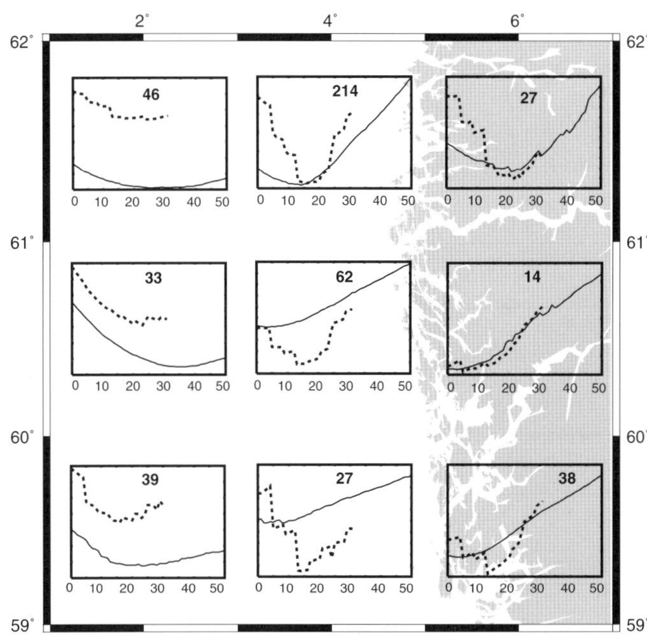

Fig. 5. Depth distribution for earthquakes in the northern North sea calculated by minimization of travel time residuals. Curves show the travel time residuals for two different velocity models, and the numbers indicate number of events behind the calculations in each bin (2° × 1°). Horizontal scales indicates: focal depth (km); vertical scale indicates r.m.s. of travel residuals.

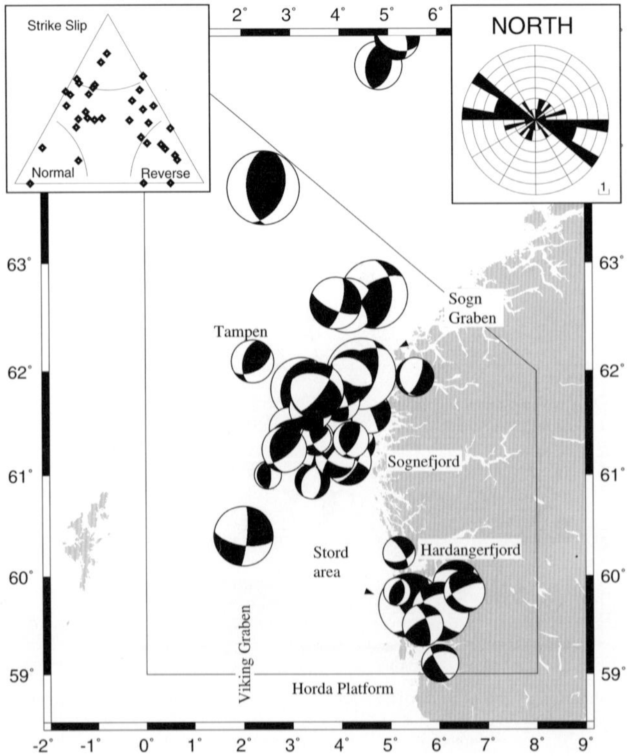

Fig. 6. Focal mechanisms in the northern North Sea. The rose diagram shows the directions of the principal horizontal compresions, and the triangle plot the sense of faulting involved. (See also Fig. 3 for detailed presentation of the focal mechanism beachballs.)

The results are shown in Fig. 5 where for each bin the sum of the r.m.s. travel time residuals are plotted as function of focal depth. It is seen that the best fitting earthquake depths (lowest sum of r.m.s.) within each bin are clearly different, with shallow depths (5–15 km) on the Horda Platform and eastward into the coastal areas, and with deeper events (20–30 km) at the western flank of the Viking Graben and northward towards the Tampen area. The summed r.m.s. values v. depths in Fig. 5 are averaged and weighted numbers from large populations, and therefore do not exclude the possibility of deep events in a shallow zone and vice versa. When evaluating Fig. 5 it is also important to bear in mind that the depth resolution (and location precision in general) is higher closer to the coast (where the stations are) than further offshore.

Earthquake focal mechanisms. Figure 6 shows the focal mechanisms with a triangle plot and a stress rose diagram as insets. The figure shows a very distinct geographical distribution, illustrating the Stord–Hardangerfjord cluster (which has long been recognized as an area with few but notable earthquakes), and the area northwest of the Sognefjord mouth, which is possibly the most seismically active region in Fennoscandia at present, with earthquakes greater than or equal to magnitude 4.0 nearly every year.

From the inset rose diagram (Fig. 6) it is clearly seen that the principal horizontal stress direction is WNW–ESE, which compares well with the expected direction if ridge push is assumed as the principal stress-generating mechanism. Lindholm *et al.* (1995) investigated this area in more detail, also including borehole breakout measurements, and concluded that there is a weak anomalous trend in the Tampen–Sogn Graben area, where σ_H seems to be rotated 90° normal to the regional trend shown in Fig. 6. This stress rotation was originally noted by Bungum *et al.* (1991).

The type of faulting involved in this area can be seen from the triangle inset in Fig. 6. Reverse and oblique strike-slip faulting is clearly dominant, also indicating an overall compressive stress regime. However, this conclusion requires

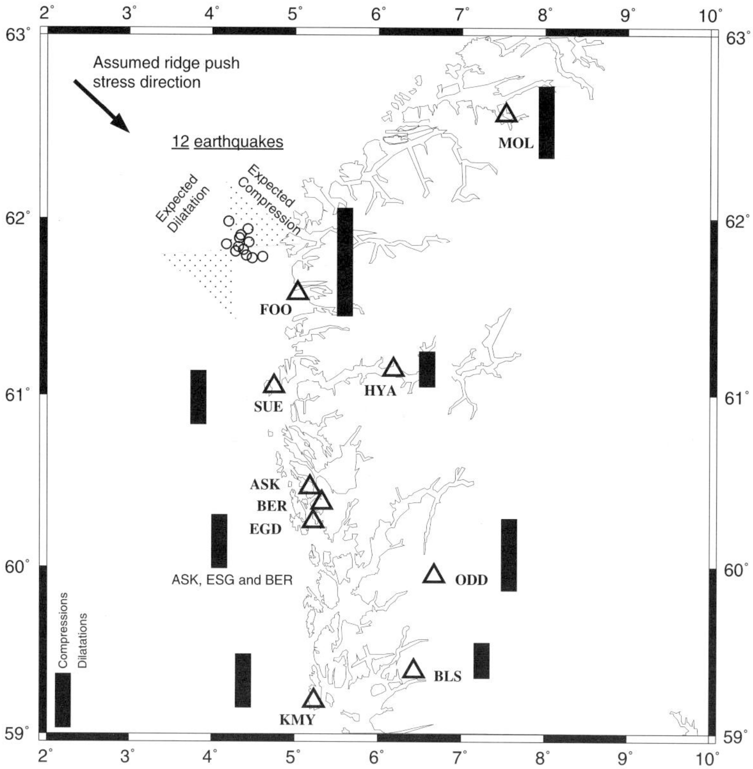

Fig. 7. First motion distribution from a selected group of 12 earthquakes in the Sogn Graben area.

a qualification, as both Bungum *et al.* (1991) and Lindholm *et al.* (1995) showed that there are indications that activity in shallow Stord–Hardanger is dominated more by normal faulting processes than in the offshore region, implying the possibility of greater glacial rebound effects in this region. A stress model supporting this was presented by Stein *et al.* (1989), and has been further discussed by Fejerskov & Lindholm (this volume).

Earthquake first motion distribution. A further attempt to map stress directions was done by selecting earthquakes from a small region as shown in Fig. 7 and read the polarity of first P-wave motions from all the stations whenever possible (Hicks 1996). The quadrant in Fig. 7 only shows the expected direction for compressions and dilatations from a hypothesized strike-slip focal mechanism with the principal horizontal compression in a NW–SE direction. As can be seen from Fig. 7, the relative dominance of observed dilatations supports this assumption in general, but with the FOO (station at Florø, Fig. 7) being anomalous. This is expected if the earthquakes in this region are sub-Moho events, as evidenced also for the M_s 5.1 earthquake of 23 January 1989, at 62°N, 4.4°E, depth 26.2 km (Hansen *et al.* 1989). Even though individual focal mechanisms cannot be calculated for most of these events, the statistics of the first motions are indicative of compression in the upper left quadrant, i.e. as expected from ridge push. There are reasons to believe that many different types of faulting do occur in the Sogn Graben region, but the data in Fig. 7 indicate that there is a dominance of tectonic activity leading to horizontal compression in the NW quadrant.

Mid-Norway

Thirteen earthquakes, generally along the continental margin, are shown in the boxed area in Fig. 8. The principal horizontal stress direction is relatively consistent and slightly clockwise rotated compared with the northern North Sea area (Fig. 6) and southeastern Norway. The bulk of the data follow the main tectonic

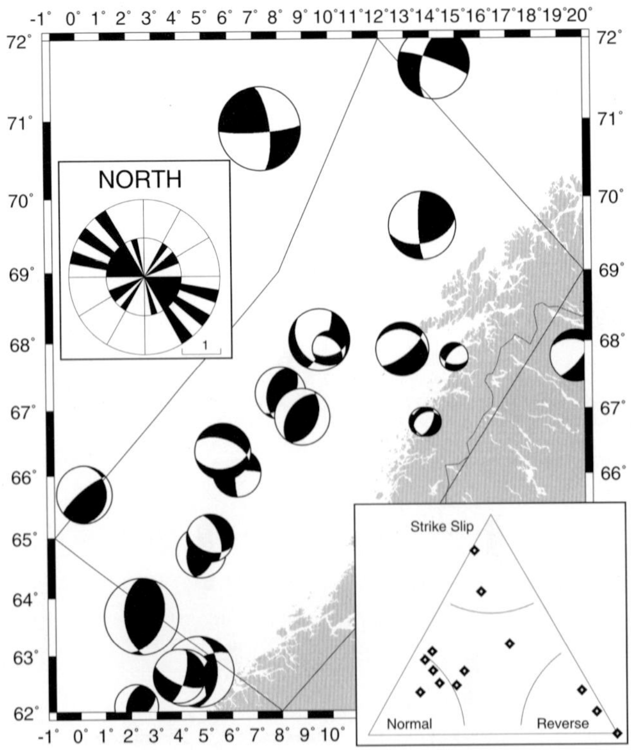

Fig. 8. Earthquake focal mechanisms in the mid-Norway region. The rose diagram shows the directions of the principal horizontal compressions.

structures along the Norwegian margin (Fig. 2), and both normal and reverse faulting is observed in the area.

The combination of both reverse faulting and normal faulting yielding the same σ_H direction (NW–SE) is unique. A tentative explanation (as pointed out by Byrkjeland *et al.* (in press)) may be connected to the fact that a clear correlation exists between rapidly deposited post- Miocene sediments and the seismicity distribution on the mid-Norwegian shelf. A subsiding basin will cause gravity faults to develop at basin margins, and if this process is placed in a regional NW–SE compressional stress regime, only those gravity faults that have a strike near NW–SE will develop.

Northern Norway and the Southern Barents Sea Region

In the Barents Sea region the seismicity is generally very low except for the activity along the main fracture zones; also, the station coverage is poor in this region, and these factors together explain the few (five) focal mechanisms in Fig. 9. Also, the solutions shown in Fig. 9 are not as well constrained as one would normally require (Bungum & Lindholm 1997). Nevertheless, the principal horizontal stress direction (weakly indicated) seems to be clockwise rotated compared with Fig. 8 and three of the four Finnmark events that occurred on or near the Stuoragurra neotectonic fault are indicating reverse faulting in good agreement with surface expressions and borehole results (Bungum & Lindholm 1997).

Discussion

Even if detailed comparative studies are now possible, using the present database, we are in this study focusing only on the regional picture. The rose diagrams of the principal horizontal stress for each region are plotted in Fig. 10 together with stress trajectories generated by the ridge push force, assuming a spreading pole at 47.3°N and 123.0°E for the mid-Atlantic spreading axis.

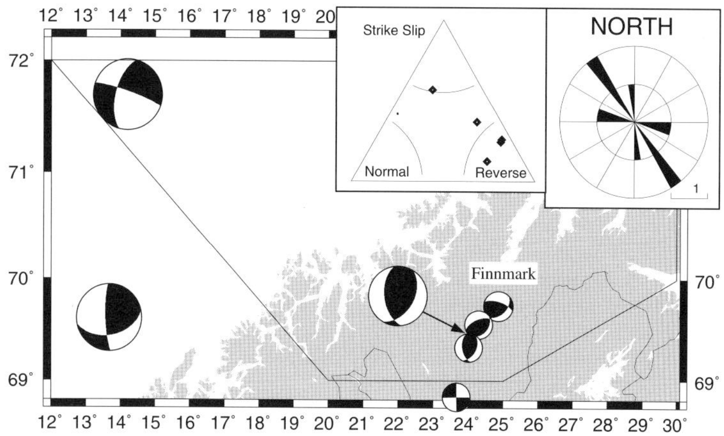

Fig. 9. Focal mechanisms in the Norwegian Sea. The rose diagram shows the directions of the principal horizontal compressions.

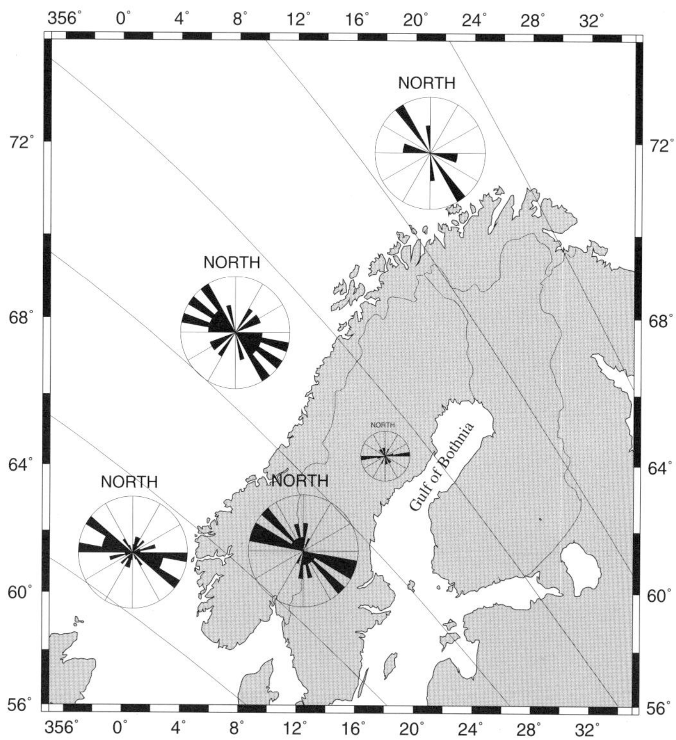

Fig. 10. Stress trajectories and rose diagrams for each of the four regions discussed in the text. The smaller rose dirgam is from nine northern Sweden earthquake focal mechanisms.

We know from the data that considerable uncertainty in the individual focal mechanisms exists, but even then the extracted principal stress directions are surprisingly stable. Moreover, there seems to be a systematic clockwise rotation when moving from southwestern to northern Norway, which matches the rotation that would be expected if the ridge push force is the principal stress-generating mechanism (see Bott 1991). Figure 10 demonstrates that ridge push should, also in terms of stress direction, be regarded as a major (first-order) source for the observed stresses, yet without underestimating possible (and likely) contributions from other (higher-order) stress sources.

An interesting detail is shown in Fig. 10 by the rose diagram of stress directions extracted from nine Swedish earthquakes (Arvidsson & Kulhanek 1994). The E–W direction found for this region is nearly perpendicular to the Fennoscandian uplift centre in the inner Gulf of Bothnia, indicating that this and possibly other more local sources of stress in this case are obscuring the more regional ridge push correlation generally found elsewhere.

With respect to the faulting process involved in the database a significant difference between the earthquakes offshore and onshore southern Norway is found, with normal faulting being more abundant in the onshore areas (comprising those in southeastern Norway and those in the Stord area). In the Stord area the events seem to occur at systematically shallower depths than offshore, and are also more shallow than the average depth in the Oslo Graben area. In the mid-Norwegian area normal and reverse faulting is found side by side, but with a consistent stress NW–SE direction. In the offshore areas (and in Finnmark), reverse and strike-slip faulting is the more common process.

The correlation between seismic activity and main faults on a regional scale is generally high, and can be seen when comparing the maps of Larsen *et al.* (in prep.) and Lindholm *et al.* (in prep.).

Conclusions

From the current study the following can be concluded:

- The database of 109 Fennoscandian earthquake focal mechanisms confirms the overall stress picture that was established through the WSM project (Gregersen 1992; Müller *et al.* 1992).
- A clockwise rotation of the stress direction when moving from south to north is indicated by the data. This can be explained as a result of ridge push; however, more data (especially in northern Norway) are needed before more firm conclusions in this direction can be drawn.
- Earthquake focal depths in the northern North Sea seem to be deeper than those onshore, and difference in depth may be correlated with a difference in tectonic regime.
- Deeper northern North Sea events more often show reverse faulting, whereas more shallow foci onshore (Stord area) earthquakes indicate preferences for normal faulting. One geodynamic model explaining this observation is the glacial rebound of Fennoscandia.
- Earthquakes in southeastern Norway exhibit a preference for strike-slip and normal faulting regimes.

This paper has focused on the presentation of data from earthquakes that can reveal the crustal state of stress in Norwegian regions. Fejerskov & Lindholm (this volume) have given a more detailed discussion of possible geodynamic models, with incorporation also of stress data from borehole breakouts.

The authors are grateful to B. T. Larsen and M. Fejerskov for the many comments and suggestions, and we thank E. Bogoslowski for the linguistic comments. This work was supported by the Research Council of Norway, as part of the IBS (Integrated Basin Studies) project under the Joule II research programme funded by the Commission of European Communities (Contract JOU2-CT 92-0110).

References

AHJOS, T. & USKI, M. 1992. Earthquakes in northern Europe in 1375–1989. *Tectonophysics*, **207**, 1–23.

AKI, K. & RICHARDS, P. G. 1980. *Quantitative Seismology; Theory and Methods*. Freeman, San Francisco, CA.

ARVIDSSON, R. & KULHANEK, O. 1994. Seismodynamics of Sweden deducted from earthquake focal mechanisms. *Geophysical Journal International*, **116**, 377–392.

BLYSTAD, P., FÆRSETH, R., LARSEN, B. T., SKOGSEID, J. & TØRUDBAKKEN, B. 1989. Nomenclature of tectonic units in the Norwegian Sea between 62° North and 69.5° North. *In*: *Structural and Tectonic Modelling and its Application to Petroleum Geology*. Norwegian Petroleum Directorate Conference, 18–20 October 1989, Stavanger.

BOTT, M. H. P. 1991. Ridge push and associated plate interior stress in normal and hot spot regions. *Tectonophysics*, **200**, 17–32.

BREKKE, H., FÆRSETH, R., GABRIELSEN, R., GOWERS, M. B. & PEGRUM, R. M. 1989. Nomenclature of tectonic units in the Norwegian North Sea south of 62 degrees north. *In*: *Structural and Tectonic Modelling and its Application to Petroleum Geology*. Norwegian Petroleum Directorate Conference, 18–20 October 1989, Stavanger.

BUNGUM, H. & LINDHOLM, C. D. 1997. Seismo- and neotectonics in Finnmark, Kola, and in the southern Barents Sea, Part 2: Seismological analysis and seismotectonics. *Tectonophysics*, **270**, 15–28.

——, ALSAKER, A., KVAMME, L. B. & HANSEN, R. A. 1991. Seismicity and seismotectonics of Norway and nearby continental shelf areas, *Journal of Geophysical Research*, **96**, 2249–2265.

BYRKJELAND, U., BUNGUM, H. & ELDHOLM, O. 2000. Seismotectonics of the Norwegian Continental Margin. *Journal of Geophysical Research*. In press.

ENGELL-SØRENSEN, L. & HAVSKOV, J. 1986. Recent North Sea seismicity studies. *Physics of the Earth and Planetary Interiors*, **45**, 37–44.

FEJERSKOV, M. & LINDHOLM, C. 2000. Crustal stress in and around Norway: an evaluation of stress-generating mechanisms. *This volume*.

——, ——, LARSEN, B. T. & NØTTVEDT, A. 2000. Stress Map. North Atlantic Area. In preparation.

FROHLICH, C. & APPERSON, K. D. 1992. Earthquake focal mechanisms, moment tensors, and the consistency of seismic activity near plate boundaries. *Tectonics*, **11**, 279–296.

GABRIELSEN, R., FÆRSETH, R., JENSEN, L. N., KALHEIM, J. E. & RIIS, F. 1990. *Structural Elements of the Norwegian Continental Shelf. Part 1: The Barents Sea Region*. Norwegian Petroleum Directorate Bulletin, **6**.

GREGERSEN, S. 1992. Crustal stress regime in Fennoscandia from focal mechanisms. *Journal of Geophysical Research*, **97**, 11 821–11 827.

HANSEN, R. A., BUNGUM, H. & ALSAKER, A. 1989. Three recent large earthquakes offshore Norway. *Terra Nova*, **1**, 284–295.

HAST, N. 1958. *The Measurement of Rock Pressure in Mines*. Sveriges Geologiska Undersøkning, Series C, Årsbok, **52**(3).

——1969. The state of stress in the upper part of the earth's crust. *Tectonophysics*, **8**, 169–211.

HICKS, E. 1996. *Crustal stresses in Norway and surrounding areas as derived from earthquake focal mechanism solutions and in situ stress measurements*. Cand.Scient. thesis, University of Oslo.

——, BUNGUM, H. & LINDHOLM, C. 1999. Stress inversion of earthquake focal mechanism solutions from onshore and offshore Norway. In preparation.

HUSEBYE, E. S., BUNGUM, H., FYEN, J. & GJØYSTDAL, H. 1978. Earthquake activity in Fennoscandia between 1497 and 1975 and intraplate tectonics. *Norsk Geologisk Tidsskrift*, **58**, 51–68.

LARSEN, B. T., NØTTVEDT, A., VÅGNES, E. *et al.* 1999. Tectonic map. North Atlantic area. *This volume*.

LINDHOLM, C. D., BUNGUM, H., BRATLI, R. K., AADNØY, B. S., DAHL, N., TØRUDBAKKEN, B. & ATAKAN, K. 1995. Crustal stress in the northern North Sea as inferred from borehole breakouts and earthquake focal mechanisms. *Terra Nova*, **7**, 51–59.

——, ——, HICKS, E., FEJERSKOV, M., OLESEN, O., LARSEN, B. T. & NØTTVEDT, A. 1999. Seismicity, earthquake focal mechanisms and neotectonics, North Atlantic area. In preparation

MUIR WOOD, R. 1989. The Scandinavian earthquakes of 22 December 1759 and 31 August 1819. *Disasters*, **12**(3), 223–236.

MÜLLER, B., ZOBACK, M. L., FUCHS, K. *et al.* 1992. Regional patterns of tectonic stress in Europe. *Journal Geophysical Research*, **97**, 11 783–11 803.

STEIN, S., CLOETHINGH, S., SLEEP, N. H. & WORTEL, R. 1989. Passive margin earthquakes, stresses and rheology. *In*: GREGERSEN, S. & BASHAM, P. W. (eds) *Earthquakes at North Atlantic Passive Margins: Neotectonics and Postglacial Rebound*. Kluwer, Dordrecht, 231–259.

STEPHANSON, O., DAHLSTRØM, I., BERGSTRØM, K. *et al.* 1987. *FRSDB–Fennoscandian Rock Stress Database*. Department of Mining, Norwegian Institute of Technology, Trondheim.

TURCOTTE, D., & SCHUBERT, G. 1982. *Geodynamics; Applications of Continuum Physics to Geological Problems*, Wiley, New York.

WESSEL, P. & SMITH, W. H. F. 1991. Free software helps map and display data. *EOS Transactions, American Geophysical Union*, **72**, 441, 445–446.

ZOBACK, M. L. 1992. First and second order patterns of stress in the lithosphere: the World Stress Map Project. *Journal of Geophysical Research*, **97**, 11 703–11 728.

Crustal stress in and around Norway: a compliation of *in situ* stress observations

MORTEN FEJERSKOV,[1,3] CONRAD LINDHOLM,[2] ARNE MYRVANG[1] & HILMAR BUNGUM[2]

[1] *Norwegian University of Science and Technology, 7034 Trondheim, Norway*
[2] *NORSAR, 2007 Kjeller, Norway*
[3] *Present address: Saga Petroleum ASA, P.O. Box 490, 1300 Sandvika, Norway (e-mail: Morten.fejerskov@saga.com)*

Abstract: *In-situ* rock stresses yield information about geodynamic processes in the crust, and are important input data for almost all kinds of geomechanics work. Compilations of rock stress measurements, such as the World Stress Map, have been used to characterize the regional stress field. On the basis of new data from the Norwegian region, a map of the maximum horizontal stress (σ_H) direction, for both on- and offshore Norway, has been established. Four main stress provinces with different regional stress trends are identified. The Barents Sea and northern Norway province exhibits a very consistent N–S σ_H direction, with high horizontal stresses. In the Norwegian Sea and mid-Norway province the σ_H direction has rotated towards NW–SE. Also here, high horizontal stresses are measured and a compressional regime appears. In the northern North Sea and western Norway province the σ_H direction is more scattered. A WNW–ESE direction dominates, but a NNE–SSW trend, parallel to the major structures in the area, also appears. The stress regime is primarily compressional, but in the Stord area normal faulting earthquakes are observed. The southern North Sea province yields a very scattered σ_H direction. However, a NW–SE trend, which aligns with the stress system in western Europe, is identified. The high variation, both laterally and with depth, indicates near-isotropic horizontal stresses and influence by local features. The different stress determination techniques applied (overcoring, borehole breakouts and earthquake focal mechanisms) yield approximately the same stress orientation, and regional and internal variation, although they cover different depth intervals. They further manifest high horizontal stresses in the Norwegian area, generally exceeding the vertical stress. This indicates the presence of compressional tectonic stresses penetrating deep into the crust. The first-order stress pattern in Norway and adjacent offshore regions show a rotation in σ_H direction from N–S in the Barents Sea to WNW–ESE in the North Sea. This rotation can be explained by the ridge push effect alone, but other effects may also be contributing factors. The increased scatter in σ_H direction in the south is believed to be due to changes in the tectonic stress magnitudes, or related to relative changes in angle between the continental margin and the ridge push force.

A knowledge of the *in situ* rock stress field is important for almost all kinds of geomechanics work. Design of underground openings for mining and civil engineering purposes is often based on stress magnitude and orientation data. In petroleum exploration and production, *in situ* rock stresses are important for petroleum retention, reservoir behaviour and borehole stability. The *in situ* stress field is linked to plate tectonics, and knowledge of the rock stresses thereby enhances our understanding of geodynamic mechanisms in the crust.

The first *in situ* stress determination methods were developed in the 1950s (Hast 1958). Before then, rock stresses had only been regarded as induced by gravity. Hast's measurement showed anisotropic horizontal stresses exceeding what was expected from gravitational effects, and often the horizontal stress exceeded the vertical stress. Today, high horizontal stresses are commonly accepted, as verified by many workers and measured by a variety of techniques.

This paper compiles all recent stress studies encompassing on- and offshore Norway, and discusses the observations derived from various techniques.

Compilation of *in situ* rock stress data

Ranalli & Chandler (1975) compiled stress data from Scandinavia to determine the regional

From: NØTTVEDT, A. *et al.* (eds) *Dynamics of the Norwegian Margin*. Geological Society, London, Special Publications, **167**, 441–449. 1-86239-056-8/00/$15.00

stress field. On the basis of Hast's measurements they concluded that the maximum horizontal stress direction was E–W in southern and N–S in the northern Fennoscandia. Overcoring data (Myrvang 1976) and earthquake focal mechanisms (Bungum & Fyen 1979; Bungum *et al.* 1991) from Norway seemed to fit well into this pattern. Klein & Barr (1986) used stress data from all over Europe to identify a regional NW–SE trend in western Europe, but characterized the stress field in Fennoscandia as complicated. On the basis of overcoring and hydraulic fracturing measurements, mainly done for engineering purposes, a Fennoscandian Rock Stress Database (FRSDB) was compiled (Stephansson *et al.* 1987). Today, the FRSDB contains more than 500 measurements from more than 120 locations in Finland, Norway and Sweden. These data were available for the World Stress Map (WSM) project (Zoback *et al.* 1989; Zoback 1992), which, when finished in 1992, contained more than 7300 stress data from 18 countries. The primary goal of the WSM was to establish the global stress field within the plates, excluding data thought to be influenced by local conditions. From the WSM data a well-defined NW–SE trend in the central parts of the European plate was identified, whereas this trend is weaker and more dispersed in southern Fennoscandia and the North Sea (Müller *et al.* 1992; Müller 1993).

On the basis of data from the FRSDB and WSM, together with new borehole breakout and focal mechanism data from the North Sea, a new stress map has been established (Fig. 1). Four main provinces with different regional stress trends have been identified. These trends have then been related to different possible stress-generating mechanisms.

Stress provinces along the coast of Norway

Northern Norway and the Barents Sea

The Barents Sea (Fig. 1), extending between the north of Norway and the Svalbard archipelago, has for the last decade been subjected to oil and gas exploration. More than 55 exploration wells has been drilled by 1996. Eighteen of these have been analysed for borehole breakouts. The occurrence of breakouts is high and occurs over depths of 700–4500 m. The maximum horizontal stress (σ_H) direction defined by the breakout data is consistently N–S, a trend which is confirmed by nine overcoring measurements in northern Norway.

Overcoring measurements in Norway are primarily carried out in connection with mining and civil engineering activities, and for this purpose a modified 3D Leeman cell is used (Myrvang 1976). For use in this compilation the measurements have been subjected to a strict quality control and only data without significant topographical effects or disturbances from nearby openings are presented. In Finnmark, overcoring data reveal very high tectonic stresses (15–25 MPa) at a shallow depth (0–200 m) in both the N–S and E–W directions (Myrvang *et al.* 1993).

The Barents Sea is characterized by very low seismic activity (Lindholm *et al.* 1995*b*) and no focal mechanisms have been derived. However, in Finnmark some seismic activity has been recorded, primarily traced back to post-glacial activity on the Stuoragurra fault zone. These data represent compressional stress regimes and reverse faulting with the maximum horizontal stress trending N–S or E–W. Another seismically active region, the Senja fracture zone, also exhibits reverse faulting with maximum horizontal principal stress striking parallel to the fracture zone.

Mid-Norway and the Norwegian Sea

This area comprises data from the Norwegian Shelf and mainland Norway (62°–70°N) (Fig. 2). Oil and gas exploration activities are concentrated at Haltenbanken and the Trøndelag Platform. More than 100 exploration wells have been drilled in this area. Fifteen wells have been analysed for borehole breakouts, and breakouts were encountered from 1100 to 4800 m, yielding a very consistent NW–SE σ_H direction.

Overcoring measurements have been performed at several locations along the coast, and the dominant stress direction is NW–SE. This confirms the directions derived from borehole breakouts. The measurements indicate very high horizontal stresses, occasionally exceeding 30 MPa (Hanssen & Myrvang 1986; Myrvang 1993) close to the surface.

The Norwegian margin is subject to relatively high seismic activity, with two interesting earthquake swarms in the coastal region at Meløy and Steigen (Bungum & Fyen 1979; Atakan *et al.* 1994). This has made it possible to compute ten earthquake focal mechanism solutions for the area, which primarily indicate compressive reverse faulting with a consistent NW–SE σ_H direction. Two events, however, give normal faulting. One, on the shelf edge, fits well into the model of passive continental margins (Stein *et al.

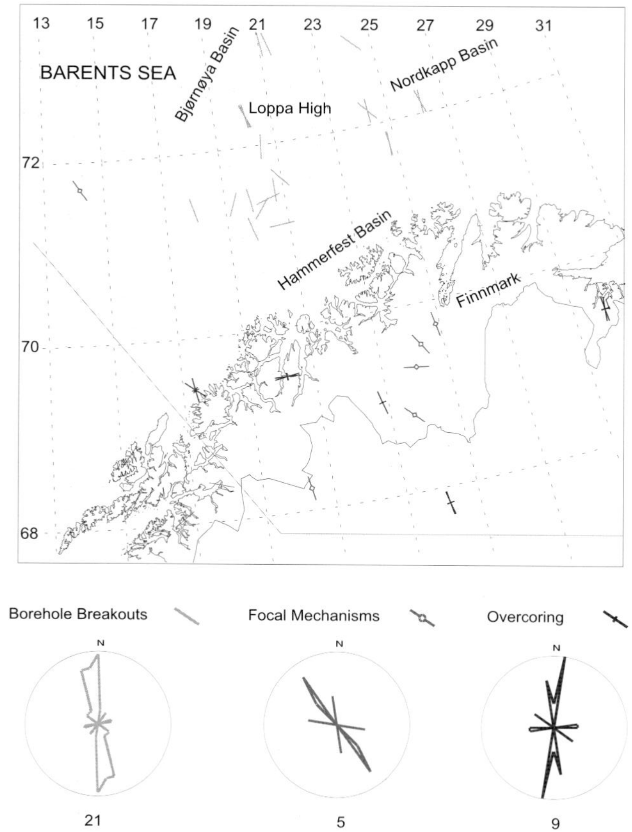

Fig. 1. Stress data from northern Norway and the Barents Sea. Rose diagrams show the compiled results for each measuring technique. The number of data points is given below each rose diagram.

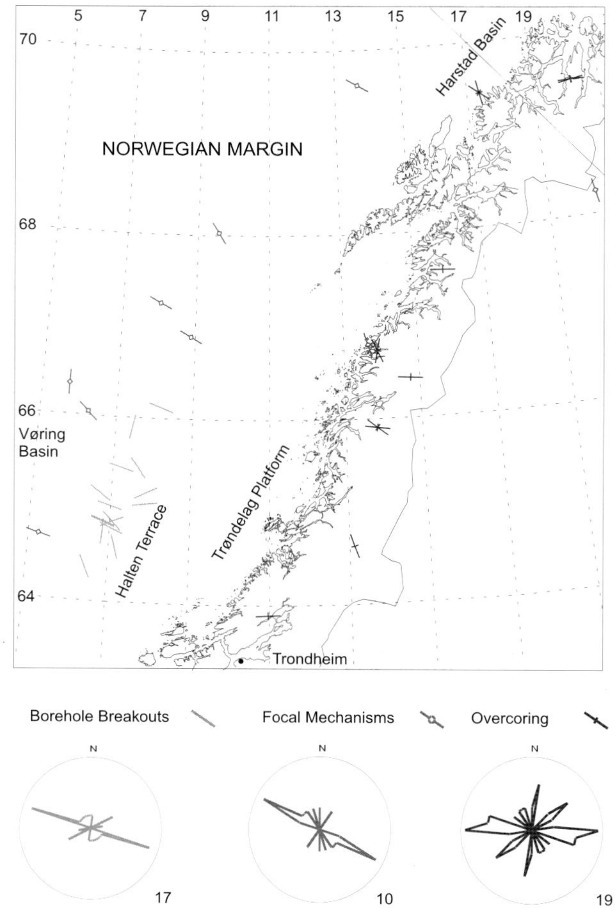

Fig. 2. Stress data from mid-Norway and the Norwegian Sea. Rose diagrams show the compiled results for each measuring technique. The number of data points is given below each rose diagram.

1989) and the other is connected to a highly elevated and glaciated onshore region close to Meløy and Steigen.

Western Norway and the northern North Sea

The major area of investigation in this study has been the northern North Sea, comprising the Tampen Spur area, the Viking Graben and the Horda Platform (Fig. 3). The area has been subjected to extensive oil and gas exploration, and many oil and gas fields are at present in production. The number of exploration wells in the Norwegian sector exceeds 370. A total of 125 wells have been examined with respect to borehole breakouts, of which 73 exhibited breakouts. From the UK sector an additional 30 wells with breakouts are reported. Breakouts are encountered at depths from 1500 to 5000 m. Breakout data from the northern North Sea yield a more dispersed stress orientation than in the Barents Sea and on the Norwegian margin. However, even though a blurred NE–SW trend is observed, the major direction is WNW–ESE. This bimodal distribution is also prominent in the other datasets from this province.

Overcoring measurements in this region are primarily made in connection with road tunnelling and construction of hydropower plants. The relief is high, with deep fjords, and the measurements thus contain large topographical contributions. Topography often forces the principal stresses to align with the fjord and each measuring site has to be examined carefully. Even though overcoring data that reflect only local topographical effects have not been included, the rose diagram looks diffuse. However, an overall NW–SE trend is obtained, and by correcting for the topographical contribution tectonic stresses of the order of 10–15 MPa are calculated.

The seismic activity in this region is relatively high and mainly concentrated in the Tampen Spur, the Sogn Graben and the Stord area. An indicative focal depth distribution from more than 500 earthquakes in this region revealed different characteristic focal depths within subregions (Lindholm *et al.* 1995*b*). Earthquakes seem to be more shallow (*c.* 5 km) in the coastal region of Stord, whereas in the Tampen Spur (*c.* 15 km) and the western side of the Viking Graben (*c.* 30 km) they appear deeper in the crust. Focal mechanisms in the same region reveal consistent trends, where the shallower earthquakes in the Stord area are mostly related to normal faulting but in the Tampen Spur and the Sogn Graben they are related to reverse and strike-slip faulting. Evaluation of the stress direction derived from the focal mechanisms shows that the major trend is WNW–ESE, with a weaker NE–SW trend also present. This correlates well with borehole breakout data from the same area (Lindholm *et al.* 1995*a*).

To explain the bimodal distribution it has been necessary to look at some of the *in situ* measurements in more detail. Three major oil and gas fields (Troll, Visund and Snorre) have been chosen for detailed breakout studies, and the results reveal a rather uniform pattern. Breakouts, appearing in wells situated centrally within the fault blocks, yield a relatively consistent WNW–ESE trend. The NE–SW stress direction is then primarily connected to breakouts encountered in wells close to major fault zones, and thus reflects a stress reorientation, where the maximum horizontal stress aligns with the tectonic structures. Furthermore, borehole breakout data indicate an increase in tectonic stress with depth, which is believed to be related to formation stiffness. Triassic and Jurassic strata exhibit near-lithostatic stresses, possibly favouring a slight strike-slip faulting regime. The Cretaceous and Tertiary sediments, however, exhibit low tectonic stress, yielding a normal faulting regime (Fejerskov 1996).

Southwestern Norway and the central North Sea

The area south of 59°N is referred to as the central North Sea in this paper. The major structural elements are the Central Graben and the Norwegian–Danish Basin (Fig. 4). This was the first area where hydrocarbons were encountered, and the area has since been subjected to extensive oil and gas activities. In the Norwegian sector alone, more than 300 exploration wells have been drilled, and Norway, the UK, Denmark and the Netherlands have strong interests in this area. In this compilation, 45 wells in the Norwegian sector and 38 wells in the UK sector exhibited breakouts. The maximum horizontal stress direction changes vigorously, both between wells as well as within single wells. However, a vague pattern may be identified, as wells situated centrally in the graben reveal a NW–SE stress orientation parallel to the graben system, and wells on the flanks or near large fault zones more often show a NE–SW stress direction normal to the graben structure. Still, breakouts do not seem to appear with the same consistency as in the northern area. This may be explained by the extensive salt diapirism in this area, which has resulted in large anticlines and

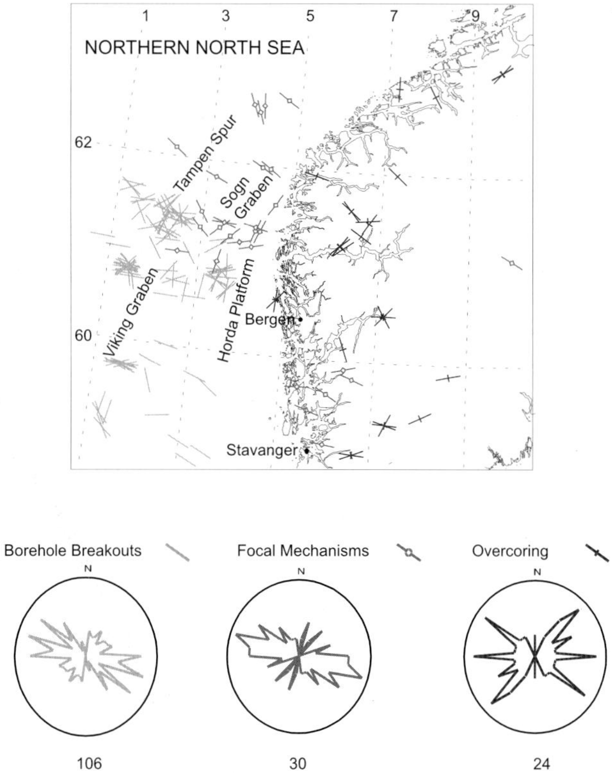

Fig. 3. Stress data from western Norway and the northern North Sea. Rose diagrams show the compiled results for each measuring technique. The number of data points is given below each rose diagram.

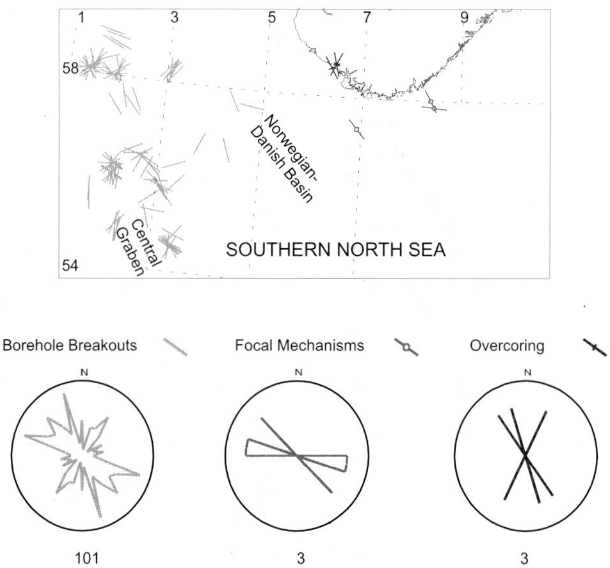

Fig. 4. Stress data from southwestern Norway and the central North Sea. Rose diagrams show the compiled results for each measuring technique. The number of data points is given below each rose diagram.

doming structures. Near-equal horizontal stresses may also explain the observations.

Very few overcoring measurements have been carried out in the coastal regions of southern Norway. The database is so sparse that no overcoring data were used for comparison. Seismic activity is also low in the southern North Sea, and only three focal mechanisms have been obtained from the region.

Discussion

When comparing stress data from the areas in question, it is particularly interesting to note the coinciding stress directions that are obtained from different stress determination methods (Fig. 5). Although the methods cover different depth intervals (0–1 km for overcoring, 1–5 km for borehole breakouts and 5–30 km for focal mechanisms) they yield approximately the same stress orientation, with the same regional and internal variation. This supports the assumption of a tectonic stress field penetrating large parts of the crust. It further indicates a close connection between stresses in crystalline and sedimentary rocks, for both on- and offshore Norway. This definitely supports the hypothesis of an attached basin (Bell 1993), in contrast to a detached basin where independent stress fields are observed above and below a detachment zone. At the very least, the consistent result obtained by the various techniques increases the confidence in each of the stress determination techniques.

Most of the measurements indicate extensive compressive stresses generally exceeding the vertical stress. Earthquake focal mechanisms primarily indicate a strike-slip or thrust faulting stress regime, although two minor areas with normal faulting are observed. Overcoring measurements also yield strike-slip or thrust faulting stress regimes with substantial tectonic stresses. The highest tectonic stress is measured in massive gneiss regions, whereas the more fractured Caledonian rocks exhibit somewhat lower tectonic stress. This is primarily believed to be due to the stiffness of the rock masses and thereby their ability to conduct tectonic stress. Borehole breakouts yield no direct information about stress magnitudes, but, in the northern Viking Graben, where other wellbore stress information has been utilized together with the breakout data, a possible strike-slip stress regime may be inferred in Triassic and Jurassic strata (Fejerskov 1996). Cretaceous and Tertiary strata do not show extensive signs of tectonic stresses. This is primarily believed to be due to the soft sediments low ability of conduct tectonic stress.

Although some areas exhibit scattered data, the general trend indicates a rotation of the stress field from N–S in the Barents Sea, to NW–SE on the Norwegian margin and to WNW–ESE in the North Sea. This is supported by all the stress measuring methods. To identify the reason for this stress rotation, several possibilities have to be investigated. An observed horizontal stress field may be thought of as being composed of different components. The gravity component yields isotropic horizontal stresses, whereas tectonic components are directional and thus form an anisotropic horizontal stress field, as is experienced in both on- and offshore Norway. For convenience, tectonic stresses are often further divided into plate-wide, regional and local, or first-, second- and third-order stress fields, according to their area of influence (Zoback 1992). Observed tectonic stresses may then be explained by a plate-wide stress field, which is possibly overprinted by regional and local effects. For a first-order approximation it is therefore essential to examine plate-wide stress-generating mechanisms. Here only a brief discussion of the possible stress generating mechanisms is made. For details, the reader is referred to the following paper on stress-generating mechanisms (Fejerskov & Lindholm 1999).

The dominant source of tectonic stress in western Europe is a combination of collision-related forces at the Euroasian–African and ridge push along the Mid-Atlantic and Arctic spreading ridge (Müller 1993). The vicinity to the spreading ridge, and the fact that the ridge push force may contribute significantly to intraplate tectonic stresses (Forsyth & Uyeda 1975; Dahlen 1981), has made the ridge push effect a prime suspect for the tectonic stress observed in the Norwegian area. Using, as an approximation, the pole of rotation as a basis for determining the direction of ridge push, traces indicating the direction of plate movement have been superimposed on the stress map. Figure 5 shows that most of the stress rotation can be explained by the flowlines. A more precise way to describe the ridge push force is to use the ridge geometry and bathymetry data. In the arctic region the Lomonosov Ridge is very prominent, and hence could contribute significantly to a rotation of the ridge push force from NW–SE in the Atlantic to more N–S in the Arctic. Therefore, for a more detailed study of the ridge push force the actual 3D geometry (bathymetry, lateral variation and plate thickness) has to be modelled. Although the ridge

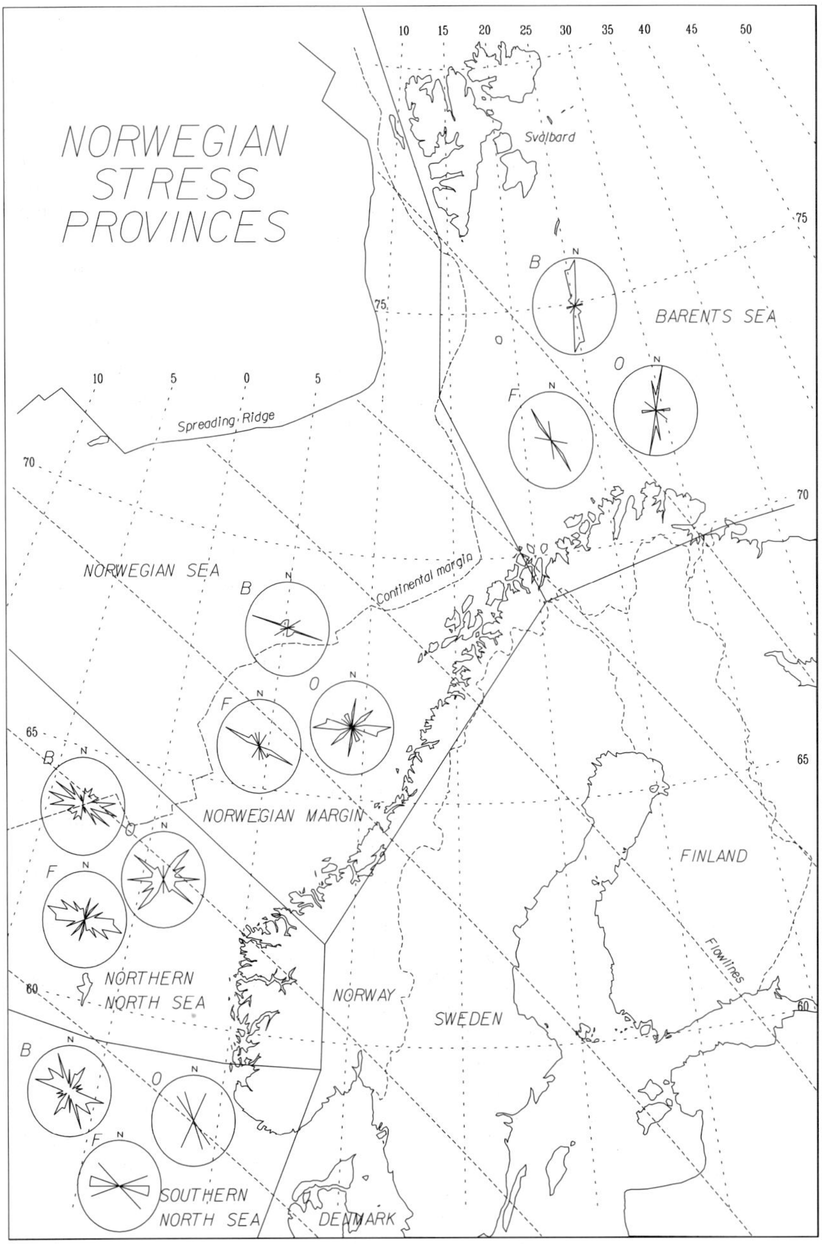

Fig. 5. Compilation of Norwegian stress data. Rose diagrams indicate maximum horizontal stress directions from borehole breakouts (B), focal mechanisms (F) and overcoring (O) within each of the four stress provinces. Geological features such as the spreading ridge and the continental margin are included. Flowlines (dashed line), representing the direction of spreading and indicating the direction of ridge push as a first-order approximation correlate well with the stress trends.

push effect can explain most of the stress observations, other possible causes for the tectonic stress, such as density variations, sediment load and deglaciation, also have to be considered. These effects, however, are of much smaller lateral extent, and will primarily affect the stress field only locally.

The southern areas, encompassing the North Sea, yield more scattered data than do the Barents Sea and the Norwegian Sea. This may be a result of the available number of data, but is rather anticipated to be caused by variations in the horizontal stress anisotropy. If the North Sea exhibits lower stress anisotropy than the Barents Sea, measurements in North Sea will more easily be influenced by local tectonic and geological structures, yielding a more scattered stress direction. The change in stress anisotropy may be caused by changes in the magnitude of the ridge push force, but can also be related to the relative angle between the continental margin and the ridge push force. In the western Barents Sea the ridge push force acts near parallel to the continental margin, whereas in the Norwegian Sea it acts perpendicular to it. As the continental margin is known to generate deviatoric tension in the continental crust (Stein *et al.* 1989), the continental margin effect may counteract the ridge push force in the Norwegian Sea simultaneously as the horizontal stress anisotropy in the western Barents Sea is enhanced.

Conclusion

The different techniques for stress determination applied on the Norwegian continental shelf yield very consistent results, although they cover different depths. This increases the confidence in each stress determination method, supports the assumption of tectonic stresses penetrating large parts of the crust, and indicates a close connection between stresses in both crystalline and sedimentary rocks, thereby supporting the hypothesis of an attached basin.

The first-order stress pattern in Norway and adjacent offshore regions seems to be characterized by high compressive horizontal stresses. A rotation of the maximum horizontal stress, from N–S in the Barents Sea, to NW–SE on the mid-Norwegian margin and to WNW–ESE in the North Sea, is observed. It is considered that most of this rotation can generally be explained by the stresses generated from ridge push, but it is likely that other effects may also contribute.

The stress directions are more scattered in the North Sea compared with the well-defined trends in the Barents Sea and the Norwegian Sea. This is believed to be caused by changes in stress anisotropy, indicating a stronger stress anisotropy in the northern areas. Change in stress anisotropy may be due to changes in the magnitude of the ridge push force, or related to a change in angle between the continental margin and the ridge push force.

This work was supported by the Research Council of Norway, as a part of the IBS (Integrated Basin Studies) project under the Joule II research programme funded by the Commission of European Communities (Contract JOU2-CT-92-0110). Data for the study were supported by Norsk Hydro, Saga Petroleum, Statoil and the PROFIT-project. The authors especially want to thank M. D. Zoback, R. Gabrielsen, A. Mascle, B. Tørudbakken, and J. Palmer for valuable input during the writing.

References

ATAKAN, K., LINDHOLM, C. D. & HAVSKOV, J. 1994. Earthquake swarm in Steigen, Northern Norway: an unusual example of intraplate seismicity. *Terra Nova*, **6**, 180–194.

BELL, J. S. 1993. Global sedimentary stress project of the international lithosphere program. Poster P71, 55th Meeting of Association of Exploration and Geophysics, Stavanger, 7–11 June.

BUNGUM, H. & FYEN, J. 1979. Hypocentral distribution, focal mechanisms, and tectonic implications of Fennoscandian earthquakes 1954–1978, *Geologiska Føreningens i Stokholm Forhandlingar*, **101**(4), 261–273.

——, ALSAKER, A., KVAMME, L. B. & HANSEN, R. A. 1991. Seismicity and seismotectonics of Norway and nearby continental shelf areas. *Journal of Geophysical Research*, **96**, 2249–2265.

DAHLEN, F. A. 1981. Isostasy and the ambient state of stress in the oceanic lithosphere. *Journal of Geophysical Research*, **86**, 7801–7807.

FEJERSKOV, M 1996. *Determination of* in situ *rock stresses related to petroleum activities on the Norwegian Continental Shelf.* Dr. Ing. thesis, Norwegian Institute of Science and Technology, Trondheim.

—— & LINDHOLM, C. 1999. Crustal stress in and around Norway: an evaluation of stress generating mechanisms. *This volume*.

FORSYTH, D. W. & UYEDA, S. 1975. On the relative importance of the driving forces of plate motion. *Geophysical Journal of the Royal Astronomical Society*, **43**, 163–200.

HANSSEN, T. H. & MYRVANG, A. 1986. Rock stresses and rock stress effects in the Kobbelv area, northern Norway. *Proceedings of the International Symposium on Rock Stress and Rock Stress Measurements*. Stockholm, 1–3 September, 625–634.

HAST, N. 1958. *The Measurement of Rock Pressure in Mines.* Swedish Geological Survey, Series C, **560**.

KLEIN, R. J. & BARR, M. V. 1986. Regional state of stress in western Europe. *Proceedings of the International Symposium on Rock Stress and Rock Stress Measurements*, Stockholm, 1–3 September, 21–32.

LINDHOLM, C., BUNGUM, H., BRATLI, R. K., AADNØY, B. S., DAHL, N., TØRUDBAKKEN, B. & ATAKAN, K. 1995*a*. Crustal stress in the northern North Sea as inferred from borehole breakouts and earthquake focal mechanisms. *Terra Nova*, **7**, 51–59

——, ——, VILLAGRAN, M. & HICKS, E. 1995*b*. Crustal stress and tectonics in Norwegian regions determined from earthquake focal mechanisms. *Proceedings from Workshop on Rock Stresses in the North Sea*, Trondheim, 13–14 February, 77–91.

MÜLLER, B. 1993. *Tectonic stress in Europe – borehole guided waves affected by stress-induced anisotropy*. PhD thesis, University of Karlsruhe.

——, ZOBACK, M. L., FUCHS, K., MASTIN, L., GREGERSEN, S., PAVONI, N., STEPHANSSON, O. & LJUNGGREN, C. 1992. Regional patterns of tectonic stress in Europe. *Journal of Geophysical Research*, **97**(B8), 11 783–11 803.

MYRVANG, A. 1976. Practical use of rock stress measurements in Norway. *Symposium on Investigation of Stress in Rock – Advances in Stress Measurements*, Sydney, 11–13 August, 92–99.

——1993. Rock stress and rock stress problems in Norway. *In*: HUDSON, J. A. (ed.) *Comprehensive Rock Engineering, Vol. 3*. Pergamon, Oxford, 461–471.

——, HANSEN, S. E. & SØRENSEN, T. 1993. Rock stress redistribution around an open pit mine in hardrock. *International Journal of Rock Mechanics, Mining Science, and Geomechanics Abstracts*, **30**, 1001–1004.

RANALLI, G. & CHANDLER, T. E. 1975. The stress field in the upper crust as determined from *in situ* measurements. *Geologische Rundschau*, **64**, 653–674

STEIN, S., CLOETINGH, S., SLEEP, N. & WORTEL, R. 1989. Passive margin earthquakes, stresses and rheology. *In*: GREGERSEN, S. & BASHAM, P. (eds) *Earthquakes at North-Atlantic Passive Margins: Neotectonics and Postglacial Rebound, Nato ASI Series C*. Kluwer, Boston, MA, 231–259.

STEPHANSSON, O., DAHLSTROEM, L. O., BERGSTROEM, K. *et al.* 1987. *Fennoscandian Rock Stress Data Base – FRSDB*. Luleå University, Research report, Luleå **1987:06**.

ZOBACK, M. L. 1992. First- and second-order patterns of stress in the lithosphere: the World Stress Map project. *Journal of Geophysical Research*, **97**(B8), 11 703–11 728.

—— *et al.* 1989. Global patterns of tectonic stress. *Nature*, **341**, 291–298.

Crustal stress in and around Norway: an evaluation of stress-generating mechanisms

MORTEN FEJERSKOV[1,3] & CONRAD LINDHOLM[2]

[1] *Norwegian University of Science and Technology, 7034 Trondheim, Norway*
[2] *NORSAR, 2007 Kjeller, Norway*
[3] *Present address: Saga Petroleum ASA, P.O. Box 490, 1300 Sandvika, Norway (e-mail: Morten.fejerskov@saga.com)*

Abstract: Recent stress observations from *in situ* measurements and earthquake focal mechanisms in the Norwegian onshore and offshore areas are evaluated with the aim of characterizing the most important mechanisms for the present stress field in and around Norway. The evaluation is based on a set of stress indicators that include both shallow measurements (overcoring and borehole breakouts) and deep data (earthquake focal mechanisms). Computer simulations of simple stress-generation models were used to constrain the relative importance of different stress-generating mechanisms. The ridge push force associated with sea-floor spreading in the North Atlantic is considered to be the primary source of the compressional stress field observed in Norway. Regional influences from the continental margin density contrast, topography and flexure induced by sediment loading are of limited lateral extent, but are important in reorienting the stress field in certain areas. The observed tectonics and stresses are generally also in accord with tectonics expected from Fennoscandian uplift.

The orientation of the stresses in the crust is among the principal indicators of current dynamic processes. Earthquake focal mechanisms together with *in situ* measurements (overcoring, hydraulic fracturing and borehole breakout analysis) are important sources of information on stress orientation and magnitude.

The first *in situ* stress determination methods were developed in the 1950s (see Hast 1958), and were primarily related to mining activity. Before that time, rock stresses had only been regarded as gravity induced, but in the 1950s the new measurements indicated that horizontal stresses exceeded what could have been expected from only gravitational forces, and since the 1960s stress data were interpreted in a wider geophysical context. Since the first measurements were obtained in Norwegian areas, the stress database has increased dramatically, primarily as a result of three scientific programmes: the Fennoscandian Rock Stress Data Base (Stephansson *et al.* 1987), the World Stress Map (Zoback *et al.* 1989; Müller *et al.* 1992; Zoback 1992), and the Dynamics of the Norwegian Margin (Fejerskov *et al.* 1995; Lindholm *et al.* 1995*b*). In the mapping of crustal stresses in northwestern Europe and Norway, the following contributions can be considered as milestones: Hast (1969), Ranalli & Chandler (1975), Bungum & Fyen (1979), Klein & Barr (1986), Clauss *et al.* (1989), Bungum *et al.* (1991), Müller *et al.*, (1992) and Müller (1993).

Parallel to the compilation of new data on observed stresses, analytical and numerical models and computer simulations were applied to investigate the characteristics of different stress-generating mechanisms. Models ranging from global analysis of different plate driving forces (Forsyth & Uyeda 1975; Bott & Kusznir 1984; Richardsson 1992) to regional and local features such as density inhomogeneities, and flexural stresses from sediment loading, glacial rebound and topography (Stephansson 1988; Stein *et al.* 1989; Spann *et al.* 1991) were developed. The present paper reviews available data and modelling results with the goal of characterizing the most important stress-generating processes behind the horizontal stress field in Norway and adjacent offshore areas.

Data

Ranalli & Chandler (1975) were the first to compile stress data from Scandinavia. On the

From: NØTTVEDT, A. *et al.* (eds) *Dynamics of the Norwegian Margin*. Geological Society, London, Special Publications, **167**, 451–467. 1-86239-056-8/00/$15.00

basis of Hast's measurements (largely overcoring) they concluded that the direction of the principal horizontal stresses was E–W in southern Scandinavia and N–S in northern Scandinavia. Although this picture has become more differentiated with the acquisition of additional data, the main conclusion of this early investigation is still valid. The database now at hand has been compiled and quality checked, and has been presented by Fejerskov *et al.* (1995) and Lindholm *et al.* (1995*b*). A synthesis of all the data is shown in Figs 1 and 2, together with the regionalization used in this paper. Within the area defined in Fig. 1 altogether 351 data points give the azimuth of the largest principal horizontal stress component, and the observed stress directions are summarized as follows:

- The maximum horizontal stress direction (σ_H) rotates from N–S in northern Norway and the Barents Sea to WNW–ESE in western Norway and the northern North Sea.
- The stress directions at shallow depths determined by *in situ* techniques, and deeper observations from the middle to lower crust based on earthquake focal mechanisms, are similar and indicate that the tectonic stresses are homogeneous in direction over a large depth range.
- In offshore areas reverse and strike-slip faulting dominates, indicating compressive stress regimes.
- Shallow earthquakes, particularly found in an area around Stord, but also in the Oslo region and in the Meløy and Steigen sequences (Bungum *et al.* 1979; Atakan *et al.* 1994) indicate tensional stress regimes with normal faulting.
- No change in horizontal stress direction is observed between sedimentary and crystalline rocks.
- Overcoring measurements show a more scattered stress direction compared with borehole breakouts and earthquake focal mechanisms.

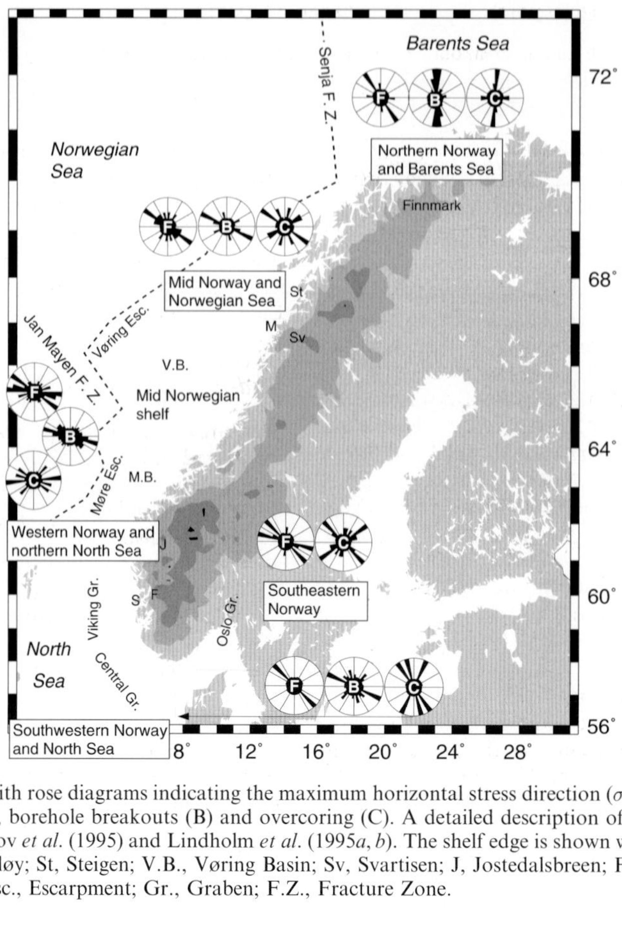

Fig. 1. Five regions with rose diagrams indicating the maximum horizontal stress direction (σ_H) determined from focal mechanisms (F), borehole breakouts (B) and overcoring (C). A detailed description of the data used has been given by Fejerskov *et al.* (1995) and Lindholm *et al.* (1995*a*, *b*). The shelf edge is shown with the dashed line. S, Stord area; M, Meløy; St, Steigen; V.B., Vøring Basin; Sv, Svartisen; J, Jostedalsbreen; F, Folgefonna; M.B., Møre Basin; Esc., Escarpment; Gr., Graben; F.Z., Fracture Zone.

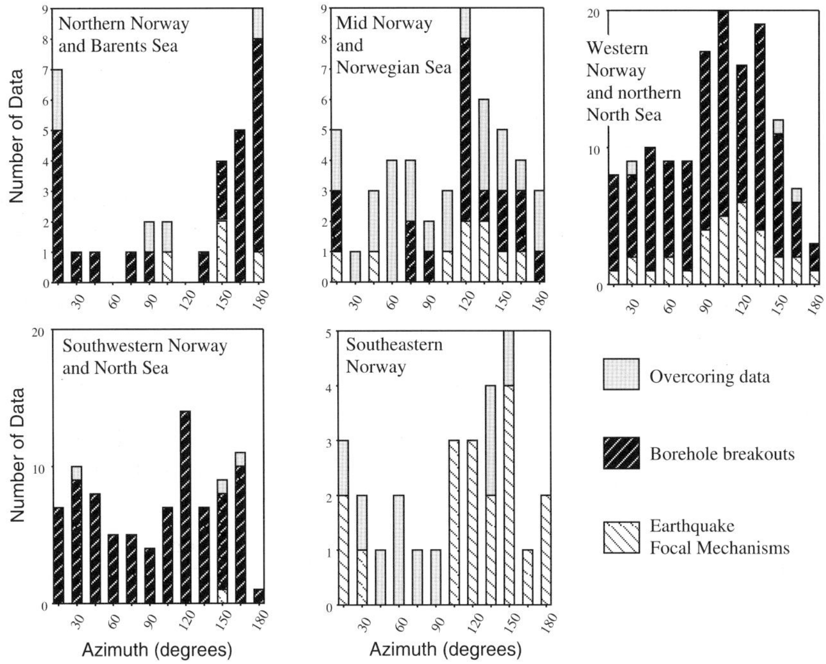

Fig. 2. Histograms of borehole breakout, earthquake focal mechanism and overcoring data for each region.

They also reveal high stress magnitudes (10–20 MPa tectonic stress), particularly in massive gneiss areas of mid-Norway.

Regions

Five main regions with different regional horizontal stress trends have been identified in Norway. The region boundaries do not represent abrupt changes in the stress field, but a continuous lateral variation.

Northern Norway and the Barents Sea

As seen from Fig. 1 the area comprises large offshore areas and the northernmost part of Norway (Finnmark). *In situ* measurements in the Barents Sea reveal a clear N–S stress direction (Dart *et al.* 1995; Fejerskov *et al.* 1995). Overcoring measurements from onshore Finnmark are consistent with the offshore observations and reveal very high tectonic stresses (15–25 MPa) both in N–S and E–W directions at shallower depths (0–200 m) (Myrvang *et al.* 1993). In contrast to the low seismic activity in the Barents Sea, neotectonic movements associated with large prehistoric earthquakes have been mapped in Finnmark and northern Sweden. Displacements on NE–SW and NNW–SSE trending structures indicate a compressional stress regime with a NNW–SSE stress direction (Olesen *et al.* 1992; Bungum & Lindholm in prep.).

Mid-Norway and the Norwegian Sea

This areas comprises a large continental platform and mid-Norway. *In situ* stress observations on the continental platform and in mid-Norway indicate a NW–SE stress direction, which is consistent with earthquake focal mechanisms. The seismic activity in the region is generally high, with two recent interesting earthquake swarms recorded in the coastal region at Meløy and Steigen (Bungum *et al.* 1979; Atakan *et al.* 1994). Earthquake focal mechanisms reveal a compressional stress regime offshore, whereas seismic onshore activities close to highly elevated and glaciated regions, such as the Meløy–Steigen area, indicate a shallow tensional stress regime. Overcoring measurements in mid-Norway reveal high tectonic stresses (up to 30 MPa), especially in massive gneiss regions (Hanssen & Myrvang 1986).

Western Norway and the northern North Sea

In situ stress observations in the northern North Sea yield a WNW–ESE regional stress direction, but a NNE–SSW direction is also weakly indicated both by *in situ* measurements and by earthquake focal mechanisms (Lindholm *et al.* 1995*a*). The seismic activity offshore is high, especially centred around 61.5°N and 3.0°E, and a compressional stress regime is indicated. Onshore, seismic activity is concentrated on the seaward side of a highly elevated and glaciated region just east of the island of Stord. Focal mechanisms indicate a small area with shallow tensional stress regime in this area. Stress direction and magnitudes from overcoring measurements in western Norway are often strongly influenced by topography, however, indicating high tectonic stresses (10–15 MPa).

Southwestern Norway and the central North Sea

Borehole breakouts in the central North Sea yield a complex stress pattern. The NW–SE European regional trend is observed in the dataset, but also other stress directions frequently appear. The observations are of very low quality and probably also largely reflect local features such as faults and salt diapirs (Cowgill *et al.* 1994; Ask 1996). As relatively few focal mechanisms and overcoring measurements have been obtained for this area, it has been difficult to accomplish a regional interpretation. The seismic activity in this region is low.

Southeastern Norway

This region is centred around the Permian Oslo Graben, and because of the proximity to seismic monitoring stations focal mechanisms could be calculated for many minor events. The data indicate both relatively deep and shallow seismic activity with a dominance of normal faulting in the upper crust. The focal mechanisms indicate a NW–SE compressional (NE–SW tensional) stress direction. This is only partly confirmed by overcoring measurements, which generally yield a NE–SW trend more parallel to the graben structure, and with stresses up to 20 MPa at shallower depths.

Stress-generating mechanisms

Stresses in the Earth's crust may be distinguished with respect to their origin and lateral extent (Engelder 1974; Sykes & Sbar 1994), and can be divided into continental, regional and local stress fields (Table 1). The observed stress is then composed of a plate-wide continental stress field overprinted by regional and local effects (Zoback 1992). To identify the stress sources in the Norwegian areas, it is important to examine and identify the most important stress-generating mechanisms and to quantify the stress magnitude, lateral extension and depth variation. This has been achieved primarily by analytical or numerical stress modelling, where the plate has been regarded as pure elastic. This is a coarse simplification of the crust's brittle–ductile behaviour. In the simplified models the computed stress directions and stress regimes will be close to reality, whereas absolute stress magnitudes are probably exaggerated. Hence the models may be used only qualitatively, and to derive appropriate stress magnitudes the effects of rheology have to be evaluated (Cloetingh & Burov 1996). In cases where we have not performed the modelling a review of other modelling results and a qualitative evaluation has been made.

The generally accepted explanation for the observed first-order intraplate stresses are plate tectonic forces acting at convergent and divergent plate boundaries (Zoback 1992). Several workers, who have modelled the significance and relative magnitude of different plate tectonic

Table 1. *Stress-generating mechanisms*

Stress field	1st-order, continental	2nd-order, regional	3rd-order, local
Lateral extent	>1000 km	100–1000 km	<100 km
Stress-generating mechanisms	Plate tectonic forces: Ridge push Slab pull Basal drag	Large-scale density inhomogeneities: Continental margin Flexural stresses: Deglaciation Sediment loading Wide topographical loads	Topography: Fjords and mountain ranges Geological features: Faults Hard and soft inclusions

forces, have concluded that ridge push, slab pull and collisional resistance forces are the principal contributors both to plate kinematics and for the global stress field (Forsyth & Uyeda 1975; Chapple & Tullis, 1977; Richardsson 1992). Müller (1993) analysed the European stress field and concluded that the dominant source of regional tectonic stress in western Europe is a combination of collision-related forces at the southern plate boundary and ridge push along the western and northern boundary. This is based on a match between ridge push and collisional torque directions and stress directions (Richardsson 1992), and a similarity between stress orientations and relative plate motion directions between the African–Eurasian and North American–Eurasian plates. Because of the Norwegian region's proximal position relative to the Mid-Atlantic and Arctic spreading ridge, the major first-order tectonic stresses in this region are attributed to the ridge push force.

Second-order stress fields arise from regional density inhomogeneities, topographical loads and plate flexure related to deglaciation or sediment loading. Some local effects, as indicated in Table 1 are below the resolution of our stress observations and are therefore not discussed in detail in this paper.

In addition to stress-generating features the paper also focuses on the effect of crustal thickness as a stress-modifying feature that has importance in Norwegian regions. As tectonic stresses are concentrated in the brittle part of the crust, a local or regional crustal thinning will significantly enhance the stress magnitude and vice versa (Hasegawa *et al.* 1985).

The ridge push force

Mid-ocean ridges are areas of shallow bathymetry in approximately isostatic equilibrium because the elevated crust is compensated at depth by hot, low-density material. As the

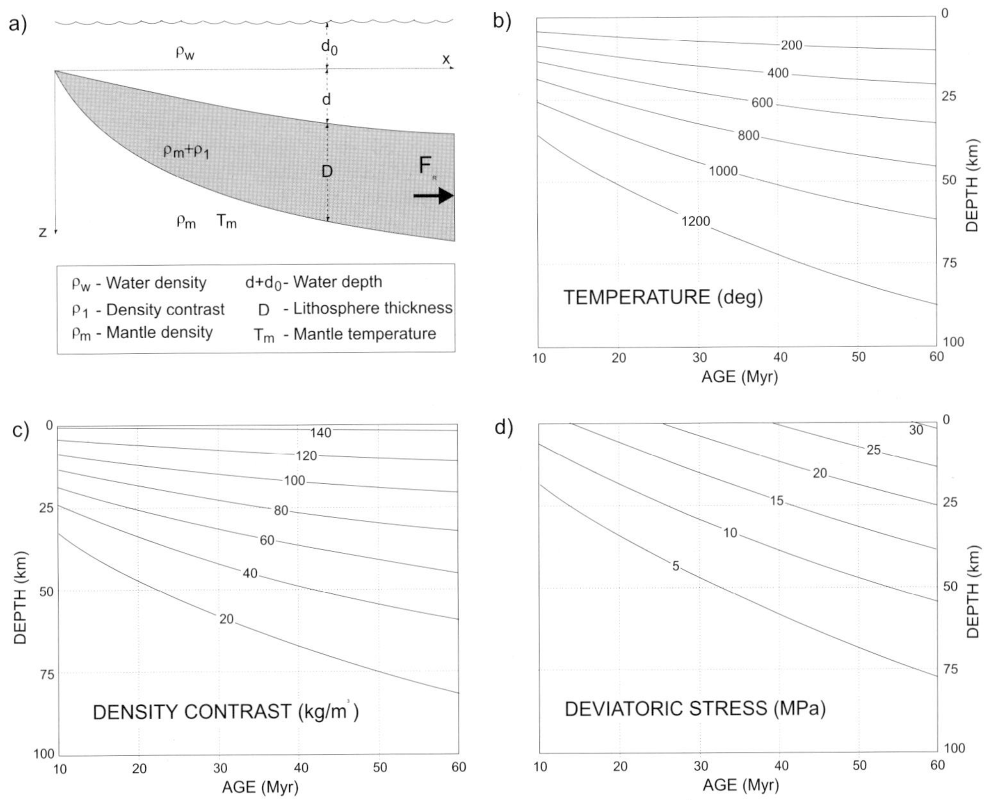

Fig. 3. Schematic cross-section of the spreading ridge model (**a**). Variation in temperature (**b**), density (**c**) and deviatoric stresses (**d**) as a function of age and depth. Assumed temperature and density of the mantle are 1350°C and $3300\,kg\,m^{-3}$, respectively; a thermal diffusivity of $0.008\,cm^2\,s^{-1}$ and coefficient of thermal expansion of $3.2 \times 10^{-5}\,°C^{-1}$ are anticipated.

Table 2. *Stress magnitudes associated with ridge push reported by various researchers*

Reference	Stresses related to ridge push
Bott & Kusznir (1984)	Average tectonic stress of the order of 20–30 MPa
Stein *et al.* (1989)	Average tectonic stress of a few tens of MPa
Bott (1991)	Maximum deviatoric stress of about 40 MPa in the upper lithosphere 200 km from the crest
Dahlen (1981)	Maximum deviatoric stress of 31 MPa at the surface of 60 Ma oceanic lithosphere
Fleitout & Froidevaux (1983)	Average tectonic stress of 25 MPa for an 80 km thick 60 Ma oceanic crust

oceanic lithosphere ages and moves away from the mid-ocean ridge it cools and subsides. The ridge elevation creates an outward compressional, gravity-generated force perpendicular to the crest. As a result of the rather simple ridge geometry and the well-constrained boundary conditions the ridge push force is quantitatively well understood and has been analysed both analytically and numerically (Lister 1975; Parsons & Richter 1980; Dahlen 1981).

By applying a cooling half-space model, Dahlen (1981) was able to estimate the stresses as a function of both depth and age of the oceanic lithosphere, and on the basis of his work, Fig. 3 shows the distribution of temperature, density and deviatoric stresses in the oceanic lithosphere. (Deviatoric stress (S) is the difference between total stress (σ) and mean Earth pressure (P).

$$\begin{bmatrix}\sigma_V\\ \sigma_H\\ \sigma_h\end{bmatrix} = \begin{bmatrix}P & 0 & 0\\ 0 & P & 0\\ 0 & 0 & P\end{bmatrix} + \begin{bmatrix}S_V\\ S_H\\ S_h\end{bmatrix}$$

where $P = (\sigma_V + \sigma_H + \sigma_h)/3$, and $S_H = -S_V$ and $S_H = 0$ is assumed.)

The ridge push force (F_R) will be zero at the ridge crest and increase linearly with age. The deviatoric stresses increase with age, but will decrease with depth. For a 60 Ma oceanic crust, a maximum deviatoric stress of 31 MPa is obtained. This corresponds well to values reported by various workers listed in Table 2.

Density contrast at the continental margin

The transition from denser and thinner oceanic crust to lighter and thicker continental crust will create a tensional stress state in the continental crust where it tends to 'spread out' over the oceanic lithosphere (Artyushkov 1973), whereby tensional deviatoric stresses will be generated in the continental crust and compressional deviatoric stresses in the oceanic crust. Stein *et al.* (1989) computed maximum deviatoric stresses of 40–50 MPa (tension in the continental crust and compression in the oceanic crust) close to the margin slope, decreasing with both depth and distance from the margin. The effect is not expected to penetrate more than about 100 km into the continental plate.

As long as the continental effect is evaluated alone, stresses will always act perpendicular to the margin, but Gölke *et al.* (1995) showed that stress direction and stress anisotropy near the continental margin are dependent on the angle between the applied ridge push force and the continental margin. Figure 4 illustrates the effect of a change in orientation of the continental margin, where margins orientated normal to the tectonic far-field stress direction were found to exhibit lower stress anisotropy (close to zero) in the continental crust compared with margins orientated parallel to the far field.

Glacial rebound and flexural stresses related to deglaciation

Glaciers can form a significant load on the lithosphere, and deglaciation, or removal of the ice load, introduces stress changes in the upper crust. Two models, both considering the crust as elastic, have been proposed to estimate horizontal stresses caused by glacial rebound

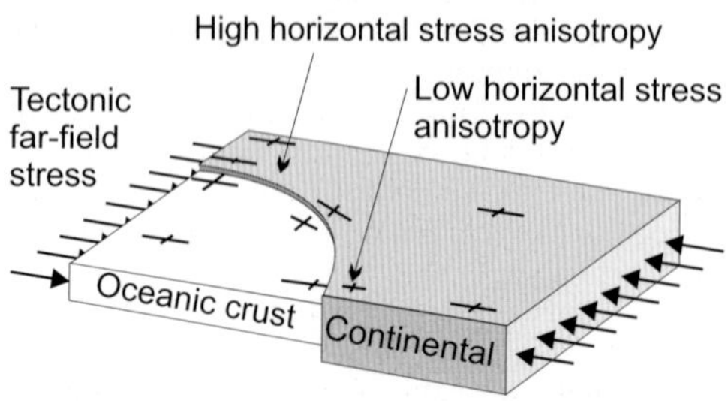

Fig. 4. Schematic model of a continental margin perturbed by a tectonic far-field stress. Tectonic stresses applied parallel to the margin yield a higher horizontal stress anisotropy in the continental crust compared with when the margin strikes normal to the far-field stress direction.

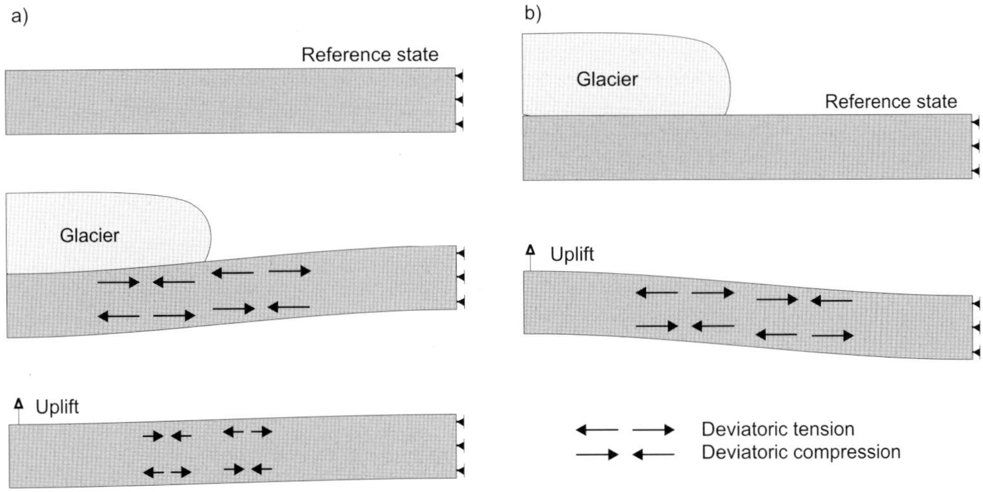

Fig. 5. Two models for the estimation of stresses associated with post-glacial rebound. The model of Stephansson (1988) (**a**) predicts compressional horizontal stresses of the order of 2–3 MPa beneath an ice sheet of 2 km thickness with 140 m of uplift remaining. The model of Stein *et al.* (1989) (**b**) predicts tensional stresses of a few tens of MPa for an ice load of 2 km thickness.

(Fig. 5). Stephansson (1988) assumed the lithosphere to be in equilibrium before ice loading and computed compressional stresses beneath the ice, so that when the ice is removed horizontal stresses slowly diminish until the uplift ceases and equilibrium is reached. Stein *et al.* (1989) assumed the plate to be in equilibrium with the ice load applied and thus derived tensional stresses associated with the uplift.

The choice of model depends on glacial history and the preferred plate relaxation time. If the plate relaxation time is long compared with the glacial history, the plate will not have time to adjust its equilibrium and stresses from deglaciation are best calculated by using an undeformed plate (Stephansson's model). For short relaxation times some viscous deformation would appear in the lithosphere, changing the state of equilibrium, which makes the model of Stein *et al.* more appropriate.

Close to the former ice margin the model of Stein *et al.* (1989) computed deviatoric stresses in the order of 30 MPa for a 2 km thick ice sheet (Fig. 6). The model of Stephansson predicts much smaller stress magnitudes (3–4 MPa) and an opposite stress pattern. In reality, the result is assumed to lie somewhere between the two models, probably favouring slight tension (P. Johnston, pers. comm.).

Flexural stresses from sediment loading

Subsiding sediment basins are driven by a combination of tectonic crustal stretching and thinning and the increasing sediment load. As in the case of ice loading, the sediment load will introduce flexural stresses with compressional stresses in the upper crust underneath the sediment basin and tensional stresses outside the basin (Fig. 7). Stein *et al.* (1989) computed stresses of several hundred MPa for a 10 km thick sedimentary basin. This is an obvious artefact of the elastic model, and by introducing viscoelastic and brittle–ductile plate behaviour stresses decrease to some tens of MPa. By varying lithosphere thickness and sedimentation rates, Stein *et al.* further concluded that the highest stresses are related to basins with high sedimentation rates underlain by a thin lithosphere. Very few places in the world show sufficiently high sedimentation rates to generate strong seismic activity, which leads to the conclusion that sediment loading generally has only a minor influence on the regional stress field. Recent observations from the Norwegian area, where areas of extraordinary high sedimentation rates occur, however, show a correlation between seismicity and sedimentary basins (Byrkjeland 1996).

Topography

Topography represents a load on the lithosphere, which has to be supported by stresses in the crust. Three main types of topographical surface loading situations may be identified (Bott 1971), and are qualitatively illustrated in

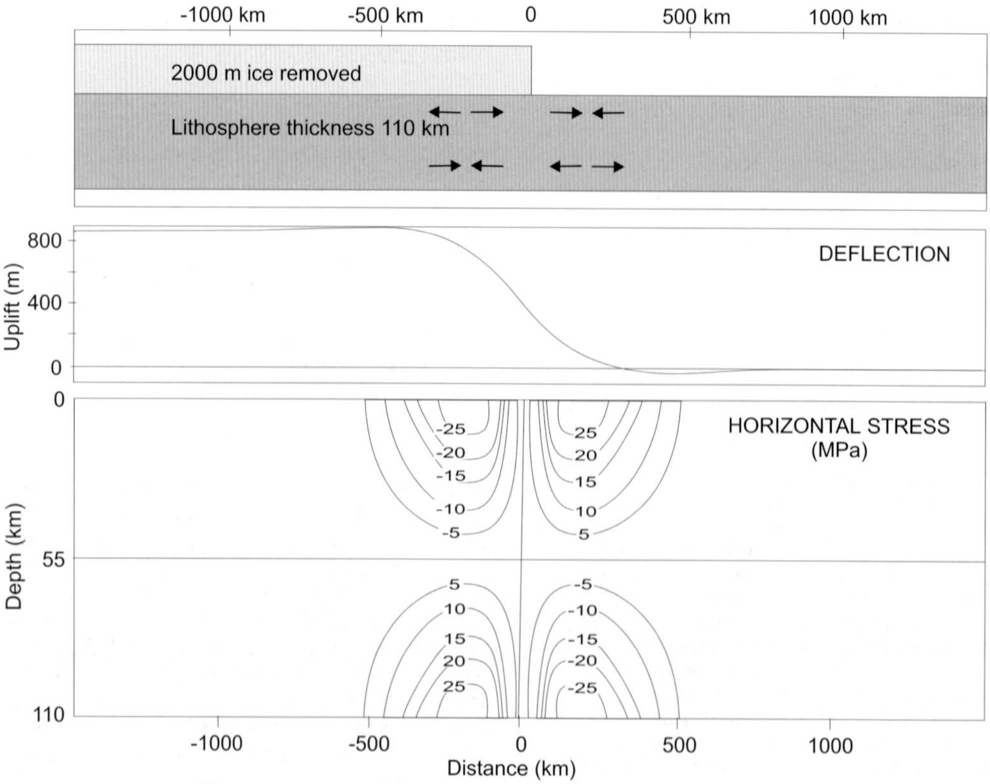

Fig. 6. Deformation and horizontal stress magnitudes resulting from the removal of 2000 m of ice from an elastic crust floating on a fluid. Modelling parameters used are: flexural rigidity ($D = \times 10^{25}$), flexural parameter ($\alpha = 205$ km), restoring density ($\rho_a = 2300\,\mathrm{kg\,m^{-3}}$), plate thickness ($T = 110$ km).

Fig. 8: narrow load; wide uncompensated load; wide compensated load.

For small surface loads less than 50 km wide, the lithosphere will not bend significantly, and stresses are adequately modelled by applying the load on an elastic half-space (Jaeger & Cook 1969; Bott & Kusznir 1984). Beneath the applied load deviatoric tension develops, whereas beyond the edge of the load, slight compression appears. More general models of symmetrical ridges in both two dimensions (Savage *et al.* 1985; Pan & Amadei 1993) and three dimensions (Liu & Zoback 1992) arrive at similar conclusions, but indicate slightly lower stresses compared with a constant load. In all cases the stresses are believed to be too small to cause tectonic activity, but may be of local importance and trigger earthquakes in areas with enhanced regional stress or pre-existing zones of weakness.

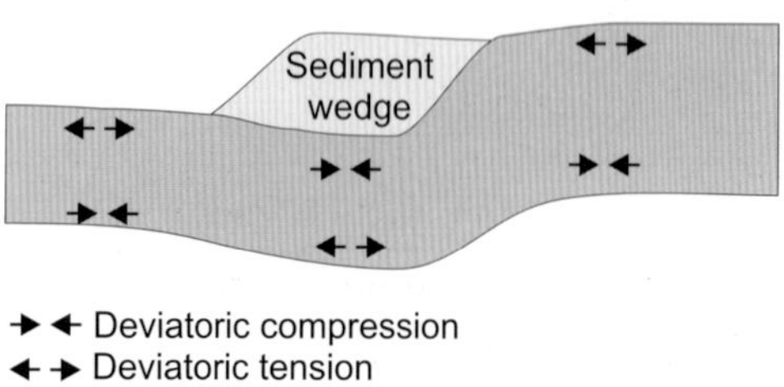

Fig. 7. Schematic representation of the bending stresses associated with sediment loading. Beneath the sediment basin compressional stresses appear, whereas tensional stresses dominate on the flanks.

For surface loads significantly wider than 50 km, which are not compensated at depth, the same type of bending as caused by sediment loading will appear. The flexure associated with bending creates compression on the concave side of the plate and tension on the convex side, and the stress magnitudes can be very high (several hundreds of MPa) if elastic models are applied (Bott 1971). Brittle–ductile behaviour, however, reduces the stress magnitudes significantly and entails stress dissipation by transient creep over a relative short geological time scale.

If the surface load is compensated at depth, the load will be counterbalanced by an upthrust and no bending will occur. The vertical stresses

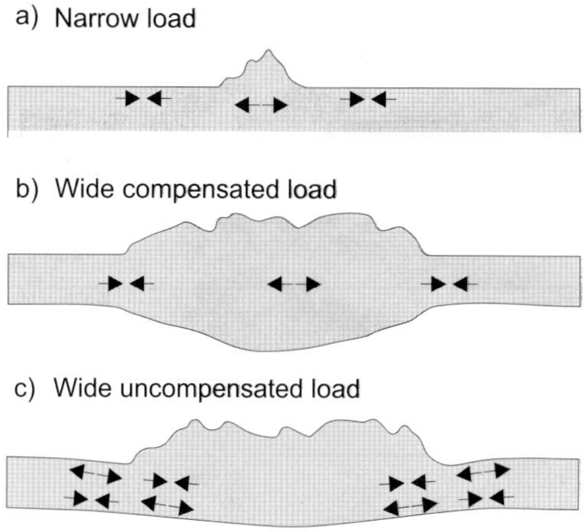

Fig. 8. Topographical load models investigated in this study. (**a**) Narrow topographical loads cause tensional deviatoric stresses beneath and compressional stresses adjacent to the load. The maximum stress difference is 0.64 times the loading pressure and defines a semicircular arc containing the ends of the load. (**b**) Large compensated loads will also cause deviatoric tension underneath the load, but here the maximum stress difference will be higher (approximately equal to the loading pressure). (**c**) Uncompensated wide loads result in bending, where the upper part of the lithosphere beneath the load is subjected to compression and the flanks to tension. The opposite situation appears at depth. The maximum bending stresses appear for loads that are about 4.4 times the flexural parameter and will be about six times the loading pressure.

in the crust will then increase by an amount proportional to the excess topography. Horizontal stresses are relatively unaffected and deviatoric tensional stress will therefore appear in the region between load and compensation and compression at the load edges. Deviatoric stresses close to the weight of the excess surface load are thus calculated, which means that for a 1500 m elevated region deviatoric stresses of the order of 40 MPa may be assumed (Table 3).

Sources of observed stresses in the Norwegian provinces

On the basis of the described sources of crustal stress and the observational data it is now time to link the stress-generating mechanisms to the defined stress provinces. The effect of each stress-generating mechanism in each stress province is discussed in view of the stress observations, and the source models are evaluated in order of lateral extent (plate-wide to local) and ranked in order of importance. Figure 9 provides an overview of the regional variations and Table 4 contains a summary of the results.

Ridge push

Since late Paleocene time (60 Ma) spreading has occurred from both the mid-Atlantic ridge and its northward continuation, the Arctic mid-ocean ridge (Jackson & Gunnarsson 1990). In the polar region the ridge geometry is well defined and the Barents Sea is expected to be subjected to a SSE-directed ridge push from the Arctic mid-ocean ridge (Fig. 9a). In the northern Atlantic Ocean, the ridge (Knipovitch and Molloy Ridges) turns N–S and is cut by several transform zones, and these zones complicate the ridge geometry and reduce the ridge push effect. It should therefore be anticipated that the Barents Sea is influenced by NNW–SSE tectonic stresses, which, for a 60 Ma crust, generate maximum deviatoric stresses of the order of 20–30 MPa. The observed stress direction (N–S) in the Barents Sea is rotated *c.* 20° clockwise from the expected ridge push direction and hence ridge push alone cannot explain the observations in this region.

Further south, the Mohns Ridge, a relatively straight ENE–WSW striking ridge bounded by the Jan Mayen and Senja fracture zones, generates a clear NNW–SSE tectonic stress. This is consistent with the observed stress direction in the Norwegian Sea and northern Norway, and

Table 3. *Predicted patterns and magnitudes of stress associated with topographical loading*

Type of loading	Maximum stress magnitude (MPa)	Lateral variation	Variation with depth	Zone of influence
Narrow load	25	Compression outside load Tension underneath	Monotonic	Local
Wide load, uncompensated	~240	Tension outside load Compression underneath	Reverses	Insignificant
Wide load, compensated	40	Compression outside load Tension underneath	Monotonic	Regional

All values are computed for a 1500 m elevation.

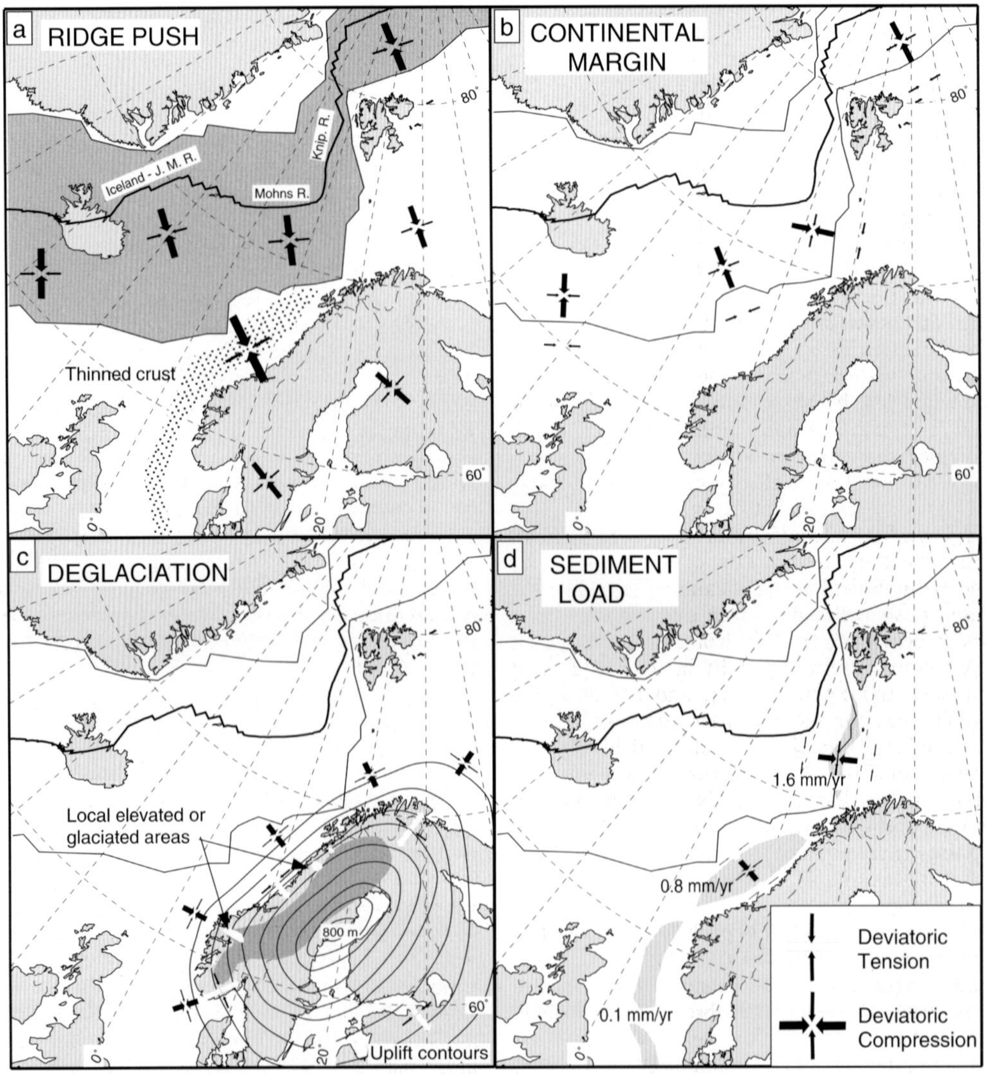

Fig. 9. Four important stress-generating mechanisms and their regional effect on Norway and adjacent regions. (**a**) The mid-Atlantic spreading ridge induces a NW–SE deviatoric stress field of the order of 20–30 MPa in the oldest oceanic crust. The magnitude is lowered somewhat in the thick continental crust, but might show local maxima in areas with thinned crust. (**b**) The continental margins are characterized by a deviatoric tensional stress field in the continental crust, and deviatoric compression in the oceanic crust, normal to the margin. (**c**) Post-glacial uplift contours indicate substantial vertical movement of Fennoscandia since deglaciation of the region. Deviatoric stresses, tension beneath the former ice, and compression beyond the ice edge, are associated with deglaciation in accordance with the model of Stein *et al.* (1989). Maximum stresses occur close to the former ice margin, and will reverse both with depth and laterally. (**d**) Sediment loading causes bending stresses to develop in the lithosphere, with compression beneath the load and tension on the flanks with stress reversal at depth. Bending stresses are dependent on a high sedimentation rate.

modelling also here indicates maximum deviatoric stresses of the order of 20–30 MPa close to the continental margin.

South of Jan Mayen, the Iceland–Jan Mayen Ridge defines the mid-Atlantic spreading ridge. Although the ocean-floor geometry in this region is complicated by the Jan Mayen microcontinent, the Ægir Ridge and the oceanic swell beneath Iceland, a net ridge push force is generated with a NW–SE direction. The stress direction may show a radial orientation around Iceland, as pointed out by Bott (1991), but the first-order stress pattern is expected to be oriented normal to the ridge (NW–SE). The stress magnitude is difficult to assess from a simple cooling model, but it is expected to be similar to or slightly less than that computed for the original 60 Ma oceanic crust (20–30 MPa).

The tectonic stresses set up by the ridge push are known to penetrate the entire north European plate (Müller *et al.* 1992; Müller 1993) and will therefore be active both in the northern and southern North Sea, as well as in southwestern Norway. In the northern North Sea and western Norway the predicted NW–SE direction fits the observations fairly well. In the central North Sea and southwestern Norway a complex stress pattern appears, but also here a NW–SE direction is clearly present in the data.

The compressional stress magnitude from the ridge push is affected by crustal thickness and plate rheology. Because the crustal thickness of the Baltic Shield is significant, a force like the ridge push will have to be distributed over a wider depth range. This will eventually reduce the tectonic stress regionally in central parts of Norway and Fennoscandia, and enhances the possibilities for second-order stress generators to dominate the observed stress field. On the other hand, the thinned crust underneath offshore rifted basins (mid-Norwegian margin, Viking and Central Grabens) enhances tectonic stresses and explains the compressional focal mechanisms and high seismic activity observed in this regions.

In summary, the ridge push explains most of the stress observations in the Norwegian area and no area seems to be outside the influence of this stress. However, the observed stress direction in the Barents Sea, as well as local onshore tensional regions and possible lateral variations in horizontal stress anisotropy, imply that second-order effects are also present. Other stress-generating mechanisms, such as the continental margin effect and flexural stresses associated with sediment loading or deglaciation, therefore need to be taken into consideration.

Continental margin

To the west, the Barents Sea is bounded by a N–S trending continental margin, which generates extensional horizontal deviatoric stresses perpendicular to the margin in the continental crust (Fig. 9b). When superimposing this margin effect on the stresses from the ridge push a stress rotation can appear. Depending on the ratio between the stresses set up by the margin effect and the stresses generated from ridge push, the continental margin may rotate the stresses so that the stress direction will be more or less parallel to the margin. Close to the margin, where the deviatoric stresses from the continental margin are at their maximum, the continental margin effect will be most significant and hence may explain the clockwise rotation from the NNW–SSE ridge push direction to the observed N–S direction.

In the Norwegian Sea the continental margin is defined by the Møre and Vøring escarpments and the shelf edge off the Lofoten islands. The margin here is approximately parallel to the spreading ridge, which means that the continental margin effect introduces deviatoric tensional stresses in the continental crust parallel to the compressional stresses caused by the ridge push. The overall effect will be a more isotropic stress state in the continental crust near the margin.

Further south the continental margin bends around the United Kingdom and is relatively distant from the region under focus. Its effect in the North Sea and southern Norway will thus be insignificant.

Sediment loading

During Tertiary and Quaternary time the Barents Sea region was uplifted and extensively eroded, and the eroded sediments were redeposited as a wedge of up to 3 km thickness covering the continental margin (Reemst *et al.* 1994). The sedimentation rate in Pliocene time (1.6 mm a^{-1}) was sufficiently high to set up significant deviatoric stresses in the underlying rock. Computations indicate that for a sediment wedge of 2–3 km thickness covering the continental margin, tensional deviatoric stresses of the order of 10–20 MPa can be set up in the continental crust close to the margin. The stresses align with the continental margin effect and may be significant both in reorienting the stress field and in increasing the stress anisotropy close to the margin in the Barents area.

In the mid-Norway and Norwegian Sea province a strong differential tilt and uplift of the mainland initiated a strong erosion and deposition of a thick prograding sequence (about 1500 m) over the mid-Norwegian Shelf in late Neogene time. The high sedimentation rate in Pliocene time (up to 0.8 mm a^{-1}) may have been large enough to cause flexural stresses on the Mid-Norwegian Shelf. Byrkjeland (1996) also found a clear correlation between seismic activity and Pleistocene and post-Miocene sediment

Table 4. *Stress-generating mechanisms and their importance for the stress field in the Norwegian area*

Mechanism	Theoretical effects and magnitudes	Effects in Fennoscandia	Comments
Ridge push (continental)	20–30 MPa for 60 Ma old oceanic crust Normal to ridge Compression in oceanic and continental crust Monotonic with depth	The basic stress generator throughout Fennoscandia	Yields a WNW–ESE to NNW–SSE stress field, affecting the whole region
Crustal thickness (regional)	Modifies the present stress field A force will generate higher tectonic stress in thin crust compared with thick crust	Areas where the crust has been extensively thinned or in areas of extraordinary thick crust	High compressional stresses in the thinned crust under the offshore basins of mid- and central Norway may be reflected by the high seismic activity The thick Fennoscandian crust dampens the tectonic stresses in central parts of Norway and Sweden
Continental margin (regional)	10–50 MPa Normal to margin Tension in continental crust and compression in oceanic crust Monotonic with depth	Probably important in the western Barents Sea and on the Norwegian continental margin	In the western Barents Sea, the continental margin effect supplements the tectonic stress from ridge push; at the Norwegian continental margin this mechanism counteracts the ridge push; this accounts for the observed reduction in stress anisotropy from north to south
Sediment loading (regional)	160 MPa for 2 km of sediments (decreases with time) Compression beneath basins and tension at margins Compression is strongest in the centre and decreases outwards Stress direction reverses with depth High sedimentation rates required	Western Barents Sea and tentatively the Norwegian margin, Viking and Central Grabens	The Western Barents Sea was the locus of moderate sedimentation rates in Paleocene time; bending stresses in the Barents Sea may reinforce the tectonic stresses from ridge push; on the Norwegian margin and in the Viking and Central Grabens the sedimentation rates are lower and the effect is less clear
Deglaciation (regional)	20–30 MPa for 2–3 km ice load (decreases with time) Tension beneath uplifted areas and compression beyond Most prominent close to the former ice margin Stress reverses with depth	Tentatively affects whole region	In Norway the effect will be small compared with other mechanisms; observed fault types in northern Norway may be explained by this mechanism, as also the normal faulting mechanisms observed in coastal areas in mid- and west Norway

Topographical wide load (regional)	40 MPa for 1500 m elevated region Tension beneath the ice load and compression beyond Monotonic with depth	The mountain ranges in western Norway may support this mechanism	Topography is moderate (*c.* 1000 m elevation); underneath the elevated regions in northern and southwestern Norway tectonic stresses from ridge push will be reduced, whereas in the surrounding (both coastal and landward) areas the tectonic stresses are expected to increase slightly; in offshore regions to the northwest the direction of compression will be similar to that of ridge push
Topographical narrow load (local)	25 MPa for 1500 m elevated region Tension beneath the elevated regions and compression beyond	Uncertain	Centres of seismicity with normal faulting mechanisms are located close to the glaciers Svartisen and Folgefonna (remnants from the glaciation?); this may be explained by local areas of high topography and possibly strengthened by local uplift connected to the glaciers
Surface relief (local)	10–30 MPa Compressional stress acts parallel to the surface gradient Only important very close to the surface	Throughout region, but particularly in high-relief areas of northern and southwestern Norway	The large variation in stress direction from shallow onshore measurements along fjords and valleys in the coastal regions of northern and western Norway may be explained by local, shallow topographical effects; depending on the orientation of fjords and valleys, topographical stress may or may not interact with the regional stress
Faults (local)	Modifies the present stress field Stress rotates normal or parallel to the faults	Throughout the region, but most clearly observed in the North Sea basins	Large variation in stress direction (sometimes rotation?) in the North Sea basins may be explained by low horizontal stress differences favouring local influence from faults
Inclusions (local)	Modifies the present stress field (up to 50% change in stress magnitude) Stress bends around soft inclusions such as salt diapirs and is directed into areas of high stiffness such as massive rock masses	Salt diapirs beneath the southern North Sea Gneiss windows in mid-Norway	The complex stress pattern in the Central Graben may possibly be explained by low horizontal stress differences favouring local influence from salt and other doming structures The very high tectonic stresses measured in the gneissic areas compared with surrounding Caledonian areas may be explained by variations in fracture systems, and stiffness contrasts

depocentres on the mid-Norwegian shelf, substantiating the idea of sedimentation-induced stresses. Analysis of focal mechanisms underneath the loaded area yields a compressional regime whereas onshore mechanisms indicate a tensional regime. It is, however, difficult to assess the quantitative importance of the sediment loading effect, as it acts constructively and in the same direction as ridge push.

In Pleistocene time, the Viking and Central Grabens subsided and the sedimentation rate was relative low. Hence, bending stresses are relatively small and the sediment loading effect is expected to be negligible here.

Deglaciation

The deglaciation models indicate maximum deviatoric stress concentrations close to the former ice edge, with a shift in polarity across the edge (largely along the coast) (Fig. 9c). This is also to some extent observed in the region, with compressional offshore focal mechanisms and some shallow onshore mechanisms (Stord, Oslo and Meløy–Steigen) indicating a tensional stress regime. The observations thus support the deglaciation model of Stein *et al.* (1989) well. However, the onshore faulting may also be explained by sediment loading and topography effects.

Finnmark and the southern Barents Sea have also been covered by thick ice sheets (Kjemperud & Fjeldskaar 1992), possibly inducing tectonic stresses in connection with deglaciation. The observed shallow reverse faulting in Finnmark, with a NNW–SSE compression reported by Olesen *et al.* (1992) and Bungum & Lindholm (in prep.), may favour Stephansson's uplift model, but no lateral shift in stress regime is identified from the available data. The deglaciation effect in the northern part of Norway is therefore believed to be either negligible or masked by other regional effects. The observed compressive stress regime is more probably connected to effects such as the compression from ridge push.

If the deglaciation effect is an important source for the observed stress field in Norway the observations surely favour the model of Stein *et al.* (1989).

Topography

The bathymetry of the Barents Sea is relatively shallow and the onshore relief is low with gentle slopes. The topographical effects, both on a regional and on a local scale, are therefore not expected to play an important role in northern Norway.

Mid-Norway is characterized by locally high-altitude mountains, especially in the northern part of the region, and topography clearly influences the near-surface stresses. The mountains can be regarded as narrow loads and modelling has shown that tensional deviatoric stresses extend downward to a depth comparable with the width of the load. In the Meløy–Steigen areas there are examples of locations relatively near high mountain complexes. Here earthquake swarms, indicating inhomogeneous stress conditions, are observed; however, the shallow normal faulting focal mechanisms are contrary to what should have been expected from topographical loading (Fig. 8).

Western Norway is also characterized by high mountains cut by deep fjords and valleys, and hence, near *in situ* measurements are very much influenced by local topographical features. The mountain range also contribute on a more regional scale, as the large topographical features act as a wide compensated load, introducing tensional deviatoric stresses underneath elevated onshore areas. The shallow tensional focal mechanisms observed in southwestern Norway can be attributed to topographical effects.

The relief in southern and eastern Norway is lower and more gentle than in western Norway. Hence, topographical effects will not be of the same importance as in the high-relief and high-altitude areas and may therefore be neglected.

Discussion

The N–S stress direction observed along the western margin of the Barents Sea is well explained by the ridge push, when modified in azimuth by the continental margin effect and flexure associated with rapid sedimentation. The ridge push is also believed to cause the shallow reverse faulting and high horizontal stress magnitudes measured in northern Norway. The consistent stress direction in this region indicates a relatively high stress anisotropy. In contrast, the low seismic activity should indicate low stress anisotropy. This behaviour is difficult to explain, and as yet is not fully understood.

The observed regional stress direction on the Norwegian margin is consistent with the direction of ridge push, and the high tectonic stresses observed are also attributable to this source. Because of the local thinning of the continental crust, the tectonic stresses may be amplified, and this may explain the reverse faulting and high seismic activity observed offshore, which in a broad sense is concentrated in areas of thinned crust. Sediment loading effects also induce offshore compressional deviatoric stresses underneath the basins. The observed stress anisotropy seems to be lower on the Norwegian margin than in the Barents Sea, presumably because of the continental margin effect counteracting the ridge push effect. As three effects (sediment loading, topography and deglaciation) produce more or less the same stress pattern in this area it is difficult to evaluate their relative importance.

The regional stress direction in western Norway and the northern North Sea correlates well with the direction of ridge push. Local thinning of the crust is also assumed to be an important factor in amplifying tectonic stresses in the continental crust underneath the graben. This explains the large amount of earthquakes (many with reverse focal mechanisms) observed in this region. Sediment loading fits partly with the observed stresses, and may also possibly be an important factor in the areas of extreme sedimentation rates in recent time. The observed local stress reorientation in the northern North Sea may indicate that stress anisotropy is lower in this region compared with other areas, or that local features are prominent and deflect the regional stress field. Onshore, the shallow normal faulting activity is attributed to regional and local topography, but regional deglaciation effects may also be important. Possible important local stress modifiers are faults, as well as local topographical features such as fjords and valleys.

The major stress direction in southwestern Norway and the central North Sea coincides with the 'European trend' (Müller *et al.* 1992), indicating that ridge push is an important source here as well. None of the other stress-generating mechanisms examined in this paper seem to be of great importance in this region. The complex stress pattern observed from borehole breakouts in the Central Graben area is primarily believed to be a result of low-quality data, but may also tentatively be caused by low horizontal stress anisotropy and significant local influence by geological structures such as faults and salt diapirism.

The NW–SE stress direction observed in southeastern Norway may be attributed to the ridge push effect, but as crustal thickness increases eastward the region will probably be subjected to somewhat lower tectonic stresses. Of the other effects analysed in this paper only the deglaciation effect was found to be able to influence the stress field in this region. Very tentative indicators (reversal of stress regime with depth indicated by focal mechanisms and N–S to NE–SW stress directions from overcoring) favour the deglaciation model Stein *et al.*It should be noted that also in this region the zones of enhanced seismic activity coincide with areas of thinned crust (the Oslo Graben) and that near-surface overcoring measurements often yield directions parallel to the major faults in this region.

Conclusions

All stress-generating mechanisms examined in this paper are summarized in Table 4 together with a brief description of their importance for the Norwegian area. The internal ranking has been based on lateral extension or regional influence, as most mechanisms are capable of generating high deviatoric stresses. Some of the mechanisms, known as non-renewable, relieve with time, and the magnitudes reported may exaggerate their importance, as they are maximum values computed for a perfectly elastic crust. Other, more local mechanisms, which are not formally stress generators, but more precisely described as modifiers, are also included in Table 4.

The only continental stress-generating mechanism that can be responsible for the high compressional stresses observed throughout the Norwegian region is the ridge push effect. Modelling indicates that horizontal deviatoric stresses from ridge push in the oceanic crust are of the order of 20–30 MPa. In the continental crust the changes in tectonic stresses depend on crustal thickness and ability of the crust to accommodate differential stresses. Regions with thinned crust, such as the Norwegian margin, the North Sea Graben system and the Oslo Graben, will have higher tectonic stresses. This is also clearly demonstrated by the enhanced seismic activity in these regions.

The other stress-generating mechanisms analysed here are found to be of variable regional importance. They modify the tectonic stresses generated by the ridge push and are responsible for locally observed stress reorientation, horizontal stress anisotropy and variations in seismicity.

Close to the western margin of the Barents Sea both the continental margin and recent high sedimentation rates act constructively with the ridge push and play an important role in rotating and strengthening the tectonic stress field. In the Norwegian Sea the continental margin acts against the ridge push, causing a lower stress anisotropy in this region.

Along the coast of southern Norway local and regional topography plays an important role in modifying stress magnitudes and explaining the shallow normal focal mechanisms observed.

The deglaciation of Fennoscandia has been regarded as an important stress-generating mechanism. The modelling did not indicate this as a source of high stress anisotropy, but all along the coastal areas and in southeastern Norway the observed stress directions and tectonics are in reasonable accord with predictions from deglaciation.

Stress modelling based on an elastic plate behaviour is a coarse simplification of the crust's brittle ductile behaviour; nevertheless, the models can be applied qualitatively. For more accurate determination of stress magnitudes viscoelastic modelling is required, where also the yield strength is accounted for (Cloetingh & Burov 1996). Further research should therefore be based on more realistic models. A quantitative evaluation of the contribution from different stress-generating mechanisms can then be conducted under otherwise identical mechanical and boundary conditions. This will shed more light on the mutual contribution from topography, sediment loading and deglaciation, taking both loading history (duration and time since maximum loading) and lateral influence into consideration.

This work was supported by the Research Council of Norway, as a part of the IBS (Integrated Basin Studies) project under the Joule II research programme funded by the Commission of European Communities (Contract JOU2-CT 92–0110). Many individuals have directly and through discussions and reviewing improved the manuscript. The authors are grateful to all of them, and especially want to thank M. D. Zoback, R. Gabrielsen, E. VÍgnes and J. Palmer for valuable input during the writing.

References

ARTYUSHKOV, E. V. 1973. Stresses in the lithosphere caused by crustal thickness inhomogeneities. *Journal of Geophysical Research*, **78**, 7675–7708.

ASK, M. V. S. 1996. In situ *stress from borehole breakouts in Denmark*. Lic. thesis, Royal Institute of Technology, Stockholm.

ATAKAN, K., LINDHOLM, C. D. & HAVSKOV, J. 1994. Earthquake swarm in Steigen, Northern Norway: an unusual example of intraplate seismicity. *Terra Nova*, **6**, 180–194.

BOTT, M. H. P. 1971. *The Interior of the Earth*. Edward Arnold, London

——1991. Ridge push and associated plate interior stress in normal and hot spot regions. *Tectonophysics*, **200**, 17–32.

—— & KUSZNIR, N. J. 1984. The origin of tectonic stress in the lithosphere. *Tectonophysics*, **105**, 1–13.

BUNGUM, H. & FYEN, J. 1979. Hypocentral distribution, focal mechanisms, and tectonic implications of Fennoscandian earthquakes, 1954–1978. *Geologiska Føreningens i Stockholm Forhandlingar*, **101**, 261–273.

—— & LINDHOLM, C. 1999. Neotectonics in Finnmark (in preparation).

——, ALSAKER, A., KVAMME, L. B. & HANSEN, R. A. 1991. Seismicity and seismotectonics of Norway and nearby continental shelf areas. *Journal of Geophysical Research*, **96**, 2249–2265.

——, HOKLAND, B. K., HUSEBYE, E. S. & RINGDAL, F. 1979. An exceptional intraplate earthquake sequence in Meløy, Northern Norway. *Nature*, **280**, 32–35.

BYRKJELAND, U. 1996. *Seismotectonics of the Norwegian Continental Margin, 62–71° North*. Cand.Scient. thesis, University of Oslo.

CHAPPLE, W. M.'& TULLIS, T. E. 1977. Evaluation of the forces that drive the plates, *Journal of Geophysical Research*, **82**, 1967–1984.

CLAUSS, B., MARQUART, G. & FUCHS, K. 1989. Stress orientations in the North Sea and Fennoscandia, a comparison to the central European stress field. *In*: GREGERSEN & BASHAM (eds) *Earthquakes at North Atlantic Passive Margins: Neotectonics and Postglacial Rebound*. Kluwers, Dordrecht.

CLOETINGH, S. & BUROV, E. B. 1996. Thermomechanical structure of European continental lithosphere: constraints from rheological profiles and EET estimates. *Geophysical Journal International*, **124**, 695–723.

COWGILL, S. M., MEREDITH, P. G. & MURRELL, S. A. F. 1994. Heterogeneous stress orientations in the North Sea Basin. *Proceedings of ISRM International Conference*, 29–31 August, Delft, 963–970.

DAHLEN, F. A. 1981. Isostacy and the ambient state of stress in the oceanic lithosphere. *Journal of Geophysical Research*, **86**, 7801–7807.

DART, C. J., INDERHAUG, O. H., KLØVJAN, O. & OTTESEN, C. 1995. The present stress regime in the Barents Sea from borehole breakouts. *Proceeding of the Workshop 'Rock Stresses in the North Sea'*, 13–14 February, Trondheim, 179–190.

ENGELDER, T. 1994. *Stress Regimes in the Lithosphere*. Princeton University Press, Princeton, NJ.

FEJERSKOV, M., MYRVANG, A. M., LINDHOLM, C. & BUNGUM, H. 1995. *In situ* rock stress pattern on the Norwegian continental shelf and mainland. *Proceeding of the Workshop 'Rock stresses in the North Sea'*, 13–14 February, Trondheim, 191–201.

FLEITOUT, L. & FROIDEVAUX, C. 1983. Tectonic stress in the lithosphere. *Tectonics*, **2**, 315–324.

FORSYTH, D. & UYEDA, S. 1975. On the relative importance of the driving forces of plate motion. *Geophysical Journal of the Royal Astronomial Society*, **45**, 163–200.

GØLKE, M., COBLENTZ, D., CLOETINGH, S. & FEJERSKOV, M. 1995. Stress system of the Norwegian continental margin – Part 1. *Proceeding of the Workshop 'Rock Stresses in the North Sea'*, 13–14 February, Trondheim, 250–274.

HANSEN, T. H. & MYRVANG, A. 1986. Rock stress and rock stress effects in the Kobbelv area, northern Norway. *Proceedings of the International Sympo-*

sium on Rock Stress Measurements. Stockholm, 625–627.

HASEGAWA, H. S., ADAMS, J. & YAMAZAKI, K. 1985. Upper crustal stresses and vertical stress migration in eastern Canada. *Journal of Geophysical Research*, **90**(B5), 3637–3648

HAST, N. 1958. *The Measurement of Rock Pressure in Mines*. Swedish Geological Survey, Series C, **560**.

——1969. The state of stress in the upper part of the earth's crust. *Tectonophysics*, **8**, 169–211.

JACKSON, H. R. & GUNNARSSON, K. 1990. Reconstruction of the Arctic: Mesozoic to present, *Tectonophysics*, **172**, 303–322.

JAEGER, J. C. & COOK, N. G. W. 1969. *Fundamentals of Rock Mechanics*. Chapman and Hall, London, p. 585.

KJEMPERUD, A. & FJELDSKAAR, W. 1992. Pleistocene glacial isostasy – implications for petroleum geology. *In*: LARSEN, R. M., BREKKE, H., LARSEN, B. T. & TALLERAAS, E. (eds) *Structural and Tectonic Modelling and its Application to Petroleum Geology*. Norwegian Petroleum Society Special Publication, **1**, 187–195.

KLEIN, R. J. & BARR M. V. 1986. Regional state of stress in western Europe. *Proceedings of the International Symposium on Rock Stress and Rock Stress Measurements*. Stockholm, 1–3 September.

LINDHOLM, C., BUNGUM, H., BRATLI, R. K., AADNØY, B. S., DAHL, N. & TØRUDBAKKEN, B. 1995*a*. Crustal stress in the northern North Sea as inferred from borehole breakouts and earthquake focal mechanisms. *Terra Nova*, **7**, 51–59.

——, ——, VILLAGRAN, M. & HICKS, E. 1995*b*. Crustal stress and tectonics in Norwegian regions determined from earthquake focal mechanisms. *Proceeding of the Workshop 'Rock Stresses in the North Sea'*, 13–14 February, Trondheim, 77–91.

LISTER, C. R. B. 1975. Gravitational drive on oceanic plates caused by thermal contraction. *Nature*, **257**, 663–665.

LIU, L. & ZOBACK, M. D. 1992. The effect of topography on the state of stress in the crust: application to the site of the Cajon Pass Scientific Drilling Project. *Journal of Geophysical Research*, **97**, 5095–5108.

MÜLLER, B. 1993. *Tectonic stress in Europe – borehole guided waves affected by stress-induced anisotropy*. PhD thesis, University of Karlsruhe.

——, ZOBACK, M., FUCHS, K. *et al*. 1992. Regional patterns of tectonic stress in Europe. *Journal of Geophysical Research*, **97**, 11 783–11 803.

MYRVANG, A., HANSEN, S. E. & SØRENSEN T. 1993. Rock stress redistribution around an open pit mine in hardrock. *International Journal of Rock Mechanics, Mining Science, and Geomechanics Abstracts*, **30**, 1001–1004.

OLESEN, O., HENKEL, H. LILE, O. B., MAURING, E. & RØNNING, J. S. 1992. Geophysical investigations of the Stuoragurra postglacial fault, Finnmark, northern Norway. *Journal of Applied Geophysics*, **29**, 95–118.

PAN, E. & AMADEI, B. 1993. Gravitational stresses in long asymmetric ridges and valleys in anisotropic rock. *International Journal of Rock Mechanics and Mining Science, and Geomechanics Abstracts*, **30**, 1005–1008.

PARSONS, B. & RICHTER, F. M. 1980. A relation between the driving force and geoid anomaly associated with mid-ocean ridges. *Earth and Planetary Science Letters*, **51**, 445–450.

—— & CHANDLER, T. E. 1975. The stress field in the upper crust as determined from *in situ* measurements. *Geologische Rundschau*, **64**, 653–674.

REEMST, P., CLOETINGH, S. & FANAVOLL, S. 1994. Tectonostratigraphic modeling of Cenozoic uplift and erosion in the south-western Barents Sea. *Marine and Petroleum Geology*, **11**, 478–490.

RICHARDSSON. R. M. 1992. Ridge forces, absolute plate motions, and the intraplate stress field. *Journal of Geophysical Research*, **97**, 11 739–11 748.

SAVAGE, W. Z., SWOLFS, H. S. & POWERS, P. S. 1985. Gravitational stresses in long symmetric ridges and valleys. *International Journal of Rock Mechanics and Mining Science, and Geomechanics Abstracts*, **22**, 291–302.

SPANN, H., BRUDY, M. & FUCHS, K. 1991. Stress evaluation in offshore regions of Norway. *Terra Nova*, **3**, 148–152.

STEIN, S., CLOETHINGH, S., SLEEP, N. H. & WORTEL, R. 1989. Passive margin earthquakes, stresses and rheology. *In*: GREGERSEN & BASHAM (eds) *Earthquakes at North Atlantic Passive Margins: Neotectonics and Postglacial Rebound.* Kluwers, Dordrecht.

STEPHANSSON, O. 1988. Ridge push and glacial rebound as rock stress generators in Fennoscandia. *Bulletin of the Geological Institute of the University of Uppsala*, **14**, 39–48.

——, SÄRKKÄ, P. & MYRVANG, A. 1987. State of stress in Fennoscandia. *Proceedings of the International Symposium on Rock Stress Measurements*, Stockholm, 21–32.

SYKES, L. R. & SBAR, M. L. 1974. Intraplate earthquakes, lithospheric stresses, and the driving mechanism of plate tectonics. *Nature*, **245**, 298–302.

THORESEN, M. K. 1990. *Kvartærgeologisk kart over Norge, Tema: jordarter*. Norwegian Geological Survey.

ZOBACK, M. L. 1992. First- and second-order patterns of stress in the lithosphere: the World Stress Map project. *Journal of Geophysical Research*, **97**, 11 703–11 728.

——, *et al*. 1989. Global patterns of tectonic stress. *Nature*, **341**, 291–298.

Index